Lecture Notes in Civil Engineering

Volume 73

Lecture Notes in Civil Engineering (LNCE) publishes the latest developments in Civil Engineering—quickly, informally and in top quality. Though original research reported in proceedings and post-proceedings represents the core of LNCE, edited volumes of exceptionally high quality and interest may also be considered for publication. Volumes published in LNCE embrace all aspects and subfields of, as well as new challenges in, Civil Engineering. Topics in the series include:

- Construction and Structural Mechanics
- Building Materials
- Concrete, Steel and Timber Structures
- Geotechnical Engineering
- Earthquake Engineering
- Coastal Engineering
- Ocean and Offshore Engineering; Ships and Floating Structures
- Hydraulics, Hydrology and Water Resources Engineering
- Environmental Engineering and Sustainability
- Structural Health and Monitoring
- Surveying and Geographical Information Systems
- Indoor Environments
- Transportation and Traffic
- Risk Analysis
- Safety and Security

To submit a proposal or request further information, please contact the appropriate Springer Editor:

- Mr. Pierpaolo Riva at pierpaolo.riva@springer.com (Europe and Americas);
- Ms. Swati Meherishi at swati.meherishi@springer.com (Asia - except China, and Australia, New Zealand);
- Dr. Mengchu Huang at mengchu.huang@springer.com (China).

All books in the series now indexed by Scopus and EI Compendex database!

More information about this series at http://www.springer.com/series/15087

Volodymyr Onyshchenko ·
Gulchohra Mammadova ·
Svitlana Sivitska · Akif Gasimov
Editors

Proceedings of the 2nd International Conference on Building Innovations

ICBI 2019

Springer

Editors
Volodymyr Onyshchenko
Poltava National Technical Yuri Kondratyuk University
Poltava, Ukraine

Svitlana Sivitska
Poltava National Technical Yuri Kondratyuk University
Poltava, Ukraine

Gulchohra Mammadova
Azerbaijan University of Architecture and Construction
Baku, Azerbaijan

Akif Gasimov
Azerbaijan University of Architecture and Construction
Baku, Azerbaijan

ISSN 2366-2557 ISSN 2366-2565 (electronic)
Lecture Notes in Civil Engineering
ISBN 978-3-030-42938-6 ISBN 978-3-030-42939-3 (eBook)
https://doi.org/10.1007/978-3-030-42939-3

This Springer imprint is published by the registered company Springer Nature Switzerland AG
The registered company address is: Gewerbestrasse 11, 6330 Cham, Switzerland

Contents

Building Constructions, Technologies and Technics

Features of Public Spaces Development in the Context of Shopping Complexes Formation

T. Apatenko, O. Bezlyubchenko, T. Lytvynenko, and T. Lvovska

Abstract The public-urban spaces, including shopping complexes, which bring together many different groups living in urban environments, have changed in parallel with the development of the city today. When the cities were first founded, the shopping action was carried out in open spaces. In modern cities, public spaces look different, because they are a platform for communication that in the era of the Internet and social networks makes the shopping center a powerful center of attraction. This study specifically is attributed to the public-urban characteristics of shopping centers which are interpreted as characteristics of urban space as the public space part.

Keywords Public space · Shopping center · Shopping complex · Shopping mall

1 Introduction

The creation of opportunities for social interactions is one of the most important aspects of the vast expanses. In these interactions, people will establish a greater silence with their own desires and suspensions. At the public establishments, there is the possibility of special interactions, part-time, open-minded activities, and participation in large-scale visits [1]. Shopping centers (complexes) are an integrant part of the vast expanse; a shopping center for ordinary people is being re-organized in a special way, not so much of a "shopping row," but a large number of commercial areas, to spend all their homeland. The trading function attracts to itself the

T. Apatenko · O. Bezlyubchenko
O.M. Beketov National University of Urban Economy, Kharkiv, Ukraine
e-mail: apatenkotanya82@gmail.com

O. Bezlyubchenko
e-mail: elen4iksokol@gmail.com

T. Lytvynenko · T. Lvovska (✉)
Poltava National Technical Yuri Kondratyuk University, Poltava, Ukraine
e-mail: lvovska.tetiana@ukr.net

T. Lytvynenko
e-mail: litta2510@gmail.com

V. Onyshchenko et al. (eds.), *Proceedings of the 2nd International Conference on Building Innovations*, Lecture Notes in Civil Engineering 73,
https://doi.org/10.1007/978-3-030-42939-3_1

rest of functions, in such a way, concentrating around itself the whole world, where everybody can find leisure activities according to the interests [2].

Thus, we see currently in the world practice of engineering and construction, there is a tendency to the enlargement of consumer services, their concentration in big complexes in combination with other institutions and businesses in the service sector and in close communication with major urban highways.

2 Trade Establishment Formation

2.1 *New Concept*

For the modern consumer, with accelerated rhythm of living, which is characteristic of modern lifestyles of most people, one of the main requirements is the concentration of more functions in one place, the ability to solve several problems at the same time. New needs define a conceptually new approach to the factors of competitiveness of trading enterprises. Although the shopping process has become a popular activity, it is not enough to continue shopping in the mall. This causes the need to fill it with additional functions that allow you to get consumer services, cultural, educational, recreational, entertainment, and other services. According to research by the American Urban Land Institute (ULI): "A shopping mall is a group of retail and commercial outlets located in one specially designed location, organized according to a specific concept and united by a single architecture, with common areas for all public utilities that work as one unit and are under single management" [3].

In general, malls are closed or open spaces that include a number of stores that may be of one or more types. These places are trading centers and create a direct link between supply and demand. According to the classified division, the variety of shopping complexes includes a gallery, a passage, a shopping and entertainment complex, a shopping complex, a shopping mall.

Throughout history, the process of commercial establishments formation was considered a complicated task. Even today, such facilities are one of the most powerful leisure activities in the world, and consumption is also defined as leisure. Jacobs, using the typology of dependency and connection between purchases, leisure and the place of purchase, stated that the "mall" is a place to buy for leisure [4].

2.2 *Research of Public Spaces Development Features in the Shopping Centers Formation*

Shopping center planning is one of the most important architectural and town-building challenges that can solve many urban problems. With the trade development, economic, and cultural relations with other countries, the need to strengthen the presentation of industrial products, with a sharp increase in urban automation, with changing lifestyles, cultural and economic priorities, consumers prefer, as research has shown, large shopping complexes where they can get a wider range of services. Therefore, in recent years, intensive development has become the creation of online multifunctional stores that represent all the needs and requirements of customers, and the competition between them leads to the emergence of new ideas and the formation of new approaches to their design. This requires research and consideration of public spaces development features in the context of shopping complexes formation:

1. Changes in the trading system require a new approach to the design of shopping centers, the need to fill them with additional functions, their competent architectural formation based on a kind of "magnet-blocks";
2. The transition to modern economic relations has shown that the most flexible, capable of a high adaptation level, even in rough conditions of Ukrainian reality, is the trade, for this reason the shopping center becomes one of the most popular types of buildings;
3. The most successful in terms of economic feasibility multifunctional shopping and entertainment centers that absorb many objects of different purpose on the basis of trade "magnets," kind of anchors-centers of gravity, are difficult in their planning structure and functional technological objects;
4. At the moment, there is a tendency to mix integrated and suburban shopping centers—modern large, multifunctional complexes can involve both features at the same time.

Multifunctional shopping centers become in the minds of people the center of social life, which can also carry a socio-cultural function [1].

2.3 *Functional Solution of Shopping Center Space*

In the initial process of preparation for functional solutions, first of all, it is necessary to decide on flexible modular systems, a feature that directly affects the sustainable space of the shopping center or the composition of its premises. The modular principle should be developed with the possibility of multiplying it several times in order to form a functional model. Therefore, a convenient geometric module does not mean a static ideal unit and will continue to change over time. However, taking into account the trends and initial design concepts, the modular design should be flexible in size, focusing on a sales area that would fit more typical design decisions. Moreover, as a rule, referring to approved standards and regulations is aimed at achieving international standards of conceptual design. Consequently, it is possible that modular mall units will achieve interconnected functional hybridization [5].

One of the main means of analyzing, adopting, adapting and improving pedestrian traffic in shopping areas should be based on empirical observation of the number of persons in the most frequently visited areas of sale (Fig. 1).

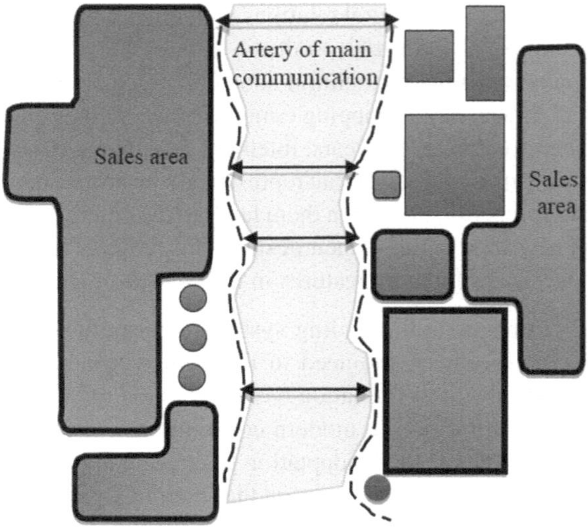

Fig. 1 Functional scheme for points of interest determining

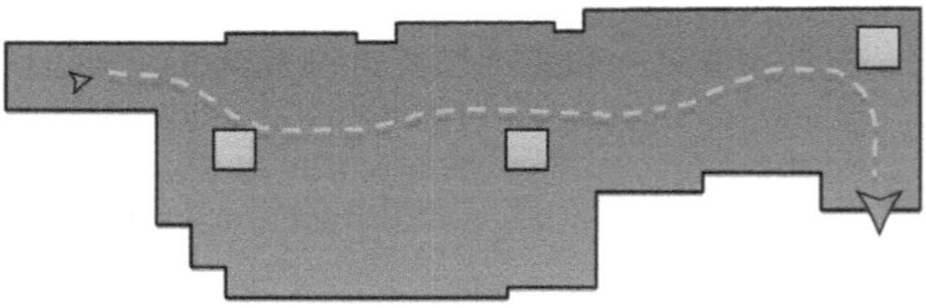

Fig. 2 Pedestrian circulation determines the mall geometry

Retail business advantageous zones are the ones that are perceived as the most important for sale: brand, exclusivity and price, or checkpoints as the most attractive outlets with other major leisure destinations (Fig. 2).

Usually, the most important areas are found with the main stores, while the interest in shopping comes through communications, unless a new attractiveness or the most interesting new object (point) appears [6].

3 Determination of Dimensions and Communication Spaces Area

An integral part of the functional process (Fig. 3) is the construction of a stream of people rational movement, compliance with fire regulations in the context of visitors evacuation, taking into account the issues of servicing the population with various disabilities. While designing shopping complexes, it is necessary to create a comfortable and convenient environment, aesthetic spaces of interiors (Fig. 4) [7].

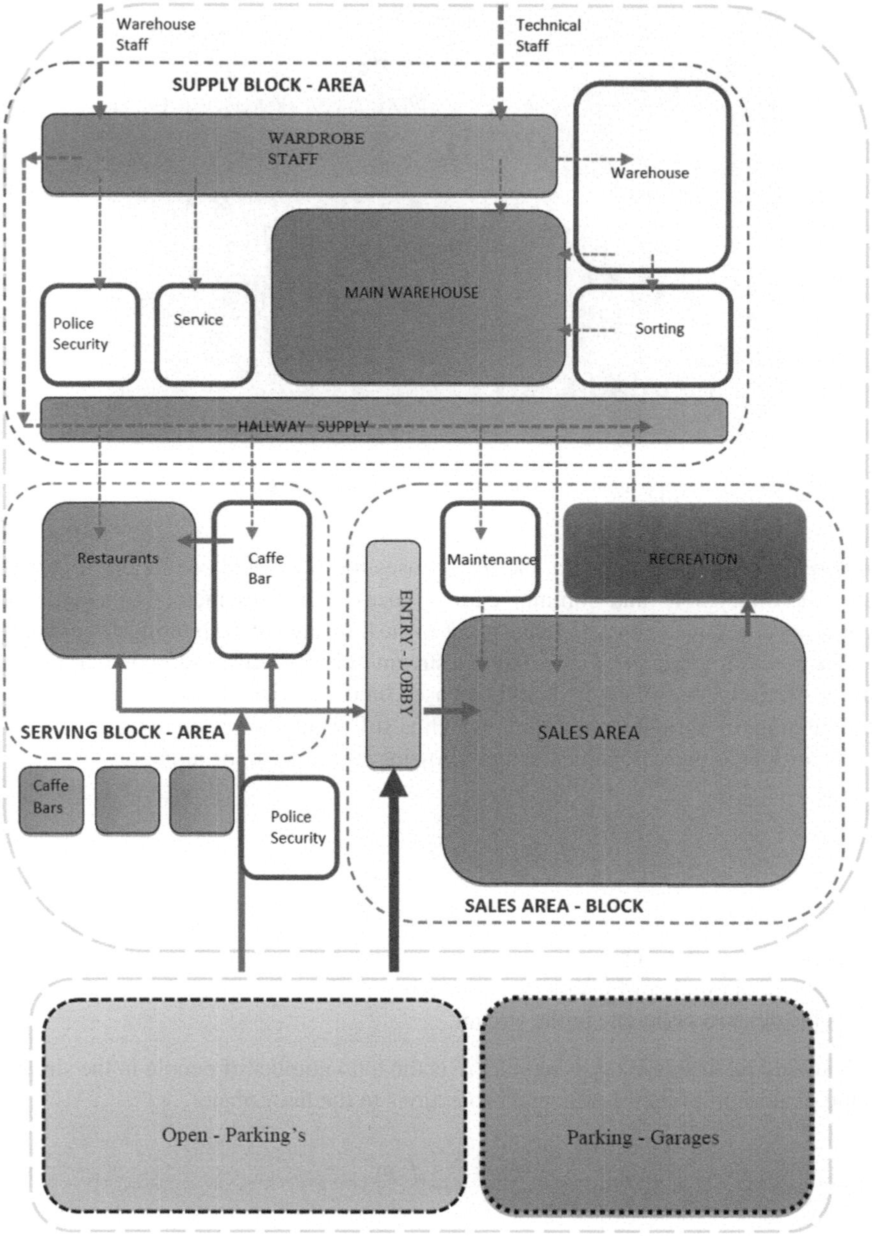

Fig. 3 Scheme of shopping malls functional organization

Fig. 4 Comfortable shopping malls formation for all segments of people

Foreign experience (Austria, Germany) uses the floor area coefficient to determine the dimensions and communication spaces area. This factor is the ratio of communication spaces area and transitions to the floor area of the building (complex).

The domestic practice of calculating a stream of people distribution in communication spaces is absent, but it should be based on the Soviet experience as well as a large amount of normative literature, which is still valid documentation [8].

According to these documents, the following formula can be applied:

$$S = \frac{N}{D}, \mathrm{m}^2 \tag{1}$$

where

S is the area of the path,
N is the total number of people in the stream,
D is the density of people in the stream.

It should be further established that N is the total number of people in the stream as the plane sum of their horizontal projections to the floor plane:

$$N = \sum f, \mathrm{m}^2 \tag{2}$$

where

f is the area of one person horizontal projection, m^2

Using the statistics of DBN data [9], we determine the sum of the horizontal plane of people projections involved in a stream of people, taken, according to the norms, comfortable flow density for business purposes 0.2–0.3 persons/m^2, for the purpose of purchases 0.12–0.2 persons/m^2, recreational movement 0.05–0.15 persons/m^2, and determine the communication space necessary parameters.

Analyzing the calculation results, the next step is to determine the ratio of the total area of the shopping centers to the communication areas, which will allow in the initial stages of design to provide economic and functional feasibility evaluation about chosen variant quality of object planning decision.

$$\frac{S_{k.p.}}{S_{m.k.}}, \text{m}^2 \tag{3}$$

where

- $S_{k.p}$ the total area of the multifunctional complex,
- $S_{m.k.}$ the total area of all communication zones, which is determined by following formula:

$$S_{m.k.} = \frac{N_m}{D_m} + \frac{N_a}{D_a} + \frac{N_0}{D_0} + \frac{N_r}{D_r} + \cdots + \frac{N_n}{D_n}, \tag{4}$$

- N_m simultaneous amount of retail and service enterprise visitors;
- N_a administrative agency visitors amount;
- N_o simultaneous amount of office and business areas visitors;
- N_r simultaneous amount of recreation area visitors;
- N_n the number of visitors to other areas of shopping complex.

Simultaneous visitors number is setting individually for each trading company, based on design, capacity and pass-through function.

D_m, D_a, D_o, D_r, D_n—stream of people density for each area.

The basis for calculating the evacuation of people should be based on Formula (5) [8], which accepts the following coefficients $\eta - 1$, the deaf, hearing impaired people and people with severe speech impairment η – 0.85, people with mental disabilities η – 0.75, people with visual impairment η – 0.62, people with limb defects (two additional artificial supports) η – 0.35, etc.

$$t = \sum \frac{l}{\nu \mu \eta} + \sum \frac{N}{\mu \eta}\left[\frac{1}{Q_{n+1}} + \frac{1}{Q_n}\right] \leq t_d, \text{ min} \tag{5}$$

where

- t evacuation time;
- l length movement, m/min;
- v speed movement, m/min;
- N estimated amount of people in a stream, m^2;
- Q pass-through function, m^2/min.
- μ coefficient of traffic conditions (emergency μ = 1.2–1.5; normal μ = 1, comfortable $\mu = 0.8$)

η coefficient of traffic conditions depending on physical aspects of particular groups of people (youth, children, disabled, etc.)

Pass-through function (Q)—the number of people passing per unit of time across the intersection of paths (the segment of the path with the smallest length).

$$Q = D_{\nu\sigma}, \text{ m/min} \tag{6}$$

where

δ the width of the path section, m.

In buildings, it is advisable to avoid crowds and ensure uninterrupted movement by

$$Q_n = Q_{n+1}, \text{ m}^2/\text{min} \tag{7}$$

Then

$$t = Q \sum \frac{l}{\nu\mu\eta} \leq t_d, \text{ min} \tag{8}$$

It is advisable to use such a shortened formula for variable architectural design [10].

Insufficient study of typological and architectural and spatial public spaces organization problem in the context of shopping complexes formation: necessitated the consideration of new trends in their design, functional zoning, planning and architectural and compositional modeling. Public space in large shopping malls is a matter of individual design and its composition is entirely author-determined. But in the process of designing, it should be guided by a system of general compositional principles, specific examples of practice, based on generalization. *Research methods* and research were obtained by reviewing the literature consisting of empirical observation through spatial planning with a focus on shopping mall structures.

The *aim* of this study was to emphasize the necessity for architectural design as a creative artistic process that must take into account all "sides": good architectural design with optimized space use, aimed at profitability and a uniquely comfortable, convenient, safe environment.

4 Conclusion

As a conclusion, we state that shopping center (complex) for the modern man becomes a place where not just "strip mall," but also various entertainment and socio-cultural zones, where you can spend your leisure time are concentrated. Trading function as a kind of "magnet" attracts other functions, thus, concentrating around the space where each person can find a certain occupation. Thus, modern types of shopping malls should have flexible, easy-to-transform architectural solutions, be able to grow in size due to the growing number of goods and services, multifunctional, and accessible in any area of the city requiring the addition of small, medium or large public and commercial areas. "Trade civilization" became one of the components of entire urban space reorganization, the logic of which is derived from the need to increase urban environment quality.

References

1. Rahimi, A., & Khazaei, F. (2018). Designing shopping centers: The position of social interactions. *Journal of History Culture and Art Research, 7*(2), 239–248. https://doi.org/10.7596/taksad.v7i2.1588.
2. Гослинг, Д. (1979). Проектирование торговых комплексов: пер. с англ./ Д. Гослинг, Б. Мэйтленд; Ред. И. Р. Федосеева, Пер. Д.Г. Копелянский. М.: Стройиздат, 136 с.
3. Малиборская, Г., & Иванов, С. (2008). Торговые центры и их оценка./ Под общ. ред. Я. И. Маркуса. К., 52 с.
4. Jacobs, J. (1993). The city unbound: Qualitative approaches to the city. *Urban Studies, 30*(№ 4/5), 827–848.
5. Alfeld, E. (1995). *System dynamics review* (Vol. 11(3), pp. 199–217). New York: Wiley. CCC 0883-7066/95/030199-19.
6. Bujar, B. (2016). Architectural conceptual design—The sustainable shopping malls structures. *European Journal of Technology and Design, 14*(4), 136–143. https://doi.org/10.13187/ejtd.2016.14.136 www.ejournal4.com.
7. Elif, Ö., & Pelin C. (2016). Public space characteristics of contemporary shopping centers: The case of Mersin forum shopping center. *International Refereed Journal of Design and Architecture*, 55–72. doi: 10.17365/TMD.2016716507.
8. Предтенский, В., & Малинский, А. (1984). Проектирование зданий с учетом движения людских потоков. М., Стройиздат, с. 375.
9. ДБН В.2.2-10:2017 Будинки і споруди. Заклади охорони здоров'я. К.: Мінрегіон України, с. 215 (2017).
10. Lytvynenko, T., Tkachenko, I., & Gasenko, L. (2017). Principles of the road beautification elements placing. *Periodica Polytechnica Transportation Engineering, 45*(2), 94–100. https://doi.org/10.3311/PPtr.8592.

Proposals Design of Steel Storage Tanks for Gas and Oil Products

Kseniia Chichulina and Viktor Chichulin

Abstract The article presents general recommendations for the calculation of steel tanks for gas and petroleum products storage. The analysis of the existing structural forms of tanks is carried out and their classification is provided. The paper analyzes the main types and features of tanks. Optimal dimensions and parameters of steel tanks are given. The article presents coefficients of working conditions for the calculation of vertical cylindrical tanks elements. The article details calculation of the tank wall and negative tolerances on the thickness of the steel sheet. The stress–strain state of the tanks is estimated according to the theory of shell calculation. This assessment takes into account only the ring and meridional stresses, operational snow and wind loads, temperature and aggressive effects. Negative trends of reducing the indicators of bearing capacity, operational reliability and durability of steel tanks were revealed. A variant of calculated diagram of the fluid pressure on the tank wall is presented. Calculation proposals based on the experience of designing steel tanks are formed. An example of calculation of a steel vertical tank is considered. The formula of calculation of the tank wall belt from condition of strength provision (according to the first group of limit states) is presented. The formula for checking the stresses in the lower zone of the tank wall taking into account the action of the boundary moment is also obtained.

Keywords Steel tanks · Gas · Oil products · Calculation · Classification · Stress–strain state · Boundary moment

1 Introduction

Modern market of building structures is saturated with innovative design methods and new forms of design solutions. But today, the requirements for building structures are growing. Considering objects of oil and gas complex, it is very important for design engineer's calculation of storage tanks for gas and petroleum products. It should be noted that there are certain difficulties during the storage of hydrocarbons

K. Chichulina (✉) · V. Chichulin
Poltava National Technical Yuri Kondratyuk University, Poltava, Ukraine
e-mail: Chichulinak@ukr.net

V. Onyshchenko et al. (eds.), *Proceedings of the 2nd International Conference on Building Innovations*, Lecture Notes in Civil Engineering 73,
https://doi.org/10.1007/978-3-030-42939-3_2

due to the qualities of such substances. It is known that they are flammable and explosive, highly sensitive and there is no possibility of changing hydrocarbons characteristics. Choosing the main criteria for creating optimal conditions for oil and gas storage, an important factor is the design of tanks according to modern standards and requirements. The type of placement and design features of tanks are important. A general factor is the physical and chemical characteristics of the tank material. Existing types of tanks can be divided into: method of location in space, namely vertical and horizontal, surface and underground. The shape of tanks is divided into rectangular, cylindrical and teardrop; thickness: single-walled and double-walled. It is also important to identify key metrics that will help you keep track of products. In particular, the mass of oil in the tank is also important. In general, there are several ways to determine the mass. In traditional cases, the tank volume and density of petroleum products are taken into account. Let us consider the options for safe operation of tanks. One such option is regular cleaning and monitoring of the existing tank condition. It provides an opportunity to prevent possible accidents and fire situations, as well as the detection of defects at the initial stage. Thus, the design of steel tanks for gas and oil products storage and their further maintenance is very relevant and meets modern challenges.

General trends in design, classification and characteristics of tanks are presented in the works [1–10]. In the article [1], storage tanks characterization based on conventional science and engineering analysis are presented. It is well known that new science has numerous contradictions and unfounded premises that make it difficult to include new synthesis. This article [1] introduces a fundamental science of mass and energy that enables readers to grasp the concept of natural energy and mass. This forms a background for characterizing unconventional gas reserves and ranks them according to their sustainability and environmental integrity. When conventional oil and gas reservoirs are reconsidered based on their geology, they can be linked to added resources of unconventional oil and gas. It creates opportunities for new developments with little additional cost of reservoir development. The paper [2] reviews 242 accidents of storage tanks that occurred in industrial facilities over last 40 years. Fishbone diagram is applied to analyze the causes that lead to accidents. Corrective actions are also provided to help operating engineers handling similar situations in the future. The results show that 74% of accidents occurred in petroleum refineries, oil terminals or storage. Fire and explosion account for 85% of the accidents. There were 80 accidents (33%) caused by lightning and 72 (30%) caused by human errors including poor operations and maintenance. Other causes were equipment failure, sabotage, crack and rupture, leak and line rupture, static electricity, open flames, etc. Most of those accidents would have been avoided if good engineering has been practiced. In [3], oil storage tanks play a very important role in economic and social development; however, there are numerous cases of damage caused by wind and earthquake. Considering fluid-structure interaction, wind interference and material nonlinearity, firstly, wind fields are simulated by computational fluid dynamics and solved by SIMPLEC method under the conditions of single tank and double tanks; second, the wind pressures in the two cases are obtained and applied to the oil storage tanks, 3D numerical calculation model is established.

Wind disturbance effect on dynamic responses is discussed, and influence of storage ratio and seismic wave type on dynamic responses under wind–earthquake action is further investigated. Results show that wind disturbance effect has a great influence on the dynamic responses of liquid storage tank, especially in the empty state. When the liquid storage ratio is large, traditional oil storage tank is easy to be damaged under the action of wind–strong earthquake. Wind interference effect should be considered in the design and application of oil storage tanks; at the same time, shock absorption and strengthening measures should be taken for oil storage tanks under wind–strong earthquake. The article [4] revealed that aboveground storage tanks are usually empty during construction, inspection and repair periods; they are more vulnerable to buckling due to wind loading than they are in use filled with liquid product. The uplift effect of the bottom plate of empty storage tanks due to wind pressure is investigated in this work. The buckling behaviors of two sets of tanks were studied using finite element analysis. The article [5] describes how tanks used for storage of process fluids have a record of vulnerability to a number of triggering events and their consequences: overfilling, leaks, overpressurization, underpressurization, spills, fires and explosions. Qualified professionals must be involved in the design, layout, maintenance and operation of storage tanks and ancillary equipment. The study [6] investigates the buckling behavior of aboveground storage tanks (ASTs) subjected to storm surge and wave loads during hurricane events and explores the importance of dynamic effects on the buckling behavior. First, a computational fluid dynamics model is developed to estimate water pressures on ASTs subjected to wave loads. The modeling assumptions of this model are also validated with experimental data. Next, a methodology is presented to perform dynamic buckling analysis of ASTs subjected to surge and wave loads by adapting procedures commonly used for ASTs subjected to wind or seismic loads; the methodology is illustrated with a case study AST located in the Houston Ship Channel. For comparison, static buckling analyses are also performed to determine the significance of dynamic effects. Lastly, design of experiments principles and regression analyses is employed to investigate the effects of varying AST and loading parameters on the buckling behavior and the relative importance of dynamic effects. Results indicate that wave loads can significantly affect the buckling behavior of ASTs subjected to storm surge and need to be considered, while the dynamic effects induced by waves have a negligible influence on the buckling strength of ASTs. Simpler and computationally inexpensive static buckling analysis provides reasonable estimates for ASTs subjected to surge and waves. However, dynamic buckling analysis might still be required if the objective is to assess the post-buckling behavior of ASTs subjected to waves, rather than only to estimate the critical load. The article [7] noted that with the development of hydrogen fuel cell vehicles, the on-board hydrogen storage technology with safety, efficiency and economy has become a fundamental part. Low cost, lightweight and good safety performance are required for the on-board hydrogen storage tanks. The composite high-pressure hydrogen storage tank has been recognized as an efficient solution that could address these problems. However, the complex working environment of hydrogen-thermo-mechanism presents challenge to the failure analysis and predictive model establishment of the composite hydrogen storage tanks. The crucial parameters or indicators for tank's failure analysis include burst pressure, damage state and fatigue lifetime, etc. So, this paper gives a comprehensive review on the failure behavior analysis methods and prediction models of composite high-pressure hydrogen storage tanks from the literature. First, the failure analysis methods of composite high-pressure hydrogen storage tanks are summarized. Second, the latest literature regarding failure mode predictive methods and models of type III and type IV tanks are reviewed. The different failure criteria are compared and summarized, including some new failure criteria. These criteria enable failure analysis methods to obtain the interaction information on the interaction between the microscopic and macroscopic aspects of the composite. Damage evolution model and constitutive model are summarized. The post-initial failure behavior of the composite laminates structure is simulated by the material property degradation method (MPDM), especially the continuum damage mechanics (CDM) in conjunction with commercial finite element (FE) analysis method. The process of progressive failure analysis of composite tank is summarized as a reference for subsequent failure analysis. In the work [8], computational fluid dynamics (CFD) is used to investigate the pressurization behavior of cryogenic storage tanks by applying the volume-of-fluid method and taking into account vaporization–condensation phenomena. The boundary conditions are estimated from a 1-dimensional model to solve the heat transfer through the tank insulation layers, eventually taking into account accidental damages. The tank CFD model is preliminary validated against small-scale experimental data obtained for cryogenic nitrogen and then extended to the simulation of an industrial cylindrical tank, whose volume is 100 m^3. The effect of fluid, i.e., ethylene and LNG (modeled as pure methane), filling level and possible insulation damage, on natural convection driving liquid stratification and ultimately tank pressurization is analyzed. Specific performance indexes are proposed to compare efficiently the different scenarios. The chapter [9] covers both atmospheric and pressurized storage tanks, as well as LNG tanks. It covers the codes and regulations required in the design of these tanks. The types of tanks, as well as their sizing, are discussed along with details of their construction. Spill containment for these tanks and the design of berms is explained and how to calculate the berm containment area. Tank heaters and relief vents are explained along with piping examples. Possible risks of living near petroleum storage tanks are presented in [10–13].

2 Main Body

In the course of the study, it was revealed that the tanks are called vessels. These vessels are designed for receiving, storing, processing and dispensing of various liquids, oil, petroleum products, liquefied gases, water, aqueous ammonia, industrial alcohol and the like. Vertical cylindrical tanks (see Fig. 1) are used at overpressure in a steam–air zone to 2 kPa and vacuum to 0.25 kPa. These tanks have flat bottoms made of steel sheets 4–6 mm thick and the walls in the form of a series of belts. The

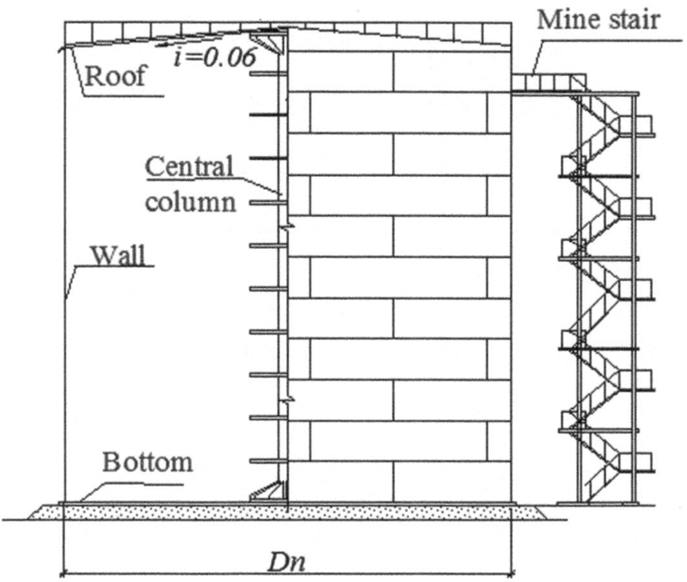

Fig. 1 Facade and section of vertical cylindrical tank

thickness of these belts increases in proportion to the increase in fluid pressure as it approaches the bottom.

The main elements of such a tank are the wall (body), bottom and roof (coating), which are made of sheet steel. The operational equipment of the tank consists of fittings (devices for filling, measuring and releasing liquid, safety valves) and devices for cleaning and inspection (stairs, light and measuring hatches, manholes). Let us consider the dimensions of the tank.

The optimum height of the tank body is determined by the Formula (1):

$$H_{\text{opt}} = \sqrt{\Delta \cdot R_{wy} \cdot \gamma_c / (\gamma_{f,\ell} \cdot \gamma_g)}, \quad (1)$$

where

$\Delta = 0.9$ cm—the sum of the given thickness of the bottom and roof (Table 1);
$R_{wy} = R_y = 24$ kN/cm^2—calculated resistance of the weld [14], taking into account that the welds are performed with physical quality control of the weld;
γ_g—the specific gravity of the fuel is determined by Table 2;

Table 1 Thicknesses for tanks are given

Volume V, Thous. m	1	2	3	4	8	12	16	20
The sum of the given thickness of the bottom and roof Δ (cm)	0.8	0.9	1.05	1.2	1.4	1.6	1.7	1.8

Table 2 Density and specific gravity of liquids stored in tanks

The name of the liquid	Density (kg/m^3)	Specific gravity (kN/cm^3)
Acetone	800	7.85×10^{-6}
Gasoline (light)	700	6.87×10^{-6}
Benzene	880	8.63×10^{-6}
Glycerin	1260	1.24×10^{-5}
Diesel fuel	1000	9.81×10^{-6}
Kerosene	800	7.85×10^{-6}
Rectified alcohol	830	8.14×10^{-6}
Ethanol	790	7.75×10^{-6}
Essential	720	7.06×10^{-6}
Black oil	960	9.4×10^{-6}

$\gamma_{f,\ell} = 1.1$—reliability coefficient on the load from the hydrostatic pressure of the liquid;
γ_c—coefficient of working conditions [14].

This height should be a multiple of the width of the standard sheet, taking into account the edge guard. The desired sweep length is:

$$L = 2\pi\sqrt{\frac{V}{\pi H_0}}, \tag{2}$$

where H_0—filling height of the tank product.

In the arrangement of the tank wall, the radius of the shell is:

$$r = (L - n \cdot \delta)/(2\pi), \tag{3}$$

where

n—number of rolls of the tank wall;
δ—the total length of the blending in one installation the junction (taken in the range 140–200 mm);
$L = -2\pi\sqrt{\frac{V}{\pi H_0}}$—the length of the tank wall.

Actual tank volume is:

$$V_{\text{fact}} = \pi r^2 H_1 \tag{4}$$

The difference with the specified volume is $(V_{\text{fact}}/V - 1)/100\%$.

We present in detail the calculation of the tank wall. The calculation is based on the Formulas (5), (6):

Table 3 Coefficients of working conditions for the calculation of elements of vertical cylindrical tanks

Type and place of calculation	The coefficient of working conditions γ_c
Walls in the calculation of strength	
Lower belt	0.6
Other belt	0.7
Pairing wall with a bottom	1.2
Tank walls in the calculation of stability	1.0
Spherical and conical cover spacer structures calculated according to the membrane theory	0.9

$$t_w = \frac{N_{r,1}\gamma_n}{\gamma_c R_{wy}} + c_1 + c_2 = \frac{[\gamma_g \gamma_{f,\ell} h_p + p_0 \gamma_{f,p} \cdot r]}{\gamma_c R_{wy}} \gamma_n + c_1 + c_2, \tag{5}$$

where

t_w—wall thickness at level x;
$N_{r,1}$—internal force on the belt (when calculating the strength);
γ_c—the coefficient of working conditions according to Table 3;
$h_p = h - h_\ell \cdot (n_i - 1) - x_\ell$—the height of the calculated level;
h—distance to calculated liquid level;
h_ℓ—constructive height of belt (1490, 1790, 1990 mm);
n_i—sequence number of the belt at the bottom of the account;
p_0—characteristic value of overpressure in steam–air medium;
$r = R_2$—radius of tank;
$\gamma_{f,\ell} = 1.1$ and $\gamma_{f,p} = 1.2$—accordingly, the load reliability coefficients for hydrostatic pressure and overpressure of the steam–air mixture:

$$y = \frac{N_{r,2}}{Et_w} r = \frac{[\gamma_g (h - h_\ell \cdot (n_i - 1) - x_\ell) + p_0] \cdot r}{Et_w} r, \tag{6}$$

where

$E = 2.06 \times 10^4$ kN/cm^2—modulus of elasticity of steel;
$N_{r,2}$—the internal stress in the belt (when calculating stiffness);
t_w—wall thickness at the stiffness check point.

It is taken into account that the wall will consist of n belts (variable $n_i \in [1 - 8]$). All belts are recruited from sheets 6000 × 1500 mm when processing the edges of the sheets on each edge of the sheet is removed by 5 mm. Additive corrosion is taken $c_1 = 1.0$ mm; negative tolerances for hire are taken according to Table 4. In addition, the thickness of the belt should not be less than 5 mm. Maximum fill level of the tank is assumed equal to its design height H.

Table 4 Negative tolerances on steel sheet thickness according to [14]

Sheet thickness (mm)	Sheet width (mm)		
	1000–1500	1500–2000	2000–2300
From 3.9 to 5.5	−0.50	−0.50	−0.50
From 5.5 to 7.5	−0.60	−0.60	−0.60
From 7.5 to 25.0	−0.80	−0.80	−0.80
From 25.0 to 30.0	−0.90	−0.90	−0.90
From 30.0 to 34.0	−1.0	−1.0	−1.0

For the first (lower) belt, the wall thickness will be determined as follows. The height of the design level is:

$$h_p = h - h_l(n_i - 1) - x_l. \tag{7}$$

Internal force on the belt (when calculating the strength) is:

$$N_{r,1} = \left[\gamma_r \gamma_{f,l} h_p + p_o \gamma_{f,p}\right] \cdot r. \tag{8}$$

The internal stress in the belt (when calculating stiffness) is:

$$N_{r,2} = \left[\gamma_g h_p + p_o\right] \cdot r. \tag{9}$$

The calculated wall thickness without consideration of tolerances c_1 and c_2 is:

$$t_{w,p} = \frac{N_{r,1}\gamma_n}{\gamma_c R_{wy}}, \tag{10}$$

where R_{wy}—calculated resistance of welds (assumed to be equal to the calculated resistance beyond the yield point R_y at the physical quality control of seams or $0.85R_y$ without physical quality control of seams).

Calculated wall thickness including corrosion tolerance and negative tolerances for rolled products is:

$$t_w = t_{w,p} + c_1 + c_2. \tag{11}$$

For the assortment of plate steel [14], take the thickness of the first (lower) belt t_w.

The actual stress in the body wall with the accepted thickness t_w is:

$$\sigma_2 = \frac{N_{r,1}}{t_w} < \frac{R_{w,y}\gamma_c}{\gamma_n}. \tag{12}$$

Deflection of the body (radial displacement) at the location of the first (lower) belt according to the Formula (13) is:

$$y = \frac{N_{r,2}}{Et_w} r. \quad (13)$$

The calculation of the second and subsequent zones is generally performed in accordance with the above algorithm, but has two features. The first is the need to specify the correction x_ℓ according to the Formula (14), which affects the determination of the height of the calculated level h_p, the second is the use of the coefficient of the working conditions with the value $\gamma_c = 0.7$ (instead of $\gamma_c = 0.6$).

$$x_\ell = 0.6\sqrt{r \cdot t_w}, \quad (14)$$

where

r—radius of tank;
t_w—belt thickness.

Steel tanks for storage of oil and oil products are used as technological objects of oil depots and main oil products. Quite an important element of the design is to ensure the necessary level of reliability. The tank is a vertical shell with a bottom, so the analytes of this type of objects calculation are associated with the momentary theory of shells. But the presence of geometric features, namely: changing the cross sections of the elements, the presence of hatches and holes, tie-in fittings and other elements significantly changes the symmetry of the structure, the distribution of stresses and strains in local areas. Such features complicate analytical calculations.

The traditional approach to assessing the residual life is to process the data of measurements of the tank wall thickness and identify the average and maximum values of corrosion failure. Further, a simple model based on the hypothesis of linear (in time) nature of corrosion failure is used to obtain the value of the resource.

To determine the residual resource the following calculations are performed: (1) test calculations of a tank wall for strength; (2) test calculations of a tank wall for resistance; (3) calculate the residual resource according to the criterion of corrosion; (4) calculation of residual resource according to the criterion of little cyclic fatigue (if necessary).

The residual life of the tank is determined on the basis of the analysis of operating conditions, the results of technical diagnostics and criteria of the limit state. Steel tanks are subjected during operation to periodic loads and corrosion–erosion fracture; therefore, the limiting condition can occur or low cycle fatigue or due to loss of bearing capacity caused by wall. These criteria are the basis for calculating the residual life of steel tanks. The residual resource is determined separately for each criterion, and the minimum of the two values found is assigned as the final one.

The advantage of traditional approaches to the assessment of residual life is the simplicity of calculations and adequate assessment of residual life with the correct use of techniques. The calculations do not take into account the nature of the distribution of destruction on the surface of the tank, which in some cases leads to serious errors, and is the main drawback. From the foregoing, it can be concluded that the evaluation of the stress–strain state of the reserve acres type RVS is based on the membrane theory of shell analysis, and the calculation according to this theory it only considers circular and meridional stresses, and therefore, it is quite difficult to assess the real picture of distribution of all stresses.

Considering the experience of designing steel tanks, we note that the main objects of design are vertical cylindrical single-wall pressure and horizontal multi-faceted non-pressure double-wall tanks. The most common material used is: ordinary structural steel, stainless steel. Common are tanks with a volume of 0.5–250 m^3, a diameter of 1.0–4.5 m, a maximum length of 30 m, weight—up to 30 tons.

In the course of the study, it was revealed that the most popular applications are tanks of gas stations; tanks for chemical, naphtha-chemical industries; tanks for bitumen industry; tanks for oil and oil industry; tanks for wastewater treatment plants; tanks for energy and ecology; tanks for wagons.

Let us consider an example of calculating a steel vertical tank (see Fig. 2).

The belt of the tank wall from the condition of ensuring the strength (according to the first group of limit states) is calculated by the formula:

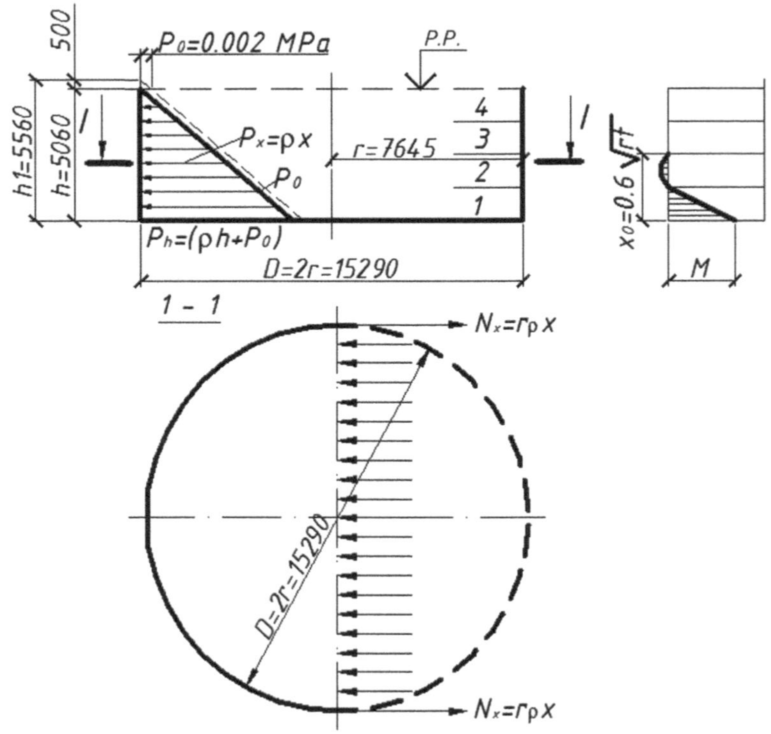

Fig. 2 Variant of the fluid pressure on the wall of the tank: **a** fluid pressure plot; **b** moment plot at the junction of the bottom with the wall (edge effect)

$$\frac{(\gamma_{f1} \cdot \rho \cdot x + \gamma_{f2} \cdot \rho_o) \cdot r}{t \cdot \gamma_c \cdot R_{wy}} \leq 1, \quad \text{or } t \geq \frac{(\gamma_{f1} \cdot \rho \cdot x + \gamma_{f2} \cdot \rho_o) \cdot r}{\gamma_c \cdot R_{wy}}, \tag{15}$$

where

γ_{f1}—load reliability factor for hydrostatic pressure;
γ_{f2}—the reliability coefficient for the load to excess pressure;
$R_{wy} = 0.85 \cdot R_y$—calculated tensile resistance of butt weld; physical quality control of the weld $R_{wy} = R_y$ equal to the yield strength of steel;
γ_c—coefficient of working conditions for the tank wall;
ρ—density of oil products;
ρ_o—the excess pressure of the tank.

We checked the voltage in the lower zone of the tank wall, taking into account the action of the edge moment M_1.

$$\frac{(T_1/t + 6M_1/t^2)}{\gamma_c \cdot R_y} \leq 1, \tag{16}$$

where $M_1 = (\gamma_{f1} \cdot \rho \cdot x_o + \gamma_{f2} \cdot \rho_o) \cdot r \cdot t$; $T_1 = (\gamma_{f1} \cdot \rho \cdot x_o + \gamma_{f2} \cdot \rho_o) \cdot r$.

3 Conclusion

As a result of the conducted research, the calculation general algorithm of steel tanks for hydrocarbons storage was presented. The classification of steel tanks is given, the main requirements for the storage of gas and petroleum products are identified. The study presents the typology and characteristics of existing hydrocarbon storage tanks.

References

1. RafiqulIslam, M. (2014). *Unconventional gas reservoirs. Chapter 6—Scientific characterization of unconventional gas reservoirs* (pp. 337–485). Copyright, Elsevier. https://doi.org/10.1016/B978-0-12-800390-9.00006-2.
2. Chang, J., & Cheng-Chung, L. (2006). A study of storage tank accidents (Internet). *Journal of Loss Prevention in the Process Industries, 19*(1), 51–59. https://doi.org/10.1016/j.jlp.2005.05.015.
3. Jing, W., Wang, J., & Cheng, X. (2019). Dynamic responses of oil storage tank considering wind interference effect. *Engineering Failure Analysis, 104,* 1053–1063. https://doi.org/10.1016/j.engfailanal.2019.06.040.
4. Hu, W. Bohra, H., Azzuni, E., & Guzey, S. (2019). The uplift effect of bottom plate of aboveground storage tanks subjected to wind loading. *Thin-Walled Structures, 144*. https://doi.org/10.1016/j.tws.2019.106241.

5. Kletz, T., Amyotte, P. (2014). *What went wrong? Chapter 11—Storage tanks* (pp. 210–234). Copyright, Elsevier. https://doi.org/10.1016/B978-0-12-810539-9.00011-2.
6. Bernier, C., & Padgett, J. E. (2019). Buckling of aboveground storage tanks subjected to storm surge and wave loads. *Engineering Structures, 197*. https://doi.org/10.1016/j.engstruct.2019.109388.
7. Zhang, M., Lv, H., Kang, H., Zhou, W., & Zhang, C. A. (2019). Literature review of failure prediction and analysis methods for composite high-pressure hydrogen storage tanks. *International Journal of Hydrogen Energy*. https://doi.org/10.1016/j.ijhydene.2019.08.001.
8. Ovidi, F., Pagni, E., & Landucci, G. (2019). Numerical study of pressure build-up in vertical tanks for cryogenic flammables storage. *Applied Thermal Engineering, 161*. https://doi.org/10.1016/j.applthermaleng.2019.114079.
9. Barker, G. (2019). *The Engineer's guide to plant layout and piping design for the oil and gas industries. Chapter 15—Storage tanks* (pp. 361–379). Copyright, Elsevier. https://doi.org/10.1016/B978-0-12-814653-8.00015-1
10. Zusman, M., Dubnov, J., Barchana, M., & Portnov, B. A. (2012). Residential proximity to petroleum storage tanks and associated cancer risks: Double Kernel Density approach vs. zonal estimates. *Science of The Total Environment, 441*, 265–276. https://doi.org/10.1016/j.scitotenv.2012.09.054.
11. Piskunov, V. G., Goryk, A. V., Lyakhov, A. L., & Cherednikov, V. N. (2000). High-order model of the stress-strain state of composite bars and its implementation by computer algebra. *Composite Structures, 48*(1), 169–176. https://doi.org/10.1016/S0263-8223(99)00091-4.
12. Piskunov, V. G., Goryk, A. V., & Cherednikov, V. N. (2000). Modeling of transverse shears of piecewise homogeneous composite bars using an iterative process with account of tangential loads. 1. Construction of a model. *Mechanics of Composite Materials, 36*(4), 287–296. https://doi.org/10.1007/bf02262807
13. Chichulin, V., & Chichulina, K. (2017). Deriving a function of the bending axis of a profiled wall in the form of orthotropic plate. *Eastern-European Journal of Enterprise Technologies, 5*(7–89), 30–37. https://doi.org/10.15587/1729-4061.2017.109687.
14. DBN V.2.6-163:2014. (2014). *Construction of buildings and structures. Steel structure. Standards of design, manufacture and installation*. Kyiv, 196.

On Clarification of the Application Area of the Concrete Plasticity Theory to the Strength Problems Solutions

O. O. Dovzhenko, V. V. Pohribnyi, V. F. Pents, and M. V. Pents

Abstract To substantiate plasticity theory application to the strength problems solution of structural concrete, concrete plates under different load schemes, truncated wedges over a dangerous inclined crack, as well as Gvozdev samples as a basis for defining shear resistance, have been studied. The virtual velocities principle has been applied. Concrete is regarded as a rigid-plastic body. Plastic strain is considered to be localized in thin layers on the elements failure surface. The function of the virtual velocities principle is dealt at stationary state. The failure kinematic schemes and the calculated dependencies for the specified samples strength evaluation have been given. The ultimate estimation of the upper load is used. The stresses are defined at the characteristic points of the concrete strength condition, which is considered as plastic potential. The stress levels at the boundary of the shear and breaking-off failure forms have been obtained. The concrete plasticity measure is established as the ratio of the compressed zone height in the normal section in the failure stage to its height in the initial stage. Taking into account the plasticity measure, the stresses on the failure surface compressed area have been determined. The interval of the stress states in the tension-compression area is narrowed compared to the interval for plastic materials. The established boundary between the shear and the breaking off is confirmed experimentally. The plasticity theory application area for the strength problems solution is specified.

Keywords Strength condition · Virtual velocities principle · Shear · Plasticity measure · Stress interval · Implementation boundary

1 Introduction

While calculating concrete and reinforced concrete elements strength, empirical dependencies have been widely used, which does not always provide the required accuracy. This is due to the narrow enough area of their possible application, limited by the experimental conditions. Therefore, the extension of these dependencies to

O. O. Dovzhenko (✉) · V. V. Pohribnyi · V. F. Pents · M. V. Pents
Poltava National Technical Yuri Kondratyuk University, Poltava, Ukraine
e-mail: o.o.dovzhenko@gmail.com

V. Onyshchenko et al. (eds.), *Proceedings of the 2nd International Conference on Building Innovations*, Lecture Notes in Civil Engineering 73,
https://doi.org/10.1007/978-3-030-42939-3_3

constructively close elements is not always justified. Concrete and reinforced concrete strength theory should be generalizable, taking into account all the determining factors, and should be based on the modern mechanics of deformable solids. Recently, significant results have been achieved in its development [1, 2]. The application of brittle failure mechanics [3] is promising at breaking off, and the plasticity theory is promising at the shear [4–10].

In most cases, it is difficult to distinguish the shear from the breaking off. Meanwhile, the mechanisms of these failure forms are radically different [10–20]. For the shear implementation, a prerequisite is the presence of plastic strain on the failure surface, which is possible only when the compression stresses are dominating. Plastic properties of concrete under different stress states have significantly different effect on the strength. Establishing a concrete plasticity measure will enable reasonably to limit the area of the strength condition under the shear.

The strength problems study is carried out by using a variational method in the ideal plasticity theory, with the application of the virtual velocities principle, discontinuous solutions, and upper estimation of ultimate load level.

The purpose of this research is to clarify the area of justified application of the plasticity theory to solve concrete strength problems by setting stress states intervals and taking into account concrete plasticity measure.

2 Strength Problem Solving

To solve the tasks mentioned, the Balandin–Geniyev condition as a strength condition in a two-dimensional stress state is used [4], which has a simple recording form, is experimentally confirmed, generalizes the theory of Mises–Genk on brittle materials, and is written as

$$\sigma_1^2 - \sigma_1\sigma_2 + \sigma_2^2 - (\sigma_1 + \sigma_2)(f_c - f_{ct}) - f_c f_{ct} = 0, \quad (1)$$

where f_c and f_{ct} areaxial compressive and tensile strengths of concrete.

In the coordinates τ_n–σ_n in the area of actually existing sliding planes, it is the Mohr circles envelope and is written as

$$|\tau_n| = \varphi(\sigma_n) = \sqrt{(f_c^2 - f_c f_{ct} + f_{ct}^2)/3 - 0.25(\sigma_n - f_c + f_{ct})^2}. \quad (2)$$

The tangent to the envelope at each point extends at an angle ψ. This angle uniquely determines:

$$\text{normal } \frac{\sigma_n}{f_{c,\text{prism}}} = 1 - f_{ct}/f_c - \frac{4}{\sqrt{3}}\sqrt{1 - f_{ct}/f_c + (f_{ct}/f_c)^2}\frac{\tan\psi}{\sqrt{1 + 4\tan^2\psi}}; \quad (3)$$

$$\text{tan gent } \frac{\tau_n}{f_{c,\text{prism}}} = \frac{1}{\sqrt{3}}\sqrt{\frac{1 - f_{ct}/f_c + (f_{ct}/f_c)^2}{1 + 4\tan^2\psi}}; \tag{4}$$

$$\text{main } \sigma_1 = \sigma_n + \tau_n(\tan\psi + \sqrt{1 + \tan^2\psi}), \tag{5}$$

$$\sigma_2 = \sigma_n + \tau_n(\tan\psi - \sqrt{1 + \tan^2\psi}) \tag{6}$$

stresses in the element failure zone according to the accepted strength condition.

It is proposed to determine the application boundary of the plasticity theory for concrete in the zone of mixed stress states that is corresponding to the condition of axial tension for plastic materials ($\psi = 19.47°$). Stress values at characteristic points of the strength condition are given in Table 1.

To confirm the area of plasticity theory application, different strength problems of the concrete elements are considered: prism, truncated wedge, Gvozdev samples, and plate with local load application. All these elements according to experimental studies [15–17, 19] can be destroyed both by the shear as well as by the breaking off. The limit of these failure cases is the boundary of plasticity theory application to their calculations. It is important to take into account the features of the ultimate state development in the presence or absence of tensile zones.

Table 1 Relative stress values at characteristic points of the Balandin–Geniyev strength condition on the area of actually existing planes of sliding

f_{ct}/f_c	Relative stresses	Boundary in the area of mixed stress states	Uniaxial compression	Maximum tangential stresses	Boundary in the area of biaxial compression
0.15	σ_1/f_c	0.311	1	1.39	1.78
	σ_2/f_c	−0.229	0	0.311	0.85
	σ_n/f_c	−0.229	0.383	0.85	1.47
	τ_n/f_c	0	0.486	0.539	0.44
0.1	σ_1/f_c	0.349	1	1.45	1.85
	σ_2/f_c	−0.201	0	0.349	0.9
	σ_n/f_c	−0.201	0.367	0.9	0.551
	τ_n/f_c	0	0.482	0.551	0.45
0.05	σ_1/f_c	0.387	1	1.51	1.93
	σ_2/f_c	−0.177	0	0.387	0.95
	σ_n/f_c	−0.177	0.35	0.95	1.6
	τ_n/f_c	0	0.477	0.564	0.46

When solving strength problems by the variational method in plasticity theory [10, 15–20], strength condition [4] is considered as a plastic potential and for convenience is written in a tensor form

$$F(\sigma_{i\,j}) = T^2 - m\sigma - T_{sh}^2 = 0, \tag{7}$$

where T—the intensity of the tangent stresses, σ—mean stress, $m = f_c - f_{ct}$, $T_{sh}^2 = f_c f_{ct}/3$.

Concrete is regarded as a rigid-plastic body. The virtual velocities principle is applied.

The function of the principle J under the condition of plastic strains localization in thin layers on the velocity rupture surface (failure surface) has the general form

$$J = \int\limits_{S_1} \left[T\sqrt{4\Delta V_n'^2/3 + \Delta V_t'^2} + \sigma \Delta V_n' \right] \mathrm{d}S - F_i V_i', \tag{8}$$

where F_i—loading parameter; $\Delta V_n'$ and $\Delta V_t'$—jumps in normal and tangential directions; and V_i'—velocity in the loading direction.

For a two-dimensional stress state, the function has the following form

$$J = f_c \int\limits_{S_l} \left\{ \left[\sqrt{(1 - f_{ct}/f_c + (f_{ct}/f_c)^2)/3}\sqrt{4 + (\Delta V_t'/\Delta V_n')^2} - 1 + f_{ct}/f_c \right] \Delta V_n' \right\} \mathrm{d}S - F_i V_i'. \tag{9}$$

The character and kinematic failure schemes of elements and the results of their calculations by dependencies (10–13) are shown in Figs. 1, 2, and 3 and in Tables 2, 3, 4, and 5.

The results of the experiments show that in the elements of high-strength concrete with axial compression (Fig. 1a), a failure shear form is developed [19]. This indicates the destruction possibility from the shear in the mixed stresses zone when hiring concrete with less strength, which has greater plastic properties.

The criterion for applying the plasticity theory to solve the concrete strength problem at axial compression is the presence on the downward branch of the strain diagram near its top of the area, on which the deformation level growth exceeds the stress reduction level.

When $\sigma_2 \neq 0$, the stresses in concrete at the boundary of the shear and breaking off in the mixed stress states zone are given in Table 2.

$$\sigma_1/f_c = \frac{\sqrt{(1 - \chi + \chi^2)/3}\sqrt{4(\kappa - \tan\gamma)^2 + (1 + \kappa\tan\gamma)^2} - (1 - \chi)(\kappa - \tan\gamma) + \kappa\sigma_2/f_c}{\tan\gamma}, \tag{10}$$

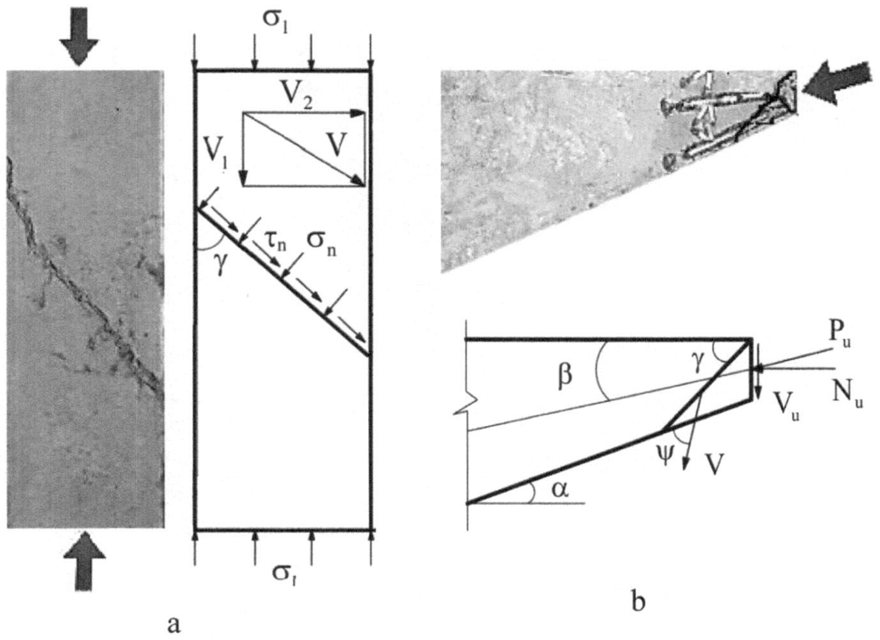

Fig. 1 Failure character and the kinematic scheme of the concrete prism (a) and the truncated wedge (b)

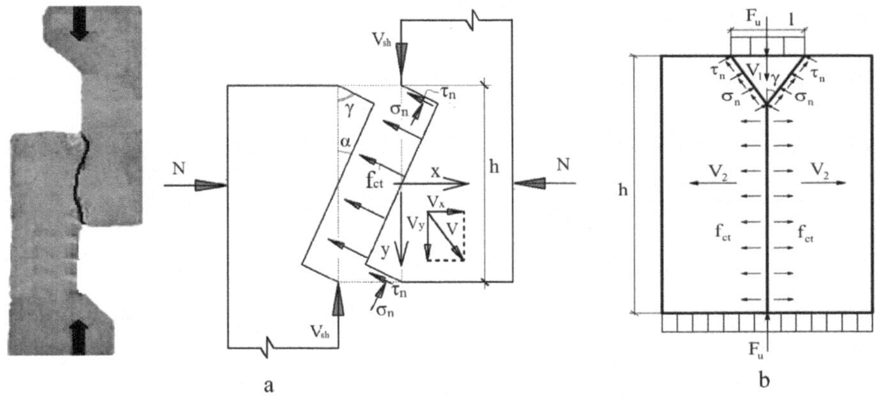

Fig. 2 Failure character and the kinematic scheme of the failure of the Gvozdev sample (a) and the plate under local load application (b)

where $\chi = f_{ct}/f_c$, $\kappa = V_2/V_1$.

Analysis of the problems solution results indicates shear development in concrete wedges (Fig. 1b, Table 3) by the condition of $\beta < \alpha + 10°$, which is confirmed by the data [16].

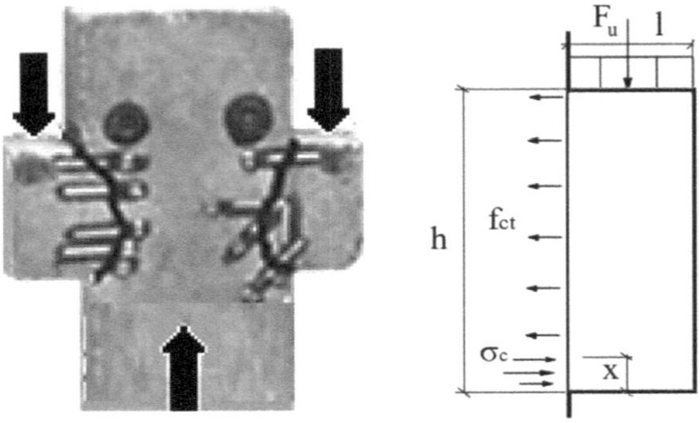

Fig. 3 Failure character of short elements and the stresses distribution in the normal section at the destruction boundary from breaking off and shear

Table 2 Values of the varying parameters and the relative stresses in the mixed stress states zone at the boundary of concrete plasticity theory application by the solution of the plate strength problem

f_{ct}/f_c	κ	tan γ	ψ (°)	σ_1/f_c	σ_2/f_c	σ_n/f_c	τ_n/f_c
0.15	1.41	0.707	19.47	0.85	−0.0841	0.227	0.44
0.1				0.9	−0.0539	0.264	0.45
0.05				0.95	−0.0259	0.3	0.46

Table 3 Values of the varying parameters and the relative stresses in the mixed stress states zone at the boundary of concrete plasticity theory application by the solution of the truncated wedge strength problem at $f_{ct}/f_c = 0.1$

α (°)	β (°)	κ	tan γ	ψ (°)	σ_n/f_c	τ_n/f_c
0	10	1.72	0.845	19.59	0.261	0.449
15	25	3.68	1.44			
30	40	281.98	2.78			

Table 4 Value of the inclination failure surface angle γ and the relative stresses in the mixed stress states zone at the boundary of concrete plasticity theory application by the solution of the Gvozdev sample strength problem

f_{ct}/f_c	γ (°)	ψ (°)	σ_n/f_c	τ_n/f_c
0.15	→ 0	18.42	0.252	0.449
0.1		18.05	0.274	0.453
0.05		17.96	0.337	0.473

Table 5 Values of the varying parameters and the relative stresses at the boundary of concrete plasticity theory application by the solution of the plate under local one-sided load application strength problem

f_{ct}/f_c	t = h/l	κ	tan γ	ψ (°)	σ_n/f_c	τ_n/f_c
0.15	0.875	1.11	0.57	18.24	0.256	0.45
0.1	0.82	1.2	0.61	18.93	0.277	0.454
0.05	0.763	1.31	0.656	19.34	0.303	0.461

$$N_u/f_c b\delta = \frac{\left[\sqrt{(1-\chi+\chi^2)/3}\sqrt{4(\kappa-\tan\gamma)^2+(1+\kappa\tan\gamma)^2}-(1-\chi)(\kappa-\tan\gamma)\right]}{(1-\kappa\tan\beta)(\tan\gamma-\tan\alpha)}, \tag{11}$$

where δ—wedge thickness.

By increasing the compression force and the ratio of l/h, respectively, in the Gvozdev sample (Fig. 2a) and the plate under local compression (Fig. 2b), the tensile areas decrease and in their absence, the breaking off is impossible.

$$\frac{f_{sh}}{f_c} = \frac{\tan\alpha}{\tan\alpha+\tan\gamma} \left[\sqrt{\frac{1-\chi+\chi^2}{3}}\sqrt{4(\kappa-\tan\gamma)^2+(1+\kappa\tan\gamma)^2}-(1-\chi)(\kappa-\tan\gamma)+\chi\tan\gamma(1+\frac{\kappa}{\tan\alpha})\right]+\kappa\frac{f_{ct}}{f_c}, \tag{12}$$

$$\frac{\sigma_1}{f_c} = \frac{1}{\tan\gamma} \left[\sqrt{(1-\chi+\chi^2)/3}\sqrt{4(\kappa-\tan\gamma)^2+(1+\kappa\tan\gamma)^2}-(1-\chi)(\kappa-\tan\gamma)\right] +\kappa\chi\left(2t-\frac{1}{\tan\gamma}\right). \tag{13}$$

3 Definition of Concrete Plasticity Measure

It is proposed to apply the ratio of the compressed zone in the normal section at the failure stage ξ_u and the initial stage of work ξ_{el} to establish the concrete plasticity measure. Taking a curvilinear plane in the compressed zone (Fig. 3) and taking into account its completeness coefficient, we have

$$\mu_{pl} = \frac{\xi_u}{\xi_{el}} = \frac{2f_{ct}}{(f_c+f_{ct})\omega}. \tag{14}$$

Then, the development area of the shear failure form at mixed concrete stress states is limited by the tensile stresses value

$$\sigma_2 = \mu_{pl}\sigma_{2,\max} = \mu_{pl}[f_c - 2\sqrt{(f_c^2 - f_c f_{ct} + f_{ct}^2)/3} - f_{ct}]. \tag{15}$$

Under compression and tension areas in the failure zone, a simultaneous existence of a limit state on the entire failure surface is required for the shear development. The specified simultaneity for flexural elements depends on the ratio of bending moment to shear force and is established experimentally. Thus, for short elements

(Fig. 3), with the dominant influence of the shear force failure from shear occur at $l/h = 0.5$[20].

Scientific novelty of the research lies in developing a sufficiently general method of calculating concrete and reinforced concrete elements strength on the basis of the ideal plasticity theory with the definition of its reasonable application area. Widespread application of structures that work on the shear forces perception marks the practical significance of the research.

4 Conclusions

1. Concrete plasticity theory can be successfully applied to solve the strength problems in uneven compression stress states and in a limited range of stresses in the tension-compression area.
2. The application limits of concrete plasticity theory in the mixed stress states zone according to the results of calculating concrete elements strength under different load schemes correspond to the main compressive stress values $\sigma_1 = f_c - f_{ct}$ and tangential stresses $\tau_n = \sqrt{2/3}\,\tau_{n,\max}$. Taking into account concrete plasticity measure, the ultimate value of the main tensile stresses is $\sigma_2 = \mu_{pl}\sigma_{2,\max}$.
3. The stress value at the shear failure boundary in the non-uniform compression zone at plastic strain localization in thin layers on the failure surface is $\tau_{n,\max} = \sqrt{(f_c^2 - f_c f_{ct} + f_{ct}^2)/3}\; i\; \sigma_n = f_c - f_{ct}$. Increasing the level of σ_n is possible with the plasticity zones expansion.
4. In case of presence of compression areas as well as tensile areas in the failure zone, the criterion for plasticity theory application is the simultaneous existence of an ultimate stress state on the entire failure surface.

References

1. Eliseyev, V. V. (2006). *Mechanics of deformable solids (Mekhanika deformiruyemogo tverdogo tela)*. Sankt-Peterburg: SPb.
2. Okopnyy, Y. A., Radin, V. P., & Chirkov V. P. (2001). Materials and structures mechanics (Mekhanika materialov i konstruktsiy). Moscow.
3. Zaytsev, Yu V. (1982). *Modeling of deformations and concrete strength by methods of fracture mechanics (Modelirovaniye deformatsiy i prochnosti betona metodami mekhaniki razrusheniy)*. Moscow: Stroyizdat.
4. Geniyev, G. A., Kissyuk, V. N., & Tyupin, G. A. (1974). Concrete and reinforced concrete plasticity theory (Teoriya plastichnosti betona i zhelezobetona). Moscow.
5. Kolmogorov, V. L. (1986). Metal forming mechanics (Mekhanika obrabotki metallov davleniyem). Moscow.
6. Mineola, L. J. (2008). *Plasticity theory*. NY: Dover.
7. Nielsen, M. P., & Hoang L. C. (2011). Limit analysis and concrete plasticity (3rd ed.). Boca Raton: CRC Press, Taylor & Francis Group.

8. Ivlev, D. D. (2001). Mechanics of plastic media. T. 1. Theory of ideal plasticity (Mekhanika plasticheskih sred. T. 1. Teoriya ideal'noj plastichnosti). Moscow: Fizmatlit.
9. Ebobisse, F., & Reddy, B. D. (2004). Some mathematical problems in perfect plasticity. *Computer Methods Applications Mechanical Engineering, 193,* 5071–5094.
10. Mitrofanov, V. P. (2006). The theory of perfect plasticity as the elementary mechanic pseudo-plastic ultimate state of concrete: Bases, imitations, practical aspects, Proceedings of the 2nd Fibre Congress (pp. 7–6). Naples.
11. Sorensen, J. H., Hoang, L. C., Olesen, J. F., & Fischer, G. (2017). Test and analysis of a new ductile shear connection design for RC shear walls. *Structural Concrete, 18,* 189–204.
12. Jorgensen, H. B, & Hoang, L. C. (2015). Load carrying capacity of keyed joints reinforced with high strength wire rope loops. In Proceedings of Fibre symposium: Concrete—Innovation and Design (13 p). Copenhagen.
13. Pedersen, R. H., & Herlev, M. E.: (2015). Shear capacity of construction-friendly element joints (Bachelor thesis). Department of Civil Engineering. Denmark.
14. Svejgaard, J. (2015). Test and analysis of keyed shear joints between precast concrete walls—Influence of indent area on the load bearing capacity (Master's thesis). Department of Civil Engineering. Denmark.
15. Pohribnyi, V., Dovzhenko, O., Karabash, L., & Usenko, I. (2017). The design of concrete elements strength under local compression based on the variational method in the plasticity theory. In MATEC Web of Conferences (116).
16. Dovzhenko, O., Pohribnyi, V., Pents, V., & Mariukha, D. (2018). Bearing capacity calculation of reinforced concrete corbels under the shear action. In MATEC Web Conferences (230).
17. Dovzhenko, O. O., Pohribnyi, V. V., & Yurko, I. A. (2018). Concrete and reinforced concrete strength under action of shear, crushing and punching shear. In IOP Conference Series: Materials Science and Engineering (Vol. 463 (1)).
18. Pohribnyi, V., Dovzhenko, O., Kuznietsova, I., & Usenko, D. (2018). The improved technique for calculating the concrete elements strength under local compression. In MATEC Web Conferences (230).
19. Dovzhenko, O., Pogrebnyi, V., & Yurko, I. (2018). Shear failure form realization in concrete, news NAS RK. *Series of Geology and Technical Science, 2*(428), 212–219.
20. Dovzhenko, O., Pohribnyi, V., & Karabash, L. (2018). Experimental study on the multikeyed joints of concrete and reinforced concrete elements. *International Journal of Engineering & Technology, 7*(3.2), 354–359.

Non-crane Method of Reconstructing Buildings with Additional Storey Erection

Evgen Dyachenko, Oleksandr Zyma, Roman Pahomov, and Oleksandr Shefer

Abstract The article focuses on the possibility of refusing from the use of cranes while reconstructing buildings with additional storey erection in the conditions of dense urban development, using monolithic construction systems erected by the floor-lifting method. Non-crane method of reconstructing buildings, preserving enclosure structures with additional storey erection, is proposed, and advantages and disadvantages of the method provided are analysed.

Keywords Reconstruction · Additional storey erection · Residential buildings · Public buildings · Replacement of floors · Embedded systems · Floor-lifting method · Monolithic structures

1 Problem Statement

Ever-increasing demands on the performance properties of residential and public buildings bring to the forefront the problem of reconstructing residential and public buildings of the old urban architectural complex, such architectural forms being, as a rule, of historical importance, creating the architectural face of the downtown. The main reasons for the need of reconstruction are the inconsistency of existing planning solutions with current conditions and the use of timber structures as overlapping structures. Due to the long service life, wooden structures are usually in poor condition and need to be replaced or cannot withstand new increased operational loads. In addition, there is often the problem of the need to increase the operating area of buildings, which can be solved by additional storey erection. In many cases, the load-bearing capacity of enclosure structures and foundations in old buildings allows this to be done (Fig. 1). One of the methods of reconstructing these buildings is the erection of embedded systems with the preservation of existing enclosure structures (Fig. 2). The peculiarities of the reconstruction work in dense urban development, especially in the central, historical parts of cities, include significantly limited size of

E. Dyachenko · O. Zyma (✉) · R. Pahomov · O. Shefer
Poltava National Technical Yuri Kondratyuk University, Poltava, Ukraine
e-mail: zymaae@gmail.com

V. Onyshchenko et al. (eds.), *Proceedings of the 2nd International Conference on Building Innovations*, Lecture Notes in Civil Engineering 73,
https://doi.org/10.1007/978-3-030-42939-3_4

Fig. 1 Reconstruction of the buildings with additional storey erection

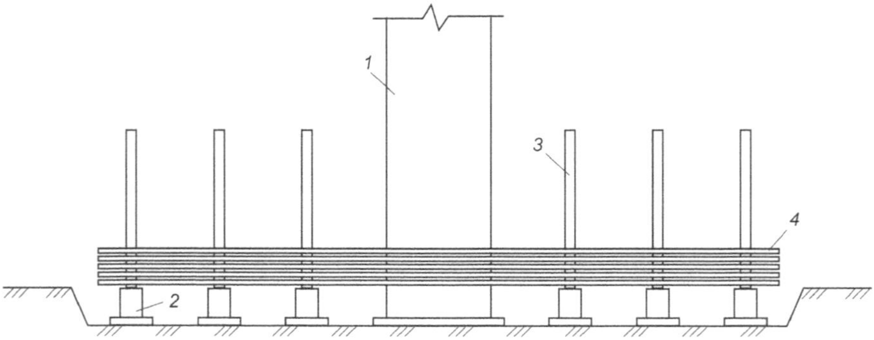

Fig. 2 Erection of a building by the floor-lifting method: 1—rigidity core; 2—foundations; 3—columns; 4—floor slab package

the construction site, located in the immediate vicinity of existing facilities (roads, sidewalks, shopping pavilions, etc.).

This condition significantly affects the ability to use large construction equipment and lifting cranes. In some cases, the use of cranes requires limiting or stopping the movement of vehicles for a long time, and in some cases the use of lifting cranes is not possible at all. It causes the refusal to use precast concrete structures as well as monolithic reinforced concrete structures as structures of the embedded frame. However, the use of monolithic structures still requires the use of lifting cranes for feeding reinforcement frames and formwork. One of the methods that enables practically complete avoidance to use lifting cranes for the erection of embedded structures is the floor-lifting method [1], even in the conditions of the need to reconstruct objects with additional storey erection.

2 Recent Research Outline

Recently, due to the urgency of the problem to reconstruct residential and public buildings of the old urban development, a large number of works, both by domestic [2–6] and by foreign [7–12] authors, have appeared. The papers [2–6] dwell on the reconstruction features of these buildings, the problems of using construction equipment in the conditions of a limited-size construction site, comparison of using precast and monolithic reinforced concrete as a material of embedded structures. It is concluded that the reconstruction of the buildings in the old city development by means of replacing overlappings and erecting embedded structures enables to preserve the architectural image of the historical parts of the cities and at the same time to adapt the buildings to the new operating conditions. The works of foreign authors are often devoted to the issues of specific object reconstruction [7, 8], and some of the works focus on strengthening the structures of historical buildings [9]. The papers [10–12] concern the general issues of historical building reconstruction. In [13, 14], the authors consider the possibility of using the floor-lifting method for the arrangement of embedded systems and analyse the advantages and disadvantages of its use.

3 Highlighting Unsolved Problems

Despite the large number of works devoted to this problem, there is still no proposed method on how to reconstruct these buildings, so that the use of lifting cranes might be completely abandoned, provided that it is necessary to stop the traffic to carry out the additional storey erection of the reconstruction object and thus significantly facilitate the organization of work in the conditions of the dense construction site. This problem still remains unsolved.

4 The Purpose of the Article

The purpose of the article is to analyse the feasibility of using floor-lifting method for the reconstruction of residential and public buildings by erecting embedded systems with the additional storey erection of the reconstruction object as well as to study the possibility of complete avoidance to use lifting cranes during the reconstruction.

5 Main Material

As is commonly known [1], the erection of frame multi-storey buildings with monolithic or precast monolithic reinforced concrete frame by the floor-lifting method is to create slab package on the surface of the earth or on the floor slab over the underground part of the entire floor. After the floor slabs have been arranged and the concrete has reached the required strength, the slabs are raised to the designed position on previously arranged columns by lifting vehicles. The specified method has the following sequence of works: (1) erection of the underground part of the building by the traditional method, which involves the arrangement of the foundations under columns, the installation of columns of the underground part, the arrangement of the enclosure structures of the underground part, floor slabs, etc. (Fig. 2); (2) arrangement of the rigidity core, which ensures the stability of the building in the longitudinal and transverse directions; as a rule, a staircase with elevator shafts is placed in the rigidity core; (3) installation or arrangement of the first-tier monolithic columns; (4) arrangement in the contours of the building of the whole floor slab package over the entire area of the building or section; (5) after the concrete slabs have reached the required strength, the slabs are lifted into the intermediate position by lifting vehicles mounted on columns; (6) installation or arrangement of columns of the next tier, movement of lifting vehicles and floor slab lifting; (7) after all the overlapping slabs are lifted in a designed position and lifting vehicles are removed, arrange a roof, enclosing structures, and perform general construction and finishing works.

The advantages of this construction method presuppose that it can significantly reduce the use of lifting cranes, sometimes completely reject their use and significantly reduce the area of the construction site; in addition, due to the lack of load-bearing walls inside the building, this method does not impose restrictions on the choice of planning decisions.

This method was intended to be used primarily for new construction; the use of this method during the reconstruction with the need for additional storey erection has not been considered yet.

From the analysis of the advantages of this method, it follows that its use in the reconstruction by constructing embedded systems would enable the reconstruction without going beyond the constraints imposed by the conditions of dense urban development. The floor-lifting method with avoiding the use of cranes would reduce the area of the construction site almost to the area of the building under reconstruction and eliminate the restriction of traffic and pedestrians and the operation of the nearby facilities.

The complex of reconstruction works of the building with additional storey erection by the embedded system construction using floor-lifting method will be as follows: (1) dismantling the building internal structures: in this case, dismantling must be carried out with strict adherence to the sequence of dismantling structures adopted in the technological maps. If necessary, reinforcement of the enclosure structures and foundations is carried out; (2) producing soil under the embedded system foundations

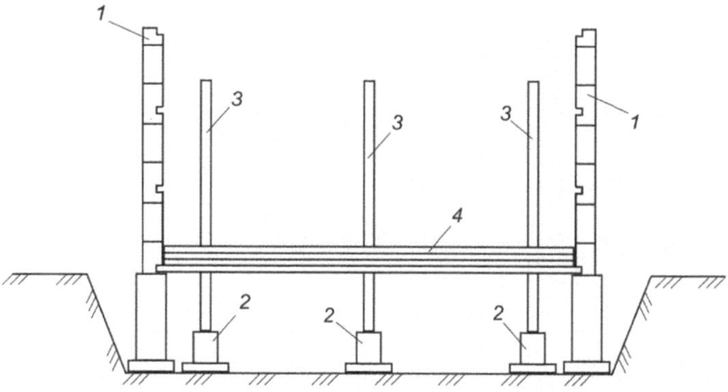

Fig. 3 Installation of the embedded frame by the floor-lifting method (stage of the overlapping slab concreting): 1—existing walls; 2—foundations for the embedded frame; 3—columns; 4—floor slab package

and arranging monolithic foundations under the embedded structures: it is possible to prepare a concrete mixture at the site and to feed it with a light concrete pump located in the dimensions of the building. After the concrete has reached the required strength, the waterproofing of the foundations and backfilling of the soil with compaction are performed. If necessary, the concrete floor of the building underground part is arranged; (3) arranging monolithic columns of the first tier: formwork and reinforcement frames are installed from scaffolding, and the concrete mixture is supplied with a lightweight concrete pump; (4) arranging a monolithic floor slab above the underground part of the building: concrete is delivered centrally, and form worked by a concrete pump is located outside the building; (5) after reaching the concrete slab above the underground part of the required strength, the whole floor slab package is arranged one after the other (Fig. 3); the plates are separated by a distribution layer; (6) after the set of the required strength with concrete, scaffolding, formwork, reinforcement frames and lightweight concrete pump for concreting the next tier of columns, materials for the erection of enclosing structures of the superstructured floor are loaded onto the last slab; (7) on the first-tier column heads, lifting vehicles are installed to raise the floor slabs [1]; slabs are raised to the intermediate position and fixed (Fig. 4); (8) concreting of the second tier of columns is performed; (9) the lifting of the slabs continues after the concrete strength of columns has been reached; (10) the erection of enclosure structures (considered enclosure structures of aerated concrete blocks) of the additional storey begins after reaching of the floor slab the level of the additional storey floor and securing it in this position (Fig. 5). The means of the floor slab fixing in the intermediate position must be designed to accommodate the loads arising during the additional storey erection. The mortar for the erection of the additional storey enclosure structures is supplied to the place of works by the mortar pump. The slab rises to the design position and fixes in steps that coincide in height with the masonry layers of the enclosure structures of additional storey (Fig. 6). Lifting to the next position is made after the completion of the tile

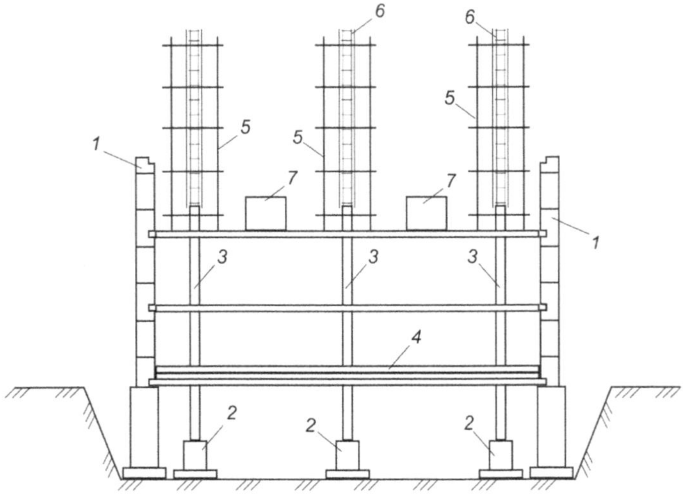

Fig. 4 Arrangement of the embedded frame by the floor-lifting method (stage of the second-tier column concreting): 1—existing walls; 2—foundations for the embedded frame; 3—columns; 4—floor slab package; 5—scaffolding; 6—column formwork; 7—materials for the enclosure structure erection

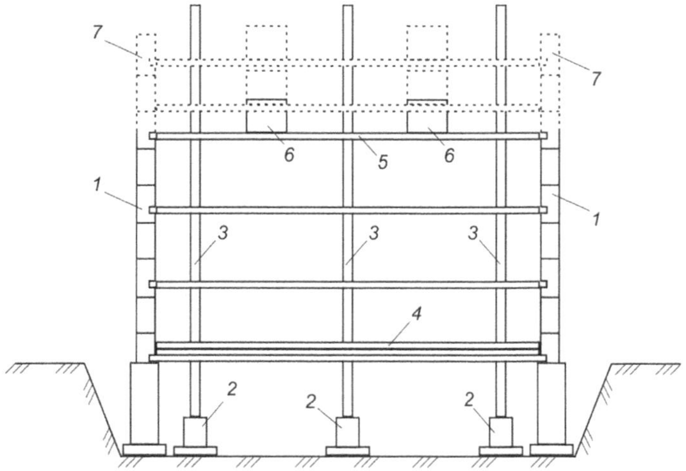

Fig. 5 Arrangement of additional storey enclosure structures: 1—existing walls; 2—foundations for the embedded frame; 3—columns; 4—floor slab package; 5—overlapping plate fixed in the intermediate position; 6—materials for the enclosure structure erection; 7—additional storey enclosure structures

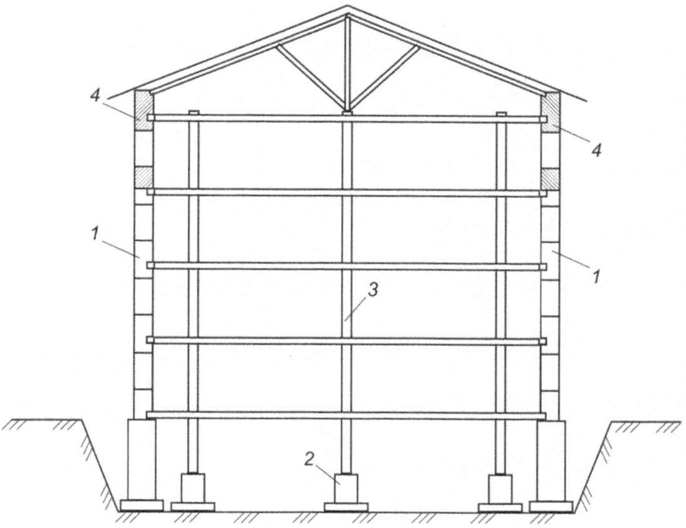

Fig. 6 Reconstruction of the building by embedded system with an additional floor erection: 1—existing enclosure structures; 2—foundations for the embedded frame; 3—embedded structures; 4—additional storey enclosure structures

laying. This eliminates the use of platforms and the need to use cranes to feed and remove them from the floor.

After the completion of laying enclosure structures and fixing all overlapping slabs in the design position, joints between columns and slabs, and between slabs and enclosure structures are made, and the concrete pump is removed from the slab surface by a crane installed for a short period.

Coverage, internal general construction works and finishing works are performed by traditional methods. As a result, after completing the above complex of works, the reconstruction object was replaced with internal structures and an additional storey was erected (Fig. 6).

As can be seen from the above-mentioned complex of works, the considered method of residential and public building reconstruction in the conditions of dense urban development in case of need for additional storey enables to practically refuse the lifting crane usage while performing the works. This increases the compactness coefficient of the building general plan to 60–80%. With traditional construction methods, this ratio is 10–40%. The proposed method eliminates the need to install slab formwork and supports under it, because formwork is pre-concreted slabs. This can significantly reduce the complexity of the work on the arrangement of slabs. In addition, the reconstruction method under consideration eliminates the need for the use of platforms in the arrangement of the enclosure structures of the additional storey due to the stepwise elevation of the floor slab to the designed position with a step coinciding with the height of the tile wall.

Fig. 7 Reconstruction of the building with an additional storey with lightweight aerated concrete blocks

However, the use of the floor-lifting method in the conditions of embedded system construction involves the manual assembly of formwork columns, the installation of reinforcing frames in formwork, the assembly and disassembly of scaffolding, the installation and removal of lifting vehicles. In addition, there is a restriction on the weight of the assembled structures on the surface of the roof slab, which is due to the bearing capacity of the embedded system framework elements during installation, the means of fastening the floor slabs in the intermediate position, the lifting capacity of the lifting vehicles that carry out the elevation of the plates in the designed position. This limitation, in some cases, implies the need to use lightweight aerated concrete blocks (Fig. 7) or lightweight thin-walled metal structures (Fig. 8), which are manually assembled, insulated and lined, as enclosure structures.

6 Conclusions

The method of reconstructing buildings, considered in the research, can be used (after economic comparison with other methods of reconstruction possible in these conditions) for the reconstruction of residential and public buildings by arranging embedded systems of monolithic reinforced concrete frame type also in the conditions of additional storey development necessity.

Reconstruction technology using the floor-lifting method allows to practically refuse the crane usage and significantly (up to 50%) reduce the required construction site area in comparison with the crane methods of embedded structure erection and enables to refuse the use of devices when erecting enclosure structures or other brick

Fig. 8 Reconstruction of the building with an additional storey with lightweight thin-walled metal enclosure structures

structure materials. This makes possible the reconstruction without restricting traffic on the adjacent streets and the provision of construction in dense urban development conditions. Significant reduction of the construction site area, the absence of the need to stop the traffic and the functioning of the adjacent objects can significantly reduce the cost of the construction site maintenance and restoration of the operational and aesthetic properties of the areas reserved for the construction site, which can amount to a total cost of up to 25%.

One criterion for the possibility of using such technology is the presence of a rigidity core, which may be the walls of staircases that are not disassembled, in the building. In addition, in some cases, the method requires the use of lightweight materials and building structures, such as aerated concrete blocks or light thin-walled metal structures, for additional storey enclosure structure erection.

References

1. Telichenko, V. (2004). *Technology of buildings and structures construction* (446 p.). Moscow: High School.

2. Osipov, A., & Akimov, S. (2002). *Features of space-planning and structural characteristics of reconstructed residential buildings in Kiev Construction and industrial safety* (Vol. 6, pp. 260–265).
3. Osipov, A., & Akimov, S. (2005). The construction of monolithic ceilings during the reconstruction of residential buildings. In *New technologies in construction* (Vol. 1, pp. 46–52).
4. Osipov, A., & Akimov, S. (2006). *Reconstruction of residential buildings. Monolithic flooring methods Dnieper science news* (Vol. 3, pp. 9–15). Dnieper.
5. Osipov, A., & Akimov, S. (2009). Development of organizational and technological models of replacement of ceilings in buildings of historical construction. In *Construction and industrial safety* (Vol. 29, pp. 101–108).
6. Voskobijnik, O., & Dyachenko, E. (2012). Features of technology of performance of works at reconstruction of buildings by means of embedded systems. In *Academic journal series: Industrial machine building, Civil Engineering* (Vol. 33, pp. 43–48). Poltava: PNTU.
7. Murgul, V. (2015). *Reconstruction of the Courtyard spaces of the historical buildings of SAINT-Petersburg with creation of atriums procedia engineering* (№117, pp. 808–818). https://doi.org/10.1016/j.proeng.2015.08.145.
8. Korkmaz, E., & Vatan, M. (2013). Retrofitting Deniz Palace historic building for reusing. *International Journal of Electronics; Mechanical and Mechatronics Engineering, 2*(3), 269–278.
9. Bhattacharya, S., Nayak, S. & Dutta, S. (2013). A critical review of retrofitting methods for unreinforced masonry structures. *International Journal of Disaster Risk Reduction, 7.* https://doi.org/10.1016/j.ijdrr.2013.12.004.
10. Kareeva, D., & Glazkova, V. (2017). Reconstruction and restoration of historical buildings of transport infrastructure. In *IOP Conference Series: Earth and Environmental Science* (90). https://doi.org/10.1088/1755-1315/90/1/012224.
11. Holland, A.G. (2011). *The reconstruction of historical buildings* (120 p). Richmond, VA: Virginia Commonwealth University.
12. Green, M. (2012). *Building codes for existing and historic buildings* (247 p). USA, Hoboken: Wiley.
13. Zyma, O., Dyachenko, E., Pahomov, R. & Zhyhylii, S. (2018). Works execution organization at reconstruction and renovation of buildings after the fire with usage of slabs lifting method. *International Journal of Engineering & Technology, 7*(2.23), 242–246. doi: https://doi.org/10.14419/ijet.v7i2.23.11951.
14. Dyachenko, E., Zyma, O., Dryizhyruk, Y., & Lazariev, D. (2018). Works execution organization at reconstruction and renovation of buildings with usage of slabs lifting method. *Academic Journal Series: Industrial Machine Building, Civil Engineering* (Vol. 1(50), pp. 246–252). Poltava: PNTU. doi: https://doi.org/10.26906/znp.2018.50.1082.

Corrosion Protection of Metal Structures in Manufacturing Conditions

O. M. Gibalenko and V. A. Gibalenko

Abstract The purpose of the research is to improve the quality of secondary protection measures during the production of sheet metal structures. The article is also aimed at maintaining the quality with guaranteed indicators of durability, taking into account the level of corrosion hazard in a coastal climate. *Methodology* Process approach principles were used to implementing the tasks of lifetime control in corrosive environments. The reliability principle of corrosion hazard level is ensured, which is described by the successive stages of assessing the survivability of sheet metal structures based on the DMAIC (define, measure, analyze, improve, control) strategy: terms, measurement, analysis, improvement and control of primary and secondary corrosion protection measures. *Results* Corrosion protection measures have been developed according to the criterion of corrosion hazard, which ensures compliance with the reliability requirements of building metal structures based on the calculated indicators. Solution of management tasks has been found. Those were concentrated on the technological readiness within determined period which is prior to the construction of the building object. *Scientific Novelty* We have been able to design maintenance procedure for sheet metal under the actual conditions, which is based on a process approach to resource management by building a system of accounting and functional control, risk analysis and management of process safety. Implementation of these principles at the object level is supposed to improve the means and the methods of corrosion protection, resource extension, taking into account indicators of vitality and justification of reliability measures program (RMP). *Practical Relevance* Based on the developed principle, the results of field and laboratory studies have been generalized. These studies are focused at substantiating the primary and secondary protection of sheet metal structures under the given level of corrosion hazard, which is proposed by RMP organizational measures. The arrangements contain evaluation of integral constructive characteristics of adaptability, technological rationality and risk analysis of corrosion hazard signs.

Keywords Metal sheet structures · Survivability · Process approach · Provision of reliability

---

O. M. Gibalenko (✉) · V. A. Gibalenko
Azov State Technical University, Universitetskaya st., 7, Mariupol 87500, Ukraine
e-mail: grin196102@gmail.com

V. Onyshchenko et al. (eds.), *Proceedings of the 2nd International Conference on Building Innovations*, Lecture Notes in Civil Engineering 73,
https://doi.org/10.1007/978-3-030-42939-3_5

1 Introduction

Improvement of regulatory and technical requirements in the field of metal structures safety in production, manufacture, operation [1, 2] is associated with a reasonable consideration of corrosion hazard level. This level is determined by the critical intervals of the availability coefficient which shows the anticorrosive protection of metal structures under the influence of factors of corrosive environments [3]. Azovstal combine features (Azovstal Iron & Steel Works) is the production and supply of steel sheet, with structures being under the influence of corrosive environments at seaside and in marine environments for quite a significant period. Such an environment belongs to the categories C4...C5 with very high corrosion aggressiveness [4]. This requires problem solution by ensuring corrosion protection taking into account the level of corrosion hazard and the requirements of primary and secondary corrosion protection [5].

Evaluation of the quality and reliability of metal constructions and temporary coatings is an important aspect of extending the resource of fixed assets, the formation of rational innovative development of new materials production and the use of resource-saving technologies. The predicted level of corrosion hazard makes it possible to formulate requirements for the provision of primary and secondary corrosion protection measures corresponding to normal operation conditions [6, 7]. Alongside, the uncertainty of the assessing criteria for the state of the metal surface and corrosion damage creates difficulties for rapid decision-making in order to ensure technological safety of objects [8].

2 Purpose

Research is aimed at ensuring the quality and reliability of primary and secondary protection measures for sheet metal structures based on signs of corrosion hazard.

Compliance with the conditions of reliability and structural safety is achieved by solving the problem of providing technological safety based on the methodology of universal quality management [9].

A significant level of wear and deadlines in adverse conditions are threats that can cause the reduction of the product quality, which, in turn, leads to the increased requirements for improving the operational properties of structures that lose quality [10–13].

The definition of structures durability indicators is based on the developed methodology that describes the reliability and availability coefficients of temporary anticorrosion protection, established according to the results of experimental studies. Those studies were focused on the properties of primary and secondary protection during definitive (accelerated or bench) tests according to the corrosive media classification features [14].

Currently, the assessment of the corrosion state of metal structures is carried out in accordance with the requirements of domestic standards [15].

The influence of corrosion damage on the durability in norms is advised to be recognized by changing geometrical characteristics of the section based on the data basis. The basis should include the size of uniform corrosion and rate of fusion cross section which enables to account uneven features (local) destruction of structural elements. The corrosion penetration depth indicator is determined without taking into account the confidence interval of the statistical measurement error. At the same time, the norms [16] allow primary protection measures (increase in rolled thickness) taking into consideration the level of corrosion hazard.

3 Methodology

The purpose of this work is to substantiate the composition and the structure of the corrosion resistance parameters to control technological safety by reducing risks, which decreases the likelihood of corrosion damage and limits possible damage during corrosion of installations and constructions.

The influence of external factors (aggressive environment) and internal parameters (structural form) on the reliability indicators of metal products building sheets is considered for location areas of groups with homogeneous structural elements (consignment), taking into account the type and intensity of corrosive influences. According to the presented approach (Table 1), the main indicator of aggressive environment is the characteristic value of the annual corrosion loss A_n, g/m^2*year, conditionally reduced to the unprotected surface of steel-class C 235 [6].

Table 1 Classification of corrosive environments

With degree aggressiveness DSTU B V.2.6-193:2013	For corrosion resistance steel K, mm/year	To corrosion losses steel C235, $A_{n,}$ g/m^2 *year	With degree of aggressiveness SNIP 2.03.11-85	Corrosion category designation ISO 12944-2 A_n, g/m^2 *year
A4 highly aggressive	0.08–0.20 0.008–0.02	650–1500	B3 medium-aggressive	C5-1 very high (industrial) 650–1500 C5-M is very high (marine) 650–1500
A5 very highly aggressive	0.20–0.50 0.02–0.05	1500–3900		–

4 Main Material

Corrosion destruction of steel structures is determined by the external influences of the operating mode and depends primarily on the degree of environment aggressiveness. Accounting of the electrochemical kinetics of corrosion damage in the strength change is based on physical models characterizing changes in the geometric parameters and properties of the material over time under aggressive environmental influences [17–19]. From the standpoint of the structural mechanics, an allowable decrease in the bearing capacity of the elements for a given system of anticorrosive protection (**SSAP**) of the structures can be taken into account in the calculations for limit states using fictitious external loads [3, 4]. The purpose of the work is to justify the conditions for monitoring and diagnosing signs of corrosion hazard during maintenance of building metal structures in actual condition.

Tolerance in operational performance of structural elements depending on responsibility category protection systems against corrosion is taken into account to reliability of corrosion protection coefficients ($\gamma\xi_{zk}$, $\gamma\xi_{zn}$). The requirements for the secondary protection durability are established with regard to the specified metal structures preparation time $T_{n\gamma,}$ of reliability index β, and the degree of aggressiveness of environmental influences. By the comprehensive indicator of maintainability—readiness coefficient (K_g), the parameters of design and technological measures of primary and secondary protection are characterized:

$$K_g = \frac{T_{k\gamma} + T_{z\gamma}}{T_{k\gamma}} \tag{1}$$

where T_k—the service life (year) of steel structures in terms of corrosion resistance (primary protection); T_z—the estimated service life (year) of temporary protection with a confidence probability of $\gamma = 0.95$ according to the results of accelerated tests.

Calculation of corrosion resistance, durability and maintainability of structures is carried out taking into consideration the parameters of corrosion resistance (Table 2) for the limiting conditions.

Ensuring reliability and structural safety at the pre-production stage includes substantiating design situations based on conditions of survivability of corrosion protection systems. Setting the service life of structures according to the characteristics of primary protection (corrosion resistance of the structure) is done taking into account the reserve of the bearing strength properties and the degree of impact aggressiveness on the calculation basis. The service life is set, based on the analysis of the quality indicators of protection systems.

Table 2 Conditions for compliance of protective measures with project quality requirements

The scale of resistance of metals and coatings				Design responsibility category	Protection reliability factors	
Resistance group according to GOST 13819-68	Strength rating, score	Depth of lesions, mm/year	Primary and secondary protection class according to SNiP (DSTU)		Primary, γ_{zk}	Secondary, γ_{zn}
Unstable (iv)	8 7	1–5 0.5–1.0	I	C4	From 0.80 Up to 0.85	From 0.85 Up to 0.90
Low resistant (III)	6 5	0.1–0.5 0.05–0.1	II	C3	≫0.85 ≫0.90	≫0.90 ≫0.95
Satisfactory persistent (II)	Four 3	0.01–0.05 0.005–0.01	III	C2	≫0.90 ≫0.95	≫0.95 ≫0.99
Persistent (I)	2 One	0.001–0.005 Less than 0.001	IV	C1	≫0.95 ≫0.99	≫0.99 ≫1.00

5 Results

In accordance with the stated approach, the requirements are formulated for determining the actual values of the corrosive effects according to the monitoring of the control standard K_p of operation of objects which determines the measure effectiveness of the reliability assurance program based on monitoring the quality indicators of **SSAP** during maintenance in the actual conditions. Representative values of the control standard indicators K_p are used to calculate the reliability coefficients of material corrosion resistance (γ_{mk}) and the characteristic values of annual corrosion losses ($A_{n,}$ g/m^2 *year)—the main indicator of the environment aggressiveness. The calculation and experimental assessment of the protective coatings durability is based on the simulation of physico-chemical effects during accelerated testing of samples with protective coatings designed for long-term corrosion protection. Practical implementation of the methodical approach was performed while monitoring output process at Azovstal Iron & Steel Works, while producing, storing and transporting sheet metal structures on the basis of the corrosion hazard (see Fig. 1).

The strategy of temporary anticorrosive protection of metal sheet structures in a production enterprise under the influence of environmental factors of the marine climate has been developed. Activities include the process approach to resource management by building a system of accounting and functional control, risk analysis and

Fig. 1 The study of corrosion protection during field studies: **a** the boundaries of the plots; **b** conditions of application; **c** places of control, **d** the appearance of the control sites (exposure 2 days)

regulation of technological safety of production assets of enterprises that is taking into account the level of corrosion hazard. The implementation of the approach to the management of technological safety at the object level is aimed at improving the **SSAP**, extending the resource with regard to survivability indicators and substantiating measures of **RMP**.

6 Conclusions

Assignment of corrosion protection measures according to the criterion of corrosion hazard allows to provide reliability and compliance with the normative requirements of the sheet metal structures quality [16] at the premises and to solve the problems of technological safety management within the established term:

- identify deviations from the requirements of current regulatory documents on corrosion protection;

- evaluate the conformity of quality indicators, constructive fitness and technological rationality of anti-corrosion protection solutions to a given level of corrosion hazard;
- determine the requirements for the selection of materials and systems for protective coatings of metal structures in accordance with the classification features of regulatory requirements;
- develop proposals for the use of anti-corrosion protection solutions based on the examination of new materials and technologies according to the conditions for assessing compliance with the requirements of reliability and efficiency.

References

1. Paton, B. E. (2009). Resource and safety issues for the operation of structures, structures and machines. Target comprehensive program of NAS of Ukraine. *Coll. Sciences. Art. according to the results obtained in 2007–2009.* NAN Ukraine, 709.
2. Shimanovskyj, O. V. (2008). Conceptual bases of system of technical regulation of reliability and safety of building structures. *Industrial Construction and Engineering, 1,* 4–9.
3. Gibalenko, A. N. Monitoring of residual life of steel structures in corrosive media. In *Coll. Sciences. ave. Series: Industrial engineering, building* (Vol. 3(45), pp. 110–116). Poltava: PoltNTU.
4. Corrosion of metals and alloys. Corrosivity of atmospheres—Classification, determination and estimation. 15. ISO 9223:2012.
5. Korolev, V. P., Filatov, U. V., & Selutin, U. V. (2014). Development of corporate management system: technological safety of production facilities. In *Coll. Sciences works of Ukrinstalkon them. V.M. Shimanovsky* (Vol. 14, 136–149). Publishing house Stal.
6. Sumi, Y. (2008). Strength and deformability of corroded steel plates estimated by replicated specimens. *Journal of Ship Production., 24*(3), 161–167.
7. Nadim, M. A. (2015). *Numerical analysis of pit shape effect on corrosion diminution and structural integrity parameters of ship* (Bachelor thesis).
8. Guedes Soares, C., Garbatov, Y., Zayed, A., & Wang, G. (2009). Influence of environmental factors on corrosion of ship structures in marine atmosphere. *Corrosion Science, 51,* 2014–2026.
9. Korolov, V. P., Vysotsky Y. V., & Gibalenko, O. M. (2010). Estimation of steel structure corrosion risk level. In *The European Corrosion Congress "From the Earth's Depths to Space Heights"* (534 p).
10. Gibalenko, A. N. (2015). Monitoring of residual life of metal structures in corrosion media. In *Collection of scientific works. Series: Industry engineering, construction* (Vol. 3(45), pp. 110–116). Poltava: PoltNTU.
11. Pichugin, S. F. (2018). Reliability estimation of industrial building structures. *Magazine of Civil Engineering, 83*(7), 24–37. https://doi.org/10.18720/MCE.83.3.
12. Pichugin, S. F., & Makhin'Ko, A. V. (2009). Calculation of the reliability of steel underground pipelines. *Strength of Materials, 41*(5), 541–547. https://doi.org/10.1007/s11223-009-9153-0.
13. Pichugin, S. (2017). Probabilistic description of ground snow loads for Ukraine. *Snow engineering 2000: Recent advances and developments* (pp. 251–256) https://doi.org/10.1201/9780203739532.
14. Korolov, V. (2014). Design criteria of reliability and safety in the design of corrosion protection of structural steel. In *The European Corrosion Congress «Improving materials durability from cultural heritage to industrial applications»* (Vol. 88).

15. Ministry of Regional Development of Ukraine. (2014). *Corrosion protection of metal structures. Design requirements*. 74. B.2.6-193:2013.
16. Ministry of Regional Development of Ukraine. (2014). *Steel structures. Norms of designing*. 199. B.2.6-198:2014.
17. Zienkiewicz, O. C. (2005). *The finite element method for solid and structural mechanics sixth* (Vol. 648). Oxford.
18. New York: American Institute for Steel Classification. (2002). *American rust standard guide*, 1, 12.
19. Bondar, V., Shulgin, V., Demchenko, O., & Bondar, L. (2017). *Experimental study of properties of heavy concrete with bottom ash from power stations*. Paper presented at the MATEC Web of Conferences, 116. doi: 10.1051/matecconf/201711602007.

Ways to Improve the Combined Steel Structures of Coatings

Myron Gogol, Tetiana Kropyvnytska, Tatiana Galinska, and Mukhlis Hajiyev

Abstract To increase the efficiency of traditional beam steel structures, more extensive use of combined steel structures is required. When using combined steel structures, it is possible to use a calculated method of regulating their stress deformation state (SDS). This method is easy to implement and low in complexity, and does not require complexity of structures. Analysis of the structural form development and the methods of calculating combined steel structures demonstrated that such structures are the most promising and have great potential. The most rational constructive form is the truss with support units at upper belt level and lower tightening. Numerical and analytical calculations showed, that while working in the elastic-plastic area and taking into account its own weight ensures a load capacity increase only for scheme (b) up to 20%, and scheme (a) is better for increasing the rigidity of the structure also when working in a critical area. Based on the principle of minimizing the plot area of moments, an improved design of a steel-combined truss is proposed. We have developed a constructive solution for a combined rational steel rafter truss, the number of nodes and elements being reduced several times.

Keywords Combined structures · Calculation method · Constructive form · Inclined racks · Rational structures · Combined trusses

1 Introduction

Among the numerous ways to solve the problem of reducing metal consumption and improving the efficiency of steel structures is the ability to select effective schemes for the most common types of structures—beams and trusses.

M. Gogol · T. Kropyvnytska
National University "Lviv Polytechnic", Lviv, Ukraine

T. Galinska (✉)
Poltava National Technical Yuri Kondratyuk University, Poltava, Ukraine
e-mail: galinska@i.ua

M. Hajiyev
Azerbaijan University of Architecture and Construction, Baku, Azerbaijan

V. Onyshchenko et al. (eds.), *Proceedings of the 2nd International Conference on Building Innovations*, Lecture Notes in Civil Engineering 73,
https://doi.org/10.1007/978-3-030-42939-3_6

These classical structural forms will take a long time to play a prominent role in steel structures, and they will remain the main coating systems for a long time [1–4]. Design means an interconnected set of schematics with well-defined parameters, types of cross sections, structure decisions of nodes and connections with technological techniques and features of manufacturing, installation and operation.

One of the possible ways to increase the efficiency of traditional beam steel structures is a wider use of combined steel structures [4, 5]. When using composite steel structures instead of traditional ones, it provides the possibility of using the calculated, not labor-intensive power, method of regulating their stress deformation state (SDS). This method is easy to implement and does not require complexity of the structures [1, 6]. Although known for a long time, it was not practically used in the framework of buildings and structures [1] and did not receive proper distribution until the last decade.

Analysis of the development of structural forms and methods of calculation of combined steel structures enables to conclude that such structures are the most promising and have great potential, but to date they have not been widely used [5, 7].

The purpose of the research is to improve the structural forms of combined steel structures based on established new scientific principles and factors for their design, calculation and search for efficient rational values of geometric parameters.

1.1 Analysis of Recent Research and Publications

From a multicriteria perspective, variants of the rational steel structure of the truss were explored for the currently popular 24 m long run [8]. Eleven alternatives were analyzed and proved that the most rational type is the truss with support units at upper belt level and the lower tightening, which certainly applies to combined truss.

Estimation of lattice types for combined beam systems in terms of material costs and load-bearing capacity was conducted in the studies [1, 9], where it was shown that sloping rack bars are rational, with optimum tilt angles of 30–60°, which further increases the rigidity of the structure under asymmetrical loads and the maximum number of elastic supports, which are oblique racks—3–4 pieces.

The work of combined truss in two types of racks (Fig. 1): a—with the slant of the sprengel, which are bent to the bottom, and b—inclined at the bottom from the center—was studied in the field of elastic–plastic work as self-reinforcing systems [10].

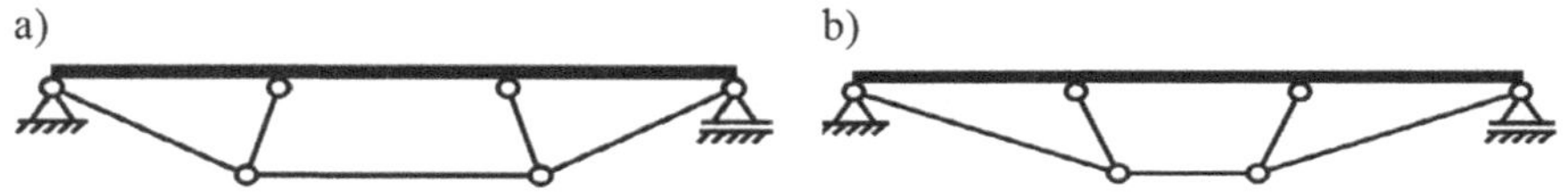

Fig. 1 Schemes of the combined sprengel beam

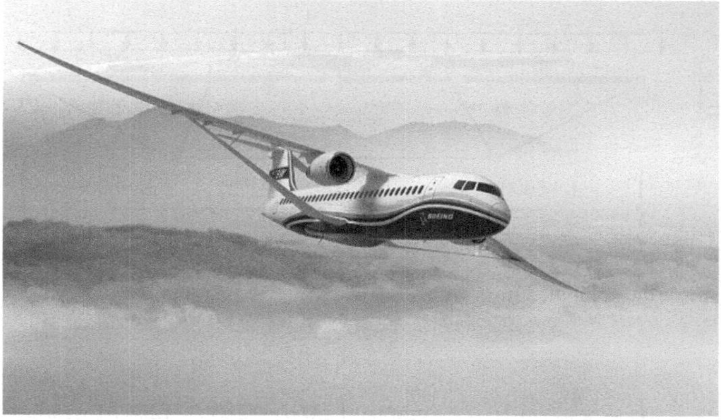

Fig. 2 Combined truss in aircraft construction: lightweight, resilient, ultra-thin sloping wing supports—Boeing and NASA, USA

Numerical and analytical calculations showed [10] that working in the elastic-plastic area (in the noncritical behavior) and taking into account its own weight, increases the load-bearing capacity only within the scheme (b) by up to 20%. The scheme (a) is better for increasing the rigidity of the structure also when working in a critical area, as evidenced by recent studies of Boeing and NASA, USA, Fig. 2 [11].

2 Main Body

Based on the principle of minimizing the plot area of moments, developed by Prof. M. S. Moskalev [3], we have proposed an improved structure of a steel-combined truss (Fig. 3a). The essence of the improvement lies in the reduced consumption of steel on the truss by adjusting the bending moment in the upper belt—in a beam of rigidity of the system by welding plates in the supporting parts of the posts. The essence of the principle is to redistribute the bending moments in the stiffness beam from the run to the supports—if 75% of the running torque is transmitted to the supports, the running mass will be minimal [3]:

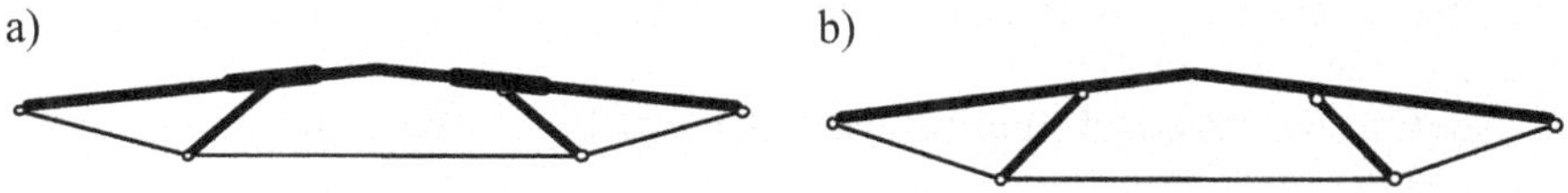

Fig. 3 Schemes of rational combined truss

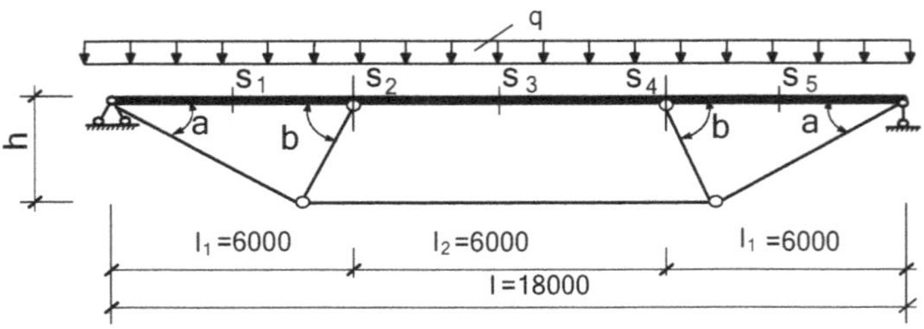

Fig. 4 Scheme of the combined truss with a run of 18 m

$$G = \frac{6\gamma}{R}\int_0^1 \frac{M}{(3-2\alpha)}\mathrm{d}x, \qquad \text{where} \int_0^1 M\mathrm{d}x = \Omega - \text{plot of moments.}$$

On the basis of the method of rational design of structures on the criterion of the minimum cost of steel [1], a constructive solution of the combined rational steel truss has been developed and patented, the number of nodes and elements of which being reduced several times (Fig. 3b). This is a form of construction adapted to real SDS and has a 17–32% lower cost and material consumption compared to typical structures, and equilibrium state is provided in the calculated sections of the stiffness beam due to the calculated regulation SDS [5]. Such combined truss can be manufactured as spatial systems by combining the two. Therefore, the **scientific novelty** of the work is to find a rational constructive form of a combined truss.

Based on numerical studies of rational parameters of the structural form [7], that are used for combined steel sprung truss with a span of 18 m (Fig. 4), the symmetrical linear load is q = 24 kN/m, Ry = 235 MPa, and the rational angle β is 54°.

Accordingly, for this value of angle β, it is necessary to analyze the influence of the lengths of the panels of the beam stiffness ($l_1 = 5$, 6 and 7 m) and K_i—the coefficients of rigidity of the elastic supports, which model the system of reinforcement—sprengel ($K = 3000$, 4000 and 5000 kN/m), and to determine by or close to σ_1, σ_2 and σ_3 rational stress states of the stiffness beam, or close to it and the corresponding geometric and physical structure parameters. This stressful state—with "equal" values of stresses in the calculated sections—will provide the maximum efficiency and rationality of the structures [2, 4], where σ_1 and σ_3 are the stresses in the calculated section of the first and second panels of the stiffness beam in MPa and σ_2 is the stress on the first support in MPa. For this purpose, the calculation of the sprung truss on a computer in the PC "Lira" has been carried out and diagrams compiled (Figs. 5, 6 and 7).

Therefore, the **practical significance** of the work lies in a rational solution for the combined steel sprung truss with a span of 18 m and is characterized by the following optimal parameters: (Figs. 5, 6 and 7) $K = 5000$ kN/m, $l_1 = 6$ m, with almost equal stresses in the calculated sections.

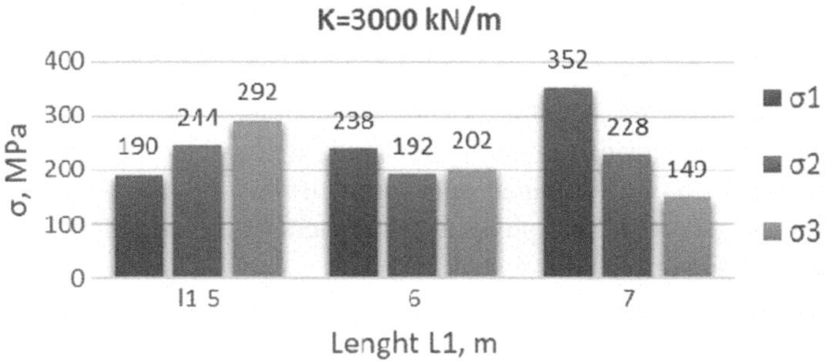

Fig. 5 Tension in the beam of the combined truss stiffness at $K = 3000$ kN/m

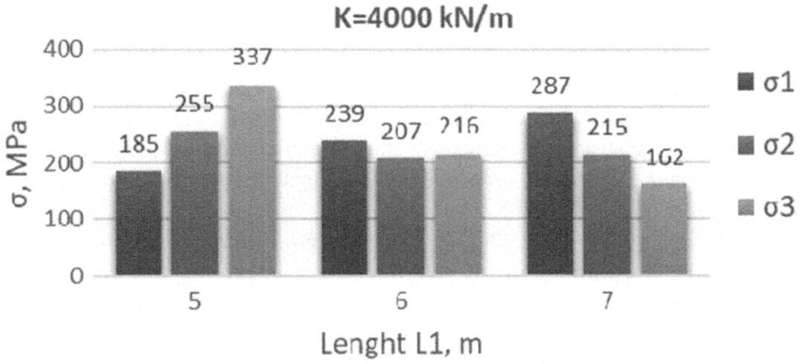

Fig. 6 Tension in the beam of the combined truss stiffness at $K = 4000$ kN/m

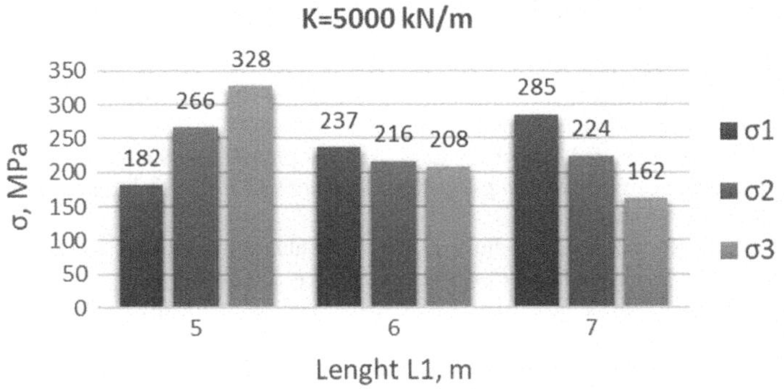

Fig. 7 Tension in the beam of the combined truss stiffness at $K = 5000$ kN/m

3 Conclusions

1. The analysis of numerical studies and publications, as well as the calculation of the minimum plot of moments, has made it possible to identify a rational constructive form of a combined steel sprung truss.
2. The rational geometric and physical parameters of the combined steel sprung truss with a span of 18 m have been found.

References

1. Gogol, M. V.(2018). *Voltage regulation in steel combined structures: Monograph.* Kyiv: Publishers of "Steel".
2. Misiunaitea, I., Juozapaitisa, A., & Daniunasb, A. (2013). Evaluation of requirements for global analysis in EC3 for the structural analysis of the simple-span under-deck cable-stayed bridge. In *11th International Conference on Modern Building Materials, Structures and Techniques. Procedia Engineering, 57*, 781–788.
3. Moskalev, N. S., & Popova, R. A. (2003). *Steel structures of light buildings.* ASV: Moskva.
4. RuizTeran, A. M., & Aparicio, A. C. (2010). Developments in under-deck and combined cable-stayed bridges. *Proceedings of the ICE. Bridge Engineering, 163*(2), 67–78.
5. Shimanovsky, O. V., & Gogol, M. V. (2018). A new approach to designing effective combined steel trus. In *International Scientific and Practical Conference «Technology, Engineering and Science–2018»*. London.
6. Juozapaitis, A., Šaučiuvenas, G., & Nagevičius, J. (2007). Strut-framed beam structure for reconstruction of pedestrian bridges. *Technological and Economic Development of Economy, 13*(2), 126–133.
7. Gogol, M., Galinska, T., & Kropyvnytska, T. (2018). Innovative combined truss: Experimental and numerical research. *International Journal of Engineering and Technology, 7*(4.8), 84–90. doi: https://www.doi.org/10.14419/ijet.v7i4.8.27219.
8. Janusaitis, R., Keras, V., & Mockiene, J. (2003). Development of methods for designing rational trusses. *Journal of Civil Engineering and Management, 9*(3), 192–197. https://doi.org/10.3846/13923730.2003.10531325.
9. Goremikins, V., Rocens, K., & Serdjuks, D. (2010). Evaluation of rational parameters of trussed beam. *Scientific Journal of RTU. Construction Science, 11,* 21–25.
10. Aliawdin, P., & Urbańska, K. (2013). Limit analysis of geometrically hardening rod systems using bilevel programming. *Procedia Engineering, 57,* 89–98.
11. Trussed-boeing-capable-going-jet-speeds-our-future, http://www.myatg.aero/, Last Accessed 2019/09/12.

Mathematical Model for Clarifying Low-Concentration Suspension by Dissolved Air Flotation

O. Haiduchok, O. Syrovatsky, A. Karahiaur, and S. Kostenko

Abstract A mathematical model for removing low-concentration suspension from raw water by dissolved air flotation method for drinking water is described. *Methods* The mathematical model is based on the known equations and dependencies of mass transfers. It takes into account the effect of the delayed bubbles on their rise velocity, as well as technological and design parameters. *Results* The numerical value of the flotation process constant has been obtained for waters with low concentrations of suspended solids. *Scientific novelty* The mathematical model which, for the first time, works with regard to the effect of bubbles delayed on rise velocity, as well as technological and design parameters such as water rise velocity, suspended solid concentration and particle size of containments, size of air bubbles and their concentration, is demonstrated. *Practical significance* The developed mathematical model enables in the future to study the design parameters of flotation tank and find the optimal dimensions with a high degree of clarification.

Keywords Clarification · Drinking water supply · Suspended solids · Dissolved air flotation · Mathematical model · Bubble · Flotation process constant

1 Introduction

Nowadays, a big problem is supplying water to consumers in the required quantity and quality according to the Ukrainian standards. Primarily, this is due to increasing level of contamination in surface water as a result of the discharge of inadequately treated wastewater from industrial and municipal wastewater treatment plants or from private households. Increasing urban population reinforces loads on wastewater treatment plants, and as a consequence, there appears a high probability for the pollutants to get into surface water. Due to the regulations of most open water bodies in Ukraine, water quality is characterized by low concentrations (up to 50 mg/l) of suspended

O. Haiduchok · O. Syrovatsky · A. Karahiaur
Kharkiv National University of Civil Engineering and Architecture, Kharkiv, Ukraine

S. Kostenko (✉)
Poltava National Technical Yuri Kondratyuk University, Poltava, Ukraine
e-mail: sergiy_kostenko@ukr.net

V. Onyshchenko et al. (eds.), *Proceedings of the 2nd International Conference on Building Innovations*, Lecture Notes in Civil Engineering 73,
https://doi.org/10.1007/978-3-030-42939-3_7

solids. These are colloidal substances, silt, sand and clay [1]. The particle size of suspended solids is usually not bigger than 10^{-6}–10^{-3} m, which complicates their removal at traditional facilities [2].

The traditional technological scheme of clarifying surface water for drinking water supply in Ukraine is a two-stage sedimentation filter scheme [3] or one-stage scheme with contact illuminators. In current conditions, the main disadvantages of these schemes are the low-efficient detention of solid particles of suspended particles, as well as the low relative productivity of plants, even when using reagents. At present, very perspective method for removing suspended solids is flotation method [4–6], but it is not widespread in Ukraine. The experience of the existing plants, conducting dissolved air flotation [7, 8] for clarification of the low-turbidity waters, shows that there are no methods for determining the most rational mode of operation, the calculation of the optimal design and technological schemes of such facilities. In addition, the mutual influence of the main factors of dissolved air flotation process and their rational values have not been defined yet.

2 The Purpose of the Research

The research is aimed at studying the key impact factors of dissolved air flotation on removing suspended solids from surface water and getting their rational values.

3 Methodology of the Theoretical Studies and Their Results

To achieve the aforementioned purpose, it is necessary to create a mathematical model of dissolved air flotation process for removing suspended solids from the water. To develop the mathematical model, we have made the following assumptions:

(1) The water flow rate is constant.
(2) The distribution of the concentration of suspended solids and bubbles over the cross-sectional area is equal.
(3) The complete separation of bubbles from the water is carried out at the beginning of the flotation chamber.
(4) The size of bubbles does not change during floating.
(5) Bubbles have a spherical shape.

Figure 1 shows a calculation scheme of a mathematical model for removing suspended solids by dissolved air flotation method.

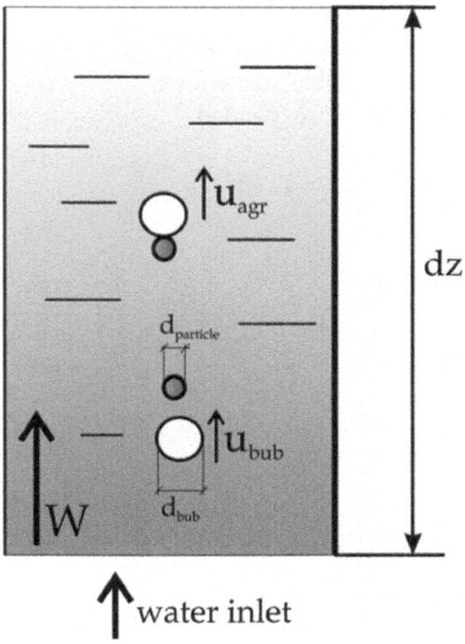

Fig. 1 Scheme of a mathematical model for removing suspended solids by dissolved air flotation method: d_{bub} diameter of bubble; $d_{particle}$ diameter of solid particle; u_{agr} 'bubble–particle' rise velocity; u_{bub} bubble rise velocity

Our mathematical model consists of the following equations:

1. The mass transfer equation (determining distribution of suspended solid concentration by height in working area of a flotation tank):

$$\frac{\partial C(z,t)}{\partial t} + W\frac{\partial C(z,t)}{\partial z} = D_C\frac{\partial^2 C(z,t)}{\partial z^2} - J(z,t), \tag{1}$$

where t is the time, s;

z is the vertical coordinate, m,

W is the rise velocity, m/s;

$C(z, t)$ is the suspended solid concentration in water, kg/m^3;

D_C is the diffusion coefficient, m^2/s;

$J(z, t)$ is the flow of suspended solid concentration from the liquid phase into the gaseous phase, kg/(m^{3}s).

According to the ideas on the kinetics of flotation process [9], the flow J is proportional to bubble (air) concentration and suspended solid concentration in water:

$$J(z,t) = b(z,t)KC(z,t), \tag{2}$$

where b is the volumetric concentration of bubbles (air) in water;

K is the flotation process constant.

The numerical value of the flotation process constant (K) is difficult to determine. A lot of scientists have developed techniques to obtain its numerical value for different conditions [10], but for low-turbidity waters similar researches have not been conducted. Therefore, it has been suggested that this coefficient is equal to the

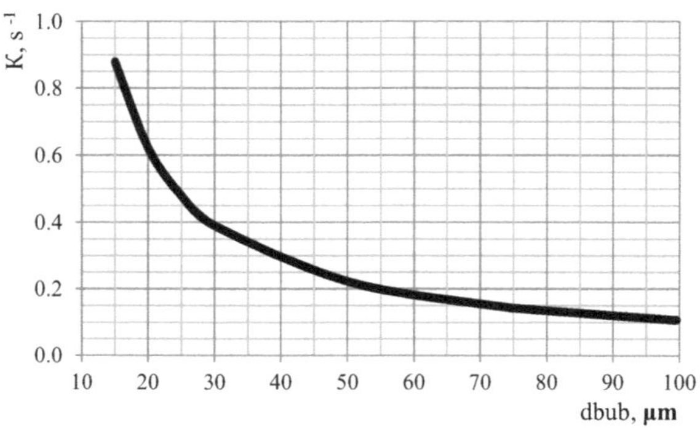

Fig. 2 Dependence of the flotation process constant (K) on the bubble size with a particle size of suspended solids—2 µm

probability of unit-time collision of one particle with bubble contained in a unit of volume (P_{CN}) [11].

Using equations [12], we obtained the numerical value of flotation process constant for the contaminant particles which are contained in surface water and which cause the turbidity of water. Figure 2 shows the dependence of the flotation process constant (K) on the bubble size with a particle size of suspended solid 2 µm.

The dependence shown in Fig. 2 can be described by the empirical equation:

$$K = a_1 \cdot (2R)^{-a_2}, \tag{3}$$

where $2R = d_{bub.}$ For particles which have size of about 2 µm, these coefficients are: $a_1 = 17{,}1$; $a_2 = 1{,}106$.

2. Concentration transfer equation (distribution of bubble concentration on height of working area in a flotation tank):

$$\frac{\partial b(z,t)}{\partial t} + \frac{\partial((W + u_{bub}(z,t))b(z,t))}{\partial z} = D_b \frac{\partial^2 b(z,t)}{\partial z^2} \tag{4}$$

where $u_{bub}(z, t)$ is the bubble rise velocity in stationary water, m/s;

D_b is the diffusion coefficient of bubble concentration, m^2/s.

The initial bubble rise velocity can be found by the Stokes equation:

$$u_{bub} = \frac{g(\rho_w - \rho_{air})d_{bub}^2}{18\mu}, \tag{5}$$

where ρ_w is the water density, kg/m^3;

ρ_{air} is the air density, kg/m^3;

μ is the dynamic viscosity, kg/m•s.

In the process of floating, particles are attached to bubbles. As a result, the density of 'bubble–particle' increases. Changing density unit in the process of floating can be estimated using our equation:

$$\rho_{agr}(z,t) = \frac{\rho_{air} + \frac{S(z,t)}{b(z,t)}}{1 + \frac{S(z,t)}{b(z,t)}\frac{1}{\rho_{ss}}}, \tag{6}$$

where S is the concentration of suspended particles on bubbles, kg/m^3;

ρ_{ss} is the density of particle suspension, kg/m^3.

To further determine the 'bubble–particle' rise velocity in Eq. (5), instead of the air density ρ_{air}, we need to substitute unit density ('bubble–particle') ρ_{agr}.

3. Suspended particle concentration on bubbles is determined by the mass transfer equation:

$$\frac{\partial S(z,t)}{\partial t} + \frac{\partial((W + u_{bub}(z,t))S(z,t))}{\partial z} = KC(z,t) \tag{7}$$

Equation (7) contains the convective component $\frac{\partial((W+u_{bub}(z,t))S(z,t))}{\partial z}$. It takes into account the fact that the change of parameter S at a given point with the coordinate z occurs not only because of the transfer from liquid, but also as a result of moving bubbles with velocity $W + u_{bub}$.

4 Scientific Novelty

Scientific novelty of the research lies in the development of a mathematical model for clarifying low-concentrated suspension by dissolved air flotation. This suspension is the characteristic of the qualitative composition for surface water sources. We have obtained the mathematical model that takes into account for the first time the effect of bubbles delayed on rise velocity, as well as technological and design parameters such as water rise velocity, suspended solid concentration and particle size of containments, size of air bubbles and their concentration.

5 Practical Significance

Our mathematical model enables to improve existing or to develop new designs of a flotation tank. This will ensure high-quality clarification of natural water from surface sources for drinking water supply.

6 Conclusion

Thus, the proposed mathematical model for clarifying low-concentration suspension by dissolved air flotation takes into account: 'bubble–particle' unit rise velocity, suspended solid concentration, particle size, bubble concentration and their sizes. As a result, our next step is to explore the design parameters for the quality of cleaning and to select the optimal design of a flotation tank.

References

1. Syrovatsky, O., Babenko, S., Haiduchok, O., & Rubachyk, Yu. (2015). Ways to improve the efficiency of clarification of natural low turbidity colored water by dissolved air flotation method. *Scientific Bulletin of Civil Engineering, 80*(2), 209–213.
2. Epoyan, S. M., Kolotilo, V. D., & Drushliak, O. G. (2010). *Vodopostachannia ta ochystka pryrodnykh vod: Navchalnyi posibnyk Kharkiv: Factor.*
3. Epoyan, S., Karahiaur, A., Volkov, V., & Babenko, S. (2018). Research into the influence of vertical drainage elements on the operational efficiency of rapid filters. *Eastern-European Journal of Enterprise Technologies, 1/10*(91), 62–69. doi:https://www.doi.org/10.15587/1729-4061.2018.123559.
4. Haarhoff, J. (2008). Dissolved air flotation: Progress and prospects for drinking water treatment. *Journal of Water Supply: Research and Technology, 57*(8), 555–567. doi:https://www.doi.org/10.2166/aqua.2008.046.
5. Edzwald, J. K. (2008). *Develompents in dissolved air flotation, water quality technology conference and exposition, 2008*. New York, USA: American Water Works Association.
6. Zhu, H. Y., Zhou, M., & Liu, X. (2014). Dissolved air theory in water process research overview. *Advanced Materials Research*, 971–973, 2035–2038. doi:https://www.doi.org/10.4028/www.scientific.net/AMR.971-973.2035.
7. London, S. (2013). New water treatment plant features first dissolved-air flotation systems in Illinois. *Journal–American Water Works Association, 105*, 18–21.
8. Sohn, B.-Y., Park, T.-J., Oh, B. S., Kwon, S.-B., & Kang, J.-W. A case study of the DAF-based drinking water treatment plant in Korea. *Separation Science and Technology, 43*(15), 3873–3890. doi:https://www.doi.org/10.1080/01614940802286008.
9. Rubynshtein, Y. B., & Fyllypov, Y. A. (1980). *Kynetyka flotatsyy*. Moscow: Nedra.
10. Hoden, A. M. (1959). *Flotatsyia*. Moscow: Hosudarstvennoe nauchno-tekhnycheskoe yzdatelstvo lyteraturы po hornomu delu.
11. Karahiaur, A. S., Syrovatsky, O. A., Titov, A. A., & Haiduchok, O. G. (2018). Determining constants flotation process for particles which case turbidity natural water. *Scientific Bulletin of Civil Engineering, 94*(4), 177–182. doi:https://www.doi.org/10.29295/2311-7257-2018-94-4-177-182.
12. Deriahyn, B. V., Dukhyn, S. S., & Rulev, N. N. (1986). *Mykroflotatsyia: Vodoochystka, obohashchenye*. Moscow: Khymyia.

Features of Operation and Design of Steel Sloping Roof Purlins

Serhii Hudz, Leonid Storozhenko, Grygorii Gasii, and Olena Hasii

Abstract The main stages of steel-restrained beams–purlins of the sloping roof for the typical metal frame with the initial geometric imperfections under the deformed scheme are considered. The restraint can occur, for example, by means of connecting to the steel beam light enclosing structures of warehouse space, namely profiled flooring. On the basis of a new approach to the analysis of internal power factors, it was aimed to find and describe the peculiarities of beam operation conditions under the joint action of the transverse bending in two planes and the warping torsion, which arises due to the eccentric laying of the load. The existing theoretical model has been improved by eliminating the shortcomings present in it. There has been carried out the comparison of techniques for determining bearing capacity by stability for flexible elements that are affected by bending and bending with torsion. Their advantages and disadvantages have been revealed according to the principles of simplicity and accuracy. It is proposed to increase material savings through a detailed calculation.

Keywords Bimoment · Restraint · Buckling · Torsion · Curvature

1 Introduction

The profiled flooring in the restrained structure increases the rotational stiffness of the purlin, so the actual work of the bearing elements of the sloping roof often does not correspond to most of the classical theoretical notions on the warping torsion of thin-walled rods. Existing processes are not always reflected in modeling, which leads to inconsistencies in the model to real picture of a stress–strain state. With sufficient shear stiffness, it is believed that the compressed belt is fully fixed against

S. Hudz · L. Storozhenko
Poltava National Technical Yuri Kondratyuk University, Poltava, Ukraine

G. Gasii (✉)
Sumy National Agrarian University, Sumy, Ukraine
e-mail: gasiigm@gmail.com

O. Hasii
Poltava University of Economics and Trade, Poltava, Ukraine

V. Onyshchenko et al. (eds.), *Proceedings of the 2nd International Conference on Building Innovations*, Lecture Notes in Civil Engineering 73,
https://doi.org/10.1007/978-3-030-42939-3_8

the transverse shifts and the axis of rotation, which in this case is called restrained, and passes through the beam top. Due to the high convergence of results of calculation and modeling, it has been revealed that for channel with sufficient stiffness of the decking the rotation axis is placed above the shear center at the level of the upper flange, where the load sloping component is applied, which hardly has eccentricity relative to this point. Therefore, warping stresses will be formed to a greater degree from the main component, and the sloping one is practically leveled. Tangible savings can be achieved by reducing the impact of the load sloping component on the bending stresses.

2 The Main Body

2.1 Overview of Recent Research Sources and Publications

Current theoretical and experimental–theoretical research of the stress–strain state and stability of thin-walled steel beams, without restrained and with restrained enclosing structures, are presented in [1–3] and other relevant studies [4–6]. The main sources, which indicate the research results of torsion and stability of beams and other thin-walled rods, include the scientific works of the classical representatives of the technical school [7–10]. The application of the energy approach to stability tasks has been rapid development and is accompanied by the implementation of computational methods [11, 12]. The concept of the lateral torsional buckling theory is found in source [13]. The definition of equivalent loads for stabilizing structures is analyzed in the dissertation [14] and article [15]. The calculation of the beams at the bend and torsion by one option of the second-order theory is found in work [16]. Theoretical principles and preconditions for the calculation of the restrained elements of the decking construction under difficult conditions of download are provided in article [17–19].

2.2 Purpose

Based on the experience of the analysis of internal forces, taking into account the constructive features of the conditions for the structure working in the joint action of transverse bending and torsion, it was decided to find and describe the differences in the operation of the decking, which distinguishes it from the work of unrestrained beam.

2.3 Methodology

Stress–strain state of a pre-curved and torsioned thin-walled beam loaded with evenly distributed transverse load in two planes, taking into account the beam fixing from transverse displacements, as well as the height of the load application in the general form, can be described by means of the spatial model for the description of stabilizing equivalent loads with the widespread system of differential equations of rod equilibrium taken from work [14, 20] (but without the longitudinal force influence, also the system is different by adding an equation describing the deflection of the beam and the part of the third equation describing the torsion of the beam, which depends on the application place of the sloping load component and deflection):

$$\begin{cases} EI_y w^{\mathrm{IV}} + (M_z(\vartheta_0 + \vartheta))'' = q_z; \\ EI_z v^{\mathrm{IV}} + \left(M_y(\vartheta_0 + \vartheta)\right)'' = q_y; \\ EI_\omega \vartheta^{\mathrm{IV}} - GI_t\vartheta'' + c_\vartheta \vartheta + q_z z_P(\vartheta_0 + \vartheta) + M_y(v_0 + v)'' + \\ + q_y y_P(\vartheta_0 + \vartheta) + M_z(w_0 + w)'' = m_x + q_y \dfrac{h_s}{2}, \end{cases} \quad (1)$$

where $\vartheta_0,\ v_0,\ w_0$ are initial rotation, curvature and deflection of the beam;

$\vartheta,\ v,\ w$ are angle of torsion, curvature and deflection of the beam under load;

$E,\ G$ are elasticity and shear modules, respectively;

$I_y,\ I_z$ are moments of inertia of cross section relative to the y- and z-axes, respectively;

$M_y,\ M_z$ are usual design bending moments relative to the y- and z-axes, respectively;

I_ω is the warping constant of the beam cross section;

I_t is the torsional constant of the beam cross section;

C_ϑ is the rotational stiffness of the profiled flooring;

$Z_P,\ Y_P$ are coordinates of the load laying point relative to the shear center of the beam;

$q_z,\ q_y$ are evenly distributed transverse loads operating along the z- and y-axes, respectively (for a purlin—projections of the design evenly distributed the load q on the axes z and y: $q_z = q\ \cos\ \alpha$; $q_y = q\ \sin\ \alpha$, where α is a cross-sectional slope angle); when matching direction of load action with direction of axis, the sign is taken positive;

m_x is the evenly distributed torsion load.

The linear m_x torsion load on the beam with sufficient shear stiffness will be determined by the formula excluding the load sloping component:

$$m_x = q_z e_z, \quad (2)$$

e_z is the shoulder of load q_z, which can be applied in the flange middle or in the web middle at the upper belt level, relative to the shear center, taking into account the rule of the signs at torsion (for channel beam $e_z = -(e - 0,\ 5t_w + 0,\ 5b_f)$, where e is the distance from the shear center to the web center; t_w is the beam web thickness; and b_f is the beam flange width, or $e_z = -e$, that is not fully elucidated), and it is better to take in the safety margin of the first bigger value, which is more disadvantageous in determining the bearing capacity.

As the deflection in the first equation and the curvature in the second equation of system Eq. (1) are independent from each other, then the system can distinguish the last two equations. Warping torsion of the beam can be represented from the equilibrium conditions in the form of belt bending. Then, the torsional load can be decomposed into a pair of conditional forces, applied at the level of flange weight centers, opposite aimed along the belts and operating in the plane of less stiffness profile.

Knowing deformations ϑ, v, w and their derivatives ϑ', ϑ'', v', v'', w', w'', stabilizing equivalent loads and internal forces required can be defined to determine the bearing capacity of the element. When entering the precondition for the existence of a restrained axis of rotation due to the presence of sufficient shear stiffness of connected structures, deformations v and θ are no longer independent from each other (see Fig. 1). Therefore, to reduce the number of unknown variables in the equation, it can be simplified, fair at small angles of torsion (sign «–» means fixing in the upper belt plane, and the linear deformation is counted from the center of the profile weight):

$$v = -\frac{h_s}{2}\vartheta. \tag{3}$$

On the other hand, there are additional deformations of torsion from the load sloping component with a part of the equivalent load, which loads or unloads the bent beam in the opposite direction from the effect of the pliability with uneven lengthwise eccentricity, equal to its design deflection (see Fig. 1). This factor can be ignored, as was done in [14, 15], because it does not significantly affect the final result, but, however, enables to increase its accuracy.

In case the bending moment is changed along the beam length, the differential equation of the problem of compatible action of transverse bending and torsion has

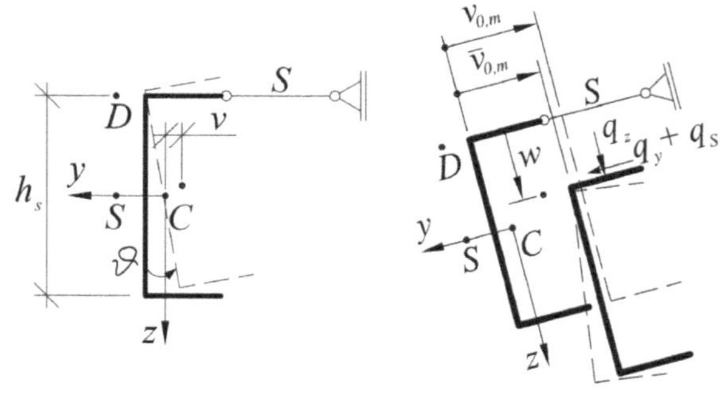

Fig. 1 Interrelation of linear and angular deformations; load and deformation for the restrained channel beam

no closed, relatively simple solution. Therefore, the definition of stabilizing loads is possible only with the help of approximate solutions.

Loads q_z and q_y are transmitted to the beam connected to the upper belt structures. To find the necessary deformation let's apply the condition of equality zero the total elementary possible (virtual) work based on the principle of possible displacements and system of differential equations of the rod equilibrium to the beam with attached at the upper belt level transverse load. Making pre-conversion and simplification when opening brackets for the second derivative from the product of functions, replacing a part of the torsional load with an expression from the second equation of the system (1), we will get the formula to virtual work:

$$\delta W = -\int_0^L \left(\left(EI_z\frac{h_s^2}{4} + EI_\omega\right)\vartheta''\delta\vartheta'' + GI_t\vartheta'\delta\vartheta' + \left(c_\vartheta + q_y y_P\right)\vartheta\delta\vartheta\right)dx - \\ -\int_0^L \left(M_y h_s \vartheta'\delta\vartheta'\right)dx + \int_0^L (m\delta\vartheta)dx = 0. \quad (4)$$

Rewriting the first equation of the system Eq. (1) in the form of the main differential equation of the bent axis (elastic line) of beam for small deformations in space staging $EI_y w'' = -M_y - M_z(\vartheta_0 + \vartheta)$ and having expressed from it a second derivative from the function of deflection, a torsional load m in Eq. (4), assuming that the curve for initial curvature is taken by sine (it is possible to take the curve for initial curvature by the square parabola), and in the absence of the initial rotation and deflection of the beam will be equal in disregard of the value of higher order of the smallness:

$$m = -M_y v_0'' - M_z w'' + m_x = M_y \bar{v}_{0,m}\frac{\pi^2}{L^2}\sin\frac{\pi x}{L} + \frac{M_z M_y}{EI_y} + m_x, \quad (5)$$

where $\bar{v}_{0,m} = v_{0,m} + v_{oc,m}$ is the maximum initial curvature of the beam, taking into account the maximal displacement of upper belt beam restrained by flooring, and $v_{0,m}$ is the maximum initial curvature of the beam, which is determined depending on the type of curve for curvature, cross section and beam span, as well as the type of calculation (e_0 according to notation EN 1993-1-1); when guiding the curvature against the direction of the y-axis, its value is accepted negative by the logic.

Strain state of beam at torsion with sufficient precision can be described with the help of the expression for the torsion angle with seven selected for the convenience of operations of differentiation and integration parameters, three of which prove to be zero:

$$\vartheta = \sum_{j=1}^{7} \vartheta_j \sin\left(\frac{j\pi x}{L}\right). \quad (6)$$

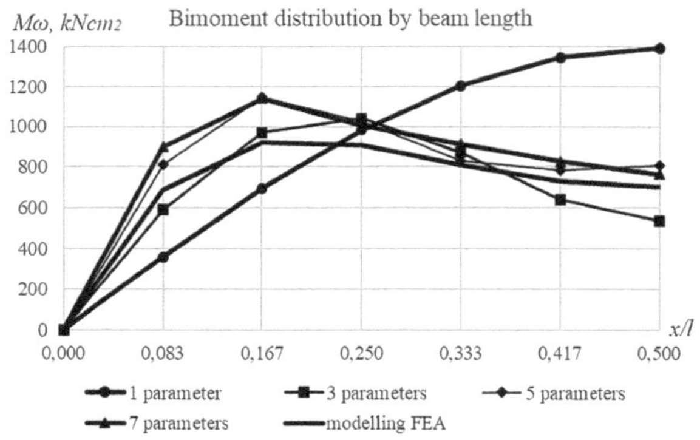

Fig. 2 Bimoment distribution by beam length

A smaller number of parameters result in a too high error in the determination of the bimoment in the middle of the beam span compared to the modeling FEA in the software module FE-LTB of complex Dlubal RSTAB 8.13 (Fig. 2), although it gives satisfactory result in determination of deformations and equivalent stabilizing load.

After entering the expressions for function of the torsion angle and its derivatives in the equation for virtual work and execution of the integration, it becomes possible to obtain the matrix equation to determine the angle of torsion with the unknown parameters, which are sought by the method of inverse matrix. Considering the expressions cumbersome, the formulas to determine the angle of torsion by the matrix method are not given. To determine the stabilizing equivalent loads q_s, let us apply to our case equations of the system Eq. (1). Taking into account the adopted simplifications and after transformations, the formula will get the following look:

$$q_s = -V_z \vartheta' - \frac{M_y}{h_s} v_0'' + \frac{GI_t}{h_s} \vartheta'' + \left(q_z - \frac{c_\vartheta + q_y y_P}{h_s} \right) \vartheta + \frac{M_z M_y}{EI_y h_s} + \frac{m_x}{h_s}, \quad (7)$$

where $V_z = -q_y x + \frac{q_y L}{2}$ is the transverse force along the beam length in the y-axis direction.

By continuous lateral restraint with some shear stiffness, the upper belt shifts from the external and stabilizing equivalent loads. Deformation of the upper flange, which arose as a result of the attachment pliability, can be determined by the formula, taking into account the direction of the action:

$$v_{oc}(x) = \frac{M_{z,s}(x)}{S}, \quad (8)$$

where $M_{z,s}(x) = -\int\int \left(q_y - q(x)\right) \mathrm{d}x \mathrm{d}x$ is the bending moment relative to the z-axis along the beam length from external and stabilizing equivalent loads;

S is the general shear stiffness of the fixing constructions (profiled flooring or profiled flooring and discrete bracing), attributable to one beam.

After completing the integration, we obtain an expression for the finding of the specified maximum value deformation of the beam upper belt by a deformed scheme considering the angle of torsion. Design bending moments M_y^{II}, M_z^{II} relative to the y- and z-axes by the second-order theory are determined based on the angle of the rod torsion:

$$M_y^{II} = M_y^I \cos(\vartheta) + M_z^I \sin(\vartheta); \tag{9}$$

$$M_z^{II} = M_z^I \cos(\vartheta) - M_y^I \sin(\vartheta), \tag{10}$$

where M_y^I, M_z^I are design bending moments relative to the y- and z-axes by the first-order theory; M_z^I is determined by taking into account the influence of equivalent load, which can be considered approximated due to the adoption of bending moment diagram from external and stabilizing equivalent loads in the sinusoid form with the maximum value equal to the product of shear stiffness and displacement of the beam, in the middle of the span $\left(M_y^I\right) = M_y$, $M_z^I = M_z + S\upsilon_{oc,m} \sin\left(\frac{\pi x}{L}\right)$.

2.4 Results

Developed on the basis of the extended system for differential equations of rod equilibrium and possible displacement principle of spatial model for the description of stabilizing loads and method of calculating steel thin-walled beams on its basis by the second-order theory by taking into account the compatible work with the profiled flooring and/or discrete bracing, the pliability of fixing structures, the deformed scheme and the curvature of the beam in the plane of less stiffness have been implemented in the software environment Mathcad. The given technique was tested by comparing the results of the calculation of internal forces and normal stresses for sufficiently restrained rolling beams (channels and I-beams) of the same area of cross section with the simulation data by finite element analysis in the program Dlubal RSTAB 8.13.01 (additional module FE-LTB), discovering at the same time optimal cross section for the sloping roof purlins.

2.5 Scientific Novelty

The comparison of the results suggests that the own method of calculating the beam by the second-order theory is more appropriate than a numerical experiment. Slight discrepancy of the results is due to simplified prerequisites for calculation; besides, it in most cases goes into a slight margin of strength, in contrast to the theory of thin-walled rods, which is limited and suitable only for unrestrained beams due to the adoption of too simple preconditions, which do not take into account the effects of the second order and the stiffness of the connected structures and substantially overestimate the result. The second-order theory convincingly agrees with the numerical experiment. The topical theoretical problem, which requires further research, is the definition of bearing capacity of not sufficiently restrained steel beams with various cross sections, considering the existing reserves of steel plastic work and stiffness of structures during special operating conditions with imperfections.

2.6 Practical Significance

The mentioned factors that characterize the peculiarities of the steel beam working in the decking with complex resistance enable to determine more accurately the values of normal and warping stresses, which affect the general stress–strain state of the structure and determine the calculated ratio. Thanks to the execution of detailed calculation for the above algorithm selection of the smaller cross-sectional steel profile, increase in material savings and facilitation of the construction of the decking become possible.

3 Conclusions

While calculating and researching, it is suggested to pay attention to the factors given below. The angle of the rod torsion significantly influences the values of the design bending moments in two planes, so in the absence of the formula for determining the total stresses of the reduction factor for LTB, which is deducted for the system deformed state, the cross-sectional rotation phenomenon should be considered. In the presence of lateral restraint and the load sloping component, the effect of the latter on the stresses is significantly reduced by decreasing the deformations in the plane of less stiffness (beam curvature). When taking the restrained axis of rotation, warping stresses in beams of sloping roof from load sloping component are not practically formed. The reason for the emergence of warping stresses can be not only the transverse load, but also the presence of geometric imperfection, namely the initial curvature, which can be considered by increasing the uniformly distributed torsional load, which increases the bimoment. The distribution angle of torsion along the beam length, even with hinged support at the ends and applying even uniformly distributed torsion load, is not recommended to accept sinusoid in the case of roof significant stiffness. I-beam cross section next to the channel can be considered the optimal form of cross section of the rolling purlin for sloping roof.

References

1. Balázs, I., & Melcher, J. (2015). Stability of thin-walled beams with lateral continuous restraint. *Technical University of Ostrava, Civil Engineering Series, 15*(1), 1–10.
2. Pavlenko, A., Rybakov, V., Pikht, A., & Mikhailov, E. (2017). Non-uniform torsion of thin-walled open-section multi-span beams. *Magazine of Civil Engineering, 67*(07), 55–69.
3. Put, B. M., Pi, Y.-L., & Trahair, N. S. (1999). Bending and torsion of cold-formed channel beams. *Journal of Structural Engineering, 125*(5), 540–546.
4. Gasii, G. M. (2017). Structural and design specifics of space grid systems. *Наука и техника, 6*, 475–484. https://doi.org/10.21122/2227-1031-2017-16-6-475-484.
5. Gasii, G., Hasii, O., & Zabolotskyi, O. (2017). Estimate of technical and economic benefits of a new space composite structure. In *MATEC Web of Conferences*. Vol. 116, p. 02014). EDP Sciences. https://doi.org/10.1051/matecconf/201711602014.
6. Storozhenko, L. I., & Hasii, H. M. (2015). The new composite designs for mine tunnel support. *Naukovyi Visnyk Natsionalnoho Hirnychoho Universytetu, 4*, 28–34.
7. Bychkov, D. (1962). *Stroitel'naya mekhanika sterzhnevykh tonkostennykh konstruktsiy*. Moscow: Gosstroyizdat.
8. Kindmann, R. (2008). *Stahlbau*. Berlin: Ernst.
9. Kuhlmann, U. (2009). *Stahlbau-Kalender 2009: Schwerpunkt—Stabilitat*. Berlin: Wiley.
10. Vlasov, V. (1959). *Tonkostennyye uprugiye sterzhni*. Moscow: Fizmatgiz.
11. Kindmann, R., & Kraus, M. (2011). *Steel structures: Design using FEM*. Berlin: Wiley.
12. Pilkey, W. D. (2002). *Analysis and design of elastic beams: Computational methods*. New York: Wiley.
13. Galambos, T. V., & Surovek, A. E. (2008). *Structural stability of steel: Concepts and applications for structural engineers*. Hoboken, NJ: Wiley.
14. Krahwinkel, M. (2001). *Zur Beanspruchung stabilisierender Konstruktionen im Stahlbau*. Düsseldorf: VDI-Verl.
15. Kindmann, R., & Krahwinkel, M. (2001). Bemessung stabilisierender Verbände und Schubfelder. *Stahlbau, 70*(11), 885–899. https://doi.org/10.1002/stab.200102860.
16. Frickel, J. (2002). Bemessung von Trägern unter Biegung und Torsion nach Th. *II. Ordnung. RUBSTAHL-Bericht, 2*, 89–94.
17. Hudz, S., Gasii, G., & Pents, V. (2018). The problem of consideration torsion emergence in beams. *International Journal of Engineering & Technology*, *7*(3.2), 141–148. https://doi.org/10.14419/ijet.v7i3.2.14392.
18. Storozhenko L., Butsky V., & Taranovsky O. (1998). Stability of compressed reinforced concrete composite tubular columns with centrifuged cores. *Magazine structural steel research, 46*(1–3), 484. https://doi.org/10.1016/S0143-974X(98)80098-9.
19. Piskunov V. G., Goryk A. V., Lyakhov A. L., & Cherednikov V. N. (2000). High-order model of the stress-strain state of composite rods and its implementation with using computer algebra. *Composite Structures, 48*(1), 169–176. https://doi.org/10.1016/S0263-8223(99)00091-4.
20. Piskunov V. G., Goryk A. V., & Cherednikov V. N. (2000). Modeling the transverse shear of piecewise homogeneous composite rods using iterative process taking into account tangential loads. 1. Building a model. *Mechanics of Composite Materials, 36*(4), 287–296. https://doi.org/10.1007/BF02262807.

Improvement of Residential Buildings Walls Operation Thermal Mode

Alla Kariuk, Victoria Rubel, Victor Pashynskyi, and Stanislav Dzhyrma

Abstract Temperature fields are analyzed for a wide range of external air temperatures in the contact node point of the window block adjoining the brick wall of a typical residential building constructed in the second half of the twentieth century. Replacing windows with modern thermal efficient structures significantly reduces overall heat losses, but does not exclude condensation formation on the internal window slopes in the window blocks. Significant increase in slope temperature can be achieved by shifting the window block inside the building. According to the numerical experiment results, the dependence of critical point temperature at the place of inner slope to the window box surface adjoining from the outside air temperature and the window block position along the wall thickness is analytically substantiated and described. Depending on the temperatures of the dew point and the outside air, the position of the window block on the wall thickness is determined, which ensures condensation absence on the inner window slopes surface and heat losses reduction through the wall contact node points.

Keywords Brick walls · Window blocks · Wall temperature · Condensation · Heat loss · Dew point · Isotherms

1 Introduction

In the second half of the XX century, residential buildings with brick walls with a thickness of 510 mm and wooden windows with separate or paired shutters were massively built in Ukraine. At that time, the requirements for resistance to the heat transfer of walls (about 1 $m^2 \cdot K/W$) were three times as low as the current requirements [1], equal to 2.8–3.3 $m^2 \cdot K/W$. In addition to the high cost of heating energy for heating, in such walls, condensation is often formed due to the fall in temperature in the areas of cold bridge ways below the dew point. One of these zones is a contact window

A. Kariuk (✉) · V. Rubel
Poltava National Technical Yuri Kondratyuk University, Poltava, Ukraine
e-mail: akariuk1975@gmail.com

V. Pashynskyi · S. Dzhyrma
Central Ukrainian National Technical University, Kropyvnytskyi, Ukraine

V. Onyshchenko et al. (eds.), *Proceedings of the 2nd International Conference on Building Innovations*, Lecture Notes in Civil Engineering 73,
https://doi.org/10.1007/978-3-030-42939-3_9

node to brick walls [2–4]. Periodic soaking caused the failure of wooden window boxes and shutters. In order to increase energy efficiency, the thermo-modernization of such buildings [5] is usually carried out toward exterior facade insulation as well as the replacement of window blocks with modern polyvinylchloride constructions with fiberglass.

Heat engineering calculations and operation experience of replacing windows with modern thermal efficient structures significantly reduces overall heat losses, but does not exclude condensation formation on the internal window slopes in the window blocks. Significant increase in slope temperature can be achieved by shifting the window block inside the building. The idea was expressed in works [2, 6], but displacement expedient amount remained groundless.

The purpose of the present research is to elaborate recommendations for window installation scheme, which at a given outside air temperature will provide the temperature of the inner window slope not lower than the dew point and eliminate the possibility of condensate formation in the window zone contact to the wall.

2 Node Design and Output Data for Analysis

The contact node of the metal-plastic window block to the brick wall is considered. The brick wall is 510 mm in thickness, made with a quarter in width 120 mm. In the initial state, a wooden window block is installed close to a quarter, and the assembly gaps are sealed with felt and cement mortar. When replacing the window, the metal-plastic coupler is executed in accordance with the requirements of the standard [4].

When the window block is displaced into the wall thickness, the gap formed between the protruding part of the quarter of the wall and the window box outer surface is filled with foam polystyrene foam 4. Internal slopes can be executed from a plaster solution, drywall with insulation, sandwich panels, and so on.

Necessary thermal characteristics for the calculated values of all materials are adopted in accordance with the standard [7]. For the window box in Table 1, thermal

Table 1 Temperature dependence of the internal slope on the outside air temperature and window block position in the wall thickness

Outside air temperature T_{out} [°C]	Temperatures at the critical point of the inner slope surface T_{cr} [°C] window installation depth X [mm]				
	120	160	200	240	280
−25	6,7	7,7	8,8	9,8	10,3
−20	8,2	9,0	10,0	10,9	11,4
−15	9,6	10,4	11,3	12,0	12,5
−10	11,1	11,8	12,5	13,2	13,6
−5	12,6	13,1	13,8	14,3	14,6

conductivity equivalent coefficient, established due to the total thermal resistance and thickness of the three-chamber polyvinylchloride profile, is given.

The temperature of the dew point $T_d = +10{,}7$ °C below, which the temperature of the inner slope surface should not be lowered, are determined by psychrometric tables in accordance with the norms [1] of the normative value of the air temperature in the living space $T_{\text{in}} = +20$ °C and relative humidity 55%.

Atmospheric air temperatures in Ukraine have been analyzed in detail in the monograph [8]. Taking into account the considerable thermal inertia of the wall, as the estimated values of outside air temperature during the winter period, the coldest five-day temperatures with probability 0,92 are selected. According to [8, 9], these temperatures decrease from south to northeast of Ukraine and change from $T_{\text{out}} = -6$ °C in Yalta to $T_{\text{out}} = -25$ °C in Lugansk and Sumy. In order to fully cover the climatic conditions of the territory of Ukraine for numerical simulation, the calculated values of the outside air temperature are $T_{\text{out}} = -5, -10, -15, -20, -25$ °C.

Internal air temperature in all calculations is assumed to be equal to $T_{\text{in}} = +20$ °C in accordance with the requirements of standards [1] applied to residential premises.

3 Research Methodology

Condensate formation possibility due to the occurrence of thermal failure can be detected by comparing the temperature of the inner slope surface with the above or another dew point temperature. To do that, the temperature field in a fairly complex contact node window block to the wall should be analyzed. The analysis of two-dimensional temperature fields can be accomplished with the help of special software complexes that implement the finite element method simulation.

One of such programs is the THERM software, developed at the Lawrence Berkeley Laboratory (LBNL) of the University of California, USA and available in free access [10]. By simulating the processes of heat transfer in fencing structures of complex configuration by the finite element method, the program enables to construct two-dimensional thermal fields, to identify the problem areas and the local temperatures in separate points of the design.

Examples of temperature fields derived from the calculation in the THERM environment are shown in Fig. 1. They represent calculation results characteristic for most of Ukraine's areas, with the temperature of the outside air $T_{\text{out}} = -20$ °C and two values of the distance from the outer surface of the wall to the window box: $X = 120$ mm (close to a quarter of the brick wall) and $X = 240$ mm (with a width of 120 mm foam polystyrene insert).

From the diagrams shown in Fig. 1, it is evident that at a distance from the node more than 1–1.5, the wall thickness of the isotherms obtained is practically parallel, indicating a heat flow perpendicular to wall surface area. While approaching the window the parallel isotherms approach and bend toward the window box. The coldest zone of the wall inner surface is located at the intersection of the inner slope with the surface of the window box. From Fig. 1, it can be seen that when the window

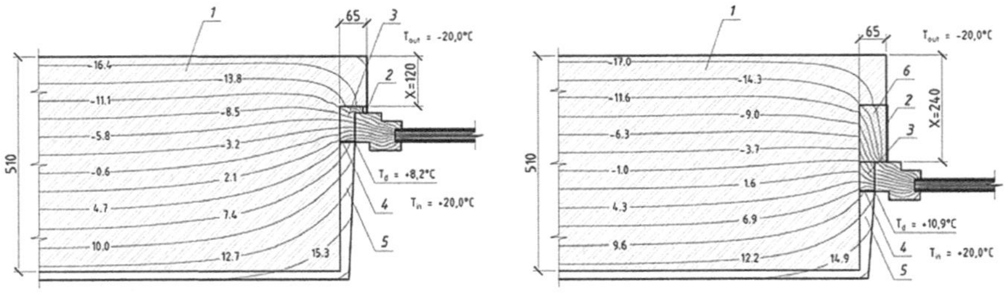

Fig. 1 Temperature fields in node window block contact to the wall. 1—Ceramic bricks common on cement-sandy mortar; 2—pre-compressed sealing tape (PSUS) or waterproofing vapor permeable tape; 3—mounting foam, 4—vapor barrier; 5—plaster solution; 6—foam polystyrene tab

is installed close to the quarter of the wall at $X = 120$ mm, the temperature at this critical point is equal to $T_{cr} = +8.2$ °C, and at a distance $X = 240$ mm increases to $T_{cr} = +\ 10.9$ °C. Consequently, at the above temperature, dew point $T_d = +$ 10,7 °C, window displacement inside the room outputs a critical point from the zone of possible thermal failure and prevents the condensation formation on the inner window slope. Similar technique was used in [3] to study the possibility of condensation formation in the contact nodes windows for buildings located in one of the cities of Ukraine.

4 Numerical Experiment Results

The possibility of thermal failure is predicted by the temperature dependence at the critical point of the T_{cr} node from the outside air temperature T_{out} and the position of the window block on the thickness of the wall *X*. To construct this dependence, a numerical experiment was carried out, the processing in the THERM program environment, 25 variants of the above-mentioned node of the window block adjoining to the brick wall were performed at the following variables parameters:

- 5 values of external air temperature $T_{out} = -25, -20, -15, -10, -5$ °C;
- 5 values of the installation depth of the window $X = 120$ mm, 160 mm, 200 mm, 240 mm, 280 mm.

Typical examples of simulation results with nodes constructed by isotherms are shown in Fig. 1. Critical points of temperatures at the point of intersection of the internal slope with the window box surface for all 25 variants of the simulated nodes are given in Table 1.

The table demonstrates that the temperature at the critical node point T_{cr} increases with the rise of the outside air temperature T_{out} and the increase in the depth of the window block installation *X*.

Dependence on both factors has a character close to the linear one. It enables to describe the general temperature dependence at the critical point T_{cr} from the design

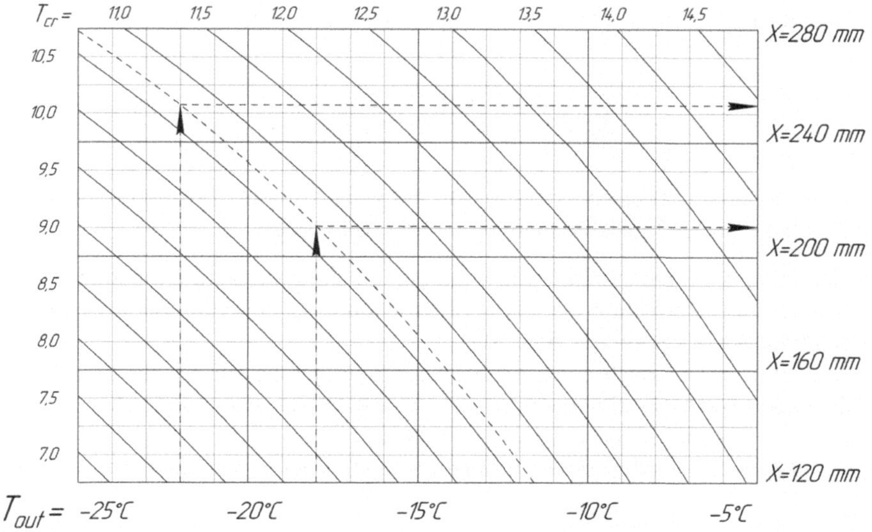

Fig. 2 Temperature dependence at the critical node point from the external air temperature T_{out} and window block installation depth X

temperature of the outside air T_{out} and window block installation depth X by the analytical expression

$$T_{cr} = 12,67 + 0,358\ T_{out} + 0,0115\ X - 0,00054\ T_{out}\ X, \qquad (1)$$

coefficients are obtained by least squares mhod.

The deviation of Formula (1) from the actually calculated temperatures from Table 1 does not exceed 1.1%, which is quite a sufficient precision for heat engineering calculations.

In Fig. 2, the dependence Eq. (1) is represented in the form of a surface depicted by temperature isotherms at the critical node point T_{cr}. The drawing clearly displays the temperature dependence at the critical node point T_{cr} from the values T_{out} and X and can be used instead of Formula (1).

5 Recommendations for Determining Window Block Position in the Brick Wall Thickness

The obtained data enable to determine window block position, which excludes the condensation possibility, taking into account the geographical area of construction and the microclimate of the premises. It can be calculated by substituting (1) the temperature at a critical point, equal to the temperature of the dew point $T_{\text{cr}} = T_d$, and solving the equation obtained with respect to the installation depth of the window X:

$$X = \frac{12,67 + 0,358\ T_{out} - T_d}{0,00054\ T_{\text{out}} - 0,0115}. \tag{2}$$

It is advisable to perform operations to the window blocks position in the brick wall thickness when replacing them in the following order:

- According to the standard [9] the coldest five-day estimated temperature Tout is determined providing 0.92 or 0.98 depending on the permissible frequency condensation formation 12.5 or 50 years;
- According to the psychrometric tables, dew point temperature T_d is determined, depending on the specified temperature values [1] Tin values and the humidity of the air in the appropriate destination premise;
- Recommended window block installation depth is determined by Formula (2) or Fig. 2.

For example, at a dew point temperature for residential premises $T_d = 10{,}70$ °C in Odessa city conditions with the calculated outside air temperature $T_{\text{out}} = -18$ °C, it is advisable to install a window at a distance $X = 211$ mm from the external surface of the brick wall, in Kropyvnytskyi at $T_{\text{out}} = -22$ °C—at a distance of $X = 253$ mm, and in Sumy with $T_{\text{out}} = -25$ °C—at a distance of $X = 279$ mm. Similar results can be obtained using the graph in Fig. 2, where the dependence Eq. (2) is represented by the level lines.

6 Conclusions

1. Temperature fields analysis method and temperatures determination at critical contact node points of translucent structures to the buildings walls, which takes into account the site design, as well as internal and external air temperatures, are developed.
2. Graphical and analytical temperature dependences at the window block critical contact node point to the brick wall of a typical residential building of the twentieth century from calculated outside air temperature and distance from the outer surface of the wall to the window block are obtained.
3. The resulting dependencies enable, taking into account the temperatures of the dew point and the outside air, to determine the window blocks position on the outer wall thickness, which ensures condensation absence on the surface at the inner window slopes.
4. The method elaborated can be used for other design nodes analyses, made from other materials, under other external air temperatures influences and room microclimate parameters.

References

1. DBN B.2.6-31. (2016). Thermal insulation of buildings. Kiev (2016). Ministry of Construction of Ukraine.
2. Boriskina, I., Boriskina, G., & Plotnikov, A. (2005). Design of modern window systems for civil buildings.
3. Pashynskyi, V., Nastoyashiy, V., & Dzhyrma, S. (2017). The influence of the position of the window blocks on the wall thickness on the thermal characteristics of the site of their adjoining. *Sciences of Europe, 3*(21), 8–13. Praha: Global Science Center LP.
4. DSTU B V.2.6-79. (2009). Construction of buildings and structures. The seams are the connecting points of the adjoining window blocks to the wall structures. Kiev (2009). National standard of Ukraine.
5. Blumberga, A., Cilinskis, E., Gravelsins, A., Svarckopfa, A., & Blumberga, D. (2018). Analysis of regulatory instruments promoting building energy efficiency. *Energy Procedia,* 258–267. https://doi.org/10.1016/j.egypro.2018.07.090.
6. Conditions, C. Impact of Window configuration on the overall building energy consumption under specific. https://doi.org/10.1016/j.egypro.2017.05.016Ge.
7. DSTU B V.2.6-189. (2013). Methods of choosing thermal insulation material for building insulation. Kiev (2013). Ministry of Construction of Ukraine.
8. Pashynskyi, V., Pushkar, N., & Kariuk, A. (2012). Temperature effects on building enclosures. Odesa.
9. DSTU-N B V.1.1. (2011). Protection against dangerous geological processes, harmful operational impacts, fire. Building climatology. Kiev (2011). Ministry of Regional Development and Construction of Ukraine.
10. Windows and Daylighting Group, Lawrence Berkeley National Laboratory, http://windows.lbl.gov. Last Accessed 2019/08/21.

Design of Effective Statically Indeterminate Reinforced Concrete Beams

D. Kochkarev, T. Azizov, and T. Galinska

Abstract The present article is concerned with the peculiarities of redistribution of forces in statically indeterminate reinforced concrete beams. Traditional continuous framing schemes allow for redistribution of bending moments within fairly narrow limits, what severely restricts their efficiency, especially under high loading. Thereupon, new effective construction solutions of continuous reinforced concrete beams have been proposed in this article. These effective construction solutions are obtained on the basis of a scheme with equal moments by changing the stiffness of the beams along their length. Utilizing the method of design resistance of reinforced concrete, there has been offered to search for effective design parameters and enforcement of continuous beam constructions. The method of design resistance of reinforced concrete is based on the nonlinear deformation model, which uses deformation curves for concrete and reinforcing steel. In reliance on currently accepted hypotheses and prerequisites, the authors of the article have laid down the geometric parameters, which allow to reduce calculation sets of equilibrium equations to the clear delineation of geometrical and physical–mathematical cross-sectional parameters. This allows tabulation of the necessary parameters and significantly simplifies all calculations of deflections of reinforced concrete elements. This paper provides a calculation example of a continuous double-span beam with central prop based on the proposed calculation procedure. The paper shows the effectiveness of the proposed construction solutions of beam systems with equal moments in comparison with the traditional ones.

Keywords Redistribution of forces · Statically indefinite beams · Deformation model · Method of design resistance of reinforced concrete · Diagram · Deflection of beam

D. Kochkarev
National University of Water and Environmental Engineering, Rivne, Ukraine

T. Azizov
Pavlo Tychyna Uman State Pedagogical University, Uman, Ukraine

T. Galinska (✉)
Poltava National Technical Yuri Kondratyuk University, Poltava, Ukraine
e-mail: galinska@i.ua

V. Onyshchenko et al. (eds.), *Proceedings of the 2nd International Conference on Building Innovations*, Lecture Notes in Civil Engineering 73,
https://doi.org/10.1007/978-3-030-42939-3_10

1 Basic Concepts of Redistribution of Forces in Axial Elements

Statically indeterminate reinforced concrete constructions are quite common in modern construction industry. Their stress–strain state depends on axial, flexural and torsional stiffness, which for their part depend on the formation of normal cracks [1–4]. Bending statically indeterminate beams are characterized by a two-digit bending moment curve. The values of this curve depend on stiffness properties of the cross section of the beam along its length. The vast majority of bending reinforced concrete elements work with cracks in tension region, which significantly affects the stiffness values of their cross sections [5–10]. Almost all design standards, including [11], allow for limited redistribution of forces in continuous beams and frames. For the most part, it becomes apparent through the reduction of bending moments at point of fixation and the corresponding increase in beam spans. The fundamental research of continuous structures, undertaken by a number of scientists [1, 2], indicates the possibility of reducing the moments at point of fixation to 30%, provided that the permissible crack width is abided. It is worth pointing out that it is not always possible to obtain redistribution of forces, providing that the conditions of standard crack growth resistance are observed. In addition, such redistribution of forces is only possible with rather low percentage of reinforcement of the reinforced concrete beam sections with longitudinal reinforcement ($\rho_f = 1\text{–}1.5\%$) [1, 2]. A more conservative variant of design of continuous beam constructions determines the calculation of reinforcement of the beam cross sections in accordance with the obtained bending moment curve at the elastic stage without redistribution of forces. But sometimes it is necessary to impose beam depth limitations according to architectural and planning standards. In this case, it is also critical to align bending moments on beam seats and in beam spans. One of many variations of this alignment may be the change in stiffness of continuous beams.

2 Effective Statically Indeterminate Reinforced Concrete Beams and Their Design Baselines

The alignment of bending moments occurring on seats and spans of continuous beams significantly improves the efficiency of material utilization in structures, reduces the labor intensity of mounting assemblies and allows to reduce the height of beam constructions.

As pointed out above, the reduction of bending moments can be done through redistribution of forces due to the formation of cracks on the beam seat and significant decrease in the stiffness of beam seat sections. This method allows only for limited and moderate decrease in the moments at point of fixation, and therefore it is not suitable for obtaining schemes with equal moments. In addition, development of cracks on the beam seat contravenes with the psychological and aesthetic requirements for structures. The second way to decrease the moments at point of fixation is based on changing the stiffness of continuous beams. Increasing the size of the beams in the span and reduction in their size on the beam seat leads to significant redistribution of forces.

Let us take a closer look at a beam span 1 of a two-span reinforced concrete beam with uniform load intensity q (Fig. 1).

In order to obtain an effective structural design with equal moments, it is necessary to increase the stiffness of the beam section in the span and to reduce it on the beam seat. There are several ways to do it for the rectangular cross-sectional cut of beams. The first is to increase the section width of the beam cross section, and the second is to increase their height (Fig. 2).

Heightening of beams complicates to a great extent the erection of both prefabricated reinforcement cages and its application. Therefore, in the authors' opinion, it is more acceptable to increase the width of its application (Fig. 3). Let us consider redistribution of forces in continuous beams with rectangular beam section during

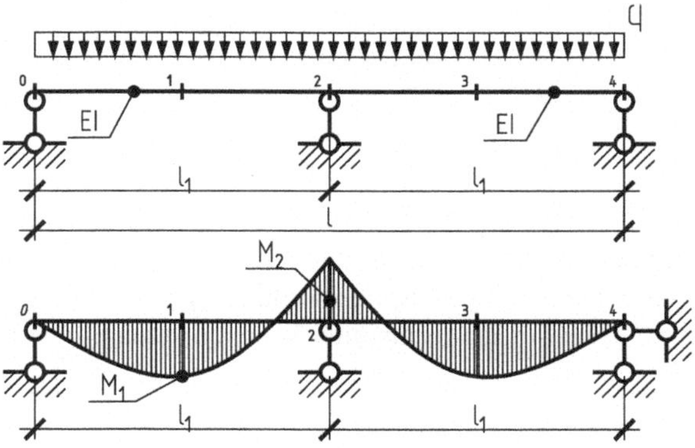

Fig. 1 Design model and the bending moment curve of a two-span continuous beam loaded with spread load intensity

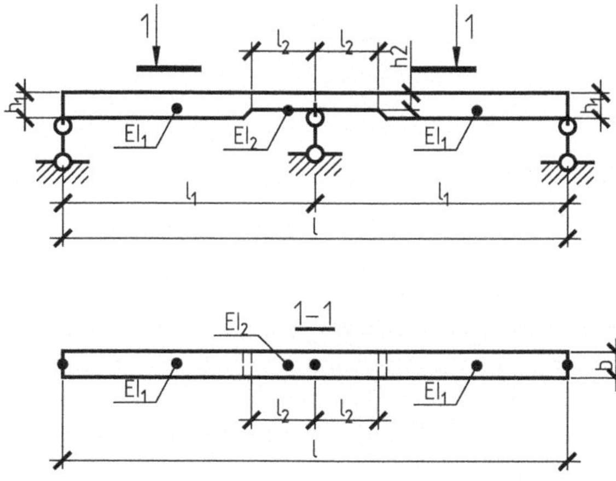

Fig. 2 Structure chart of a continuous double-span beam structure with the change in beam depth over the length

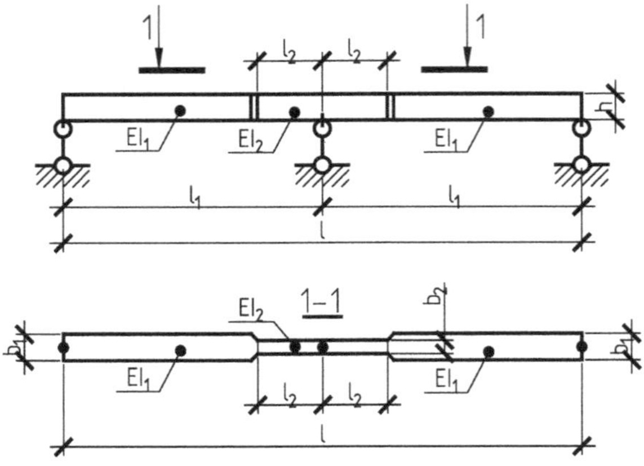

Fig. 3 Structure chart of a continuous double-span beam structure with the change in beam width over the length

elastic work of materials, as exemplified in the beam shown in Fig. 1. For such a beam, the span and seat moments at point of fixation will be:

$$M_1 = \frac{ql_1^2}{16}; \tag{1}$$

$$M_2 = \frac{ql_1^2}{8}. \tag{2}$$

If we increase the stiffness of the beam span approximately two (1.5–2) times, it will lead to the alignment of support and span moments at point of fixation. In such a case, the bearing of beam should be approximately $l_2 \approx l_1/5$. For the elastic stage, the stiffness of the rectangular beam cross section will be the following:

$$EI = E\frac{bh^3}{12}. \tag{3}$$

Then, to obtain a two-span continuous uniform beam, the ratio of width or height must be within such limits:

$$\frac{EI_1}{EI_1} = \frac{b_1}{b_2} \approx 2; \tag{4}$$

$$\frac{EI_1}{EI_1} = \frac{h_1}{h_2} \approx \sqrt[3]{2} = 1.26. \tag{5}$$

The relations (4), (5) give the opportunity to estimate approximately the possibility of designing uniform beams at the elastic stage. It is common knowledge that reinforced concrete is elastoplastic material, and therefore the work of structures made from it can be significantly different from the work of structures made of elastic materials. In order to take into account nonlinear properties of concrete and the formation of cracks in the tension region, we will apply the method of design resistance of reinforced concrete.

The peculiarities of the method of design resistances [12–14] are as follows:

1. The bases of calculation are experimentally and theoretically substantiated prerequisites and hypotheses: The credence is given to the plane section hypothesis (Bernoulli's assumption)—linearity of axial deformations in the cross section of the element. Nonlinear material deformation curves are also considered to be dependable.
2. Geometric parameters, which reduce the calculation systems of equilibrium equations to the clear distinction between geometric and physical–mechanical parameters of the section, are laid down.
3. The basic design parameters are tabulated by performing sample calculations.

In this method, the main parameters characterizing the load level are engineering stresses, which are tabulated by the basic parameters, namely

$$\sigma_{zM} = M/W_c = f(\Sigma\varepsilon, \rho_f, C, f_y). \tag{6}$$

In the expression (6), M is the bending beam moment in the corresponding beam section, kN m; $\Sigma\varepsilon$ are the total deformations of the most compressive zone of the concrete and tension steel $\varepsilon_c + \varepsilon_s$; C is the strength grade of concrete, which is characterized by the parameters of the deformation curve under tension and compression, i.e., E_s, E_c, ε_{c1}, ε_{cu}, f_c, ε_{ct1}, ε_{ctu}; ρ_f is the reinforcement percentage of the cross section of the element by longitudinal reinforcement $\rho_f = A_s/(bd)\cdot 100\%$; W_c is the modulus of net section of concrete $W_c = bd^2/6$; and f_y is the yield point of reinforcement.

The dependency (6) is determined in the following order:

1. Concrete and reinforcement deformation curves are preliminarily taken. As a rule, they prefer functional relations that meet commonly accepted deformation criteria (buckling criterion). The most general is the deformation function presented in [15].
2. The plane section hypothesis (Bernoulli's assumption) is accepted, with its correction after the formation of cracks in the tensile zone of the concrete.
3. They write down the system of equilibrium equations for cross sections with cracks and without them, from which at different levels of loading it is possible to determine engineering stresses σ_{zM} in accordance with the expression (6). In this case, after the formation of cracks it is better to use the parameters obtained for the midsection. They are defined as the average values of the parameters obtained for the section with cracks and without them.

Dependence (6) has been determined by the above-mentioned method, and for some classes of concrete it has been illustrated in Table 1.

Further calculation is performed according to the iterative method, by specifying continually the real stiffness of the cross sections of the beam.

Table 1 Parameters of strain–stress state of cross sections of bending elements with reinforcing steel of class-A500

Strength grade of concrete	Load level	Percentage of reinforcement					
		1		2		3	
		σ_{zM} (MPa)	$\sum\varepsilon$ ($\times 10^4$)	σ_{zM} (MPa)	$\sum\varepsilon$ ($\times 10^4$)	σ_{zM} (MPa)	$\sum\varepsilon$ ($\times 10^4$)
C16/20	M_{W1}	3.60	2.99	4.80	3.26	5.96	3.53
	M_{W2}	3.60	5.22	4.80	4.32	5.96	4.20
	0.4	11.01	16.77	12.70	12.09	13.55	10.12
	0.6	16.51	26.43	19.04	19.27	20.33	16.25
	0.8	22.02	37.90	25.39	28.07	27.10	23.90
	1.0	27.52	59.96	31.74	46.48	33.88	40.67
C20/25	M_{W1}	4.00	3.02	5.25	3.27	6.45	3.51
	M_{W2}	4.00	5.61	5.25	4.50	6.45	4.30
	0.4	12.84	18.80	14.99	13.55	16.11	11.33
	0.6	19.26	29.45	22.48	21.45	24.16	18.05
	0.8	25.68	41.87	29.98	30.95	32.22	26.28
	1.0	32.10	64.74	37.47	49.92	40.27	43.51
C25/30	M_{W1}	4.62	3.18	5.94	3.41	7.23	3.65
	M_{W2}	4.62	6.30	5.94	4.91	7.23	4.60
	0.4	14.59	20.70	17.19	14.91	18.58	12.44
	0.6	21.89	32.30	25.79	23.48	27.87	19.73
	0.8	29.18	45.67	34.39	33.67	37.16	28.53
	1.0	36.48	69.52	42.98	53.38	46.46	46.37

It is suggested that the final design of schemes with equal moments will be performed as follows:

1. The calculation scheme of the continuous beam structure is established.
2. They perform static analysis during elastic work of materials and set the values of bending moments over the beam length.
3. They establish necessary increase in the dimensions of the beam span, upon condition that span and seat moments at point of fixation in the beam are equal, as well as during elastic work of materials.
4. Necessary reinforcement for span and seat cross sections of the beam is established.
5. They refine the bending moment curve by the iteration method using the method of design resistance of reinforced concrete. To that end, the beam over the length is divided into a certain number of sections. In the center of each section, at the known value of the bending moment, we have to determine the engineering stresses σ_{zM} and then the total deformation of the most compressive zone of

the concrete and tensile reinforcement $\Sigma\varepsilon$. The angle of section flexure of the corresponding segments can be determined by the expression

$$1/r_{mi} = \frac{\Sigma\varepsilon}{d}. \quad (7)$$

The stiffness of cross sections is refined by following expression:

$$D_i = \frac{M_i}{1/r_{mi}}. \quad (8)$$

where D_i is the bending stiffness, M_i is the bending moment in the corresponding section of the beam over the length of the reinforced concrete element, and $1/r_{mi}$ is the mean curvature at a particular element to the corresponding cross section.

We will exemplify the proposed method by designing a two-span continuous reinforced concrete beam.

3 Design Example of an Effective Statically Indeterminate Reinforced Concrete Beam

Example It is necessary to design a two-span continuous beam with a span $l = 12$ m (see Fig. 1), which should assume uniform load with intensity $q = 90$ kN/m. The beam has to satisfy space-and-planning height requirements. The height of the beam must be no more than 400 mm. It is advisable to use for the beam strength grade of concrete C20/ 25 and reinforcing steel of class-A500.

The solution

1. Let us perform the static calculation of the beam under conditions of elastic work of materials with a previously accepted cross section $b \times h = 400 \times 400$ mm. The corresponding bending moments are equal to:

$$M_1 = ql_1^2/8 = 90 \times 6^2/8 = 405\,\text{kN} \times \text{m};$$
$$M_2 = ql_1^2/16 = 90 \times 6^2/16 = 202.5\,\text{kN} \times \text{m};$$

2. When aligning the span and seat moments at point of fixation, we will get the following:

$$M_1 = M_2 = ql_1^2/12 = 90 \times 6^2/12 = 270\,\text{kN} \times \text{m};$$

3. Let us take ad interim the cross section on the beam seat as $b_1 \times h_1 = 400 \times 400$ mm, in the span $b_2 \times h_2 = 600 \times 400\,\text{mm}$, $d = 400 - 30 = 370\,\text{mm}$;

Table 2 Calculation bending strength of reinforced concrete f_{zM} at singular reinforcement (MPa)

Class of concrete	The coefficient of reinforcement (ρ_f)							
	0.05	0.50	1.00	1.50	1.75	0.02	0.025	0.03
	Reinforcement class A-500C							
C8/10	1.32	10.90	14.57	15.35	15.60	15.79	16.07	16.27
C12/15	1.33	11.66	19.40	20.70	21.13	21.48	21.99	22.35
C16/20	1.33	12.13	21.53	26.50	27.16	27.71	28.53	29.13
C20/25	1.33	12.41	22.65	30.69	32.67	33.42	34.57	35.43
C25/30	1.34	12.57	23.28	32.12	35.82	37.89	39.33	40.40
C30/35	1.34	12.69	23.74	33.18	37.28	40.93	43.80	45.10
C32/40	1.34	12.78	24.11	33.99	38.39	42.40	48.01	49.53
C35/45	1.34	12.86	24.44	34.75	39.42	43.78	51.43	54.52
C40/50	1.34	12.92	24.67	35.26	40.11	44.68	52.94	58.52
C45/55	1.34	12.96	24.86	35.68	40.69	45.43	54.10	61.40
C50/60	1.34	13.01	25.04	36.10	41.25	46.17	55.26	63.37

4. We will establish the necessary reinforcement, having used the method of design resistances of reinforced concrete [12–18].

   The required design resistance will be:

$$f_{zM1} = 6\,M/\left(b_1 d^2\right) = 29.58\,\text{MPa};$$
$$f_{zM2} = 6\,M/\left(b_2 d^2\right) = 19.72\,\text{MPa};$$

5. According to Table 2 and using interpolation, we will find:

$$r_{f1} = 143\%;\ r_{f1} = 0.86\%.$$

6. Now, we can define the necessary reinforcement of sections:

$$A_{s1} = r_{f1} \times b_1 d/100\% = 1.43 \times 400 \times 370/100 = 2116.4\,\text{mm}^2;$$
$$A_{s2} = r_{f1} \times b_1 d/100\% = 0.86 \times 600 \times 370/100 = 1909.2\,\text{mm}^2.$$

   We will accept symmetrical reinforcement 4Ø28 A500 $A_{s1} = A_{s2} = 2463\,\text{mm}^2$.

6. Using the method of design resistances, we can determine the final bending moment diagram (see Tables 3 and 4).
7. The performed calculations made it possible to establish the true bending moment diagram. Its validation is the obtained value of beam deflection equal to zero on the midspan bearing of beam (see Table 4).
8. The final form of the beam is shown in Fig. 4.

Table 3 Bending moment diagram and reinforcement over the beam length

No section number	Reinforcing steel diameter d (mm)	Quantity, item	M (кN m)	b (mm)	h (mm)	The area of reinforcement (mm^2)	Percentage of reinforcement (ρ_f)
0	28	4	0.00	60	40	24.63	1.080
1	28	4	205.13	60	40	24.63	1.080
2	28	4	280.67	60	40	24.63	1.080
3	28	4	226.60	60	40	24.63	1.080
4	28	4	42.93	60	40	24.63	1.080
5	28	4	−270.33	40	40	24.63	1.620
6	28	4	42.93	60	40	24.63	1.080
7	28	4	226.60	60	40	24.63	1.080
8	28	4	280.67	60	40	24.63	1.080
9	28	4	205.13	60	40	24.63	1.080
10	28	4	0.00	60	40	24.63	1.080

Table 4 Refinement of the bending moment diagram of a continuous beam

No section number	The distance from t.0 (m)	W_c (cm^3)	Full load			
			σ_{ze} (MPa)	$\Sigma\varepsilon \times 10^4$	$1/r_{mi}$, ($\times 10^4$ cm^{-1})	$a_i \times 1/r_{mi}$, ($\times 10^4$ cm^{-1})
0	0.00	14,440	0.0	0.0	0.000	0.0000
1	1.20	14,440	14.2	18.6	0.491	0.0245
2	2.40	14,440	19.4	27.3	0.719	0.0719
3	3.60	14,440	15.7	20.9	0.550	0.0825
4	4.80	14,440	3.0	2.1	0.056	0.0113
5	6.00	9627	−28.1	−90.3	−2.377	−0.5944
6	7.20	14,440	3.0	2.1	0.056	0.0113
7	8.40	14,440	15.7	20.9	0.550	0.0825
8	9.60	14,440	19.4	27.3	0.719	0.0719
9	10.80	14,440	14.2	18.6	0.491	0.0245
10	12.00	14,440	0.0	0.0	0.000	0.0000
					$f = \Sigma a_i \times 1/r_{mi}$=	0.0000

4 Conclusion

We have considered the basics of designing uniform beam structures. It is proposed to redistribute bending moments by changing the stiffness of the span and seat moments at point of fixation of continuous beam structures. The above-mentioned calculation

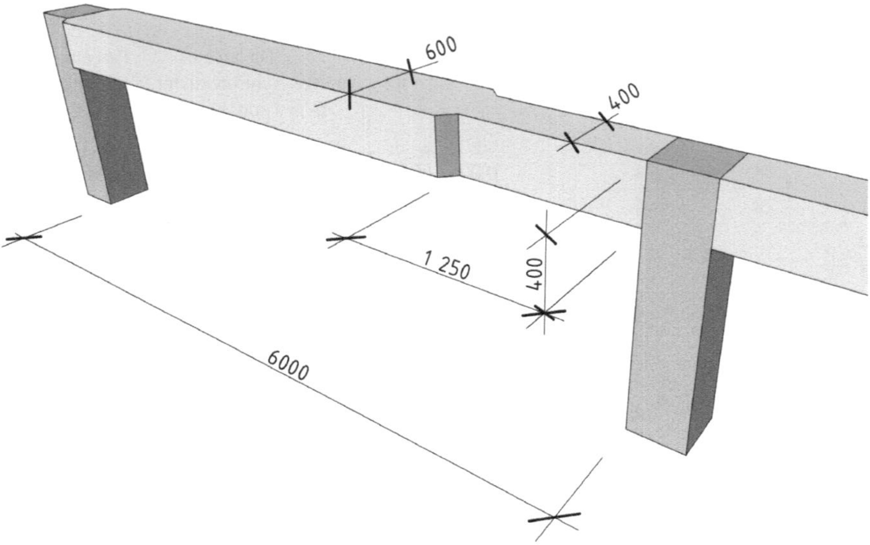

Fig. 4 Continuous uniform reinforced concrete beam

example confirms the effectiveness of the proposed structures and the possibility of their widespread use in modern construction practice. The proposed structures can be effectively used in the reconstruction and reinforcement of the existing continuous beams. In order to confirm all conducted calculations, one can use the following online resource http://sciencehunter.net/Services/Apps/Concrete.

References

1. Krylov, S. M. (1964). *Redistribution of forces in statically indefinable reinforced concrete structures* (121 p). Gostroyizdat.
2. Mayilyan, L. R. (1983). Approximation method for calculating non-slab beams taking into account the redistribution of efforts. *Concrete and Reinforced Concrete, 8*, 35–36.
3. Babich, E. M., Filipchuk, S. V., & Ilchuk, N. I. (2012). Work and calculation of frames under the action of repeated loads: Monograph. -Exactly: NSUPP, 176 p.
4. Kochkarev, D., Azizov, T., & Galinska, T. (2018). *Bending deflection reinforced concrete elements determination*. Paper presented at the MATEC Web of Conferences, 230 doi:10.1051/matecconf/201823002012.
5. Dem'yanov, A., Kolchunov, V. l., Iakovenko, I., & Kozarez, A. (2019). Load bearing capacity calculation of the system "Reinforced concrete beam—Deformable base" under torsion with bending. In *XXII International Scientific Conference "Construction the Formation of Living Environment" (FORM-2019) E3S Web Conference* (Vol. 97). https://doi.org/10.1051/e3sconf/20199704059.
6. Azizov, T., Jurkowska, N., & Kochkarev, D. (2019). Basis of calculation on torsion for reinforced concrete structures with normal cracks. In *Fib Symposium 2019 Concrete Innovations in Materials, Design and Structures* (pp. 489–490). Cracow 27–29 May 2019. Book of Abstracts.

7. Salnikov, A., Kolchunov, V. I., & Yakovenko, I. (2015). The computational model of spatial formation of cracks in reinforced concrete constructions in torsion with bending. *Applied Mechanics and Materials, 725–726*, 784–789.
8. Azizov, T., Derkowski, W., & Jurkowska, N. (2019). Consideration of the torsional stiffness in hollow-core slabs' design. In *Materials science forum* (Vol. 968, pp. 330–341). Submitted: 2019-05-28. ISSN: 1662-9752, Accepted: 2019-05-29.
9. Babych, E. M., & Andriichuk, O. V. (2017). Strength of elements with annular cross sections made of steel-fiber-reinforced concrete under one-time loads. *Materials Science, 52*(4), 509–513. https://doi.org/10.1007/s11003-017-9983-z.
10. Słowik, M., & Błazik-Borowa, E. (2011). Numerical study of fracture process zone width in concrete members. *Architecture Civil Engineering Environment Journal, 2,* 73–78.
11. ENV 1992-1. (1993). Eurokode- 2. Design of concrete structure. Part 1, General rules and rules for buildings, GEN.
12. Kochkarev, D., & Galinska, T. (2017). Calculation methodology of reinforced concrete elements based on calculated resistance of reinforced concrete. In *Matec Web of Conferences 116, 02020, Materials Science, Engineering and Chemistry*, Transbud–2017, Kharkiv, Ukraine, 19–21 April, 2017.
13. Pavlikov, A., Kochkarev, D., & Harkava, O. (2019). Calculation of reinforced concrete members strength by new concept. In *Proceedings of the fib Symposium 2019: Concrete—Innovations in Materials, Design and Structure*s. Krakow, Poland.
14. Azizov, T., Kochkarev, D., & Galinska, T. (2020). Reinforced concrete rod elements stiffness considering concrete nonlinear properties. *Lecture Notes in Civil Engineering, 47*(2020), 1–6. https://doi.org/10.1007/978-3-030-27011-7_1.
15. Kochkaryov, D. V. (2015). Nonlinear resistance of reinforced concrete elements and structures to force effects: Monograph. Exactly: O. Zen, 384 p. ISBN 978-617-601-125-5.
16. Storozhenko, L., Butsky, V., & Taranovsky, O. (1998). Stability of compressed steel concrete composite tubular columns with centrifuged cores. *Journal of Constructional Steel Research, 46*(1–3), 484. https://doi.org/10.1016/S0143-974X(98)80098-9.
17. Piskunov, V. G., Goryk, A. V., Lyakhov, A. L., & Cherednikov, V. N. (2000). High-order model of the stress-strain state of composite bars and its implementation by computer algebra. *Composite Structures, 48*(1), 169–176. https://doi.org/10.1016/S0263-8223(99)00091-4.
18. Piskunov, V. G., Goryk, A. V., & Cherednikov, V. N. (2000). Modeling of transverse shears of piecewise homogeneous composite bars using an iterative process with account of tangential loads. 1. construction of a model. *Mechanics of Composite Materials, 36*(4), 287–296. https://doi.org/10.1007/bf02262807.

Time Measurement of Ultrasonic Vibrations Extension in Concrete of Different Compositions

Victor Kolokhov, Mykola Savytskyi, Artem Sopilniak, and Grygorii Gasii

Abstract The identification of the physical and mechanical characteristics of the material of the structures used by non-destructive methods of determination has significant limitations. They are caused by differences in concrete composition, structural structure parameters and measurement conditions. A significant factor in the reliability of non-destructive methods of determination is the personal factor that depends on the expert performing the determination. Ultrasonic vibration propagation time measurements were performed using a Pulsar 1.1 and Novotest IPSM-U instrument for changes in the pressing force of the device to the sample concrete within the limits recommended by the device manufacturers and smaller. The time of propagation of ultrasonic vibrations during the change of the force of pressing the device to the surface of concrete of different compositions has been determined. It has been established that the dependence "time of propagation of ultrasonic vibrations—force of pressing the device" is approximated linearly with the highest accuracy if this dependence is divided into three intervals. The first and the third intervals are characterized by a slow change of properties, the second—by a sharp one. Studies have confirmed that the results of the measurements were influenced by the force of pressing the device to the surface, the composition of concrete and the conditions of structure construction. Improving the accuracy of determining the physical and mechanical characteristics of the material of structures appears to be possible taking into account the above-mentioned factors.

Keywords Physical–mechanical characteristics · Non-destructive control · Ultrasound

V. Kolokhov · M. Savytskyi · A. Sopilniak
State Higher Education Establishment "Pridneprovsk State Academy of Civil Engineering and Architecture", Dnipro, Ukraine
e-mail: kolokhovdnepr@i.ua

G. Gasii (✉)
Sumy National Agrarian University, Sumy, Ukraine
e-mail: gasiigm@gmail.com

V. Onyshchenko et al. (eds.), *Proceedings of the 2nd International Conference on Building Innovations*, Lecture Notes in Civil Engineering 73,
https://doi.org/10.1007/978-3-030-42939-3_11

1 Introduction

The determination of the physico-mechanical characteristics (PMC) of concrete by the use of ultrasonic vibrations (UVs) has become popular due to the simplicity of the method used. Together with the ease of the use, ultrasonic method has high degree of uncertainty [1, 2]. This uncertainty is due to a high degree of heterogeneity of PMC of the concrete, which, in turn, depends on a number of technological operations and environmental conditions [3]. The standards, governing the application of this method, [4, 5] suggest the use of different devices and grading dependencies with their adjustment for each new concrete composition. Modern ultrasonic vibrations propagation time measuring devices (UVPT) mainly use the so-called dry contact between the surface and the device [4, 6]. But, as shown in [7], the conditions of the use of the devices recommended by their manufacturers [6] do not always ensure unambiguous results.

In order to improve the accuracy of PMC determination of concrete, it is necessary to take into account the influence of the force of pressure of the device on the surface, the structure and composition of the material, the conditions of formation and hardening of concrete.

The purpose of the study, therefore, is to evaluate the effect of differences in the structure of concrete on the time of ultrasound propagation in concrete for changes in measurement conditions.

2 Outline of the Material

The experiments were performed with samples made of heavy and fine-grained concrete. Granite crushed stone of fraction of 2–20 mm, screenings of fraction of 0–5 mm of the fishing quarry and Dnieper annual sand were used as a filler. The samples are made in a 100 × 100 × 400 mm metal formwork. For the production of specimens of the first batch (four specimens), heavy concrete of industrial composition (fillers: crushed stone of fraction of 5–20 mm and Dnipro sand) was used. Concrete for the second batch (five specimens) was different from the first batch by 25% increase in the amount of cement. The third batch (four specimens) is made of fine-grained concrete (filler is a 0–5 mm section of the fishing quarry). For series four (five specimens), fine-grained concrete with Dnipro annual sand was used. Within the fourth series, the composition of the concrete was changed by the size of the aggregate. For one concrete composition, the aggregate was used in its natural state. For the other four, partial residues on sieves of 0.14, 0.315, 0.63, and 1.25 mm were obtained by sieving the small filler through a standard set of sieves. For the third and fourth series, the ratio between cement and aggregate was 1/3. Samples of the first three series were tested at 28 days of age. Samples of the fourth series were made in advance (the age at the test time was six years).

During the study, the time of propagation of ultrasonic vibrations (UVTP) was measured on a constant basis, which is caused by the design of the devices. There were applied strength meters for building materials "Novotest IPSM-U" and "PULSAR–1.1". The measurement base was 120 mm. The determination was performed on two lateral surfaces of the samples. One surface was in contact with the metal formwork during the molding and curing of the sample but the other one—wasn't (in contact). For each sample, four series of measurements were performed on each surface. The measurement areas were selected at random.

The interval of the pressing force change of the device to the sample is chosen from conditional zero (weight of the device only) to the level recommended by the manufacturers (5–10 kg). The change step was 1.0 kg at all stages, except the first one, for which the effort was changed from the unit weight to 1 kg. Statistical processing of the results was performed using EXEL tools. There have been determined: the average value of the UVPT at each stage; maximum and minimum value; standard deviation; coefficient of variation, students' test and confidence interval limits with 0.95 assurance.

Figure 1 shows the results of determining the dependence "UVPT—pressing force" for sample 2 of the first series. Determination was carried out on the surface in contact with the formwork.

It is easy to see that there are three intervals within the dependency definition for which linear approximation is possible. At the beginning and at the end of the boundaries of the dependence determination, there are areas with a small value of the change in the UVPT from the change in the pressing force ("slow intervals"). Between them, there is a zone with a significant value of the change of the UVPT from the change of the pressing force ("sharp interval").

Other samples in this series have similar dependencies. Only the limits of the intervals with "slow" and "sharp" changes are different. Attempts to construct a trend line for the entire definition domain have resulted in significantly nonlinear dependencies

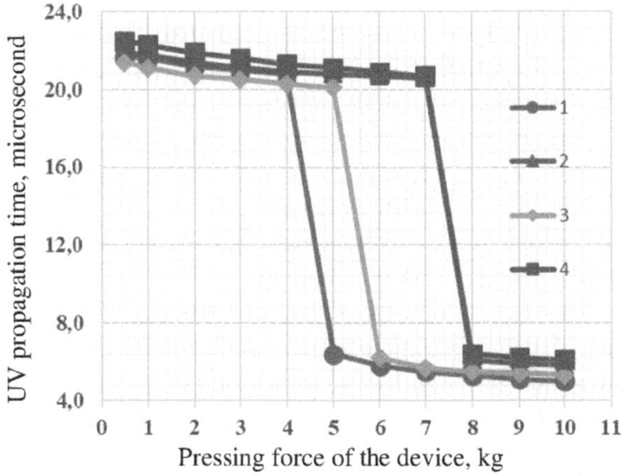

Fig. 1 Results of determining the dependence "UVPT—pressing force" for sample 2 of the first series; the determination was carried out on the surface in contact with the formwork

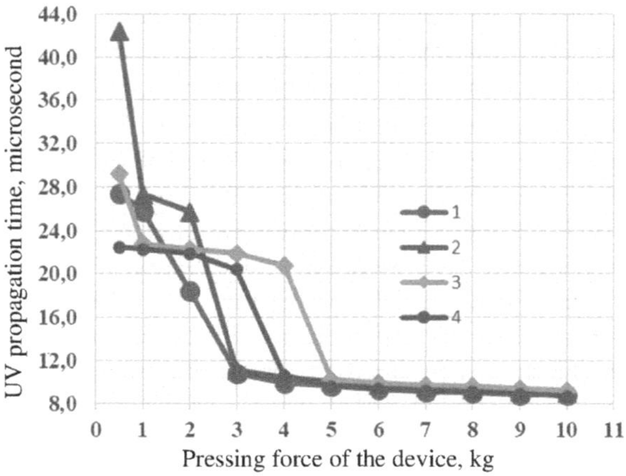

Fig. 2 Results of determining the dependence "UVPT—pressing force" for sample 2 of the first series; the determination was carried out on the surface that did not contact with the formwork

with low approximation reliability. When performing the approximation linearly at three intervals, the reliability increases significantly.

It should be noted that in only one case of the four boundaries of the second "slow interval" do they match the declared device manufacturers before determining the interval.

A slightly different nature of the dependence "UVPT—pressing force" on the surface was observed, which did not come into contact with the formwork during molding and curing. Figure 2 shows the results of determining the same dependence of the same sample (No. 2) of the first series. As in the previous case, only part of the samples demonstrates the possibility of using a "factory interval." In addition, in many cases, the first "slow interval" has a vague initial limit. We compare the change of CNCW when changing the force of the device from 1 to 10 kg. In the case of a surface that has not come in contact with the formwork, such a change is on average a quarter higher than the change in the surface that has come in contact with the formwork.

Figures 3 and 4 show the results of determining the dependence "UVPT—pressing force" for sample 5 of the second series. Determination was carried out on a surface that did not contact and contact the formwork (respectively).

The second series of samples demonstrates a similar behavior of defined dependencies. The increase in cement in the composition of the concrete makes it possible to determine within the pressure interval, which is recommended by manufacturers of devices. It should be added that the coefficient of variation within the second "slow" interval is almost halved.

That is, for concrete with a high amount of cement, it is possible to use the intervals of determination recommended by manufacturers of measuring instruments. But the actual concrete formulations used in the construction do not provide this opportunity. Figures 5 and 6 show the results of determining the dependence "UVPT—pressing force" for sample 1 of the third series. Figures 7 and 8 show the results of "UVPT—pressing force" dependence determination for sample 1 of the fourth

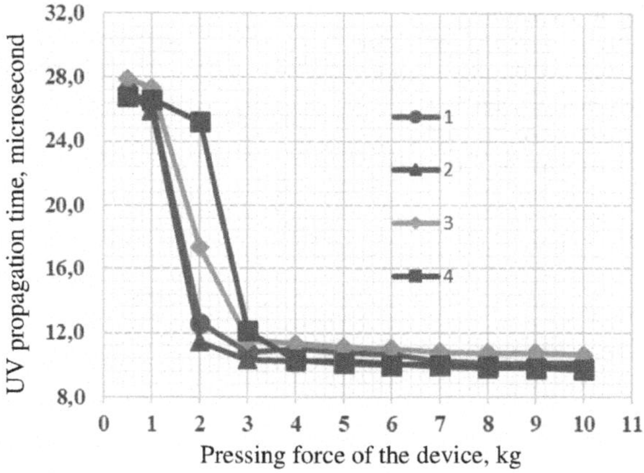

Fig. 3 Results of determining the dependence "UVPT—pressing force" for sample 5 of the second series; the determination was carried out on a surface that did not contact with the formwork

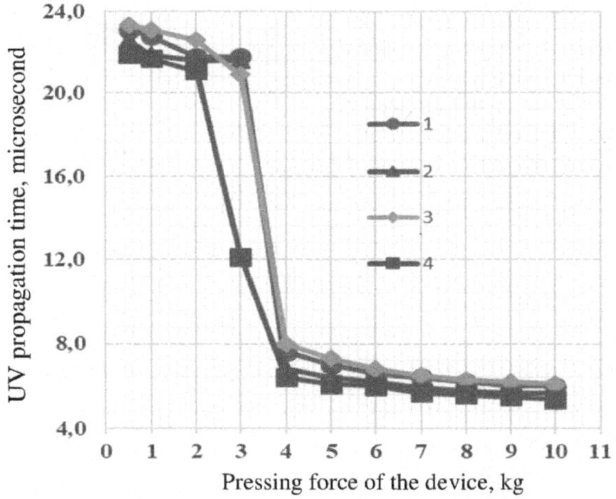

Fig. 4 Dependence "UVTP—pressing force" for sample 5 of the second series; the determination was carried out on the surface in contact with the formwork

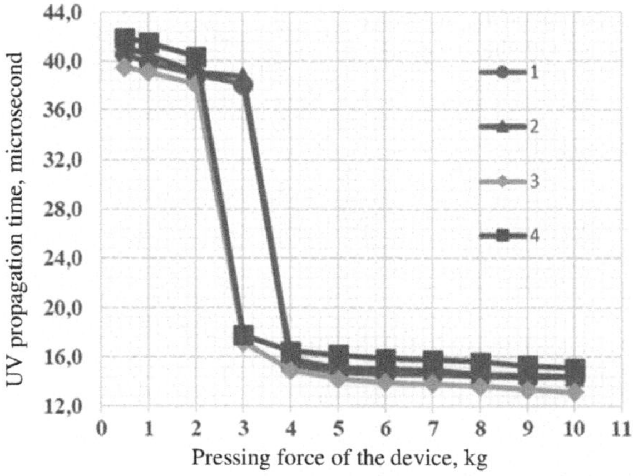

Fig. 5 Dependence "UVPT—pressing force" for sample 1 of the third series; the determination was carried out on the surface that did not come in contact with the formwork

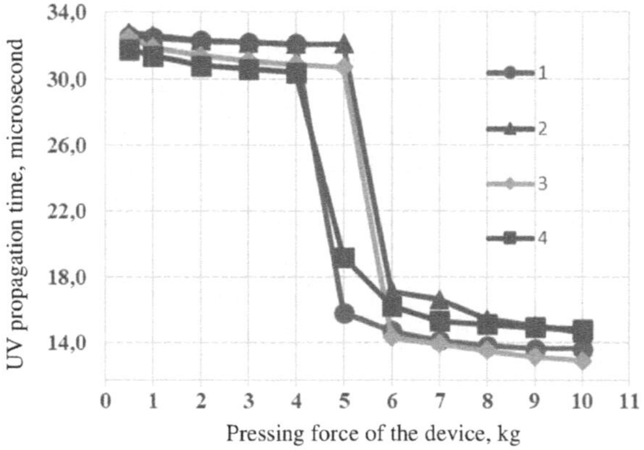

Fig. 6 Dependence "UVPT—pressing force" for sample 1 of the third series; the determination was carried out on the surface in contact with the formwork

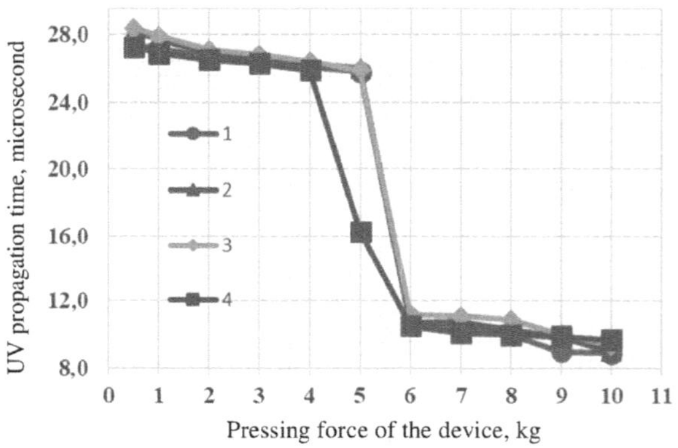

Fig. 7 Dependence "UVPT—pressing force" for sample 1 of the fourth series; the determination was carried out on the surface that did not contact with the formwork

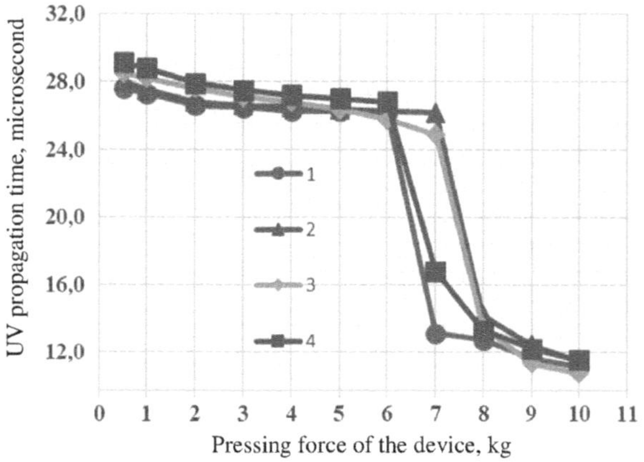

Fig. 8 Dependence "UVPT—pressing force" for sample 1 of the fourth series; the determination was carried out on the surface in contact with the formwork

series. Determination was carried out on the surface that did not contact and contact the formwork (respectively).

The above dependencies are qualitatively more similar than analogous ones for heavy concrete [8], but quantitative dependencies are also difficult to establish.

For specimens of fine-grained concrete, the effect of curing conditions (presence or absence of formwork adjacent to the concrete surface) was confirmed, but there is no direct correlation between contact with formwork surface and UVPT. In some cases, the variation of the UVPT for the second "slow" interval (as opposed to heavy concrete) becomes significantly nonlinear. In addition, for fine-grained concrete samples, the interval with drastic changes in the UVPT is between 10 and 20% of the total measurement interval.

According to the results of the statistical processing of measurements (for all series), specimens of fine-grained concrete show more uniformity than specimens of heavy concrete. The boundaries of the "sharp" interval in the vast majority did not exceed 10–20% of the entire measurement range, and for a third of the samples were even smaller. The coefficient of variation is also lower for the entire interval. But, as for heavy concrete, the boundaries of the "sharp" zone do not have a clear fixed place but migrate. The beginning of this zone is within 2–6 kg (pressing force) and the end—4 to 8 kg (pressing force). Only for 40% of the samples, the second "slow" interval is within the limits recommended by the manufacturers of the measuring instruments.

During the co-processing for all samples, the maximum and minimum values of the UVPT are beyond the confidence interval. This characterizes the individual dependencies (separately for each concrete sample) as not belonging to the same general type.

The use of concrete PMC determination data for the verification of structural calculations [8–10] is possible only if the UVPT measurement technique is improved, and subsequently to evaluate the technical conditions of the building structures.

3 Conclusions

The analysis of the obtained results has confirmed the following influence of UVPT on the results of measurements in the case of "dry" contact between the device and the concrete:

- efforts to press the appliance into the concrete;
- the shape and the size of the filler
- uniformity of the concrete.

It is necessary to improve the existing technique for determining the PMC of concrete using ultrasonic devices. Enhancing the accuracy of the technique can be achieved by reducing the impact of heterogeneity of concrete composition, conditions of its manufacture and operation.

References

1. Komlos, K., Popovics, S., Nürnbergerová, T., Babal, B., & Popovics, J. S. (1996). Ultrasonic pulse velocity test of concrete properties as specified in various standards. *Cement & Concrete Composites, 18*(5), 357–364.
2. Breysse, D. (2012). Nondestructive evaluation of concrete strength: An historical review and a new perspective by combining NDT methods. *Construction and Building Materials, 33*, 139–163.
3. Kolokhov, V., Sopilniak, A., Gasii, G., & Kolokhov, A. (2018). Structure material physic-mechanical characteristics accuracy determination while changing the level of stresses in the structure. *International Journal of Engineering &Technology, 7*(4.8), 74–78.
4. Concrets. Ultrasonic method for determining strength: DSTU B V.2.7-226: 2009. Effective from 2010-09-01. Kyiv: DP Ukrarahbudinform, 2010, 27 p.
5. Concretes. Ultrasonic strength determination method: GOST-17624-2012. (Date of introduction 2014-01-01), Standartinform, 2014, 16 p.
6. Measurement of strength of concrete and building materials Novotest IPSM. Novotest. Control and quality devices: catalog. Scientific and industrial center "Industrial equipment and technologies", Novomoskovsk, 2012, 26 p.
7. Kolokhov, V. V., & Kolokhov, O. V. (2019). Changing the time of ultrasonic oscillation propagation in concrete for changing conditions of measurement. *Bulletin of Prydniprovsk State Academy of Civil Engineering and Architecture, 2*, 92–101.
8. Kolokhov, V. V. (2012). Some aspects of the application of methods for non-destructive testing of concrete properties. In *Theoretical Foundations of Civil Engineering. Polish–Ukrainian Transactions (Conference)* (Vol. 20, pp. 443–448). Warsaw.
9. Kolokhov, V. V. (2013). Formalization of the procedure for determining the physicomechanical properties of concrete and its hardware. *Construction, Materials Science, Engineering, 69*, 231–236.
10. Storozhenko, L., Butsky, V., & Taranovsky, O. (1998). Stability of compressed steel concrete composite tubular columns with centrifuged cores. *Journal of Constructional Steel Research, 46*(1–3), 484. https://doi.org/10.1016/S0143-974X(98)80098-9.

Hydraulic Single Pump with Combined Higher Volume Compensator Operation Analysis

Bogdan Korobko, Inna Khomenko, Mykola Shapoval, and Viktor Virchenko

Abstract The article dwells on the operation peculiarities of a mortar pump with a hydraulic drive, ball suction and spring-loaded discharge valves, a special insert in the suction chamber and an increased volume compensator. A comparative analysis of the mortar pump operating modes with various drives is presented. Based on theoretical studies of determining the productivity and volumetric efficiency, it is proved that a hydraulic drive mortar pump with a combined compensator of increased volume works more efficiently in relation to a mortar pump with a combined compensator of increased volume. The rational operating modes areas of the hydraulic drive mortar pump are determined.

Keywords Mortar pump with a combined compensator of increased volume · Hydraulic drive · Capacity · Volumetric efficiency · Power · Solution mobility

1 Introduction

The analysis of modern mortar pumps operation indicates the search for ways to improve their structures, ensure high reliability and improve technical performance.

Therefore, the main directions for the development of new mortar pump designs are identified, and a new one-piston mortar pump with a combined expansion joint compensator is proposed. Implementation of hydraulic drive mortar pumps is relevant and has several advantages over electromechanical ones. First of all, the use of a hydraulic actuator will provide a constant speed of the working body, which will positively affect the reduction of backflow through the suction and discharge valves by stabilizing their speed of lifting and lowering, especially with pumping solutions of reduced mobility *P* 7–9 cm; due to the constant speed of the solution movement during the injection, the solution ripple pressure will be reduced to a minimum; there is a possibility of a smooth solution flow regulation during its pumping, which will positively affect the finishing works quality, used in conjunction with hydraulic drive plastering stations or kits. These advantages will significantly increase the

B. Korobko · I. Khomenko · M. Shapoval · V. Virchenko (✉)
Poltava National Technical Yuri Kondratuyk University, Poltava, Ukraine
e-mail: virchenko.viktor@gmail.com

V. Onyshchenko et al. (eds.), *Proceedings of the 2nd International Conference on Building Innovations*, Lecture Notes in Civil Engineering 73,
https://doi.org/10.1007/978-3-030-42939-3_12

technical and economic indicators of the hydraulic actuator use in mortar pumps both individually and in the composition of plaster units.

Therefore, it is necessary to theoretically study the workflows of a single-piston mortar with a combination of a larger volume compensator using a hydraulic drive and to determine the specifications.

Review of the latest research sources and publications indicates that not all mortar pumps satisfy the low-impulse solution flow, as well as reliability and versatility [1–3]. Therefore, it is necessary to create a mortar pump that will provide a high level of technical characteristics and a solution pressure fluctuations low-level supply.

Establishing unsolved aspects of the problem: It is necessary to make a comparative performance analysis of technical characteristics pertaining to mortar pumps with electromechanical and hydraulic actuators with combined expansion joints and prove the feasibility of introducing a hydraulic actuator in the mortar.

Problem statement: Hydraulic actuator should be used in a single-piston pump with a combination of a larger volume compensator to increase the performance level and reduce the fluid pressure pulsations level.

The purpose and the tasks of the study: The research is aimed at increasing the single-piston mortar pump with the combined expansion valve efficiency due to the use of the hydraulic drive in the rational modes of operation.

To achieve this aim, the following tasks must be accomplished:

1. To theoretically investigate the hydraulic pump working body-driven motion law solution.
2. To determine the main indicators of the advanced design hydraulic pump effective operation with regard to the combined compensator of the increased volume offered: productivity, volume efficiency and capacity; prove the benefits of using a hydraulic drive.
3. Compare the specifications of the hydraulic and electromechanical actuators with the combined expansion joint: performance, volumetric efficiency and power.

The object of research is the single-piston hydraulic-driven mortar pump with a compensator for increased efficiency workflows of mortar transportation taking into account the pressure ripples that occur during their movement through the pipeline.

The subject of research is a single-piston hydraulic drive mortar pump with a combined expansion valve.

Research methods: The following methods were used in the research: basic principles of hydraulics, hydrodynamics, methods of mathematical physics, physic-mathematical modeling by methods of applied mechanics, statistical processing of experimental data, methods of planning experiment, computer programming Microsoft Office, Compass 3D, MathCAD 14, Maple 17.

Research results: The motion law of a hydraulic-driven pump working body has been theoretically investigated, and its technical characteristics have been determined: performance, volumetric efficiency taking into account the rheological properties of the solution, the supply pressure of the solution when using the combined compensator of the increased volume.

A comparative analysis of the performance and technical characteristics of the solution pump with electromechanical and hydraulic drive has been performed.

Basic material and results: The existing mortar pumps work analysis shows that it is necessary to create a mortar of a simple design for stable mortars pumping of especially low mobility on pipelines at moderate supply pressure fluctuations, high reliability of their work and due to the use of the closed-type combined compensator.

Poltava National Technical Yuri Kondratyuk University's single-piston mortar with a large volume compensator, which has an electromechanical actuator, has proved to be a reliable and highly efficient bulk machine [4–6]. But pumping with solutions of reduced mobility does not fully meet modern requirements below the level [7–9].

However, to improve the mortar pump, such technical characteristics must be introduced in the mortar pump design, which will provide a piston constant speed during reciprocating motion, both in the suction stroke and in the stroke cycle [10]. This will have a positive effect on the suction capacity of the pump, especially with pumping solutions of reduced mobility *P* 8–9 cm, and will reduce backflow through suction and discharge valves due to faster lifting and balls lowering near "dead" points, which, in turn, will positively affect the pressure level.

Therefore, we propose the design of a single-acting mortar pump with a combined larger volume compensator (Fig. 1), which contains suction nozzles 1 and discharge 12, suction chamber 3, in the middle of which there's placed a special cylindrical insert with a cutted part of the segmental shape, tangent to the chord of which is at an angle of 45° to the horizontal, ball valves suction 2 and discharge spring 4, working cylinder 6 with a piston and a slider, which is flushed in the rod cavity cooling-grease moving fluid (soap–oil–water emulsion). The hydraulic pump of the solution pump

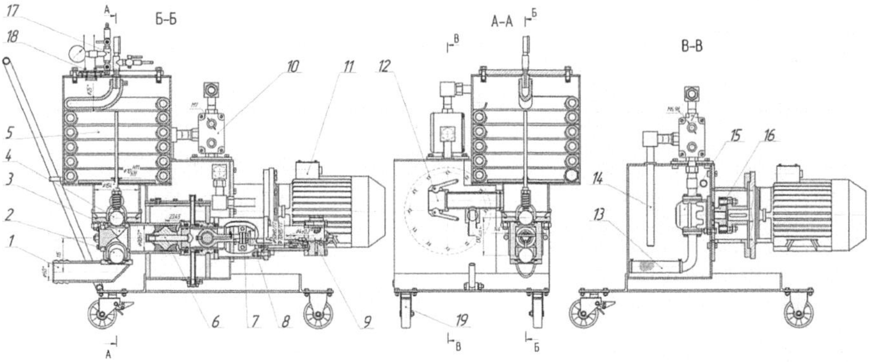

Fig. 1 Design diagram of a single-piston hydraulic drive mortar pump with combined expansion: 1, 12—suction and discharge pipe; 2, 4—suction and discharge spring ball valves; 3—suction chamber; 5—combined compensator; 6—piston with a guide plunger; 7—clamp hydraulic drive cylinder with distributor; 8—hydraulic cylinder with a piston and a rod; 9—spool valve; 10—hydraulic fluid supply regulator; 11—electric motor; 13—filter of lubricating fluid; 14—discharge pipe hydraulic fluid; 15—hydraulic pump gear; 16—coupling sleeve finger; 17—air pump reducer; 18—glass window with illumination

is equipped with a hydraulic cylinder 8 with a piston and a rod having a cavity with an inner flange. The piston of the hydraulic cylinder 8 is mounted opposite (horizontally) with respect to the piston and divides the hydraulic cylinder into the piston and rod cavities. The spool valve 9 is located in the rightmost cavity of the hydraulic cylinder and contains two double-belt differential spools—the main and the control ones. The control spool is mounted coaxially with the piston of the hydraulic cylinder and contains a shank. The connection of the working piston rod to the hydraulic piston rod is carried out by means of a special clamp 7. The rod cavity of the hydraulic cylinder 8 is permanently connected to the line of pressure, and the piston cavity through the main valve is alternately connected to the lines of pressure or discharge, whereby reciprocating motion of piston 6 with stem is provided. The control spool is securely held in one of the working positions throughout the stroke of the piston 6 due to the high or low pressure of the oil in the piston cavity (which is connected with the line of pressure or discharge) and near the dead points is transferred by the inner collar through the shank to the next working position. The reciprocating speed of the reciprocating stroke of the piston is ensured by the fact that the piston area of the hydraulic cylinder is twice as big as the area of intersection of the stem. The flow of hydraulic fluid through the highways is supplied by means of a hydraulic pump gear 15 from the motor 11.

Figure 2 presents a sump pump with combined air pressure compensator drive hydraulic system diagram which consists of two chambers: cylindrical, which is connected to the discharge chamber, and closed, which is an elastic rubber tissue hose, which is attached to the nozzle of the chamber of the air pump and installed with special limiters. The closed chamber is equipped with a nipple for injection of air at a pressure of 0.5–0.7 MP by means of a compressor. In the center of the cylindrical chamber on the, guide a rod float limiter is installed, which separates the compressed air with the solution and thus provides minimal removal of air from the cylindrical chamber.

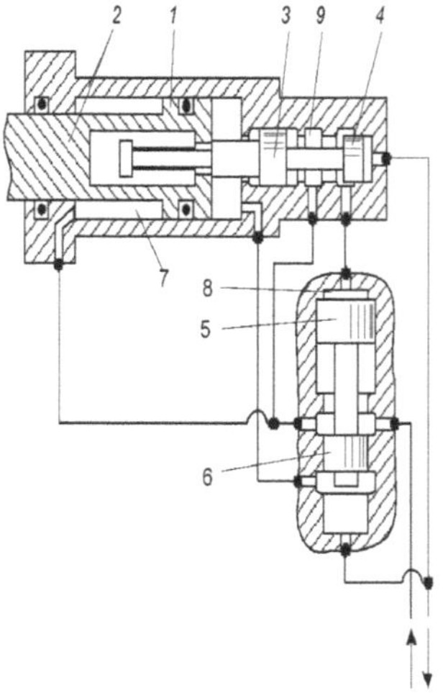

Fig. 2 Scheme of hydraulic drive system of the mortar pump: 1—piston of the hydraulic cylinder; 2—piston rod; 3—lower belt spool control; 4—upper belt spool control; 5—left girdle of the main spool; 6—right belt of the main spool; 7—rod cavity of the main hydraulic cylinder; 7—piston cavity of the main spool; 8—rod cavity of control spool

To increase the compensating volume of the compensator in the upper lid, there is provided an air pump reducer 17 in a cylindrical chamber with a pressure monitoring manometer to the pressure in the closed compensator chamber. Also, to control the volume of air in the cylindrical chamber, a glass window with a bulb 18 is installed in the upper lid for inspection and visual control of the volume of air in the cylindrical chamber.

There is a mathematical model [11, 12] of pump drive hydraulic cylinder column piston work which allows to understand better the suction power of the mortar pump, the nature of ball valves operation for opening and closing, the mechanism of return solution leaks formation during the closing of valve, loads of valve balls during their operation on closing and noise level from this phenomenon and speed of wear of valve sockets, the mechanism of forming volumetric hydraulic pump efficiency level and the level of supply pressure fluctuations.

The scheme of mortar pump automatic drive hydraulic system used in the description of the mathematical model is shown in Fig. 2.

The system of differential equations describing the first phase is the acceleration of the piston from the lower dead point:

$$\begin{cases} m \cdot \frac{\mathrm{d}\upsilon}{\mathrm{d}t} = -(F_m + m \cdot g) + P(t) \cdot (S - S') \\ \beta \cdot \frac{\mathrm{d}P}{\mathrm{d}t} = -\upsilon(t) \cdot \left(S - S'\right) + Q_0 \end{cases} \tag{1}$$

where m is the mass of moving metal parts; F_m—force of displacement of oil on discharge from the piston cavity of the hydraulic cylinder; g—acceleration of gravity; S and S'—accordingly, the area of the piston and the cylinder rod intersection; β—coefficient of volume deformation of oil; Q_0—oil supply.

The complete solution will be in the form of sum $\upsilon_{\text{odn}}(t)$ and υ_{ch}, that is

$$\upsilon(t) = \upsilon_{\text{odn}}(t) + \upsilon_r = C_1 \cdot \cos\left(\frac{S - S'}{\sqrt{\beta \cdot m}} \cdot t\right) + C_2 \cdot \sin\left(\frac{S - S'}{\sqrt{\beta \cdot m}} \cdot t\right) + \frac{Q_0}{S - S'}. \tag{2}$$

The initial conditions for this solution are as follows:

$$\begin{cases} \upsilon(0) = 0 \\ \upsilon'(0) = \frac{P(0)\cdot(S-S')}{m} - \frac{F_{\text{c}}+m\cdot g}{m}. \end{cases}$$

From Eq. (2) for $t = 0$ $C_1 = -\frac{Q_0}{S-S'}$, and after differentiation of Eq. (2) at $t = 0$ $C_2 = \frac{\upsilon'(t)\cdot\sqrt{\beta\cdot m}}{S-S'}$.

Substituting the value C_1 in C_2 Eq. (2) is an equation that describes the rate of piston acceleration in the first

$$\upsilon(t) = -\frac{Q_0}{S - S'} \cdot \cos\left(\frac{S - S'}{\sqrt{\beta \cdot m}} \cdot t\right) + \upsilon'(0) \cdot \frac{\sqrt{\beta \cdot m}}{S - S'} \cdot \cos\left(\frac{S - S'}{\sqrt{\beta \cdot m}} \cdot t\right)$$

$$+\frac{Q_0}{S-S'}\text{phase}. \tag{3}$$

Substituting into Eq. (3) instead of the $v(t)$ magnitude $\frac{Q_0}{(S-S')}$, that is, the final piston acceleration velocity, and after completing some transformations, we find piston acceleration time

$$t_p = \frac{\sqrt{\beta \cdot m}}{S-S'} \cdot \operatorname{arctg}\left(\frac{Q_0}{v'(0)\sqrt{\beta \cdot m}}\right). \tag{4}$$

The piston acceleration time determined by Formula (4) is 0.00214 s, which is about a magnitude order less than the discharge valve closing time (0.0225 s). The piston is accelerated when the discharge valve is fully open; therefore, the separation of the piston from the solution is not possible, since under the piston pressure close to the nominal discharge pressure is maintained.

From Eq. (1), it is also possible to determine the oil pressure at the end of the piston acceleration $P(t) = \frac{F_{\text{‹‹}} + m \cdot g + m \cdot v'(t)}{S-S'}$, which is 4.22 MP.

In the second phase, the constant speed of the piston will be

$$v = \frac{Q_0}{S-S'} = \frac{6.93 \times 10^{-4}}{38.5 \times 10^{-4} - 19.6 \times 10^{-4}} = 0.367\,\text{m/s},$$

and the oil pressure before closing the pressure valve is 3 MP and after closing—8.3 MP at a pressure of 2.5 MP.

In the third phase, the piston movement speed will increase due to the displacement of oil from the differential control spool cavity into the pressure line and will be

$$v = \frac{Q_0}{S-S'-S_1+S_1'} = \frac{6.93 \times 10^{-4}}{(38.5 - 19.6 - 3.8 + 2.1) \times 10^{-4}} = 0.403\,\text{m/s},$$

where S_1 and S_1'—accordingly, the area of the larger and smaller control spool belts.

The oil pressure will increase in the same proportion and will be equal to 9.12 MP.

The time of this phase will last $t_3 = \frac{y_1}{v} = \frac{6 \times 10^{-3}}{6.403} = 0.0149$ from where the y_1 control spool moves until the gap opens.

In the fourth phase, the oil from the oil pump is divided into two streams—into the rod cavity of the cylinder and into cavity 8. The flow separation equation of oil is as follows

$$v(t) \cdot (S - S' - S_1 + S_1') + \mu \cdot S_{shch} \cdot \sqrt{\frac{2g}{\gamma}\Delta P} = Q_0, \tag{5}$$

where μ—the coefficient of oil consumption; S_{shch}—the opening gap of the spool; γ—oil density; ΔP—differential pressure of oil before the gap.

Solution of Eq. (5) according to the method adopted for the first phase gives the formula for the speed of braking

$$\upsilon = y'(t) = \frac{Q_0}{S - S' - S_1 + S_1'} \cdot \mathrm{e}^{-\alpha \cdot \sqrt{P} \cdot t}. \tag{6}$$

The deceleration time $\upsilon \cdot 10^{-3}$ is $t_{\mathrm{zup}} = \frac{\ln 1000}{\alpha \cdot \sqrt{P}} = \frac{2.3 \cdot 3}{1.91 \cdot \sqrt{P}} = \frac{3.61}{\sqrt{P}}$, i.e., the time until the piston is completely stopped, inversely proportional to the root square of the oil pressure. At operating oil pressure in the hydraulic system within 5–10 MP, the braking time will be from 0.0011 to 0.0016 s.

After stopping the piston, all the oil coming from the oil pump is spent on moving the main valve. Calculations show that the residual displacement time of this spool is 0.0069 s.

The piston speed, oil pressure and duration for 5–8 phases are calculated using a similar method.

On the basis of the mathematical model, the dependence of the change in the speed of the piston (1) and the pressure of the oil (2) during the cycle of the hydraulic cylinder is determined (Fig. 3).

Quick transition of the piston from the state of braking and acceleration in "dead" points to the operating speed ensures that much of the progress piston moves at a constant speed, but, in turn, has a positive effect on the uniformity of supply of the pumped solution and pressure pulsations reduction (Table 1).

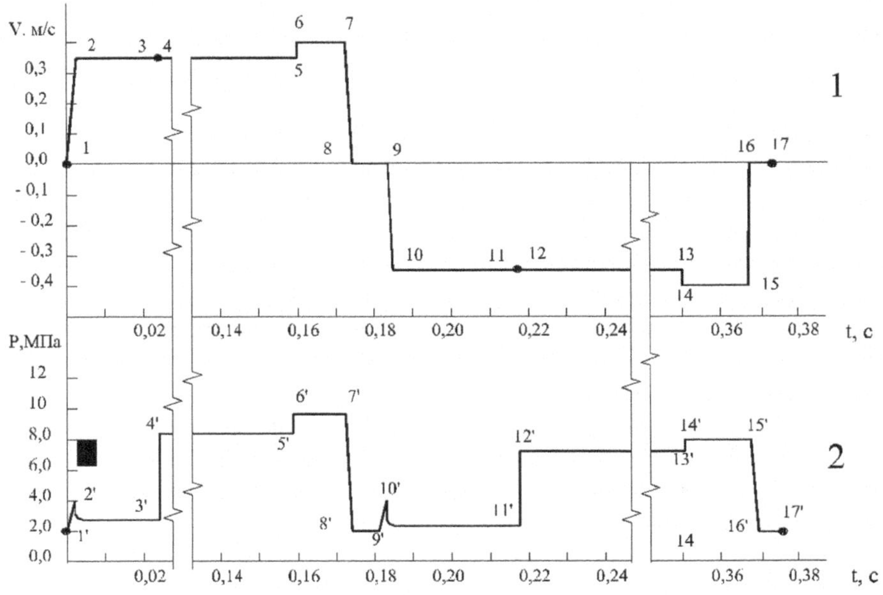

Fig. 3 Velocity change dependence of the piston (1) and oil pressure (2) during the cylinder cycle

Table 1 Summary data for all phases of piston travel speed and oil pressure during the cycle of the cylinder

Section	Section name	Duration	υ (m/s)		P, MP of	
			Start	End	Start	End
1–2	Phase 1, overclocking	0.00214	0.000	0.367	2.00	4.22
2–3	Phase 2, part 1	0.0225	0.367	0.367	3.00	3.00
4–5	Phase 2, part 2	0.1347	0.367	0.367	8.30	8.30
6–7	Phase 3	0.0149	0.403	0.403	9.12	9.12
7–8	Phase 4, deceleration	0.0015	0.403	0.000	9.12	2.00
8–9	Switching of main spool	0.0069	0.000	0.000	2.00	2.00
9–10	Phase 5, acceleration down	0.00302	0.000	0.354	2.00	4.02
10–11	Phase 6, part 1	0.0307	0.354	0.354	2.29	2.29
12–13	Phase 6, part 2	0.1351	0.354	0.354	7.20	7.20
14–15	Phase 7	0.0157	0.389	0.899	7.92	7.92
15–16	Phase 8, braking	0.0020	0.389	0.000	7.92	2.00
16–17	Switching base spool	0.0038	0.000	0.000	2.00	2.00

Mathematical analysis of the hydraulic actuator made it possible to determine that the long piston stops at the dead points are caused by the oil consumption for switching the main spool. In this case, the total stoppage time for switching this spool will be $0.0069 + 0.0038 = 0.0107$ compared to the time of one cycle of the solution pump $\frac{60}{161} = 0.373$. Thus, the stopping time of the piston at dead points will be 2.9%, while the total duration of all accelerations and brakes of the piston in one cycle is only 2.32%. To reduce the time of piston switching in further design, it is necessary to reduce the diameters of the main spool belts, as well as the stroke length of this spool, although these are also high rates of stability of the spools at dead spots.

Theoretical dependences (Fig. 4) of solution supply pressure at the outlet of the discharge pipe during the cycle of pump operation indicate a decrease in the level of pressure pulsations when used in the design of a hydraulic actuator. The decrease in the degree of fluctuation in the solution pressure is accounted for by the constant speed of piston movement in the half cycle of injection ($1'$–$7'$ phase), the decrease in the return leakage of the solution through the suction valve. The pressure stabilization of the solution in the suction half cycle is due to the rational volume of the air in the closed and cylindrical compensator chambers.

Dependence (7) allows, depending on the nature of the pressure change in the working chamber, the mode of working body movement, taking into account the rheological and elastic properties of the pumped solution to calculate the volumetric efficiency of the pump.

$$\eta_{\text{ob}} = \frac{V_{p.f}}{V_r} = \frac{\begin{pmatrix} \left[V_{\text{povn}} \cdot (1 \pm \varepsilon) - \Delta V_{vs.kl}\right] \cdot \left(1 - \varepsilon_{st2}\right) - \Delta V_{n.kl} \cdot \left(1 + \varepsilon_{st2}\right) - V_{shk} - \\ -\left(F_p \cdot \left[\left(x_p - \frac{h_p}{2}\right) - \frac{h_p}{2\pi} \cdot (\varphi_{ng} - \pi)\right] \cdot (1 - \delta)\right) \end{pmatrix}}{V_r \cdot \left(1 - \varepsilon_{st2}\right)} \tag{7}$$

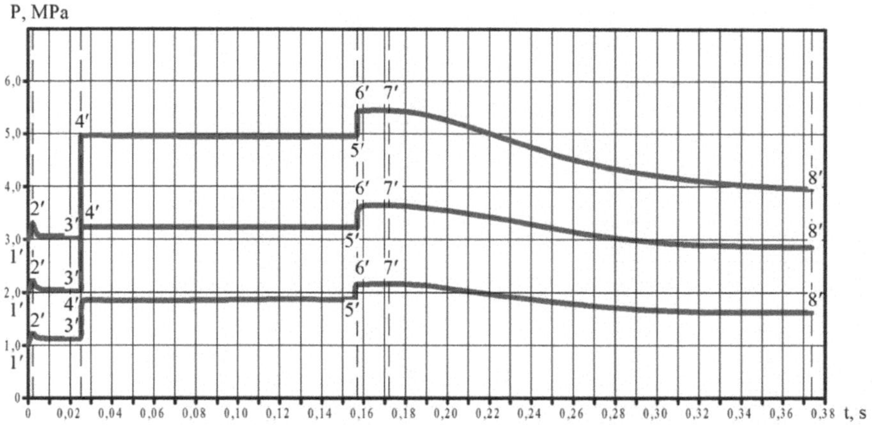

Fig. 4 Dependences of pressure change of a solution supply on an exit from a discharge branch pipe during a working cycle of a solution pump

where V_r—working volume; $V_{p.f}$—actual volume of solution supplied to the discharge pipeline and reduced to normal conditions; $V_{\text{povn}} = V_r + V_{shk}$—full volume, as the sum of harmful and working volumes; $\Delta V_{vs.kl}$, $\Delta V_{n.kl}$—the volume of losses at the closing of the suction and discharge valves, respectively; ε_{st}—coefficient of volume compression at pressure ε_{st}; $\varepsilon_{st\,2}$—the relative compression of the solutions acquires the maximum value due to the complete dissolution of the bubble air and with increasing pressure. Depending on the mobility of the pumped solution, these values are for *P* 8 cm—$\varepsilon_{st2} = 3.3 - 3.5\%$ for *P* 10 cm—$\varepsilon_{st\,2} = 2.0\%$ for *P* 12 cm—$\varepsilon_{st\,2} = 1.5\%$ [13].

The graphical dependences (Fig. 5) show that in the hydraulic-driven pump and the combined volume compensator the volume efficiency increased by 8%.

The obtained graphical dependences show that the theoretical (dashed lines in Fig. 5) agree with the actual experimental ones (solid lines). The increase in the actual volumetric efficiency compared to the theoretical for solutions of *P* 8–12 cm is explained by the increase in pressure in the working chamber and the decrease in the expansion of the solution at the end of the suction stroke according to theoretical calculations.

Dependences (Fig. 5) indicate a significant effect of the solution mobility on the volumetric loss of the solution, which significantly reduces the volumetric efficiency of the pump. This is especially noticeable when pumping solutions with the mobility of *P* 8–12 cm.

The results of the analysis show the dependence suitability (7) for the volumetric efficiency calculation of the solution pump taking into account the elastic properties of the pumped medium.

The experimental dependences (Fig. 5) are consistent with the theoretical ones for determining the volumetric efficiency from the work process factors.

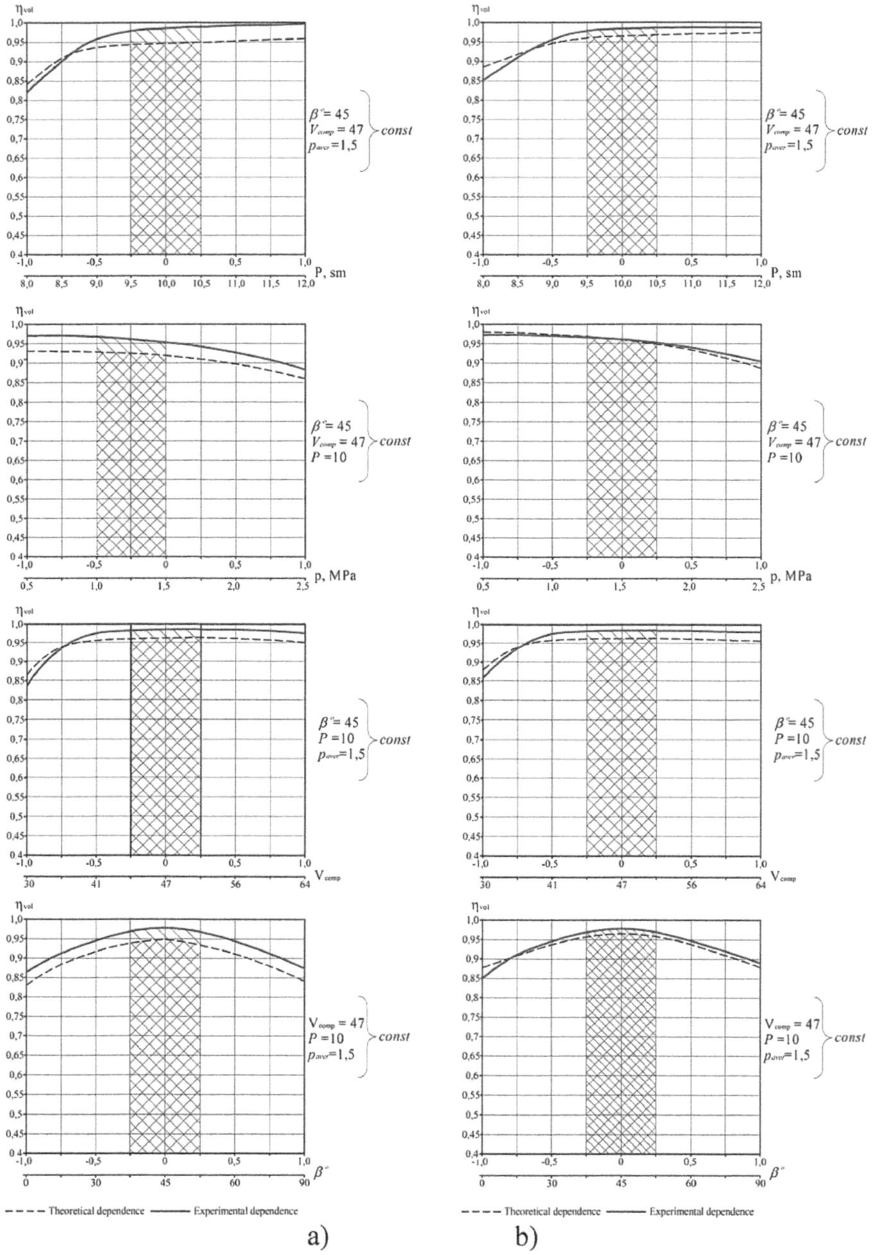

Fig. 5 Graphic dependences of the solution pump volumetric efficiency: **a** with the increased volume combined compensator; **b** with hydraulic actuator and combined compensator of increased volume

The nature of the graphical dependences, shown in Fig. 5, confirms the special insert rational parameters, which increases the value of mortar pumps volumetric efficiency with different drives by 12–15%.

The results of the influence studies (Fig. 3.25) of the accepted factors on the pump volumetric efficiency show that all the indicators are higher in the hydraulic pump-driven pump and the combined compensator of the increased volume, for which the level of pressure ripple (by 8%) is significantly reduced and for which the obtained ranges of rational technological parameters are $X_1 = 9.5–10.5$ cm; $X_2 = 1.0–1.5$ MP; $X_3 = 43–50$ DM^3; $X_4 = 37.5–52.5°$ (Fig. 5).

Therefore, the hydraulic drive mortar under similar operating conditions has better technical characteristics compared to the mortar with the combined compensator of the increased volume:

- solution supply pressure pulsation degree decreased by 8%;
- increased productivity by 14%;
- reduced power costs by 11%.

2 Conclusions

The working body motion law influence mechanism at the hydraulic drive expense and factors on the productivity, volumetric efficiency and capacity of the pump is theoretically established. Also, the maximum value of 88% volumetric efficiency of the mortar pump was reached when pumping mortar with a mobility of P 8 cm with a special insertion at an angle of inclination $\beta = 45°$ in the suction chamber and spring-loaded pressure valve. Due to such design decisions, the value of pump volumetric efficiency when pumping the solution $P = 8$ cm increased by 8%.

The rational mode of operating hydraulic-driven pump occurs at the following ranges of factors: pumping the solution with the mobility $P = 9–10$ cm, the supply pressure $p = 1.0–1.5$ MP, the reduced volume of the compensator $V_{comp} = 44–50$ DM^3, installation in the suction chamber of a special insert at an angle of 45°.

Moreover, the use of a mortar pump is appropriate in the composition of hydraulic drive plastering units and plastering stations, which will ensure their high-energy efficiency. The use of a hydraulic actuator in the mortar makes it possible to smoothly adjust the flow during the operation of the mortar pump, which will have a positive effect on the finishing process.

References

1. Parfyonov, E. P. (1972). Determination of the performance of piston mortars. Mekh. tools and finishing machines: *Information Scientific-Technical Sat*, (Vol. 4, pp. 12–13) TsNIITEstroy-mash.

2. Mortelpumpen und ihre Entwicklung// "Fordern und Heben". 1969. - No. 15. (Germany).
3. EP 0200026, INT. Cl. 4 F 04 B 43/12, 15/02. Neumuller pumps Walter, Sturmer Gerhard. 10.12.1986. Patentblatt 86/45.
4. Emelyanova, I. A. & Shapoval, M.V. (2017). Determination of productivity and volumetric efficiency of the mortar pump, depending on the geometric parameters of the suction chamber and the compensators of various design solutions. *Scientific Bulletin of Construction, 88*(2), 195–203.
5. Korobko, B., Virchenko, V., & Shapoval, M. (2018). Feed solution in the pipeline with the compensators mortar pump of various design solutions pressure pulsations degree determination. *International Journal of Engineering & Technology, [S.l.], v.7*(n.3.2), 195–202. https://doi.org/10.14419/ijet.v7i3.2.14402.
6. Emelyanova, A., & Shapoval, M. V. (2017). Determination of performance and volumetric efficiency of the mortar pump depending on the geometric parameters of the suction chamber and the compensators of different design solutions. *Scientific Bulletin of Construction, 88*(2), 195–203.
7. Korobko, B., & Vasyliev, I. e. (2017). Test method for rheological behavior of mortar for building work *Acta mechanica et automatica, 11/3*(41), 173–177. https://doi.org/10.1515/ama-2017-0025.
8. Korobko, B., Vasiliev, A., & Rogozin, I. (2015). The analysis of mixture kinematics in the mixer body frame with a screw elevator with variable generatrix. *Eastern-European Journal of Enterprise Technologies, 3*(7), 48–52. doi: 10.15587/1729-4061.2015.43053 48-52.
9. Rohozin, I., Vasyliev, O., & Pavelieva, A. (2018). Determination of building mortar mixers effectiveness. *International Journal of Engineering & Technology, 7*(3.2) (S.I. 2), 360–366. https://doi.org/10.14419/ijet.v7i3.2.14553.
10. Shapoval, M. V., Virchenko, V. V., Skoryk, M. O., & Shpilka, A. M. (2019). Improving the efficiency of the pump by using a hydraulic actuator. In *Collection of Scientific Papers of the II International Ukrainian-Azerbaijan Conference "Building Innovations—2019"* (202–205 p), May 23–24, 2019. Poltava: PoltNTU.
11. Kukoba, A. T., & Vasilyev, A. V. (2000). Investigation of the volumetric efficiency of a hydraulic drive mortar. In *Collection of scientific works (branch mechanical engineering, construction)* (Vol. 5, pp. 19–24). Polt. state. tech. Yuriy Kondratyuk. Poltava: PDTU.
12. Kukoba, A. T., Vasilyev, A. V., & Yakubtsov, O. M. (2000). The impact of the law on the piston against capacious efficiency pump. In *Collection of scientific works (branch engineering, construction)* (Vol. 6, pp. 12–17). Polt. state. tech. Yuriy Kondratyuk. Part 1. Poltava: PDTU.
13. Kukoba, A. T., Korobko, B. O., & Vasilyev, A. V. (2000). Changing the volume of the mortar mixture during pumping with a solution pump. *Mechanization of Construction, 3*.

Modern Possibilities of Management of Technogenic-Natural Systems of Heat-Energy Objects of Industrial and Construction Industry

P. M. Kulikov, N. Y. Zhuravska, and A. M. Savchenko

Abstract According to the strategies of state economic policy of Ukraine, industry and construction should develop without active transformation of the environment. Consequently, the methodological principles of operating production-technogenic system of heat-energy objects (TEO) in the context of ensuring their stability in the application of directed artificial technogenesis (for electromagnetic fields) are substantiated. The main scientific principle of ensuring the stability of the feasibility systems is the equivalence of the parameters (established by us) of the obtained magnetized water with respect to their specific indicators. The evolutionary nature of coherent relationships and interactions (the term of research—five years) is scientifically consistent with quantitative values. They include: hydrochemical, electrophysical and thermophysical potentials. From a practical point of view—the prolonged nature of the material flows and the final result of the technology of non-reagent water treatment in the systems of thermal power facilities of industrial and construction fields.

Keywords Production systems of construction · Organizational and management decisions · Electromagnetic fields

1 Introduction

With the development of market relations, the term "ecological-economic management" (environmental management) is increasingly used, being often confused with the term "administrative environmental management," but they have a certain distinction. Administrative environmental management is closely linked to public administration in the field of environmental management and environmental protection, that is, public administration of environmental policy implementation. Environmental policy, according to N. A. Malash, is a set of tools and measures related to the impact of society on nature and is aimed at ensuring environmentally balanced development and civility [1]. Environmental management, in its turn, is an initiative and productive

P. M. Kulikov · N. Y. Zhuravska (✉) · A. M. Savchenko
Kyiv National University of Civil Engineering and Architecture, Kiev, Ukraine
e-mail: nzhur@ua.fm

V. Onyshchenko et al. (eds.), *Proceedings of the 2nd International Conference on Building Innovations*, Lecture Notes in Civil Engineering 73,
https://doi.org/10.1007/978-3-030-42939-3_13

activity of economic entities, directed at achieving their own environmental goals, projects and programs, developed on the basis of the principles of eco-efficiency and environmental justice.

The main goal of Ukraine's environmental policy is to improve the quality of life with the optimal use and reproduction of natural resources. Environmental policy is to reconcile social and environmental objectives of the society as a basis for solving the problem of global environmental crisis. The main principles of Ukraine's environmental policy are set out in Art. 3 of the Law of Ukraine "On Environmental Protection". On the basis of the generalization of domestic and foreign experience in the formation of mechanisms for solving social and environmental problems, the basic principles of environmental policy of Ukraine during the transition period have been defined as follows: (a) priority of ecology over the economy, gradual and consistent transition to new mechanisms of public administration in the sphere of implementation of environmental policy; (b) conformity of the mechanism of public administration in the sphere of implementation of environmental policy to the actual state of society development; (c) generalization of positive domestic and foreign experience in order to solve environmental problems; (d) participation of the state in financing environmental problems; e) ensuring sustainable and "friendly" use of nature; (e) continuous environmental education and the acquisition of environmental information [2].

State policy in the field of nature protection is also formed by the relevant regulatory documents, integrated management, financial, economic and other management mechanisms in the field of protection, use and reproduction of natural resources. Based on this, Ukraine has previously developed "Main directions of state environmental policy until 2020." This document identifies the purpose and the priorities of the environmental policy, the mechanisms for their implementation, including the financial mechanism for environmental issues and environmental quality improvement.

Financial mechanism for the implementation of the country's environmental policy includes the following main parts:

1. Based on the application of different financial methods: financial planning, financial management, financial control, financial security and financial regulation of environmental activities;
2. Contains certain instruments that regulate the activities of the state in establishing mandatory laws and legal rules (norms) of the behavior of subjects of law;
3. Contains financial levers (instruments) that most comprehensively support the development of "greening" the economy. As for the fees for special use of natural resources, fines for violations of environmental legislation and financial resources received in the form of compensation for environmental damage, their fiscal purpose is justified by the practice of modern management, based on a high level of anthropogenic environmental burden. In recent years, the number of types of payments, both for the use of natural resources and for the pollution of the environment, has significantly increased. At the same time, the rate of payments and penalties for violation of environmental legislation has been raised.

4. Contains organizational tools, the main of which are: eco-sponsoring, eco-tagging, eco-outsourcing, eco-technology, information and technical support.

The normative level of ecologically safe development of the environment, today, is possible only with the exception of negative technogenic influences from production processes (Law on the Basic Strategies of the State Environmental Policy of Ukraine for the Period up to 2030; No. 8328 of 28.02.2019). The Law sets priorities for improvement and development of the state system of environmental management.

According to this Law, it is stated that the economy in the industrial sphere should develop in such a way, that active transformation of the environment will not occur. Such technologies include non-reagent water preparation, in the systems of thermal power facilities of industry and construction, when applying electromagnetic fields. Taking into account the fact that in the near future the problems of modern energy in providing the population with heat will be realized at the expense of thermal energy, the present research is relevant and timely. At the same time, the issues of non-reagent water treatment in heat supply systems have a limited nature of its implementation [3] and do not fully reveal the benefits of non-reagent water preparation [4]. Therefore, the paper shows that the environment and the socio-economic components of industrial objects cannot be represented separately from each other. It is then possible to achieve simultaneous scientific and technological success for the development of industrial facilities (ensuring the innovative nature of technological production processes) and, thanks to modern tactical management intentions, reduce or eliminate anthropogenic pressure on the atmosphere.

2 Materials and Methods

The study applies a systematic approach to conducting research using factor analysis of the systematization and formalization of passive monitoring data regarding the assessment of the ecological status of heat supply systems [5]. Due to the fact that the main system-forming unit is the material flows of heat-energy objects of industry and construction, then all attention was drawn to them.

The organizational and managerial orientation of the step-by-step mathematical formalization of the technological process made it possible to determine the basic interconnections and interactions of the functional components of material flows, and thereby to determine the basic regulation levers of these processes [6]. Figure 1 represents our block diagram of implementing factor analysis in the experimental work. It is designed to put into practice the factual features of passive monitoring results. According to this block diagram, the general direction of the research is to reach the level of probabilistic (predicted) functioning of the feasibility study systems.

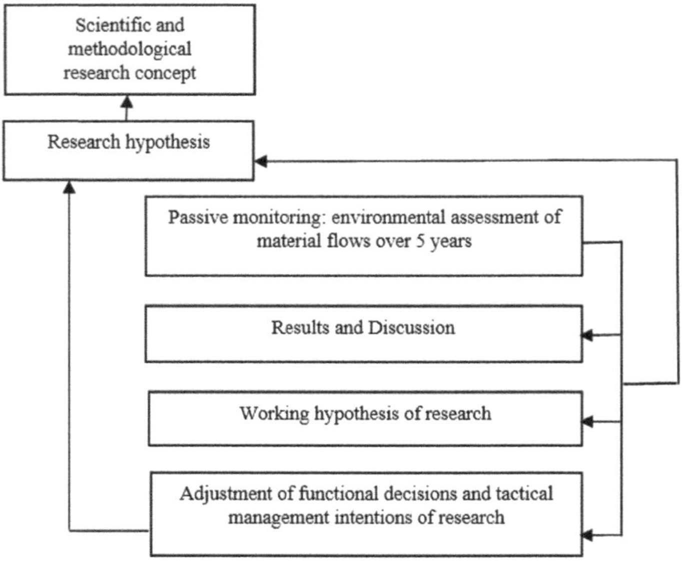

Fig. 1 Flowchart of the practical implementation of the factorial features of the passive monitoring results

3 Results and Discussion

The integral mathematical manifestation of the action of magnetized water in the operation of the systems of heat-energy objects (HEO) in electromagnetic fields (EMF) is the specific composition of material flows. In that, the formation of complex heterogeneous inorganic systems (the primary basis—simple inorganic systems) can be observed. That is, the complexity of the systems due to their thermodynamic and synergistic restructuring has occurred. It can be noted that artificial targeted technogenesis (obtaining the authors of magnetized water in the "Illios-M" apparatus with clearly defined optimal parameters) led to the production of magnetized water. What is the prolonged nature of the action due to the increase in the catalytic activity of material flows? The decisive role at this important stage of research was played by organizational and managerial decisions. This is the choice of apparatus for obtaining non-reagent (b/g) water preparation, the choice of control parameters, the creation of a two-level control system and the like.

In our developed, scientific and applied concept of experimental work, it is established that the main basis for making organizational decisions is engineering and technological issues of the proposed nanotechnology under the conditions of reliable management of them.

To ensure the innovative nature of the proposed nanotechnology is to determine the functional features of b/g water preparation in the EMF for the HEO systems that determine the working hypothesis of the research. The working hypothesis consists in a factor sign of determining estimation of technical updating tendencies of technological processes and their forecast of formation for now and prospect. In order to establish the theoretical and applied aspects, we have proposed such generalized tendencies-principles that ensure the development of a scientific and methodological concept of experimental works, namely:

1. The principle of equivalence (consistency) of optimal parameters of the obtained magnetized water and specific indicators and potentials (according to indicators and potentials) that characterize the material flows of the feasibility study system;
2. The principle of evolution with respect to the transformation of simple inorganic systems within a heterogeneous system, as a factor of complication of the ecological and structural state of material flows of the HEO systems under the influence of EMF and in the process of thermodynamic and synergetic rearrangements (self-organization), term of study of technogenic-natural systems;
3. The principle of optimality, as a factor trait that excludes the over-normative formation of technogenic-hazardous risks for the operation of the created nanotechnology itself, and the formation of ecologically hazardous risks for the environment (resource and energy costs; technological emissions into the atmosphere);
4. The principle of prolonged nature of the action of magnetized material flows, as a result of organizational and management decisions regarding production processes in heat supply systems;
5. The principle of specific conditions for the combination of functional features, in the form of certain structures, which are interconnected by links and interactions (system-forming components of material flows) allows to create a mechanism for regulation of technological process of water preparation in EMF in the HEO systems with the integral indicator control mechanism, etc.

Figure 2 presents a structural and graphical conceptual model.

The figure demonstrates a number of features that are typical only of natural-technogenic systems (energy costs and exergy costs; prevention of thermal pollution of the atmosphere at the stage of theoretical studies, when the optimal economic optimum of these pollution is found, a graphical conceptual model for the conservation of the assimilation potential of the atmosphere is created, etc.) [7]. Data analysis in Fig. 2 shows that technogenic-natural (production) systems, with the use of EMF in heat supply systems, have factor characteristics that are unique to them. In addition, it should be noted that we first proposed the term “passive monitoring” as a type of impacted anthropogenic monitoring to determine the dynamics of changes in the state of material flows to implement the principles of equivalence and optimality in practical conditions. The establishment of regulated values-gradations made it possible to ensure the prolonged nature of the action of the magnetized water of material flows.

At the same time, it can be stated with confidence that the leading role of this state of material flows is fully linked to perfect organizational and managerial decisions of ensuring production processes. That is, the innovative nature of the proposed nanotechnology (predictive-evolutionary principle) is constantly maintained. The

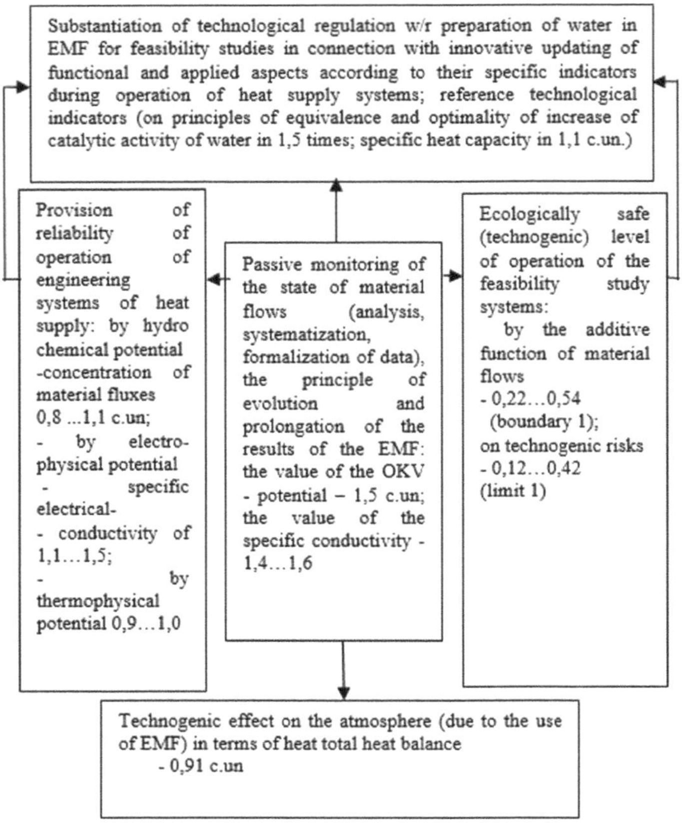

Fig. 2 Structural and engineering components of heat supply systems, which determine the coherent interconnections and interactions of systemic factors of material flows

decisive role in ensuring the prolonged nature of the action is related to a two-level integrated control system, which allows timely signaling of changes in the ecological and technological state of material flows (critical situation) in the hit of their specific indicators, certain potentials in the danger zone, which shows the general direction of algorithms in the algorithms of material flows. It can be noted that the application of the proposed indicator control allows analytical control at all stages of operation of the HEO systems.

Thus, the conditions and criteria (theoretical and methodological concept of the research) with the use of quantitative determination of the level of functioning of technogenic-natural production systems are associated with the development of theoretical bases (principles) and the creation of appropriate tactical intentions for their implementation (management decisions), which are mutually beneficial, of course.

References

1. Malysh, N. A. (2011). *Effective mechanisms of formation of state environmental policy* (348 p)/ Kyiv: KIS.
2. Zarzhitsky, O. S. (2003). *Legal aspects of regional environmental policy* (pp. 26–27). Dnepropetrovsk: Science and Education.
3. Malkin, E. S., Furtat, I. E., & Zhuravska, N. Y. (2017). *Special issues of heat and mass transfer* (288 p). Kiev: KNUBA.
4. Malkin E. S. (2014). Prospects for creation of resource-saving technologies by magnetic treatment of water and aqueous solutions. In E. S. Malkin, I. E. Furtat, V. P. Usachev, & N. Y. Zhuravska. (Eds.), *Ventilation, lighting and heat and gas supply: NTZ* (17, pp 120–127). Kyiv: KNUBA.
5. Zhuravska, N. (2017). Ecological scientific aspects of technogenic safety with nonchemical water treatment for a technical water-supply. In *III Міжнародна науково-практична конференція «Underwater technologies»* (p. 8). Kyiv: KNUCA.
6. Zhuravska, N. (2018). *Ensuring technological reliability of energy-saving technologies with a reagent-free water treatment* (Vol. 2(2), pp. 1–8). USEFUL. SVP4U-KYIV-1-FUND LLC. Maimi.
7. Kulikov, P. M., Zhuravska, N.Y., & KNUBA. (2018). *Intellectual property—copyright for a work.* Scientific and methodological structure of man-made hazard and risk management in the preparation of technical water in the heat supply system. Literary written work of a scientific and technical nature. Application No. 86018. Kiev: Ministry of Economic Development of Ukraine. 12. 12. 2018. 11 sec.

Deformations of Soil Massifs Under the Existence of Saline Solutions with Different Concentration and Temperature

Mykola Kuzlo, Yuriy Vynnykov, Volodymyr Ilchenko, and Nataliya Zhukovska

Abstract The regularities of the influence of saline solutions' concentration and their temperature on the deformational properties of soils have been experimentally investigated and determined. At the background of experimental research and its statistical processing, nonlinear dependences in the form of polynomials of the deformation module and Lame coefficients from the concentration of saline solutions and their temperature which allowed to improve the mathematical model of the stress–strain state of soil, taking into account nonlinear filtration and deformation processes occurring in soil masses under the condition of presence and filtration of saline solutions, have been obtained.

Keywords Saline solutions · Deformational properties · Mathematical model · Experimental research

1 Introduction

Till nowadays, quantitative and qualitative estimations of soil massifs deformations, saturated with saline solutions, are still not discovered. Usually, data that are developed for soil mechanics, which are subject to the action of natural groundwater, are used in the analysis of deformations of saline soils and their physical interpretation [1]. However, as shown before, the compressibility of soils, which are subject to saline solutions, depends both on the degree of loading and on the concentration of saline solutions [2].

M. Kuzlo · N. Zhukovska
National University of Water Management and Environmental Engineering, Rivne, Ukraine
e-mail: kuzlo-@ukr.net

Y. Vynnykov · V. Ilchenko (✉)
Poltava National Technical Yuri Kondratyuk University, Poltava, Ukraine
e-mail: znpbud@gmail.com

V. Onyshchenko et al. (eds.), *Proceedings of the 2nd International Conference on Building Innovations*, Lecture Notes in Civil Engineering 73,
https://doi.org/10.1007/978-3-030-42939-3_14

2 Defining the Problem

Experimental research has been carried out in the geotechnical laboratory of the National University of Water Management and Environmental Engineering in order to determine the effect of the saline solutions' concentration and their temperature on the deformation characteristics of the soil. Experiments have been carried out on a compression–filtration device according to the standard method [3].

The design of the device has provided the application of the solution to the soil sample from below (upstream flow scheme) or from above (flow chart); the removal of the solution that has been filtered through a soil sample; tightness of the main parts of the device; preserving the set temperature of the filtering fluid.

3 Research Results

The soil samples for disturbed structures with specified values of density and humidity have been used for the test. As soil for research, sand clays with a plasticity number $I_r = 7.0\%$ and porosity coefficient $e = 0.55$ have been taken. To conduct the experiment, ground soils' pastes have been prepared, which have been saturated with saline solutions at the concentration of 0.0; 2.0; 4.0; 6.0; 8.0; 10.0; 12.0; 14.0; 16% for 2 days. Each sample has been tested at load stages 0.1; 0.2; 0.4 and 0.6 MPa and the given temperature of saline solution.

After applying each degree of loading, the deformation indicator data have been recorded through 0.025; 0.5; 1.0; 2.0; 5,0; 10; 20; 30 and 60 min, and then hourly until conditional stabilization of the deformation of the soil is achieved.

For conditional stabilization of soil deformations at a given pressure level, deformations of not more than 0.01 mm have been taken for the last 16 h of observations.

The magnitude of the absolute deformation of the soil (Δh) has been calculated as the average arithmetic value of the indicators; the value of the relative deformation of the soil sample $\Delta h/h$, where h is the initial height of the soil sample has been determined in the process of soil testing in the compressor.

The deformation module E (c, T), the Lame parameter λ (c, T), μ (c, T) have been determined on the basis of the research results.

The results of these experiments are shown in Tables 1, 2 and 3.

The processing of the data obtained from experimental studies of the elastic modulus $E(c, T)$, the Lame parameters $\lambda(c, T)$ and $\mu(c, T)$ depending on the concentration of C saline solutions and their temperature has been fulfilled using the software complex for the processing of the experiment data [4].

The choice of approximating functions $E = f(c, T)$, $\lambda = f(c, T)$ $\mu = f(c, T)$ for the best approximation of experiment data in the sense of the mean-square deviation has been made automatically using this complex. Approximating functions $E = f(c, T)$, $\lambda = f(c, T)$ $\mu = f(c, T)$ have the form:

Table 1 Value of the deformation module $E\ (c,\ T)$

C (g/l)	Meaning of the module deformation $E_{(c,\ T)}$ (kPa)			
	t = 16 °C	t = 25 °C	t = 50 °C	t = 75 °C
0	5700	5220	4950	4450
20	4800	4600	4500	4125
40	4350	4250	4120	3850
60	4400	4150	3850	3600
80	4490	4240	3900	3500
100	4650	4280	3975	3575
120	4760	4400	4100	3750
140	4940	4550	4275	3975
160	5250	4750	4500	4200

Table 2 Meaning of the parameter Lame $\lambda_{(c,\ T)}$

C (g/l)	Meaning of the parameter Lame $\lambda_{(c,\ T)}$ (kPa)			
	t = 16 °C	t = 25 °C	t = 50 °C	t = 75 °C
0	4925	4511	4277	3845
20	4148	3975	3888	3564
40	3759	3672	3560	3327
60	3802	3586	3327	3111
80	3880	3664	3370	3025
100	4018	3698	3435	3089
120	4113	3802	3543	3240
140	4269	3932	3694	3435
160	4537	4104	3888	3629

Table 3 Meaning of the parameter Lame: $\mu_{(c,T)}$

C (g/l)	Meaning of the parameter Lame $\mu_{(c,\ T)}$ (kPa)			
	t = 16 °C	t = 25 °C	t = 50 °C	t = 75 °C
0	2111	1933	1833	1833
20	1777	1704	1666	1527
40	1611	1574	1526	1425
60	1630	1537	1425	1333
80	1663	1570	1444	1296
100	1722	1585	1472	1324
120	1762	1630	1518	1388
140	1830	1685	1583	1472
160	1944	1759	1666	1555

$$\lambda(c, T) = a_5^1 \cdot c^2 + a_4^1 \cdot c + a_3^1 \cdot c \cdot T + a_2^1 \cdot T^2 + a_1^1 \cdot T + a_0^1, \quad (1)$$

where $a_5^1 = 0.11658$, $a_4^1 = -19.13338$, $a_3^1 = -0.01433$, $a_2^1 = 0.07664$, $a_1^1 = -18.29209$, $a_0^1 = 4875.65995$.

$$\mu(c, T) = a_5^2 \cdot c^2 + a_4^2 \cdot c + a_3^2 \cdot c \cdot T + a_2^2 \cdot T^2 + a_1^2 \cdot T + a_0^2, \quad (2)$$

where $a_5^2 = 0.05347$, $a_4^2 = -8.56879$, $a_3^2 = -0.01451$, $a_2^2 = -0.04571$, $a_1^2 = -8.02227$, $a_0^2 = 2098.68892$.

$$E(c, T) = a_5^3 \cdot c^2 + a_4^3 \cdot c + a_3^3 \cdot c \cdot T + a_2^3 \cdot T^2 + a_1^3 \cdot T + a_0^3, \quad (3)$$

where $a_5^3 = 0.13495$, $a_4^3 = -22.19182$, $a_3^3 = -0.01556$, $a_2^3 = 0.08929$, $a_1^3 = -21.30479$, $a_0^3 = 5646.9363$.

The calculated Fischer criterion for determination with the help of Formula (1) is $F_{cal} = 28.0117$, and its tabular value with the probability $P = 0.95$ is $F = 5.7558$. Since $F_{cal} > F$, then with probability $P = 0.95$ it can be argued that Formula (1) describes the functional dependence on the parameter definition.

Fischer's calculated criterion for the determination according to the Formula (2) is $F_{cal} = 21.0136$, and its tabular value with the probability $P = 0.95$ is $F = 5.1158$. Since $F_{cal} > F$, then with probability $P = 0.95$ it can be affirmed that Formula (2) describes the functional dependence on the parameter definition.

The calculated Fischer criterion for determination of $E(c, T)$ according to the Formula (2) is $F_{cal} = 24{,}01517$, and its tabular value with a probability of $P = 0.95$ is $F = 5.4858$. Since $F_{cal} > F$, then with probability $P = 0.95$ it can be confirmed that Formula (2) describes the functional dependence on the parameter $E(c, T)$.

The results of the mathematical processing of data on experiments with an optimal scale are shown on Figs. 1, 2 and 3.

As our experiments have shown, the concentration of saline solutions affects the soil compression. So, when the soils are saturated with saline solutions with the concentration from 0 to 16%, their post compaction is 15% more than compaction of the soils saturated with distilled water. At the same time, most of the post compaction occurred when the soil was saturated with saline solutions at the concentration of 60 g/l, the same 6%. Previously, it was found that the filtration coefficient also reaches its maximum value at a given concentration of soils' saturation with saline solutions [5]. Thus, the effects of mineralized water on soil compression should be found in the effect of electrolytes on mineral particles of soil.

Such regularity can be explained on the basis of physical and chemical processes occurring between mineral particles (solid phase) and saline solutions (liquid phase). In this case, we have a disperse system. The superficial energy of such a system is measured by the surface tension that occurs at the interface of the disperse phase with the dispersive environment and the magnitude of the total surface of all parts of the disperse phase. Any disperse system tries to reduce its surface energy. In the system of "mineral particles + solution," it can be reduced both by reducing the size of the total surface and the surface tension of the liquid. The latter factor leads to the

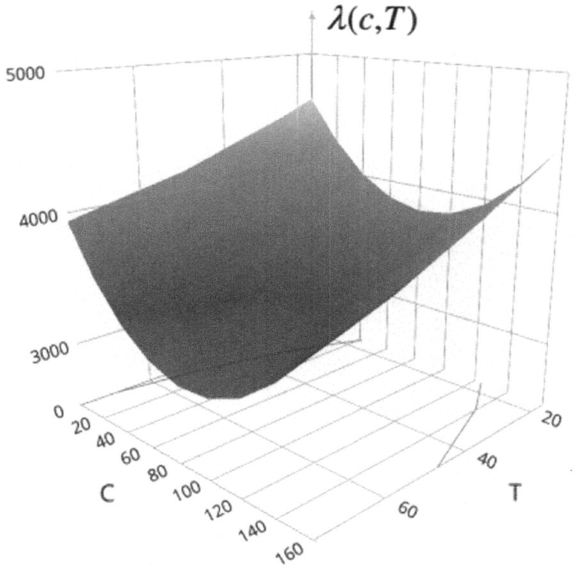

Fig. 1 Graphic of Lame coefficient dependency $\lambda(c, T)$

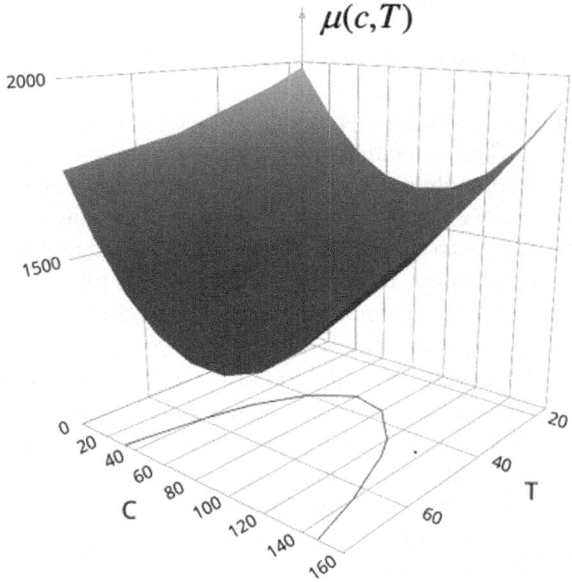

Fig. 2 Graphic of Lame coefficient dependency $\mu(c, T)$

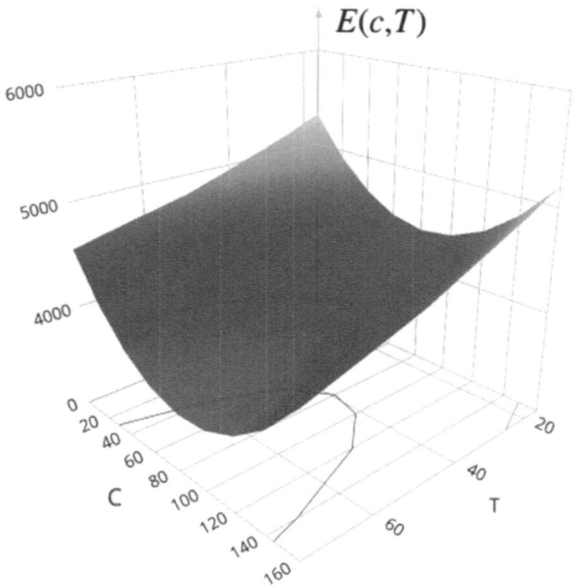

Fig. 3 Graphic of Lame coefficient dependency $E(c, T)$

compression of the diffusion layers of the liquid, and, consequently, to the additional compaction of soils.

Significant influence on the compressibility of soils is made by the composition of exchange cations. Thus, saturation of soils with sodium ions causes a noticeable decrease in their compressibility. The sharp decrease in the compressibility of the soil from the presence of exchange cation is due to its dispersing effect. As a result, pore sizes and formation of a significant amount of bound water in the soil decrease, which reduce the permeability and compressibility of soils [6].

In addition, there is always the third component, the electrolyte, which is the saline solution under the presence of saline solutions in dispersed systems such as soils, in addition to the two components, the solid disperse phase and the liquid disperse medium. When the salts are dissolved in water, the mineral particles of the soil do not remain passive to the molecules or ions of soluble salts. Dipole water molecules, orienting in the force field of molecules or soluble salt's ions, form around it a densified hydration layer. They are hydrating, or as it is said in such cases, mineral particles become wet and solid. So, clay soils are very hydrophilic. Depending on the degree of hydrophilicity, the surface retains a different amount of hydration, or, as it is called, bound water. The content of bound water in clay reaches 25%, which leads to a decrease in soil compressibility [7].

The dependence of the Lame coefficients $\lambda(c, T)$ and $\mu(c, T)$ on the concentration of the filter solution and its temperature is of scientific and practical interest and can be used later in the construction of mathematical models of the underground hydromechanics tasks and analysis of the filtering properties of soil environment.

In particular, in the general case, the mathematical model describing the stress–strain state of soil environments, taking into account the dependences of the Lame coefficients and the Yung's module on the concentration of the filter solution and its temperature in the two-dimensional case, is as follows:

$$\frac{\partial\sigma_x}{\partial x}+\frac{\partial\tau_{xy}}{\partial y}-(2\lambda(c,T)+2\mu(c,T))\alpha_T\frac{\partial T}{\partial x}+X=0,$$
$$\frac{\partial\tau_{yx}}{\partial x}+\frac{\partial\sigma_y}{\partial y}-(2\lambda(c,T)+2\mu(c,T))\alpha_T\frac{\partial T}{\partial y}+Y=0,\quad X\in\Omega, \tag{4}$$

$$\nabla\cdot(D\nabla c)+\nabla\cdot(D_T\nabla T)-(\upsilon,\nabla c)=n_p\frac{\partial c}{\partial t},\ X\in\Omega,\quad t>0, \tag{5}$$

$$\nabla\cdot(\lambda_T\nabla T)-\rho c_\rho\upsilon\nabla T=c_T\frac{\partial T}{\partial t},\ X\in\Omega,\quad t>0, \tag{6}$$

$$\upsilon=-K(c,T)\nabla h+\nu_c\nabla c+\nu_T\nabla T,\ X\in\Omega,\quad t>0, \tag{7}$$

$$\varepsilon_x=\frac{\partial U}{\partial x},\quad \varepsilon_y=\frac{\partial V}{\partial y},\quad \varepsilon_{xy}=\frac{1}{2}\left(\frac{\partial U}{\partial y}+\frac{\partial V}{\partial x}\right),\quad X\in\Omega, \tag{8}$$

$$\sigma_x=\lambda(c,T)\varepsilon_\theta+2\mu(c,T)\varepsilon_x-(2\lambda(c,T)+2\mu(c,T))\alpha_T\bar{T} \tag{9}$$

$$\sigma_y=\lambda(c,T)\varepsilon_\theta+2\mu(c,T)\varepsilon_y-(2\lambda(c,T)+2\mu(c,T))\alpha_T\bar{T} \tag{10}$$

$$\tau_{xy}=2\mu(c,T)\varepsilon_{xy},\quad X\in\Omega, \tag{11}$$

$$c(X,0)=\tilde{C}_0(X),\ T(X,0)=\tilde{T}_0(X),\quad X\in\Omega, \tag{12}$$

$$l_1h(X)|_{\Gamma_T}=\tilde{H}_1(X),\ l_2c(X,t)|_{B_T}=\tilde{C}_1(X,t), \tag{13}$$

$$l_3T(X,t)|_{B_T}=\tilde{T}_1(X,t),\ X\in\Omega,\quad t>0, \tag{14}$$

under appropriate boundary conditions for displacements and stresses in the researched area Ω.

In the (4)–(14), it used the following symbols:

$X = (x, y)$—point of area Ω;
B_T—boundary of area Ω; $\lambda(c, T)$ and $\mu(c, T)$—Lame coefficients;
$c(X, t)$—concentration of filtrated solution;
$T(X, t)$—temperature;
$h(X)$—piezometric pressure;
υ—vector of filtration speed;
n_p—volume of saline solution of water in the soil volume unit;

D—coefficient (tenzor) of convective diffusion;
D_T—coefficient (tenzor) of thermodiffusion;
λ_T—coefficient (tenzor) of effective thermoconductivity;
ρ—density of saline solution;
c_ρ—specific heat of saline solution;
c_T—volumetric heat capacity of soil under stable volume;
$K(c, T)$—coefficient (tenzor);
v_c—coefficient (tenzor) of chemical osmosis;
v_T—coefficient (tenzor) of thermal osmosis;
$\varepsilon_x, \varepsilon_y, \varepsilon_{xy}, \sigma_x, \sigma_y, \tau_{xy}$—normal and tangent deformations and strains;
X, Y—components of mass force;
α_T—average coefficient of linear thermal expansion in the interval of temperatures (T_0, T);
t—time, $t > 0$;
$l_i, i = \overline{1, 3}$—differentiative operators that give boundary conditions according to pressure, concentrations of salt and temperature on lateral surface of B_T cylinder $Q_T = \Omega\ (0; T)$.

Existing mathematical models for today do not fully describe the stress–strain state of soil with the consideration of nonlinear deformation processes occurring in soils under the presence and filtration of saline solutions with different temperatures [8–10].

4 Conclusions

The statistical significance and adequacy of the empirical dependences of the deformation module and the Lame coefficients in the form of polynomials from the concentration of saline solutions and their temperature have been obtained and verified. It allowed us to improve the elaboration of the mathematical models of the stress–strain state of soil, taking into account the nonlinear deformation processes occurring in soils under the presence and filtration of saline solutions at different temperatures.

The data of experimental research can be used in the assessment of soil bases state and structures with the help of deformations and bearing capacity. The perspective of further research can be an assessment of the stress–strain state of soil bases and structures, taking into account nonlinear processes occurring in the presence of saline solutions with different concentrations and temperatures.

References

1. Sergeev, E. (1983). *Soil science*. Moscow: Moscow State University.

2. Kuzlo, M., & Filatova, I. (2006). Investigation of the influence of the concentration of saline solutions on the deformation characteristics of soils. *Hydromelioration and Hydrotechnical Construction, 31,* 175–181.
3. DBNBV.2.1-4-96 (GOST 12248-96). (1997). Soils. Methods of laboratory determination of durability and deformation characteristics.
4. Rogalev, N., Ochkov, V., Orlov, K., & Voloshchuk, V. (2016). Thermal engineering studies with excel. *Mathcad and Internet*. https://doi.org/10.1007/978-3-319-26674-9.
5. Vlasyuk, A., & Kuzlo, M. (2001). Experimental studies of some parameters saline solutions filtration in sandy soils. *Reclamation and Water Management, 87,* 139–145.
6. Bohnhoff, G. L., & Shackelford, C. D. (2015). Salt diffusion through a bentonite-polymer composite. *Clays and Clay Minerals, 63*(3), 145–162. https://doi.org/10.1346/CCMN.2015.0630301.
7. Chernuha, O. (2005). Admixture mass transfer in a body with horizontally periodical structure. *International Journal of Heat and Mass Transfer, 48,* 2290–2298. https://doi.org/10.1016/j.ijheatmasstransfer.2005.01.003.
8. Bonelli, S. (2009). Approximate solution to the diffusion equation and its application to seepage-related problems. *Applied Mathematical Modeling, 33*(1), 110–126. https://doi.org/10.1016/j.apm.2007.10.017.
9. Vlasyuk, A., & Fedorchuk, N. (2013). Numerical modeling of the stress-strained state of a multilayered soil mass under the presence of ground water level and the effect of heat-mass transfer in one-dimensional case. *Mathematical and Computer Modelling. Series: Technical Sciences, 8,* 31–44.
10. Kuzlo, M., Moshynskyi, V., & Martyniuk, P. (2018). Mathematical modelling of soil massif's deformations under its drainage. *International Journal of Applied Mathematics, 31*(6), 751–762. doi:http://dx.doi.org/10.12732/ijam.v31i6.5.

Experimental Researches of Concrete Ultimate Characteristics and Strength of Compressed and Bended Reinforced Concrete Elements

Dmytro Lazariev, Yurii Avramenko, Oleksandr Zyma, and Pavlo Pasichnyk

Abstract The results of experimental researches of the work of short compressed and bent elements under the action of, respectively, longitudinal loading and bending moment are given. The influence of the normal cross-section reinforcement percentage on ultimate deformation of the most compressed concrete fiber and strength of reinforced concrete elements is established. Comparative analysis of experimental data with theoretical calculations based on deformation model with an extreme criterion was performed.

Keywords Bearing capacity · Deformation model · Strength · The extreme criterion · Ultimate strain

1 Problem Latest Researches and Analysis

The deformation model with the extreme criterion [1, 2], taking into account the descending branch of the compression diagram, makes it possible to obtain analytically the ultimate deformations of compressed concrete fiber of reinforced concrete element. This reflects the process of reducing the bearing capacity of material due to increase in the degree of its destruction.

Confirmation of theoretical calculations by experimental researches and their statistical analysis plays an important role. Therefore, experimental researches of compressed concrete zone ultimate characteristics, the deformability and reinforced concrete elements strength that work on bending and compression is of an urgent task [3–15].

Theoretical researches of compressed concrete zone ultimate characteristics and strength of the above concrete elements carried out on the basis of deformation model with an extreme criterion are given [1].

D. Lazariev (✉) · Y. Avramenko · O. Zyma
Poltava National Technical Yuri Kondratyuk University, Poltava, Ukraine
e-mail: lazariev.pntu@gmail.com

P. Pasichnyk
Kyiv National University of Construction and Architecture, Kyiv, Ukraine

V. Onyshchenko et al. (eds.), *Proceedings of the 2nd International Conference on Building Innovations*, Lecture Notes in Civil Engineering 73,
https://doi.org/10.1007/978-3-030-42939-3_15

2 The Work Purpose

Experimental researches of normal cross-section reinforcement percentage influence on the ultimate strain of compressed concrete fiber and strength of reinforced concrete elements working on bending and flat non-center and central compression and comparing obtained experimental data with theoretical calculations on the basis of the deformation model is the work purpose.

3 The Main Material

Deformations of concrete and reinforcement in characteristic cross sections of reinforced concrete elements were measured over the entire load range during the experimental researches of reinforced concrete columns and beams. Particular attention was paid on determining of ultimate deformations of concrete and reinforcement in characteristic cross sections of reinforced concrete elements.

Eighteen reinforced concrete columns and six reinforced concrete beams with different reinforcement percentages were researched to achieve this goal (Table 1).

The experimental samples had rectangular cross sections along the length. The height and width of normal cross section of columns and beams were assumed to be constant: for columns -120×120 mm and for beams -180×120 mm (Fig. 1).

All reinforced concrete columns were reinforced symmetrically by four longitudinal rods of class A400C of one diameter, as well as welded clamps of reinforcement of class A240C, diameter of 6.5 mm, with a distance 80 mm (Fig. 1).

The beams were reinforced by four longitudinal rods class A400C and A500C of different diameters, as well as welded clamps of A240C class fixtures, 6.5 mm in diameter, with a distance 50 mm in the area where a sloping crack may occur, and with a distance 100 mm in the area of pure bending (Fig. 1).

The strength of concrete samples of the C-1-1…9 series was equal $f_{\text{ck,cube}} = 15,83$ MPa, and of the C-2-1…9 series and B-1-1…3, B-2-1…3—$f_{\text{ck,cube}} =$

Table 1 Reinforcement percentages of columns and beams

№	Eccentricity of load application e_0, m	Reinforcement percentages ρ_f, %		Non-overreinforced elements	Overreinforced elements
		Non-overreinforced elements	Overreinforced elements		
Compressed elements					
1	$e_0 = 0$	1,764	5,517	C-1-1…3	C-2-1…3
2	$e_0 = 0,03$			C-1-4…6	C-2-4…6
3	$e_0 = 0,12$			C-1-7…9	C-2-7…9
Bent elements					
1	–	0,844	4,281	B-1-1…3	B-2-1…3

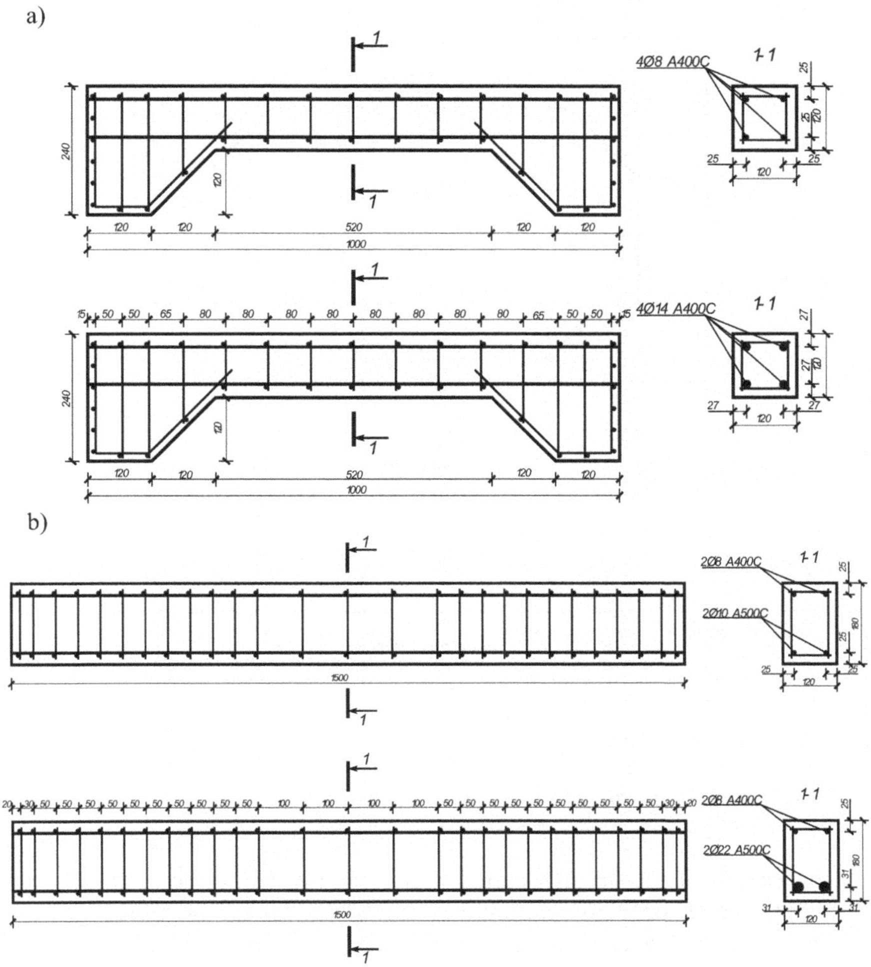

Fig. 1 Reinforcement scheme and geometric dimensions of experimental samples of reinforced concrete columns (**a**) and beams (**b**)

15, 06 MPa. According to the experimental researches, compressed concrete zone ultimate characteristics and strength of normal cross sections of the test specimens were analyzed.

The scheme of the sensors on the concrete is shown in Fig. 2.

The analysis of experimental data results allowed us to confirm the validity of theoretical researches results by the method based on deformation model with extreme criterion. The specimens were also analyzed at different levels of loading depending on compressed concrete fiber deformation of normal cross section of reinforced concrete element from reinforcement percentage.

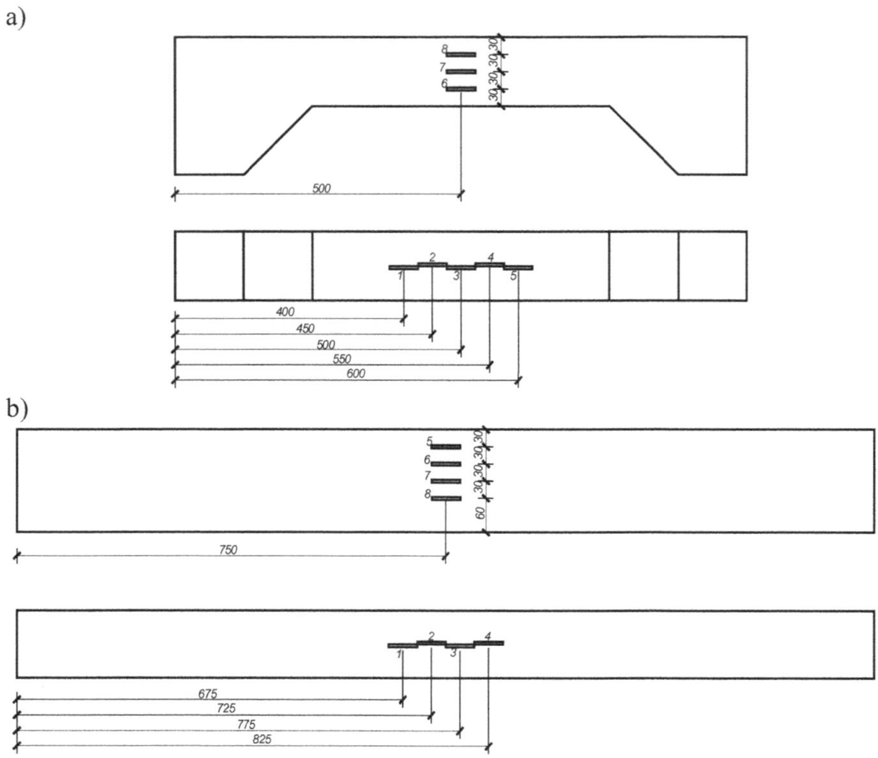

Fig. 2 Location of sensors on concrete of columns (**a**) and beams (**b**)

The researches results of non-centrally and centrally compressed and bent reinforced concrete elements of constant cross section of hinged supported elements were analyzed on characteristic samples according to eccentricity of load application, reinforcement percentage and load type (Figs. 3, 4, 5 and 6). The analysis of deformability and strength of compressed and bent reinforced concrete element normal sections was performed in accordance with Table 1.

The concrete deformation dependence of loading level sample and reinforcement percentage diagrams (Figs. 3, 4, 5 and 6) show that deformations of the most compressed concrete fiber in non-overreinforced specimens are smaller than in overreinforced both compressed and bent reinforced concrete elements. In all tested samples, the hypothesis of flat sections is confirmed, almost up to the maximum load level.

In the bending elements of concrete most compressed fiber deformation in the overreinforced elements is not significant, but the deformations in non-overreinforced elements also exceed. Excess deformation is explained by the remapping of the tension concrete zone only, which led to the destruction of overreinforced elements in compressed zone.

Similar results were obtained using a deformation model with an extreme criterion [1].

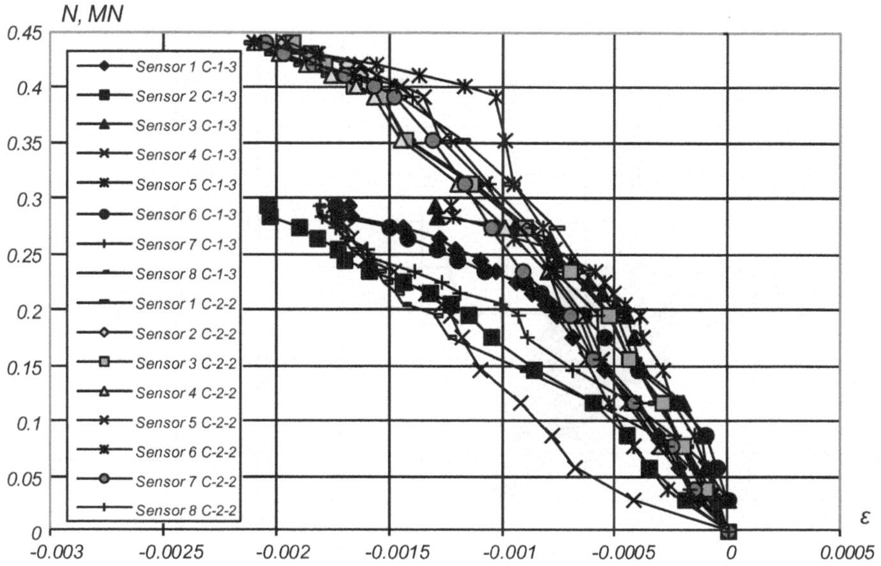

Fig. 3 Change of concrete strain in normal section of reinforced concrete element when loading the sample, depending on reinforcement percentage of centrally compressed elements

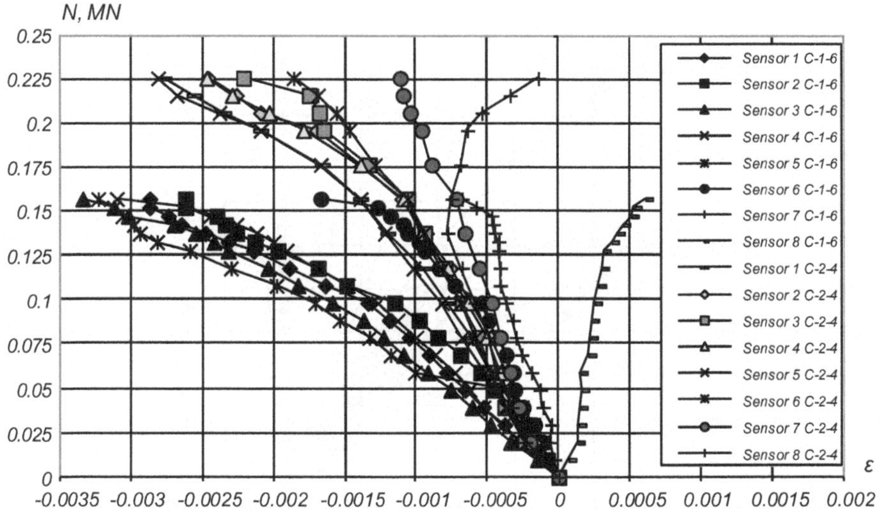

Fig. 4 Change of concrete strain in normal section of reinforced concrete element when loading the sample, depending on reinforcement percentage of non-centrally compressed elements ($e_0 = 0,03$m)

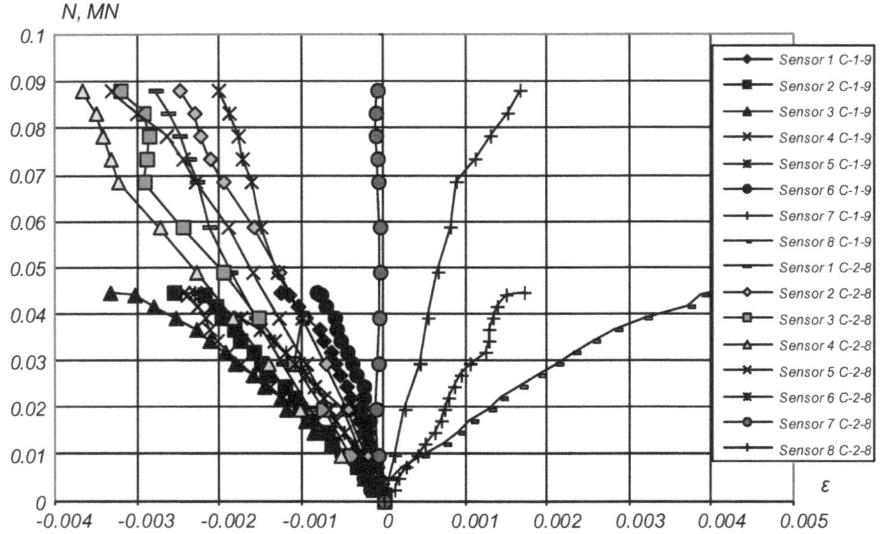

Fig. 5 Change of concrete strain in normal section of reinforced concrete element when loading the sample, depending on reinforcement percentage of non-centrally compressed elements ($e_0 = 0, 12$m)

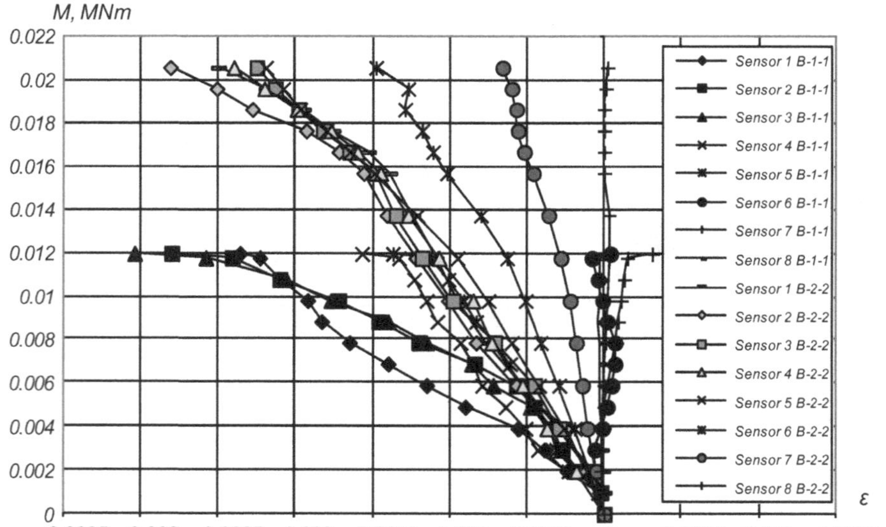

Fig. 6 Change of concrete strain in normal section of reinforced concrete element when loading the sample, depending on reinforcement percentage of bent elements

The results of theoretical (by the method of deformation model with extreme criterion) and experimental data comparison of compressed concrete fiber ultimate deformations and strength of reinforced concrete element normal cross sections with statistical processing are given in Table 2.

Table 2 Experimental and theoretical values of deformations and strength of reinforced concrete elements normal section in the stage close to destruction

№	Code sample	Maximum relative deformation of the most compressed concrete fiber		$\Delta\varepsilon_{cu1}$, %	Axial force and bending moment *N (M), MN (MNm)*		Δ, %
		$\varepsilon_{cu1,exp}$	$\varepsilon_{cu1,theor}$		N_{exp} (M_{exp})	N_{theor} (M_{theor})	
1	2	3	4	5	6	7	8
Compressed elements							
1	C-1-1	−0,0020	−0,001924	0,962	0,284	0,299769	1,056
2	C-1-2	−0,0021	−0,001924	0,916	0,255	0,299769	1,176
3	C-1-3	−0,00205	−0,001924	0,939	0,294	0,299769	1,020
4	C-1-4	−0,0029	−0,002646	0,912	0,167	0,150186	0,899
5	C-1-5	−0,0028	−0,002646	0,945	0,162	0,150186	0,928
6	C-1-6	−0,0033	−0,002646	0,802	0,157	0,150186	0,957
7	C-1-7	−0,0031	−0,002942	0,949	0,046	0,042523	0,924
8	C-1-8	−0,0039	−0,002942	0,754	0,046	0,042523	0,924
9	C-1-9	−0,0033	−0,002942	0,892	0,045	0,042523	0,945
10	C-2-1	−0,0021	−0,001993	0,949	0,451	0,459868	1,020
11	C-2-2	−0,0021	−0,001993	0,949	0,441	0,459868	1,043
12	C-2-3	−0,0026	−0,001993	0,767	0,451	0,459868	1,020
13	C-2-4	−0,0028	−0,002883	1,030	0,226	0,234095	1,036
14	C-2-5	−0,003	−0,002883	0,961	0,255	0,234095	0,918
15	C-2-6	−0,0029	−0,002883	0,994	0,255	0,234095	0,918
16	C-2-7	−0,0034	−0,003561	1,047	0,093	0,084235	0,906
17	C-2-8	−0,0037	−0,003561	0,962	0,088	0,084235	0,957
18	C-2-9	−0,0035	−0,003561	1,017	0,088	0,084235	0,957
Arithmetic average value				0,930			0,979
Root-mean-square value				0,116			0,087
Standard error				0,078			0,068
Variation coefficient, %				8,393			6,926
Bent elements							
19	B-1-1	−0,0030	−0,003017	1,006	0,012	0,011693	1,026
20	B-1-2	−0,0034	−0,003017	0,887	0,0122	0,011693	1,043

(continued)

Table 2 (continued)

№	Code sample	Maximum relative deformation of the most compressed concrete fiber		$\Delta\varepsilon_{cu1}$, %	Axial force and bending moment *N (M), MN (MNm)*		Δ, %
		$\varepsilon_{cu1,exp}$	$\varepsilon_{cu1,theor}$		N_{exp} (M_{exp})	N_{theor} (M_{theor})	
21	B-1-3	−0,0028	−0,003017	1,077	0,013	0,011693	1,112
22	B-2-1	−0,0028	−0,002897	1,035	0,0218	0,020297	1,074
23	B-2-2	−0,0028	−0,002897	1,035	0,0216	0,020297	1,064
24	B-2-3	−0,0027	−0,002897	1,073	0,0216	0,020297	1,064
Arithmetic average value				1,019			1,064
Root-mean-square value				0,024			0,004
Standard error				0,059			0,025
Variation coefficient, %				5,787			2,311
Sum of all elements							
Arithmetic average value				0,953			0,999
Root-mean-square value				0,175			0,125
Standard error				0,084			0,071
Variation coefficient, %				8,792			7,068

4 Conclusions

(1) Made experimental researches allow making conclusion about the reliability of deformation model with an extreme criterion, which allows to analyze the complete set of normal section ultimate parameters in the stage of their destruction, to identify the elastic or plastic state of the reinforcement and to use corresponding calculated equations.

(2) Consideration of real work of reinforcement has significant effect on normal section ultimate characteristics and its strength. Overreinforcing of reinforced concrete element normal sections significantly increases compressed concrete fiber ultimate deformation (ε_{cu1}). The same effect ε_{cu1} causes decrease in the strength of concrete with constant reinforcement [1].

References

1. Shkurupiy, O., Lazariev, D., & Davydenko, Y. (2018). Strength design of compressed reinforced concrete elements by deformation method based on extreme criterion. *International Journal of Engineering & Technology, 7*(3.2), 334–338. https://doi.org/10.14419/ijet.v7i3.2.14430.
2. Mitrofanov, V. P. (2000). Optimization strength theory of reinforced concrete bar elements and structures with practical aspects of its use. *Bygningsstatiske Meddelelser. Copenhagen: Danish Society for Structural Science and Engineering, 71*(4), 73–125.
3. Noghabai, K. (2000). Beams of fibrous concrete in shear and bending: Experiment and model. *ASCE Journal of Structural Engineering, 126,* 243–251. https://doi.org/10.1061/(ASCE)0733-9445(2000)126:2(243).
4. Bambura, A. N., & Gurkovskiy, A. B. K. (2003). postroeniyu deformatsionnoy teorii zhelezobetonnyih sterzhnevyih sistem na eksperimentalnoy osnove. Budivelni konstruktsii: zbirnyk naukovyh prats. K.: Budivelnyk. Vyp. 59. pp. 121–130.
5. Hwang, J.-H., Lee, D. H., Ju, H., Kim, K. S., Seo, S.-Y., & Kang, J.-W. (2013). Shear behavior models of steel fiber reinforced concrete beams modifying softened truss model approaches. *Materials, 6,* 4847–4867. https://doi.org/10.3390/ma6104847.
6. Azizov, T. N. (2008). Sposob opredeleniya granichnyih deformatsiy betona na nishodyaschey vetvi. Resursoekonomni materialy, konstruktsii, budivli ta sporudy: Zbirnyk naukovyh prats. Rivne: NUVGP. Vyp. 16. Ch. 2. pp. 3–7.
7. Bencardino, F., Rizzuti, L., Spadea, G., & Swamy, R. N. (2008). Stress-strain behavior of steel fiber-reinforced concrete in compression. *Journal of Materials in Civil Engineering, 20,* 255–263. https://doi.org/10.1061/(ASCE)0899-1561(2008)20:3(255).
8. Babych, Y. M., & Zarechanskyi, O. O. (2006). Eksperymentalni doslidzhennia hnuchkykh stysnuto-zihnutykh zalizobetonnykh elementiv pry poperechnykh povtornykh navantazhenniakh riznykh rivniv. Budivelni konstruktsii: Zbirnyk naukovykh prats. K.: NDIBK. Vyp. 65. pp. 253–259.
9. Galinska, T., Ovsii, D, & Ovsii, M. (2018). The combining technique of calculating the sections of reinforced concrete bending elements normal to its longitudinal axis, based on the deformation model. *International Journal of Engineering & Technology, 7*(3.2), 123–127. https://doi.org/10.14419/ijet.v7i3.2.14387.
10. Mitrofanov, V. P., Shkurupiy, A. A., & Lazarev, D. N. (2007). Metodika izmereniya predelnoy deformatsii betona na szhatoy grani zhelezobetonnyih elementov. *Bashtovi sporudy: Materialy, konstruktsii, tehnologii: zbirnyk naukovyh prats. Makiivka: DNABA, 6*(68), 96–100.
11. Masiuk, G., Yushchuk, O., & Paschenko, A. (2018). Experimental investigations of the stress and strain state of continuous reinforced concrete beams under the action of low-cyclic repetitive and alternating loads. *International Journal of Engineering & Technology, 7*(3.2), 236–238. https://doi.org/10.14419/ijet.v7i3.2.14410.
12. Dovzhenko, O., & Pohribnyi, V., Karabash, L. (2018). Experimental study on the multikeyed joints of concrete and reinforced concrete elements. *International Journal of Engineering & Technology, 7*(3.2), 354–359. https://doi.org/10.14419/ijet.v7i3.2.14552.
13. Barashykov, A. I., & Zadorozhnikova, I. V. (2005). Sproshcheni rozrakhunky nesuchoi zdatnosti normalnykh pereriziv zghynalnykh zalizobetonnykh elementiv za deformatsiinoiu modelliu. Resursoekonomni materialy, konstruktsii, budivli ta sporudy: zbirnyk naukovykh prats. Rivne: NUVHP. Vyp. 12. pp. 109–115.
14. Zheng, L., & Wang, S. (2002). Experimental investigational the failure patterns and mechanical properties for plain concrete with crocks. Rapair, Rejuvanifion and enhancement of concrete. In *Proceedings of the International Seminar held at the University of Dunolec, Scotland, UK*. 5–6 Sept. 2002. London: Thomas Telford. pp. 257–266.
15. Krus, Y. A., & Krus, A. Y. (2007). Vliyanie rezhima nagruzheniya na ochertanie diagrammyi mehanicheskogo sostoyaniya betonu, tip i strukturu approksimiruyuschey funktsii. Kommunalnoe hozyaystvo gorodov: sbornik nauchnyih trudov. K.: Tehnika. Vyp. 76. pp. 89–95.

Application of the Universal Design Principles in the Improvement of Street and Urban Road Environment

Tetyana Lytvynenko, Iryna Tkachenko, Viktoriia Ivasenko, and Tetiana Lvovska

Abstract The work is devoted to the application of the universal design principles (UDP) in design of the street and urban road environment improvement. The authors suggested a new method of street and urban roads improvement elements placing on the basis of spatial corridor principle modeling. This method ensures the implementation of seven principles of universal design: equitable use; flexibility in use; simple and intuitive use; perceptible information; tolerance for error; low physical effort; size and space for approach and use. To apply the proposed method, street and urban roads beautification elements have been distributed by importance level into four groups and elements of each level have been placed in defined spatial corridor. Authors have designed the examples of application of the UDP in the street and urban road environment improvement and examples of inclusive street and urban road environment formation based on the principles of universal design by suggested method.

Keywords Universal design principles · People with limited mobility · Improvement · Streets and urban roads environment

1 Introduction

Every settlement should be comfortable for all residents, without considering their health, knowledge, or physical form. Population consists of different categories of citizens, such as: children; pedestrians with strollers; middle-aged pedestrians; old pedestrians; pedestrians with disabilities (their number in Ukraine is constantly increasing, see Fig. 1); pedestrians, who came to this city for the first time; pedestrians, who live in this city; drivers of cars or other vehicles; drivers of individual environmental vehicles. None of these categories should be uncomfortable while moving the city and all the city's objects should be equally accessible to everyone.

T. Lytvynenko (✉) · I. Tkachenko · T. Lvovska
Poltava National Technical Yuri Kondratyuk University, Poltava, Ukraine
e-mail: litta2510@gmail.com

V. Ivasenko
O.M. Beketov National University of Urban Economy, Kharkiv, Ukraine

V. Onyshchenko et al. (eds.), *Proceedings of the 2nd International Conference on Building Innovations*, Lecture Notes in Civil Engineering 73,
https://doi.org/10.1007/978-3-030-42939-3_16

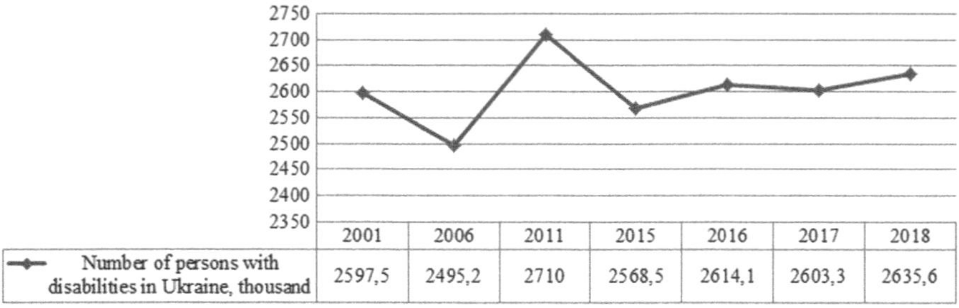

Fig. 1 Number of persons with disabilities in Ukraine, thousand

Many scientists from different countries are interested in application of the universal design principles in the street and urban road environment improvement. Jane Bringolf and Margaret Louise Ward from *Australia* are researching Universal Design in Housing [1, 2]. They are writing that voluntary guidelines were developed jointly by industry and disability advocacy groups but mandating these guidelines has been shunned at every turn. Sweta Byahut, Anagha Mujumdar, Anand Patel, Amit Sheth, Arindam Mitra, Raajesh Moothan, and Yatin Pandya from *India* in 2005 have developed Design Manual for a Barrier-Free Built Environment India [3]. In *Canada* in 2017, there was publication of Barrier-Free Design Guide by Alberta Municipal Affairs and Safety Codes Council [4]. Mai Eid Khalil Ahmed and Aslı Sungur Ergenoglu from *Turkey* have done an assessment of street design with UDP: case in Aswan [5]. In 2012, *Norway* hosted the largest conference on universal design held in Europe until then. The conference gathered researchers, students, users, planners, public officers, and other practitioners from 44 countries. Sigrid Skavlid, Hans Petter Olsen, and Ase Kari Haugeto have collected the most interesting information from conference in Trends in Universal Design [6]. Kelly Carr, Patricia L. Weir, Dory Azar, and Nadia R. Azar from *Canada* think that it is necessary to provide environments designed to suit the needs of older adults [7]. Paul Jones from *United Kingdom* is trying to respond to growing calls for a theoretical unpacking of UD [8]. Bangun Harsritanto, Indriastjario, Asri Nabila Wijayanti from *Indonesia* have examined several design characteristics of themed streets in several countries from three different continents using UDP for giving proper directions to develop more user-friendly streets and they resumed that design direction can be suggested universal along with the richness of local aspects [9, 10]. Studies of all these scientists are very interesting and show that the UDP should be applied in design of the street and urban road environment improvement.

The work is related to the research work of Highways, Geodesy, Land management and Rural buildings department of Poltava National Technical Yuri Kondratyuk University (Ukraine)—"Improvement of highways and street-road network" (Project ID: 0114U000354).

2 Research Goal and Objectives

The *goal* of the work is searching for new means and elements that provide a universal design (UD) of streets and urban roads. The intent of "UD" is to simplify life for everyone by making products, communications, and to build more usable environment for as many people as possible at low or no extra cost.

The main *objectives* of this research:

- To find the requirements for a high-quality accessible and a legible street that directs and assists the most casual of users in finding their way around;
- To lead a way in achieving an attractive and inclusive street for all people;
- To draw attention of professionals, which cover architects, city planners, industrial designers, lecturers, academicians, administrators, and investors to the concept of universal design.

3 Research Methods

The following research methods were selected for this scientific research: *comparative analysis* and *generalization* of streets and urban roads improvement and existing theoretical developments regarding the application of the UDP; *full-scale survey* of the existing condition of streets and urban roads of Ukraine (measurement of height and distance of improvement elements placement, width of traffic lanes, slopes, excesses, photo-fixation, etc.); *systematic analysis* of criteria for a comprehensive assessment of street and road environment maintenance level, taking into account the needs of limited mobility population groups; a *hypothetical method* for improve the principles for placement of streets and urban road beautification elements; *modeling* of the spatial corridor, elements of street improvement; *experimental design* of street improvement, *comparative analysis* of the obtained results.

4 Research Results

As a part of streets and urban roads environment research for the purpose of adapting it to the needs of people with limited mobility (PLM), an experiment "survey of the existing state streets and urban roads environment in Poltava and other settlements of Ukraine regarding the conditions of accessibility for people with disabilities" was conducted. In general, the following problems of using the street and urban roads environment of the people with limited mobility were identified:

- The non-possibility or difficulty of overcoming the high-altitude differences between the roadway and the sidewalk;
- Difficulty in maneuvering in a confined space;
- Difficulty into overcome the intersection, obstacles on the sidewalks;

- Impossibility of free movement through a dense movement and different speeds of pedestrian flows along sidewalks and transit roadways;
- Difficulty in identifying obstacles; lack of access to underground and overland pedestrian crossings, to public transport stops; lack of parking zones;
- Lack of territories marking; lack of time for the movement during the passage through the pedestrian crossing;
- Incompatibility of the pedestrian crossing width, information support, number of rest places;
- Inconsistency of pavement quality and lack of specialized surface and equipment.

The use of UDP as a roadmap design strategy aims to ensure equal rights for all actors in society and encourages the full participation of PLM in public life by ensuring the accessibility and ease of built environment use, thus improving the "quality of life" of the entire settlement community. According to State Building Regulations B.2.2-40: 2018 [11, p. 5] PLM are people, who experience difficulty moving independently, providing services, information, or orienting in space. PLM includes persons with disabilities, people with temporary disabilities, pregnant women, the elderly, and people in wheelchairs [12]. Also, PLM can be attributed to completely healthy people who are visiting this city for the first time and need detailed, clear and visible indications for orientation in space. The term "universal design" originated in the USA and was first used in the mid-1980s by American architect Ronald L. Macy [13]. Universal design is the design and composition of an environment that is accessible, understandable, and used by all people, regardless of age, size, ability, or disability.

The *principles* of UD include: equitable use; flexibility in use; simple and intuitive use, perceptible information; tolerance for error; low physical effort; size and space for approach and use [10].

Particularly important and urgent is the search for effective techniques for the application of UD in the design of street and road environment improvement, which should become inclusive (universal). This certainly requires improvement and change of approaches to the technical, architectural, and design the implementation of measures to ensure it.

Authors suggested criteria for a comprehensive assessment of street and road environment maintenance level, taking into account the needs of limited mobility population groups [12], which are divided on four groups: safety, accessibility, comfort, and informativity. *Safety* include: exclusion of false effects, a viable pedestrian network, tolerance for mistakes, and rationality. Safety provides: stairs, appropriate lighting, ramps, traffic lights, a number of road accidents, island of security, separation of transport, and pedestrian streams. *Accessibility* include: versatility, equality, "smart device," diversity, and hierarchical sequence. Accessibility provides: the corresponding bias, sidewalk width, and crosswalk, availability of the necessary space, border, specialized playgrounds, ramp raids, parking area, minimum transit distances, and ac elements placement complishment. *Comfort* include: appearance, aesthetic appeal, application of small physical forms, and optimal placement. Comfort provides: surface drainage, availability of a recreation area, condition of coverage, environmental data, pedestrian zone, and priority of movement. *Informativity* include: understandable ability, continuity, read ability, interactivity, and variety. Informativity provides: information support, tactile means, optimal placement, distance, angles of visibility, reduction of time, and effort.

Authors suggested the system of beautification elements of streets and urban roads environment in settlements [14] that consists of three systems beautification elements: pedestrian—sidewalks—environment; bicycle—bikeway—environment [15]; automobile—roadway—environment.

Authors suggested using a new design principle—*street and urban road spatial corridor modeling*, which provides the principles of universal design. To apply this, authors have recommended street and urban roads beautification elements distributed by importance level into four groups in accordance with purpose. Beautification elements, ensuring continuous movement function suggested to include to the I importance level, the function of safe movement—to the II importance level, the function of ecological—to the III importance level, the function of easy movement, which depends on the satisfaction of physiological and psychological needs of movement and environment—to the IV importance level. Each beautification element has placement in the corresponding space sub-corridor. Authors have defined parameters of sub-corridors. Street spatial corridor includes all sub-corridors. *User—transit way—environment* system designing should be done according to the laws of harmonious unity, when favorable conditions for user's optimal psychological state are established. The geometric generalization of this system can be created, if its elements are considered in interconnectedness, interpenetration, and mutual agreement. Street space elements harmony consists of correspondence factors and equilibrium, rhythm, contrast, magnitude, symmetry and asymmetry, and the regulatory element of which is the proportionality. The method of designing an integrated road and street beautification was based on street and urban road spatial corridor modeling principle which authors have proposed in the previous article [16].

Authors have designed the *examples of application* of the universal design principles in the street and urban road environment improvement using the proposed method (Figs. 2 and 3).

Authors have designed the *examples of inclusive street and urban road* environment formation using the principles of universal design using the proposed method (Fig. 4).

5 Scientific Innovation

The main scientific innovations of this work are:

- Authors conducted comparative analysis and generalization of streets and urban roads improvement and existing theoretical developments regarding the application of the UDP;

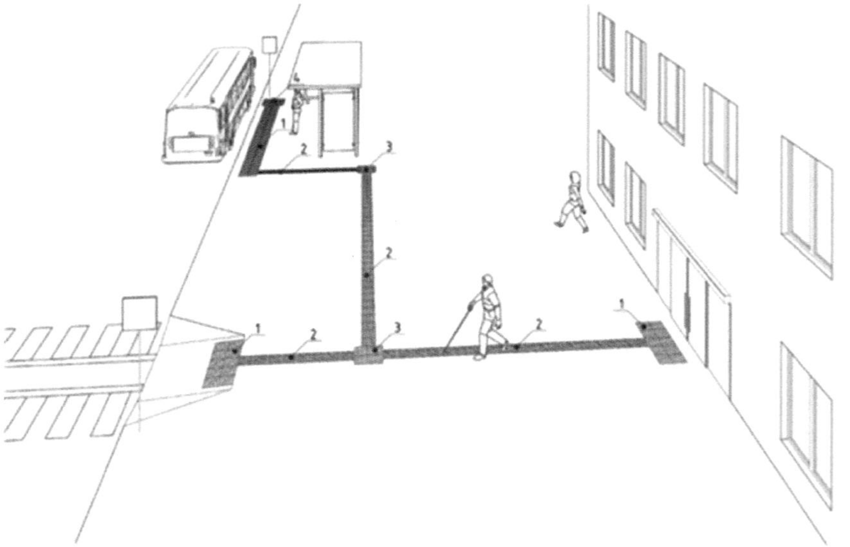

Fig. 2 Example of tactile strips (TS) paving: 1 – warning TS; 2 – guiding TS; 3 – information TS

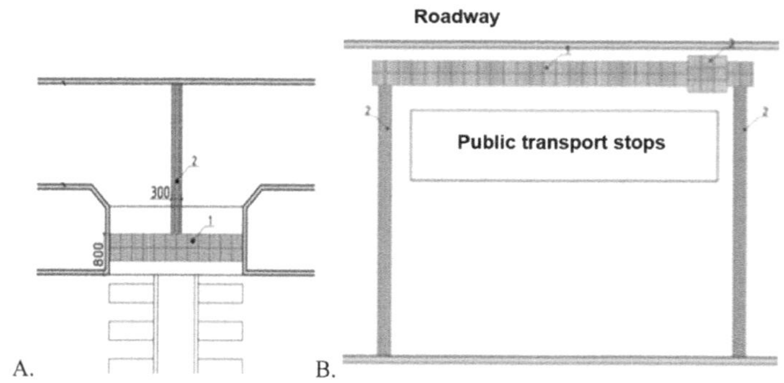

Fig. 3 a—example of tactile strips paving on land pedestrian crossings: 1—warning TS; 2—information TS; b—example of tactile strips paving at public transport stops: 1—warning TS; 2—information TS; 3— information TS, which indicates the place of entering in urban transport

- Systematic analysis of criteria for a comprehensive assessment of street and road environment maintenance level, taking into account the needs of limited mobility population groups;
- Authors improved the principles for streets and urban road beautification elements placement; modeling the spatial corridor and elements of street improvement.

Fig. 4 Example of inclusive street and urban road environment formation using the principles of UD

6 Practical Value

The main practical values of this research are:

- The design method of streets and urban roads complex improvement, taking into account the UDP, was developed;
- Authors have designed the examples of UDP application in the street and urban road environment improvement and examples of inclusive street and urban road environment formation based on the principles of universal design using the proposed method;
- The results of the study can be used in the design of street improvements and urban roads;
- The results of the research have found practical application in the educational process, in particular, in course and diploma design, masters programs and special courses on Department of Highways, Geodesy, Land Management and Rural Buildings, Poltava National Technical Yuri Kondratyuk University and on Department of Urban Construction, O. M. Beketov National University of Urban Economy in Kharkiv.

7 Conclusions

The work is devoted to the application of the universal design principles in design the street and urban road environment improvement. There are main results of this research:

- The authors have shown the relationship between universal design, accessible, adaptable, lifespan, barrier-free design, and seven universal design principles;
- The authors have excreted the criteria for barrier-free street and urban road environment to meet the needs of all people;
- The authors have proposed a new method of placing street and urban roads improvement elements on the basis of the modeling spatial corridor principle.

This method ensures implementation of seven principles of universal design: equitable use; flexibility in use; simple and intuitive use; perceptible information; tolerance for error; low physical effort; size and space for approach and use. To apply the proposed method, street and urban roads beautification elements have been distributed by importance level into four groups and elements of each level have placed in defined spatial corridor;
- The authors have proposed means and elements that provide universal design of streets and urban roads;
- The authors have proved the need for future search means and elements that provide universal design of streets and urban roads.

References

1. Bringolf, J. (2008). Universal design: Is it accessible? *Multi: The RIT Journal of Plurality and Diversity in Design, 1*(2), 45–52.
2. Ward, M., Bringolf, J. (2019). Universal design in housing in Australia: Getting to yes. *Transforming our World Through Design, Diversity and Education, Publisher: Open Access by IOS Press,* pp. 209–355. https://doi.org/10.3233/978-1-61499-923-2-299.
3. Byahut, S., Mujumdar, A., Patel, A., Sheth, A., Mitra, A., Moothan, R., Pandya, Y. (2005). Design manual for a barrier-free built environment.
4. Alberta Municipal Affairs and Safety Codes Council Barrier-free design guide. (2017). p. 176.
5. Khalil, M. (2016). An assessment of street design with universal design principles: Case in Aswan/As-Souq. MEGARON/Yıldız Technical University, Faculty of Architecture E-Journal. https://doi.org/10.5505/megaron.2016.98704.
6. Skavlid, S., Olsen, H. P., Haugeto, A. K. (2013) Trends in universal design, Norway, First Edition, Norwegian Directorate for children, youth and family affairs, The Delta Centre.
7. Carr, K., Weir L., Azar, D., Azar, N. (2013). Universal design: A step toward successful aging. *Journal of aging research*, (3), 324624. https://doi.org/10.1155/2013/324624.
8. Jones, P. (2014). Situating universal design architecture: Designing with whom? *Disability and Rehabilitation, 36,* 1–6. https://doi.org/10.3109/09638288.2014.944274.
9. Harsritanto, B., Wijayanti, A. (2017). Universal design characteristic on themed streets. In *IOP Conference Series: Earth and Environmental Science*. Vol. 99(1), pp. 012025. https://doi.org/10.1088/1755-1315/99/1/012025.
10. Harsritanto, B. (2018). Urban environment development based on universal design principles. In *E3S Web of Conferences*. Vol. 31. doi:09010. https://doi.org/10.1051/e3sconf/20183109010.
11. State Building Regulations B.2.2-40. (2018). Inclusivity of buildings and structures. substantive provisions.
12. Ivasenko, V., Lytvynenko, T., Zhidkova, T., Tkachenko, I. (2018). Integrated assessment level of service street and road environment to the needs of persons with disabilities. *International Journal of Engineering & Technology, 7*(4.8), 824–830. https://doi.org/10.14419/ijet.v7i4.8.28131.
13. Anderson, R. (2018). Universal design learning (UDL) in Higher Education.
14. Lytvynenko, T., Tkachenko, I., Ilchenko, V. (2018). Principles of street and urban road space formation in modern cities. *International Journal of Engineering & Technology, 7*(3.2), 642–648. https://doi.org/10.14419/ijet.v7i3.2.14606.
15. Lytvynenko, T., & Gasenko, L. (2015). Peculiarities of infrastructure designing for the movement of individual environmental friendly vehicles. *Periodica Polytechnica Transportation Engineering, 43*(2), 81–86. https://doi.org/10.3311/PPtr.7593.
16. Lytvynenko, T., Tkachenko, I., & Gasenko, L. (2017). Principles for road beautification elements placing. *Periodica Polytechnica Transportation Engineering, 45*(2), 94–100. https://doi.org/10.3311/PPtr.8592.

Investigation of the Mechanical Properties of Pipes for Long-Term Cooling Systems

Valerii Makarenko, Yuriy Vynnykov, and Andrii Manhura

Abstract The results of experimental studies of mechanical properties of cooling systems pipes, which are exploited at the enterprises of fermentation productions for a long time, are presented in the article. In particular, it has been established that with the service life extension of both pipelines from the transport of refrigeration mixture and the pipes of heat-exchange apparatuses of cooling systems, there is a flood of metal with an increase in the structure of non-metallic inclusions, which leads to its exfoliation and softening. It leads to decrease of crack resistance, as evidenced on reducing of metal resistance to the origin of cracks that lead to destruction of the structure.

Keywords Corrosion · Crack resistance · Flood · Vibrations · Strength

1 Introduction

Most of the metal equipment and tubular constructions of oil and gas, chemical, food and agro-processing industries operate under difficult loading conditions under the influence of acyclic working and external environments. It often causes irreversible physical and chemical changes in the material because of the occurrence of corrosion, sorption, erosion, cavitation and other processes, which lead to the loss of strength and destruction of structures [1–4]. Increase of strength (resistance to destruction under the influence of external loads) and reliability (faultless operation of the structure) under the given operating conditions is based on increased stability, i.e., resistance to the effects of corrosive media. Therefore, it is important to study the stability of metal equipment of agro-processing and food industries in the context of corrosion-aggressive gas-liquid media, which lead to electrochemical corrosion [2, 3].

The foregoing shows the importance of issues related to the corrosion-mechanical destruction of equipment, which requires an analysis of the common ways to increase the resistance of the material structures to destruction, the causes of destruction and methods for increasing the stability of metal equipment, especially shell designs

V. Makarenko · Y. Vynnykov · A. Manhura (✉)
Poltava National Technical Yuri Kondratyuk University, Poltava, Ukraine
e-mail: mangura2000@gmail.com

V. Onyshchenko et al. (eds.), *Proceedings of the 2nd International Conference on Building Innovations*, Lecture Notes in Civil Engineering 73,
https://doi.org/10.1007/978-3-030-42939-3_17

(such as vessels, containers, pipelines) in specific technological environments [2, 5]. In works [5–16], it is established that during the long-term operation, especially at temperatures below zero, there is a flood of metal, which adversely affects the impact strength and plastic properties of carbon and low-alloy steels. Moreover, hydrogen not only reduces the value of impact strength, but also increases the tendency to cold brittleness. Its influence becomes noticeable with sufficiently large content, varies for different steels ranging from 2 to 8 $cm^3/100$ g [6].

In accordance with the existing theories of hydrogen brittleness, the harmful effects of hydrogen are manifested in significant increase in the energy of cracked dislocation and the reduction of the force at which nucleus fracture is formed, as well as in the interaction of hydrogen atoms with dislocations in the process of plastic deformation [12–15]. From works [7, 8], it follows that flooding, especially at temperatures below zero (to −60 °C), decreases the number of cycles to fracture samples in the presence of stress concentrations, especially in conditions of low cycle fatigue failure. It is known [10, 11] that with a decrease in temperature (to −20...−60 °C), the impact strength and the deformation ability (ψ, δ) of metal are reduced. The author [9] found out that the destruction of samples decreased by 10–12 times when the temperature dropped to −70 °C. It was shown in [5] that cold-resistant steels with low hydrogen content ($N_{dif} < 3$ $cm^3/100$ g) are less prone to overstretching at minus temperatures than steel with higher hydrogen content.

Thus, decrease in the hydrogen content of metal, which is exploited for a long time under conditions of corrosive environments and negative temperatures (up to −50 °C), is a prerequisite for increasing the viscous-plastic properties, in particular, the buckling viscosity, which is largely responsible for operational resource in metal structure. As it is shown by the analysis of literary sources [2, 7–9, 13, 14] and practice, the long service life of pipeline systems, including those corrosion resistance of the metal determines working in extreme temperature-baric conditions of chemical aggressive productions.

However, for the scientific solution of practical tasks of corrosion protection of pipeline networks of cooling systems, additional complex investigations of the influence of negative temperatures in the conditions of prolonged operation in aggressive environments on corrosion damage of steel are needed.

Work objective is to study mechanical properties of cooling systems pipes in the process of prolonged operation in a wide range of minus temperatures.

The change in stress level of steel pipes with a different service life was determined on standard (flat) samples, which were manufactured with working part size of 35 × 4 × 1,5 mm according to the procedure [17]. The samples were cut from the billets of industrial steels and subjected to one-piece stretching at a deformation rate of ε = 10–5 s-1 [2]. Mechanical tests were performed on the universal "Instron-1251" (Great Britain) machine in the range of temperature variations from +20 to −50 °C.

According to the results of experimental tests for each sample, standard parameters of mechanical properties of steels were determined: strength (σv) and yield strength ($\sigma_{0,2}$), transverse (ψ) and longitudinal (δ) deformation [2]. Moreover, it is considered [2, 11] that e-mechanical characteristics—$\sigma_{0,2}$ and ψ are most susceptible to the overstretching of metal structure. One of the characteristic indicators of resistance to brittle fracture of steels is the impact strength. For its study, samples with a V-shaped incision—KCV (Charpy) were made. In this case, the direction of the samples and the placement of the incisions were the same. The samples were tested for dynamic bending over a wide temperature range in accordance with GOST9454-78 and GOST1497-84. To determine the impact viscosity, we used samples 55 mm long and cross-sectional area of 8 × 8 mm^2 with a V-notch 2 mm long and a 450 rosette angle cut from tubular steel in various states. The tests were carried out according to known method [14]. The test results are averaged at least 3 studies.

In addition, as a criterion of crack resistance, known from the fracture mechanics of metal structures, the indicator R_{ms} was used—microscopic steel resistance, which depends on the structural state of the steel, the thickness of the grain size and the thickness of the plates of cementite inclusions. It does not depend on external factors, unlike other integral criteria mechanical properties (temperature, strain rate, the geometric dimensions and shape of the sample, the type of stress state) [1]. The determination of R_{ms} was carried out in accordance with the method [11].

Also, the characteristics of crack resistance were determined—the parameters of the destruction viscosity of K_{1s} and δ_s, samples of standard sizes were prepared for them [1, 2, 6]. The fatigue cracks in the samples were created using a hydraulic pulsator CDM-10 (Germany) at a load frequency of 10… 15 Hz and a coefficient of asymmetry of the cycle r = 0,1…0,2. The tests for parameters of the viscosity of destruction of K_{1s} and δ_s determining were carried out at UME-10 and “Instron” (UK) installations using standard methods [1, 2].

The objects of research were samples from steel grades 3, 10 and 20, made from pipelines for transferring the refrigerant mixture (service life—from 0 to 30 years), and also from steel grades 3, 10, 20 shell-and-tube heat exchangers of refrigeration systems (service life from 0 up to 12 years) operated in fermentation plants. The basis for the preparation of samples fragments of pipes served, and cut out in the process of forced or planned repair from emergency pipelines or tubes of heat exchangers cooling systems. The results of tests for determining the parameters of crack resistance in a wide range of changes in minus temperatures are shown in Figs. 1, 2, 3 and 4.

Figure 1 shows that with increasing lifetime over the entire range of temperature changes from +20 to −50 °C, the magnitude of the stresses of steel microcleaving R_{ms} 3 decreases, and this is noticeable for samples of pipes with a lifetime of 10 years or more and temperatures of −35 − 50 °C. So, for samples made of emergency reserve pipes, the parameter R_{ms} at temperatures of +20 and −20 °C is equal to 580 and 560 MPa, respectively, and at −50 °C this parameter has the following values: 400 and 460 MPa, respectively, that is, R_{ms} decreases in 1,2–1,45 times. At the same time, the parameter R_{ms} at temperatures of −35…−50 °C for non-extruded steel samples 3 has the following values: 530 and 500 MPa, and after 30 τ of operation, respectively, 250 and 200 MPa, that is, the value of R_{ms} decreases by 1,6–2,0 times. Moreover, such tendency persists throughout the range of changes in minus temperatures and operating periods of pipe constructions.

The negative effect of the test temperature reduction from +20 to −50 °C on the parameters of crack resistance K_{1s} and $\delta_{s,}$ is observed throughout the term of long

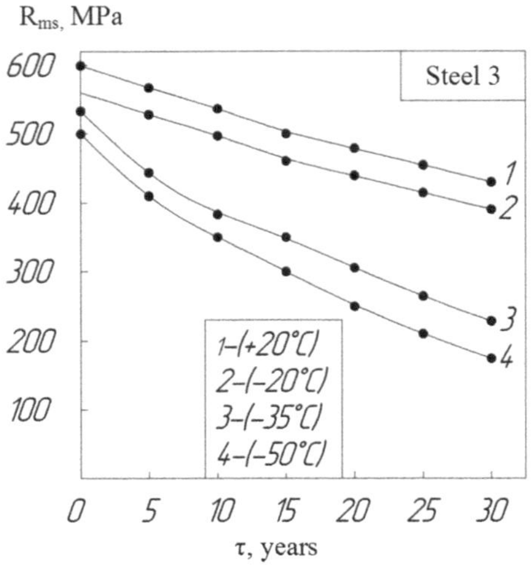

Fig. 1 Grapical dependence of steel microcleaving 3 on the lifetime of pipelines for pumping the refrigeration mixture at the enterprises of fermentation industry. Indication of the test temperature (in °C): 1–(+20); 2–(−20); 3–(−35); 4–(−50)

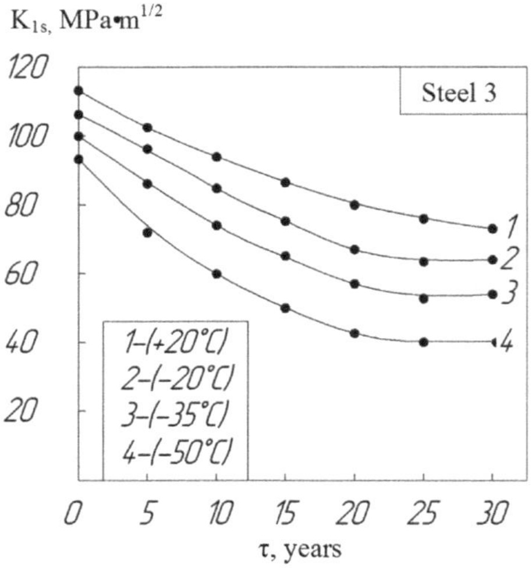

Fig. 2 Graphical dependencies of K_{1s} stress intensity factor from operation period of pipelines for pumping refrigeration mixture at the enterprises of the fermentation industry. Indication of test temperature (°C): 1–(+20); 2–(−20); 3–(−35); 4–(−50)

operation of the pipelines. Analysis of the data shown in Figs. 2 and 3 shows that the values of $K_{1s,}$ and δ_s for unexploited steel at temperature +20 °C are, respectively, 115 MPa•m$^{1/2}$ and 0,79 mm, and for steel pipe operation, for example, for 25 years the same parameters at the same temperature have the corresponding values of: 74 MPa•m$^{1/2}$ and 0,39 mm. That is, 2,2 times (K_{1s}) and 2 times (δ_s) decrease the crack resistance of steels. For unexploited steel at -50 °C, the values of K_{1s} and δ_s are 95 MPa•m$^{1/2}$ and 0,67 mm, while for steel with a lifetime of, for example, 30 years, the same parameters have the following values: 41 MPa•m$^{1/2}$ and 0,17 mm, respectively, i.e., they decrease by 2,3 times (K_{1s}) and ≈4 times (δ_s) That is, at minus temperature −50 °C, the resistance of steel 3 against cracking significantly decreases

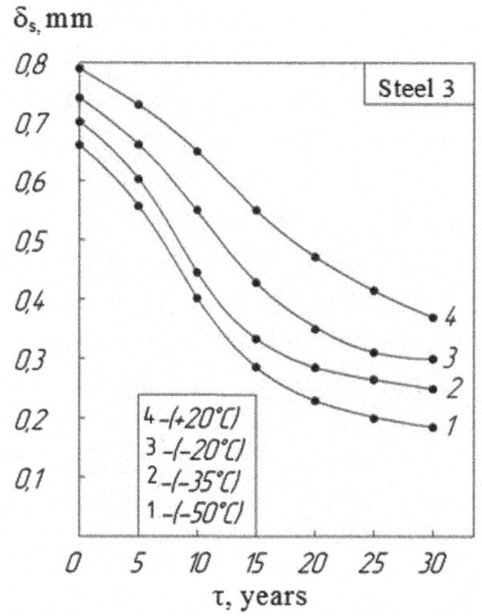

Fig. 3 Graphical dependences of the coefficient of critical crack opening from the life of the pipelines for the transport of the refrigeration mixture at the enterprises of the fermentation industry. Indication of the test temperature (in °C): 4–(+20); 3–(−20); 2–(−35); 1–(−50)

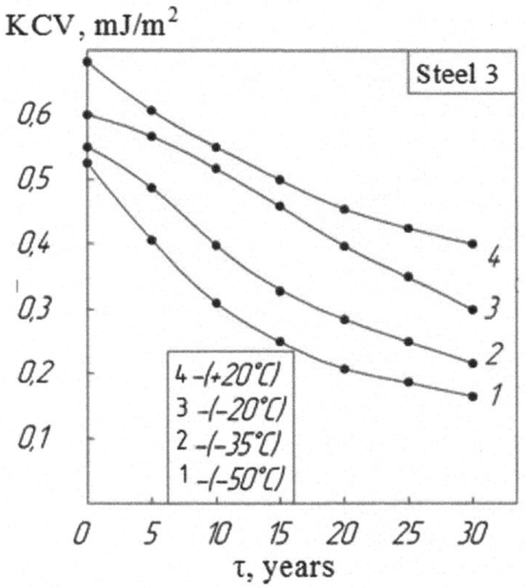

Fig. 4 Graphic dependence of the impact strength (Charpy sample) on the metal from the service life of the pipelines for the transport of the refrigeration mixture at the enterprises of the fermentation industry. Indication of the test temperature (in °C): 4–(+20); 3–(−20); 2–(−35); 1–(−50)

with the subsequent destruction of the pipe structures directly in contact with the aggressive technological environment of fermentation production.

The graphs presented in Fig. 4 indicate a significant change in the viscosity of the KCV fracture in a wide range of changes in both the lifetime of the pipe constructions and the temperature of the test samples. It can be seen that for unexploited steel at +20 °C (curve 1), the KSV parameter is 0,68 mJ/m^2, and an increase in the interval of operation, for example, up to 30 years, leads to decrease in the KCV parameter, which in this case is equal to 0,38 mJ/m^2, that is, decrease by 1,7 times. At the same time, at a test temperature of −50 °C, the KCV, value is 0,54 mJ/m^2 (unexploited steel) and 0,15 mJ/m^2 (the working service of steel is 30 years), which is reduced by 3,6 times. It is noteworthy that the same trend persists for steels with different operating periods (from 0 to 30 years) in the temperature test (from +20 to −50 °C).

Figure 5 shows that with the change of pipeline service life (steel grade 3) from 0 to 30 years, the value of the strength limit σv is almost unchanged, but the value of the yield point σt is reduced from 240 MPa (unexploited steel) to 170 MPa (30 years of operation), that is, almost 1,5 times. The values of longitudinal (δ) and transverse (ψ) deformations, respectively, vary from 33% (δ) and 75% (ψ) (non-elastic steel) to 14% (δ) and 50,5% (ψ) (30 years of operation), i.e., in 2,2 (δ) and 1,48 (ψ) times the plasticity indexes for steel with a long lifetime decreased.

Figure 6 shows the graphical dependences of the plasticity index $\sigma_{0.2}$ and the degree of metal flood [H] from the service life of steel grade 3, 10 and 20 pipelines for a refrigerated mixture transporting of fermentation production. It can be seen that the highest plastic properties are characterized by steel grade 20, while steel 10

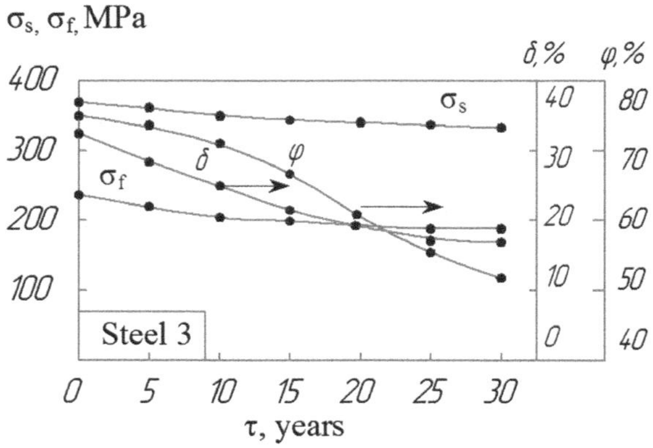

Fig. 5 Graphical dependencies of the strength (σ_s) and fluidity (σ_f), longitudinal (δ) and transversal (ψ) metal deformation from the service life of the pipelines for refrigeration mixture transporting at the enterprises of the fermentation industry

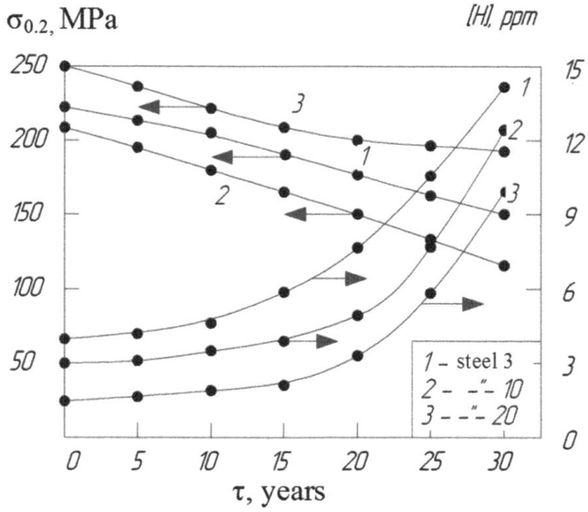

Fig. 6 Graphs of changes in plasticity and the degree of metal flooding, depending on the period of operation of pipelines for transporting refrigerated mixtures at fermentation plants

showed the lowest values of the parameter $\sigma_{0.2}$ over the lifetime of pipelines. The analysis of data in Fig. 6 shows that an increase in the life of the operation causes a decrease in the indicator $\sigma_{0.2}$, in particular, for steel 3 value $\sigma_{0.2}$ for unexploited steel is 220 MPa, and for exploited steels with a term of 30 years it is already equal to 140 MPa, that is, the plasticity is reduced by 1,57 times. The same trend is observed for steel of 10 and 20 grades. For example, in the case of unexploited steels, the value of the parameter $\sigma_{0.2}$ is 210 MPa (steel 10) and 250 MPa (steel 20), and for exploited steels with a term of 30 years the parameter $\sigma_{0.2}$ is equal to 115 MPa (steel10) and 180 MPa (steel 20), i.e., plastic properties decrease in 1,82 (steel 10) and 1,38 (steel 20) times, respectively.

The above data indicates the probable flood of tubular steels in the course of prolonged operation, which is confirmed by the results of measurements of residual hydrogen [H] in the metal of the pipelines transporting the refrigeration mixture—Fig. 6. So, for unexploited steel grades 3, 10 *i* 20 value [H] is (in ppm): 4,1 (steel 3); 3,1 (steel 10) and 1,5 (steel 20), and for these steels with 30-year lifetime, the values [H] are, respectively, equal (in ppm): 14,3 (steel 3); 12,5 (steel 10) *i* 10,2 (steel 20), that is, the flood of investigated steels with an increase in service life from 0 (emergency stock) to 30 years increases by 3,5 times (steel 3); in 4 times (steel 10) and 6,8 times (steel 20), respectively.

From the data given in Fig. 6, it follows that the values of the plasticity parameter decrease significantly less $\sigma_{0.2,}$ and steel 20 in comparison with other steel grades 3 and 10. So, the value $\sigma_{0.2}$ for unexploited steels are (in MPa): 250 (steel 20); 240 (steel 3) and 205 (steel 10), and for the exploited steels with the term, equal to 30 years, the parameter $\sigma_{0.2}$ has the following values (in MPa): 190 (steel 20); 155 (steel 3) and 105 (steel 10), that is, the value of $\sigma_{0.2}$ is reduced to 1,32 times (steel 20); in 1,55 times (steel 3) and in $\approx$2 times (steel 10), respectively. Such data, as above for the pipelines, indicate the flood of the tubular steels in the course of long-term operation, as evidenced by the results of the determination of the concentration of residual hydrogen in the steels (see Fig. 6). The analysis of the given data in Fig. 6 shows the following. For unexploited steels, the hydrogen content [H] is (in ppm): 3,2 (steel 3); 1,0 (steel 10) and 2,1 (steel 20), and for steels with a service life of 30 years, the values of [H] are (ppm): 19,3 (steel 3); 15,8 (steel 10) and 11,0 (steel 20), that is, with an increase in the lifetime of 0 (steel nonexploited) to 12 years, the flood of steels increases 6 times (steel 3); in 15,8 times (steel 10) and 5,2 times (steel 20), respectively.

In addition, as it is shown by the metallographic studies, the results of which are given in the table, which were obtained on “Kvantimet-720” apparatus, with an increase in the life of the pipelines, the size and amount of non-metallic impurities, which are known [6, 7, 19], are increasing the nuclei of the origin of microcracks, because the metal in such cases is characterized by low deformation ability in the course of interchangeable loads in conditions of corrosive environments (Table 1).

Thus, it can be assumed that as the useful life of the pipelines from the refrigerating mixture transport and pipes of the heat-exchange units of the cooling systems increases. The flood of the metal increases with the structure of the non-metallic inclusions that leads to its exfoliation and softening, which results in lowering the

Table 1 Number of non-metallic inclusions in the metal of pipelines with different useful life

Useful life, years	Total share of inclusions, %	Size of inclusions, microns						
		>1.0 Total	1–1.5	1.5–2	2–2.5	2.5–3	3–3.5	3.5–4	>4.0
0	0.055	1289	790	134	98	45	36	19	10
10	0.15	1798	1362	253	189	67	46	34	21
20	0.25	3423	2308	367	234	89	61	57	64

Note Non-metal impurities in metal were determined on "Kvantimet-720" apparatus resistance to cracking, and that indicates decrease in the resistance of the metal to the origin of cracks that leads to the destruction of the structure.

2 Conclusions

Mechanical studies have established that when the temperature regime changes from +20 to −50 °C, the indexes of crack strength of the steel tube (grade 3), in particular, the impact strength (by Charpy) KCV, are reduced by 1,5 times; stress intensity factor K_{1s}—in 2,3 times; the coefficient of crack opening width δ_s—4 times and the microcleaving R_{ms} ratio—2 times, indicating a decrease in the resistance to the origin and spread of microcracks in the steel.

It has been established that the viscous-plastic properties of the yield point $\sigma_{0.2}$, relative longitudinal (δ) and transverse deformation (ψ) are sharply reduced in the process of prolonged exploitation of steel tubes of grade 3, in particular, during operation for 30 years these parameters decrease in 1,5 times ($\sigma_{0.2}$); in 2,2 times (δ) and in 1,48 times (ψ). At the same time, there is an intense flood of metal, in particular, the hydrogen content during the 12-year operation in steel 3 increased 1,58 times; in steel 10–6 times and in steel 20—in 5,2 times.

References

1. *ANSYS Mechanical Users Guide. Release* (2013) 15.0. USA. Canonsburg, PA, p. 1832.
2. Bai, Yong. (2001). *Pipelines and risers* (p. 495). USA, Oxford: Elsevier.
3. Hrudz, Y. V. (2012). *Enerhoefektyvnist hazotransportnykh system* (p. 208). Ivano-Frankivsk: Lileia-NV.
4. ASME B31.4. (2002). *Pipeline transportation systems for liquid hydrocarbons and other liquids* (p. 126). New York: American Society of Mechanical Engineers.
5. Makarenko, V. D., Korobko, B. O., & Vynnykov, Yu L. (2018). *Eksperymentalni vyprobuvannia truboprovodiv: monohrafiia* (p. 551). Nizhyn: NDU im. M. Hoholia.
6. Makarenko, V. D., Korobko, B. O., & Vynnykov, Yu L. (2018). *Innovatsiini materialy ta tekhnolohii v naftohazovii haluzi: monohrafiia* (p. 233). Nizhyn: NDU im. M. Hoholia.
7. Onyshchenko, V. O., Vynnykov, Yu L, Zotsenko, M. L., Pichuhin, S. F., Kharchenko, M. O., Stepova, O. V., et al. (2018). *Efektyvni konstruktyvno-tekhnolohichni rishennia obiektiv transportuvannia nafty i naftoproduktiv u skladnykh inzhenerno-heolohichnykh umovakh: monohrafiia* (p. 258). Poltava: FOP Pusan A.F.
8. Sukhenko, Y. H., Lytvynenko, O. A. & Sukhenko, V. Iu. (2010). *Nadiinist i dovhovichnist ustatkuvannia kharchovykh i pererobnykh vyrobnytstv*. K.: NUKhT, p. 547.
9. ISO 13623:2009(en).(2009). *Petroleum and natural gas industries.* Pipeline transportation systems. Technical Committee ISO/TC 67, p. 173.
10. DIN EN 12007-1:2012-10. (2012). *Gas infrastructure—Pipelines for maximum operating pressure up to and including 16 bar—P. 1:* General functional requirements; German version EN 12007-1:2012, Germany, p. 298.
11. Ellenberger, J. P. (2014). *Piping and pipeline calculations manual. Construction, design fabrication and examination.* USA, Oxford: Elsevier, p. 398.

12. Code of Practice for Pipelines—Part 1. (2004). *Steel pipelines on land, PD 41 8010, British Standards Institution*, p. 52.
13. Kharchenko, M., Manhura, A., Manhura, S. & Lartseva, I. (2017). Analysis of magnetic treatment of production fluid with high content of asphalt-resin-paraffin deposits. *Journal Mining of Mineral Deposits, 11*(2), 28–33. http://ir.nmu.org.ua/handle/123456789/149572.
14. Vynnykov, Yu. L., Makarenko, V. D., Kravets, I. A. & Mynenko, I. S. (2019). *Doslidzhennia prychyn znyzhennia mitsnosti truboprovodiv TETs* Problemy tertia ta znoshennia, *1*(82), 63–68. ISSN 03702197. https://doi.org/10.18372/0370-2197.1(82).13488.
15. Makarenko, V. D., Chebotar, I. M., Petrenko, O. O., & Nohina, A. M. (2019). *Doslidzhennia mekhanichnykh vlastyvostei trub okholodzhuiuchykh system dovhotryvaloi ekspluatatsii v shyrokomu intervali minusovykh temperatur v umovakh brodylnoho vyrob-nytstva* Problemy tertia ta znoshennia, *1*(82), 69–77. ISSN 03702197. https://doi.org/10.18372/0370-2197.1(82).13489.
16. Vynnykov, Y., Manhura, A., Zimin, O., & Matviienko, A. (2019). Use of thermal-magnetic devices for prevention of asphaltene, resin, wax deposits on oil equipment surfaces. *Journal Mining of Mineral Deposits, 13*(2), 34–40. https://doi.org/10.33271/mining13.02.034.
17. Zotsenko, N. L., Lapin, N. I., & Petrash, R. V. (2008). Comparative effectiveness of bed reinforcement based on plate tests and mathematical modeling. *Soil Mechanics and Foundation Engineering, 45*(4), 138–143. https://doi.org/10.1007/s11204-008-9014-z.

Public Cadastral Maps as a Basis for a Construction of the Building General Layout

Svitlana Nesterenko, Roman Mishchenko, Vira Shchepak, and Grigoriy Shariy

Abstract The source materials for the construction master plan were analyzed. It was determined that the components of the source data for the development of the document are materials of town-planning and land cadastres, that is the cadastral data reflecting graphic and textual information about land boundaries, configurations, and locations of buildings, including linear objects, easements, or other legal restrictions. The necessity of considering the work of public cadastral maps and information portals on which the information of the State Land Cadastre was published, namely the graphic location of the land plot, its boundaries, area, cadastral number, form of ownership, purpose, in accordance with the classifier was revealed. Public cadastral maps of Ukraine, Georgia, Bulgaria, Czech Republic, Latvia, Poland, Russia, and Moldova were considered. The analysis of graphical and textual information of different cadastral maps was made, and the advantages and disadvantages of the cadastral system of the countries of Europe and Asia were stated. The main directions of improvement of the public cadastral map of Ukraine for the effective use of it for urban planning purposes were formulated, in particular for drawing up the general plan of the site with elements of displaying the terrain, the boundaries of the construction site and the location on it of all existing and planned for the construction of buildings and structures, and for construction the master plan design.

Keywords Construction master plan · Master plan · Land cadastre · Public cadastral map

1 Introduction

To draw up a construction master plan, a master plan of the site with elements for displaying the terrain (horizontal, elevation) is required, on which the limits of the

S. Nesterenko · R. Mishchenko · V. Shchepak (✉) · G. Shariy
Poltava National Technical Yuri Kondratyuk University, Poltava, Ukraine
e-mail: kanameshch@gmail.com

S. Nesterenko
e-mail: NesterenkoS2208@gmail.com

V. Onyshchenko et al. (eds.), *Proceedings of the 2nd International Conference on Building Innovations*, Lecture Notes in Civil Engineering 73,
https://doi.org/10.1007/978-3-030-42939-3_18

construction site and the location on it of all existing and planned for the construction of buildings and structures are established. According to DBN B.1.1-15 2012 “Composition and content of the general plan of the settlement,” the components of the initial data for the development of the document are the materials of urban planning and land cadastres. Cadastral data reflects graphical and textual information about the boundaries of land, the configuration, and location of buildings, including linear objects, easement, or other legal restrictions. Since the collection of cadastral information over the centuries has constantly changed qualitatively, technologically, and geographically, there is a lot of accompanying cadastral information on land ownership, schemes, and digital drawings, which were coordinated by using modern electronic measurements. Since January 1, 2013, the public cadastral map of Ukraine has become available on the official website of the State Agency for Land Management, which has become the basis for building a construction master plan.

The purpose of the article is to study the work of the public cadastral map in different countries of the world for widespread use for urban development purposes.

2 Main Material and Results

The public cadastral map is an information portal on which the information of the State Land Cadastre is published, namely the graphic location of the land plot, its borders, area, cadastral number, form of ownership, purpose, in accordance with the classifier. Anyone with access to the Internet can get acquainted with this information, but information about the owner of the land is not displayed.

The electronic layering of the public cadastral map of Ukraine contains a large amount of information: a digital map of Ukraine, its borders, the borders of regions, the borders of districts, the borders of settlements, index-cadastral maps, land plots and their borders, cadastral number of the parcel, form of ownership, purpose, area, as well as a map of Ukrainian soils [1]. A feature of the Ukrainian public cadastral map is the use of orthophotoplans as a single cartographic basis. Orthophotoplans, which are accurate in comparison with satellite images or topographic maps, are used throughout the map area, and fully covered the territory of Ukraine [2].

The path from the cadastral map to the construction master plan is complex, phased and requires the involvement of specialists from various industries, ranging from a land engineer to a civil engineer (see Fig. 1).

Public cadastral maps of different countries of the world are characterized by a set of information about land plots and buildings, their quantitative and qualitative characteristics, rules and the procedure for providing data from cadastres or registers [3].

The basis for the public cadastral map of Georgia was the Google Map service (see Fig. 2). The portal provides information in Georgian, English, and Russian. The menu of the “services” section provides information on real estate (registration, pledge, leasing, bank guarantees, various prohibitions on transactions with specific real estate, and information on taxation). The portal contains only the conditions for

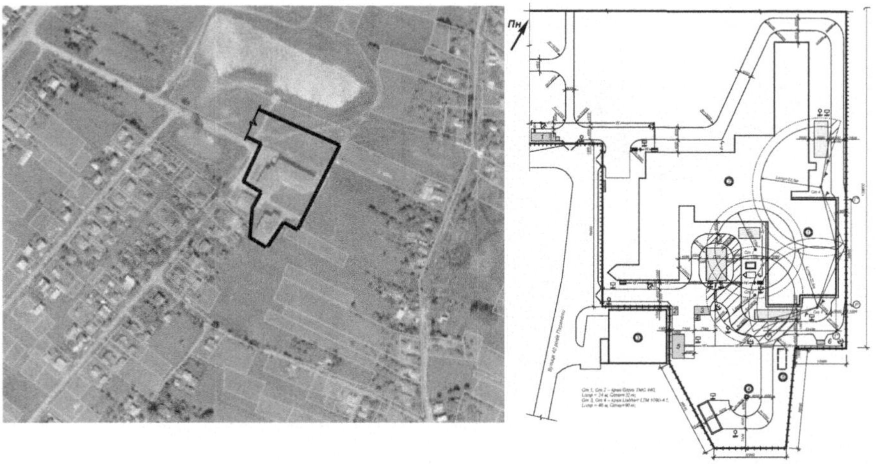

Fig. 1 An example of using a public cadastral map to create a construction master plan

the provision of such information, and instructions on how to obtain such information. The process of land registration is monitored online [4].

The home page provides links to the cadastral map of Georgia, the new cadastral map of Georgia, and the cadastral map of Tbilisi [5].

A link to the inventories of the European Union countries can be viewed on the portal www.eurocadastre.org "Standing Committee on Inventories in the European Union" [6].

The registration system of Bulgaria includes two registration systems—real estate and land. The public cadastral map contains information in Bulgarian and English on land real estate and construction (including incomplete) (see Fig. 3). It also displays information on documents confirming rights or transfer of ownership, changes (termination) of the right to real estate, or redemption of mortgages on them [7].

The portal indicates up to 70 services that can be provided electronically, and the procedure for their provision. The cadastral portal of Bulgaria also provides for the provision of archival information. The portal has a list of all legal entities and individuals who have the right to provide services for the formation of technical information on real estate.

Geoportal ČÚZK is a comprehensive Internet interface for access to spatial data obtained and updated at the Czech Office of Geodesy, Cartography and Cadastre (COSMC) [8].

The geoportal liables you to search in one place for information about the spatial data of land, as well as view or organize them (see Fig. 4).

The information given is regarding ownership of land, buildings, premises (apartments or non-residential premises), and construction rights registered in the real estate register, on the state of affairs in the cadastral office for registering property rights of persons in the Czech Republic, or for the purpose of confirming geometric plans.

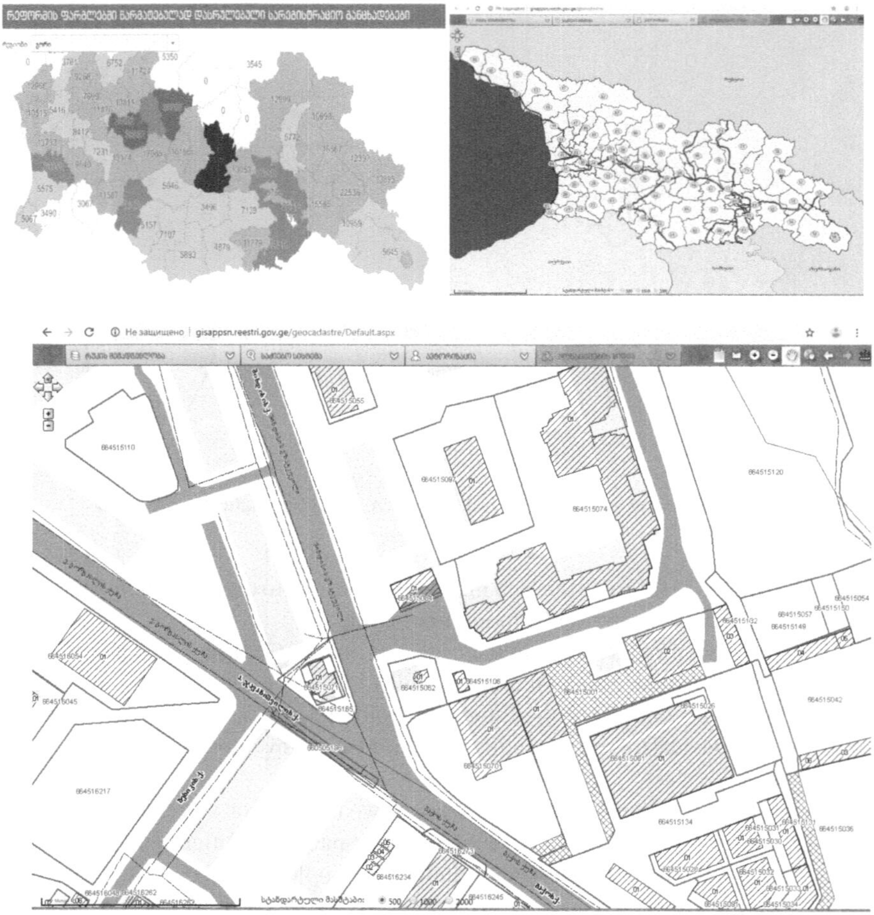

Fig. 2 Fragments of the public cadastral map of Georgia

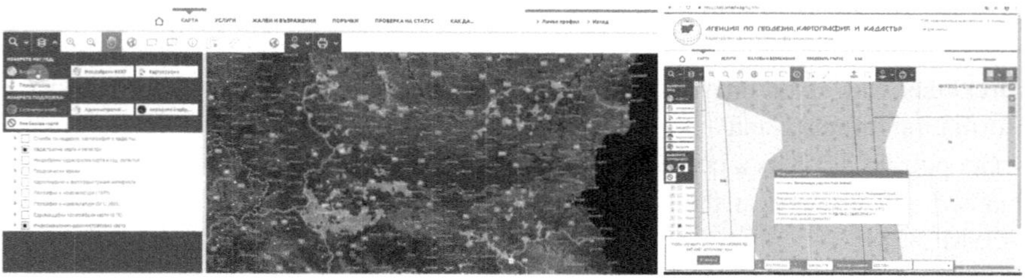

Fig. 3 Fragments of the public cadastral map of Bulgaria

Fig. 4 Fragment of the public cadastral map of the Czech Republic

Access to the portal is free for all Internet users, and does not require registration and it is free. Output options are limited to remote access. An excerpt from the inventory and some other services can be purchased in the application [9]. Payment is made through the payment portal.

The State Real Estate Cadastre of Latvia reflects information on property, extracts from the land cadastre, area of buildings and structures, real value of property, easements and restrictions, as well as data on the legal owner or user [10] (see Fig. 5).

The cadastral service of Latvia should: ensure the functioning of the State Real Estate Cadastre; ensure the functioning of the State Register of Addresses; ensure the functioning of the Information System for encumbered territories; ensure the functioning of the central database of detailed topographic information; carry out cadastral valuation of real estate; participate in the implementation of state policy in the field of land reform; and carry out cadastral surveys of structures and groups of premises. Services related to the registration of new real estate, apartment ownership; changes in data regarding joint ownership, in particular of apartments; rent—buildings, land share, forest; registration or cancellation of the ban on alienation, including the workshop of the artist; registration of data on utilities or their updating of buildings; permits for privatization [11, 12].

The Central Bureau of Geodesy and Cartography of Poland provides effective services to citizens and country development by providing access to current geodetic and

Fig. 5 Fragment of the public cadastral map of Latvia

cartographic information and creating a modern spatial information infrastructure [13, 14] (see Fig. 6).

The public cadastral map of Russia on the site http://maps-rosreestr.ru liables you to set the location of the land plot, indicating its boundaries with neighboring objects (see Fig. 7). There are available information confirming the cadastral value of the property and its unique number. The user receives information about the status of the site (previously considered temporary, permanent). The category of land and type of permitted use are indicated. The relevant information about the organization and the specialist who put the site on record is provided [15].

The map also allows you to get detailed data on the capital construction project. Information includes: cadastral number of the object, its address, date of completion of construction, and registration. The use of the resource is possible even in cases where the interested person already has a permanent or temporary cadastral passport to the land. By specifying the number of the document, you can find out: the site is registered or removed from it for not filling out title documents.

A cadastral engineer, other specialists, and legal entities are interested in using the service. You can quickly obtain information that allows you to conduct an accurate survey of the site, without violating the rights of citizens and business entities. The resource contributes to the security of real estate transactions and clarification of technical information about the object. The data allow you to determine the prospects of the site for intended use. Based on the cadastral value, land tax can be calculated.

Fig. 6 Fragment of the public cadastral map of Poland

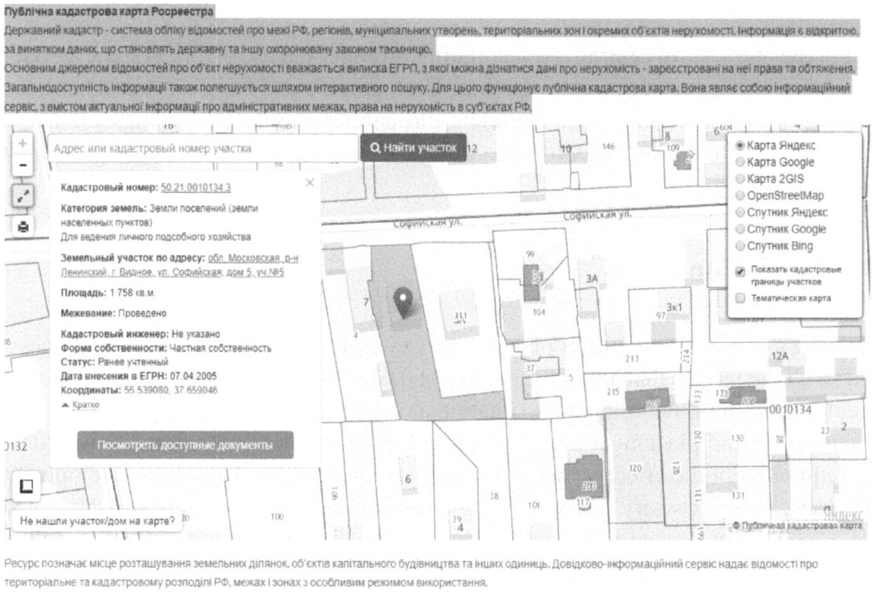

Fig. 7 Fragment of the public cadastral map of Russia

Fig. 8 Fragment of the public cadastral map of Moldova

For the purposes of centralized accounting, storage, and use of topographic, geodetic, and cartographic documents, the State Cartographic and Geodesic Fund has been created, which stores relevant materials throughout Moldova (see Fig. 8). These materials have technical, scientific, economic, historical, social, and cultural interest [16].

3 Conclusion

Building a construction master plan is a very complex and time-consuming process, and so all the factors that affect the quality of design are needed to be considered. For a real reflection of the production situation at the construction site, in addition to the organization and placement of construction facilities, reliable source cadastral data are required. For the effective use of the public cadastral map for urban development purposes, it is necessary to improve its information content, ensure uninterrupted use, save time by receiving services online, and the like. Creating an effective public cadastral map is a long-term investment in developing the infrastructure of a country's economy and increasing the competitiveness of enterprises by providing online services based on geospatial data, including cadastral data [17]. The resource will contribute to the security of real estate transactions and clarification of technical information about the object. The data will determine the prospects of the land for intended use.

References

1. Public cadastral map of Ukraine. Retrieved from http://map.dazru.gov.ua/kadastrova-karta.
2. Trevoho, I., Matishchuk, A., Ilkyv, Y., Haliarnyk, M. (2015). Public cadastres maps. *World Experience I. Journal Modern Achievements of Geodetic Science and Production*, *II*(30).
3. Mishchenko, R., Nesterenko, S., Demchenko, O. (2019). Publychny cadastrovy carty crain svytu. In *Conference « Building Innovations-2019 »* , Poltava, pp. 296–298.
4. Cadastral map of Georgia. Retrieved from http://maps.napr.gov.ge.
5. Georgia's National Registration Agency. Retrieved from http://www.reestri.gov.ge.
6. Standing Committee on Cadastre in the European Union. Retrieved from http://www.eurocadastre.org/.
7. Agency for Geodesy, Cartography and Cadastre (Bulgaria). Retrieved from https://kais.cadastre.bg/bg/Map.
8. State Administration of Geodesy and Cadastre (Czech Republic). Retrieved from https://geoportal.cuzk.cz.
9. Raveaux, R., Burie, J.-C., Ogier, J.-M. (2008). Object extraction from colour cadastral maps. In *The Eighth IAPR International Workshop on Document Analysis Systems*, Nara, Japan. Doi:https://doi.org/10.1109/das.2008.9.
10. Nagorna, O. (2013). Mandry publychnymy cadastrovymy cartamy svytu. *Land Management Newsletter, 3*, 51–54.
11. LR VALSTS ZEMES DIENESTS. Retrieved from https://www.vzd.gov.lv/lv/.
12. Kadastrs.lv. Retrieved from https://www.kadastrs.lv/#.
13. Karabin, M. (2017). A cadastral map in poland—the proposal based on analysis of cadastral maps in selected countries. In *17th International Multidisciplinary Scientific GeoConference SGEM2017*. Doi:https://doi.org/10.5593/sgem2017/22/s09.002.
14. Geoportal Infrastruktury Informacji Przestrzennej. Retrieved from http://mapy.geoportal.gov.pl.
15. Public cadastral map of the Russian registry. Retrieved from https://maps-rosreestr.ru/.
16. The Main Geoportal of Moldova. Retrieved from http://geoportal.md/.
17. Tesalovsky, A. (2016). The use of cartographic matter from open sources for mass assessment at the pre-stage study of hydropower construction. Doi:https://doi.org/10.18184/2079-4665.2016.7.3.107.111.

Analysis of Eccentrically Loaded Members of Circular Cross Section by Nonlinear Deformation Model

A. Pavlikov, D. Kochkarev, and O. Harkava

Abstract The strength analysis of columns with circular cross sections and evenly distributed reinforcement subject to an axial load and bending moment about one axis is considered. For such sections, a simple analysis technique is proposed to be applied in practice, which allows the basic formulas of nonlinear analysis to be reduced to formulas of strength of materials. The methodology is based on theoretically and experimentally substantiated assumptions, as well as on the introduced concept of the design strength of reinforced concrete. The proposed method is also grounded on modern deformation models using nonlinear stress–strain diagrams of materials. With the help of the accepted preconditions, the basic design dependencies were obtained. According to the developed method, examples of the area of reinforcement determination and definition of bearing capacity of eccentrically loaded members with circular cross section are given.

Keywords Columns · Deformation model · Design strength of reinforced concrete · Diagram · Compression · Circular section

1 Basic Concepts About Eccentrically Loaded Members

Eccentrically loaded members exist in practically every building or structure. These include columns, posts, pylons, members of arch and truss structures. Such members are characterized by the perception of bending moments and axial forces along the member. These members typically have reinforcement evenly distributed across the section, and the stresses in each reinforcement bar under certain conditions do not always reach the yield strength. This significantly complicates their analysis [1–5]. Particularly, acute questions for developing a practical calculation method arise precisely when considering elements of a circular cross section, since the specific shape of the cross section significantly increases the volume and complexity of the

A. Pavlikov · O. Harkava (✉)
Poltava National Technical Yuri Kondratyuk University, Poltava, Ukraine
e-mail: olga-boiko@ukr.net

D. Kochkarev
National University of Water and Environment Engineering, Rivne, Ukraine

V. Onyshchenko et al. (eds.), *Proceedings of the 2nd International Conference on Building Innovations*, Lecture Notes in Civil Engineering 73,
https://doi.org/10.1007/978-3-030-42939-3_19

calculation. For such cases, it is quite difficult to perform the integration, which leads to the need to use complex computer systems to solve simple design problems [6–8]. Replacing the nonlinear function of deforming the compressed area of concrete into rectangular for such cross sections results in much greater errors in the analysis than for rectangular members. Therefore, the development of a simple method of analysis of members with circular cross sections subject to an axial load and bending moment about one axis, taking into account nonlinear stress–strain diagrams of compressed concrete, is extremely necessary.

2 Effective Reinforced Concrete Columns and the Basics of Their Design

It is well known that eccentrically loaded members can work in two cases. The first case corresponds to the small eccentricities of the force application. Typically, in this case, the eccentricity of the application of force is within the height of the cross section of the member. Then, the entire cross section of the member is compressed, and the position of the neutral axis is outside the cross section (Fig. 1). The second case is characterized by a much larger eccentricity of the force application. In this case, the member works with cracks in the tensile area, and the neutral axis is within

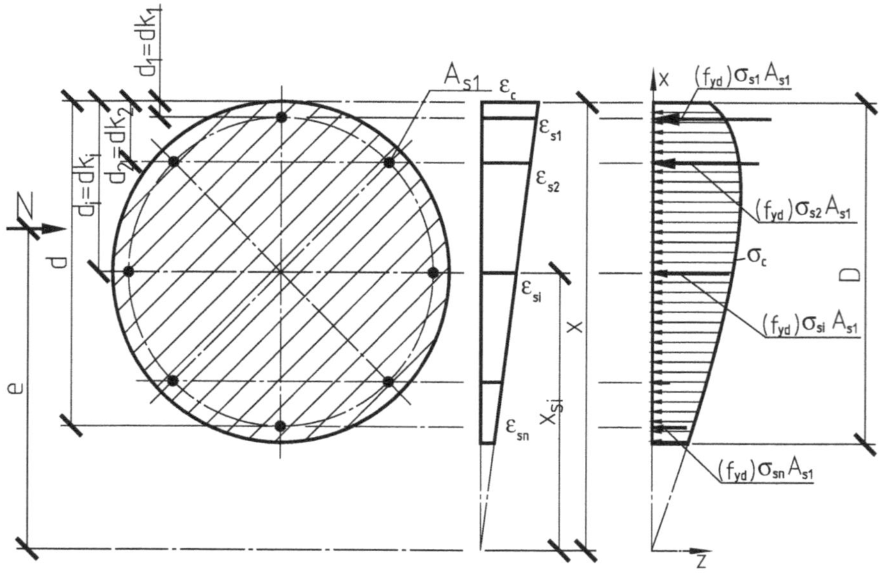

Fig. 1 Mode of deformation of the eccentrically loaded reinforced concrete member of circular cross section at the position of the neutral axis outside the cross section

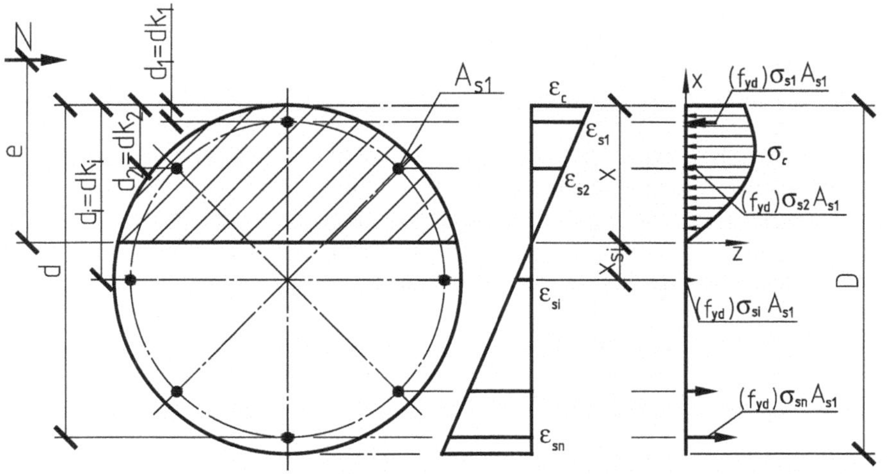

Fig. 2 Mode of deformation of the eccentrically loaded reinforced concrete member of circular cross section at the position of the neutral axis within the cross section

the cross section of the member (Fig. 2). The equilibrium equation for both cases of stress–strain state presented in the corresponding figures is written as follows:

$$\sum F_z = 0;\ N_c - \sum_{i=1}^{n_s} N_{si} = N; \tag{1}$$

$$\sum M_z = 0;\ M_c + \sum_{i=1}^{n_s} M_{si} = N \cdot e. \tag{2}$$

In expressions (1) and (2), M_c, M_{si} are the bending moments of the internal forces in the concrete and the ith reinforcement bar about the neutral axis, respectively; n_s is the number of reinforcement bars in the section of the member. For further transformations of expressions (1) and (2), it is assumed that the stresses in the reinforcement are described by the idealized Prandtl diagram and are also a valid hypothesis of plane sections. The hypothesis of plane sections is represented in this form

$$\frac{1}{r} = \frac{\varepsilon_c}{x} \quad \text{or} \quad \varepsilon_c = \frac{1}{r}x, \tag{3}$$

where x is the neutral axis depth and $1/r$ is the curvature.

After simple transformations of expressions (1) and (2), it is obtained:

$$\int_0^{A_c} \sigma_c(x, \varepsilon_c)\mathrm{d}A_c - E_s A_{s1} \sum_{i=1}^{n_s} \left(\frac{k_i d}{x} - 1\right) = N; \tag{4}$$

$$\int_{0}^{A_c} \sigma_c(x, \varepsilon_c) x \mathrm{d}A_c + E_s A_{s1} \sum_{i=1}^{n_s} \varepsilon_c x \left(\frac{k_i d}{x} - 1 \right)^2 = N \cdot e, \tag{5}$$

where $\sigma_c(x, \varepsilon_c)$ is stress–strain function in the compressed concrete area; A_{s1} is the area of one bar in section of the reinforced concrete member.

Taking into account that the eccentricity of force application according to the schemes (Figs. 1 and 2) $e = e_0 + x - D/2$ after simple transformations of Eqs. (4) and (5), using the expression for e, it is obtained that

$$\frac{\int_0^{A_c} \sigma_c(x, \varepsilon_c) x \mathrm{d}A_c + E_s A_{s1} \sum_{i=1}^{n_s} \varepsilon_c x \left(\frac{k_i d}{x} - 1 \right)^2}{\int_0^{A_c} \sigma_c(x, \varepsilon_c) \mathrm{d}A_c - E_s A_{s1} \sum_{i=1}^{n_s} \left(\frac{k_i d}{x} - 1 \right)} = e_0 + x - \frac{D}{2}, \tag{6}$$

where $e_0 = M/N$ is the initial eccentricity of applying force, M, N are the external bending moment and the external axial force.

After dividing both parts of Eq. (6) by the effective depth of the concrete cross section d, it is got

$$\frac{\int_0^{A_c} \sigma_c(x, \varepsilon_c) \frac{x}{d} \mathrm{d}A_c + E_s A_{s1} \sum_{i=1}^{n_s} \varepsilon_c \frac{x}{d} \left(\frac{k_i d}{x} - 1 \right)^2}{\int_0^{A_c} \sigma_c(x, \varepsilon_c) \mathrm{d}A_c - E_s A_{s1} \sum_{i=1}^{n_s} \left(\frac{k_i d}{x} - 1 \right)} = \frac{e_0}{d} + \frac{x}{d} - \frac{D}{2d}. \tag{7}$$

The next notation is introduced

$$\rho_f = \frac{n_s A_{s1}}{A_c}; \; A_c = \frac{\pi \cdot d^2}{4}. \tag{8}$$

Equation (7), taking into account notation (8), is looked like

$$\frac{\frac{1}{d} \int_0^{A_c} \sigma_c(x, \varepsilon_c) x \mathrm{d}A_c + A_c \frac{E_s}{n_s} \rho_f \sum_{i=1}^{n_s} \varepsilon_c \frac{x}{d} \left(\frac{k_i d}{x} - 1 \right)^2}{\int_0^{A_c} \sigma_c(x, \varepsilon_c) \mathrm{d}A_c - A_c \frac{E_s}{n_s} \rho_f \sum_{i=1}^{n_s} \left(\frac{k_i d}{x} - 1 \right)} = \frac{e_0}{d} + \frac{x}{d} - \frac{D}{2d}. \tag{9}$$

Ultimately, Eq. (9) is transformed to this

$$\frac{\frac{1}{A_c d} \int_0^{A_c} \sigma_c(x, \varepsilon_c) x \mathrm{d}A_c + \frac{E_s}{n_s} \rho_f \sum_{i=1}^{n_s} \varepsilon_c \frac{x}{d} \left(\frac{k_i d}{x} - 1 \right)^2}{\frac{1}{A_c} \int_0^{A_c} \sigma_c(x, \varepsilon_c) \mathrm{d}A_c - \frac{E_s}{n_s} \rho_f \sum_{i=1}^{n_s} \left(\frac{k_i d}{x} - 1 \right)} = \frac{e_0}{d} + \frac{x}{d} - \frac{D}{2d}. \tag{10}$$

The compressed area of circular cross section of concrete is represented with a certain number of parts (Figs. 3 and 4). Within each part, the strain values are constant and corresponding to the strain values at the center of gravity level of the resulting rectangles.

In this case, the internal forces in the concrete compressed area will be determined by the expressions:

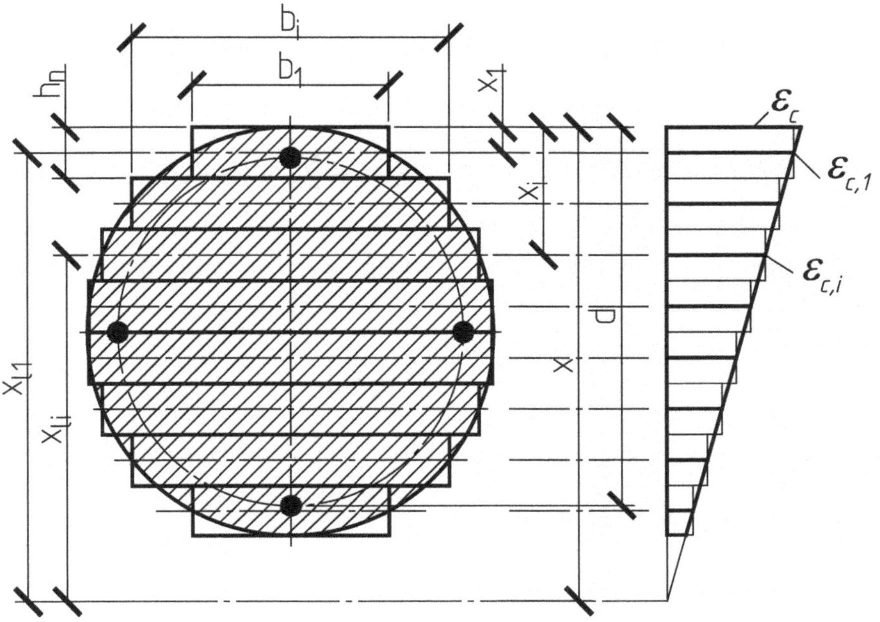

Fig. 3 To the definition of the internal forces in the concrete compressed area of the circular cross section of the member at small eccentricities

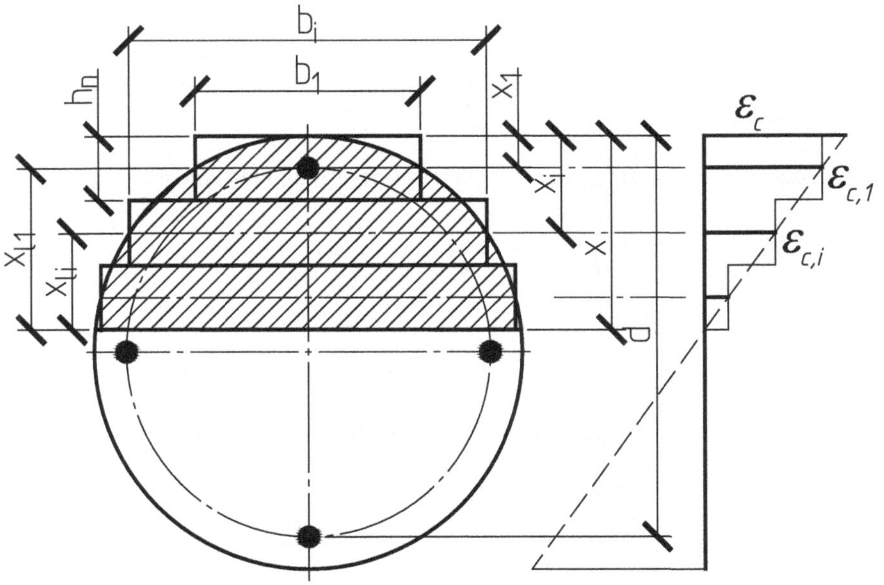

Fig. 4 To the definition of the internal forces in the concrete compressed area of the circular cross section of the member at large eccentricities

$$N_{\mathrm{c}} = \int\limits_{0}^{A_{\mathrm{c}}} \sigma_{\mathrm{c}} \mathrm{d}A_{\mathrm{c}} = \sum_{i=1}^{n_{\mathrm{c}}} \sigma_{\mathrm{c}i}(\varepsilon_{\mathrm{c}i}) b_i h_n = \sum_{i=1}^{n_{\mathrm{c}}} \sigma_{\mathrm{c}i}(\varepsilon_{\mathrm{c}i}) b_i \frac{x}{n_{\mathrm{c}}}; \tag{11}$$

$$M_{\mathrm{c}} = \int\limits_{0}^{A_{\mathrm{c}}} \sigma_{\mathrm{c}} x \mathrm{d}A_{\mathrm{c}} = \sum_{i=1}^{n_{\mathrm{c}}} \sigma_{\mathrm{c}i}(\varepsilon_{\mathrm{c}i}) b_i h_n x_{\mathrm{l}i} = \sum_{i=1}^{n_{\mathrm{c}}} \sigma_{\mathrm{c}i}(\varepsilon_{\mathrm{c}i}) b_i \frac{x}{n_{\mathrm{c}}} x_{\mathrm{l}i}. \tag{12}$$

The strains of the *i*th part of concrete compressed area are determined by the expression

$$\varepsilon_{\mathrm{c},i} = \varepsilon_{\mathrm{c}} \left(1 + \frac{1 - 2i}{2n_{\mathrm{c}}} \right), \tag{13}$$

where n_{c} is the number of sections into which the compressed concrete area is divided; $i = 1 \ldots n_{\mathrm{c}}$.

Geometric relationships have the form:

$$b_i = \sqrt{D^2 - (D - h_n)^2}; \tag{14}$$

$$x_{\mathrm{l}i} = \frac{x}{\varepsilon_{\mathrm{c}}} \varepsilon_{\mathrm{c},i}. \tag{15}$$

The following notations are introduced

$$k_x = \frac{d}{x}, \; n_a = \frac{a_{\mathrm{s}}}{d}. \tag{16}$$

In view of formulas (13) and (16), expressions (14) and (15) are as follows:

$$b_i = d \sqrt{2 \frac{1 + n_a}{n_{\mathrm{c}}} \frac{x}{d} - \left(\frac{x}{d} \right)^2}; \tag{17}$$

$$x_{\mathrm{l}i} = x \left(1 + \frac{1 - 2i}{2n_{\mathrm{c}}} \right). \tag{18}$$

Formulas (17) and (18) are substituted into expressions (11) and (12)

$$N_{\mathrm{c}} = d \cdot \sum_{i=1}^{n_{\mathrm{c}}} \left(\sigma_{\mathrm{c}i}(\varepsilon_{\mathrm{c}i}) \sqrt{2 \frac{1 + n_a}{n_{\mathrm{c}}} \frac{x}{d} - \left(\frac{x}{d} \right)^2} \frac{x}{n_{\mathrm{c}}} \right); \tag{19}$$

$$M_{\mathrm{c}} = d \cdot \sum_{i=1}^{n_{\mathrm{c}}} \left(\sigma_{\mathrm{c}i}(\varepsilon_{\mathrm{c}i}) \sqrt{2 \frac{1 + n_a}{n_{\mathrm{c}}} \frac{x}{d} - \left(\frac{x}{d} \right)^2} \frac{x^2}{n_{\mathrm{c}}} \left(1 + \frac{1 - 2i}{2n_{\mathrm{c}}} \right) \right). \tag{20}$$

In view of the above expressions, Eq. (10) can be represented by the following system:

$$\frac{N^* + \frac{E_s}{n_s}\rho_f \sum_{i=1}^{n_s} \varepsilon_c \frac{x}{d}(k_i \cdot k_x - 1)^2}{M^* - \frac{E_s}{n_s}\rho_f \sum_{i=1}^{n_s} (k_i \cdot k_x - 1)} = \frac{e_0}{d} + k_x - \frac{1+n_a}{2}.$$

$$N^* = \sum_{i=1}^{n_c} \left(\sigma_{ci}(\varepsilon_{ci}) \frac{4 \cdot k_x}{\pi \cdot n_c} \sqrt{2\frac{1+n_a}{n_c}k_x - k_x^2} \right); \tag{21}$$

$$M^* = \sum_{i=1}^{n_c} \left(\sigma_{ci}(\varepsilon_{ci}) \frac{4 \cdot k_x^2}{\pi \cdot n_c} \sqrt{2\frac{1+n_a}{n_c}k_x - k_x^2} \left(1 + \frac{1-2i}{2n_c} \right) \right). \tag{22}$$

Using (21), (22) and expression (20), the equation is obtained in which only the neutral axis depth is unknown at certain strains of the compressed area of concrete, at known cross-sectional dimensions and characteristics of materials.

Next, the basic concepts of the method of the design strength of reinforced concrete [9] are used. For circular cross sections subjected to axial load and bending moment, the design strength is determined by the dependence

$$f_{zN} = \frac{N}{A_c} = \frac{4 \cdot N}{\pi \cdot d^2}. \tag{23}$$

The first equation of equilibrium (4) is divided by the effective cross-sectional area A_c, and after simple transformations taking into account the expression (19), it is obtained

$$\sum_{i=1}^{n_c} \left(\sigma_{ci}(\varepsilon_{ci}) \frac{4k_x^2}{\pi n_c} \sqrt{2\frac{1+n_a}{n_c}k_x - k_x^2} \right) - \frac{E_s}{n_s}\rho_f \sum_{i=1}^{n_s} (k_i k_x - 1) = \frac{N}{A_c}. \tag{24}$$

In this way, the design strength for the columns of circular cross section is determined by the expression

$$f_{zN} = \sum_{i=1}^{n_c} \left(\sigma_{ci}(\varepsilon_{ci}) \frac{4k_x^2}{\pi n_c} \sqrt{2\frac{1+n_a}{n_c}k_x - k_x^2} \right) - \frac{E_s}{n_s}\rho_f \sum_{i=1}^{n_s} (k_i k_x - 1). \tag{25}$$

To find the design strength of reinforced concrete by expression (25), it is necessary to first apply an extreme criterion to expression (20). That is, such strains of the concrete compressed area need to be determined that would correspond to the maximum value of the design strength of concrete (25). In this case, the main criteria for destruction can be described by the following system:

$$\begin{cases} \frac{\mathrm{d}f_{zN}}{\mathrm{d}\varepsilon_c} = 0, \text{ when } \varepsilon_c \le \varepsilon_{cu}; \\ \sigma_{si} = \varepsilon_{si} E_s \le f_{yd}; \ i = 1 \ldots n \ ; \\ \varepsilon_{si} \le \varepsilon_{su}, \ i = 1 \ldots n. \end{cases} \tag{26}$$

In order to perform analysis by the method of design strength of concrete, it is necessary to take the form of a concrete stress–strain diagram at first and to determine the design strength depending on such characteristics

$$f_{zN} = N/A_c = f(\rho_f, e_0/d, C, f_y). \tag{27}$$

In expression (27): N is the compression force in the corresponding section, kH; C is class of concrete, which is characterized by the parameters of the stress–strain diagram at compression, i.e., E_c, ε_{c1}, ε_{cu}, f_c; f_y is the yield strength of the reinforcement; e_0/d is the eccentricity of force application.

Table 1 shows the design strength of compressed reinforced concrete members of circular cross section. As a concrete stress–strain diagram "$\sigma - \varepsilon$", Eurocode 2 function is accepted [1].

Based on the stated the condition of the bearing capacity of normal cross sections of reinforced concrete members of circular cross section subjected to axial force and bending moment has the following form

$$f_{zN} \le N/A_c. \tag{28}$$

Condition (28) allows to solve all problems of reinforced concrete members design: to check and define the strength of the section and to determine the reinforcement. This method allows to reduce the analysis of reinforced concrete members to the method of classical strength of materials, but taking into account nonlinear deformation of materials. It should be noted that the proposed analysis method is not simplified.

3 Examples of Analysis of Reinforced Concrete Columns of Circular Cross Section

Example 1 Reinforced concrete column of circular cross section $d = 40$ cm of C25/30 concrete subject to axial force $N_{Ed} = 1400$ кN with eccentricity $e_0 = 12$ cm. Calculate the reinforcement of a column of steel grade A400C.

The solution Initial eccentricity of the force and the design strength of reinforced concrete required for the perception of a given load

$$\frac{e_0}{d} = \frac{12}{40} = 0.3; \ f_{zN} = \frac{N_{Ed}}{A_c} = \frac{N_{Ed}}{0.785d^2} = \frac{1400 \times 10}{0.785 \times 40^2} = 11.15 \text{ MPa}.$$

Table 1 Design strength of reinforced concrete members of circular cross section subjected to axial force and bending moment f_{zN}, MPa, $\lambda \leq 4$

Concrete class	$n_a = 0.06 \div 0.1$ A500C $f_{yd} = 435$ MPa		Initial eccentricity of axial force application e_0/d						
			0.01	0.15	0.30	0.65	1.00	2.00	5.00
C25/30	ρ_f	0.005	19.386	14.542	8.905	3.016	1.508	0.592	0.208
		0.01	21.561	15.948	10.324	4.383	2.523	1.064	0.381
		0.02	25.911	18.848	12.856	6.588	3.979	1.833	0.697
		0.03	30.261	21.750	15.202	8.234	5.304	2.482	0.951
		0.04	34.611	24.620	17.453	9.758	6.544	3.109	1.196
C30/35	ρ_f	0.005	21.900	16.443	9.955	3.190	1.554	0.605	0.212
		0.01	24.075	17.851	11.416	4.650	2.642	1.087	0.387
		0.02	28.425	20.735	13.957	6.943	4.146	1.893	0.712
		0.03	32.775	23.653	16.306	8.642	5.496	2.552	0.974
		0.04	37.125	26.536	18.580	10.161	6.769	3.184	1.221
C32/40	ρ_f	0.005	24.284	18.334	10.995	3.337	1.595	0.616	0.216
		0.01	26.459	19.751	12.495	4.899	2.725	1.107	0.393
		0.02	30.809	22.635	15.030	7.279	4.297	1.937	0.722
		0.03	35.159	25.564	17.346	9.015	5.673	2.613	0.990
		0.04	39.509	28.458	19.572	10.535	6.970	3.252	1.242

According to Table 1, the percentage of reinforcement of the cross section is determined, which will provide the required design strength: $\rho_f = 1.45\%$. Required cross-sectional area of reinforcement $A_s = \rho_f \times A_c = 0.0145 \times 0.785 \times 40^2 = 18.21$ cm^2. 8Ø18 A400C is accepted, $A_s = 20.36$ см^2.

Example 2 Reinforced concrete column of circular cross section $d = 25$ cm has C32/40 concrete, reinforcement 8Ø16 mm of A500C steel, $A_s = 16.09$ см^2. Calculate the value axial force applied with eccentricity $e_0 = 7.5$ cm that can be perceived by the column.

The solution Initial eccentricity and reinforcement ratio

$$\frac{e_0}{d} = \frac{12}{40} = 0.3; \ \rho_f = \frac{A_s}{0.785 d^2} = \frac{16.09}{0.785 \times 25^2} = 0.0328.$$

According to Table 1, the design strength of reinforced concrete is found $f_{zN} = 17.97$ MPa.

The value of the axial force that can be perceived by the column

$$N = f_{zN} A_c = 17.97 \times 0.785 \times 25^2 \times 10^{-1} = 881.65 \,\text{kN}.$$

4 Conclusion

The method of analysis of reinforced concrete members of a circular cross section subjected to axial force and bending moment is proposed. The method uses experimentally and theoretically grounded assumptions and hypotheses [10]. It allows to solve a wide range of tasks that are common in construction practice. It permits to perform rather efficiently and quickly check the results of complex computer analysis both in the selection of reinforcement and in determining the bearing capacity of the section [8–10]. In the following, it is suggested to consider the analysis of columns of high flexibility by the proposed method.

References

1. EN 1992-1-1:2004. (2004). *Eurocode 2: Design of concrete structures—Part 1-1: General rules and rules for buildings*. fib Bulletin 34 (2006). *Model code for service life design*. Lausanne, Switzerland: Fédération Internationale du Béton (fib).
2. Pavlikov, A. M. (2007). *Neliniina model napruzheno-deformovanoho stanu kosozavantazhenykh zalizobetonnykh elementiv u za krytychnii stadii*. PoltNTU.
3. Kochkarov, D. V. (2015). *Neliniinyi opir zalizobetonnykh elementiv i konstruktsii sylovym vplyvam*. O. Zen.

4. Kochkarev, D., Azizov, T., & Galinska, T. (2018). Bending deflection reinforced concrete members determination. *MATEC Web of Conferences, 230*. https://doi.org/10.1051/matecconf/201823002012.
5. MacGregor, J. G., & Wight, J. K. (2005). *Reinforced concrete: Mechanics and design*. Upper Saddle River.
6. Hu, H. T., Huang, C. S., & Chen, Z. L. (2005). Finite element analysis of CFT columns subjected to an axial compressive force and bending moment in combination. *Journal of Constructional Steel Research, 61*(12), 1692–1712. https://doi.org/10.1016/j.jcsr.2005.05.002.
7. Kim, J. K., & Lee, S. S. (2000). The behavior of reinforced concrete columns subjected to axial force and biaxial bending. *Engineering Structures, 22*(11), 1518–1528. https://doi.org/10.1016/S0141-0296(99)00090-5.
8. Alfano, G., Marmo, F., & Rosati, L. (2007). An unconditionally convergent algorithm for the evaluation of the ultimate limit state of RC sections subject to axial force and biaxial bending. *International Journal for Numerical Methods in Engineering, 72*(8), 924–963. https://doi.org/10.1002/nme.2033.
9. Pavlikov, A., Kochkarov, D., & Harkava, O. (2019). Calculation of reinforced concrete members strength by new concept. In *CONCRETE. Innovations in Materials, Design and Structures: Proceedings of the fib Symposium 2019* (pp. 820–827), Kraków, Poland, May 27–29, 2019.
10. Pavlikov, A., Harkava, O., Prykhodko, Yu., & Baryliak, B. (2018). Experimental and theoretical testing results of reinforced concrete columns under biaxial bending. *International Journal of Engineering & Technology, 7*(4.8), 145–151. https://doi.org/10.14419/ijet.v7i4.8.27230.

Reliability Assessment of Multi-bolt Joints of Silo Capacity's Wall

Sergei Pichugin, Anton Makhinko, and Nataliia Makhinko

Abstract The paper is aimed at stochastic calculation of nodal joints of cylindrical steel silos body, formed by using bolts. The research focuses on analyzing the reliability of a multi-bolt connection as a system of parallel elements. In particular, it is found that for homogeneous joint, when all the fastening elements belong to the same sample, the connection reliability is determined exceptionally by the statistical characteristics of the strength of one bolt, the number of bolts and stochastic characteristics of the load. For quasi-homogeneous joints, neglect for the failure order of individual bolts leads to a significant reassessment of their reliability. The formula for determining the random value of critical connection factor is obtained, and the following calculations are made. The density distribution of the maximum load probabilities was assumed in accordance with Gumbel double exponential law. The calculation results are presented graphically on a critical probability scale for different combinations of coefficients of the effort variations in each bolt. The graphical representation of changes in the system reliability rate is carried out, and the factors influencing on this process are determined.

Keywords Stochastic calculation · Bolt connections · Silo · Wall · Critical factor

1 Introduction

An important aspect of steel silos design is the choice of nodal joint type for thin-walled panels of body, which should provide durable and high-tech connection. Nowadays, use of bolted joints is significant advantage of storage silos manufacturers and designers. The reason for this, first of all, is proved by the assembly operations

S. Pichugin
Poltava National Technical Yuri Kondratyuk University, Pershotravnevyj Ave, 24, Poltava, Ukraine

A. Makhinko
ETUAL LLC, Bortnytska Street 1, Petropavlivske, Boryspil District, Kiev, Ukraine

N. Makhinko (✉)
National Aviation University, Kosmonavta Komarova 1, Kiev, Ukraine
e-mail: pasargada1985@gmail.com

V. Onyshchenko et al. (eds.), *Proceedings of the 2nd International Conference on Building Innovations*, Lecture Notes in Civil Engineering 73,
https://doi.org/10.1007/978-3-030-42939-3_20

facilitation, the simplicity of assembly, the reduction of total cost and also by the possibility of constructing inventory capacities or, if necessary, swift replacement of individual elements in case of failure. In capacities of different outputs, the number of bolts in one joint may vary from 15 to 70 pieces. Depending on the design solution and the load value, the capacity metal constructions can be joined by using bolts of normal or high strength. Thus, the reliability of silo capacity, its size, weight, cost and other parameters number in many cases will depend on the characteristics of the fasteners.

Analytical calculation of bolt joints constructions is given in classical reference books on metal constructions course [1, 2] and is regulated by the normative documents and also is actively studied by modern scientists, depending on the area of chosen studies [3–7]. Particularly, we can mention the well-known works [8, 9] about thin-walled profiles. The bolt joint reliability calculations concerning thin-walled steel elements were taken into consideration. The tasks for nodes of steel constructions reliability assessing including bolt joints were initiated in our works [10–16].

2 Research Results

Bolt joints of the body sheets of cylindrical storage capacities are the most relevant elements for this type of constructions (Fig. 1).

The reason for this is the fact that, as a rule, their level of workload almost reaches the calculated one. Therefore, the failure of one or two bolts in one joint can cause an immediate destruction of the whole construction. Surely, the strength loss of one of the capacity body sheets also causes similar effects. However, in this case, there is a fundamental difference between "element" and "system" concepts [17, 18].

If we study the silos sheets, then we have one hypothetical element—the sheet and one parameter of its strength—the steel flux boundary. If we refer about a node,

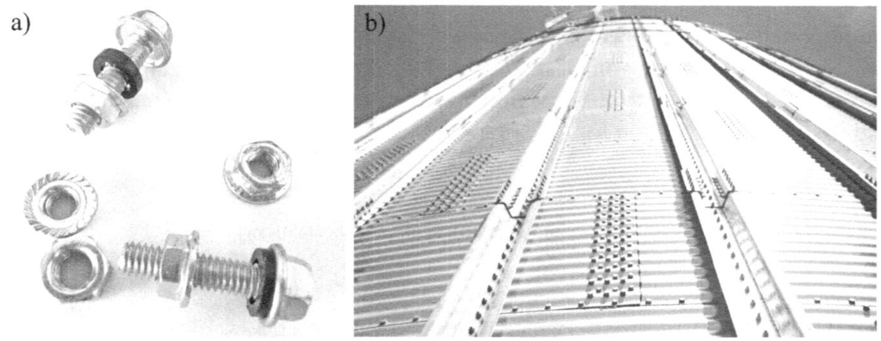

Fig. 1 Bolt joints of the silo capacities: **a** structure of bolts; **b** vertical and horizontal joints of the silo body panels

where two sheets are connected by means of several dozens of bolts, then we need to consider a number of factors. They are the bearing capacity of each of the bolts and probable failure of mechanisms, such as a bolt cut, crushing of the bolt or fracture of the base metal. It is also necessary to quantify the conditional reliability indicators [17]. These values show how the reliability of the entire joint will change when one, two or more bolts fail in the connection.

Thus, the joint of the two body sheets should be considered as a system and the bolts of this joint as its elements. In the theory of reliability, a system is any set of objects, in which the failure of one or more elements leads to a breach of the whole reliability system [17, 18].

If one of the element failures leads to the failure of the system, then we have the case of a sequential system elements connection. For example, you can refer to any statically defined construction. However, if the destruction of one or more elements does not cause a direct failure of the system, then such connection is regarded to be parallel. The parallel connection of the elements to some extent is formed by continuous beams or slabs lean upon the contour.

Let us consider the body sheet joint as a system of elements that consist of m bolts. The probability of a failure-free operation of one bolt will be denoted as p_0. Let us assume that the quantitative assessments of all bolts in the connection failure are equal. In this case, the probability of failure q_0 will be equal

$$q_0 = 1 - p_0 \tag{1}$$

For the case of sequentially connected elements, the probability of a failure-free operation of the joint will be determined by the following expression

$$P_\Sigma = \prod_{k=1}^{n} p_0 = p_0^n = \prod_{k=1}^{n} (1 - q_0) = (1 - q_0)^n \tag{2}$$

For the case of the parallel connection of the elements

$$P_\Sigma = 1 - \prod_{k=1}^{n} q_0 = 1 - q_0^n = 1 - \prod_{k=1}^{n} (1 - p_0) = 1 - (1 - p_0)^n \tag{3}$$

However, the actual work of the silo sheet joint cannot be perfectly idealized by these simplest schemes. This is due to the fact that when there are failures of individual bolts, each time in the joint, we have different working conditions, which correspond to different calculation schemes and stochastic parameters of the system. Therefore, the adequate ratings of such systems reliability should be based on the correct combination of Eqs. (2) and (3).

As an example, let us consider a failure caused by a cut of one or more connecting bolts. Let us assume that the joint is influenced by a random tensile force $\tilde{S}$, causes in each bolt accidental cutting effort $\tilde{S}_0$. Also, let us introduce the condition that the bolt failure is brittle; it means that after reaching the limiting value, the cutting effort in a separate bolt becomes equal to zero. The shear resistance $\tilde{R}_{bs}$ of individual bolt we will also assume as a random variable with a normal distribution law. In such case, several situations may occur when all the connection bolts n are the elements of the same sample or make two $n = n_1 + n_2$ or three $n = n_1 + n_2 + n_3$ independent samples. The laws of each sample strength distribution are the same. In this study, we will consider only the first two cases.

The relevance of this problem consideration is justified by two points. Firstly, from the practice of mounting, examining and analyzing the causes of failure, it is not known that heterogeneity of bolts within a single joint is common. In this case, the fastening parts of different manufacturers or bolts of different strength classes can be used. Secondly, the analysis of multi-bolt connection reliability can be carried out by the analogy with the studies of parallel work of elastic fragile elements, presented in the works [10, 18, 19]. However, the feature of these studies is to consider all elements of the system as independent random variables. Such approach has only theoretical direction and is not suitable for solving practical problems.

2.1 Stochastic Calculation of Homogeneous Bolt Connection

Under the "homogeneity" of bolt connection, all fasteners are considered to be of the same sample. The problem of stochastic calculation under this formulation has the simplest solution, because the boundary state occurs simultaneously in all elements. For a critical factor of connection $\tilde{K}_{b,i}$, which expresses the value of generalized effort ratio $\tilde{S}_i$ on the bolt connection and its generalized strength $\tilde{R}_{bs}$

$$\tilde{K}_{b,i} = \frac{\tilde{S}_i}{\tilde{R}_{bs,i} \cdot A_b \cdot n} \tag{4}$$

$$\tilde{K}_{b,i} = \frac{1.273}{d_b^2 \cdot n} \cdot \frac{m_S}{m_{bs}} \cdot \frac{1 + V_S \tilde{\gamma}_{S,i}}{1 + V_{bs} \tilde{\gamma}_{bs,i}} = m_b \cdot \frac{1 + V_S \tilde{\gamma}_{S,i}}{1 + V_{bs} \tilde{\gamma}_{bs,i}} \tag{5}$$

where n is the number of sample bolts; $\tilde{R}_{bs,i} = m_{bs}(1 + V_{bs}\tilde{\gamma}_{bs,i})$ is the calculation resistance of the i-bolt cut; $A_b = \pi d_b^2/4$ is the bolt area; d_b is the diameter of the bolt; m_S is the expected value of the generalized effort $\tilde{S}_i$; m_{bs} is the expected value of the generalized strength $\tilde{R}_{bs}$; V_S is the variation coefficient of the generalized effort $\tilde{S}_i$; V_{bs} is the variation coefficient of the generalized strength $\tilde{R}_{bs}$; $\tilde{\gamma}_{bs,i}$ is the value of normalized random variable of the bolt strength; $\tilde{\gamma}_{S,i}$ is the value of normalized random variable of external efforts maximums; m_b is the expected value of critical factor of bolt joint.

Equation (5) corresponds to the simultaneous failure of all bolts in case of accidental effort overrun $\tilde{S}_i$ of the accidental level of bearing capacity connection $\tilde{R}_{bs} \cdot A_b \cdot n$. The reliability of the entire connection is determined exceptionally by statistical characteristics of the strength of one bolt, their number and stochastic load characteristics.

2.2 Stochastic Calculation of the Connection Formed by Two Independent Samples of Bolts

This bolt connection is schematized by the case when one or more bolts of a nonproject strength class, but of the same diameter, accidently appear in the mounting joint. For this case, we will introduce such term as quasi-homogeneous bolt connection.

For the critical factor of quasi-homogeneous bolt connection, we will write the formula similar to Eq. (5)

$$\begin{aligned}\tilde{K}_{\mathrm{b},i} &= \frac{\tilde{S}_i}{A_\mathrm{b}\cdot(\tilde{R}_{\mathrm{bs}1,i}\cdot n_1+\tilde{R}_{\mathrm{bs}2,i}\cdot n_2)}\\ &= \frac{1.273}{d_\mathrm{b}^2}\cdot\frac{m_\mathrm{S}\cdot(1+V_\mathrm{S}\tilde{\gamma}_{\mathrm{S},i})}{[m_{\mathrm{bs}1}\cdot n_1\cdot(1+V_{\mathrm{bs}1}\tilde{\gamma}_{\mathrm{bs}1,i})+m_{\mathrm{bs}2}\cdot n_2\cdot(1+V_{\mathrm{bs}2}\tilde{\gamma}_{\mathrm{bs}2,i})]}\end{aligned} \tag{6}$$

Since $\tilde{\gamma}_{\mathrm{bs}1}$ and $\tilde{\gamma}_{\mathrm{bs}2}$ are independent random variables, we can apply Eq. (5) using new denotation for the values m_b and V_bs

$$m_\mathrm{b}=\mathrm{M_b}\cdot\frac{1}{1+\eta_\mathrm{bs}},\quad V_\mathrm{bs}=\frac{\sqrt{V_{\mathrm{bs}1}^2+V_{\mathrm{bs}2}^2\eta_\mathrm{bs}^2}}{1+\eta_\mathrm{bs}} \tag{7}$$

$$\mathrm{M_b}=\frac{1.273}{d_\mathrm{b}^2}\cdot\frac{m_\mathrm{S}}{m_{\mathrm{bs}1}n_1},\quad \eta_\mathrm{bs}=m_{\mathrm{bs}2}n_2/m_{\mathrm{bs}1}n_1 \tag{8}$$

where η_bs is the ratio of characteristics and number of bolts.

However, accordingly to Eq. (6), it is possible to get the correct result, only for values of the random variable $\tilde{K}_{\mathrm{b},i} < 1$, which will show the connection failure absence. Otherwise, ($\tilde{K}_{\mathrm{b},i} > 1$) the performed calculations will not be reliable, as part of the bolts will not be influenced by the external load.

As an example, we will consider the process of connection fracture, which consists of two bolts of the same size, but of different strength. This system is schematized in Fig. 2.

For destruction, the strength of one of the bolts should be less than $\tilde{S}/2$ and strength of the other less than $\tilde{S}$. The effort is divided equally between the bolts if none of bolts is failure or the effort affects only one of joint bolts in case of failure other.

If the random strength of the first bolt in the effort space is marked as $\tilde{S}_{01}$ and the second as $\tilde{S}_{02}$, then the formula of the critical factor will be like this

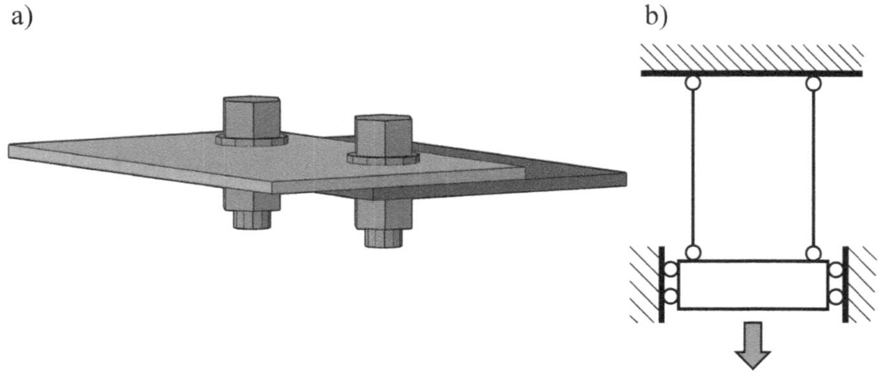

Fig. 2 Real (**a**) and idealized (**b**) schemes of the connection system of two bolts of different strength classes

$$\tilde{K}_{b,i} = \left| \begin{array}{ll} \tilde{S}_i / \left(\tilde{S}_{01,i} + \tilde{S}_{02,i} \right), & \text{if } \tilde{S}_i / \tilde{S}_{01,i} < 2 \wedge \tilde{S}_i / \tilde{S}_{02,i} < 2; \\ \tilde{S}_i / \tilde{S}_{01,i}, & \text{if } \tilde{S}_i / \tilde{S}_{01,i} < 2 \wedge \tilde{S}_i / \tilde{S}_{02,i} \geq 2; \\ \tilde{S}_i / \tilde{S}_{02,i}, & \text{if } \tilde{S}_i / \tilde{S}_{01,i} \geq 2 \wedge \tilde{S}_i / \tilde{S}_{02,i} < 2; \\ \tilde{S}_i / \left(\tilde{S}_{01,i} + \tilde{S}_{02,i} \right) & \text{if } \tilde{S}_i / \tilde{S}_{01,i} \geq 2 \wedge \tilde{S}_i / \tilde{S}_{02,i} \geq 2. \end{array} \right. \quad (9)$$

Formula (9) leads to the calculation of the critical factor when one of the four conditions is done:

1. when the effort $\tilde{S}_i$, the strength of both bolts is ensured;
2. the effort $\tilde{S}_i$ causes the failure of the second bolt, and the first continues to perform its functions;
3. the effort $\tilde{S}_i$ causes the failure of the first bolt, the second continues to perform its functions;
4. the strength of both bolts is depleted simultaneously.

Figure 3 shows the calculation results of the critical factor accordingly to Eq. (9) and its distribution on the critical stochastic scale, when the argument of double exponential Gumbel distribution is plotted along the abscissa axis. Gumbel used the similar scale [20–22] by such term as «extreme probability paper», and it was also used in the work of the author [10, 11] as "the general form of loads' presentation"

$$y = -\ln[-\ln(F_\gamma)] \quad (10)$$

where F_γ is the probability of a failure-free operation.

The probability of the failure-free operation of bolted connection, according to our long-term compatible studies, can be presented as [15]

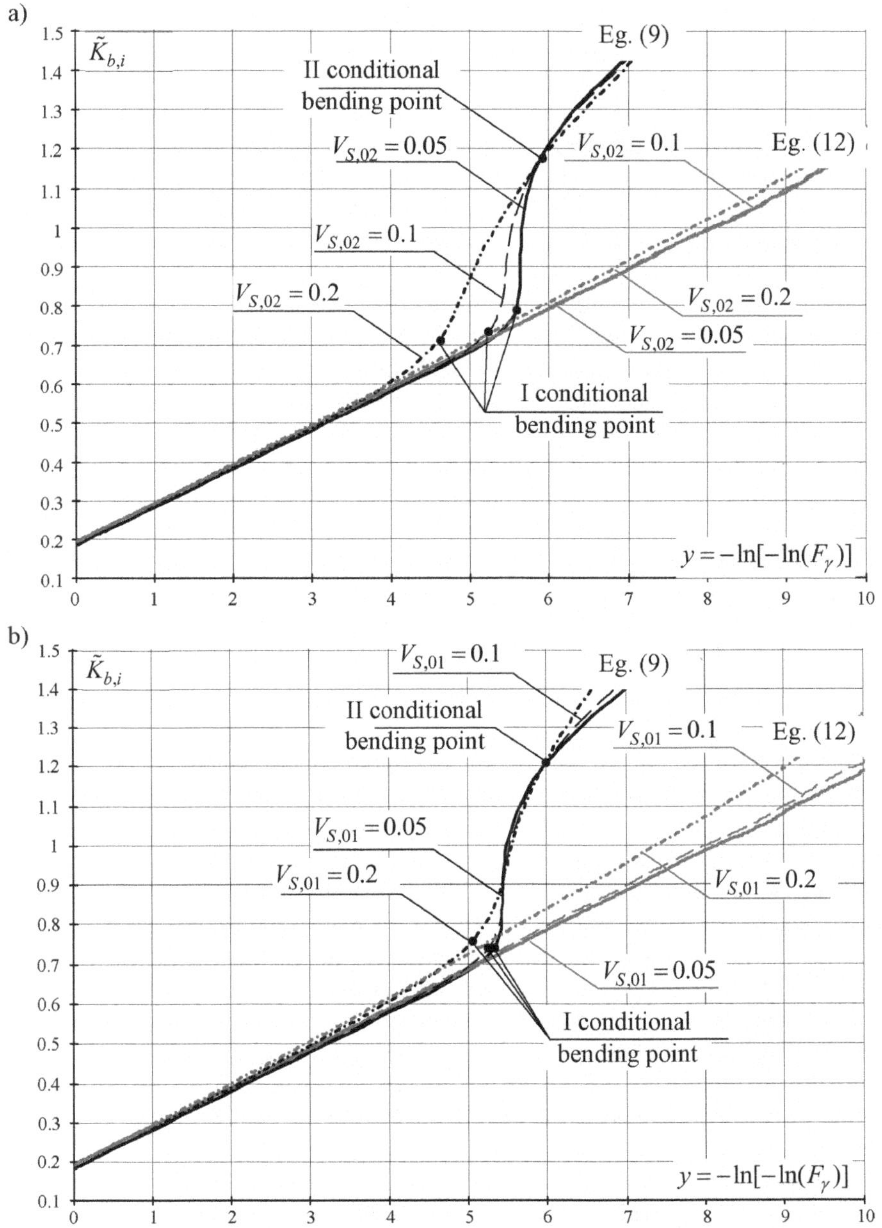

Fig. 3 Probability of the failure-free operation of the two bolts' joint on the critical stochastic scale when $V_{S,01} = 0.1$ (**a**) and $V_{S,02} = 0.1$ (**b**)

$$F_{\gamma} = \exp\left[-\exp\left(\frac{\alpha_B - \sqrt{\alpha_B^2 - 4\alpha_A^2(1 - \alpha_C V_S - 1/m_K)}}{2\alpha_A}\right)\right] \tag{11}$$

where α_A, α_B and α_C are the dimensionless coefficients, which consider the influence of the variation coefficient of the bearing capacity and the maximums of the random load. Their values are assumed to be equal: for the load distribution under the normal distribution law $\alpha_A \approx -0.02$, $\alpha_B \approx 0.65$, $\alpha_C \approx 0.18$ and for the double exponential Gumbel law $\alpha_A \approx 0$, $\alpha_B \approx 0.84$, $\alpha_C \approx 0.57$.

While calculating, the following output data is applied: The connection has a random power with mathematical expectation $m_S = 100$ and variation coefficient $V_S = 0.5$. The density of probability distribution of maximum loads was taken in accordance with the double exponential Gumbel law. The bearing capacity of both bolts was assumed to be distributed accordingly to the normal law and with different statistical characteristics: The mathematical expectation of the first bolt is equal $m_{S,01} = 250$ kN, and the other is equal $m_{S,02} = 150$ kN. For the coefficients of variation $V_{S,01}$ and $V_{S,02}$, three values were considered 0.05, 0.1 and 0.2.

Critical factor curves are constructed for two variants. According to Eq. (9) and considering the assumption that the values of the critical factor are calculated without the ratio of the bolts' strength and effort in them

$$\tilde{K}_{b,i} = \tilde{S}_i / \left(\tilde{S}_{01,i} + \tilde{S}_{02,i}\right) \tag{12}$$

When analyzing Fig. 3, we can see that ignoring the sequence of failures of individual bolts in the stochastic evaluation of critical factor values Eq. (12) leads to significant reassessment of their reliability. For the curves of the critical factor, constructed according to Eq. (9), there are two distinct points of inflection where the angle of inclination of the curves to the axis abscissa is changed, i.e., the speed of change of the reliability system. The greater the inclination of the curves between these points is, the greater the decrease speed of the reliability is. The inclination of the curves depends on the coefficient of variation, i.e., the extent to which the external force is changing. After the second point of inflection, the inclination of the curves decreases. Thus, in the case, when the coefficient of variation of the first bolt effort is stable $V_{S,01} = 0.1$ (Fig. 3a), this deviation practically does not depend on the variation of the load; otherwise, $V_{S,02} = 0.1$ (Fig. 3b); such tendency is not observed.

The occurrence of the I–II site from a mechanical point of view is explained by sharp increase in the loading of one of the bolts when the other has a failure, thus, a sharp decrease in the reliability of the entire joint.

From a mathematical point of view, it is explained by lack in the sample of critical factor coefficient $\tilde{K}_b$, which have size N_b (where $i = 1, 2, \ldots, N_b$) of elements, where the strength of one of the connection bolts is not provided. So, when one of the conditions is done $\tilde{S}_i/\tilde{S}_{01,i} \geq 1$ or $\tilde{S}_i/\tilde{S}_{02,i} \geq 1$, the value of the critical factor is calculated both for connection with one bolt and it is found to be much smaller than the value obtained for the connection with two bolts. Therefore, the most accurate is the stochastic model constructed for Eq. (9).

3 Conclusion

Formula is obtained, and conditions are defined for calculating random variable of the critical factor for homogeneous and quasi-homogeneous bolt joints.

The reliability of homogeneous connection with *n*—number of bolts that make up one sample—is determined exceptionally by statistical characteristics of the strength of one bolt, their number and the stochastic characteristics of the load.

Ignoring the sequence of failures of individual bolts of quasi-homogeneous connection in the stochastic evaluation of critical factor values leads to a significant reassessment of their reliability.

Graphical comparison of the calculation results of critical factors values for quasi-homogeneous bolt joint is presented on critical probability scale.

The analysis of graphical representation of the speed of changes in the system reliability is carried out, and the factors influencing this process are determined.

References

1. Anurev, V. (2001). *Spravochnik konstruktora mashinostroitelia*. Moskva: Mashinostroenie.
2. Orlov, P. (1977). *Osnovy konstruirovaniia*. Moskva: Mashinostroenie.
3. Oldfield, M., Ouyang, H., & Mottershead, J. E. (2005). Mottershead simplified models of bolted joints under harmonic loading. *Computers & Structures, 84*(1–2), 25–33.
4. Maggi, Y., Gonçalves, R., Leon, R., & Ribeiro, L. (2005). Parametric analysis of steel bolted end plate connections using finite element modeling. *Journal of Constructional Steel Research, 61*(5), 689–708.
5. Liu, F., Shan, M., Zhao, L., & Zhang, J. (2018). Probabilistic bolt load distribution analysis of composite single-lap multi-bolt joints considering random bolt-hole clearances and tightening torques. *Composite Structures, 194,* 12–20.
6. Elliott, M., Teh, L., & Ahmed, A. (2019). Behaviour and strength of bolted connections failing in shear. *Journal of Constructional Steel Research, 153,* 320–329.
7. Farahmand, B. (2001). *Fracture mechanics of metals, composites, welds, and bolted joints: Application of LEFM, EPFM, and FMDM theory*. New York: Springer Science + Business Media.
8. Kim, T., Yoo, J., & Roeder, C. (2015). Experimental investigation on strength and curling influence of bolted connections in thin-walled carbon steel. *Thin-Walled Structures, 91,* 1–12.
9. Yu, C., & Panyanouvong, M. (2013). Bearing strength of cold-formed steel bolted connections with a gap. *Thin-Walled Structures, 67,* 110–115.
10. Pichugin, S. (2009). *Nadezhnost stalnykh konstruktsii proizvodstvennykh zdanii*. Poltava: ASMI.
11. Pichugin, S. (1995). Veroiatnostnoe predstavlenie nagruzok, deistvuiushchikh na stroitelnye konstruktsii. *Izvestiya VUZov. Stroitelstvo i arkhitektura, 4,* 12–18.
12. Pichugin, S. F., & Makhin'Ko, A. V. (2009). Calculation of the reliability of steel underground pipelines. *Strength of Materials, 41*(5), 541–547. https://doi.org/10.1007/s11223-009-9153-0.

13. Makhinko, N. (2019). Imovirnisnyi rozrakhunok koefitsiientu krytychnoho faktoru dlia tsentralno stysnutykh elementiv. *Zbirnyk naukovykh prats Ukrainskoho derzhavnoho universytetu zaliznychnoho transportu, 183,* 80–86.
14. Makhinko, N. (2019). Imovirnisne predstavlennia koefitsiienta krytychnoho faktoru v zadachakh nadiinosti budivelnykh konstruktsii. *Nauka ta budivnytstvo, 2,* 56–61.
15. Makhinko, N. (2018). Do rozrakhunku nadiinosti elementiv stalevykh yemnostei zberihannia. *Visnyk PDABA, 6*(247–248), 71–76.
16. Pichugin, S., & Makhinko, N. (2019). Otsinka nadiinosti bahatoboltovykh ziednan stinky sylosnoi yemnosti. In *Building innovations—2019* (pp. 176–178). PoltNTU.
17. Perelmuter, A. (2000). *Izbrannye problemy nadezhnosti i bezopasnosti stroitelnykh konstruktsii.* Kiev: UkrNIIproektstalkonstruktsiia.
18. Raizer, V. (1995). *Raschet i normirovanie nadezhnosti stroitelnykh konstruktsii.* Moskva: Stroiizdat.
19. Rzhanitsyn, A. (1978). *Teoriia rascheta stroitelnykh konstruktsii na nadezhnost.* Moskva: Stroiizdat.
20. Gumbel, E. (1965). *Statistika ekstremalnykh znachenii.* Moskva: Mir.
21. Pichugin, S. F. (2018). Reliability estimation of industrial building structures. *Magazine of Civil Engineering, 83*(7), 24–37. https://doi.org/10.18720/MCE.83.3.
22. Pichugin, S. (2017). Probabilistic description of ground snow loads for ukraine. In *Snow engineering 2000: Recent advances and developments* (pp. 251–256). https://doi.org/10.1201/9780203739532.

Accidents Analysis of Steel Vertical Tanks

S. F. Pichugin and L. A. Klochko

Abstract This work presents the accident statistics of steel vertical tanks with a detailed review of the identified causes and consequences of their occurrence. Information on accidents was collected on the basis of the Internet sources, literature on this issue, as well as scientific publications of the previous years. It is also considered the prerequisites for the necessity to study this issue; the focus is paid on the most large-scale accidents of the last century, which served as an impetus for the introduction of methods for predicting accidents. Accidents materials statistics of steel vertical tanks over the past years have been worked over in the paper. The need to create an algorithm for modeling the probability of an accident at a construction site of the highest level of danger, which is based on the commercial need for such calculations on the construction customer side, has been formulated. Based on collected and studied material, a chart of the accidents types' percentage at the enhanced safety facility for 2009–2019 has been compiled. As a result of the carried work, the most common accidents types in steel vertical tanks have been identified and appropriate conclusions have been made. Particular attention is paid to the analysis of the reservoir destruction from precipitation, which indicates a decrease in the cases of geometric shape defects detection and uneven draft. The reasons for referring steel vertical tanks to one of the most dangerous industrial facilities have been outlined.

Keywords Steel vertical tanks · Tanks · Construction · Refusal · Accident · High-risk object · Accident type · Accident statistics · Destruction of buildings and structures

1 Introduction

The causes of accidents at high-risk facilities have been studied for dozens of years, but the need to improve accident statistics and data processing methodology is also

S. F. Pichugin · L. A. Klochko (✉)
Poltava National Technical Yuri Kondratyuk University, Poltava, Ukraine
e-mail: lina.dmitrenko@gmail.com

S. F. Pichugin
e-mail: pichugin.sf@gmail.com

V. Onyshchenko et al. (eds.), *Proceedings of the 2nd International Conference on Building Innovations*, Lecture Notes in Civil Engineering 73,
https://doi.org/10.1007/978-3-030-42939-3_21

quite relevant today. The reason for this is a series of accidents, which resulted in dead and injured, and companies suffered from huge economic losses.

Studying the causes of accidents based on the methodology enables solving the most important practical issues of industrial safety. The identification of hazardous production factors and zones, their impact on residential buildings adjacent to enterprises contributes to the introduction of new safety technologies and the optimization of measures and means of suppressing the development and localization of accidents.

To improve state building codes, more and more attention is paid to accident statistics, the method of possible accident and progressive destruction predicting. Thus, the processing and the analysis of tank accidents also occupy an important position in this topic.

Nowadays, the accident prediction at the stage of construction site design begins to be introduced into government regulations in developing countries. This fact makes it clear to realize the need for developing an appropriate methodology in this matter. When studying the historical experience of creating algorithms of various types, it is necessary to be engaged in their improvement. The reasons here are technical and informational development of the construction industry as a whole, increasing the engineering tasks complexity, updating the architectural forms concept and buildings designations.

2 Main Body

2.1 The Analysis of Recent Research Sources and Publications

The topic of analyzing the reasons for the complete or partial destruction of reservoirs has been relevant for quite a long period, the reason for this is the great responsibility of this type of objects and their increased danger. Such scientists head the list of works devoted to this issue as Kondrasheva [6, 7], Konovalov and Mangushev [8], Khanukhov [4], Zemlyansky et al. [22], Konovalova et al. [9]. The analysis of such accidents was carried out using the information taken from technical literature, periodicals, personal experience of the authors, Internet sources, etc.

Completeness and quality of material processing in the works of such scientists as Mangushev R. A., Konovalov P. A., Tarasenko A. A., Galeev V. B., Zemlyansky A. A., Rosenstein I. M. is of particular note.

2.2 The Selection of Previously Unsolved Parts of the General Problem

In connection with the introduction of new technologies, the types of accidents and their number have been changed.

2.3 Problem Definition

The analysis of accidents with steel vertical tanks is based on the materials of recent publications, scientific works, Internet resources and mass media.

2.4 The Main Material and Results

2.4.1 Background to the Compilation of Statistics on Accidents of High-Risk Construction Objects

On Saturday July 10, 1976, the control system of a chemical reactor for the production of trichlorophenol, a component of several herbicides, was damaged, and the temperature rose beyond the limits (Fig. 1). The explosion of the reactor was avoided by the opening of safety valves, but the high temperature had caused a change in the reaction that led to a massive formation of 2,3,7,8-tetrachlorodibenzo-p-dioxin (TCDD), substance commonly known as dioxin, a high toxic compound.

Fig. 1 Seveso disaster, Italy, 1976

This event became internationally known as the Seveso disaster, because Seveso is the name of a neighboring municipality that was the most severely affected [10].

The catastrophic accident in Seveso (Italy) in 1976 led to the adoption of European Union legislation aimed at preventing accidents in certain industries with the use of hazardous substances and, thus, limiting the impact on workers, the population as a whole and the environment. The resulting standard was Directive 82/501/EEC [2], better known as Seveso I. This regulatory framework established that manufacturing company that used in its process hazardous substances (listed in Appendix A or stored hazardous substances listed in Appendix B, or both) should develop (among other documents) internal and external protection plans and emergency plans, including risk assessment.

With the introduction of Seveso I in Europe, more than 130 serious accidents have occurred, and as a result of technological advances, new risks have appeared. Therefore, the European Commission introduced Directive 96/82/EC (called the Seveso II Directive) in 1996. This directive classified risks as "minor," "low risks" and "high risks" depending on the amount of hazardous substances. Seveso II has been revised in Directive 2012/18/EU or Seveso III to increase the level of protection for people, property and the environment.

In Spain, in 2016, according to the Directorate General for Civil Defense [2], in accordance with the Seveso Directive, there were 422 high-risk facilities and 470 low-risk facilities. The geographical distribution is similar to the distribution of goods turnover: Catalonia was the first (23.9%), Andalusia with 70 (16.6%), the Valencian community—39 (9.2%) and the Basque Country (6.6%).

The chemical industry has implemented improvements in safety process and environmental protection through four strategies: safer design; risk assessment processes; use of instrumental security systems and the introduction of security management systems. In the risk assessment process, the HAZOP method is the method mostly used to identify risks [1]. The HAZOP study, developed with Imperial Chemical Industries (ICI) as a "critical study" method, was formulated in the mid-1960s [12]. A decade later, HAZOP was officially published as a disciplined procedure for identifying deviations in the manufacturing industry by Kletz in 1978 [5], as well as in some publications [17, 20], corporate reference books, standards (IEC 61882 [3]) and national guidelines (Nota Técnica Prevención (NTP) 238) was developed later.

2.4.2 Causes and Consequences of Accidents with Steel Vertical Tanks

Speaking about the reliability of steel vertical tanks (Fig. 2), it is provided by the following parameters:

- characteristics of sections of the main bearing structures, steel properties;
- quality of welded joints;
- tolerances in the manufacture and installation of structural elements.

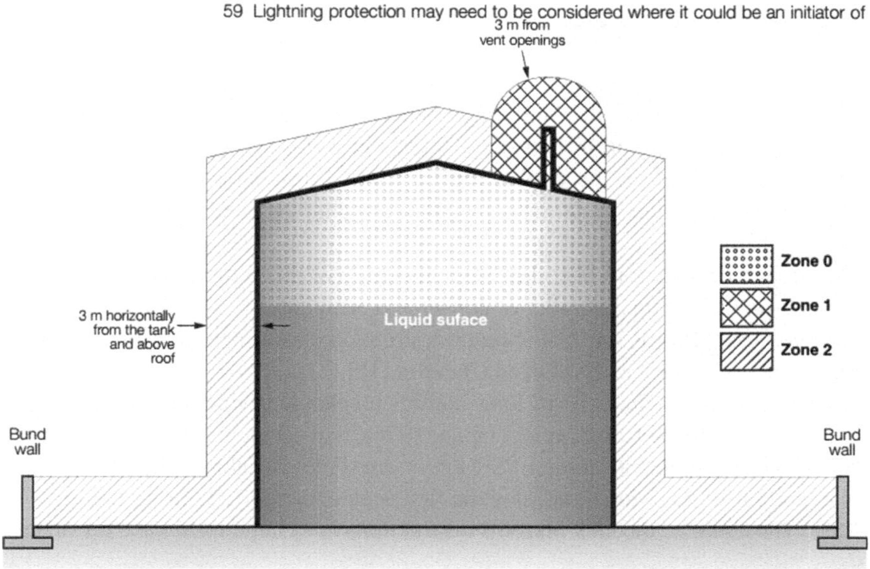

Fig. 2 Vertical storage tank—typical hazardous area classification

Meeting all standards and ensuring the reliability of the design with regulatory documents, nowadays steel vertical tanks act as one of the most dangerous industrial facilities. This is due to a number of reasons, such as

- large length of the welds of the structure, which is quite difficult to control on the full scale;
- high fire and explosion hazard produced;
- imperfect geometric shape, that occurs on the stage of reservoir hydrotesting;
- significant movement of the tank wall in the operation process as well as in the process of technological operations;
- high structural elements corrosion rate;
- low-cycle fatigue of the individual zones of the structure;
- complex nature of the load structure in the zone of the loop joint [13].

Tank accidents lead to severe material, environmental and social consequences. Among the main consequences of accidents are the following:

- full or partial destruction of the emergency tank itself, as well as other nearby tanks, buildings and structures;
- soil and water bodies pollution with oil and oil products, as well as air pollution by combustion products;
- injuries and deaths.

According to statistics in extreme cases, material damage from tank accidents is 500 times more than the initial costs of their construction [7].

2.4.3 Review of Past Research

Based on the previous perennial research, the following vertical steel tanks destruction causes have been singled out: direct and indirect. Direct causes include brittle cracks, viscous cracks, pre-rolled steel, defects in welds and uneven draft. Indirect causes, in turn, include unsuccessful rolling solutions, unsatisfactory quality of work, poor quality of materials, violation of installation technology and poor quality control of works [12].

Moreover, uneven draft becomes the main cause of this type tanks destruction. It accounts for 33% of total number of considered accidents.

Uneven base draft is one of the main destruction causes and is distinguished by such global companies as ESSO and Chevron [18].

Such accidents include 10 m long bottom breakdown with 0.15 m opening to the tank of Mitsubishi Corporation (Japan, 1974), two accidents at an oil depot near London, several accidents at the ESSO tank farm (Foley, England). It is notable that at the tank farm in the town of Foley the first accident occurred during their testing (1955). The cause of the tank bottoms destruction was a large local base subsidence [6].

In the early 1970s, three more major accidents occurred with tanks having the diameter of 53 m. Two tanks were filled with water, one—with oil. During testing, one of the damaged tanks received an average draft of 254 mm, and the peripheral draft on the bottom area in a section 2.0 m wide from the wall to the center was 150 mm, while in non-destroyed areas it was 40–50 mm. A detailed examination of tank accidents acts investigation over the past 30 years shows that in 38 cases out of 44 there was an uneven draft of the foundation, which, in combination with other factors, caused the destruction. It is difficult to judge the quantitative ratio of the influence of draft and other factors, because there is no true picture of the draft of these tanks. However, a number of cases are known when only draft was the cause of the damage. In one case, it was different in draft size of the tank body and process pipelines, which resulted in the latter breaking off from the wall and subsequent rupture of the latter; in the other—an uneven draft, which reached 320 mm, which led to the rupture of the wall and complete destruction of the reservoir in the third—an uneven draft, which led to the rupture of the bottom. It should be noted that accidents are usually caused by a complex of reasons, one of which is the uneven draft of certain base areas [11] (Fig. 3).

Every year, the accidents number on expired tanks increases because a large percentage of tanks have already developed their design resources. The wear of the operated vertical steel tanks (RVS) is 60–80%. Based on a survey by TSNIIPSK [19, 14], it has been established that the total accidents number is 3–5 times more than that recorded. The intensity level of emergencies remains quite high amounting over the last 30 years, to about 0.0003 tank destruction per year [15, 16].

The analysis of the destruction risk dynamics [21] showed that the actual accidents risk is two orders of magnitude higher than the standard value and is 1.6 10^{-3}. The accidents danger is estimated by the amount of damage, which depends on how the accident manifests itself: in the form of explosions and fires from the spilled

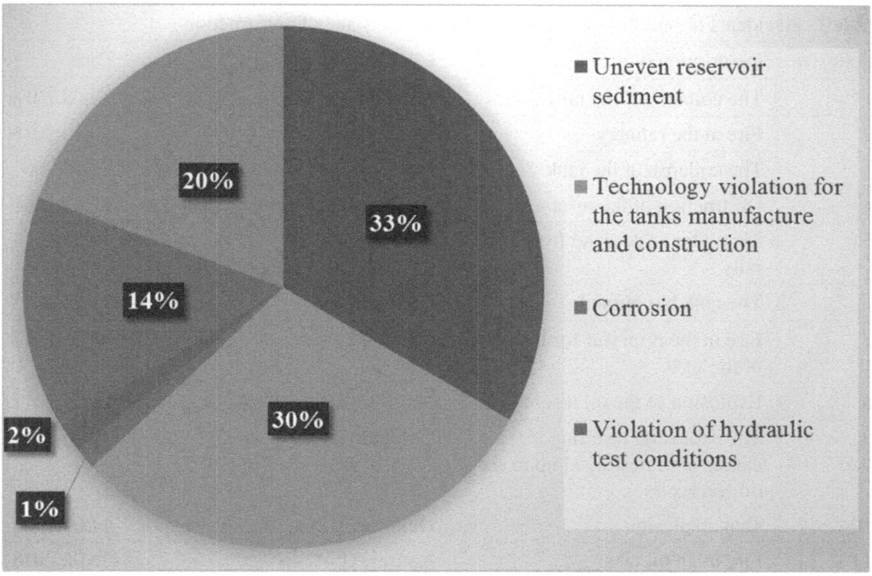

Fig. 3 Main destruction reasons VST

product or in the form of fragile damage or local failure of tanks. As practice shows, RVS accidents in most cases are accompanied by a significant loss of products, poisoning of the area and the deaths [21]. In extreme cases, according to statistics, the total material damage exceeds by 500 times or more the initial costs for reservoirs construction [11].

2.4.4 Statistics of Accidents on Steel Vertical Tanks

Based on the conducted research, Table 1 of steel vertical tanks accidents statistics has been compiled. The data were retrieved using the Internet sources, research papers and other media resources.

According to Table 1, a percentage rate of tanks destruction by accident type from 2009 until 2019 has been made up. The analysis shows that fire occurrence or a sudden explosion (75%) accounts for the highest failure percentage of the normal operation of the structure. Such accidents include the ignition of the internal vapors of reservoir, the explosion, the reservoir ignition from neighboring objects enveloped in a fire, the ignition of the gas–air mixture during dry cleaning, etc. (Fig. 4).

The tanks collapse covers 14% of the total accidents number, indicating an improvement in the methods of this type construction. Only 11% is accounted for by accidents, caused by faulty tanks, their depressurization and subsequent inability to operate as intended (Fig. 5).

Table 1 Accidents of buildings and constructions at the stage of construction

Font size	Description of accident	City, country	Date
1	The collapse of the tank	Muravlenko, Russia	17.08.2018
2	Fire at the refinery	Ugra, Russia	30.10.2018
3	The collapse of the tank	India	28.08.2006
4	Destruction of the reservoir	Chesapeake, Virginia, USA	12.11.2008
5	Six oil tanks descend from the rails	California, USA	09.05.2014
6	The tank has burned	Colorado, USA	17.04.2015
7	Fire in the reservoir for bursting of the well	Colorado, USA	08.05.2017
8	Explosion of the oil reservoir	North Colorado, USA	25.05.2017
9	The explosion resulting in damaged or destroyed up to six oil reservoirs	Colorado, USA	19.06.2018
10	Tank explosion	Delaware, USA	17.07.2001
11	Fire in an oil tank	Kansas, USA	26.06.2018
12	Tank burst	Mississippi, USA	31.10.2009
13	The explosion, which led to 36 oil reservoirs	New Mexico, USA	11.07.2016
14	Several oil reservoirs burned	South dakota	18.08.2018
15	Tank explosion	Texas, USA	17.10.2017
16	Fire in an oil reservoir in Cherokee	Texas, USA	27.12.2017
17	Fire in an oil tank in Madison County	Texas, USA	03.01.2018
18	Fire in a crude oil storage tank across the US pipeline east of Wichita Falls	Texas, USA	28.08.2018
19	Several oil storage tanks were involved in a fire in Campbell's northern constituency	Wyoming, USA	16.03.2018
20	Due to a malfunction of the reservoir at the winemaking plant, 30,000 L of prosecco spilled	Veneto, Italy	28.09.2018
21	Overpressure of one of the reservoirs	Bila Tserkva, Ukraine	28.08.2017
22	Sulfuric acid spilled out of the reservoir	Oberhausen, Germany	16.02.2017
23	The tank was lit when cleaning	Samara, Russia	31.08.2018

(continued)

Table 1 (continued)

Font size	Description of accident	City, country	Date
24	During the drain of fuel oil into an underground reservoir an explosion with destruction of the capacity occurred	Russia	26.12.2010
25	The "dead" residue of oil was burning	Russia	24.09.2010
26	A reservoir burned during welding	Tamur district, Dagestan, Russia	23.04.2010
27	Tank explosion	Russia	28.03.2010
28	In the whitewash shop for cellulose production, a dust–gas–air mixture in the hydrochloric acid reservoir exploded	Irkutsk region, Russia	03.09.2009

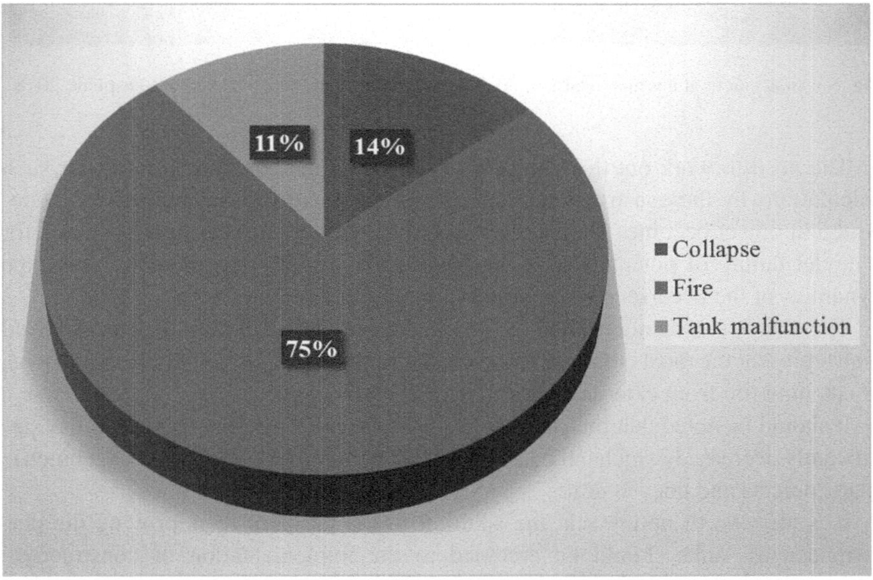

Fig. 4 Percentage ratio of the accident type at the high security facility since 2009

3 Conclusion

The prerequisites for the implementation of the methodology and algorithm model for the probability of accident at a high-risk construction site gave impetus to the development of such direction in scientific activity of construction industry as predicting the progressive destruction of structures.

Fig. 5 An accident at a winery in Italy, 30 thousand liters of prosecco leaked onto asphalt, 2018

The need to work out this algorithm is based on the commercial need for such calculations by the construction customer. The demand for this analysis of a construction object becomes the basis for introducing clear construction standards for a model failure of potential accident, based on statistics from previous years, the dynamics of the occurrence of certain types of accidents.

The analysis of accidents and causes of accidents of steel vertical tanks showed that at present the most common reservoir destructions are: fires, namely explosions, or catching fire from external objects covered by the flame.

It should be noted that the percentage of reservoir destruction from draft has significantly decreased, which indicates a decrease in the cases of detection of geometric shape defects and uneven draft.

It is also worth noting that the solution to the problem of improving the performance of tanks should be confined to the implementation of constructive-technological, operational and organizational measures. Constructive-technological measures to improve the resource of the safe operation of tanks are performed at the stages of designing, manufacturing and installation. One of the most important conditions for ensuring high reliability and safety of tanks is the use of fine-grained steels with high resistance to brittle fracture in their manufacture.

References

1. Dirección General de Protección Civil (DGPC). (2017). *¿Qué Hacemos?/Riesgos: Prevención y Planificación/Tecnológicos/Químicos/Distribución*. Accessed January 17, 2017, www.proteccioncivil.es/riesgos/quimicos/distribucion.
2. European Union Directive 82/501/EEC of the Council of 24 June 1982 on the major accident hazards of certain industrial activities. (1982). *The Official Journal of the European Union, 1*, 1–18.
3. International Electrotechnical Commission (IEC). (2016). IEC 61882:2001. *Hazard and operability studies (HAZOP studies)—Application guide*. Geneva, Switzerland: IEC.
4. Khanukhov, K. M., & Alipov, A. V. (2011). Regulatory, technical and organizational support for the safe operation of reservoir structures. *Electronic Journal "Prevention of Accidents of Buildings and Structures"*.
5. Kletz, T. A. (1999). *Identifying and assessing process industry hazards* (4th ed.). Rugby, UK: IChemE.
6. Kondrasheva, O. G., & Nazarova, M. N. (2004). Causal analysis of accidents of vertical steel tanks. *Oil and Gas Business, 2*, 36–43.
7. Kondrashova, O. G., & Nazarova, M. N. (2014). Causal analysis of vertical steel tanks accidents. *Electronic Scientific Journal Oil and Gas Business, 2*.
8. Konovalov, P. A., Mangushev, R. A., Sotnikov, S. N., Zemlyansky, A. A., & Tarasenko, A. A. (2009). *Foundations of steel tanks and deformations of their bases* (p. 336). Publishing House Association of Construction Universities.
9. Konovalova, O. P. (2002). *Accounting for consolidation in improving the operational reliability of steel tanks of large capacity* (p. 176). Tyumen.
10. Major Industrial Accidents. (2018). *Tosco refinery fire flawed management supervision*. Accessed January 17, 2018, http://accidentsoilandgas.blogspot.com/2013/01/seveso-disaster-dioxin-crisis-icmesa.html.
11. Michele, B. G., & Bartoli, M. O. (2010). Evaluation study on structural fault of a Renaissance Italian palace. *Engineering Structures, 7*(32), 1801–1813.
12. National Institute of Health and Safety at Work (NIHSW). (2015). *Papers prevention. N° 238: HAZOP at processing facilities*. Accessed July 13, 2015, http://www.insht.es/InshtWeb/Contenidos/Documentacion/FichasTecnicas/NTP/Ficheros/201a300/ntp_238.pdf.
13. Perel'muter, A. V. (2000). *Selected problems of reliability and safety of building structures* (p. 182). V. Shimanovsky UkrRDISteelconstruction.
14. Pichugin, S. F. (2018). Reliability estimation of industrial building structures. *Magazine of Civil Engineering, 83*(7), 24–37. https://doi.org/10.18720/MCE.83.3.
15. Pichugin, S. F., & Makhin'Ko, A. V. (2009). Calculation of the reliability of steel underground pipelines. *Strength of Materials, 41*(5), 541–547. https://doi.org/10.1007/s11223-009-9153-0.
16. Pichugin, S. (2017). Probabilistic description of ground snow loads for ukraine. In *Snow engineering 2000: Recent advances and developments* (pp. 251–256). https://doi.org/10.1201/9780203739532.
17. Planas, E., Arnaldos, J., Darbra, R. M., Muñoz, M., Pastor, E., & Vílchez, J. A. (2014). Historical evolution of process safety and major-accident hazards prevention in Spain. Contribution of the pioneer Joaquim Casal. *Journal of Loss Prevention in the Process Industries, 28*, 109–117. https://doi.org/10.1016/j.jlp.2013.04.005.
18. Salnikov, A. P. (2016). *Assessment of the stress-strain state of reservoirs according to the results of ground-based laser scanning* (p. 167).
19. Solomon, E., Enebe, E., & Onoh, F. E. (2016). Accidents in building construction sites in Nigeria; A case of Enugu State. *International Journal of Innovative Research and Development, 5*(4), 244–248. ISSN 2278–0211.
20. Tiago, P., & Júlio, E. (2010). Case study: Damage of an RC building after a landslide—Inspection, analysis and retrofitting. *Engineering Structures, 7*(32), 1814–1820.

21. Williams, S. O., Hamid, A. R., Misnan, M. S., Abimaje, J., Seghier, E. T., & Aminu, Y. D. (2017). Review of building construction accidents: Concept, cases, causes, consequences and control measures. In *Conference Proceeding of the 3rd International Conference on Sciences, Engineering and the Social Sciences (ICSESS 2017). Promoting Innovative Multidisciplinary Research for Sustainable Development*, May 17–18.
22. Zemlyansky, A. A. (2006). *Design principles and experimental-theoretical studies of large tanks* (p. 417). Balakovo.

Experimental Investigation of Masonry and Reinforced Masonry Walls Under Local Loading

Nataliia Pinchuk and Volodymyr Byba

Abstract In modern construction, the brick is quite widely used as a material for wall structures. The masonry buildings are durable and have beautiful architectural forms and high heat transfer resistance. The practice of masonry buildings construction dates has several thousand years, and some of the architectural monuments have survived to these days. But construction experience shows that the thickness of the walls in these buildings is considerable—sometimes up to 1 m. Such material overruns are unjustified. One of the reserves for reducing material consumption, energy consumption, labor costs, and cost of the above elements is to improve the methods of their calculation and construction. The results of experimental investigations of masonry and reinforced masonry walls under one-sided central and edge loading are presented. According to the experimental data, the strength of masonry and reinforced masonry structures under local loading is influenced by: case (place) of loading application; geometric sizes of elements and loading platforms; strength characteristics of masonry; the number and nature of the reinforcement fabrics.

Keywords Masonry walls · Reinforced masonry walls · Local loading · Experimental investigations

1 Introduction

Local compression is one of the most common types of masonry work. It occurs in masonry under the influence of loading on a limited area. The ends of runs, beams, trusses, slabs of overlapping, crane beams are almost always supported not along the section, but on a part of wall or a pillar. Local compression also occurs near the upper cross-sections of the foundations, as they are usually larger than the lower cross-sections of walls and columns supporting the foundations.

Professor Onishchik [1] conducted experimental studies of masonry under local compression. He considered the case of central unilateral local compression with different platform sizes.

N. Pinchuk (✉) · V. Byba
Poltava National Technical Yuri Kondratyuk University, Poltava, Ukraine
e-mail: natali.pinchuk.pntu@gmail.com

V. Onyshchenko et al. (eds.), *Proceedings of the 2nd International Conference on Building Innovations*, Lecture Notes in Civil Engineering 73,
https://doi.org/10.1007/978-3-030-42939-3_22

Sementsov [2] tested specimens of solid and hollow bricks for cornering. It was found that when the surface area of the sample to the area of less than two fracture occurred by the cut. With a ratio of more than two, the nature of the fracture varied dramatically; at loads that were 60–70% of the boundaries, vertical cracks developed, and by the time of fracture, the column with the dimensions of the die section was completely separated from the masonry array. This column collapsed as it did with axial compression.

Shapoval [3] investigated the work of brick masonry with local compression in the case of loading in the entire thickness of the wall in the middle and at the edge of the specimen. In the case of central local compression, depending on the size of the ratio of the length of the loading pad l_{loc} to the length of the sample L, there were two characteristic types of destruction. With smaller ratios, the specified fracture parameter occurred by the formation of a wedge from which a vertical crack departed, and the specimen was splitting almost in half. At larger ratios, an asymmetrical fracture occurred; an asymmetrical wedge was formed, from which the slopes cracked. In the case of angular crushing, there were also two forms of destruction: chamfering of the corner of the element and crushing of the column with the cross-sectional dimensions of the die separated from the sample (as in [2]).

2 Research Methodology

According to the norms [4], the determining factors of the strength of masonry in local compression are the calculated resistance of the masonry to compression and the ratio of the so-called "estimated" area to the area of local compression. The formula of norms is approximate, and it does not allow to cover all cases occurring in the practice of construction and in some cases gives a significant difference with the data of the experiment [2, 3]. In addition, the reinforced brick elements do not take into account the different nature of the work of the reinforcement in the areas of separation and shear, and in addition to the tension, there is a bend, which the norms neglect.

The purpose of the experimental investigations was to study the behavior of the specimens during loading [5], deformed state of the mesh reinforcement (for reinforced brick walls), to determine the shape of the fracture and the boundary load depending on the ratio $\alpha = l_{loc}/h$, and the amount and nature of the reinforcement.

To achieve this goal, two series of samples were made: the first consisted of six walls with dimensions $h \times b \times l = 1050 \times 250 \times 1050$ mm; three of which were brick, and three reinforced wire mesh Ø 4 mm Bp-I and cell size 50 mm. Grids were laid in three rows in height of masonry. Each pair of specimens was tested for central unilateral erasure by a stamp of the appropriate size: $l_{loc1} = 100$ mm, $l_{loc2} = 150$ mm, $l_{loc3} = 250$ mm, in accordance $\alpha_1 = 0.01$, $\alpha_2 = 0.14$, $\alpha_3 = 0.24$.

Within the second series, the dimensions of one brick and eight reinforced brick samples ($h \times b \times l = 630 \times 250 \times 630$ mm) and loading platforms ($l_{loc} = 150$ mm) were constant, with the diameter of the mesh reinforcement (Ø 3 and Ø 5 mm Bp-I) and the nature of their placement varied: during the specimen height (1, 2, 3 grids in each row directly in the zone of confusion [Fig. 1a] or evenly through three rows of masonry height of the specimen [Fig. 1b]).

After testing the brick and reinforced brick specimens for one-sided local loading, it was planned to test them with the load applied to the edge of the specimen.

The surface of the specimens was plastered to better observe their loading behavior. To measure deformations of the longitudinal mesh reinforcement, 5 mm base strain gauges were placed on the rod, which were placed in pairs on two diametric sides of the reinforcing rod (Fig. 1). The readings on the strain gauges were taken by an electronic strain gauge of the AVD-4 type. The load was applied in steps of one-tenth of the expected destructive, with a holding time of 5–6 min at each stage. At this time, the sample was reviewed, and the appearance and development of cracks were recorded.

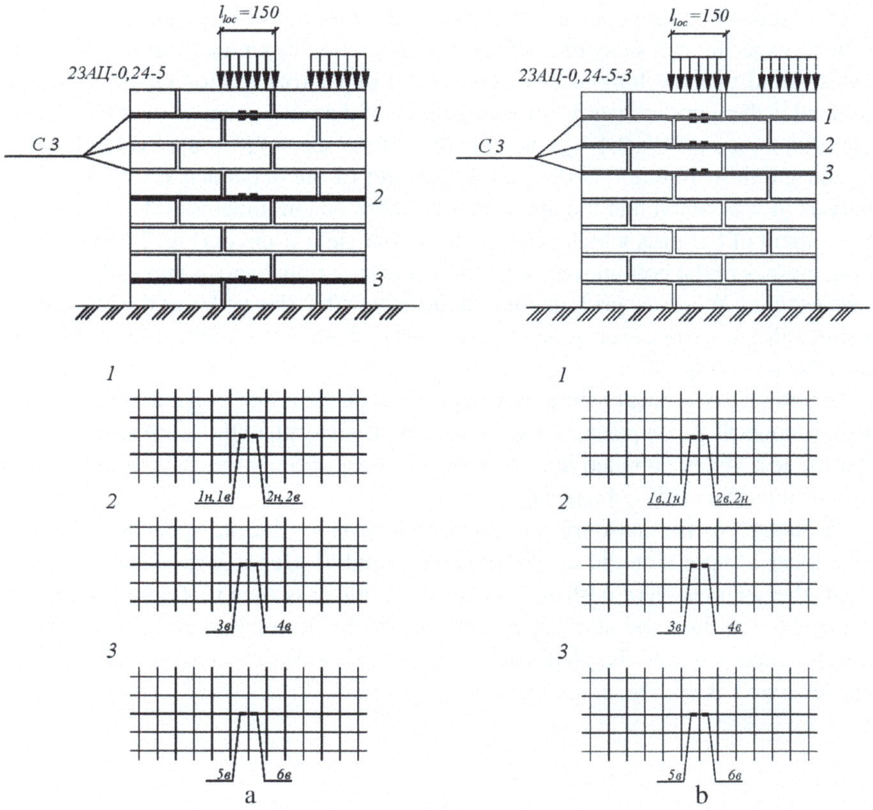

Fig. 1 Scheme of strain sensors placement on the longitudinal reinforcement of the sample meshes: **a** 23АЦ-0.24-5, **b** 23АЦ-0.24-5-3

3 Research Results

At one-sided central local compression, a wedge of seal was formed in all samples under the loading platform, and its dimensions were determined by the dimensions of the loading platform. Vertical splitting cracks began to spread from the top of the wedge. In the specimens with the loading site $l_{\mathrm{loc1}} = 100$ mm, at the last moment, an inclined crack was formed and immediately spread, which connected the top of the wedge with the lower corner of the specimen. In our opinion, the asymmetric nature of the destruction of the sample during symmetrical loading can only be explained by the uneven loading.

Reinforcement did not fundamentally change the nature of the destruction of the walls, but only increased the ultimate load.

The nature of the destruction of the brick and reinforced brick samples is presented in Figs. 4 and 5, respectively.

The data on the limit load and characteristics of the samples are given in Table 1.

In the experimental study of reinforced brick walls with central local compression, it was found that the longitudinal rods of the upper row reinforcement mesh are included in the work with uniform arrangement of the reinforcement over the sample height (at a load of 0.3–0.4 N_{u}) then the bars of the second row at a load of 0.4–0.5 N_{u} and the third row at 0.6–0.7 N_{u}. As a result of the dependence «$N/N_{\mathrm{u}} - \varepsilon_{\mathrm{s}}$» analysis, it was found that the upper mesh reinforcement undergoes deformation at the moment of fracture, which corresponds to the yield strength (Fig. 2). Rebar rods in the meshes of the bottom row at the time of destruction do not completely exhaust their resource. When the grids are placed directly within the wedge of the seal during loading, all grids work evenly, and the stresses in them at the moment of destruction reach the limit (Fig. 3).

According to the data of strain sensors located on the reinforcement of the meshes on two diametrically opposite sides of the rod and as a result of the samples inspection after the test, it can be noted that the reinforcement of the upper nets of the sample has a tilting effect (Figs. 4 and 5).

The nature of the masonry and reinforced masonry failure specimens in one-sided local compression did not differ (Figs. 6 and 7). Depending on the ratio $\alpha = l_{\mathrm{loc}}/h$, the samples were destroyed differently: at α less than 0.16, the destruction occurred by cutting, and at α larger ones during the destruction process, a column with the dimensions of the platform was separated from the sample surface, which was destroyed as with axial compression.

Table 1 Test results of masonry and reinforced masonry walls

Sample code	Sample sizes, m		d, mm	l_{loc}	α	μ, %	N_u^{test} кN	f_{loc}^{test} MPa
	Length, l	Height, h						
1 set (breadth, b = 0.25м, mark of mortar M50 mark of brick M75)								
13ЦЦ-0.01	1.05	1.17	–	0.10	0.10	–	248	9.9
13ЦЦ-0.14	1.04	1.16	–	0.15	0.14	–	286	7.5
13ЦЦ-0.24	1.03	1.15	–	0.25	0.24	–	600	9.6
13АЦ-0.01	1.04	1.19	4	0.10	0.10	0.2	478	19.1
13АЦ-0.14	1.04	1.18	4	0.15	0.14	0.2	556	14.6
13АЦ-0.24	1.05	1.18	4	0.25	0.24	0.2	1000	15.9
13ЦК-0.01	1.05	1.17	–	0.10	0.10	–	201	8.0
13ЦК-0.14	1.04	1.16	–	0.15	0.14	–	284	7.5
13ЦК-0.24	1.03	1.15	–	0.25	0.24	–	316	5.0
13АК-0.01	1.04	1.19	4	0.10	0.10	0.2	232	9.3
13АК-0.14	1.04	1.18	4	0.15	0.14	0.2	488	12.8
2 set (breadth, b = 0.25м, mark of mortar M100 mark of brick M100)								
23ЦЦ-0.24	0.64	0.62	–	0.15	0.24	–	346	9.1
23АЦ-0.24-3	0.63	0.64	3	0.15	0.24	0.1	390	10.3
23АЦ-0.24-3-1	0.63	0.62	3	0.15	0.24	0.2	438	11.5
23АЦ-0.24-3-2	0.63	0.63	3	0.15	0.24	0.4	500	13.2
23АЦ-0.24-3-3	0.63	0.64	3	0.15	0.24	0.1	598	15.7
23АЦ-0.24-5	0.63	0.64	5	0.15	0.24	0.3	667	17.6
23АЦ-0.24-5-1	0.63	0.62	5	0.15	0.24	0.5	390	10.3
23АЦ-0.24-5-2	0.63	0.62	5	0.15	0.24	1.0	644	16.9
23АЦ-0.24-5-3	0.63	0.64	5	0.15	0.24	0.3	718	18.9
23ЦК-0.16	0.64	0.62	–	0.10	0.16	–	132	5.3
23ЦК-0.24	0.64	0.62	–	0.15	0.24	–	170	4.5
23АК-0.24-3	0.63	0.64	3	0.15	0.24	0.1	262	6.9
23АК-0.24-3-1	0.63	0.62	3	0.15	0.24	0.2	212	5.6
23АК-0.24-3-2	0.63	0.63	3	0.15	0.24	0.4	274	7.2
23АК-0.24-3-3	0.63	0.64	3	0.15	0.24	0.1	284	7.5
23АК-0.24-5	0.63	0.64	5	0.15	0.24	0.3	268	7.1
23АК-0.24-5-2	0.63	0.62	5	0.15	0.24	1.0	250	6.6
23АК-0.24-5-3	0.63	0.64	5	0.15	0.24	0.3	394	10.4

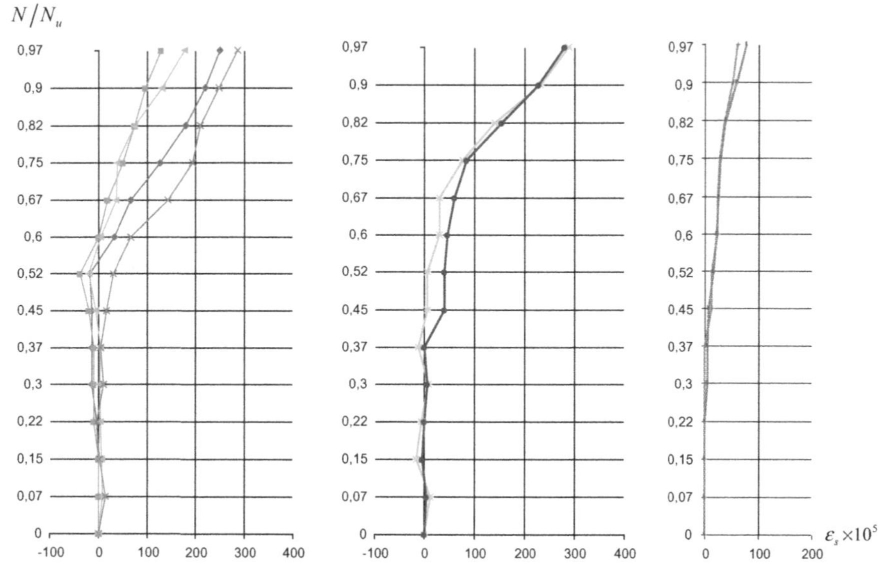

Fig. 2 Deformation diagrams in the longitudinal bars of the sample meshes: 23АП-0.24-5
1в 1н 2в 2н 3в 4в 5в 6в

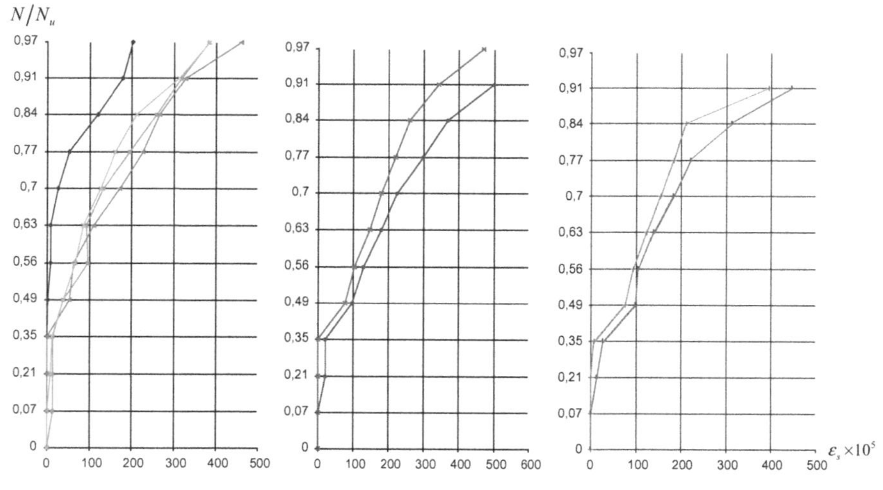

Fig. 3 Deformation diagrams in the longitudinal bars of the sample meshes: 23АП-0.24-5-3
1в 1н 2в 2н 3в 4в 5в 6в

4 Conclusions

The nature of the destruction of brick and reinforced brick walls in local compression does not differ fundamentally. Under central loading, a wedge is formed under the

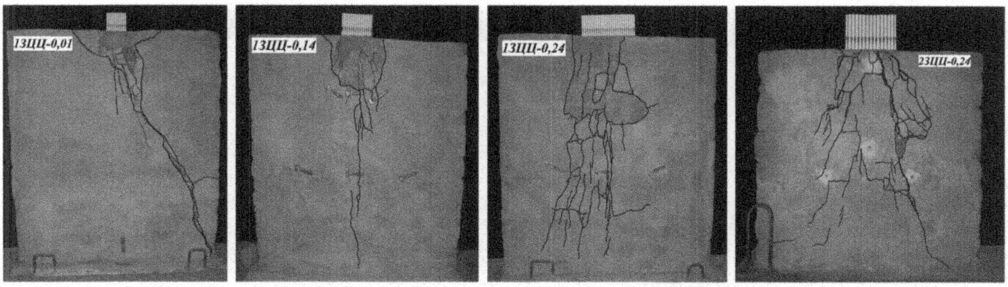

Fig. 4 Nature of the masonry specimens destruction under central local compression

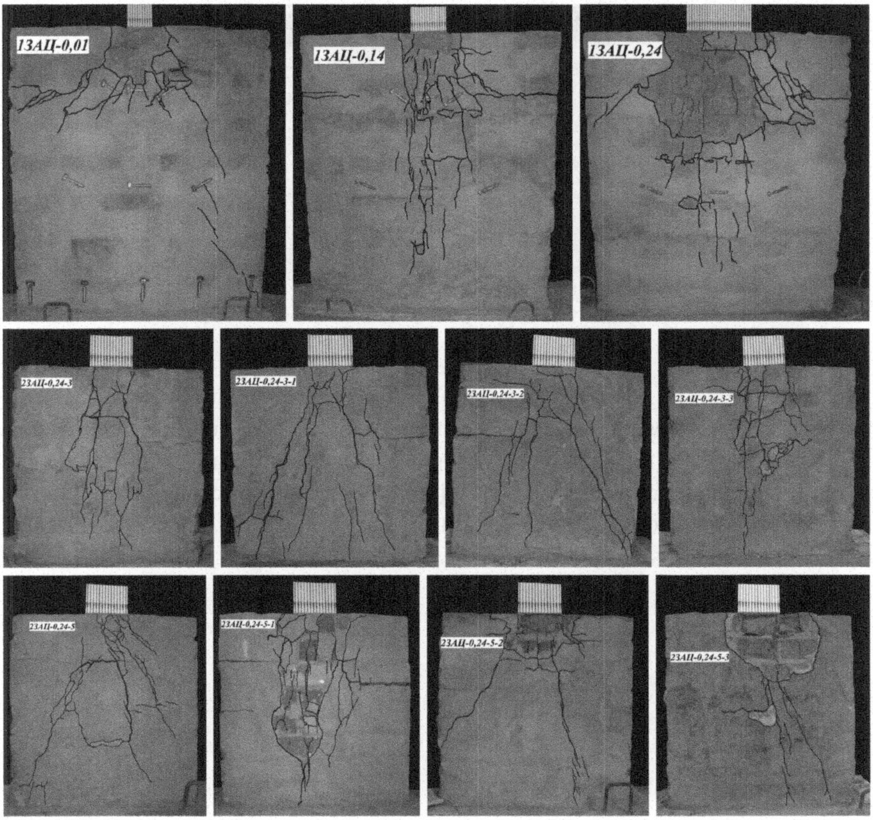

Fig. 5 Nature of the reinforced masonry specimens destruction under central local compression

surface of the platform, from the top of which the splitting cracks begins. As the ratio $\alpha = l_{\text{loc}}/h$ increases, the bearing capacity of the element decreases and the size of the compression zone under the loading platform (wedge) increases. In one-sided local compression on the edge of the wall, there are two types of destruction depending on $\alpha = l_{\text{loc}}/h$: with smaller α failure occurs by cutting, with larger ones due to

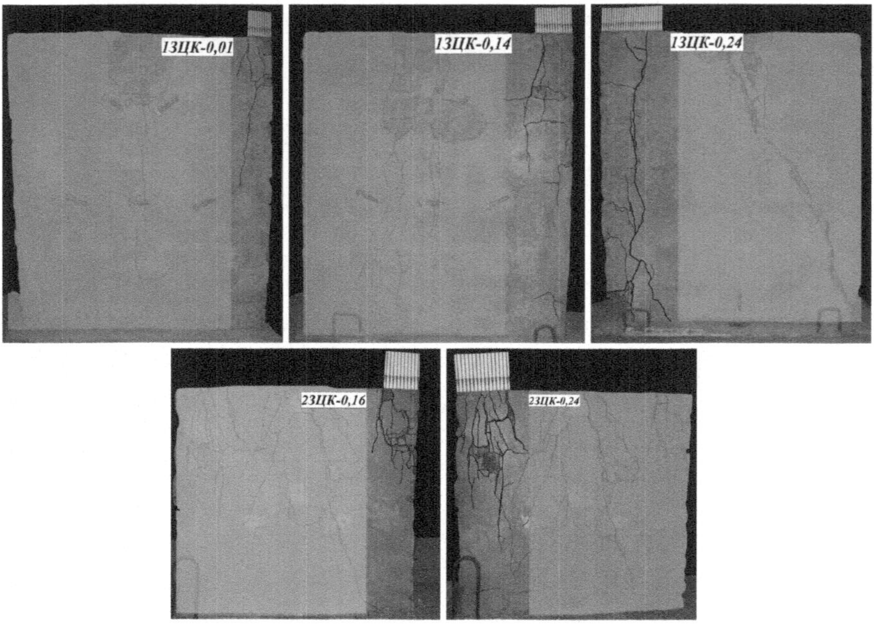

Fig. 6 Nature of the masonry specimens destruction under edge local compression

the destruction of the column below the platform surface. Reinforcement of a brick wall increases its bearing capacity. The strength under local compression depends on the masonry and reinforcement strength characteristics, the parameter α, and the reinforcement amount and location. The analysis of reinforcement deformations allows to confirm about the dowel effect arising in the zone of the wedge in the reinforcement meshes.

Fig. 7 Nature of the reinforced masonry specimens destruction under edge local compression

References

1. Onishchik, L. I. (1939). *Stone structures of industrial and civil buildings*. Moscow: Gosstroyizdat.
2. Sementsov, S. A. (1959). Local and extra-center compression of concrete and masonry. *Structural Mechanics and Calculation of Structures, 1*, 11–19.
3. Shapoval, S. L. (2005). *Stress-strain state and strength of brickwork at local compression* (Dis. … Cand. Tech. Sciences: 05.23.01). Poltava National Technical Yuri Kondratyuk University, PNTU, Poltava.
4. EN 1996-1-1. (2005). *Eurocode 6: Design of masonry structures—Part 1-1: General rules for reinforced and unreinforced masonry structures*.
5. Pinchuk, N., Fenko, O., & Kyrychenko, V. (2018). The stress-strain state computer modelling of reinforced masonry under local loading. *International Journal of Engineering & Technology, 7*(3.2), 316–321. https://doi.org/10.14419/ijet.v7i3.2.14427.

Research of Acoustic Properties of Modern Building Structures

P. Sankov, Y. Zakharov, V. Zakharov, and B. Hvadzhaia

Abstract The use of acoustic methods of noise reduction in residential and non-residential areas of modern cities is a time-consuming task, which is not solved sometimes at the proper quality level. The complexity and importance of this problem lies in the quantitative determination of the magnitude of the noise emission from the sources of noise in the path of its propagation. Practical sound insulation issues are often misinterpreted without their scientifically sound analysis and practical confirmation. The article deals with the issues of elimination of errors arising from the application of one of the most common architectural and design methods of protection against noise—modern metal plastic window fillings. The complex approach to solving the "acoustic purity" of these structures is offered at the stage of their application and operation. The analysis of evaluation of acoustic properties of window fillings, the essence of which is to consider some typical errors in solving the issues of protection against external people noise, can lead to a sharp decrease in the quality and safety of modern person's life in the premises during work or rest. It is established that a qualitative solution to the problem of security for human habitation places requires the obligatory acoustic certification of building materials, buildings, and structures made of them.

Keywords Double-glazed windows · Acoustic efficiency · Sound insulation

1 Introduction

1.1 Analysis of Recent Studies on the Effects of Environmental Factors on Humans

Nowadays, there are many dangerous factors of environmental impact on humans. One of the most common and dangerous factors that accompanies a person everywhere is noise. It is difficult to name now the places of rest, production, and everyday

P. Sankov (✉) · Y. Zakharov · V. Zakharov · B. Hvadzhaia
Prydniprovska State Academy of Civil Engineering and Architecture, Dnipro, Ukraine
e-mail: petrsankov5581@gmail.com

V. Onyshchenko et al. (eds.), *Proceedings of the 2nd International Conference on Building Innovations*, Lecture Notes in Civil Engineering 73,
https://doi.org/10.1007/978-3-030-42939-3_23

life, where no noise would be present in the sound spectrum that is an annoying mix of sounds, which prevents us to have a good rest, with maximum efficiency to work, and to feel calm in the home environment. High power sounds and noises affect the hearing aid and nerve centers and can cause pain and shock [1–3]. The article [4] deals with the physiological and psychological effects of noise on the human body and its impact on the quality of life. The authors of the article [5] consider the impact of noise on quality of life in several ways:

1. as the absence of an effective national noise policy;
2. changes in noise impact on the US population;
3. changes in the population in one or other places as they wish;
4. changes in the number of noise sources, their location, and acoustic power;
5. inadequate knowledge of the effect of noise on the human body.

The authors state [6] that the environmental impact, including traffic-related noise, can be attributed to risk factors for type 2 diabetes. In the article [7], it was found that the quality of the workplace in the office and their satisfaction of employees are influenced by the interior design, the level of openness, the number and variety of jobs, and accessibility to the building. The secondary factors of employee satisfaction are the influence of the physical characteristics of the work environment.

1.2 Analysis of Recent Studies on the Effects of Environmental Factors on Humans

Issues of preservation and improvement of living conditions and health of people have been and remain one of the most important directions of socio-economic policy of any state and require finding their effective solution, both at the state level and at the level of branches, regions, economic entities. Literature analysis has confirmed that a lot of attention is paid to human health in the world. One of the reasons for the deterioration of human health is found to be noise from any source in any city. Scientists are considering different ways to reduce the impact of noise on human habitation. However, the direction of qualitative consideration of the passage of noise through the fence (walls, windows, etc.) remains open. Therefore, in this study, we will try to expose this topic qualitatively.

The use of acoustic methods of noise reduction in residential and non-residential areas of modern cities is a time-consuming task, which is not solved at the proper quality level. The complexity and importance of this problem lie in the quantitative determination of the magnitude of the noise emission from the sources of noise in the path of its propagation. Practical sound insulation issues are often misinterpreted without their scientific sound analysis and practical confirmation.

At the Prydniprovska State Academy of Civil Engineering and Architecture (PSACEA), work is performed to protect places of residence from the noise [8–10]. The reverberation chamber of the Academy tests the acoustic properties of any building materials [11]. We present the first part of an analysis of typical errors in solving problems of applied acoustics in the field of noise protection. The article is constructed in the form of a list of common mistakes and their scientific sound exposition.

2 Purpose

We can reach the compliance with standards and improvement of acoustic conditions of people's stay in residential, public, and industrial buildings in urban development by setting the acoustic efficiency of window translucent fillings and modern wall materials by theoretical means and methods of in situ observations by predicting noise pollution design of buildings and structures and their operation.

To achieve this goal, we must solve the following tasks:

- to conduct research on the organization of noise protection in premises of residential, public, and industrial buildings from external or internal sources of noise;
- to perform theoretical and practical studies of the acoustic properties of these materials in the field conditions and in the conditions of the PSACEA reverberation chamber;
- to undertake research into the effectiveness of the implementation of protective measures aimed at reducing noise pollution in the middle of premises, such as the outpatient clinic in the city of Kamianske, which is being built in the area of noise from the railway and highway.

3 Methodology

In this work, the acoustic properties of the enclosure structure made of layers of deciduous material (USB sheets, drywall, and glass) are analyzed.

The studies were conducted in a certified PSACEA reverberation chamber according to DSTU B B.2.6-86: 2009.

3.1 Acoustic Properties of Enclosing Structures

To obtain the acoustic properties of the enclosing structures, the chamber recorded the sound pressure levels at each high and low levels in the tert-octave bands with geometric frequencies in the range 100–3150 Hz (6 and 9 measurement points, respectively). The airborne noise insulation index R_w of the enclosure with the known measured airborne noise insulation R is determined by comparing this frequency characteristic with the standard airborne noise insulation frequency rating R_N. The magnitude of the R_w index is the numerical value of the ordinate displaced downward or upward of the standard estimation characteristic at the geometric mean frequency of 500 Hz.

3.2 Practical Mitigation of Major Errors in Noise Shielding by Enclosing Structures

The basic errors are corrected by careful analysis of the interpretation of the error itself and practical justification of the state of the actual noise protection capabilities of a particular noise protection measure.

4 Results

4.1 Discussion of Error No 1

Error No 1 The physical properties of the material sound insulation and sound absorption are of the same concept.

New information: sound absorption decreases in the energy of the sound wave reflected from the enclosing structure. Such structures include walls, partitions, floors, and ceilings. The process is carried out by dissipating energy, excitation of vibrations, and the transition of sound energy into thermal energy. By placing noise sources in the middle of the room with the help of sound absorbing materials, it is possible to reduce actually the noise level in the room from 3 to 8 dBA.

Soundproofing reduces the sound level when sound passes through a fence. Three variants of sound transmission are considered: (1) from one room to another, (2) from the room to the outside, and (3) from the outside into the room. The sound insulation performance is evaluated by the airborne noise isolation index R_w (averaged over the range of most typical auditory frequencies 100–3150 Hz), and the inter-story floorings are also the index of the reduced noise level under the floor L_{nw}. The more R_w and less L_{nw} is the higher the sound insulation is. Both values are measured in dB (decibels).

Practical recommendations: (1) More powerful and thicker enclosures should be used to increase sound insulation. Decorating of the room with sound absorbing materials alone is inefficient and does not significantly increase the sound insulation between rooms. (2) To eliminate the word ISOVER on their banners by phrase: "… effective heat and sound insulation" is the word SOUND.

4.2 Discussion of Error No 2

Error No 2 The greater the value of the noise isolation index R_w is, the higher the soundproofing of the fence is. Basics for it is the airborne noise insulation index R_w constructions. The enclosure is a characteristic that is only applicable to the frequency range 100–3150 Hz and is designed to estimate noise of household origin (spoken language, radio, television). Therefore, it is considered that the greater the value R_w is, he higher the isolation from sounds of this type is. And in practice, we have to take into account not only sounds of this type, but more often, sources of external harmful noise are transport and industrial noise sources with new technological equipment.

There may be a situation where a lightweight partition wall made of plasterboard has an index R_w higher than that of a brick wall of similar thickness. In such case, the partition wall is much better at isolating the sounds of voice, TV, phone or alarm clock, but subwoofer sound (this is an acoustic system that plays low-frequency sounds from about 20 to 65 Hz) will reduce the brick wall more effectively. (The same situation is possible with window designs. It will be a challenge for further research.)

Practical recommendations: Before using modern window fills indoors, it is necessary to analyze the frequency characteristics of all potential sources of noise. When choosing window designs, we recommend comparing their sound insulation in tert-octave frequency bands, not indices R_w. This recommendation becomes more expedient after the entry into force of the new State Sanitary Standards in Ukraine [12], which introduces normalization of the permissible sound pressure levels depending on the NC criterion. In addition, these standards are more tightly regulated singing, music, or other noise of biogenic origin. The correction to the permissible noise levels in Appendix 3 of these standards is 5 dB(A).

4.3 Discussion of Error No 3

Error No 3 Windows with double-glazed windows, where three windows are used, clearly have better sound insulation characteristics than windows with single-glazed windows (where only two windows are used). Due to the acoustic connection between the glasses and the occurrence of resonance in small air spaces (these intervals are 8–10, maximum 16 mm), double-glazed windows, as a rule, do not provide significant sound insulation from external noise in comparison with single-glazed windows of similar width and total thickness of glass. With the same thickness of double glazing and the total thickness of glass in them, a single-chamber double-glazed window (the total thickness of the glass in both cases is the same) will have a higher value of the noise insulation index R_w, compared with two-chamber.

5 Scientific Novelty

1. It is established that the use of certain technologies that have smoothly transitioned into our construction practice, very often not only does not fit our construction conditions, but also causes economic, moral, and environmental damage. For example, energy-saving enclosures, from the practice of countries with milder climates, lead to a significant reduction in bearing capacity, durability, and the use of materials that insulate with dubious "environmental cleanliness." Ecological cleanliness is being challenged, for example, by having plates based on ISOVER. The mineral component of the latter is able to penetrate through the drywall at the molecular level, spreading to other rooms. In addition, rooms and corridors treated with drywall are transformed into music boxes with drum membranes. It is known that minerals do not dissolve and, once in the human body, one small glass particle can subsequently cause cancer—malignancies.
2. A comprehensive approach is proposed to address the "cleanliness" of materials, services, and goods from the stage of raw material extraction to the stage of their use or operation.

6 Practical Importance

6.1 Practical Application of Research Results

As a practical application of our scientific work results, theoretical calculations were carried out to protect the premises of the clinic from the noise of transport sources. The construction site of the outpatient clinic is located in the area of noise from the railway and the motorway (Kamyanske, Dnipropetrovsk region). To address the issue of noise protection of the outpatient staff and its clients, the following studies were conducted:

1. Noise characteristics of the railway and motorway were established.
2. The noise power loss of the sources is calculated taking into account the distance to the protection object for each source separately.
3. The total noise level, which will be observed within 2 m from the most subordinate noise of the facade of the outpatient clinic, has been determined.
4. The calculations of noise penetration into the room through the construction of the wall and the window were made.
5. A minimum value for sound insulation from window traffic noise is set. The value was $R_{\text{Атран}} = 35$ дБА.

   The maximum size of sound insulation of a metal plastic window according to the formula 6-20-6 is 30 dB, which does not correspond to the required sound insulation.

6. Acoustic measurements were carried out in the reverberation chamber of the PSACEA, and calculations were made for the windows of separate ones. The following results were established:
   - with the distance between the glass −75 mm, sound insulation at the action of transport sources in a scale adjusted "A" (used to assess the personal perception of noise by the human body) $R_{\text{Atran}} = 24.9$ dBA;
   - with the distance between the glass −160 mm, sound insulation at the action of transport sources in a scale adjusted "A" $R_{\text{Atran}} = 38.4$ dBA.

Thus, we recommend the use of a window with a distance between the glass—160 mm, sound insulation when exposed to transport sources $R_{\text{Atran}} = 38.4$ dBA.

6.2 Overall Practical Results

1. The results of the study are aimed at improving the current system of control and management of the industrial-construction process in terms of ensuring environmental safety in places of human habitation (by the formula: work, life, rest), which does not meet the requirements of today.
2. It is practically proved in the work that complete and effective solution of the problem of safety for human habitation can be provided only by a comprehensive approach in combination with environmental protection measures in general at all stages of design and construction, as well as at the stages of operation of already existing buildings and structures.
3. It is established that a qualitative solution of the problem of security for places of people's residence requires necessary acoustic certification of building materials, buildings, and structures made of them. This certification should become an integral part of the integrated management system of the process of organization and operation of industrial and structural structures, including the integrated quality management system of building materials, structures, and construction objects as a whole.

7 Conclusions

1. To increase the sound insulation of the window, it is recommended to use double-glazed windows of the maximum possible width (much larger than common in modern window structures: 36 mm), consisting of two glasses, preferably of different thickness (for example, 6 and 4 mm) and the widest possible spacer.
2. However, using double-glazed windows, contractors are recommended to use air gaps of different widths between glass and glass of different thickness. The profile system should provide three contour seals of the sash around the perimeter

of the window. In real conditions, the quality of the doorway affects the sound insulation of the window even more than the structural formula of the double glazing.

3. Given sound insulation is a frequency-dependent characteristic, and it is necessary to analyze carefully the performance of the packet in the frequency bands and the same frequency range of the noise source itself, because it may be the case when a double-glazed window with a higher index value may be more efficient than a double-glazed window with a higher index R_w в in some frequency ranges.
4. The main recommendation is that manufacturers in the specifications of their products need to add the following characteristic as: R_{Atran}, dBA. Size is required R_{Atran} (at the request of DSTU BV.2.6-85: 2009) to evaluate the insulation by external enclosures of air noise generated by urban traffic flows.

References

1. Hänninen, O., et al. (2014). Environmental burden of disease in Europe: Assessing nine risk factors in six countries. *Environmental Health Perspectives, 122*(5), 439–446. https://doi.org/10.1289/ehp.1206154.
2. WHO and JRC. (2011). *Burden of disease from environmental noise—Quantification of healthy life years lost in Europe*. Geneva, Switzerland: World Health Organization. Accessed May 5, 2014.
3. WHO. (2018). *WHO environmental noise guidelines for the European region*. Copenhagen: World Health Organization, Regional Office for Europe. Accessed December 7, 2018.
4. Seidman, M. D., & Standring, R. T. (2010). Noise and quality of life. *International Journal of Environmental Research and Public Health, 7*(10), 3730–3738. https://doi.org/10.3390/ijerph7103730.
5. Harris, A. S., Fleming, G. G., Lang, W. W., Schomer, P. D., & Wood, E. W. (2003). Reducing the impact of environmental noise on quality of life requires an effective national noise policy. *Noise Control Engineering Journal, 51*(3), 151–154. https://doi.org/10.3397/1.2839708.
6. Jørgensen, J. T., Bräuner, E. V., Backalarz, C., Laursen, J. E., Pedersen, T. H., Jensen, S. S., et al. (2019). Long-term exposure to road traffic noise and incidence of diabetes in the Danish nurse cohort. *Environmental Health Perspectives, 127*(5). https://doi.org/10.1289/EHP4389.
7. Brunia, S., De Been, I., & van der Voordt, T. (2016). Accommodating new ways of working: Lessons from best practices and worst cases. *Journal of Corporate Real Estate, 18*(1), 30–47. https://doi.org/10.1108/JCRE-10-2015-0028.
8. Sankov, P. M. (2016). Organization of safe working and resting conditions for citizens by the factor of noise pollution. In V. I. Bolshakov (Ed.), *Construction, materials, engineering: A collection of scientific papers* (Vol. 90, pp. 158–163). Dnipro: DVNZ PDABA.
9. Sankov, P. M., Tkach, N. O., Dikarev, K. B., Blyzniuk, A. M., & Hvadzhaia, B. D. (2018). Effect of motor transport on the working places in the service infrastructure (by noise factor and urban air pollution in the city center of Dnipro). *Science and Innovation, 14*(3), 59–66. Retrieved from https://doi.org/10.15407/scin14.03.067.
10. San'kov, P. N., Tkach, N. A., Gorb, A. V., Miheenko, Y. Y., & Chechuro, A. V. (2015). Razrabotka razdela proekta OVOS dlya ob"ekta rekonstrukcii v gorode Dnepropetrovske. *Mezhdunarodnyj nauchnyj zhurnal, 6*, 78–83.

11. Sankov, P., Makarova, V., Tkach, N., & Hvadzhaia, B. (2016). Analysis of noise protective properties of sheet material composite structures. *Technology Audit and Production Reserves, 6*(2(32)), 24–28. Retrieved from https://doi.org/10.15587/2312-8372.2016.83814.
12. Derzhavni sanitarni normy dopustymykh rivniv shumu v prymishchenniakh zhytlovykh ta hromadskykh budynkiv i na terytorii zhytlovoi zabudovy (nakaz MOZ Ukrainy vid 22.02.2019 No. 463).

Modern Smart City Concept Considering Population Safety Issues

P. Sankov, K. Dikarev, Y. Kushnir, and N. Tkach

Abstract The article considers the questions of the new approach introduction to urban policy based on wide application of information technologies that provide economical and environmentally friendly use of urban systems of life and at the same time, to maintain continuous sustainable development. The practical ideas on the foundations of the Smart City platform in Dnipro city are reflected. A thorough analysis of examples and principles of the Smart City concept implementation in the world, modern technologies, and practical ideas was carried out. The recommendations given on the use of a Smart City technology can provide further research and justification for their implementation in the city's real-time development. The integrated approach to the issues of protection from external noise the people in the territories is offered, in the premises during work or leisure. The subject of the research is a detailed analysis of modern Smart City technologies, theoretical, methodological and practical problems, and taking into account the transport infrastructure, develop directions for improving the development system of the Smart City concept for Dnipro city. The subject of the research is a detailed analysis of modern Smart City technologies, theoretical, methodological and practical problems, and directions for improving the development system of the Smart City concept for Dnipro city, taking into account the transport infrastructure.

Keywords «smart city» · Noise reduction · Landscaping systems

---

P. Sankov (✉) · K. Dikarev · Y. Kushnir · N. Tkach
Pridneprovsk State Academy of Civil Engineering and Architecture, Dnipro, Ukraine
e-mail: petrsankov5581@gmail.com

V. Onyshchenko et al. (eds.), *Proceedings of the 2nd International Conference on Building Innovations*, Lecture Notes in Civil Engineering 73,
https://doi.org/10.1007/978-3-030-42939-3_24

1 Introduction

1.1 Analysis of Recent Studies on the Concept of Smart City

Construction is a complex, but at the same time, extremely important industry, which is one of the main indicators of the development and stability of the country's economy. Scientific, pedagogical, research, and practical activities in the field of construction are a real contribution to the renewal of cities, towns, and villages of our independent Ukraine. To support continuous sustainable development, each city is urgently required to introduce a new high-quality approach to urban planning policy based on the widespread use of information technologies that ensure the economic and environmental use of urban life systems.

The Smart City concept involves upgrading the city's infrastructure with fundamentally new centralized management capabilities, a new level of service and security. This strategic development combines the diverse factors of urban development into a single system. "Smart City" is characterized by a combination of urban development areas, which require the use of IT tools, as well as special "smart" technologies, methods, and calculations. This concept recognizes the role of artificial intelligence, information and communication technologies, social and environmental potential as a resource for the development and competitiveness of the city.

A Smart City is smart management, smart living, smart people, smart environment, smart economy, and smart mobility. The list of areas that require the introduction of smart technologies covers practically all areas of the urban economy and urban infrastructure, including analytics, banking, buildings, commerce, e-government, communications, education, energy, emergency situations, catering, healthcare, manufacturing and the field services, transportation, retail, public safety, ecology and environmental monitoring, water and gas supply, and much more [1]. Among the reputable international ratings of "Smart City" are the following most significant [2]: (1) IBM's annual Smarter Cities Challenge; (2) European Smart Cities—model of European smart cities; (3) Intelligent Community Forum—rating of intellectual communities; (4) AT Kearney Global Cities Index and Emerging Cities Outlook-index of global cities; (5) Smart City Expo World Congress, Barcelona-Fira; (6) Siemen's Green Cities Index-ecocities; (7) IDC Smart Cities Index +—rating of smart cities in Spain; (8) PwC's «Cities of Opportunity» —rating of satisfaction of the population; (9) Innovation Cities Index—rating of innovative cities; (10) Broadband Communities' Magazine Annual Conference and Cornerstone Awards; (11) Digital Governance in Municipalities—innovative use of ICT; (12) Center for Digital Government's Digital Cities Survey; (13) Mercer's Quality of Living Survey—Quality of Life Rating; and (14) IESE Cities in Motion Index (CIMI).

In the city of Dnipro, there is a very strong team of IT volunteers, which the City Council listens to, and who themselves create "smart" solutions for urban infrastructure [3]. The Dnipro Smart City Group has created the concept of infrastructure modernization for D. Jawornytsky Avenue and Pryvokzalnaya Square. There was an American foundation that offered grant support to gather public opinion on these projects. Here is a brief description of the three most suitable for implementation for the city of Dnipro [4–6]. The Intelligent Community Forum (ICF) rating gives priority to communities that, through a crisis or anticipation of the future, have come to understand the potential challenges of a broadband economy (globalization) and taken deliberate steps to create an economically viable economy. The ICF rankings take into account what results in cities have achieved in improving living standards, quality of education, improving working conditions, and developing infrastructure. "Smart City Expo World Congress, Barcelona-Fira" in Smart City highlights not only new technologies that are emerging in the context of urban life to help manage the city more efficiently but also change the thinking of people and processes that promote the activity and activity of the city. The "Siemen's Green Cities Index" favors the most environmentally sustainable city that is developing and solving problems using efficient and smart technologies.

1.2 Problem Statement

One of the main, according to the authors, in the concept of "Smart City" is the intelligent transport system. Transport sources of noise (motorways, parking places, railways: tram and city lines) are the closest to people's places of residence (residential and office buildings, recreation places, and jobs in general). One of the most important components of Smart City is the intelligent transportation system (ITS), which optimizes traffic by displaying traffic on street dashboards and users' smartphones showing them the optimal route and carrying many other useful features.

In order to manage the road network, the intelligent transport system is very closely linked to the geoinformation system (GIS). It is the basis for intelligent transport systems. It is a "single window" at all levels of ITS: from mobile solutions to the situation center. The main tasks of geoinformation systems in this field are as follows: (1) collection of detailed with precise binding to the terrain of information about such objects of road economy as road surface, signs, bridges, lighting, etc.; (2) transport modeling, analysis of transport networks, planning of traffic flows; (3) display of the operational traffic situation on the 3D-digital map of the city; (4) public transport dispatching and emergency response; determining the location of objects and tracing; (5) planning and routing, including between different types of transport networks; (6) ATCS, highway maintenance management; (7) management of the road rail transport; (8) power supply, communication, signaling equipment; (9) event analysis; (10) demographic analysis and change of transport routes; and (11) meteorological support, environmental monitoring.

A striking example of the implementation of the intelligent road transport system is the system deployed in Tokyo [7]. It is based on tens of thousands of fairly simple sensors, a small number of intelligent camcorders (about 400), and provides information from moving objects and informing drivers about the situation ahead on roads, alternative routes, etc. The system is a great computing resource, combined with telecommunications, and detailed road GIS. One of the major problems with the transportation system is the fight against traffic jams. This problem is devoted to work [8]. This article uses ARIMA-based modeling to study some of the factors that significantly affect traffic intensity

The object of the research is the development of the Smart City concept for Dnipro city.

The subject of the research is a detailed analysis of modern Smart City technologies, theoretical, methodological, and practical problems and directions of improvement of the Smart City concept development system for Dnipro city in view of the transport infrastructure.

2 Purpose

The purpose of the study is to reflect practical ideas for building the foundations of the Smart City platform in a city with respect to transportation infrastructure by analyzing world-renowned examples and principles for the implementation of the Smart City concept in many existing cities. The purpose of the Smart City concept is to provide a system of basic insights into the principles, approaches, goals, priorities, and objectives of implementing Smart City technologies. The implementation of this concept is capable of improving the quality of life of the population and reducing the cost of operating the infrastructure by automating routine processes for managing the region.

3 Methodology

The theoretical and methodological basis of the study is the dialectical method of cognition, a systematic approach to the study of economic phenomena and processes, scientific works of domestic and foreign scientists on the problems of urban development. In the course of the work the following basic methods and means of approach were used: abstract science (theoretical generalizations and formulation of conclusions), analysis of electronic resources, bibliographic search, systematization of the obtained results. The theoretical and empirical basis of the research is the scientific works of domestic, foreign scientists, and personal scientific achievements of the authors of real work on the problems of development and improvement of the urban structure, materials of scientific-practical conferences, reference, and information publications.

4 Results

For many years, scientific research on the effects of noise of transport sources on humans has been carried out at PSACEA [9–12].

We have found that one of the most important components of Smart City is the intelligent transport system (ITS). In order to manage the **road network**, the intelligent transport system is very closely linked to the geoinformation system (GIS). The components of ITS are such important indicators as the state of noise pollution on highways, gas pollution, and the presence of traffic jams. In [10], the authors proposed a method for determining the noise characteristics of two main linear noise sources in cities. Namely (1) the sources of acoustic pollution closest to the objects of protection (residential areas and residential buildings located on them) are the noise of automobile transport moving along the main streets and roads; (2) noise of railway transport—as the most powerful source of acoustic pollution. And the work of the authors [11] allows us to assess the dynamics of the calculated indicators of the equivalent sound level and CO concentration when organizing the removal of transit traffic from the city. The value of the increase in the equivalent noise level on the highways of city traffic taking into account transit traffic was obtained. This value (0.6 dBA) does not affect the estimated noise class of the main street. However, the calculation of the increase in the concentration of exhaust gases on the main street (3.77 mg/m^3), taking into account transit vehicles, leads to the transition of the considered line into a more dangerous class of gas. Based on theoretical calculations, the change of noise characteristics of the main streets in the city is shown and the value of increase of the calculated concentration of the exhaust gases for the same highways is determined, taking into account the traffic intensity of Dnipro city. In the study [12], the authors gave a scientific justification and analysis of the possibility of improving life safety in residential and industrial areas of modern cities using mobile landscaping systems. Scientists in the field of construction and architecture pay much attention to such systems. In the research [13], the Boston-based architecture firm Höweler + Yoon and Los Angeles-based Squared Design Laboratories propose to cover the Boston building (which is not completed) with robotic environmental capsules to turn it into an autonomous farm! Bold decision! We propose a more realistic method for the organization of safe working conditions and recreation of citizens by the factor of noise pollution through the reconstruction of the greening system of modern cities. It is recommended that as one of the main directions in the development of the architecture of the city to apply a modern way of forming zones of ecological comfort in the conditions of compacted development of residential areas and reconstruction of industrial zones. The method is based on: landscaping roofs of buildings, the use of vertical landscaping facades, the construction of ecological parking lots, mobile landscaping systems. Figure 1 shows an example of the possible use of mobile landscaping systems to protect residential areas from noise, dust and exhaust gases from vehicles moving down the street Velyka Diivska in Dnipro city. Figure 2 clearly demonstrates the possibility of increasing comfort in the areas adjacent to the parking areas of vehicles with a sharp increase in the level of motorization.

a)

b)

Fig. 1 Velyka Diivska St before (**a**) and after (**b**) installation of screens with tilt and modular systems, landscaping

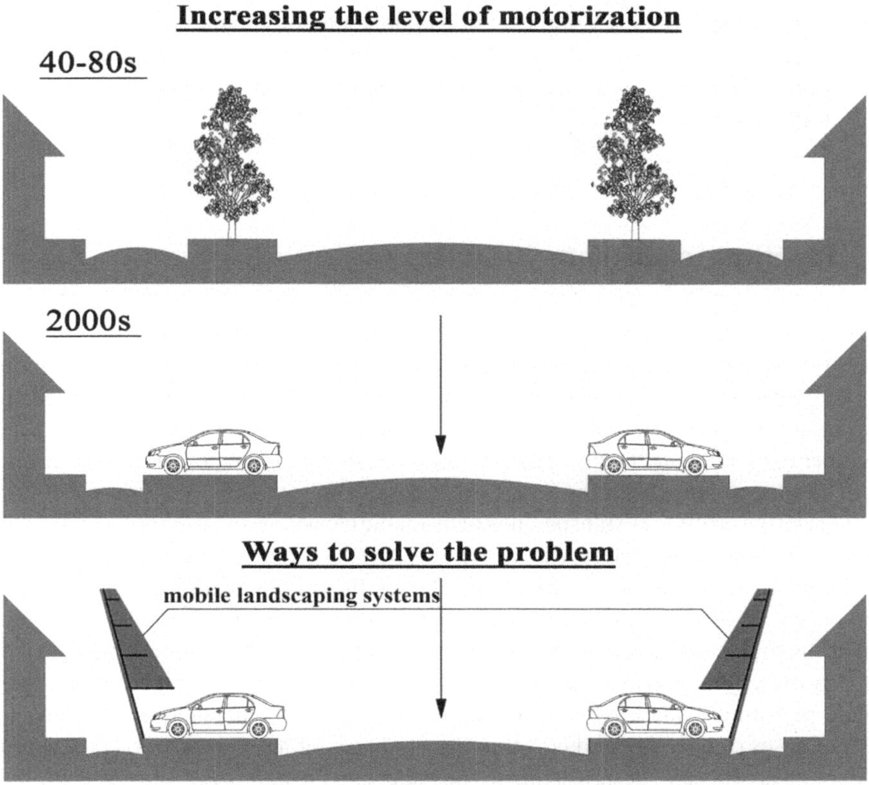

Fig. 2 Dynamics and ways of solving unfavorable environmental situation in residential development due to the use of mobile landscaping systems in cases of sharp increase in the level of motorization

The latter is recommended to be used as green noise shields along noisy cars and railways, around car parking places in residential areas of cities and in areas of noisy industries or businesses, areas, and workplaces requiring protection against noise and dust caused by neighboring businesses.

5 Scientific Novelty

(1) A thorough analysis of the examples and principles of the implementation of the Smart City concept in the world, based on modern technologies and practical ideas.
(2) Practical ideas on the foundations of the Smart City platform in the city of Dnipro are reflected.

(3) Have found further development of recommendations on the use of a particular Smart City technology that can be implemented in the development of the city in real time in view of the state of noise pollution on the highways, gas, and traffic jams.

6 Practical Importance

The results of the work are aimed at creating safe living conditions in the territories of different functional zones of modern cities;—reducing the amount of pollution of the territories of cities by components, aimed at protecting the population from the harmful effects and consequences of pollution;—increase of the territory and number of recreational facilities in the conditions of reconstruction of modern cities.

The expected effects of the Smart City project are:

(1) Design errors. (2) Transparency of available capacities of all components of the city's planning infrastructure, especially transport. (3) Improving the quality of optimal specifications for the construction and operation of any facilities. (4) Reducing the coordination of optimization schemes for all components of city life management, including optimization of the city transport scheme. (5) Improvement of the quality of examination of project documentation and carrying out of the author's care while performing any kind of works according to this documentation. (6) Prevention of cases of unauthorized construction of objects of any ownership. (7) A sharp reduction is expected: design estimates on paper and the number of approvals for the implementation of design decisions.

7 Conclusions

1. It is important to understand that the resources of the city currently available will never be enough to solve all the problems that have accumulated over the long hours of simple consumption of environmental and infrastructure problems. In this regard, it is necessary to select priority areas for concentration of financial and administrative resources aimed at shaping sustainable development of the city.
2. The promotion of a "Smart City" format for Ukraine is a policy that offers a new quality of life. And the success of its implementation largely depends on the willingness of the Ukrainians themselves to break stereotypes [14].
3. The authors' further work looks at the problem of qualitative assessment of life safety issues, the beginning of which is given in [15].
4. A smartly integrated system consisting of many subsystems is required for Smart City to work effectively. It does not make much sense to implement such projects

in part, it is necessary to develop a common concept of "smart city," which will take into account both the current needs of different urban services and the prospects for development, taking into account all external factors. Therefore, Smart City projects require an integrated approach, which is a major obstacle to the concept.

5. The authors of the most favorable international rating for the city of Dnipro are the Siemen's Green Cities Index. This conclusion is based on the fact that the priority for solving the problem in the city is to improve the quality and safety of life of the population for the transition to environmentally sustainable way of Dnipro.

References

1. Yashchyshyna, I. V., & Konovalova, M. V. (2018). Smart city concept and its implementation in Ukraine (2018) Economy of the city and urban, International scientific and practical Internet conference, KNEU 2018. pp. 236–239. Retrieved from http://ir.kneu.edu.ua/bitstream/2010/24388/1/236-239.pdf.
2. Smart City Expo World Congress. (2019). Date: 11/19/2019—11/21/2019 Venue: Fira de Barcelona Gran Via, Barcelona, Spain. Retrieved from https://www.showsbee.com/fairs/58696-Smart-City-Expo-World-Congress-2019.html.
3. Smart start of the Dnieper: how the city develops "smart" infrastructure. (2016). Retrieved from https://delo.ua/econonomyandpoliticsinukraine/smart-start-dnepra-kak-gorod-razvivaet-umnuju-infrastrukturu-324342/.
4. *International Journal of Open Information Technologies*. (2017). (Vol. 5, no.3). ISSN: 2307-8162.
5. The Top7 Intelligent Communities of the Year. (2019). Retrieved from https://www.intelligentcommunity.org/top7.
6. Smart cities challenge. (2019) Retrieved from https://www.smartercitieschallenge.org/about.
7. Myhal, V. D. (2018). *Intelligent systems in the technical operation of cars: A monograph Kharkiv*. UA: Majdan.
8. Alghamdi T., Elgazzar, K., Bayoumi, M., Sharaf, T., & Shah, S. (2019). Forecasting traffic congestion using ARIMA modeling. In *15th International Wireless Communications & Mobile Computing Conference (IWCMC)*. Tangier, Morocco, Morocco. doi: https://doi.org/10.1109/IWCMC.2019.8766698//. Retrieved from https://ieeexplore.ieee.org/document/8766698.
9. Sankov, P. M., Tkach, N. O., Dikarev, K. B., Blyzniuk, A. M., & Hvadzhaia, B. D. (2018). Effect of motor transport on the working places in the service infrastructure (by noise factor and urban air pollution in the city center of Dnipro). *Science and Innovation, 14*(3), 59–66. Retrieved from doi: https://doi.org/10.15407/scin14.03.067.
10. Sankov, P., Tkach, N., Gvadzhaia, B., & Tyoshina, L. (2017). Quality of acoustic forecasting of noise characteristics of motor vehicles and railways. *Norwegian Journal of Development of the International Science, 4*(2), 7–9. ISSN 3453-9875.
11. Sankov, P. M. (2017). Accounting for transit traffic on highways of the city in order to improve the environmental safety of the population. In V. I. Bolshakov (Ed.), *Construction, materials, engineering: A collection of scientific papers* (Vol. 96, pp. 128–132). DVNZ PDABA: Dnepr.
12. Sankov, P. M. (2016). Organization of safe working and resting conditions for citizens by the factor of noise pollution. In V. I. Bolshakov (Ed.), *Construction, materials, engineering: A collection of scientific papers* (Vol. 90, pp. 158–163). DVNZ PDABA: Dnepr.
13. Vertical farms concept of urban robots. (2019). Retrieved from http://www.robotbuzz.fr/inclassable/concept-de-fermes-verticales-robotisees-urbaines/.

14. Skidanova, V. (2009). Ethnic stereotypes as components of the linguistic picture of the world. Scientific notes. *Philological Sciences, 81*(4), 27–31. Kirovohrad. Kirovohrad State Pedagogical University named after Volodymyr Vynnychenko.
15. Sankov, P., Trifonov, I., Tkach, N., Hilov, V., Bakharev, V., Tretyakov, O., & Nesterenko, S. (2017). Development of the method of evaluation the level of environmental safety of housing accommodation and its approbation. *Eastern-European Journal of Enterprise Technologies, 4/10*(88). ISSN 1729-3774 2017. pp. 61–69, 79–80. doi: https://doi.org/10.15587/1729-4061.2017.108443.

The Rational Parameters of the Civil Building Steel Frame with Struts

Oleksandr Semko, Anton Hasenko, Volodymyr Kyrychenko, and Vitaliy Sirobaba

Abstract The article deals with the field of construction, in particular the design of load-bearing steel frame structures, which can be used in the construction of industrial and civil buildings. It is known the analogues of such frames as the typical designs of frames consisting of racks and beams. The stiffness of such frames is provided either with the installation of a system of vertical and horizontal knits (in this case the joint model of supporting beams on the columns is simple), or the arrangement of stiff joints between the beams and columns. Stiff connections form a frame system which is capable to keep horizontal loads and reduce the design torque in beams. The disadvantage of such joints is the complicated fabrication technology and the considerable steel costs required to ensure the local stability of the supporting sections of the column members by setting the stiffening ribs or increasing the thickness of these members. The basis of the material given below is the task to design a building cross steel frame with the struts between the columns and the continuous floor beams with simple joint model, while optimizing the dimensions and locations of the struts. In this case, the design torque in the beams is reduced compared to the standard frames, which allows their section to be reduced. The article also shows the drawings of the standard joints between columns and beams that can be used in the development of work projects of civil building steel frames with struts.

Keywords Steel frame with struts · Optimization · Joints

1 Introduction

The variant design of building frames always aims to optimize the parameters of their members [1]. It concerns the development of the optimal frames' scheme of the bearing framework, and the effective construction of cross-sections of structures [2].

O. Semko · A. Hasenko (✉) · V. Kyrychenko
Poltava National Technical Yuri Kondratyuk University, Pershotravnevyi Avenue, 24, 36011 Poltava, Ukraine
e-mail: gasentk@gmail.com

V. Sirobaba
Sumy Building College, Petropavlivska Street, 108, 40000 Sumy, Ukraine

V. Onyshchenko et al. (eds.), *Proceedings of the 2nd International Conference on Building Innovations*, Lecture Notes in Civil Engineering 73,
https://doi.org/10.1007/978-3-030-42939-3_25

These factors entail the reduction of material costs for the construction of buildings' frames.

2 Analysis of Recent Research and Publications

This optimization problem can be successfully solved using steel elements for the bearing framework of both industrial and civil buildings due to the wide range of rolling metal profiles and the ability to create the favorable welded sections [3].

A standard solution of the buildings' framework with small and medium-sized spans is the use of identical cross frames, which are installed with the same pitch [4]. It is almost always the cross frames consist of racks and beams [5]. The stiffness of the frames is provided either with a system of vertical and horizontal knits (in this case, the joint model of supporting beams on the columns is simple), or the arrangement of stiff joints between the beams and columns [6]. Hinge joints between beams and columns are easy to manufacture, as they transmit only normal and cross forces [7]. Stiff connections form a frame system which is capable to keep horizontal loads and reduce the design torque in beams [8]. The disadvantage of such joints is the complicated fabrication technology and the considerable steel costs required to ensure the local stability of the supporting sections of the column members by setting the stiffening ribs or increasing the thickness of these members [9].

Due to the development of computer technology and its wide technical capabilities, it is convenient to investigate the stress–strain state of joints using computer programs of finite element analysis [10].

One of the solutions to ensure the cross stiffness of the frame using simple joints is to install the struts between the column and the beam (see Fig. 1) [11].

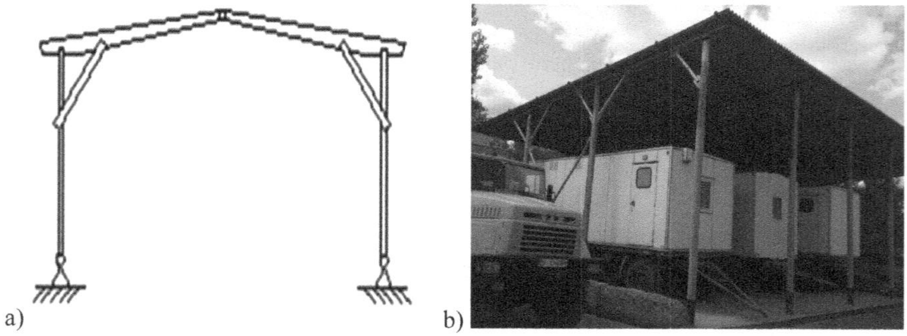

Fig. 1 installation of struts between columns and beams to ensure the cross stiffness of the building frames: **a** A design diagram; **b** a real frame with struts

Item No.		Operation name
1	→	The determination of the optimal angle of the strut's inclination
2	→	The determination of the strut position about the beam and column
3	→	The selection of optimal spans of the building's cross frame
4	→	The design of standard joints for struts

Fig. 2 Structural and logical scheme of the search for rational parameters of building steel frame with struts

3 Formulating the Article's Goals

The purpose of the work is to design the building cross steel frames with the struts between the columns and the continuous floor beams with simple joint model, while optimizing the dimensions and locations of the struts. In this case, the design torque in the beams is reduced compared to the standard frames, which allows their section to be reduced.

4 The Main Material with the Justification of New Scientific Results

4.1 The Procedure for Solving the Optimization Problem

Structural and logical scheme of the search for rational parameters of building steel frame with struts is shown on Fig. 2. According to this scheme, the techniques and results of the research work are given below.

In solving this problem, the costs of materials for one cross frame of the load-bearing framework of a building 36 m wide were compared. The distance between the columns of the frame (spans) was different: 6, 9, and 12 m (see Fig. 3). To simplify the calculation, the load on the beams was equally distributed.

4.2 Determination of the Optimal Slope Angle

The minimum stress in the strut will be with the minimum value of internal forces in it and the maximum value of radius of gyration [12]. A larger value of radius of gyration is necessary to reduce the flexibility of the strut. Only the axial force of the six possible internal forces in the strut with its simple joint model appears. To find the axial force in the strut, a building two-span cross frame should be considered

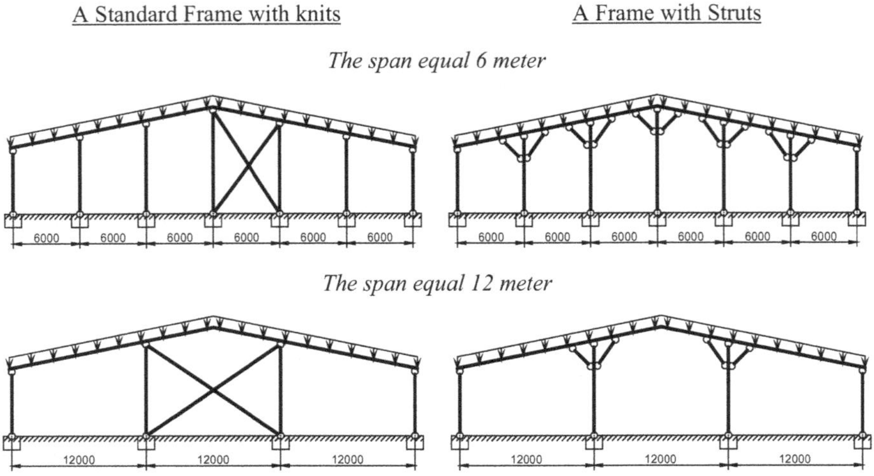

Fig. 3 Separate design schemes of the analyzed building cross frames

(see Fig. 4a). The number of spans is two in order to reduce the additions in the static equilibrium equations. The struts are installed only on the middle column, conditionally attaching them to the beams at a distance of ¼ span from the middle column. The vertical load on the frame is distributed over the beams.

We consider a continuous floor two-span beam (see Fig. 4b) showing the supporting reactions in the locations of the posts and struts. We write the equation of equilibrium of the sum of moments about the right point C Eq. (1). The horizontal components of the axial forces in the struts in Eq. (1) are not recorded because they balance each other.

$$\sum M_C = 0;\ R_A \cdot 2L + R_B \cdot L + N_S \cdot \cos\alpha \cdot \frac{3}{4}L + N_S \cdot \cos\alpha \cdot \frac{5}{4}L - q \cdot 2L \cdot L = 0. \quad (1)$$

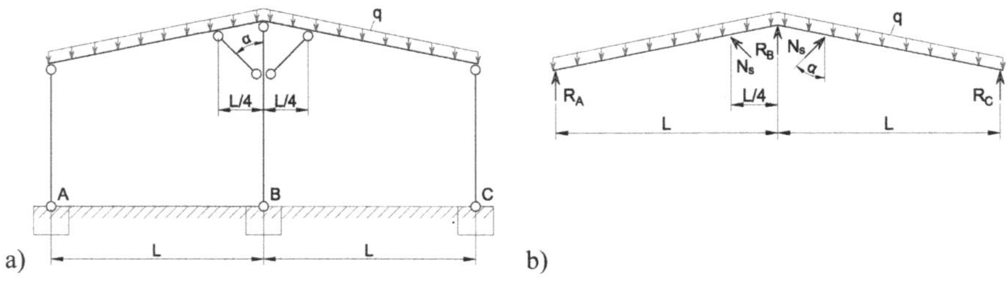

Fig. 4 Two-span frame with struts on the middle column: **a** A design scheme; **b** before the axial force in the struts

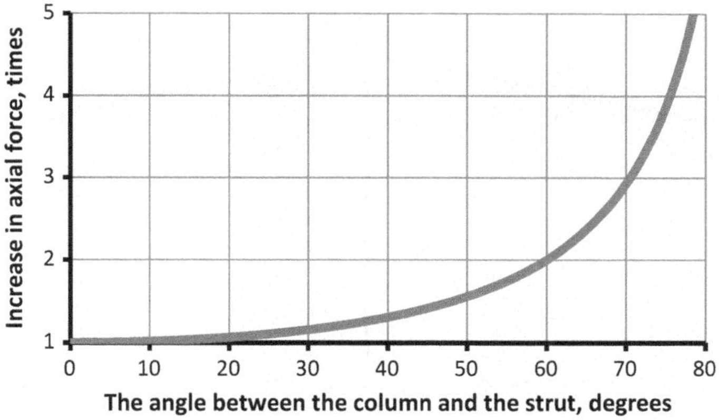

Fig. 5 increase of the axial force in the strut, depending on the increase of the slope angle α

Substituting the value of the supporting reactions in Eq. (1), after the corresponding transformations and some simplifications, we obtain an approximate Formula (2) for determining the axial force in the struts.

$$N_S = \frac{q \cdot L}{2 \cdot \cos \alpha}. \tag{2}$$

When plotting the increase of the axial force in the strut depending on the increase of its slope angle α (see Fig. 5), it is easy to see the following. The axial force in the strut is practically not increased at a slope angle of20° from the vertical. At an angle of 60°, the axial force is doubled compared to the vertical position of the strut. At the greater angle, the axial force increases several times. Thus, the optimal slope angle related to the vertical is 40...50°. At this angle, the axial force increases 1.3...1.6 times.

4.3 Determining the position of the strut reletive to the beams and columns

In addition to ensuring the stiffness of the building frame in the area of the frames, the struts lead to a decrease of the bending moment in the coating beams by reducing the spans of the latter [13]. The optimal locations of the struts will be in the case when the spans and bearing bending moments are equal, $M_{sp} = M_{sup}$ (see Fig. 6). However, then the joints of attachment of the struts to the beam should be arranged with an equal pitch, which at the slope angle of the strut 40...50° lowers them down the column too low into the working space of the premise.

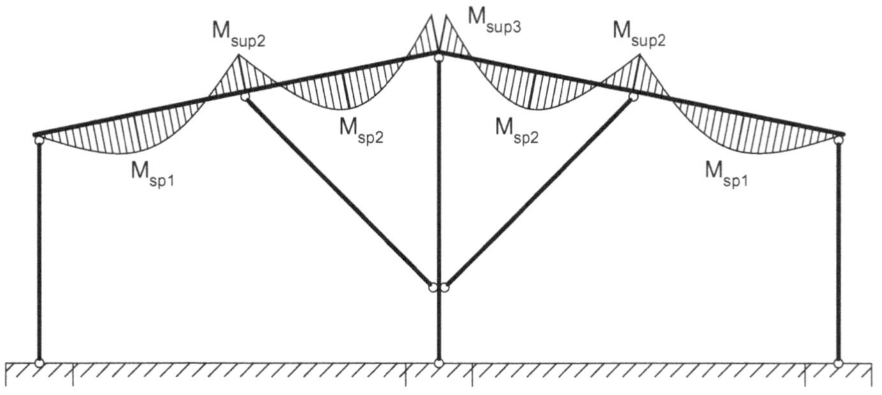

Fig. 6 Optimum plot of bending moments in a continuous floor beam

4.4 Choosing the optimum spans of the building frame

During the development of this section of the work, the internal efforts were determined and cross-sections of the supporting members (columns, beams, struts, as well as structurally vertical knits, if they are) of the building cross frames with the spans of 6, 9, and 12 m were selected (see Fig. 3). The sizes of the struts for the different spans of frames were assumed to be the same. The frames' pitch is 6 m. The load on the frame is equally distributed in accordance with the weight of the roof and the snow load. The weight of the spans and their knits are not given in the comparison of the frame variants, because the frame pitch was taken as a constant and the cost of the spans was not changed. The calculation results are summarized in Table 1.

The results of the calculation are also shown graphically in Fig. 7. It shows the comparison only the cost of metal, because the costs of concrete foundations for the three analyzed options are almost the same (see Table 1). According to the graph, the frame with strut is cheaper than a standard vertical frame. The rational span of the frame with strut is 9 m.

4.5 Designing Typical Joints of Strut Arrangement

The problem of the construction of strong, stiff, and durable joints of frame members that reliably transfer effort is relevant for load-bearing structures made of any building materials, including steel. At the same time, modern design requires increasing the reliability, economy, and manufacturability of building structures as a whole and their joints. Therefore, it is necessary to design and investigate new types of joints connecting struts with columns and beams. Figure 8 shows the solution of standard joints of strut arrangement to columns and beams.

Table 1 Comparison of material costs for different designs of cross frames

Parameter	A standard frame with knits			A frame with struts		
	The column	The beam	Knits	The column	The beam	The strut
The span equal 6 m						
#profile	2 channel 14	I-beam 22	tube 60 × 4	2 channel 14	I-beam 22	tube 80 × 4
Consumption, kg	994	868	124	994	868	158
In all, kg	**1986**			**2020**		
Foundation (concrete, m^3)		**9.3**				
The span equal 9 m						
# profile	2 channel 14	I-beam 36	tube 80 × 4	2 channel 14	I-beam 30	tube 80 × 4
Consumption, kg	704	1761	211	704	1323	120
In all, kg	**2676**			**2147**		
Foundation (concrete, m^3)		**9.4**				
The span equal 12 m						
# profile	2 channel 16	I-beam 40	tube 80 × 4	2 channel 16	I-beam 40	tube 80 × 4
Consumption, kg	636	2066	258	636	2066	81
In all, kg	**2960**			**2783**		
Foundation (concrete, m^3)		**10.1**				

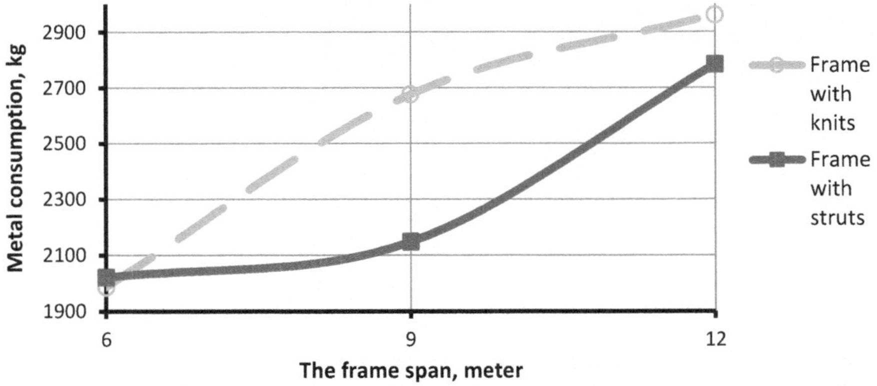

Fig. 7 Comparison of the cost of metal for one building cross frame at different spans

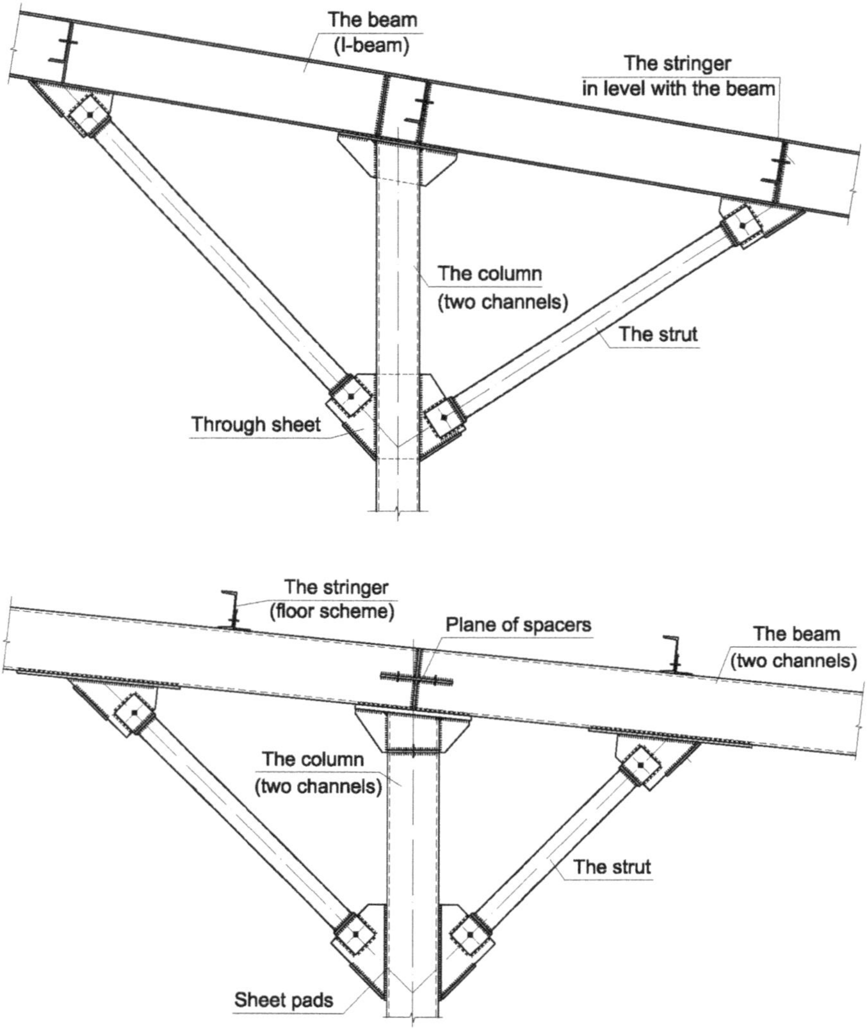

Fig. 8 Standard joints of strut arrangement to columns and beams

5 Conclusion

The frame with struts is cheaper than a standard vertical frame. The rational span of the frame with strut is 9 m. In addition to ensuring the stiffness of the building frame in the area of the frames, the struts lead to a decrease of the bending moment in the coating beams by reducing the spans of the latter. The optimum slope angle of the struts about the vertical is 40…50°.

References

1. Pavlikov, A. M., Mykytenko, S. M., & Hasenko, A. V. (2018). Effective structural system for the construction of affordable housing. *International Journal of Engineering & Technology: Publisher of International Academic Journals, 7*(3.2), 291–298. doi: https://doi.org/10.14419/ijet.v7i3.2.14422.
2. Liew, A., & Gardner, L. (2015). Ultimate capacity of structural steel cross-sections under compression, bending and combined loading. *Structures, 1,* 2–11. https://doi.org/10.1016/j.istruc.2014.07.001.
3. Morhun, A. S., & Soroka, M. M. (2017). Rozv″yazannya zadach parametrychnoyi optymizatsiyi budivel′nykh konstruktsiy v prohramnomu kompleksi ANSYS. *Visnyk Vinnyts′-koho politekhnichnoho instytutu, 5,* 18–23.
4. Permyakov, V. A., Perel'muter, A. V., & Yurchenko, V. V. (2008). *Optimal'noye proyektirovaniye stal'nykh sterzhnevykh konstruktsiy*. Kyiv, K.: TOV "Steel Publishing House".
5. Kolesnichenko, S. V., & Mironov, A. N. (2002). Issledovaniye napryazhenno-deformirovannogo sostoyaniya i kontsentratsii napryazheniy v K-obraznykh uzlakh ploskikh reshetchatykh konstruktsiy. *Avtomobíl'ní dorogi í transportne budívnitstvo, 64,* 127–130.
6. Hasenko, A. V. (2012). Konstruktyvni rishennya vyrishyly poyednuvaty mizh soboyu stale-betonnyy kolon iz perekryttyam hromads′kykh ta promyslovykh budivel′. *Zbirnyk naukovykh prats′. Seriya: Haluzeve mashynobuduvannya, budivnytstvo, 3*(33), 55–61.
7. Korkh, O. O. (2009). *Napruzheno-deformovanyy stan i nadiynist′ zvarnykh vuzliv stalevykh ram* (Abstract of dissertation).
8. Semko, O. V., & Hasenko, A. V. (2006). *Patent of Ukraine 25636*. Kyiv: State Patent Office of Ukraine.
9. Storozhenko, L. I., Semko, O. V., & Pents, V. F. (2005). *Stalezalizobetonni konstruktsiyi: Navchal′nyy posibnyk.* Poltava: PoltNTU.
10. Hasenko, A. V., & Novytskyi, O. P. (2018). Numerical experiment for the determination of the stress-strain condition of the system "Basis—Vibroreinforced soil-cement pile". *International Journal of Engineering & Technology: Publisher of International Academic Journals, 7*(4.8), 41–47. doi: https://doi.org/10.14419/ijet.v7i4.8.27211.
11. Cai, M., Yu, J., & Jiang, X. (2018). Stress and strength analysis of non-right angle H-section beam. *Periodica Polytechnica Civil Engineering, 62*(3), 612–619. https://doi.org/10.3311/PPci.11280.
12. Megson, T. H. G. (2014). Normal force, shear force, bending moment and torsion. *Structural and Stress Analysis, 3,* 38–78.
13. Komarov, V. A., & Mrykin, S. V. (2006). Optimizatsiya razmeshcheniya opor staticheski opredelimykh balok. *Vestnik Samarskogo gosudarstvennogo aerokosmicheskogo universiteta, 1,* 86–92.

Investigation of the Temperature–Humidity State of a Tent-Covered Attic

O. V. Semko, O. I. Yurin, O. I. Filonenko, and N. M. Mahas

Abstract The results of theoretical and experimental studies of the temperature–humidity state of a tent-covered attic are presented. The investigations were carried out in the attics of administrative buildings in Poltava. It was performed an analysis of the possibility of formation the ice area on the surface in the winter–spring period, which prevents the flow of meltwater from the roof. The accumulation of meltwater on the roof leads to leakage of water through the leaky joints of metal sheets on the attic floor. This phenomenon occurs in the case of simultaneous presence on the cover of the positive and negative temperatures. The range of external temperatures at which this phenomenon occurs is revealed. The attic air temperatures at which the formation of ice on the surface is minimized are found. The analysis of the possibility of condensation on the inner surface of the roof at the current temperature–humidity state of air in the attic is performed. The air temperature in the attic at which moisture condensation does not occur is revealed.

Keywords Tent cover · Ice formation on the roof · Moisture condensation

1 Introduction

Due to the long service life of office buildings, built in the early twentieth century, reduction of roofing metal sheets joints density and restructuring of the ventilation systems in the attic, meltwater flows through the roof and condenses on its inner surface.

Experimental studies performed in the attics of administrative buildings in Poltava showed that the average air temperature in the attic is 8 °C, and relative humidity is 40%. At this temperature in the winter–spring period, two areas with negative and positive temperatures are observed on the outer surface of the roof. Due to the rise of warm air into the upper part of the attic, the positive temperature section is at the top of the cover. The negative temperature area is above the eaves of the roof. With the simultaneous existence of these areas on top of the roof, the snow is melting there.

O. V. Semko · O. I. Yurin · O. I. Filonenko · N. M. Mahas (✉)
Poltava National Technical Yuri Kondratyuk University, Poltava, Ukraine
e-mail: mahasnataliia@gmail.com

V. Onyshchenko et al. (eds.), *Proceedings of the 2nd International Conference on Building Innovations*, Lecture Notes in Civil Engineering 73,
https://doi.org/10.1007/978-3-030-42939-3_26

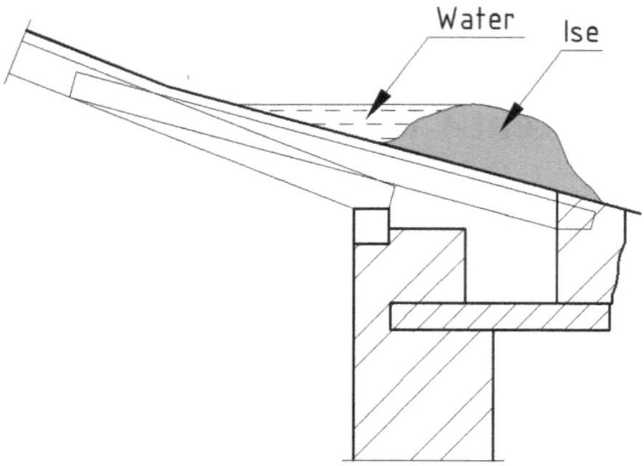

Fig. 1 Ice formation and water accumulation on the roof

The meltwater flows down, enters into area with a negative temperature and creates an ice barrier that prevents the flow of meltwater from roof (see Fig. 1).

The water that accumulates before the ice flows through the non-sealed joints of metal sheets and moisturizes the wooden roof structures, the attic heater and the exterior walls.

After the restructuring of the ventilation system, the air from the rooms is not removed on the outside, but into the attic. This leads to an increase in the temperature and relative humidity of the attic, which, in turn, promotes the formation of condensation on the inner surface of the cover. At low outside temperatures, the condensing moisture turns to frost. As the outside air temperature rises, the frost begins to melt; water flows down to the eaves and moisturizes the roof structure and the exterior walls.

The listed above bringing the attic temperature and humidity to the values that prevent or minimize the negative effects is an urgent task.

Thermophysical studies of enclosure structures at PoltNTU were considered in [1–6], and researchers in this field worked abroad [7–10].

The purpose of this work was to determine the range of air temperatures in the tent-covered attic of administrative buildings in which occurs or does not occurs the ice formation, meltwater accumulation and moisture condensation on the inner surface of the metal roof.

2 Basic Material and Results

Experimental studies performed in the attics of administrative buildings in Poltava showed that the average air temperature in the attic is 8 °C, and relative humidity is 40%. Such a high temperature is explained by the fact that after the restructuring of the ventilation system, the air from the rooms is not removed on the outside, but into the attic. In addition, the heat transfer resistance of the attic floor is $R_{\sum} = 1.233$ м^2

К/Вт which is much less than the normalized value, which according to [11] is $R_{q\,min}$ = 4.95 м^2 К/Вт.

Theoretical studies of the temperature distribution over the roof surfaces were performed on the basis of the calculation of the temperature fields. The attic cross section accepted for the study is shown in Fig. 2.

Design model for the calculation of temperature fields is shown in Fig. 3.

The results of calculating the size of the area with a negative temperature at different outside temperature and the temperature in the attic 8 °C are shown in Table 1.

At a temperature from −15 °C and below, the entire roof surface has a negative temperature. At a temperature from 0 °C and above, the entire roof surface has a positive temperature. At a temperature from −4 to −1 °C, there will be two areas

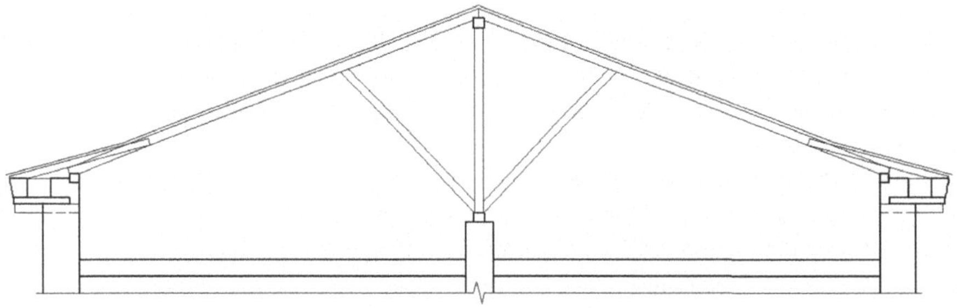

Fig. 2 Attic cross section

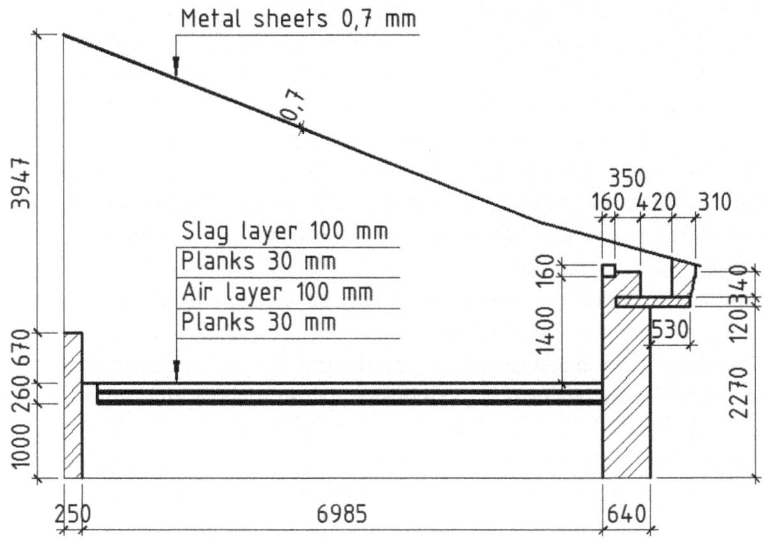

Fig. 3 Design model for the calculation of temperature fields

Table 1 The segment with a negative temperature length depending on the outside air temperature

№	The temperature of the outside air, °C	The length of the outside surface, m, roofs with		The percentage of length of the roof outside surface with negative temperature, %
		Positive temperature	Negative temperature	
1	−5	0	9.09	100.0
2	−4	7.64	1.45	16.0
3	−3	8.52	0.57	6.3
4	−2	8.72	0.37	4.1
5	−1	8.83	0.26	2.9
6	0	9.09	0	0.0

with positive and negative temperature on the roof, that is, ice will form and meltwater will accumulate on the roof. As the outside air temperature decreases, the size of the negative temperature area decreases. Similar studies at other attic air temperatures are shown in Table 2.

Table 2 shows that at an air temperature in the attic of 1 °C, the temperature on the entire outside surface of the roof is either negative, with snow melting does not occur, or positive—snow melts on the entire surface. In the second case, nothing prevents the flow of meltwater from the roof. Figure 4 shows the temperature field of the eaves at the air temperature in the attic 1 °C and the outside temperature −1 °C.

Also from Table 2, it is seen that as the air temperature decreases in the attic, the temperature range of the outside air at which ice is formed on the roof decreases. That is, the time of ice formation and its quantity decreases. Thus, at an air temperature in the attic of 8 °C, the ice on the roof is formed at an outside air temperature of −4 to −1 °C, and at a temperature of 2 °C only at an outside air temperature of −1 °C.

Ventilation system in which the air is removed from the premises not outside, but in the attic leads to an increase in temperature (up to 8 °C) and relative humidity in the attic up to 40%. This contributes to the formation of condensate on the inner surface of the cover, which at frosty temperatures on this surface forms a frost. As the surface temperature rises, the frost melts and dampens the wooden roof structures, attic flooring and exterior walls.

To analyze the conditions under which water vapor condensation occurs, calculations of the temperature fields of the roof cross section at the air temperature in the attic from 1 to 8 °C with a step of 1 °C and the outside air temperature of −5, −10, −15 and −20 °C were performed. At these temperatures, the minimum temperature on the inner surface of the cover and the length of the section on that surface with a temperature below the dew point was determined. The dew point temperature in the attic air temperature from 1 to 8 °C and relative humidity of 40% is shown in Table 3.

Table 2 The segment with a negative temperature length depending on the outside air temperature at different air temperatures in the attic

№	The temperature of the outside air, °C	The temperature of the inside air, °C	The length of the outside surface, m, roofs with		The percentage of length of the roof outside surface with negative temperature, %
			Positive temperature	Negative temperature	
1	7	−4	0	9.09	100.0
		−3	7.76	1.33	14.6
		−2	8.68	0.41	4.5
		−1	8.8	0.29	3.2
		0	9.09	0	0.0
2	6	−4	0	9.09	100.0
		−3	7.61	1.48	16.3
		−2	8.51	0.58	6.4
		−1	8.77	0.32	3.5
		0	9.09	0	0.0
3	5	−3	0	9.09	100.0
		−2	7.75	1.34	14.7
		−1	8.72	0.37	4.1
		0	9.09	0	0.0
4	4	−3	0	9.09	100.0
		−2	7.58	1.51	16.6
		−1	8.51	0.58	6.4
		0	9.09	0	0.0
5	3	−2	0	9.09	100.0
		−1	7.75	1.34	14.7
		0	9.09	0	0.0
6	2	−2	0	9.09	100.0
		−1	7.51	1.58	17.4
		0	9.09	0	0.0
7	1	−1	0	9.09	100.0
		0	9.09	0	0.0

The results of the study of the minimum temperature on the inner surface of the cover and the length of the area on this surface with a temperature below the dew point are shown in Table 4.

Table 4 shows that the length of the area with a temperature below the dew point with decreasing air temperature in the attic decreases. Thus, at an attic temperature of 8 °C and an outside air temperature of −15 °C, the length of the condensation area is

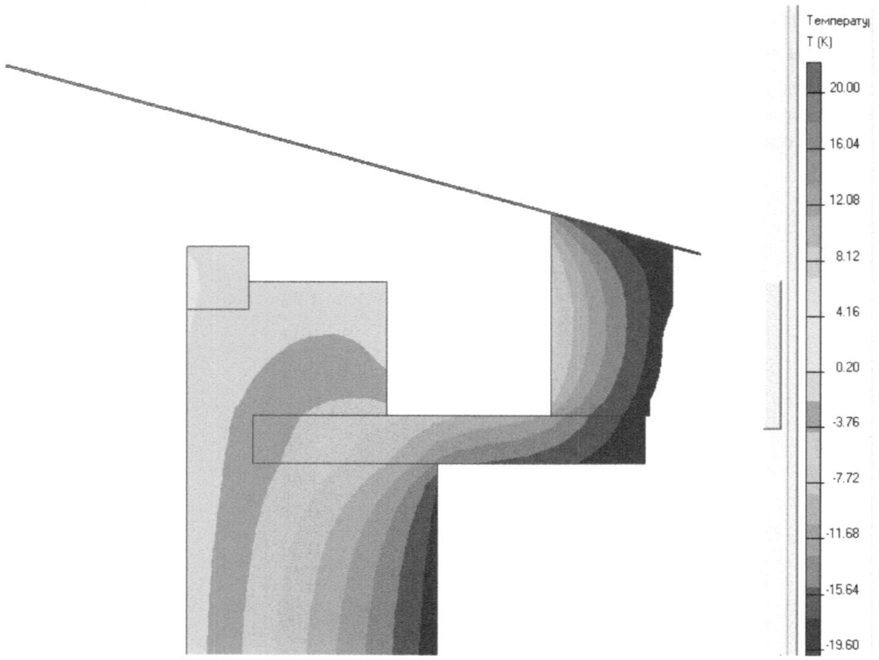

Fig. 4 The temperature field of the eaves at the air temperature in the attic 1 °C and the outside temperature −1 °C

Table 3 Dew point temperature

Dew point temperature, °C, at relative humidity of 40% and its temperature							
1	2	3	4	5	6	7	8
−11.1	−10.2	−9.3	−8.4	−7.5	−6.6	−5.7	−4.8

Table 4 The minimum temperature on the inner surface of the cover and the length of the area with a temperature below the dew point

№	Attic air temperature, °C	The minimum temperature on the inner surface of the cover, °C, and the length of the area, m, with a temperature below the dew point, at outside air temperature			
		−5	−10	−15	−20
1	1	−3.9/0	−7.6/0	−11.2/0.07	−14.8/8.69
2	2	−3.6/0	−7.3/0	−10.9/0.1	−14.6/8.69
3	3	−3.4/0	−7/0	−10.5/0.92	−14.3/8.69
4	4	−3.1/0	−6.7/0	−10.4/8.69	−14/8.69
5	5	−2.8/0	−6.5/0	−10.1/8.69	−13.7/8.69
6	6	−2.6/0	−6.2/0	−9.8/8.69	−13.5/8.69
7	7	−2.3/0	−5.9/0.02	−9.6/8.69	−13.2/8.69
8	8	−2/0	−5.6/0.125	−9.3/8.69	−12.9/8.69

8.69 m, and at a temperature in the attic of 1 °C only 0.07 m. Also with decreasing the attic temperature decreases the outside air temperature range at which condensation occurs. So at a temperature in the attic of 8 °C, the condensation occurs at an outside air temperature of −10 °C and below, and at 1 °C from −15 °C and below.

3 Conclusions

1. Restructuring of the attic ventilation system, in which the air from the rooms is not output outside but into the attic and there is insufficient resistance to the heat transfer of the attic floor, leads to an increase in the attic temperature.
2. An increasing the attic temperature in the winter–spring period promotes the formation of areas with a negative and positive temperature on the outer surface of the roof. With the simultaneous existence of these areas on top of the roof the snow melting occurs. The meltwater flows down, enters the area with a negative temperature and creates an ice barrier that prevents the flow of meltwater from roof. The water that accumulates before the ice flows through the non-sealed joints of metal sheets and moisturizes the wooden roof structures, the attic heater and the exterior walls.
3. Raising the temperature in the attic promotes condensation on the inner surface of the cover. At low outside temperatures, the condensing moisture turns to frost. As the outside air temperature rises, the frost begins to melt; the water flows down to the eaves and moisturizes the roof structure and the outer walls.
4. To prevent these effects, it is necessary to lower the attic temperature to 1 °C. At this temperature, the simultaneous existence of areas with different temperatures does not occur. Also at this temperature, a large area of water vapor condensation on the inner surface is observed only when the outside temperature is −20 °C. The time period with such outside temperature for the climatic conditions of Poltava region is insignificant.
5. In the case where the attic temperature exceeds 4 °C, the eaves of the roof section must be warmed up in accordance with paragraph 12 [12].
6. To reduce the attic temperature, it is necessary that the air from the premises of the building is removed by the ventilation system not into the attic but directly outside. You must also provide a reduction of the relative humidity in the attic by bringing it ventilation in compliance with rules.

References

1. Semko, O., Yurin, O., Avramenko, Yu., & Skliarenko, S. (2017). Thermophysical aspects of reconstruction of cold roof spaces. *MATEC Web of Conferences, 116,* 02030. https://doi.org/10.1051/matecconf/201711602030.

2. Yurin, O., & Galinska, T. (2017). Study of heat shielding qualities of brick wall angle with additional insulation located on the outside fences. *MATEC Web of Conferences, 116,* 02039. https://doi.org/10.1051/matecconf/201711602039.
3. Filonenko, O. (2018). Definition of the parameters of thermal insulation in the zone of building foundation according to the ground freezing depth. *Energy Efficiency, 11*(3), 603–626. https://doi.org/10.1007/s12053-017-9600-x.
4. Leshchenko, M. V., & Semko, V. O. (2015). Thermal characteristics of the external walling made of cold-formed steel studs and polystyrene concrete. *Magazine of Civil Engineering, 8,* 44–55. https://doi.org/10.5862/MCE.60.6.
5. Pichugin, S. F., Semko, V. O., & Leshchenko, M. V. (2017). Probabilistic analysis of thermal performance of the wall from light-gauge thin-walled steel structures. *Collection of Scientific Papers. Series: Haluzeve mashynobuduvannia, budivnytstvo. Poltava. Issue 1*(48), 144–155.
6. Avramenko, Y., Filonenko, O., Leshchenko, M., Mahas, N., Maliushitskii, O., Semko, V., et al. (2017). *Uteplennia, remont ta rekonstruktsiia ploskykh pokrivel tsyvilnykh budivel: posibnyk.* Poltava: Astraia. ISBN 978-617-7669-00-4.
7. Murgul, V. (2016). Methodology to improving energy efficiency of residential historic buildings in St. Petersburg. *MATEC Web of Conferences, 53,* 01046. https://doi.org/10.1051/matecconf/20165301046.
8. Diakaki, C., Grigoroudis, E., Kabelis, N., Kolokotsa, D., Kalaitzakis, K., & Stavrakakis, G. (2010). A multi-objective decision model for the improvement of energy efficiency in buildings. *Energy, 35*(12), 5483–5496. https://doi.org/10.1016/j.energy.2010.05.012.
9. Garay, R., Uriarte, A., & Apraiz, I. (2014). Performance assessment of thermal bridge elements into a full scale experimental study of a building façade. *Energy and Buildings, 85,* 579–591. https://doi.org/10.1016/j.enbuild.2014.09.024.
10. Naji, S., Çelik, O. C., Alengaram, U. J., Jumaat, M. Z., & Shamshirband, S. (2014). Structure, energy and cost efficiency evaluation of three different lightweight construction systems used in low-rise residential buildings. *Energy and Buildings, 84,* 727–739. https://doi.org/10.1016/j.enbuild.2014.08.009.
11. DBN V.2.6-31:2016. (2016). *Teplova izoliatsiia budivel* (70s). Kyiv: Minrehion Ukrainy.
12. DBN V.2.6-220:2017. (2017). *Pokryttia budynkiv i sporud* (129s). Kyiv: Minrehion Ukrainy.

One-Piston Mortar Pump with Increased Volume Combined Compensator Working Processes Analysis

Mykola Shapoval, Viktor Virchenko, Maksym Skoryk, and Anatolii Kryvorot

Abstract The mortar pump with various combined compensators design solutions proposed, namely with a combined pressure compensator and a combined compensator of increased volume. The solution pumps design features and the principles of their operation are presented. The analysis of single-piston increased volume combined compensator pump working processes presented, which in the studies proved to be more effective.

Keywords Solution pump · Suction chamber · Suction and discharge valves · Pressure in the injection nozzle

1 Introduction

For the mechanization of construction work during finishing work, the common use of the mortar pump has become common. The working pressure in the pipelines, the specific consumption of electricity for pumping, the convenience and quality of solutions mechanized application on the surface to be treated, the amount of losses during joining, the life of the pipelines and the parts of the mortar pumps depend on the level of pulsation when supplying building solutions.

When creating a modern meltdown pump with moderate pulsation, it is necessary to be guided by the appropriate factors, namely using a simple design of the drive, which will provide the working bodies with constant speed during smooth dynamic loads during the transition through the extreme points and endurance over time; construction of the hydraulic part of the pump in such way that a stable soluble mixture flow is ensured, taking into account its rheological properties, both in the half-cycle of suction and in the semi-cycling of the injection, which in turn will increase the volume efficiency of the pump; using high-performance and simple pressure pulsation compensators.

M. Shapoval · V. Virchenko (✉) · M. Skoryk · A. Kryvorot
Poltava National Technical Yuri Kondratyuk University, Poltava, Ukraine
e-mail: virchenko.viktor@gmail.com

V. Onyshchenko et al. (eds.), *Proceedings of the 2nd International Conference on Building Innovations*, Lecture Notes in Civil Engineering 73,
https://doi.org/10.1007/978-3-030-42939-3_27

In order to reduce the pulsation in modern single-piston rigs, pressure compensators are used in the form of different volumes air caps. However, the air compensators, along with the advantages, have significant disadvantages: The compressed air is in direct contact with the pumped solution and therefore removed intensively from the compensator during the operation of the solenoid pump, especially accelerated air removal at high pressure (above 1.5 MPs). In such case, the efficiency of the compensator is significantly reduced, namely productivity and volume efficiency decrease, while the pulsations are increasing.

Therefore, it is necessary to create a single-piston pump with high efficiency, reliable with an advanced cylinder-propeller group and drive, which will provide moderate pulsations of supply of solution, smoothness of the drive and increased volumetric efficiency. The new compensators designs development and the improvement of existing ones will ensure the efficient operation of the mortar pumps.

2 Review of the Latest Research Sources and Publications

There are designs of both domestic and foreign production for pumping solutions of various mobilities.

Domestic piston mortar pumps are used for pumping mortars with a minimum of 5–7 cm and a fraction size of not more than 5–12 mm.

Pumping of the solution is carried out by the piston reciprocating motion with a direct influence on the solution during its suction and injection.

Piston pumps of the mark CO [1] consist of an actuator, cylinder-piston group, working and valve chambers with suction and pressure spherical self-acting valves, an air cap (except two pistons) for pressure pulsations smoothing, a control panel and a frame, where all the nodes are mounted.

The main characteristics of the piston mortar are as follows: solution-independent flow from the developed head, good suction power and high resource cylinder-piston group (about 2000 mash-year). Piston pumps are as uniform as possible and are intended for the complete set of plaster aggregates and stations. They are mounted, as a rule, on wheelchairs, which ensures their high mobility.

To the most significant single-piston mortars disadvantage of the brand CO [1], the flow pulsation of the pumped environment can be attributed, which is caused by a single work action. The use of air caps pumps constructions will provide pressure pulsations degree smoothing.

Differential mortar pumps with one piston used to provide a reduced level of pulsations of pumped solution [2].

The differential scheme of the pump operation ensures the process of suction periodically, and injection—continuously, which makes it possible to reduce significantly the pressure pulsations level of the pumped solution. However, in the given pump, the working chambers have very complex design and the solution motion direction in them.

In differential solenoid pumps with a flow piston, despite the tangible advantages in the simplicity and reliability of such design, it was not possible to get rid of the pulsations of supplying solution completely, which is due to the use of drives that provide uneven motion of the piston.

Famous firms involved in the development and implementation of production are the German firm Putzmeister Werk Maschinenbau GmbH and the Italian firm Turbosol [3].

In 1965 in Germany, the Putzmeister [4] patented a two-cylinder differential solenoid pump K-139, in which periodic suction process ensures the reduction of the pulsations, and the process of injection is carried out continuously.

Plastering machine of this model line with piston pumps KA-139 [4] is a universal machine for preparing solutions and plaster of almost all kinds. These machines are characterized by high performance, even when operating in the most difficult cases, and moderate pressure pulsations. Due to the powerful mechanical drive use, the mortar pumps, in contrast to other types of pumps, transport even hard-to-mix materials.

Mortar pumps of the brand UNI 30 (manufactured by Turbosol [3]) with an engine power of 7–13 kW, with a feed pressure of 0.45–0.5 MPs are ones of the most powerful piston pumps, ideal for mixing, feeding, drawing and extruding with conventional or special cement solutions with a fraction of application up to 10 mm. They are supplied with an electric or diesel engine of standard configuration (S) or with built-in mixer (MF).

Mortar pump can be used for a wide range of works, when performing Shrekot works for application of general and special liquid mixtures, for pumping, delivery of both special solutions and ready-mixed premixed mixtures in difficult conditions, as well as injections with regulated pressure.

Mortar pumps are equipped with an automatic device for protection against excess pressure, recirculation device and pneumatic remote control (start/stop).

The disadvantage of Turbosol solenoid pumps [3] is the increased pressure pulsation due to single operation, low volume efficiency and increased metal capacity.

3 Definition of Unsolved Aspects of the Problem

Analysis of solution pumps existing designs has shown that it is necessary to develop a solution pump simple in design and effective when pumping building solutions with improved parameters, such as increased productivity, volume efficiency and reduced energy consumption.

Use of solution pump composition in a combined compensator will provide a solution pulsations flow low degree.

4 Problem Statement

In order to achieve an increase in the parameters of the solution pump, as well as to reduce the flow pulsations degree, it is necessary to investigate and analyze:

- the performance of one-piston mortar with an increased volume combined compensator;
- to investigate the working processes of the solution pump under conditions of operation close to the production;
- to analyze the feasibility of using design changes in the design of the mortar pump to ensure its efficient operation.

5 Basic Material and Results

The melting pump [5–9] proposed for consideration has a principle scheme that allows the use of each of the combined compensators, the constructive solutions of which are presented in Figs. 1 and 2.

The results of theoretical and experimental studies (Figs. 3, 4 and 5) indicate that the solution pump with the presence of an increased volume combined compensator in relation to the solution pump with the pulsation pressure compensator has considerably improved technical characteristics: performance, pressure pulsations, volume efficiency and energy consumption, therefore it is the actual design consideration.

The pressure influence in the suction chamber and the nozzle, displacements of the injection balls, suction valves and the piston on the solution pulsation during the solenoid pump operation with combined compensator of enlarged volume was experimentally determined.

The diagrams show the general physical patterns of characteristic phases of work processes (indicated by points) and the influence of elastic elements on the operation of valves when pouring solutions of different mobility.

When the piston moves to the extreme "dead" point in the suction chamber, the suction process begins (point A, Figs. 3, 4 and 5). In this period, there is a sharp drop in pressure, both in the suction chamber and in the nozzle (section A–B, Figs. 3, 4 and 5, f, g). In the suction chamber, there is a pressure below the atmospheric pressure (0.05–0.1 MPs), which affects the opening of the suction valve ball. In this case, the suction valve ball, both spring loaded and free acting (point B, Figs. 3, 4 and 5, b, c), begins to open at the same time at the movement of the solution P 8 cm. This is due to the fact that the suction valve elastic element in thick solutions loses their properties (the less mobile solution resistance forces exceed the elastic element resistance strength). On more mobile solutions of P 10–12 cm, the spring-loaded suction valve operates with a slight delay in relation to free moving (section A–B, Figs. 3, 4 and 5, b, c), the period of valve operation with both design features is less, and the pump has a higher suction capacity on more mobile solutions.

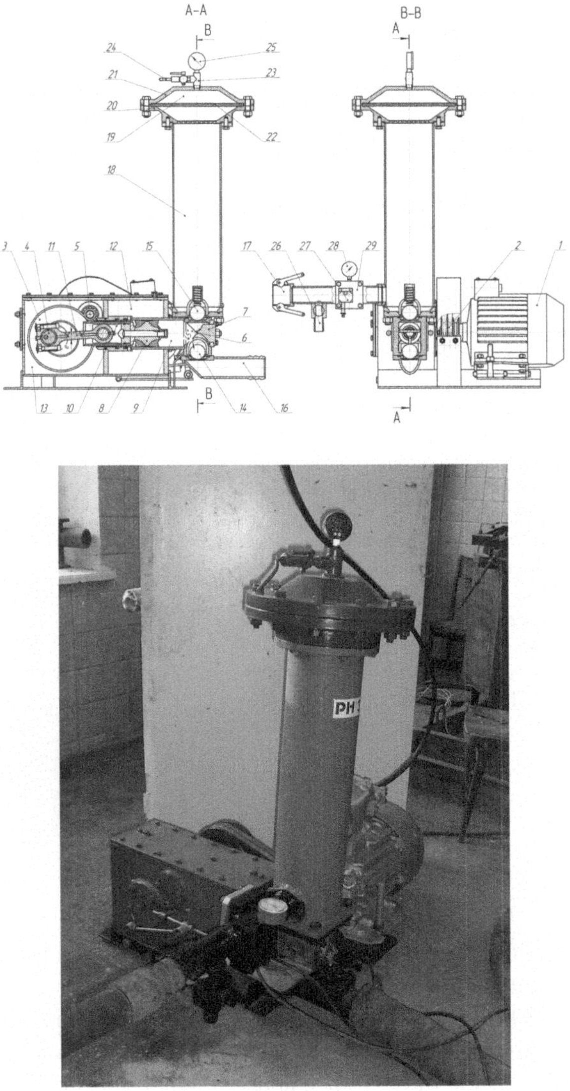

Fig. 1 Single pump with pressure pulsation compensator: 1—electric motor; 2—Cape Harbor transmission; 3—gear wheel; 4—crankshaft; 5—gear shaft; 6—suction chamber; 7—special insert; 8—piston; 9—working cylinder; 10—a slider; 11—a connecting rod; 12—camera; 13—gearbox housing; 14, 15—suction and pressure spring-loaded ball valves; 16—suction pipe; 17—injection nozzle; 18—cylindrical chamber; 19—locked camera; 20—base; 21—upper cover; 22—rubber-coated diaphragm; 23—nozzle; 24—nipple; 25, 28—pressure gauge; 26—pressure drop valve; 27—pressure relay; 29—diaphragm chamber

Then, the further absorption process goes, which reduces the pressure in the suction chamber (point B, Figs. 3, 4 and 5, d), and the flow of the solution suction well chamber under the dilution influence moving.

The drop in pressure in the suction chamber is associated with a decrease in the speed of the working body. At the same time, the opening of the ball of the suction valve and its subsequent freezing (section B–G, in Figs. 3, 4 and 5, b, c) and the pressure drop in the suction chamber to the minimum value (point 2, Figs. 3, 4 and 5, g) are observed. Moreover, when pumping a solution with a movement of P 8 cm,

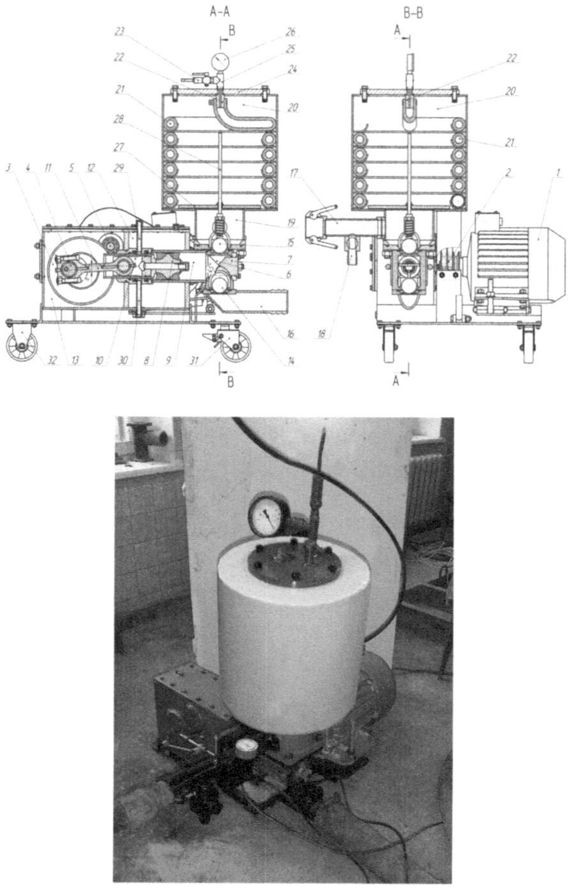

Fig. 2 Single pump solvent pump with combined compensator for increased volume: 1—electric motor; 2—Cape Harbor transmission; 3—gear wheel; 4—crankshaft; 5—gear shaft; 6—suction chamber; 7—special cylindrical insert; 8—piston; 9—working cylinder; 10—a slider; 11—a connecting rod; 12—camera; 13—gearbox housing; 14, 15—suction and pressure spring-loaded ball valves; 16, 17—suction and discharge nozzles; 18—pressure drop valve; 19—injection chamber; 20—cylindrical chamber; 21—closed camera; 22—the connection of the unit of air swapping; 23—a nipple; 24—cover; 25—crane; 26—pressure gauge; 27—a float limiter; 28—guide rod; 29, 30—channel branch pipes; 31, 32—a pair of wheels

both freezing and operation with a spring-loaded valve are longer in relation to free running.

When the piston moves in the direction of the injection stroke and the passage of the extreme "dead" point (point B, Figs. 3, 4 and 5, a), there is a reverse flow of the solution through the suction valve in the suction chamber, which is located in the upper open position and is delayed. Especially, the largest reverse turning occurs when pumping a solution with a mobility of P 8 cm.

In further piston movement, the pressure in the suction chamber begins to increase (section B–G, Figs. 3, 4 and 5, g), and the solution contained in it is compressed. In the discharge nozzle, the pressure drops to a minimum (section B–G, Figs. 3, 4 and 5, g). At the moment when the pressure in the suction chamber reaches the pressure level in the nozzle (point G, Figs. 3, 4 and 5, f, g), the pressure valve starts to open and almost simultaneously closes the suction valve.

The injection spring-loaded valve ball opening is carried out with some delay in relation to the free ball, which is due to the elastic element resistance strength of the

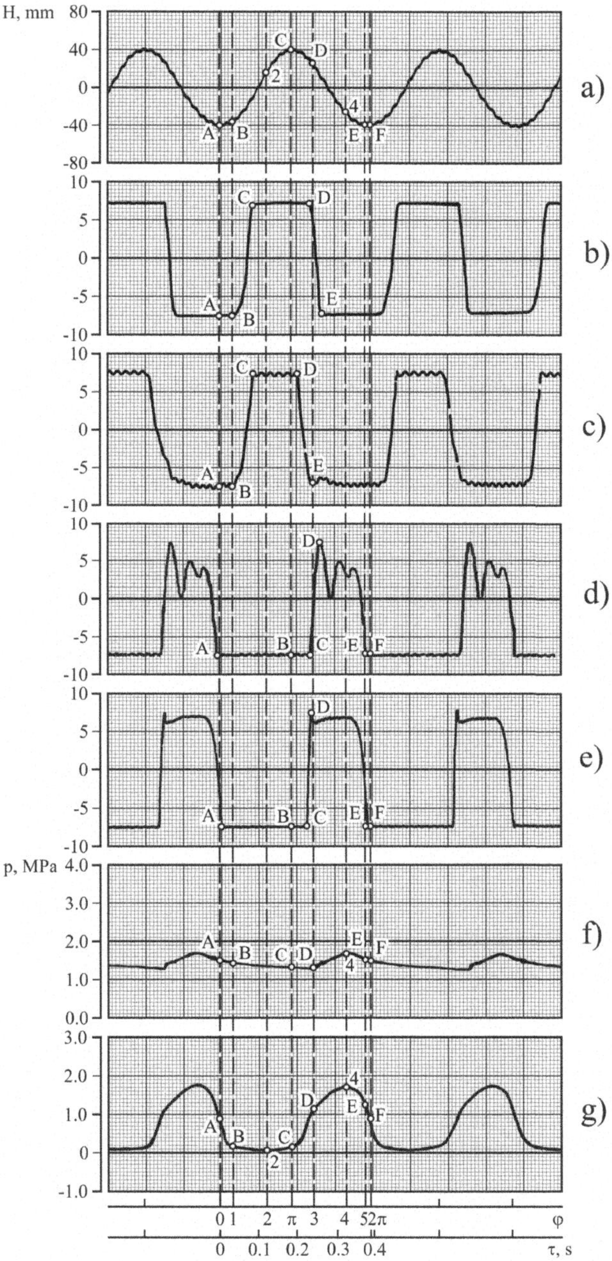

Fig. 3 Dependencies: displacement of the piston (**a**), moving the ball of the suction spring valve (**b**), moving the ball of the suction free-acting valve (**c**), moving the ball of the injection spring-loaded valve (**d**), moving the ball of the injection free-wheel valve (**e**), changing the pressure in the discharge nozzle (**f**) and the suction chamber (**g**) when pumping the solution P 8 cm from the angle of rotation of the crank or time for the cycle of the solenoid pump

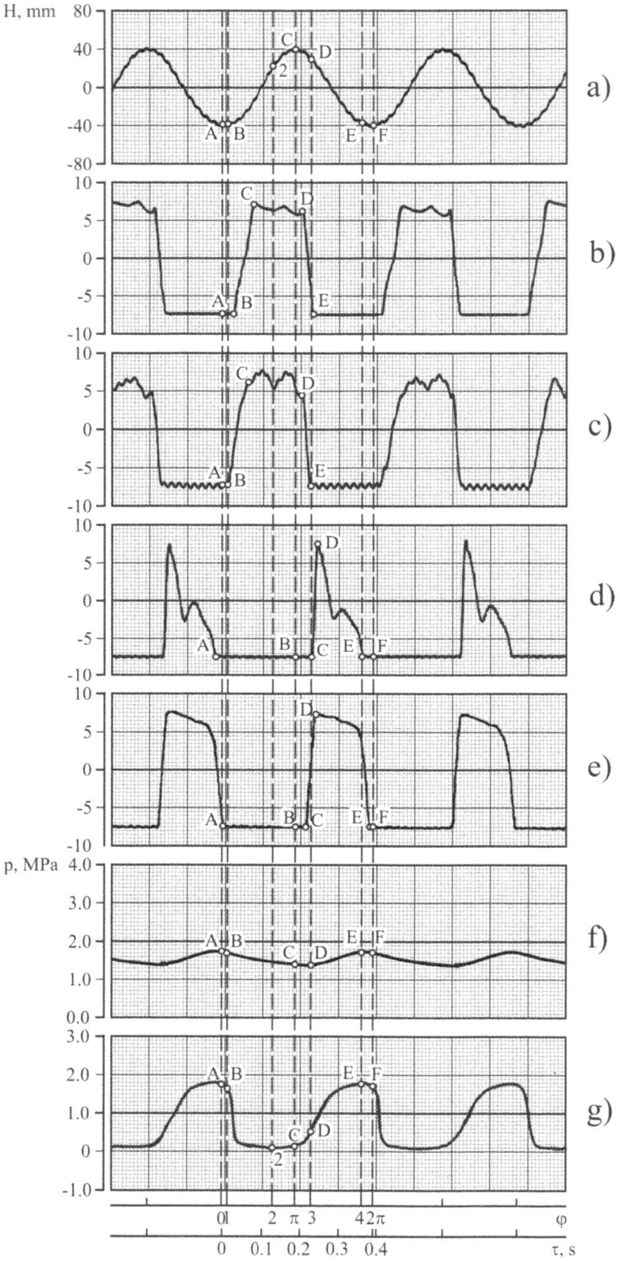

Fig. 4 Dependencies: displacement of the piston (**a**), displacement of the ball of the suction pylon (**b**), the displacement (**c**) of the ball of the injection spring-loaded valve (**d**), the displacement of the ball of the injection free-running valve (**e**), the pressure change in the discharge nozzle (**f**) and the suction chamber (**g**) when the solution is pumped L 10 cm from the angle of rotation of the crank time or for a series of pump

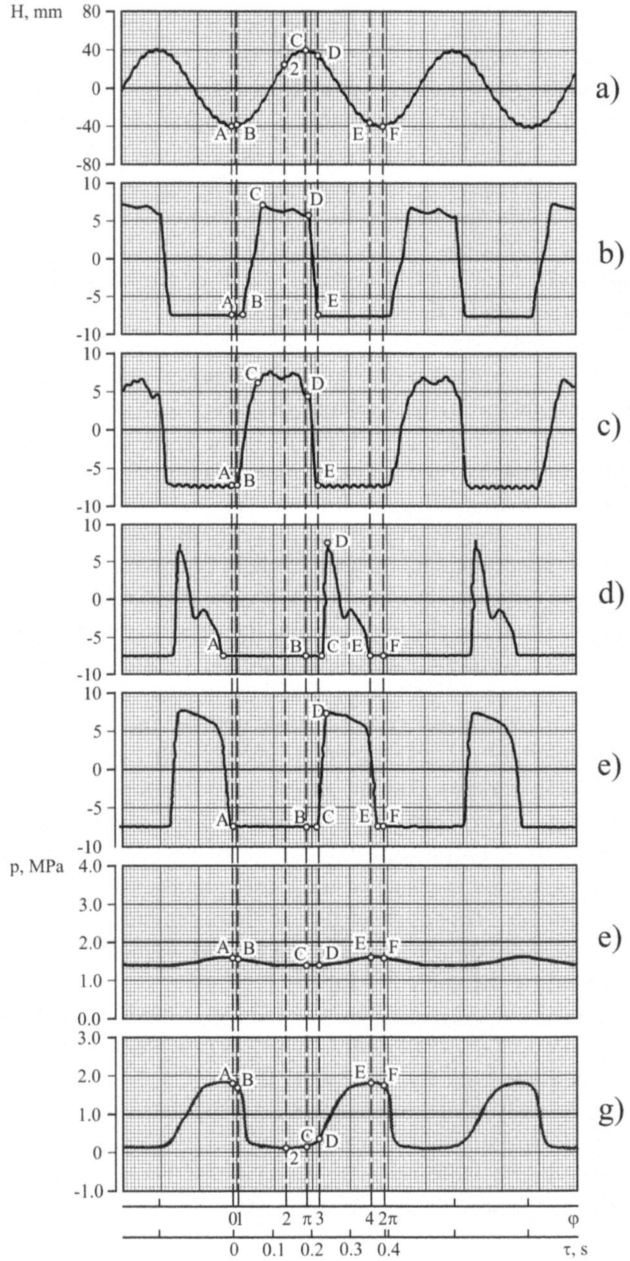

Fig. 5 Dependencies: displacement of the piston (**a**), moving the ball of the suction spring valve (**b**), moving the ball of the suction free-acting valve (**c**), moving the ball of the injection spring-loaded valve (**d**), moving the ball of the injection free-acting valve (**e**), changing the pressure in the injection nozzle (**f**) and suction chamber (**g**) when pumping a solution P 12 cm from the angle of rotation of the crank or time for the cycle of the solenoid

valve (point B, Figs. 3, 4 and 5, g, f). During the piston subsequent movement, the pressure in the suction chamber and the nozzle continues to increase to its maximum value simultaneously (point 4, Fig. 3, point D of Figs. 3, 4 and 5, g, f).

It is advisable to note that when pumping a solution of P 8 cm, the maximum pressure is reached somewhat earlier than when pumping solutions of P 10 and 12 cm. This is due to the greater resistance of the P 8 cm fluid solution flow.

Further, there is a decrease in pressure in the suction chamber. This is typical for pumping P 10–12 cm solutions—the reduction in pressure almost coincides with the injection spring-loaded valve ball lowering, which is caused by the movement of pumped solutions (section D–E, Figs. 3, 4 and 5, f).

The analysis of the diagrams shows that, due to the elastic element efficiency, the injection valve ball lowering is more likely to be carried out in relation to the free acting, notwithstanding the later, its relative discovery.

After the shut off valve (point D, Figs. 3, 4 and 5, a), the pressure drop (and, accordingly, the expansion of the solution) in the suction chamber continues to the pressure level of the suction valve opening and subsequent filling with its solution. In the discharge nozzle, the pressure drops to a certain level of pressure due to the operation of the compensator, and then the cycle of operation repeats.

In addition to the peculiarities of the work process, it should be noted that for all the given diagrams, both the closing time and the time of pressure valve, pressing regardless of its spring load (the segment from point 5 to point $\varphi = 2\pi$, the segment from point 4 to the point $\varphi = 2\pi$) (Figs. 4 and 5), is much less, but is ahead of the suction (segment $\varphi = \pi$ to point 3). In addition, with the increase in the mobility of the solution, the length of these segments for the suction valve is reduced, and for the injection valve, it increases. Taking into account that the length of the segment of the diagram along the axis between the "dead" points of the same name is 40 ± 0.5 mm (one cell of the diagram corresponds to approximately 9° turn of the crank), this allows an approximate estimate of the closing angle valves by measuring the corresponding segments. Thus, the lengths of the segments $5 - 2\pi$ in Fig. 4 are 1.5 mm, that is, it is approximately equal to 13.5°, and segments $\pi - 3$ are 4–5 mm, that is, it corresponds to the angle 36–45°.

In Figs. 3, 4 and 5, the pressure diagrams indicate a moderate pulsation, respectively, at P8 26%, P10 22.8%, P12 16.9%, which is provided by the construction of an enlarged compensator with a volume of 47 dm^3.

The results of the analysis of the dependencies of the characteristics of the working process of the prototype single-piston mortar with a combined compensator of the increased volume confirm and summarize the results of theoretical and experimental studies to determine the performance of the work.

6 Conclusions

According to the results of experimental studies, it is seen that the pressure of the pressure valve due to spring load is much faster, and the springing of the suction valve did not give positive results. Therefore, the installation of the special insert in the suction cup reduced its harmful volume and ensured a quicker suction of the suction valve, which positively influenced on the growth of productivity and volume efficiency and reduced the level of the pulsations of the pressure of the solution. Also, the results of studies on the influence of accepted factors on pump performance and volume efficiency indicate that all the indicators are higher in a solution pump with a combined compensator of the increased volume, as well as a significant decrease in the level of pulsation of pressure (by 25%) for which the ranges of rational technological parameters are the mobility of the solution $\Pi = 9.5$–10.5 cm, pressure of the solution $p = 1.0$–1.5 MPs; the air volume of the compensator brought to atmospheric pressure $V_{\text{komp}} = 43$–50 dm^3; angle of inclination of a special insert in the suction chamber $\beta = 37.5$–52.5°.

References

1. Solvent Pumps CO—167 Passport. (1986). *Dneprorudnensky plant of construction and finishing machines*. Vasilyevka: Vasilyevskaya Printing House.
2. Ustyantsev, V. V., Onischenko, A. G., Sofiychenko, V. F., & Litvinenko, P. S. (1988). AC 1390434 MCI 4 F 04 B 5/00, 11/00. Place of publication: in BI, 7.05.88. № 17. USSR.
3. Mastyayev, V. M., Parfyonov, E. P., & Baulin, E. G. (1970). *Testing of the plastering unit of the Italian company Turbosol Meh. Tool and finishing machines. Issue 1: Inform. scientific-technical Sat.* Moscow: TsNIITEstroymash.
4. Mortelpumpen und ihre Entwicklung. (1969). *Fordern und Heben*. № 15. (Germany).
5. Emelyanova, I. A., & Shapoval, N. V. (2016). An analysis of the effective operation of solute pumps with a combined pressure surge compensator and a combination of a larger volume compensator. In *Materials of the XII International Scientific and Practical Conference, Science and Civilization, Volume 16. Ecology. Geography and geology. Agriculture. Construction and architecture* (Vol. 80, pp. 60–67). Sheffield: Science and Education LTD.
6. Onishchenko, A. G., Shapoval, N. A., & Vasil'ev, A. V. (2001). One-piston mortar pump with combined compensator for pressure. *Mekhanizatsiya Stroitel'Stva, 4,* 4–6.
7. Shapoval, M. V. (2012). Influence of parameters of work of the combined compensator on the level of pressure ripple. In *Construction. Material science. Mechanical engineering. Intensification of work processes of construction and pre-harvesting machines. Sat. scientific research ser.: Lifting, transporting, building road machines and equipment* (Vol. 66, pp. 204–211). Dnepropetrovsk: PSU.
8. Shapoval, M. V. (2013). Analysis of the effect of flow pressure on the volumetric pump efficiency. In *Collection of scientific works sectoral mechanical engineering, construction* (Vol. 1, No. 36, pp. 19–204). Poltava: PoltNTU.
9. Yemelyanova, I. A., & Shapoval, M. V. (2017). Determination of performance and volumetric efficiency of a mortar pump depending on the geometric parameters of the suction chamber and the compensators of various design solutions. *Scientific Bulletin of Construction, 88*(2), 195–203.

Calculation of Bending Composite Steel and Concrete Elements with Glutinous Connection of Concrete and Steel According to Theory of Compound Rods

Oleksandr Skurupiy, Yurii Davidenko, Oleksandr Horb, and Pavlo Mytrofanov

Abstract The article describes methods of calculation of bearing capacity of composite elements in which the joint work of steel and concrete is provided through bonding, according to the theory of compound rods. This method of calculating structures with adhesive joints takes into account the physical and mechanical characteristics of the joint material. We compare experimental results with theoretical calculation.

Keywords Composite steel and concrete element · Bonding · Compound rod · Loading · Bearing capacity

1 Problem Statement

Recently, due to the difficult economic situation most areas of industry are developing by reducing the cost, lower cost of materials and the complexity of the basic technological processes, and the construction industry is no exception. The creation of new progressive structural elements for the erection and reconstruction of buildings and structures in the context of strategic national and world trends is increasingly in need.

Nowadays, the steel-concrete structures in which steel and concrete are combined the most rationally satisfy these requirements. However, researchers of these structures increasingly face with the problem of ensuring the interoperability of their components. The use of adhesives has become widespread in the manufacture and restoration of load-bearing structures, as well as in the anchoring of joints to ensure the concrete and steel joint work. The most effective in these cases were acrylic adhesives. Therefore, a large number of traditional steel-reinforced concrete structures have been used instead anchors adhesives based on these compounds' acrylates.

O. Skurupiy · Y. Davidenko · O. Horb · P. Mytrofanov (✉)
Poltava National Technical Yuri Kondratyuk University, Poltava, Ukraine
e-mail: Mytrofanov.P@gmail.com

O. Horb
e-mail: olhorb@gmail.com

V. Onyshchenko et al. (eds.), *Proceedings of the 2nd International Conference on Building Innovations*, Lecture Notes in Civil Engineering 73,
https://doi.org/10.1007/978-3-030-42939-3_28

However, the question of how to calculate the load-bearing capacity of a new type of structural element, namely considering the material properties of the adhesive joint, was unresolved.

2 Analysis of Recent Research and Publications

For a long time, the Zolotov Scientific School [1], which was established at the Kharkiv National Academy of Urban Economics, was studying the physical and mechanical characteristics of adhesive joints and were applying them in construction practice. In recent years, established schools of the National Aviation University (Kyiv) [2] and Poltava National Technical Yuri Kondratyuk University have contributed to the process [3]. They proved a significant advantage of acrylic adhesives among all others used in construction. This glue is durable, reliable and easy to prepare, fits well. And the latest developments of traditional anchoring tools [4–8] are too time consuming, their installation has high material intensity and requires high skills of workers [9–14]. These developments were used for the combination of steel and concrete in the manufacture of steel-reinforced concrete slabs on profiled flooring [15]. It is suggested to arrange them instead of usual anchoring of the connection based on acrylic adhesives. The experience of using the theory of compound rods [16], which was developed and tested by A. Rzhanitsyn at the beginning of the last century, is taken into account. We decided to use it to develop a method of calculating structures with adhesive joints, which will allow to take into account the physical and mechanical characteristics of the joint material.

3 The Purpose of the Article

The purpose of this article is to publish an adapted technique and an algorithm for determining the load-bearing capacity of steel-reinforced concrete structural elements, in which the joint work of steel and concrete is ensured by compound rod theory bonding using.

4 The Main Material

One of the main options for determining the load-bearing capacity of steel-reinforced concrete bending structural members is the calculation according to the theory of compound rods. Folded is a rod whose cross-section consists of two or more separate parts. In most cases, these parts are rigidly interconnected throughout the length, so they are considered monolithic. These package rods are considered like this, even when their components are made of different materials. However, in most cases, it is not possible to connect rigidly the individual rods sufficiently, and therefore there is a need to consider the bonding flexibility. Such rods are not monolithic and should be considered as a separate group, which is called compound rods.

The most common are metal and wooden folded rods, namely folded metal columns, stretched and compressed lap jointed steel package, glued plywood and board structures. Reinforced concrete and steel-reinforced concrete structures are also not difficult to divide into folded rods. There are also flat and spatial folded rods. The gap between the components of the link rod is called a seam. Among the steel-reinforced concrete structures, the most typical examples of composite rods are elements with external rigid reinforcement (beams and racks with framing steel sheets, profiled floor slabs, etc.) and steel-reinforced concrete void elements with cavities filled with concrete. As adhesive joints are increasingly used to make concrete and steel work together, there is a problem in determining the peculiarities of such structures and taking into account the physical and mechanical characteristics of the joint material during their calculation. If the theory of compound rods makes it possible to consider each component of composite structures in detail, and especially the adhesive joint, then its position should be used when calculating the above-mentioned elements.

Let us consider flat component rods. We assume that the working of each individual rod occurs according to the traditional laws of resistance of materials, in particular, the law of flat sections. Therefore, the internal stress state of each rod is considered to be determined, when it is possible to calculate the magnitudes of moments, normal and transverse forces in each section.

We consider the deformations of the rods too small in comparison with their length, so the problem can be solved with the help of linear equations according to the law of independence of action of forces. We believe that the rods are connected by continuously distributed links whose deformations are considered linear.

By purpose, the links in the composite rods are divided into two types: shear and cross-link. The main characteristic of the connection is determined by the relationship between the deformations occurring inside the compound rod and the internal forces caused by these deformations. In most cases, with small deformations, this dependence is considered linear. This stage of bonding is characterized by a coefficient of rigidity that expresses the relation of the bonding forces to their respective deformations.

For shear relationships, the stiffness factor ε is determined:

$$\varepsilon = \frac{T_3 m}{\delta_3}, \tag{1}$$

where T_3—is a one-link shifting effort, m—is the number of connections per unit length of seam, δ_3—is the shear deformation of adjacent fibers of two adjacent rods joined by shear bonds.

For cross-links, the corresponding stiffness factor η is determined:

$$\eta = \frac{S_3 m}{\Delta y_3}, \tag{2}$$

where S_3—is a one-link tensile force, m—is the number of connections per unit length of seam, Δy_3—is the magnitude of the transverse divergence of the adjacent bars connected by the cross-links. The dimension of the stiffness coefficients ε and η is [Pa].

Let us consider rods with evenly spaced joints along the length of each seam, and with varying the thickness of the joints along the length of the seam only stepwise. Even distribution of joints is also considered as placing separate identical joints along the length of the seam at an equal distance from each other.

During the operation of a composite rod formed of $n + 1$ individual rod, in each seam, forces arise which are a function of the coordinate x of the given section along the length of the rod. The ratio of these forces to the unit of length is denoted by τ [N/m]. The relative displacement of any point of the extreme fiber of one rod and the corresponding point of adjacent fiber of the adjacent rod is also a function of x and is denoted by δ [m]. As a result of the proportionality of the shear force and the responsiveness of the shear, we obtain:

$$\tau = \delta\varepsilon. \tag{3}$$

The extreme stress in the upper rod relative to the selected seam is denoted by σ_v and in the lower rod by σ_n.

The modulus of elasticity of the material of the upper and lower rods is denoted by E_v and E_n, respectively.

The increase in shear along the length dx will be equal to the difference of the elongation of the extreme fibers located on either side of the seam:

$$\mathrm{d}\delta = \left(\frac{\sigma_n}{E_n} - \frac{\sigma_v}{E_v}\right)\mathrm{d}x, \ \delta' = \frac{\sigma_n}{E_n} - \frac{\sigma_v}{E_v}. \tag{4}$$

The basic system of composite rod theory is a packet of rods that is not bonded by shear bonds. The value of δ'_κ for the kth seam arising from external loading and displacement forces that replace the action of unaccounted shear relationships is determined by:

$$\delta'_\kappa = \Delta_{\kappa 1} T_1 + \Delta_{\kappa 2} T_2 + \cdots + \Delta_{\kappa n} T_n + \Delta_{\kappa 0}. \tag{5}$$

The Δ_{ki} denotes the increment of displacement in the kth seam from the total single shear force acting in the ith seam to the left of the x-section. T_i denotes the total shear force in the ith seam, acting to the left of the cross-section x's and is determined by:

$$T_i = \int_{x_0}^{x} \tau_i \mathrm{d}x. \quad (6)$$

where $\Delta_{\kappa 0}$—is the increment of displacement in the kth seam from the action of an external load on the folded rod not secured by the bonds that perceive the shear forces on the seams.

The external load can be given by the axial force N_k^0 in the section of the x-axis of the given core of the main system and the moment M^0, which, due to the absolute rigidity of the cross-links, acts at once on all rods, causing a bend equal for all rods in the package. Z and u denote the distances from the seam to the centers of gravity of the section above and below the spaced rods, through v the sum of these distances. Thus, the stresses in the extreme fibers of the upper and lower rods from the action of moment M^0 are equal to:

$$\sigma_v = \frac{E_k z_k}{\rho} = \frac{M^0 E_k z_k}{\sum EI}, \quad (7)$$

$$\sigma_n = -\frac{M^0 E_{k+1} z_{k+1}}{\sum EI}. \quad (8)$$

Substituting the obtained values in (4) we get:

$$\delta'_\kappa = -\frac{M^0 z_k}{\sum EI} - \frac{M^0 u_k}{\sum EI} = -\frac{M^0 v_k}{\sum EI}, \quad (9)$$

Summing up the influence of axial forces and moments, we obtain the final expression for the increase of displacements from external loading:

$$\Delta_{k0} = \frac{N_{k+1}^0}{A_{k+1} E_{k+1}} - \frac{N_k^0}{A_k E_k} - \frac{M^0 v_k}{\sum EI}, \quad (10)$$

Next, we construct a system of differential equations for all seams considering (3), (5), (6) after determining the increment of displacements from the single total displacement forces and the total displacement increment for each seam:

$$\left.\begin{array}{l} \dfrac{T_1''}{\varepsilon_1} = \Delta_{11} T_1 + \Delta_{12} T_2 + \cdots + \Delta_{1n} T_n + \Delta_{10}, \\ \dfrac{T_2''}{\varepsilon_2} = \Delta_{21} T_1 + \Delta_{22} T_2 + \cdots + \Delta_{1n} T_n + \Delta_{20}, \\ \cdots\cdots\cdots\cdots\cdots\cdots\cdots\cdots\cdots\cdots\cdots \\ \dfrac{T_n''}{\varepsilon_n} = \Delta_{n1} T_1 + \Delta_{n2} T_2 + \cdots + \Delta_{nn} T_n + \Delta_{n0}. \end{array}\right\} \quad (11)$$

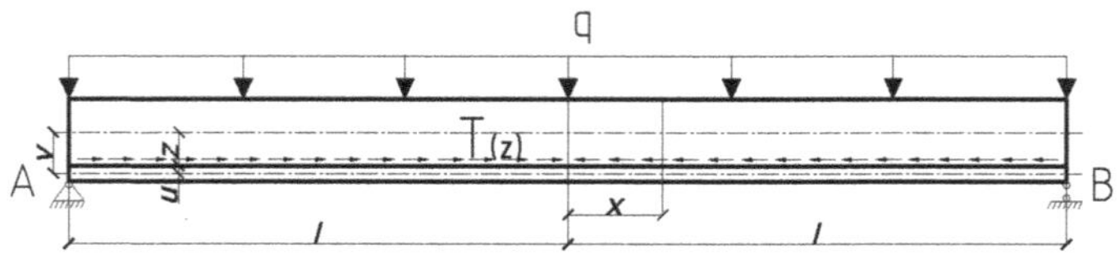

Fig. 1 Design scheme of the beam loaded with uniformly distributed load

The main unknown of equations system is T_i. Their definition is the main task. If the compound rod consists of two monolithic rods and is loaded with a uniformly distributed load (Fig. 1), then instead of system (11) we obtain one differential equation corresponding to the middle seam:

$$\frac{T''}{\varepsilon} = \gamma T + \Delta, \tag{12}$$

where the coefficient γ and the free term Δ are equal:

$$\gamma = \frac{1}{A_1 \cdot E_1} + \frac{1}{A_2 \cdot E_2} + \frac{v^2}{\sum EI}; \tag{13}$$

$$\Delta = -\frac{N_1^0}{A_1 \cdot E_1} + \frac{N_2^0}{A_2 \cdot E_2} - \frac{M^0 \cdot v}{\sum EI}; \tag{14}$$

For the case under consideration (a beam on two supports loaded with a uniformly distributed load composed of concrete and steel rods glued together by acrylic adhesive), the force is defined as:

$$T = \frac{q \cdot v}{2 \cdot \gamma \cdot \sum EI}(l^2 - x^2) - \frac{q \cdot v}{\gamma \cdot \lambda^2 \sum EI}\left(1 - \frac{\mathrm{ch} \cdot \lambda \cdot x}{\mathrm{ch} \cdot \lambda \cdot l}\right); \tag{15}$$

where λ is equal to:

$$\lambda = \sqrt{\varepsilon \cdot \gamma}; \tag{16}$$

and the coefficient of rigidity ε:

$$\varepsilon = \frac{b \cdot E_{\mathrm{sh}}}{v}, \tag{17}$$

Approximately the shear force T can be determined:

$$T = \left(\frac{M^0}{v}\right)\left(1 - \frac{K_{\mathrm{g}} \cdot M}{M^0}\right), \tag{18}$$

Total moment:

$$M^0 = M_0 + M_T = M_0 + T \cdot z_c^0; \tag{19}$$

The total moment without taking into account the forces in joints:

$$M_0 = V_0 \cdot \xi_c^2 \cdot b \cdot h_0 \cdot f_{ctk} + A_s \cdot R_s \cdot z_c^0; \tag{20}$$

Axial forces in rods:

$$N_s^0 = -N_c^0 = \frac{M^0}{z_c^0}; \tag{21}$$

Equivalent forces in the steel and concrete:

$$N_s = -N_c = N^0 - T; \tag{22}$$

Based on the values

$$x = \frac{\left(N^0 - T\right)}{b \cdot f_{cd}}; \tag{23}$$

$$z_c = h_o - 0.4 \cdot x \tag{24}$$

we check the bearing capacity of the compressed and stretched zones:

$$M \leq M_c = b \cdot x \cdot f_{cd} \cdot z_c; \; M \leq M_s = A_s \cdot R_s \cdot z_c. \tag{25}$$

Therefore, the algorithm for checking the normal sections bearing capacity of bending elements with adhesive joint of steel and concrete, which are rectangular, looks as follows:

1. Check compliance of the given section with the optimal design conditions;
2. Preliminary estimates the section bearing capacity, and if the conditions are not met, we review the parameters of the given section;
3. Refine the value of the displacement forces according to the selected design scheme taking into account the properties of the adhesive joint and determine, if necessary, the tension in the displacement ties;
4. Determine internal forces;
5. Calculate the cross-sectional area of reinforcement and check its compliance with optimal design conditions;
6. Determine the compressed zone height and the inner pair of forces shoulder;
7. Determine the bearing capacity of reinforced concrete block and external reinforcement.

According to the above procedure, the calculation of steel-concrete composite slabs on profiled flooring [15] was performed (Table 1).

Table 1 Bearing capacity of templates

Template	Series	Bearing capacity, M, k Nm			Divergence, %	
		Experiment values	Theoretical values		Two-layer	Three-layer
			Two-layer	Three-layer		
	P1	10.5	11.62	–	9.64	–
	P2	18	18.75	18.45	4	2.17

5 Conclusion

The proposed algorithm for determining the bearing capacity of bending reinforced concrete elements, in which joint work of steel and concrete is ensured by bonding, showed sufficient accuracy for engineering calculations. The discrepancy between theoretical and experimental values of bearing capacity is within 10%. Another major advantage of this technique is its simplicity and accessibility for design engineers. So, you can talk about the possibility of its application in the real construction projects design.

References

1. Zolotov, M. S. (2005). *Anchor bolts: Design, calculation, design*. Kharkov: KSAME.
2. Storozhenko, L. I., Davydenko, Yu. O., Gorb, O. G., & Gorb, O. O. (2015). Adhesive joints of steel and concrete. *Collection of Scientific Works of the Ukrainian State Academy of Railway Transport, 155*, 184–190.
3. Davydenko, Y., Horb, O., & Avramenko, Y. (2018). Adhesive-bonded joint influence on deflection of composite steel and concrete beams with strengthening by external steel reinforcement. *International Journal of Engineering & Technology, 7*(3.2), 349–353. https://doi.org/10.14419/ijet.v7i3.2.14551.
4. Storozhenko, L. I. (2006). *Reinforced concrete: A collection of scientific works*. Poltava: PNTU.
5. Storozhenko, L. I., & Lapenko, O. I. (2008). *Reinforced concrete structures in fixed formwork*. Poltava: ASMI.
6. Al-Toubat, S., Karzad, A. S., Maalej, M., & Estephane, P. (2017). Repair of reinforced concrete beams using carbon fiber reinforced polymer. *MATEC Web of Conferences, 120*(01008), 1–10. https://doi.org/10.1051/matecconf/201712001008.
7. Gua, L., & Meng, X. H. (2016). Review on research and application of stainless-steel reinforced concrete. *MATEC Web of Conferences, 63*(03003), 1–5. https://doi.org/10.1051/matecconf/20166303003.
8. Kalina, T., Sedlacek, F., & Krystek, J. (2017). Determination of the influence of adherent surface on the adhesive bond strength. *MATEC Web Conferences, 157*(05012), 1–13. https://doi.org/10.1051/matecconf/201815705012.
9. Storozhenko, L., Butsky, V., & Taranovsky, O. (1998). Stability of compressed steel concrete composite tubular columns with centrifuged cores. *Journal of Constructional Steel Research, 46*(1–3), 484. https://doi.org/10.1016/S0143-974X(98)80098-9.
10. Piskunov, V. G., Goryk, A. V., Lyakhov, A. L., & Cherednikov, V. N. (2000). High-order model of the stress-strain state of composite bars and its implementation by computer algebra. *Composite Structures, 48*(1), 169–176. https://doi.org/10.1016/S0263-8223(99)00091-4.
11. Piskunov, V. G., Goryk, A. V., & Cherednikov, V. N. (2000). Modeling of transverse shears of piecewise homogeneous composite bars using an iterative process with account of tangential loads. 1. Construction of a model. *Mechanics of Composite Materials, 36*(4), 287–296. https://doi.org/10.1007/bf02262807.
12. Mitrofanov, V., Pinchuk, N., & Mytrofanov, P. (2019). *How can the plastic shear failure of the RC elements be obtained?* (pp. 1700–1709). Paper presented at the Proceedings of the Fib Symposium 2019: Concrete—Innovations in Materials, Design and Structures.
13. Semko, O., Fenko, O., Hasenko, A., Harkava, O., & Kyrychenko, V. (2018). *Influence of external and internal cooling at solidification on strength of brittle duralumin in compression* (Vol. 230). Paper presented at the MATEC Web of Conferences. https://doi.org/10.1051/matecconf/201823002029.

14. Pichugin, S., & Severin, V. (2004). Reliability of structures under snow load in Ukraine. *Snow Engineering, V*, 67–72.
15. Storozhenko, L. I., Lapenko, O. I., & Gorb, O. G. (2010). Reinforced concrete floor structures on profile flooring with joint work of concrete and steel by means of bonding. *Bulletin of the National University of Lviv Polytechnic, 662*, 360–365.
16. Kozar, V. I. (1999). *Monolithic reinforced concrete slabs on profiled steel flooring* (Author's abstract. Dis. Candidate of Technical Sciences).

Improving the Efficiency of Road Machines During Introduction Innovative Control Systems

Nataliia Smirnova, Alexander Yefimenko, Anna Filatova, and Oksana Demchenko

Abstract A system for monitoring the life cycle of a road using innovative systems for controlling the working bodies of road machines based on GPS technology is developed. The basic principles of the intellectualization of modern road machines are given. A structural diagram and a mathematical model of the intellectual system of road machines, which provides adaptive optimization of working processes in the aspect of modern trends of BIM technologies in the road sector, are proposed. The prospects for managing the design, construction, and operation of road construction objects are considered.

Keywords Intellectualization · System · Road machines · Road · Life cycle · Road performance modeling

1 Introduction

Currently improving the network of international transport corridors of Ukraine, as one of the priority transit potentials in Europe, the urgency of solving the problem of optimizing road projects in terms of modern consumer properties is increasing [1–3]. In connection with the goal of providing a set of transport and operational indicators, timely fulfillment of large volumes of capital and current repairs is required. Consequently, an important task is to increase the efficiency of road construction by introducing innovative GPS-based systems, which provide for shortening the working cycle, minimizing the energy costs of the work process, and maximizing productivity. This task requires the unambiguous use of CAD systems as the basis for modern Building Information Modeling (BIM) technologies with the integration of roadbed model parameters at the stages of subsequent construction and repairs [2, 4–7].

N. Smirnova · A. Yefimenko · A. Filatova
Kharkiv National Automobile and Highway University, Kharkiv, Ukraine
e-mail: NataliaVS.0105@gmail.com

O. Demchenko (✉)
Poltava National Technical Yuri Kondratyuk University, Poltava, Ukraine
e-mail: khadi.alef@gmail.com

V. Onyshchenko et al. (eds.), *Proceedings of the 2nd International Conference on Building Innovations*, Lecture Notes in Civil Engineering 73,
https://doi.org/10.1007/978-3-030-42939-3_29

Thus, the problem of designing intelligent automated systems that optimize the interaction process between the working bodies of road-building machines using an innovative system for development of subgrade soil and road-building materials is relevant. Creation of such systems will improve the quality of construction technology organization under conditions of modern resource-saving technologies in the road sector.

The traditional method of designing a new construction or major repairs and reconstruction of the road does not provide the required accuracy in calculating the amount of work and building digital cartograms of technological schemes of work. The problem is complicated by the fact that geodetic survey data of a point cloud of existing relief or coverage is not processed into an acceptable format for control systems of working bodies of road-building machines [8–10].

Measurements of the roadway surface show that the irregularities of roadway, waviness of the lanes in longitudinal profile are noticeable even after a series of repairs, which significantly affects coating performance. Analysis of modern technologies shows that the applied leveling methods require a large expenditure of energy and materials. Resources cost in technological processes organization of overhaul and reconstruction of roads can be optimized by using innovative systems for managing work processes of road-building machines.

The analysis shows that Trimble and Leica companies have begun to make extensive use of positioning systems machines working bodies. For example, determining the exact location of bulldozer knife allows a significant reduction in the cost of developing and recultivating construction sites [10–14]. Analysis of data from foreign sources shows that due to the reduction of working time, efficiency increases by 30–50%, and fuel and materials are saved [15–17].

Therefore, it is of interest to further develop technologies and optimize use of GPS-based road-building machines [11]. With such an innovative comprehensive approach, a clearer algorithmization of object information model modeling processes in the BIM format is carried out. Coordinating and converting the obtained data using geo-information systems are the primary task for creating a system of adaptive management taking into account development of automated management of modern road-building machines working bodies, resulting in reduction of the execution time of the work while simultaneously increasing productivity and quality.

Specificity of modern road construction works involves situation assessment of the conditions, a prediction of the machine's behavior in normal mode, synthesis and assessment of possible operator actions, and selection of the best. It means that there is a problem of effective maintenance of road machine, solution of which is transferred to the intellect of the machine itself. A distinctive feature of intelligent systems is the ability to plan behavior and to adapt. Such system automatically changes the algorithms of its functioning and its structure in order to maintain or achieve the optimal state when external conditions change [8, 18].

The integration of algorithmic methods for managing complex objects and artificial intelligence methods is aimed at solving problems with uncertainty of initial information [19].

Solving the problem will allow to manage fully by monitoring the road functioning process as a nonlinear system with variable parameters. Such process management at the stage of the targeted search for the best road option at any stage of the life cycle will also lead to a significant reduction in the risk of loss of truly optimal design solutions.

Along with the traditional developments, the authors suggest fundamentally new decision-making methodology in terms of modeling the functioning of a highway during its life cycle. Researchers solve the problem of significant savings in the time of work and increase productivity at each technological stage of the use of road-building equipment [20, 21]. Implementation of intelligent systems indicates their prospects in the field of road construction [1].

To solve the problem, the authors suggest to improve existing CAD systems by adding algorithmic software blocks responsible for technological processes using the intellectualization of road-building machines and mechanisms based on GPS [11]. To solve these problems, it is necessary to provide IT support for the life cycle of roads by consolidating computer-aided design systems and geo-information technologies [22–25].

Consequently, analysis of existing problems showed adaptive methodology validity of management system of road operation processes during the life cycle, taking into account technological processes features and the use of modern road-building machines and its effectiveness in solving urgent problems of construction, reconstruction, and overhaul of highways.

2 The Objective and Tasks of the Study

The research objective is to increase roadwork efficiency, shorten the working cycle, minimize the energy costs of the workflow, and maximize productivity by introducing innovative systems for managing working bodies of construction equipment based on GPS and modeling the information model of the object in BIM format.

To achieve this objective, the following tasks should be solved:

- to develop a mathematical model of adaptive optimization of road machines work processes;
- to study the structure of the intellectual control system of the road machine;
- to develop a mathematical model of working processes adaptive optimization based on a set of integrated sensors indications.

3 The Method of Modern Road Machines Intellectualization for Solving Technological Problems in the Construction, Reconstruction and Overhaul of Roads

To solve this problem, we consider the modern road machine operation method more detailed through coordination via intelligent sensors suggested in [18]. The basic modules of an intelligent machine are presented in Fig. 1.

To carry out research tasks consistently, it is necessary to perform study of a road machine intellectual control system structure.

Any intelligent system must meet the requirements of openness, multitasking, resistance to conflicts, resistance to element failures, adaptability, modularity of the structure, real-time operation, and possibility of choosing the criteria for optimality of work. The structural scheme of road machine intellectual system in which the described functions are realized is presented in Fig. 2.

In accordance with the block diagram in Fig. 2, intelligent system is controlled by operator and with the microprocessor of the automatic control system (MP ACS) based on vector of data on operating status of machine $\vec{D}$.

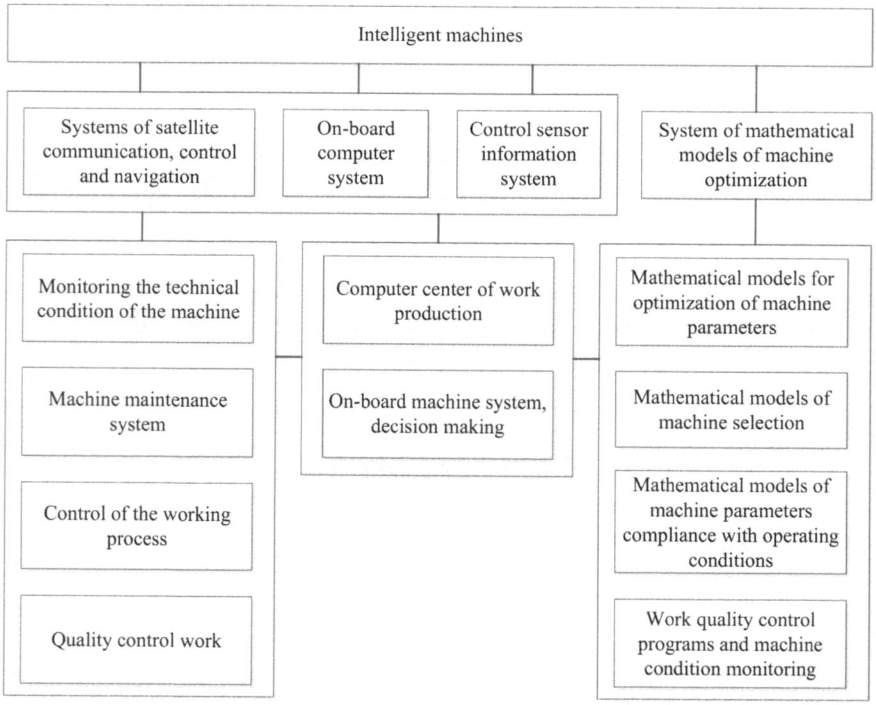

Fig. 1 Main modules of road construction machine intelligence

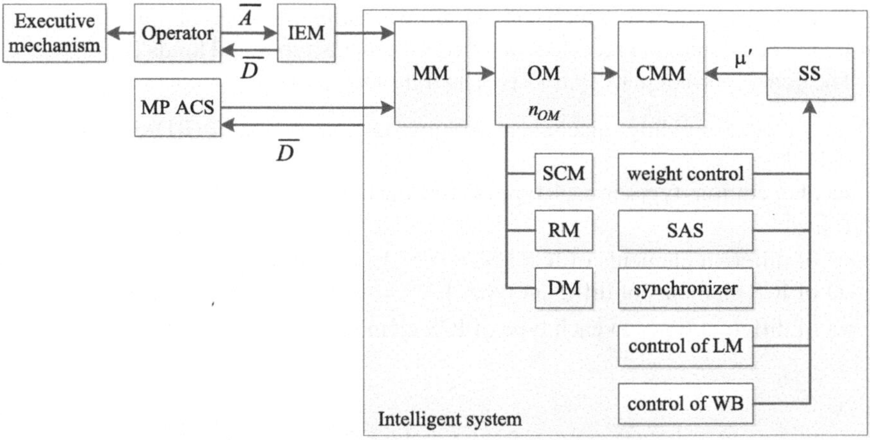

Fig. 2 Structural diagram the road machine of intellectual system

The control influence of operator is transmitted through information exchange module (IEM) to matching module (MM) in parallel with the control signals from automatic control system of microprocessor. Matching module generates a weight vector of workflow optimization criteria. The optimization module (OM) assesses the work processes effectiveness. Sensor control module (SCM) assesses performance of control elements. Reliability module (RM) performs failure control. Data module (DM) is designed to issue the stored values on requests of other modules of the system.

The sensory system (SS) controls the following subsystems of the loading machine:

- weight control;
- stabilization control (SAS—system of active stability);
- signal synchronization;
- lifting mechanism control (LM);
- working bodies control (WB).

The sensor system generates a vector μ of sensor readings and transmits it to the control and management module (CMM), which, in accordance with modules priorities, generates a control signal for power distributors and regulators of the machine power.

Let us use the structural–parametric synthesis method [4, 18] to select the element base of intelligent control system (ICS). Structural-parametric synthesis is the process by which object structure is determined and values of parameters of its constituent elements are found to meet the requirements of technical specification. Structure is a set of elements and connections between them. The task of elemental base systemic synthesis is complicated by multidimensionality of characteristics that the control system elements have.

Thus, the general task, which has a large dimension, was decomposed into partial problems of evaluation and selection of ICS elements types and kinds [4].

The general problem of synthesis is as follows [5]. Let it be known:

- set of road-building machines working processes types (RDM) $\text{RDM} = \{\text{RDM}^e\},\ e = \overline{1, e'}$;
- set of operations types in each type of working process (WP) $\text{WP}^e = \{\text{WP}_n^e\},\ n = \overline{1, n^e}$;
- set of different elements of ICS $\text{ES} = \{\text{ES}^m\},\ m = \overline{1, m'}$;
- set of ICS elements of different types $\text{ES}_o^m = \{\text{ES}_{ow}^m\},\ w = \overline{1, w^o}$;
- set of different types in each type of ICS elements $\text{ES}^m = \{\text{ES}_o^m\},\ o = \overline{1, o^m}$.

$$\overline{S}_{\text{MM}}\{\overline{C}, \overline{D}_{\text{MM}}\} = f_{\text{MM}}(\overline{J}_{\text{MM}}\{\bar{\mu}', \overline{C}_{\text{O}}, \overline{C}_{\text{ASP}}\}, \overline{N}_{\text{MM}}); \quad (1)$$

Let variables be introduced by $J_{en} = \{0; 1\}$ where $J_{en} = 1$, if the eth type working process of the nth kind is selected, $J_{en} = 0$ otherwise; $\tilde{J}_{mow} = \{0; 1\}$ where $\tilde{J}_{mow} = 1$, if the mth element of ICS is chosen of the oth type of the wth kind, $\tilde{J}_{mow} = 0$ otherwise.

It is necessary to determine the type and kind of WP, ICS elements according to selected criteria and restrictions.

The intelligent system solves the problem of adaptive machine subsystems working processes optimization. The main criteria for effectiveness of road machine are operational performance and reliability. Two typical working situations arise as follows:

- necessity to ensure maximum performance of loading and unloading operations which requires maximum productivity from the machine;
- implementation of given performance, taking into account ensuring maximum efficiency, minimizing failures due to the rational loading of power systems.

4 Development of Mathematical Model for Working Processes Adaptive Optimization Based on a Set of Integrated Sensors Indications

The mathematical model for intellectual system adaptive optimization of a road-building machine can be presented in the following form:

$$\overline{S}_{\text{OM}_i}\{\overline{N}_{\text{WP}_i}, \overline{D}_{\text{OM}_i}\} = f_{\text{OM}_i}(\overline{J}_{\text{OM}_i}\{\bar{\mu}', \overline{C}, \overline{D}_{\text{SCM}}\}, \overline{N}_{\text{OM}_i}), \quad i = \overline{1, n_{\text{OM}_j}};$$

$$\overline{S}_{\text{SCM}}\{\overline{Z}_{\text{SCM}}, \overline{D}_{\text{SCM}}\} = f_{\text{SCM}}(\overline{J}_{\text{SCM}}\{\mu'\}, \overline{N}_{\text{SCM}});$$

$$\overline{S}_{\mathrm{RM}}\{\overline{Z}_{\mathrm{RM}},\overline{D}_{\mathrm{RM}}\} = f_{\mathrm{RM}}(\overline{J}_{\mathrm{RM}}\{\mu'\},\overline{N}_{\mathrm{RM}});$$

$$\overline{S}_{\mathrm{DM}}\{\overline{D}_{\mathrm{DM}}\} = f_{\mathrm{DM}}(\overline{J}_{\mathrm{DM}}\{\mu',\overline{D},\overline{F}\},\overline{N}_{\mathrm{DM}});$$

$$\overline{S}_{\mathrm{CMM}}\{\overline{Z},\overline{D}_{\mathrm{CMM}}\} = f_{\mathrm{CMM}}\left(\overline{J}_{\mathrm{CMM}}\left\{\overline{N}_{\mathrm{WP}_i},\overline{Z}'_{\mathrm{O}},\overline{Z}_{\mathrm{SCM}},\overline{Z}_{\mathrm{RM}},\overline{Z}_{\mathrm{ACS}}\right\},\overline{N}_{\mathrm{CMM}}\right)$$

where $\overline{S}_{\mathrm{MM}},\overline{S}_{\mathrm{OM}_i},\overline{S}_{\mathrm{SCM}},\overline{S}_{\mathrm{RM}},\overline{S}_{\mathrm{DM}},\overline{S}_{\mathrm{CMM}}$ are output vectors of corresponding modules: matching module, optimization module, sensor control module, reliability module, data module, control and management module; $\overline{D}_{\mathrm{MM}},\ldots,\overline{D}_{\mathrm{CMM}}$—workflow data vectors of corresponding modules; $f_{\mathrm{MM}}(),\ldots,f_{\mathrm{CMM}}()$—function vectors that show algorithms for modules functioning; $\overline{J}_{\mathrm{MM}},\ldots,\overline{J}_{\mathrm{CMM}}$—input vectors of corresponding modules; $\overline{N}_{\mathrm{MM}},\ldots,\overline{N}_{\mathrm{CMM}}$—parameter vectors of the corresponding modules; $\overline{C}_{\mathrm{O}}$—vector of matching module object parameters; $\overline{C}_{\mathrm{ACS}}$—parameter vector of the ACS module; $\overline{Z}_{\mathrm{O}}$—parameter of control and management module; $\overline{C}$—parameter vector of matching module; $\overline{F}$—workflow data vector; $\overline{Z}$—parameter of control and management module; $\bar{\mu}'$—sensor readings vector.

Mathematical model of machine intellectual system in the integral form can be written in this way:

$$\bar{f} = f_{\mathrm{IS}}(\bar{\mu},\overline{N}_{\mathrm{IS}})$$

where $\bar{\mu}\{\bar{\mu}',\overline{C}_{\mathrm{O}},\overline{C}_{\mathrm{ACS}}\}$ is vector of the input information and control signals; $\bar{f}\{\overline{Z},\overline{D}\}$—control vector for power distribution distributors and power drive controllers; $N_{\mathrm{IS}}\{N_{\mathrm{MM}},N_{\mathrm{OM}_i},N_{\mathrm{SCM}},N_{\mathrm{RM}},N_{\mathrm{MD}},N_{\mathrm{CMM}}\}$—intelligent system parameter vector.

Cycle time and performance of road machines show that these values significantly depend on the car mass as the main parameter and, therefore, on the traction machine characteristics. Minimum time of the loader working cycle and maximum performance in certain operating conditions occur at a certain optimal value of machine mass.

Selection of criterion is carried out by the working situation, and intellectual system of the machine must ensure effective control by each of the criteria. To determine the optimal composition of diagnostic information based on certain priorities (technical and economic indicators, resource indicators, safety indicators, communication system capabilities, etc.), a matrix of diagnostic parameters is created. The diagnostic matrix is a logical model that describes relationship between diagnostic parameters and possible object malfunctions [1]. The most important value characterizing the process of diagnosing any object is the informativeness of diagnostic parameters, which is determined by sensitivity of these parameters to structural changes in the object and uniqueness in the diagnosis.

5 Conclusion

Proposed intellectual system of the road machine must be equipped with all the necessary software and equipment for monitoring its condition. Developed block diagram and mathematical model of the intellectual system of loading and unloading machine provide solution for the tasks of adaptive optimization of its work processes. It is possible to assess performance of actuators and predict trouble-free operation of the machine structural elements. The data system on the parameters of the workflow allows to monitor the system and to ensure its maximum security at any stage of operation.

When designing and developing intelligent systems for road machines, it is necessary to use a modular principle that ensures coordination of the working process of the machine with operation of other equipment. Such system allows developing of highly intelligent machines, ensures efficient use of technology in the conditions where it gives the greatest production effect and will contribute to improvement of quality.

Summarizing results of the above analysis of the modern concept of substantiating design decisions of roads from the point of view of production technology using intelligent road construction equipment, it can be concluded that this concept needs to be improved in terms of increasing reliability and detailing the basic calculations of road works performance. Thus, a clear algorithmization of information model modeling processes of the road is carried out.

References

1. Plugin, T., & Markozov, D. (2013). Designing intelligent operator stations of distributed control systems. *Bulletin of Kharkiv National Automobile and Highway University, 57,* 38–43.
2. Smirnova, N. (2015). The use of BIM-technologies in the tasks of searching for the design solutions of roads. *Avtoshliakhovyk Ukrainy, 5–2,* 47–49.
3. Filippov, V., & Smirnova, N. (2014). *Modeling traffic flow on roads of II–IV categories: Monograph*. Kharkiv: Kharkiv National Automobile and Highway University.
4. Kuzmin, I. (1981). *Fundamentals of modeling complex systems: An educational tool for university students*. Kiev: High School.
5. Pisarchuk, O. (2014). Technology of situational structural-parametric synthesis of complex information-control system. *Collection of Scientific Works of Zhytomyr Military Institute of State University of Telecommunications, 9,* 56–61.
6. Hagan, M., De Jesús, O., & Schultz, R. (2000). Training recurrent networks for filtering and control. In L. Medsker & L. Jain (Eds.), *Recurrent neural networks: Design and applications* (pp. 325–354). CRC Press.
7. Hagan, M., & Demuth, H. (1999). Neural networks for control: Invited tutorial. In *1999 American Control Conference* (pp. 1642–1656), San Diego.
8. Plugina, T., & Stotsky, V. (2014). The task of intellectualizing modern construction and road machines. *Instrumentation Technology: Special Issue,* 40–43.
9. Isermann, R. (2003). Modellgestützte präventive Diagnosemethoden (Fehlerfrüherkennung) für Dieselmotoren. In *Informationstagung Motoren* (Heft R 159). Frankfurt am Main.

10. Loffler, J., Bolz, M., & Hulser, H. (1998). *Vorrichtung und Verfahren zur koordinierten Steuerung des Antriebstrangs eines Kraftfahrzeugs wahrend Getriebeschaltvorgangen*. Patentschrift DE 198 48 520.4. Deutsches Patentamt, Munchen.
11. *AccuGrade GPS grade control system*. (2005). Caterpillar.
12. Harms, H.-H., & Seeger, J. (2002). *Energieeinsparung durch moderne Motoren-und Getriebetechnik bei Traktoren*. Helmstedt: Landwirtschaftskammer Hannover.
13. Martinus, M., & Freimann, R. (2002). Prozesssicherheit Landmaschinen Elektronik. *Gerät steuert Traktor, Agrartechnische Forschung, 8–3,* 61–69.
14. Nørgaard, M. (2000). *Neural network based control system design toolkit* (Tech. Report 00-E-892). Technical University of Denmark.
15. Voronin, A., Ziatdinov, Yu., Kharchenko, A., & Ostashevsky, V. (1997). *Complex technical and ergatic systems: Method of use*. Kharkov: Fact.
16. *Fahrzeugantrieben zur Erlangung der Würde eines Dr.-Ing. genehmigte Abhandlung*. (2000). Stuttgart, 156 S.
17. Smirnova, N., Ivanov, V., & Polyakova, T. (2017). Determination of patterns of variability of the main indicators of road transport and operating condition. In *Integration processes and innovative technologies: Achievements and prospects of engineering sciences* (Vol. 7-II, pp. 251–254).
18. Efimenko, O., Plugina, T., & Musayev, Z. (2016). Models of parametric synthesis of elemental base of control system of software and hardware complex. *Technology of Instrument Making, 2,* 10–15.
19. Nefedov, L., & Osmachko, A. (2009). Generalized model of system synthesis of automatic transmission. *Eastern-European Journal of Enterprise Technologies, 6*(42), 10–13.
20. Barabanov, N., & Prokhorov, D. (2002). Stability analysis of discrete-time recurrent neural networks. *IEEE Transactions on Neural Networks, 13*(2), 292–303.
21. De Jesús, O., & Hagan, M. (2001). Backpropagation through time for a general class of dynamic network. In *Proceedings International Joint Conference on Neural Networks* (Vol. 4, pp. 2638–2643).
22. De Jesús, O., Pukrittayakamee, A., & Hagan, M. (2001). A comparison of neural network control algorithms. In *Proceedings International Joint Conference on Neural Networks* (Vol. 1, pp. 521–526).
23. Nørgaard, M. (1997). *Neural network based system identification toolbox* (Tech. Report 97-E-851). Technical University of Denmark.
24. Schultz, R., Hagan, M., & De Jesús, O. (1999). Training multi-loop networks. In *Proceedings International Joint Conference on Neural Networks* (Vol. 3, pp. 1580–1585).
25. Shulgin, V., Yermolenko, D., Durachenko, H., Petrash, O., & Demchenko, O. (2018). Investigation of the bottom ash slags influence on the heavy concrete frost resistance. *MATEC Web of Conferences, 230*. https://doi.org/10.1051/matecconf/201823003019.

Some Physicochemical Aspects of the Preparatory Stages in the Formation of Self-cleaning Photocatalytic Active Coatings for Building Construction Materials

D. Storozhenko, O. Dryuchko, T. Jesionowski, and I. Ivanytska

Abstract A complex systematic study of the interaction of structural components in cerium subgroup lanthanide nitrate systems and representatives of the IA group (Li, Na, K) of the elements of the periodic system, precursors of modern multicomponent oxide multifunctional materials based on them, established the formation of a representative class of alkaline coordination nitrates Ln. Their composition, formation conditions, atomic-crystalline structure, forms of Ln coordination polyhedral, types of ligand coordination, a number of their properties were studied using a complex of physicochemical methods: chemical, X-ray phase, X-ray structural, IR-spectroscopic, crystal-optical, thermographic, and SHG laser radiation. The data obtained are the basis for identifying and monitoring the phase state of processing facilities in the preparatory stages when forming self-cleaning coating layers of building construction materials by using innovative technologies; using nanostructured composite systems of lanthanides and transition elements with photocatalytically active and hydrophilic properties; various combined methods of their activation; and establishing technological and functional dependencies to modify the properties of the products obtained.

Keywords Alkali coordination nitrates of lanthanides · Conditions of formation · Crystalline structure of compounds · Properties

1 Introduction

This work is aimed at finding new comprehensive solutions for the creation of a new generation of float glass for the construction industry (both external and interior), covering the designs of solar photovoltaic energy, and other specialized industries. It is known that the leading companies (Pilkington, AGC Flat Glass Europe, Saint-Gobain, and others) develop and sell lines of types of self-cleaning glass, products

D. Storozhenko · O. Dryuchko (✉) · I. Ivanytska
Poltava National Technical Yuri Kondratyuk University, Poltava, Ukraine
e-mail: dog.chemistry@gmail.com

T. Jesionowski
Poznan University of Technology, Poznan, Poland

V. Onyshchenko et al. (eds.), *Proceedings of the 2nd International Conference on Building Innovations*, Lecture Notes in Civil Engineering 73,
https://doi.org/10.1007/978-3-030-42939-3_30

based on them with a coating of complex action with photocatalytic and hydrophilic properties by developed own technologies and using compositions of titanium dioxide. Their products are in great demand when constructing high-rise buildings, glass roof structures, for glazing facades of commercial and artistic architectural objects, parking lots, greenhouses, structures with good visibility solar photovoltaic panels, sports arenas of stadiums, outdoor premises of airports [1, 2], that is, in cases of difficult access and complicated, labor-intensive cleaning. The implementation of these simple solutions also allows you to solve the problem of improving the efficiency of converters of renewable energy forms in energy, energy conservation and sun protection simultaneously, providing a spectacular appearance of the facade of buildings and creating a comfortable microclimate for people who are indoors.

1.1 Problem Statement and Solution Methods

Self-cleaning eliminates the pollution of the environment, contains particles of photocatalytic titanium dioxide (TiO_2), and is capable of forming electron–hole pairs in electromagnetic irradiation, in particular ultraviolet (UV) close to ultraviolet light and/or visible light. Preferably, the photocatalytic titanium dioxide in the anatase modification is capable of significant photoactivity when exposed to visible light. It has been found that careful regulation of the crystalline form and size of TiO_2 particles can provide photocatalysts that is capable of removing contaminants even in low-light UV light, in particular in interior light, and which have significant initial activity, even without activation by solvent washing. The degree of crystallinity and the nature of such crystalline phases are determined by X-ray diffraction methods. The average particle size of TiO_2 for use in the compositions is 5–30 nm.

Photocatalytically active coatings are precipitated by using a variety of technologies, such as sol–gel, spray pyrolysis, by means of chemical vapor deposition or gas phase, magnetron sputtering. Regulations are currently being implemented using "solution burning" methods [3].

An interesting innovative solution is the method of preparation of sol–gel method of hydrophilic self-cleaning coating with photocatalytic activity, in which butyl titanate and ethyl orthosilicate are used as precursors for the formation of nanosol TiO_2/SiO_2 particles [4]. Anatase composite film is applied to the surface of the object (window or architectural glass, solar photovoltaic panel) and naturally dried at room temperature in a natural way to obtain photocatalytic activity rapidly. Pollutants, when decomposed into inorganic substances, are easily washed off with rainwater or artificial washing.

It has also been found that titanium oxide coatings deposited on the surface of sodium–calcium glass can reduce the level of photoactivity and hydrophilicity due to the migration of alkali metal ions from such glass into the TiO_2 layer and therefore introduce barrier layers of SiO_2 more often.

Nowadays, the implementation of innovative solutions is linked to:

- the development of methods of synthesis of thin films, functional coatings, and nanosized inorganic oxide fillers with chemically modified surface;
- the development of nanochemical liquid phase and pyrolytic methods for the synthesis of thin oxide films on the surface of solids;
- the study of the peculiarities of chemical modification reactions and the formation of nanoparticles in thin oxide films and composite systems;
- the creation of film structured photocatalysts; functional, special, protective, anticorrosive, and decorative coatings on the surface of inorganic materials;
- the development of a protective superhydrophobic coating on the surface of TiO_2/SiO_2, TiO_2/Si compositions, which facilitates dust, polar, and non-polar contaminants and also protects against the adverse effects of atmospheric factors.

One of the most promising classes of photocatalytically complex rare earth oxides and titanium oxide materials that can act as alternatives to existing developments is nanostructure-layered perovskite-like compounds and solid solutions based on them. Depending on their composition and structure, they have a wide range of physicochemical properties. The perovskite-like layered titanates presented in this paper belong to the homologous series $(Me, Ln)_{n+1}Ti_nO_{3n+1}$—of the Ruddlesden–Popper phase, where Me—H, Li, Na, K, Rb, Cs; Ln—La, Nd; n—the number of nanowires of perovskite, with a thickness of about 0.5 nm; in accordance, $MeLnTiO_4$ contains one perovskite nanolayer in its structure, $Me_2Ln_2Ti_3O_{10}$—contains three. As an example, a three-layer perovskite-like oxide $K_2Nd_2Ti_3O_{10}$ was obtained and researched by the authors of the following works [5–7]. Figure 1 shows his expanded elemental cell.

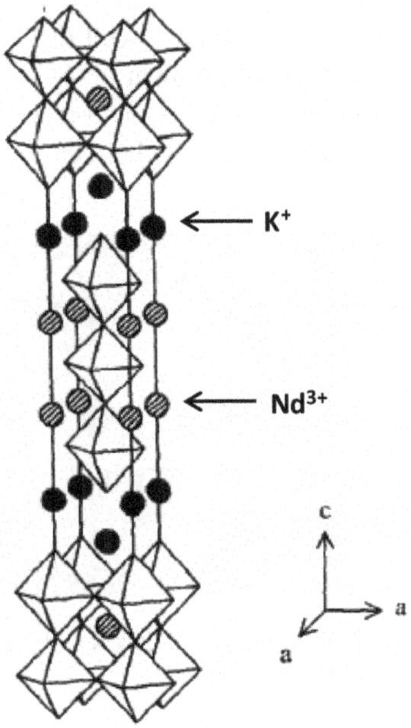

Fig. 1 Extended unit cell $K_2Nd_2Ti_3O_{10}$

Oxides $K_2Nd_2Ti_3O_{10}$ crystallize in a tetragonal structure similar to the structure $Sr_4Ti_3O_{10}$, which is one of the first synthesized phases of Ruddlesden–Popper. The space group for these compounds is defined as I4/mmm. The thickness of the layered oxides of this type is characterized by three titanium-oxygen octahedra $[Nd_2Ti_3O_{10}]$, alternating between themselves and separated by cations of alkali metals, in this case by potassium cations, between the layers. The lattice parameter c ($\approx$ 30 Å) indicates the displacement of adjacent perovskite layers by ½. The neodymium cation is located in the center of the perovskite lattice and is characterized by a 12-coordinated oxygen environment. The alkali metal cation is in the interlayer space and is usually 9-coordinated [8].

For innovative applications, it is important to study the stability of systems that have proven to be promising photocatalysts, in conditions close to their use in aqueous solution, in conditions simulating a flow reactor, and in moist air. In research [9] using methods of TGA, DSC and X-ray phase analysis regularities of the behavior of layered perovskite-like oxides $A_2Ln_2Ti_3O_{10}$ (A = Li, Na, K; Ln = La, Nd) with water for different times are established. It has been found that all alkaline forms of layered oxides are characterized by the flow of protonation processes; in the case of $K_2Ln_2Ti_3O_{10}$, this process is accompanied by the intercalation of water into the interlayer space. The dependence of the protonation process from the choice of the rare earth metal cation is revealed. For Nd-containing compounds, ion exchange for protons proceeds faster than La-containing counterparts due to the greater mobility of alkali metal cations in the interlayer space, which is associated with greater distortion of titanium-oxygen octahedra and smaller (compared to La^{3+}) radians. It has been found that $K_2Nd_2Ti_3O_{10}$ and its stable intercalated proton derivatives are not changed when interacting with atmospheric humidity, which indicates the stability of the compounds under possible operating conditions. A layered structure consisting of lamellar particles is stored for the incubation product, which is confirmed by scanning electron microscopy (see Fig. 2, [9]).

Features of application of methods of "soft chemistry" during the formation of nanosystems are the possibility of synthesis of new photocatalytically active coating materials with high chemical homogeneity of multicomponent systems due to the

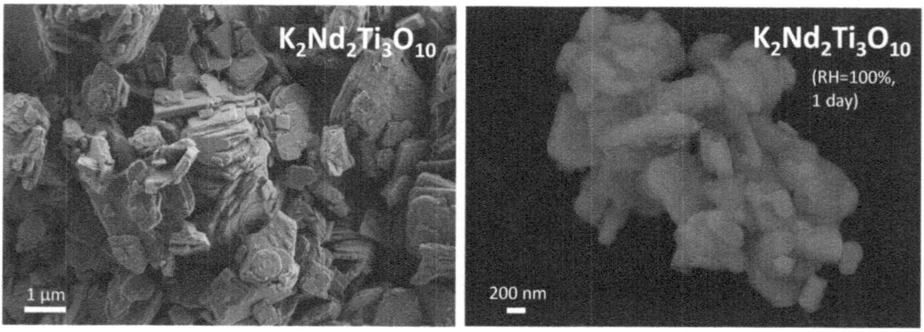

Fig. 2 Output micrographs $K_2Nd_2Ti_3O_{10}$ i $K_2Nd_2Ti_3O_{10}$ after keeping in a humid atmosphere (RH = 100%, 1 day) [9]

uniform (molecular level) distribution of components in the original solution, a significant decrease in the temperature control of their properties, the target products obtained in the synthesis process, and the simplicity and accessibility of this method contribute to its application on an industrial scale. However, as for the mechanisms of formation of nanocomposites from film-forming solutions by the above technologies, the influence of the formation conditions on the structure, physicochemical and target properties, these issues are hardly covered. Therefore, the successful use of nanomaterials in mass technology requires a systematic fundamental study of these aspects.

In [10–13], the effect of non-film-forming components of colloidal solutions (salts of inorganic REE acids, oxide filler systems, metals) on the film-forming properties of such systems is investigated. It is shown that salts of inorganic acids act as the electrolyte stabilizer of colloidal solutions and significantly extend the interval of film-forming properties of solutions, and increasing the concentration of electrolyte and increasing the charge of the nucleus on a number of REE, increase the stability of the system to aggregation and coagulation of the structure and transition of the solution to the gel state.

The processes of the formation of simple and complex oxides in thin-film and dispersed state are non-equilibrium, transient flows through a series of successive stages. They include evaporation of adsorbed solvents, the formation of complex associates, hydrolysis and condensation of hydrolysis products, the emergence of amorphous or metastasis, salts and removal of gaseous products, combustion of organic residues, crystallization processes, and other physicochemical transformations. The processes that take place in the thin layer are more energy-efficient—there is a decrease in the temperature of film formation compared to powders and a decrease in the activation energy of the processes going on the substrate surface.

The influence of the dimensional factor on the structure and properties of the fine-layer coatings is determined [10], which is to stabilize the metastable and high-temperature phases in the nanosystems. The influence of the oriented action of the substrate and the conditions of forming of obtaining films with amorphous and crystalline structures are also revealed.

For effective management of the processes of the formation of self-cleaning coatings of building structural materials using compositions based on titanium dioxide and inherent valuable set of photocatalytic and hydrophilic properties requires a deep understanding of the physicochemical processes, phenomena that occur during their manufacture. Moreover, their complex research with the application of modern physicochemical methods allows us to improve our knowledge about the characteristic features of transient processes, the stage of evolution of the structure, and microstructure of technological objects.

1.2 The Purpose and Tasks of the Study

The aim is to carry out research of cooperative processes of interaction between structural components in systems of nitrate precursors of representatives of REE of the cerium subgroup and elements of IA subgroup of the periodic system (Li, Na, K) under conditions similar to the regulatory ones when creating hydrophilic self-cleaning coatings of building structural materials using photocatalytic active TiO_2 in order to form credible ideas and obtain objective knowledge about the peculiarities of transformations and the cumulative behavior of the constituent elements in the preparatory stages of processing of technological objects with thermal activation, necessary for the improvement and development of methodologies and regulations of modern technologies of their production.

2 Experimental Part

2.1 Experiment Methodology

To evaluate the possibility of controlling, the processes of multistage formation of complex-oxide compositions with polyfunctional properties and to substantiate the mechanisms of phase formation as model we use a complex of physicochemical methods studied water–salt systems of nitrates $MeNO_3$–$Ln(NO_3)_3$–H_2O (Me—Li, Na, K; Ln—La–Sm) at 25–100 °C. The choice of the composition of the objects of study, the temperature cross sections are due to a number of factors. Among the rare earth elements, the higher complexing ability is found by the cerium subgroup; among them, the biggest changes in the composition, structure, properties of their compounds are the elements of its middle, Pr and Nd. The selected system components specify the specifications of the target product or are modifiers of its properties. In addition, the presence of large quantities for the use of potential electronic analogues (representatives of natural series of rare earths, alkaline elements) determines the considerable variability and breadth of the range of modification of their characteristics. The temperature cross sections are due to the regions of the crystallohydrate forms of the original components.

To determine the nature of the chemical interaction and phase equilibria in water–salt systems of the investigated nitrates (precursors of multicomponent oxide polyfunctional materials) in full concentration ratios in the temperature range of solutions, the additive method described in [14, 15] and based on the study was used as one of the properties of the most "sensitive" to detect phase transformations in systems, which is both a parameter of their state, and, moreover, the simplest experiment available notional methods. The method allows finding the boundaries of self-development, to which, under specific conditions, in an equilibrium state, an isolated system of a given composition goes.

Phase equilibrium was reached within 2–3 days. As starting, salts hydrated and anhydrous nitrates of these brand elements "p.f.a." were used.

Chemical analysis of the liquid and solid phases, "residues," was performed for the content of Ln^{3+} and nitrogen. The content of Ln^{3+} was determined trigonometrically; nitrogen—by distillation; Me^{+} ions—calculated by difference based on total nitrate content and partly on dry residue.

The data obtained for the individual ions were converted to salt content and plotted according to the principle of correspondence. The graphical representation of the composition of the solid phases formed in the system was carried out by Skreemakers [14, 15], confirmation of their individuality and characterization, chemical, X-ray phase, X-ray structural, IR-spectroscopic, crystal-optic, thermographic, and other methods.

Crystal-optical determination of the compounds was performed by immersion using a microscope MIN-8. Phase analysis was performed on a DRON-3M diffractometer (Cu K_{α} radiation and Ni filter) using the powder method. The diffraction patterns were decoded according to the JCPDS PDF file. Determination of symmetry, parameters of elementary cells, and measurement of the intensity of diffraction reflections from single crystals were performed on an automatic X-ray single crystal diffractometer CAD-4F «*Enraf-Nonius*» (Mo K_{α} radiation, graphite monochromator; $\omega/2\theta$—method). All calculations for the determination and refinement of atomic structures were performed using SHELX, XTL-SM, AREN crystallographic software packages. The IR absorption spectra of the synthesized compounds in the 400–4000 cm^{-1} region were recorded on a UR-20 spectrophotometer using a standard Vaseline oil suspension technique. Thermogravimetric analysis was performed on a Q-1500 D derivatograph at temperatures from 293 to 1273 K in the air with a heating rate of 10°/min and the developed device for DTA.

2.2 *Results of the Studies and Their Discussion*

Generalized and important for practical use of information on alkaline coordination nitrates of rare earth elements of the cerium subgroup—precursors of perspective modern multifunctional materials—regarding the conditions of their formation and existence, the nature of chemical bond, composition, structure, type of coordination of ligands, the existence of isotypic rows in the steppe structures, manifested properties systematized by the authors and in the most obvious form are given in Tables 1, 2 and 3. The choice of this form of data presentation is the most informative and useful in the development of innovative projects allows to predict the cause and effect fundamental patterns of behavior of structural components in similar production processes, to properly choose the modes, stages, methods of formation and obtaining of target products with reproducible structure-sensitive characteristics.

Identified patterns in the behavior of structural components in nitrate systems of rubidium, cesium and La–Sm; in similar objects based on REE of the yttrium subgroup (Y, Gd–Lu) indicate the possibility of only limited or special use of the

Table 1 Composition of the newly formed associated forms and the conditions of their crystallization in the systems of nitrate precursors REE cerium subgroup and alkali metals—lithium, sodium, potassium (wt%)

	System	La	Ce	Pr	Nd	Sm
$LiNO_3$–$Ln(NO_3)_3$–H_2O at 100 °C						
E_1	$LiNO_3$	27,11	26,84	24,09	24,03	17,56
	$Ln(NO_3)_3$	53,35	54,35	54,66	54,68	67,31
Connection	Correlation components	3:2:4	3:2:4	3:2:4	3:2:4	
	Nature solubility	Congruent				
E_2	$LiNO_3$	9,92	9,76	8,73	9,68	
	$Ln(NO_3)_3$	71,03	71,45	72,20	72,51	
$NaNO_3$–$Ln(NO_3)_3$–H_2O at 50 °C						
E_1	$NaNO_3$	17,43	17,37	17,08	16,55	8,39
	$Ln(NO_3)_3$	50,97	51,24	51,57	51,62	60,87
Connection	Correlation components	2:1:2	2:1:2	2:1:2	2:1:2	
	Nature solubility	Incongruent				
E_2	$NaNO_3$	7,43	7,51	7,87	8,60	
	$Ln(NO_3)_3$	62,66	62,55	62,62	62,58	
KNO_3–$Ln(NO_3)_3$–H_2O at 50 °C						
E_1	KNO_3	28,03	27,19	25,94	27,26	23,44
	$Ln(NO_3)_3$	42,80	43,88	48,37	51,62	54,29
Connection. I	KNO_3	35,23	35,35	35,31	35,38	29,96
	$Ln(NO_3)_3$	57,63	57,83	57,68	57,77	67,43
	Correlation components	2:1:2	2:1:2	2:1:2	2:1:2	3:2:1
	Nature solubility	Incongruent				
E_2	KNO_3	11,23	11,55	21,11	21,34	18,19
	$Ln(NO_3)_3$	62,66	62,72	53,86	54,91	56,08
Connection. II	KNO_3			30,28	30,95	21,38
	$Ln(NO_3)_3$			67,47	67,41	70,68
	Correlation components			3:2:1	3:2:1	3:2:1
	Nature solubility			Incongr.	Congr.	
E_3	KNO_3			11,58	11,49	9,67
	$Ln(NO_3)_3$			63,17	63,31	64,22

above predecessors in the studied field. There are a number of objective and economic reasons for this. These are the features of the electronic structure of their atoms, a lower manifestation of the chemical activity and complexion ability of these Ln3+ compared to elements of the cerium subgroup, a weaker effect of the considered influencing factors on the processes under study. In order to find out the general laws and to construct a holistic objective picture of the behavior of such technological

Table 2 Radiographic data of neodymium lithium, sodium, potassium coordination nitrates

$Li_3[Nd_2(NO_3)_9]\cdot 3H_2O$				$Na_2[Nd(NO_3)_5]\cdot H_2O$		$K_2[Nd(NO_3)_5(H_2O)_2]$				$K_3[Nd_2(NO_3)_9]\cdot H_2O$			
d, Å	I/I_0, %	*d*, Å	I/I_0, %	*d*, Å	I/I_0, %	*d*, Å	I/I_0, %	*d*, Å	I/I_0, %	*d*, Å	I/I_0, %	*d*, Å	I/I_0, %
8,36	30	2,013	19	7,84	63	5,42	66	2,056	24	9,48	85	2,281	21
7,64	77	1,979	19	7,54	47	5,27	90	1,993	27	7,74	92	2,249	69
6,68	13	1,944	28	7,07	73	4,94	45	1,947	26	7,65	32	2,189	54
6,00	38	1,931	26	5,18	27	4,11	72	1,777	11	5,36	39	2,108	68
5,75	79	1,855	15	4,23	100	3,88	15			5,27	40	2,082	37
5,42	98	1,778	15	3,80	57	3,80	42			4,94	30	2,058	15
5,26	74	1,726	28	3,15	23	3,66	12			4,76	46	2,012	49
4,76	51	1,708	34	3,09	67	3,53	42			4,49	87	1,909	21
4,64	100			3,02	86	3,35	38			4,26	27	1,837	27
4,35	43			2,629	20	3,18	13			4,06	100	1,757	19
4,19	47			2,391	17	3,05	73			3,89	40	1,729	16
3,94	51			2,346	17	2,873	17			3,78	22	1,714	18
3,90	40			2,307	8	2,843	33			3,73	39		
3,56	26			2,234	13	2,783	14			3,36	19		
3,32	19			2,178	27	2,750	17			3,27	26		
3,22	47			1,979	13	2,724	40			3,18	55		
2,978	34					2,664	14			3,07	16		
2,772	19					2,639	19			3,04	17		
2,617	28					2,594	100			2,844	20		
2,545	43					2,463	37			2,755	12		

(continued)

Table 2 (continued)

$Li_3[Nd_2(NO_3)_9]\cdot 3H_2O$				$Na_2[Nd(NO_3)_5]\cdot H_2O$		$K_2[Nd(NO_3)_5\ (H_2O)_2]$				$K_3[Nd_2(NO_3)_9]\cdot H_2O$			
d, Å	I/I_0, %	*d*, Å	I/I_0, %	*d*, Å	I/I_0, %	*d*, Å	I/I_0, %	*d*, Å	I/I_0, %	*d*, Å	I/I_0, %	*d*, Å	I/I_0, %
2,385	21					2,392	19			2,730	18		
2,328	17					2,374	44			2,647	55		
2,305	23					2,314	15			2,592	16		
2,226	19					2,235	25			2,508	87		
2,135	49					2,188	10			2,468	78		
2,111	43					2,099	21			2,349	32		

Note d, Å—interplanar distances; I/I_0, %—the relative intensities of the reflexes

Table 3 Temperature values of the detected effects during the heat treatment of the representatives of the established groups of alkali coordination nitrates REE

Compounds; spatial group of crystals	Representatives	Temperature interval of formation, °C	The nature of solubility	Dehydration	Melting in crystallization water	Polymorphic transitions	Melting anhydrous form	The composition of the products of conversion at 980°C
$Li_3[Ln_2(NO_3)_9]·3H_2O$ cubic.; $P2_13$	La–Sm	65–100	Congr.	65 183 216	183	–	274	$LiLnO_2$
$Na_2[Ln(NO_3)_5]·H_2O$ monocl.; $P2_1/a$	La–Sm	50–100	Congr.	81 148 236	–	271	328	$NaLnO_2$
$K_2[Ln(NO_3)_5(H_2O)_2]$ rhomb.; Fdd2	La–Nd	50–100	Incongr.	95,111	95	219	314	$KLnO_2$, Ln_2O_3
$K_3[Ln_2(NO_3)_9]·H_2O$ cubic.; $P4_332$	La–Sm	50	Congr.	126	–	–	347	Ln_2O_3
$K[Ln(NO_3)_4(H_2O)_2]$ prim. rhomb.; $P2_1cn$	Y, Gd–Lu	50–100	Congr.	138, 172	138	–	–	Ln_2O_3

precursors, the authors investigated systems by natural series Y, La–Lu, Li–Cs. The analysis of the results of the study is published in the previous works of the authors [16–23].

In ternary REE-containing systems of nitrate precursors, which are integral components of more complex multicomponent systems, exchange transformations begin from the moment of dissolution of the constituents in water. It was found that the Ln^{3+} cerium subgroup under active conditions is active complexing agents, forming anionic coordination compounds of Me^{+} of all alkali metals, and their stability and complex of inherent properties are found to be powerful technological factors that significantly affect the nature of the transformations, and the results of the processes as a whole.

The obtained data allow us to model the behavior of structural components at the preparatory stages of formation of self-cleaning photocatalytically active coatings of building structural materials according to innovative technological regulations using nitrate precursors.

Existing identified trends of phase formation in model investigated systems are thermodynamically the most probable limits of transformations in technological objects under conditions of formation and receipt of target products. In addition, the possible real deviations in such systems are due to the heterogeneity of the reaction medium in composition, the content of the reacting components, the conditions of their finding, the finiteness of the rate of transformations, diffusion peculiarities, heat capacity, viscosity, the nature of transformations at the boundaries of the formed heterophase modes, activation, and other specific factors. The revealed processes of complexation in aqueous solutions of nitrates contribute to the homogenization of systems of structural components at the molecular level in complex or combined processing.

The analysis of the obtained data indicates that systems in the immediate environment of the Ln^{3+}-complexing agent of competing processes of substitution of water molecules with nitrate ions occur in the systems. The degree of completeness of substitution depends on the nature of Ln^{3+} present in Me^{+}, the properties of electron-donor oxygen atoms and the spatial structure of the ligands, the concentration of anions, and the amount of solvent. Significant influence on these processes of temperature factor is revealed. Differences in the complexing ability of the elements of cerium and yttrium subgroups, Y, and among REEs in the middle of the first subgroup are observed. The obtained results indicate the stage of complex formation processes. The presence of certain values of the temperature of the beginning of the release into the solid phase of complex compounds—the existence of an energy barrier and the need for some activation energy to effect such transformations.

The formation of nitrate complexes largely fulfills the requirements of symmetry, and planar small-sized ligand NO_3^- is "convenient" for the formation of a highly symmetric environment of the Ln^{3+} ion. Lanthanides are characterized by a tendency to form a limited number of coordination polyhedron species (see Fig. 3) and three types of coordination NO_3^--ligands. This leads to the formation of both isolated complexes and their polymerization into dual nuclei, chains, and frames. The coordination numbers of the Ln^{3+} cerium subgroup—12 detected in low-temperature associated forms

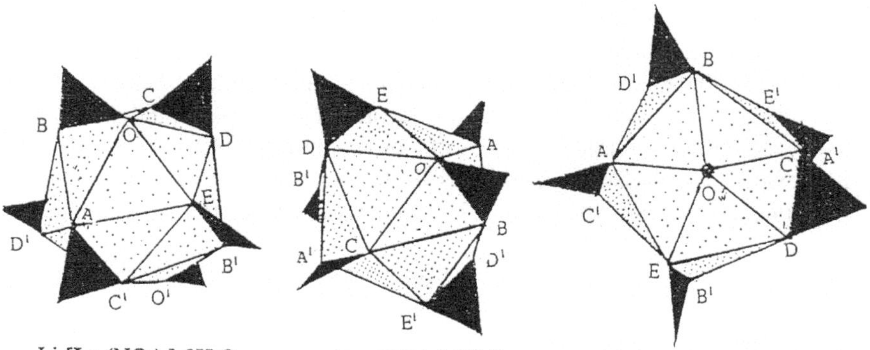

Fig. 3 Schematic representation and general view of Ln-icosahedrons found in the structures of the REEs of the cerium subgroup and lithium, sodium, and potassium

and remain unchanged until the formation of stable high-temperature multicomponent oxide phases of $MeLnO_2$, $Me_2Ln_2Ti_3O_{10}$, the influence of the nature of Me^+, on the shape of the coordination polyhedral Ln, the method of packing complexes in the spatial structure, and a number of properties of compounds.

We present below:

- in Table 2, Fig. 4—installed X-ray diffraction characteristic parameters of the newly formed phases for the possibility of their identification and detection during processing;

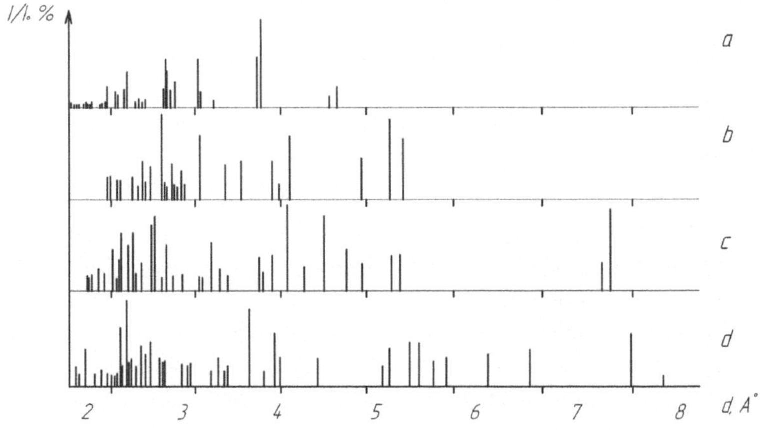

Fig. 4 Bar graphs of the starting salts of nitrates **a** potassium, **d** neodymium and identified coordination compounds **b** $K_2[Nd(NO_3)_5(H_2O)_2]$, **c** $K_3[Nd_2(NO_3)_9]{\cdot}H_2O$

- in Table 3—we explain the nature and patterns of thermal transformations of compounds in the temperature range 25–1000 °C to find out their thermal stability and model the behavior of process objects under similar conditions. For comparison and analysis, information about the low stability and thermal stability of potassium coordination nitrates Y, Gd–Lu of the composition $K[Ln(NO_3)_4(H_2O)_2]$ is provided, which causes restrictions in the use of this type of precursors in technological transformations in order to modify the properties target products.

2.3 Scientific Novelty

A complex systematic study of the interaction of structural components in rare earth nitrates and IA group of elements of the periodic system—precursors of modern multicomponent oxide functional materials based on them—revealed the formation of a wide class of alkali coordinate nitrates of lanthanides.

Identified objective laws have a fundamental and applied value, deepen understanding of:

- chemical and physical properties of Ln and their complexing ability,
- possibility of formation and existence in the associated systems of associated new phases, their atomic-crystalline structure, stability, and stability,
- the influence of the nature of lanthanides and alkali metals on the structure of complex anions and compounds in general,
- the individuality of the Ln complexes,
- the existence of isotopic groups in the composition and structure of groups of compounds by the natural series of lanthanides and alkali metals,
- the role of NO_3^--groups in the stereochemistry of this class of nitrates,
- the role of water in the formation of the closest environment of Ln^{3+} ions-complexing agents.

2.4 Practical Importance

The obtained system of knowledge about transformation processes in systems of REE-containing nitrate precursors and crystal-chemical properties of samples of coordinate nitrates Ln is of particular value in the formation of nanostructure-layered perovskite-like compounds of lanthanides and transitional elements (including titanium); solid solutions based on them; in establishing the technological and functional dependencies between the method of preparation; the variability of the method of activation of technological systems; the methodology of production of the target product and its phase composition, lattice parameters, the specific surface area, the morphology of the constituent particles, the activity of layers of self-purifying compositions with hydrophilic structures and special structural elements; in the practical implementation of innovative projects of decomposition of water for the purposes of hydrogen production (as an alternative fuel), decomposition of toxic organic substances in solutions and air, incomplete oxidation of carbohydrates; upon receipt of other perovskite-like phases by ion-exchange reactions and in other fields.

3 Conclusions

1. The results of the study show that the processes of obtaining oxide REE-containing structural and functional materials of various purposes using nitrates of elements of the different electronic structure are performed by chemical mixing of the initial components in the joint separation of products from the liquid phase by sequential or compatible deposition with subsequent heat treatment. Data on their composition, content, and behavior in each case require prior systemic empirical knowledge in full concentration ratios at a given temperature interval.
2. The differences in the behavior of structural components in the systems of lanthanides of cerium and yttrium subgroups, in their nature of interaction, stage, features, and regularities of flow are revealed.
3. The new knowledge obtained forms the basis for:
    - finding ways to increase the activity of Ln forms;
    - to find out the nature of sequential thermal transformations in nitrate REE-containing multicomponent systems of different aggregate states during their heat treatment; conditions of formation and existence, properties of intermediate phases; influencing factors; possible ways to control how your target product is obtained;
    - in case of the creation of modern advanced low-cost technologies of formation of functional materials of various purposes with reproducible properties.

References

1. https://www.pilkington.com/ru-ru/ru/products/product-categories/self-cleaning/pilkington-activ. Last accessed 2019/09/10.
2. http://www.ivit.ua/article/arhitekturnye_i_stroitelnye_stekla-obzor. Last accessed 2019/09/10.

3. Varma, A., Mukasyan, A. S., Rogachev, A. S., et al. (2016). Solution combustion synthesis of nanoscale materials. *Chemical Reviews, 116,* 14493–14586.
4. Zhang, X., Zhang, A., Zhang, H., Luo, J., & Chen, L. (2012). *Preparation and application of a hydrophilic self-cleaning coating with photocatalytic activity*. Pat. CN. 102382490A. Claimed 30.08.2011; publ. 23.03.2012.
5. Schaak, R. E., & Mallouk, T. E. (2001). $KLnTiO_4$ (Ln = La, Nd, Sm, Eu, Gd, Dy): A new series of Ruddlesden–Popper phases synthesized by ion-exchange of $HLnTiO_4$. *Journal of Solid State Chemistry, 161*(2), 225–232.
6. Zhu, W. J., Feng, H. H., & Hor, P. H. (1996). Synthesis and characterization of layered titanium oxides $NaRTiO_4$ (R = La, Nd and Gd). *Materials Research Bulletin, 31*(1), 107–111.
7. Richard, M., Brohan, L., & Tournoux, M. (1993). Synthesis, characterization, and acid exchange of the layered perovskites $A_2Nd_2Ti_3O_{10}$ (A = Na, K). *Journal of Solid State Chemistry, 112,* 345–354.
8. Gopalakrishnan, J., & Bhat, V. (1987). $A_2Ln_2Ti_3O_{10}$ (A = potassium or rubidium; Ln = lanthanum or rare earth): A new series of layered perovskites exhibiting ion exchange. *Inorganic Chemistry, 26*(26), 4299–4301.
9. Utkina, T., Chislov, M., Silyukov, O., Burovikhina, A., & Zvereva, I. (2016). TG and DSC investigation of water intercalation and protonation processes in perovskite-like layered structure of titanate $K_2Nd_2Ti_3O_{10}$. *Journal of Thermal Analysis and Calorimetry, 125,* 281–287.
10. Borilo, L. P. (2003). *Thin-film inorganic nanosystems.* Tomsk: Publishing House Tom. University.
11. Belyaev, A. V. (2003). *Methods for producing inorganic non-metallic nano-particles*. Moscow: Publishing House of Moscow Art Theater. D.I. Mendeleev.
12. Syrkova, O. V., & Tsvetkova, V. K. (1987). Obtaining composite materials by the method of interfacial polycondensation. In V. B. Aleskovsky (Ed.), *Directional synthesis of solids: Interuniversity collection*. Leningrad: Publishing house of Leningrad State University.
13. Kozik, V. V., Borilo, L. P., & Turetskova, O. V. (2002). Thin-film composite materials based on SiO_2 and REE oxides. *Condensed Matter and Interphase Boundaries, 4*(3), 231–235.
14. Anosov, V. Ya., Ozerova, M. Y., & Fyalkov, Yu. Ya. (1976). *Basis of the physical chemistry analysis*. Moscow: Nauka Publ.
15. Horoshchenko, Ya. H. (1978). *Physical chemistry analysis of the homogeneous and heterogeneous systems*. Kiev: Naukovadumka Publ.
16. Storozhenko, D. O., Dryuchko, O. G., Bunyakina, N. V., et al. (2015). Phase formation in REE-containing water-salt systems at the preparatory stages of the multicomponent oxide functional materials formation. *Innovations in Corrosion and Materials Science, 5*(2), 80–84.
17. Storozhenko, D. O., Dryuchko, O. G., Bunyakina, N. V., et al. (2013). Chemical interaction and phase formation in sulfates, nitrate, chloride water-salt systems of neodymium and alkali metals. *Bulletin of NTU "KPI", 57*(1030), 121–126.
18. Bunyakina, N. V., Storozhenko, D. O., Shevchuk, V. G., & Dryuchko, O. G. (1996). Solubility polytherm of the $Mg(NO_3)_2$–$Nd(NO_3)_3$–H_2O system. *Journal of Inorganic Chemistry, 41*(9), 1577–1579.
19. Driuchko, O. G., Storozhenko, D. O., Buniakina, N. V., et al. (2016). Peculiarities of transformation REE-containing systems of nitrate precursors in the preparatory process of formation perovskite-like of oxide materials. *Bulletin of National Technical University "KhPI", 22*(1194), 63–71.
20. Dryuchko, O. G., Storozhenko, D. O., Bunyakina, N. V., et al. (2018). Physicochemical characterization of REE coordination nitrates and alkali metals—Precursors of oxide polyfunctional materials. *Bulletin of the National Technical University "KPI", 39*(1315), 3–13. https://doi.org/10.20998/2079-0821.2018.39.01.
21. Vigdorchik, A. G., Malinovsky, Yu. A., & Dryuchko, A. G. (1992). Low-temperature X-ray diffraction study of potassium-neodymium nitrate $K_3[Nd_2(NO_3)_9]$ and $K_2[Nd(NO_3)_5(H_2O)_2]$. *Crystallography, 37*(4), 882–888.
22. Dryuchko, O. G., Storozhenko, D. O., Bunyakina, N. V., & Ivanytska I. O. (2019). Formation conditions and physico-chemical characteristics of lithium coordination nitrates of lanthanides

$Li_3[Ln_2(NO_3)_9]\cdot 3H_2O$ (Ln–La–Nd). *Bulletin of the National Technical University «KhPI», 1*, 41–48. https://doi.org/10.20998/2079-0821.2019.01.08.
23. Dryuchko, O., Storozhenko, D., Vigdorchik, A., Bunyakina, N., Ivanytska, I., et al. (2019). Features of transformations in REE-containing systems of nitrate precursors in preparatory processes of formation of multifunctional oxide materials. *Molecular Crystals and Liquid Crystals, 672*(1), 199–214. https://doi.org/10.1080/15421406.2018.1542066.

Some Technical Solutions for the Use of Aerodrome Pavements in the Soft Soil Conditions

Svitlana Talakh, Oleksandr Dubyk, Olha Bashynska, and Volodymyr Ilchenko

Abstract Reconstruction model of aerodrome pavement with the use of new design of precast concrete prestressed slab is proposed. Numerical study of stress–strain state of strip pavement was performed. The aim of the article is studying stress–strain state of the proposed design of precast concrete aerodrome pavement when interacting with soft soil conditions. The design of the jump joint, which maintains jointless connection of pavement, is offered. To study stress–strain state of aerodrome pavement constructed from precast concrete prestressed slabs, numerical technique is used in the article. In furtherance of numerical technique, it was defined the equivalent cross-section unit of the three-layer thin slab construction. The scheme with one plane of symmetry for discrete model fragment of flight strip prestressed cover slabs was constructed. Based on the results of numerical calculations, nodal displacements plot, driving bending moments and nodal reactions of soil base are made. It is established that conditions of aerodrome pavement structure strength and deformability are satisfied. Practical importance of the work is that, with sufficient feasibility study, a significant reduction in material capacity of coating structure can be achieved with the necessary durability and reliability, even for heavy-duty oversized aircraft.

Keywords Stress–strain state · Reinforced concrete slabs · Soft soil conditions · Aerodrome pavements

S. Talakh · O. Dubyk · O. Bashynska
National Aviation University, Kiev, Ukraine
e-mail: svetlanatalah@gmail.com

O. Dubyk
e-mail: saschadubik@ukr.net

O. Bashynska
e-mail: olchik01@ukr.net

V. Ilchenko (✉)
Poltava National Technical Yuri Kondratyuk University, Poltava, Ukraine
e-mail: znpbud@gmail.com

V. Onyshchenko et al. (eds.), *Proceedings of the 2nd International Conference on Building Innovations*, Lecture Notes in Civil Engineering 73,
https://doi.org/10.1007/978-3-030-42939-3_31

1 Introduction

Within infrastructure projects for Odessa International Airport realization, implementation of aerodrome total redevelopment, construction of the new passenger terminal complex and flight strip are suggested. These measures will include reconstruction of existing flight strip, which is of poor smoothness and insufficient load-bearing capacity, and arrangement of rigid pavement capable of bearing D class aircraft (AC) loads.

In accordance with engineering and geological investigations [1], the aerodrome soils belong to soft ones and are unable to bear loads originated from modern aircraft wheels.

The key reasons that cause destruction of the aerodrome pavements are as follows: increased loads due to constant aircraft takeoff weight gain; flying rate with airline fleet intensity increasing; water accumulating in the base under the pavement due to water and temperature condition disturbances, drainage and water disposal system operational disorders.

In accordance with materials of engineering and geological investigations, the aerodrome pavement soil subbase is comprised of type II subsidence powdered clay loam. Thickness of the subsiding soil equals to 7 m. The soils given are stable only in their native state. When their native state is disturbed due to watering or under stress of physical force, these soils lose their stability and deliver irregular subsidence.

On the area under investigation along the flight strip, we can observe water table rise, which is suggestive of active and unbalanced activities impact. The first water-bearing layer below the surface is of anthropogenic origin. Mainly, it is fed by infiltration waters and leakages from sewerage systems [1].

Taking into account that the first water-bearing layer below the surface lies on different depths, then soil subsiding properties appear differently. Thus, thickness of subsiding depth below the surface varies from 8.00 to 11.20 m along the axis of flight strip. The amount of subsidence when wetting, against the loads equal to the own weight of soil, is up to 6 cm. All these factors determine the necessity to carry out extra activities to get the subgrade soil stabilized.

Existing construction of the flight strip's pavement is presented with rectangular cement concrete slabs 24–26 cm thick on the compacted subgrade soil and reinforced with the double-layer asphalt concrete 16–18 cm thick.

The authors suggest the design of precast reinforced concrete pavement from prestressed slabs, which can be temporarily used to construct the aerodrome artificial flight strip for Odessa airport. In the variant offered, the precast reinforced concrete slabs of the pavement are to be laid on the base course made from the machined asphalt concrete (granulated asphalt concrete) 15 cm thick on the existing rigid concrete base 24 cm thick.

The main advantage of this method is that it is possible to carry out construction almost all year round, which greatly reduces the terms of construction [2].

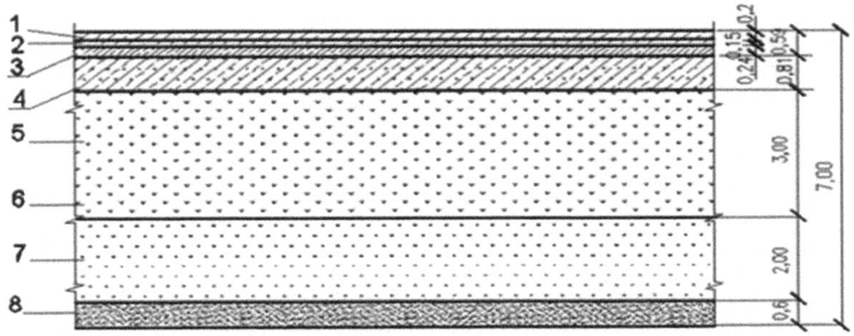

Fig. 1 Structural design of the pavement with the active zone of the subgrade soil: 1—precast reinforced concrete prestressed slabs made from B45 concrete, $R_{btb} = 3.73$ MPa (by strength), $E_b = 3.53 \times 10^4$ MPa; $R_{btb,ser} = 4.4$ MPa (by crack formation), thickness $t_1 = 0.2$ m; 2—base and levelling coarse made from the machined asphalt concrete B15, $R_{btb} = 2.26$ MPa (by strength), $E_b = 2.3 \times 10^4$ MPa, thickness $t_2 = 0.15$ m; 3—artificial subgrade made from B25 concrete, $B_{btb} = 2.75$ MPa, $E_b = 2.84 \times 10^4$ MPa, $t_3 = 0.24$ m; 4—soil layer, $E = 14.0$ MPa; 5—loess soil, $E = 14.0$ MPa; 6—low-plasticity loess soil, $E = 13.0$ MPa; 7—flowing loess soil, $E = 6.0$ MPa

In addition, with sufficient technical and engineering justification, it is possible to achieve a significant reduction in the material-bearing capacity of coating construction, providing the necessary strength and reliability even for super-heavyweight aircraft.

Design of this pavement with an active zone of subgrade soil is shown in Fig. 1. Coefficient of subgrade reaction is $K_{se} = 50.04$ MN/m^3.

2 Defining the Problem

In the suggested variant, the authors consider a new design of precast prestressed slab made from solid and heavy B45 concrete (of moderate strength) reinforced with high-strength BII wire of 5.5 in diameter and with the design tensile strength for limiting states of the second group $R_{s,ser} = 1295$ MPa. The dimensions of slab were taken as non-standard in order to consider the development of transport and lifting machines, i.e. 7.4 × 3.7 × 0.2 m, weight is 13.5 t. The prestressing force of the wiring consisting of two grids spaced at 140 mm, fastened with the additional light triangular frames from the BI common reinforcement rods 8 mm in diameter, shall be determined for a single bundle of 4-rods reinforcing wire BII 5.5 mm in diameter attributable to part of the slab's zone 300 mm in width. The stretching force N_{II} is equal to 110.72 kN, taking into account the parameters of reinforcing wires $R_{s,ser} =$ 13,200 kgf/cm^2 and wiring area $F_a = 0.95$ c/cm^2 and the coefficient of primary loss under relaxation of reinforcing wires on the abutments $\gamma_c = 0.9$.

After removing prestressing force on the stand and transferring it to slab structure within lengthwise and crosswise directions through the fixings (Fig. 2), we obtain central pressing (excluding eccentricity) of average intensity:

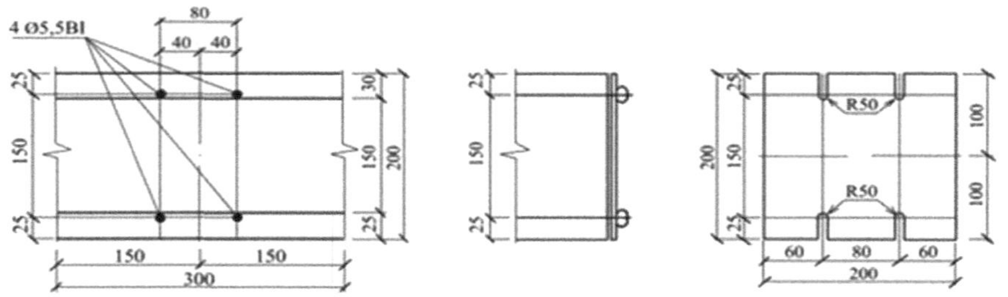

Fig. 2 Typical unit of stressed reinforcing wires from high-strength wire and its anchoring at the slab butt

$$q = \frac{\eta_c \cdot N_{II}}{h \cdot l_p} \cong 1.7 \text{ MPa}, \quad (1)$$

where η_c is the transition factor of compression on the anchor devices of the slab—the secondary loss on the concrete shrinkage and creep, $\eta_c = 0.9$; l_p is the bundle spacing, $l_p = 0.3$ m; h is the slab thickness, $h = 0.2$ m [3, 4].

In the variant of precast pavements being considered in question, the authors suggest the butt joint design which is able to ensure practically seamless pavements (Fig. 3). The design comprises BI Ø 20 mm joint clips, $l = 0.5$ m with spacing $l_s = 0.90$ m (8 pcs clips along the long sides of the slab and 4 pcs along the short ones). After placing slabs, the clips of the adjoining slabs are welded rigidly using the connecting pieces, i.e. round bar cuts 36 mm in diameter, $l = 0.38$ m, and additionally, the steel joint bars are welded to anchor plates 14 mm thick (see Fig. 3).

For further numerical calculation, parameters of reduced cross-section of the plate are determined taking into account the areas of reinforcement A_{sp} and A'_{sp} (Fig. 4).

The design fragment thanks to its continuous scheme of pavement functioning includes two precast slabs 746 × 740 cm in size with symmetrical part of the wheel landing gear of B767-300 aircraft was placed on that fragment (Fig. 5). The dimensions of the design fragment's net domain are M1 × M2 × M3—2 × 23 × 24 that

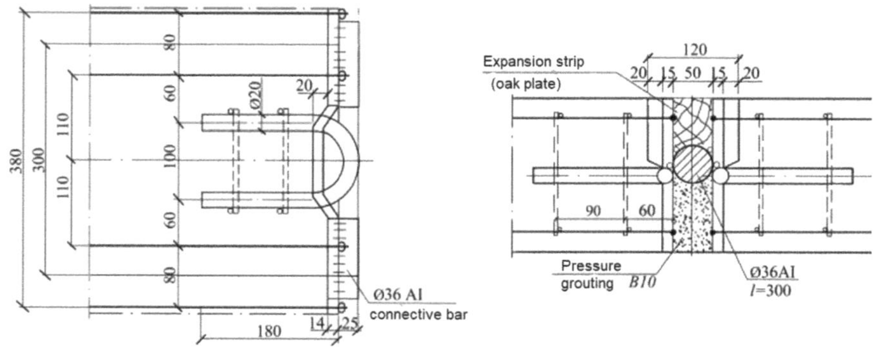

Fig. 3 Butt joint design of the precast reinforced concrete slabs

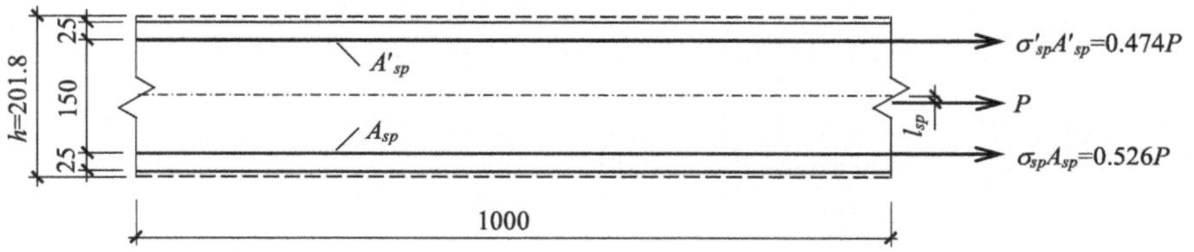

Fig. 4 Scheme of prestressing within plate cross-section unit

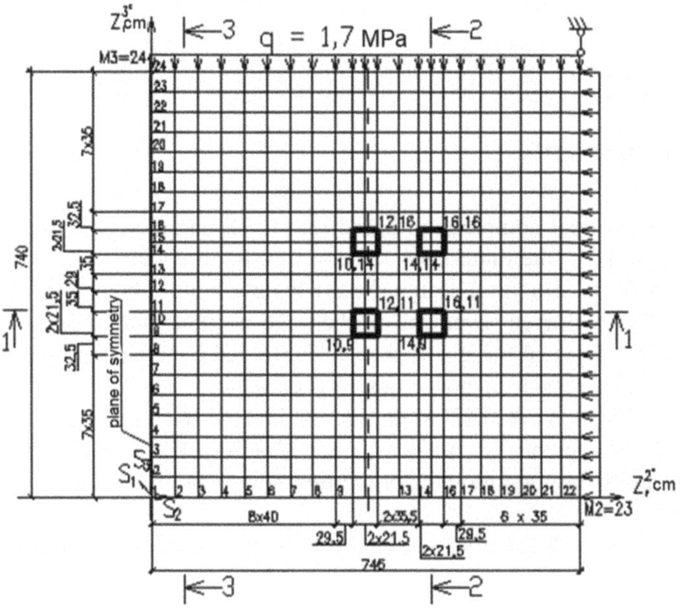

Fig. 5 Design diagram with one symmetry plane for discrete model fragment consisting of two precast reinforced concrete prestressed slabs of flight strip's pavement

corresponds to the system of resolving algebraic equations in the order $k = 2 \times 20 \times 25 \times 3 = 3312$ without imposing constraints.

Here, we present geometric and mechanical and physical properties of discrete model of flight strip's double-layer coating on solid artificial subgrade in accordance with the structural design shown in Fig. 1. Since the upper layer of the pavement is precast reinforced concrete slab, we first calculate parameters of equivalent cross-section of this slab based on the typical unit shown in Fig. 2, taking into account cross-section given Area (see Fig. 4):

- height of equivalent cross-section $h_e = 0.21$ m;
- longitudinal stiffness of cross-section: $(EA)^d_{x_c} = 7.123 \times 10^8$ kgf;

$$E_e = \frac{(EA)^{\partial}_{x_c}}{h_e \cdot 100} = 3.392 \times 10^5 \text{ kgf/cm}; \quad (2)$$

- Poisson's ratio and average unit weight:

$$\nu_e = \frac{\nu_{cm}\left(A'_{sp} + A_{sp}\right) + \nu_b A_b}{20 \times 100} = 0.2205; \ \gamma^1_{cp} = \frac{\gamma_b A_b + \gamma_s\left(A'_{sp} + A_{sp}\right)}{20 \times 100} = 0.00241. \quad (3)$$

3 Research results

To perform numerical computation based on [1, 2], parameters of the equivalent unit cross-section for the three-layer structure of the thin slab on the solid base in accordance with the structural design should be determined and shown in Fig. 1.

For this, the two layer pavement on the solid artificial subgrade is considered [5–12]. The first layer of the given precast reinforced concrete slab model has the geometrical and mechanical properties in accordance with (2): $t_1 = 0.21$ m, $E_1 = 3.392 \times 10^4$ MPa, $\nu_1 = 0.2205$; the second layer is from the granulated asphalt concrete with parameters: $t_2 = 0.15$ m, $E_2 = 2.32 \times 10^4$ MPa, $\nu_2 = 0.22$; the artificial concrete subgrade has the following properties: $t_3 = 0.24$ m, $E_3 = 2.84 \times 10^4$ MPa, $\nu_3 = 0.225$.

In accordance with scheme, shown in the Fig. 6, following parameters were calculated:

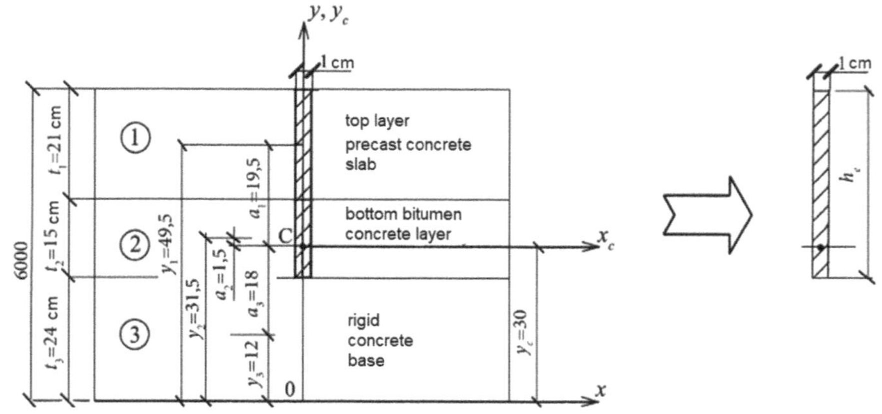

Fig. 6 Scheme to reduce single cross-section of actual thin slab structure to equivalent cross-section of simulated thin slab

- coordinates of the gravity centre for thin slab structure in frame *XOY* coordinates:

$$Y_c = \frac{t_1 y_1 + t_2 y_2 + t_3 y_3}{t_1 + t_2 + t_3} = 0.30\text{ m}; \tag{4}$$

- inertia moments for two layer structure related to its gravity centre axis—x_c and y_c:

$$I_{x_c} = I_1^o + a_1^2 \cdot A_1 + I_2^o + a_2^2 \cdot A_2, \tag{5}$$

where $I_{x_c} = I_1^{x_c} + I_2^{x_c} = 9072\text{ cm}^4$; $I_1^{x_c} = \frac{t_1^3}{12} + a_1^2 \cdot t_1 = 8757\text{ cm}^4$; $I_2^{x_c} = \frac{t_2^3}{12} + a_2^2 \cdot t_2 = 315\text{ cm}^4$.

- bending and longitudinal stiffness of actual slab structure:

$$\begin{aligned} EI &= E_1 I_1^{x_c} + E_2 I_2^{x_c} = 3.0426 \times 10^8\text{ MPa}; \\ EA &= E_1 t_1 + E_2 t_2 + E_3 t_3 = 1.739 \times 10^6\text{ kgf}; \end{aligned} \tag{6}$$

- equivalent height (thickness) of given cross-section plate:

$$h_e = \sqrt{\frac{12EI}{EA}} = 45.82 \cong 0.46\text{ m}; \tag{7}$$

- equivalent modulus of given cross-section elasticity:

$$E_e = \frac{EA}{l \cdot h_e} = 3.7802 \times 10^4\text{ MPa}; \tag{8}$$

- Poisson's ratio of cross-section equivalent of the plate:

$$v_e = \frac{v_1 t_1 + v_2 t_2}{t_1 + t_2} = 0.22;\ \gamma_{\text{half - value}} = \frac{\gamma_s' t_1^{\text{eq}} + \gamma_2 t_2 + \gamma_3 t_3}{t_1 + t_2 + t_3} = 0.0024. \tag{9}$$

Based on results of numerical calculations, diagrams for nodal displacements, linear bending moments and soil subgrade nodal reactions were constructed in 1.1 and 2.2 cross-sections.$u_N^{1'}$ and internal efforts are localized in footprint of the main gear wheels PS B767-300. Their values can be seen in the graphs of epures presented in Figs. 7 and 8: $u_{506}^{1'} = -1.08$ mm; $M_{405}^{33} = -83.93$ kNm/m; $R_{494}^{1'} = 54.02$ kN.

The maximum primary stresses can be observed in the zone of maximum linear bending moments, i.e. $\sigma_{405}^{\max} = 2.153$ MPa; $\sigma_{405}^{\min} = -2.545$ MPa.

The bending constituent has the value:

$$\sigma_{\text{bending}}^{\min} = \frac{\sigma_{405}^{\max} - \sigma_{405}^{\min}}{2} = 2.35 \text{ MPa},$$

That is, pure bending strains determine the value for tensile in outer fibre amounting to 2.35 MPa, but the pressure loads as a result of prestress reduce the tensile loads to 2.15 MPa in the lower layer of plate, where $R_{\text{btb}} = 2.26$ MPa. Thus, the conditions for strength and deformability for the pavement structure are being satisfied.

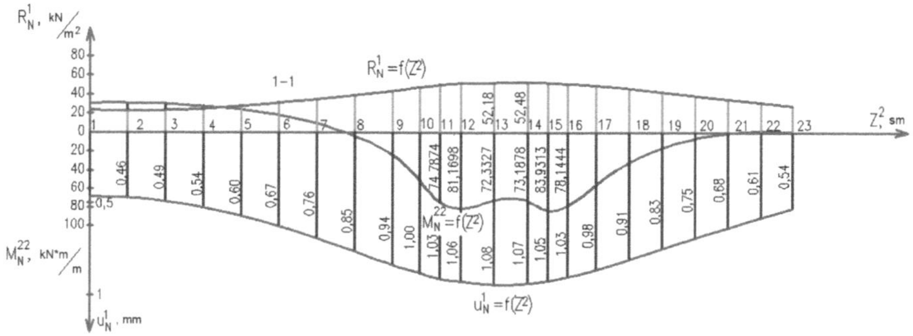

Fig. 7 Diagrams of the linear bending moments M^{22}, kNm/m and nodal responses of the base R^1, kN/m^2 and nodal displacements u^1, mm of the cross-section 1.1 from the main gear wheels of PS B767-300

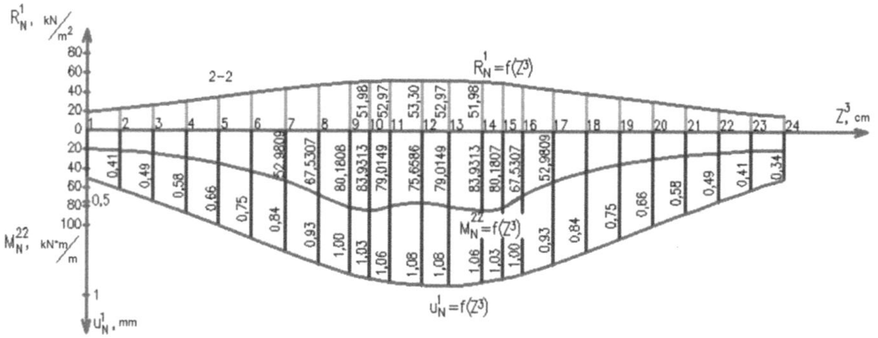

Fig. 8 Diagrams of linear bending moments M^{22}, kNm/m and nodal responses of the base R^1, kN/m^2 and nodal displacements u^1, mm of the cross-section 2.2 from the main gear wheels of PS B767-300

4 Conclusion

Model of reconstructing the aerodrome pavement using reinforced concrete prestressed slabs laid on the machined asphalt concrete is suggested.

If to consider the design as alternative variant to conventional concrete and reinforced concrete pavements with well-established construction technology, they have advantages based on reduction criterion in the input of materials and capital costs.

Primary advantage of applying the reinforced concrete prestressed slabs lies in the fact that they enable carrying out construction works actually the whole year round ensuring significant reduction in construction time [2]. Additionally, with a sufficient level of engineering and technical evaluation, sufficient reduction in the input of materials aimed at the pavement structure ensuring the required level of strength and durability even in case of the super heavy large body aircraft is achievable.

References

1. Technical report on engineering and geological conditions for the reconstruction area within the aerodrome of Odessa International Airport CE: Odessa, Engineering and Project Investigations Centre (2015).
2. Kulchitskij, V. A., Makagonov, V. A., Vasilev, N. B., et al. (2002). *Aerodrome Pavements*. Fizmatliteratura, Moscow: Modern View.
3. SNiP 2.03.01-84* (1989). Concrete and reinforced concrete structures. Gosstroy USSR.
4. Arbuzov, N. T., Trigoni, V. E., et al. (1972). *Operating instructions on design and construction of continuously seamless aerodrome pavements (CSAP)*. Moscow: ONTI.
5. Chau, K. (2013). Numerical Methods. In *Proceedings of the 18th International Conference on Soil Mechanics and Geotechnical Engineering* (pp. 647–654). Paris.
6. Henke, S., & Grabe, J. (2009). Numerical modeling of pile installation. In *Proceedings of the 17th International Conference on Soil Mechanics and Geotechnical Engineering (Alexandria, 2009)* (pp. 1321–1324). Amsterdam: IOS Press.
7. Pak, A., Shahir, H., & Ghassemi, A. (2005). Behavior of dry and saturated soils under impact load during dynamic compaction. In *Proceedings of 16th International Conference on Soil Mechanics and Geotechnical Engineering* (pp. 1245–1248). Osaka.
8. Bazhenov, V. A., Tsykhanovskii, V. K., & Kislookii, V. M. (2000). *The finite-element method in nonlinear-deformation problems for thin and soft shells*. Kyiv: KNUBA.
9. Bazhenov, V. A., Sakharov, A. S., & Tsykhanovskii, V. K. (2002). Moment scheme of the finite elements method in the problems continuous medium nonlinear mechanics. *Applied Mechanics, 38*(48), 24–63.
10. Tsytovich, N. A. (1963). *Mechanics of Soil*. Moscow: GILSAiSM.
11. Tsykhanovskii, V. K., Kozlovets, S. M., & Koryak, A. S. (2008). Calculation of Thin Slabs on the Elastic Subgrades Using the Finite Elements Method. Stal PH.
12. Development of a Computer Program—COMFAA—for Calculating Pavement Thickness and Strength (2003).

Research of Possible Methods of Increasing the Duration of the Insolation of Rooms in Residential Buildings

Oleh Yurin, Yurii Avramenko, Maryna Leshchenko, and Olesia Rozdabara

Abstract **The aim** of the work was to determine the best methods for increasing the insolation duration of rooms in residential buildings with high-density buildings using 10- and 16-story buildings. **The results** of theoretical studies on the analysis of possible methods for increasing the insolation duration of rooms of residential buildings are presented. In order to bring the insolation duration of these rooms to the required building codes, the following **methods** were considered: changing the orientation of the building, reducing the width of the loggia side screen, increasing the window width, changing the layout of the apartment, shifting the window to the north, shifting sections of the house, increasing the distance between the houses, the separation and displacement of the section of the house located closer to the south in the direction to the south, the change in the planning decisions, and reducing the height of the shading house. **Scientific novelty**. For the first time, a set of constructive methods for the design of rooms to provide insolation in rooms in residential buildings has been applied. **Practical significance**. This approach of constructive methods of effective insulation in residential buildings can be widely used in compact planning urban of microdistricts in big cities.

Keywords Building · Construction · Room insolation · Methods for increasing the duration of insolation

1 Introduction

Currently, due to the rise in price of the urban area, there is a significant consolidation of its building. Developers significantly reduce the distance between buildings and increase their number of floors. This greatly complicates the planning of comfortable

O. Yurin · Y. Avramenko · M. Leshchenko (✉) · O. Rozdabara
National University "Yuri Kondratyuk Poltava Polytechnic", Poltava, Ukraine
e-mail: mv.leshchenko@gmail.com

M. Leshchenko
Poltava Scientific Research Forensic Center of the MIA (Ministry of Internal Affairs) of Ukraine, Poltava, Ukraine

V. Onyshchenko et al. (eds.), *Proceedings of the 2nd International Conference on Building Innovations*, Lecture Notes in Civil Engineering 73,
https://doi.org/10.1007/978-3-030-42939-3_32

courtyard space, worsens the air regime of the courtyard, due to the difficulty of airing it, and reduces the insolation time of the courtyard and living rooms in the houses.

Reducing the duration of insolation of rooms adversely affects the central nervous system of a person, tone, health, and performance. Therefore, the analysis and identification of optimal methods for increasing room insolation is an urgent task.

Most publications on this topic are devoted to the method of calculating the duration of insolation and the regulation of insolation [1–6], the hygienic properties of insolation [7–11], and the advantages of using tower-type houses to increase the duration of insolation of rooms [12, 13]. In publications devoted to the study of insolation, no analysis was made of possible methods for increasing the duration of insolation of rooms.

2 The Aim

The aim of the work was to determine the best methods for increasing the insolation duration of rooms in residential buildings with high-density buildings using 10- and 16-story buildings.

3 The Methods

In order to bring the insolation duration of these rooms to the required building codes, the following methods were considered: changing the orientation of the building, reducing the width of the loggia side screen, increasing the window width, changing the layout of the apartment, shifting the window to the north, shifting sections of the house, increasing the distance between the houses, the separation and displacement of the section of the house located closer to the south in the direction to the south, the change in the planning decisions of the section as a whole, and reducing the height of the shading house. Combined methods were also considered: moving the window to the north and removing the side screen of the loggia, moving the window to the north, removing the side screen of the loggia and shifting the western wall of the section, moving the window to the north, removing the side screen of the loggia, shifting the western wall of the section, and increasing the distance between houses and others.

4 Main Material and Results

The study of the duration of insolation was carried out in accordance with the requirements of [12]. The calculations were performed using the insolation line, given in Appendix A [9]. Insolation duration was determined for equinox days.

According to clause 7.5 [12], the estimated insolation duration is determined from 7^{00} to 17^{00} as the difference between the insolation duration within the horizontal insolation angle and the shading duration of the opposite houses.

The houses are located around the low-rise residential district, which are located at a considerable distance; therefore, the study did not take into account the insolation duration.

The master plan of the site, where the houses are located, in which the insolation duration of the rooms was investigated, is shown in Fig. 1.

Window numbers in these houses are shown in Fig. 2.

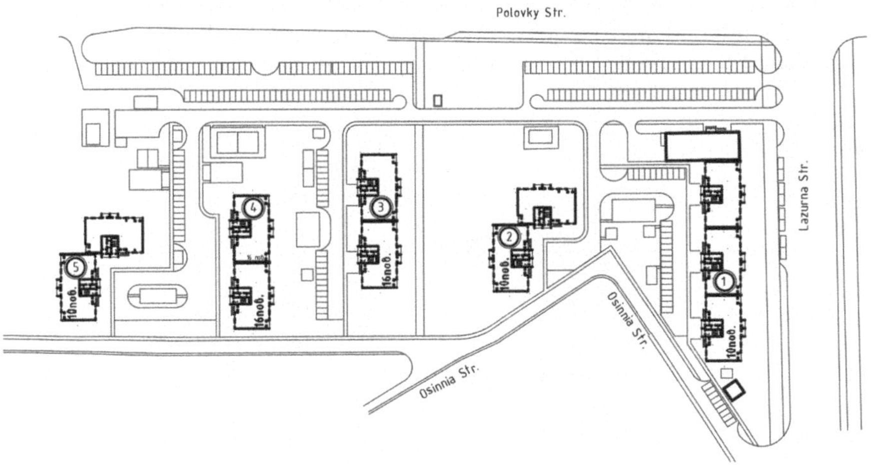

Fig. 1 Site master plan

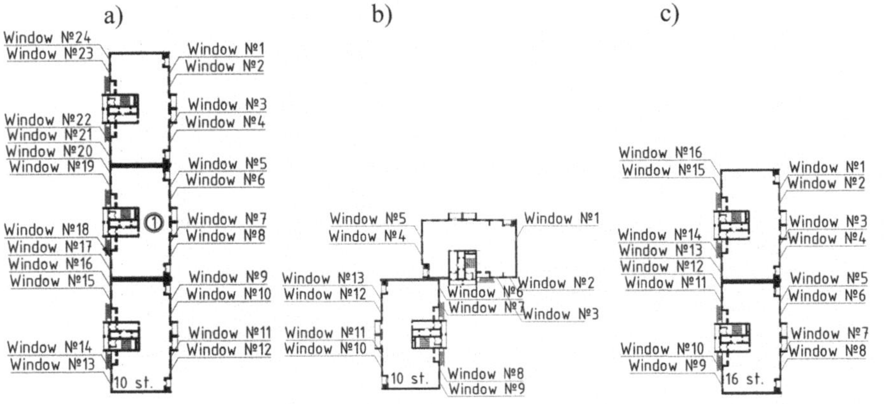

Fig. 2 Window numbers in the houses: **a** number 1; **b** number 2 and number 5; **c** number 3 and number 4

The results of the study of the duration of insolation in the rooms of houses are presented in Table 1. The table shows the windows of apartments in which the duration of insolation is less than the standard.

The location of the apartments where insolation requirements are not fulfilled is presented in Fig. 3.

From the studies performed, it can be seen that in house number 1, the insolation norms are performed in all apartments.

In house number 2 in rooms with windows number 4 and number 5, the insolation requirements are not fulfilled due to a significant shift of the meridionally located section to the west, which closes a significant horizon sector for these windows. The 16-story building located close to the west side also affects the decrease in the insolation duration. This house also leads to the non-fulfillment of insolation requirements in a meridional section in an apartment with window number 12. In

Table 1 Insolation duration

Number window	Insolation		Insolation duration, hours
	Start	The end	
House number 2			
4	–	–	0^{00}
5	14^{34}	15^{08}	0^{34}
12	14^{42}	16^{03}	1^{21}
House number 3			
1	9^{09}	10^{59}	1^{50}
2	9^{09}	9^{16}	0^{16}
3	9^{09}	11^{24}	2^{15}
4	9^{05}	11^{24}	2^{19}
6	8^{16}	9^{17}	1^{01}
11	13^{01}	14^{29}	1^{28}
12	12^{38}	14^{18}	1^{50}
13	12^{36}	14^{09}	1^{33}
14	12^{36}	14^{02}	1^{26}
15	13^{01}	13^{58}	0^{57}
16	12^{38} 16^{41}	13^{58} 17^{00}	1^{20} 0^{19}
House number 4			
1	9^{12}	10^{59}	1^{47}
2	9^{00}	9^{17}	0^{17}
House number 5			
1	9^{39}	11^{24}	1^{45}
4	15^{44}	17^{00}	1^{16}
5	14^{34}	17^{00}	2^{26}

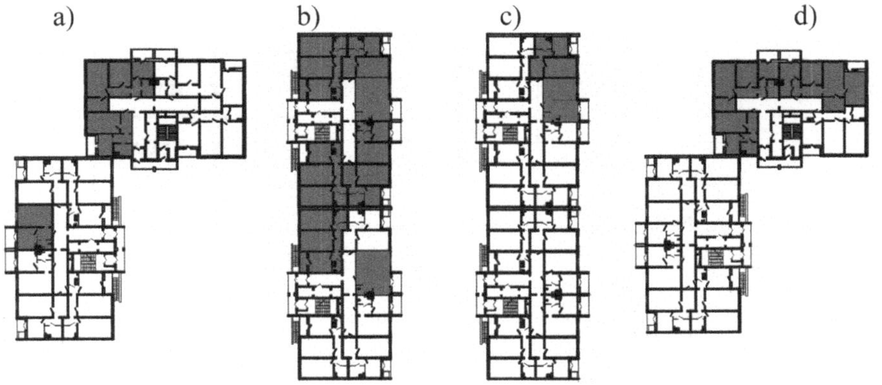

Fig. 3 Location of the apartments where insolation requirements are not fulfilled

addition, a loggia screen located to the south of the window contributes to a decrease in the insolation duration of this window.

In house number 3 in rooms with windows number 1, number 2, and number 6, the side-mounted solid loggia screen is located to the south of the windows, and the 10-story building is located on the east side. In rooms with windows number 3 and number 4, the norms are not fulfilled due to the fact that a significant part of the horizon, visible from these windows, is closed by a 10-story building. In rooms with windows number 11, number 12, number 15, and number 16, the lateral continuous loggia screen is located to the fulfillment of the norms located to the south of the windows and the 16-storied house is located on the west side. In rooms with windows number 13 and number 14, the norms are not fulfilled due to the fact that a significant part of the horizon, visible from these windows, is closed by a 16-story building.

In the house number 4 in rooms with windows number 1 and number 2, insolation requirements are not met due to the reduction of the horizon sector, which allows direct sunlight to penetrate into the apartment, due to the 16-story house closely located on the east side and the side loggia screen located south of these windows.

In the house number 5 in a room with a window number 1, insolation is not performed because a large part of the horizon sector, which can be seen from the window, closes the 16-story building located in the east. In rooms with windows number 4 and number 5, the requirements for insolation are not fulfilled, since the section meridionally located is displaced with respect to the section located meridionally, covers a considerable part of the horizon, and provides room insolation.

As can be seen from the analysis of the duration of insolation of rooms, the main reasons for non-fulfillment of insolation requirements are the presence of a continuous side screen on the loggias, insufficient distance between the houses, and their significant number of floors. In addition, it should be considered unsuccessful connections meridionally and latitudinal sections of houses number 2 and number 5. A significant displacement of the meridionally located section with respect to the

a) b)

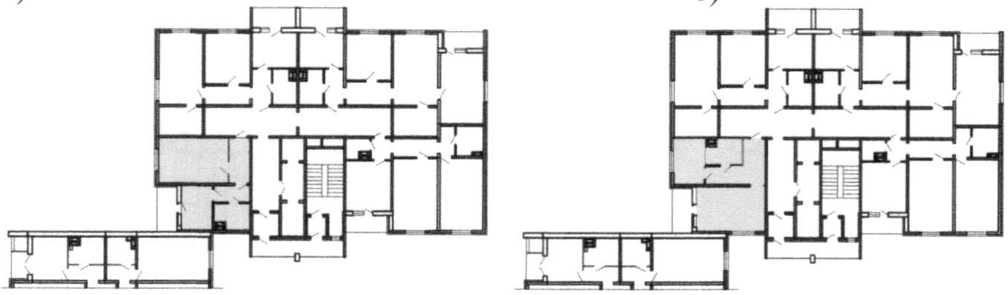

Fig. 4 Apartment plan with window number 4: **a** existing plan and **b** a new plan

section located in latitude makes it difficult to fulfill the insolation requirements for windows located in the western end of the latitudinal section.

To bring insolation of rooms in house number 2 to regulatory requirements, the following methods were considered:

(1) Turn the house. To fulfill the insolation requirements in a room with a window number 4, the house must be rotated 42°. But it becomes less normalized. The application of this method is impossible.
(2) Changing the planning solution for a room with window number 4 (Fig. 4).
As a result of rescheduling, the calculation room window was moved from the western facade to the southern one. As a result, the duration of insolation of the room was 2 h and 33 min. Rules are met.
(3) Offset of window number 5 in the north direction by 3 m. The duration of insolation was 0 h and 56 min. Rules are not met.
(4) Offset of window number 5 to the north by 3 m and removal of the side screen of the loggia located at the end of the meridional section on its western facade. The duration of insolation was 1 h and 10 min. Rules are not met.
(5) Offset of window number 5 to the north by 3 m, removal of the side screen of the loggia, and offset of the western wall of the meridional section by 1.5 m. The insolation duration was 1:00 20 min. Rules are not met.
(6) Offset of window number 5 to the north by 3 m, removal of the side screen of the loggia, offset of the western wall of the meridional section by 1.5 m, and increase in the distance between the houses by 37.5 m. The insolation duration was 2:00 30 min. Rules are met.
(7) Separation of the sections and the displacement of the lower section in the direction to the north by 80.5 m. The insolation duration of the room with window number 5 was 2:00 30 min. Rules are met.
(8) Change of the planning decision of the meridional section as a whole (Fig. 5).
As a result of redevelopment, the number of apartments in the section was reduced from four to three. This allowed the windows of the settlement rooms to be located on the eastern and northern façades. With such an arrangement of the windows of the calculation rooms, the insolation norms are fulfilled.

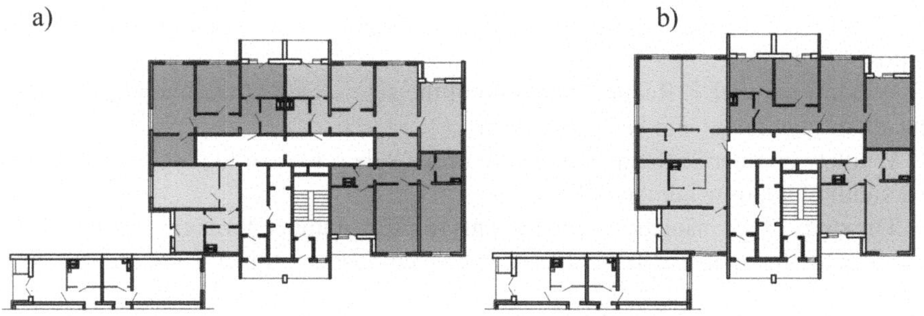

Fig. 5 Planning of the meridional sections: **a** the existing plan and **b** a new plan

(9) Reduction of the side screen of the loggia located to the south of the window number 12 by 0.6 m. The insolation duration of a room with window number 12 was 2 h and 30 min. Rules are met.

Reducing the insolation of rooms in the house number 3 to regulatory requirements is possible if:

Window number 1.

(1) Reduce the height of the meridionally located section of the house number 2 on three floors.
(2) Increase the distance between houses number 2 and number 3 by 16 m.
(3) Reduce the width of the side screen of the loggia, located to the south of window number 1 by 0.75 m, and increase the distance between houses number 2 and number 3 by 6 m.

Window number 2. Remove the side loggia screen, which is located south of window number 2, and increase the distance between houses number 2 and number 3 by 6 m.

Window number 3. Increase the distance between houses #2 and #3 by 6 m.

Window number 4. Increase the distance between houses number 2 and number 3 by 4.5 m.

Window number 6. Reduce the width of the side screen of the loggia, located to the south of window number 6, by 0.9 m.

Windows number 13 and number 17. Reduce the width of the side loggia screens located to the south of windows 13 and 17 by 0.1 m, reduce the height of house 4–6 floors (up to 10 floors), and increase the distance between houses 3 and 4–8 meters.

Windows number 12 and number 16. Reduce the width of the side screens of the loggias located to the south of windows 12 and 16 by 0.7 m, reduce the height of house 4 by 6 floors (up to 10 floors), and increase the distance between houses 3 and 4 by 8 m.

Windows number 14 and number 15. Reduce the height of house number 4–6 floors (up to 10 floors), and increase the distance between houses number 3 and number 4 by 8 m.

Reducing the insolation of rooms in the house number 4 to regulatory requirements is possible if:

Window number 1. Reduce the width of the side screen of the loggia, located to the south of window number 1, by 0.9 m.

Window number 2. Reduce the width of the side screen of the loggia, located to the south of window number 2 by 1.3 m.

The study of the insolation of apartments in house number 4 was carried out taking into account the increase in the distance between houses number 3 and number 4 by 8 m.

Reduction of insolation of rooms in the house number 5 to regulatory requirements is possible if:

Window number 1.

(1) Increase the width of the window by 1.8 m.
(2) Increase the distance between houses number 4 and number 5 by 6 m.
(3) Move the house number 5 to the north by 14 m.

The study of the insolation of apartments in house number 5 was carried out taking into account the reduction in the height of house number 4 on 6 floors.

To bring insolation of apartments with windows number 4 and number 5, it is necessary to change the planning solution of the latitudinal section as in house number 2.

The analysis of possible ways to bring the insolation duration of apartments in houses allowed us to choose the following best of them:

(1) In house number 1, reduce the width of the side loggia screen in all sections by 0.2 m (eastern façade).
(2) In the house number 2, change the planning decision of the meridional section; reduce the width of the side screen of the loggia in the meridional section of the house by 0.6 m (western façade).
(3) In house number 3, it is necessary to remove the side loggia screen in the upper section (east façade); reduce the width of the side screen of the loggia in the lower section by 0.9 m (east facade); and reduce the width of the side screen of the loggia in the upper and lower sections by 0.7 m (western façade).
(4) In house number 4, it is necessary to reduce the width of the side loggia screen in the upper section by 1.3 m (eastern façade).
(5) In house number 5, it is necessary to change the planning decision of a meridional section.
(6) Reduce the height of the house number 4 on six floors (up to ten floors).
(7) Increase the distance between the houses: number 2 and number 3 by 6 m; number 3 and number 4 at 8 m; number 4 and number 5 at 6 m.

The planning solution of the microdistrict and the constructive solution of the buildings after the changes are made are shown in Fig. 6.

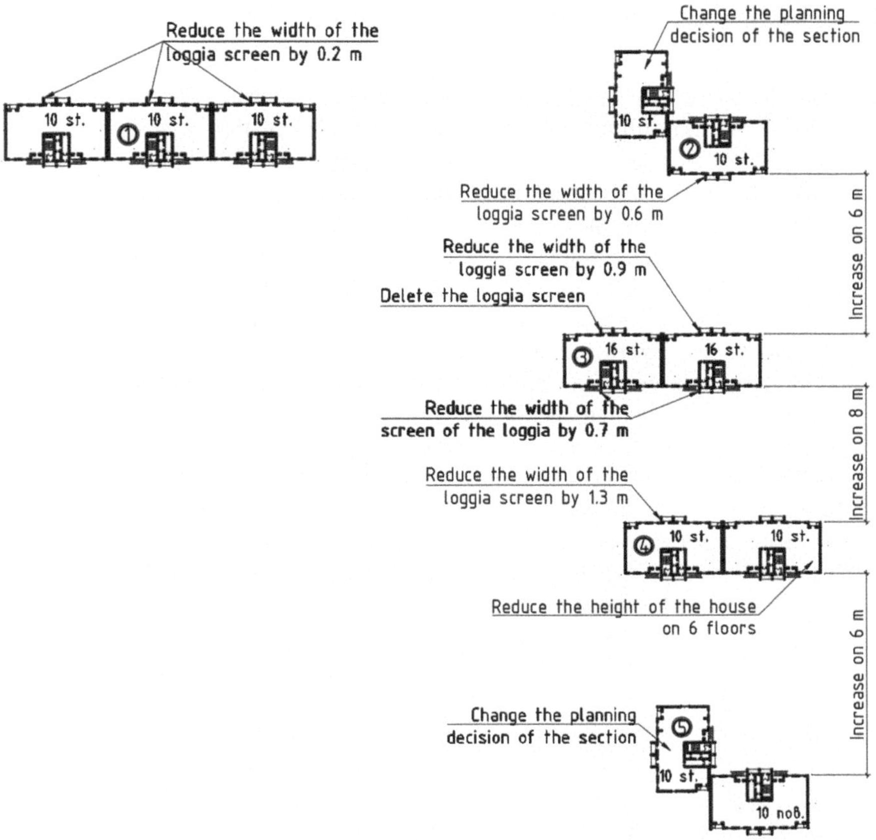

Fig. 6 Planning decision of the microdistrict and constructive solution of the buildings after the changes made

5 Scientific Novelty

For the first time, a set of constructive methods for the design of rooms to provide insolation in rooms in residential buildings has been applied.

6 Practical Significance

This approach of constructive methods of effective insulation in residential buildings can be widely used in compact planning urban of microdistricts in big cities.

7 Conclusions

1. With the modern dense development of microdistricts with high-rise buildings, it is difficult to fulfill the insolation norms in the calculation rooms.
2. Analysis of possible ways to bring insolation requirements to building standards showed that in meridionally located sections, replacing the outwardly protruding from the plane of the facade of the loggias with continuous side screens to the balconies significantly improves the insolation of rooms located to the north of them.
3. It should be considered unsuccessful connections meridionally and latitudinal sections with a significant shift (house numbers 2 and 5). It is impossible to fulfill the conditions of insolation of rooms with windows located on the western facade of the latitudinal section. In this case, it is necessary to apply such a planning solution of the section, in which the windows of the calculation rooms will face the eastern and southern facades.
4. In some cases, without decreasing the number of floors of houses and increasing the distance between houses, it is impossible to fulfill the norm for insolation.
5. It should be recognized that it is more expedient to use tower-type houses instead of multi-section houses.

References

1. Kinash, R., Yatsiv, M., Lahush, V., & Ihnatiuk, A. (2009). Problems of regulation and calculation of housing insolation. *Bulletin of the National University "Lviv Polytechnic", 656*, 118–122.
2. Darula, S., Christoffersen, J., & Malikova, M. (2015). Sunlight and insolation of building interiors. In *6th International Building Physics Conference, IBPC 2015. Energy Procedia*, vol. 78, pp. 1245–1250. https://doi.org/10.1016/j.egypro.2015.11.266.
3. Ali, A., Elnokaly, A., & Mills, G. (2017). Mind the gap; methodology discussion of the extraction and analysis of pilot phase data to generate multi-configuration household behavioural profiles. In *Proceedings of 33rd PLEA international conference design to Thrive Edinburgh, 2–5 July 2017 PLEA 2017 conference*, vol. 1, pp. 385–392. London: Network for Comfort and Energy Use in Buildings (NCEUB).
4. Valladares-Rendón, L. G., & Lo, S. L. (2014). Passive shading strategies to reduce outdoor insolation and indoor cooling loads by using overhang devices on a building. *Building Simulation, 7*(6), 671–681.
5. Kämpf, J. H., Montavon, M., Bunyesc, J., Bolliger, R., & Robinson, D. (2010). Optimisation of buildings'solar irradiation availability. *Solar Energy, 84*(4), 596–603. https://doi.org/10.1016/j.solener.2009.07.013.
6. Semko, O., Yurin, O., Avramenko, Y., & Skliarenko, S. (2017). Thermophysical aspects of reconstruction of cold roof spaces. *Paper presented at the MATEC Web of Conferences*, 116 https://doi.org/10.1051/matecconf/201711602030.
7. Sankov, P., Tkach, N., Voziian, K., & Yermolaieva, Y. (2016). Provision of full lighting and insolation of residential areas in the reconstruction conditions. *International Scientific Journal. No., 5*, 18–21.
8. Utesheva, G. (2018). Modern trends in the development of the problem of insolation and sun protection in architecture and design. Traditions and Innovations in Construction and Architecture. Urban Planning, pp. 217–220 (2018).

9. Ayoub, M., & Elseragy, A. (2017). Parameterization of traditional domed-roofs insolation in hot-arid climates in Aswan Egypt. *Energy and Environment, 29*(1), 109–130. https://doi.org/10.1177/0958305X17741285.
10. Hii, D. J. C., Heng, C. K., Malone-Lee, L. C., Zhang, J., Ibrahim, N., Huang, Y. C., & Janssen, P. (2011). Solar radiation performance evaluation for high density urban forms in the tropical context. In *12th Conference of International Building Performance Simulation Association* (pp. 2595–2602). Sydney: Proceedings of Building Simulation.
11. Yurin, O., Azizova, A., & Galinska, T. (2018). Study of heat shielding qualities of a brick wall corner with additional insulation on the brick. *Paper presented at the MATEC Web of Conferences*, 230.
12. Ling, C. S., Ahmad, M. H., & Ossen, D. R. (2007). The effect of geometric shape and building orientation on minimising solar insolation on high-rise buildings in hot humid climate. *Journal of Construction in Developing Countries, 12*(1), 27–38.
13. Yurin, O., & Galinska, T. (2017). Study of heat shielding qualities of brick wall angle with additional insulation located on the outside fences. Paper *presented at the MATEC Web of Conferences*, 116. https://doi.org/10.1051/matecconf/201711602039.

Construction Features Durable Storage of Toxic Waste in Boreholes

M. L. Zotsenko, O. V. Mykhailovska, and S. P. Sivitska

Abstract The article proposes a technological solution for the arrangement of life-long sludge storage of toxic waste using soil-cement. The authors propose to use a waterproof natural soil layer as the bottom of the sludge deposit. The walls of the storehouse shall be made of chopped vertical soil-cement elements by blending technology. It is proposed to mix the drilling waste with soil within the storage area to a rigid plastic consistency. Thickening to a rigid plastic consistency is necessary to achieve optimum humidity of the mixture and its further compaction. After filling the storage volume, the authors encourage to cover the soil-cement layer. Cover with a fertile soil layer after hardening. The article investigates the characteristics of the components of the soil mixture and sludge during drilling of the Yablunivka oil-gas-condensate field. The optimal humidity of the mixture at different ratios of drilling waste and soil was determined.

Keywords Sludge storage · Waterproof soil layer · Drilling waste · Soil · Mixture · Optimal humidity · Soil-cement · Waterproofing of Soil-cement

1 Relevance of Work

The success of the development and operation of oil and gas wells, as well as the exploration of oil and gas fields is largely determined by the efficiency of well construction. During the drilling and subsequent operation of oil and gas wells, significant amounts of waste are generated that adversely affect the environment. Wastes include a wide range of organic and inorganic substances. Concentrated solutions of acids, surfactants, inhibitors are used for the operation of wells for the intensification of hydrocarbon production. Getting drilling waste into reservoir, soil, and groundwater in large quantities is environmentally hazardous.

Well construction technology and environmental protection measures foreseen by construction projects must first be focused on preventing possible causes and pathways of pollutants entering the environment. The main potential sources of

M. L. Zotsenko · O. V. Mykhailovska (✉) · S. P. Sivitska
Poltava National Technical Yuri Kondratyuk University, Poltava, Ukraine
e-mail: MykhailovskaOV@pntu.edu.ua

V. Onyshchenko et al. (eds.), *Proceedings of the 2nd International Conference on Building Innovations*, Lecture Notes in Civil Engineering 73,
https://doi.org/10.1007/978-3-030-42939-3_33

environmental pollution during well construction are drilling waste, drilling mud, and formation fluids [1]. Only in 2017–2018, the facts of environmental pollution by drilling waste in the region were recorded in the territory of Poltava region (Velyka Bagachka, Lokhvitsa, Myrhorod District) [2]. Much publicity has been given to the export of drilling sludge to the municipal solid waste dump in the Velyka Bagachka District. Drilling waste was also found at a landfill near Myrhorod. About 100 tonnes of wellbore sludge were brought to the municipal solid waste landfill near Kovalivka. Ignoring environmental standards can lead to the fact that the chemicals that are used in the drilling can penetrate into the groundwater. Usually, it is disposed by removal from the territory of drilling operations and disposal in another territory.

Well construction technology provides the possibility of drilling in one of the following ways: "barn" or "no-barn." With "no-barn" way of drilling, the site of possible contamination is screened. The drilling waste (waste drilling mud and sludge) is made dense and neutralized, then taken to specially designated disposal sites, which are often several hundred kilometers from the site. According to the state standard, the removal of sludge from the drilling site (the so-called "no-barn" method of drilling) is provided only in the case of high groundwater levels and if drilling is carried out "on lands of reserves, national, dendrological, botanical, memorial parks, burial sites, and archeological excavations" [2]. These are about 5–10% of wells drilling in the country today. In all other cases, the accumulation, neutralization, and disposal of drilling waste occur directly in the site designated for drilling operations.

Well construction technology involves drilling operations in a "barn" way, in which the collection, accumulation, disposal, neutralization, and disposal of drilling waste are carried out in sludge bins (sludge) directly in the site of drilling operations.

The size of the barns (sludge) is determined by the project and should correspond to the volume of drilling waste. The conditions for the construction of drilling waste storage barns and their construction depend on many factors, such as engineering-geological conditions of the area allocated for the construction of wells and soils characteristics of this area (their physical-chemical composition, filtration properties, etc.). After the construction of the pit beneath the storage barn, work is carried out to equip its surface with an anti-filtration screen. The screening is arranged to protect against contamination of soil and groundwater.

As anti-filtration screens can be used as soil anti-filtration screens of poorly permeable compacted clay soils, resistant to the salts contained in the liquid phase of drilling waste; clogging anti-filtration screens of surfactants and polymers; colloidal-chemical screens based on an aqueous suspension of hydrolyzed polyacrylamide (HPAA) and bentonite clay; anti-filtration screens made of polyethylene.

This method of waterproofing is considered the best among the existing ones. However, the disadvantage of this method is that the waterproofing of the drilling wastes barns thus performed does not provide a guaranteed protection of surface and groundwater from pollution. Thus, insufficient tightness of the connection of the polyethylene material leads to geofiltration of liquid drilling waste [1].

This method of drilling waste disposal in sludge barns does not exclude the possibility of subsequent migration of polluting drilling waste components due to the ingress of rain and flood water into the barn bed. All these methods of waterproof screens are time consuming and expensive. Over time, the effectiveness of such screens to resist the chemical action of waste drilling components is reduced [1, 3].

Tymofeieva offers a method of sludge storage arrangements from soil-cement by means of production of soil-cement elements by blending technology without soil extraction [4, 5]. According to the proposed technology around the perimeter of the sludge, a protective impermeable screen is made of sloping soil-cement elements. The elements are made by the mixing method. Then, the pit is torn off, the bottom of which is arranged by the same technology [4]. Also another method is the arrangement of the pit bottom made of soil-cement which is fed separately into the concrete mixer and laid with a continuous layer on the bottom of the storage [5]. It has at least 0.5 m height and mineral wire hedges along the perimeter of the mineral soil. This method is time consuming, and the main disadvantage of this solution is that the repository is open. In this case, the possibility of harmful substances evaporation and accidents is not excluded. Thus, there is a need to arrange the coating sludge storage.

Tymofeieva also offers a sludge storage facility with a wall-to-soil anti-filtration curtain. The wall-to-soil anti-filtration curtain of soil-cement elements is immersed in a water-resistant layer of soil. This prevents the sludge tank from having an anti-filtration screen. The waterproof layer of soil [5, 6] will perform the function of the screen. This design of the bottom will be more economical. The sludge storage facility is designed as a closed type, but the sludge storage facility requires reagents to thicken the sludge, which requires considerable funds. In this case, it is necessary to remove and dispose of the soil, which is used in the construction of the sludge storage [3].

2 The Purpose of the Work

Therefore, the purpose of the work is to investigate the creation of arrangement method of sludge storage during the construction of wells. This method will provide guaranteed protection of surface, ground, and groundwater from contamination. The purpose of the article is achieved by developing a constructive solution for an efficient, economical, and safe life-long closed-type sludge with a soil-cement screen.

3 Methodology and Research

This task is solved by applying the technology of mixing drilling waste with soil of the construction site. Mix the mixture to a certain consistency and then apply a coating over the sludge from the soil-cement.

The essence of the technology is that the walls of the pit of life-long sludge are waterproofed by a vertical anti-filtration veil of soil-cement. It is proposed to arrange the soil-cement elements according to the blending technology without removing the soil according to the "wall in the soil" type. The elements are immersed in a waterproof soil layer. The sludge deposit is closed.

The construction of life-long sludge storage begins with the fact that a monolithic vertical anti-filtration curtain is being constructed around the perimeter of the planned sludge storage facility. Such an anti-filtration veil is constructed according to the "wall in soil" type of soil-cement elements (Fig. 1). The distance between the centers of adjacent elements should be 0.8*d* (*d*—diameter of soil-cement elements).

Soil-cement is a material that consists of multicomponent systems—cement and soil. The main leading factor in the conversion of soil properties is cement, which is a polydisperse and polymeric system that can form a solid after the addition of water [7, 8].

Waterproofing of soil-cement has been proved by the results of laboratory and industrial researches. For example, samples of soil-cement in laboratory and field

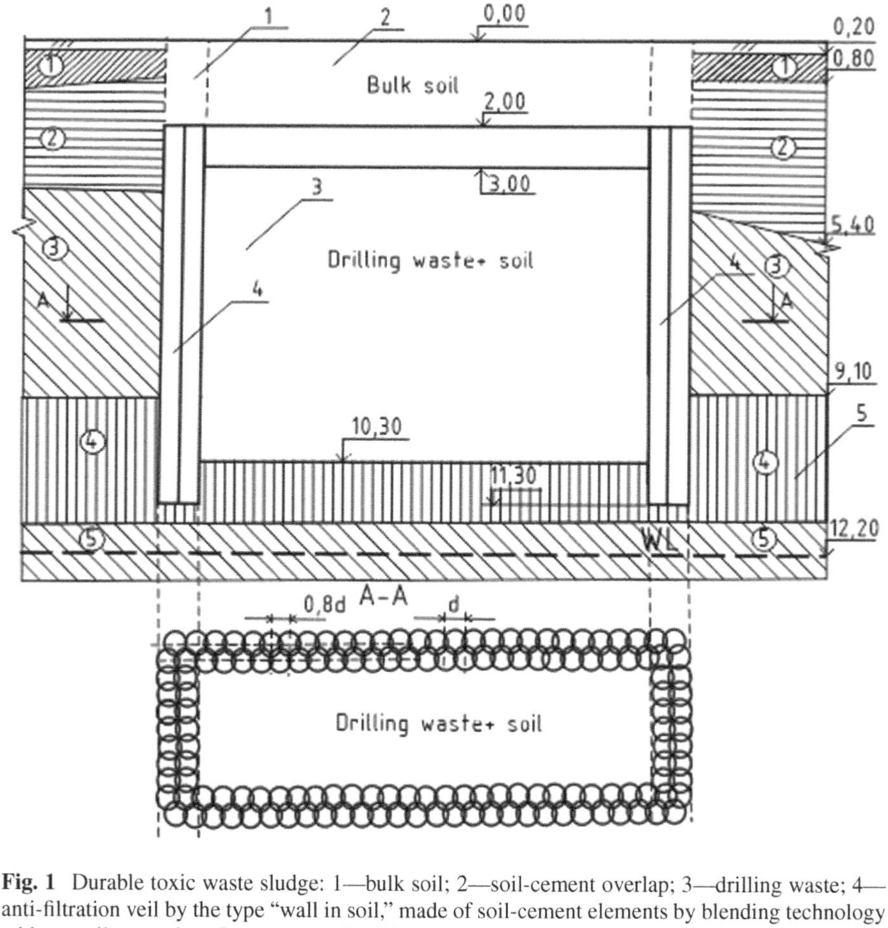

Fig. 1 Durable toxic waste sludge: 1—bulk soil; 2—soil-cement overlap; 3—drilling waste; 4—anti-filtration veil by the type "wall in soil," made of soil-cement elements by blending technology without soil extraction; 5—waterproof soil layer

conditions were made of loess soil with the addition of 20% Portland cement M400 by weight of dry soil. The samples were made at a water–cement ratio of 1, without additional compaction and hydrophobic additives, by the method of "wet spot" and the express method, which is to determine the breathability. On the basis of these results, the soil-cement grade corresponding to the W12–W16 classification of concrete for water resistance was determined [9, 10].

The soil-cement elements are made by a blending method, which consists in the fact that with the help of special equipment when the loosening of the soil is performed without its removal. At the same time, slurry is pumped into the loose soil, and mixing and compaction of the soil-cement mixture are performed. Thus, we obtain cylindrical soil-cement elements with a diameter of 0.3–0.8 m and a length of up to 30 m [7]. A wall-to-soil anti-filtration curtain of sub-soil soil-cement elements is immersed in a waterproofing to a depth not less than 1 m in order to ensure no filtration. Waterproof layer is a layer of waterproof rocks that limits the aquifer from above or below. Waterproof roof is formed by waterproof rocks, such as clays, solid limestones, and solid crystalline rocks, clay shales, crystalline shales.

After solidification of the soil-cement elements, the soil mass is extracted by the volume of the sludge [1]. The period of wetting in the moistened state lasts 28 days. Over time, the strength and water resistance of soil-cement increase. Preparation and packing work are carried out within the storage area. After filling the project volume of the storage with a compacted mixture, a soil-cement horizontal screen 50 cm thick is arranged behind its surface.

After the soil-cement coating becomes hard, it must be covered with a layer of fertile soil with a thickness of at least 1.5 m.

The size of the life sludge and its volume, profile, and depth are determined at the stage of design work in relation to a particular site, taking into account the soil category, depth of groundwater, depth of water resistance, and other characteristics. An important factor in the design of the repository is the choice of the location of the repository, provided there is a waterproof layer at the optimum depth from the surface (8–20 m).

4 Scientific Novelty

The scientific novelty of the article is that the degree of filling of the sludge is performed by mixing the soil and drilling waste in the sludge to a rigid plastic consistency. Mixing is carried out in order to thicken the waste and to arrange the soil-cement coating on top of the sludge. The soil-cement coatings are laid directly on the thickened to optimal humidity and compacted mixture of drilling waste and soil.

The results of many studies carried out by various authors confirm the opinion that the optimal characteristics of compaction depend significantly on the soil composition. Therefore, when studying the patterns of sealing, the relationship between optimal humidity and plasticity should be taken into account. Optimal moisture and plasticity are complex indicators that determine the dispersed and mineralogical composition of soil [7].

Investigation of soil characteristics and determination of optimal moisture were carried out with drilling waste of Yablunivka oil-gas-condensate field well No. 355. Drilling waste had a density of 1.49 g/cm^3, sediment volume of 1.5 ml, solids content of 3%, and pH 6.71. For the experiment, the loam loess is refractory from a depth of 2 m. The average moisture of the soil samples was about $W = 0.2$ (coefficient of variation, $\upsilon = 0.08$).

According to the results of laboratory studies, it was determined that the drilling waste had an average moisture of soil samples $W = 0.50$. Its moisture at the yield point $W_L = 0.36$, and its moisture at the rolling point $W_p = 0.21$. It was determined that this drilling waste belongs to the loam of the fluid. Scientific experiments were conducted according to standard laboratory methods of soil research according to DSTU B B.2.1-17: 2009 [11].

The authors propose to mix soil with drilling waste in different proportions and to determine its moisture and characteristics in order to determine the optimal moisture of the mixture [6]. The general appearance of the samples of the mixture of drilling waste and loamy refractory in different proportions is shown in Fig. 2.

Table 1 shows the basic physical characteristics of the objects of research, namely drilling waste and loam loess. All characteristics were determined as the mean of the six parallel shifts.

The results of calculations and laboratory experiments on the mixture of soil, namely loam loess natural humidity and drilling waste in certain proportions and according to the methods approved by state building standards, are summarized in Table 2.

Fig. 2 Type of soil samples of the mixture of drilling waste (**a**) and loam loess refractory (**b**) in proportion (a:b),%: 1–70:30; 2–50:50; 3–80:20; 4–60:40

Table 1 Indicative characteristics of the mixture components of drilling waste and loam at the coefficient of variation, υ

Objects of research	Moisture, W/υ	Plasticity limit, W_p/υ	Fluid boundary, W_L/υ	Quantity of plasticity, I_L/υ	Density, ρ/υ, g/sm^3	Density of soil skeleton, ρ_d/υ, g/sm^3
Drilling waste	1.01/0.012	0.21/0.06	0.36/0.14	0.15/0.1	1.49/0.9	0.74/0.8
Loam loess	0.20/0.08	0.25/0.07	0.37/0.12	0.12/0.11	1.72/0.7	1.43/0.5

Table 2 Indicative characteristics of mixtures of drilling waste and loam loess in different proportions

Proportion of drilling waste (a) and soil (b), %	Non-compacted mixture density, ρ/υ, g/sm^3	Moisture of the mixture W/υ	Quantity of plasticity I_p/υ	Moisture on the boundary of rolling, W_p/υ	Moisture on the boundary strength, W_L/υ	Optimal moisture, W_{opt}/υ	Optimal density, ρ_d/υ, g/sm^3
50:50	1.53/0.07	0.50/0.12	0.15/0.07	0.25/0.13	0.40/0.09	0.25/0.04	1.67/0.12
60:40	1.49/0.09	0.57/0.09	0.13/0.1	0.27/0.1	0.40/0.06	0.27/0.08	1.65/0.07

For further analysis, select the mixture at the ratio of drilling waste to loam as 60–40%, respectively. In accordance with the standard soil compaction technique [12], the values of the optimal moisture $W_{\text{opt}}/\upsilon = 0.27/0.08$ were determined with the help of a stationary mechanized mechanical soil compactor MDU-1 (Fig. 3) for the mixture under consideration. Optimal skeleton density $\rho_d/\upsilon = 1.65/0.07$ g/cm^3.

With these characteristics, samples of the compacted mixture were made to evaluate its compressibility in the K-1 compression device. According to the results of the compression tests, the modulus of deformation was $E/\upsilon = 27/0.09$ MPa. Such slight compressibility of the mixture is not necessary for the site of disposal of drilling muds. According to DBN B.2.1-10: 2018 [13, 14], soils with a deformation modulus $E > 5$ MPa can serve as a natural basis. Therefore, the modulus of deformation of the mixture at moisture $W = 0.35$ and $\rho_d = 1.45$ g/cm^3 was determined. According to the compression tests, the modulus of deformation of the mixture that was compacted at these parameters was $E/\upsilon = 27/0.09$ MPa, which can be considered sufficient.

5 Practical Importance

On the basis of the conducted researches, the method of life preservation of toxic waste of drilling of oil and gas wells by arrangement of underground storages isolated

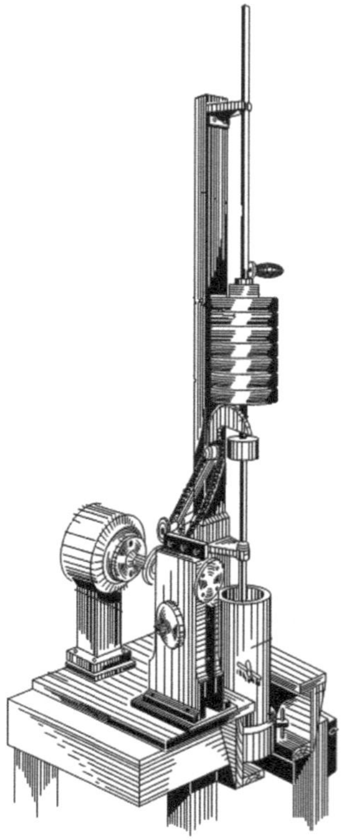

Fig. 3 Stationary mechanized device for dynamic compaction of MDU-1 soils

from the environment by soil-cement screen is proposed. A fertile soil layer suitable for any agricultural work is arranged above all the storage [15, 16].

The method is that in the place, which is selected mainly from the conditions of the geological structure of the site of the repository location, a pit of a given space is arranged. The ditch is enclosed by a solid wall in the soil, which is arranged with cut-off soil-cement elements by blending technology. After soil cementation, soil is removed from the enclosure. 60% of the excavated soil is exported for the construction of embankments and 40% remains within the site for partial drying followed by mixing with drilling waste. Production of the mixture "soil-drilling waste" is carried out directly in the pit at the specified design parameters of the mixture.

The finished mixture is leveled at the bottom of the pit with a layer of design thickness for further sealing by the rollers to the specified skeleton density of the mixture ρ_d, g/cm^3 and its modulus of deformation *E*, MPa. Then, the next layer of compacted mixture is arranged. The formation of layers of compacted mixture continues to a depth of 2.0 m from the earth's surface. Then, a solid monolithic soil-cement screen is arranged over the entire compacted surface of the mixture. After shrinking the screen into the remaining space, a fertile soil layer is placed for later use for agricultural work and for other purposes.

6 Conclusion

Thus, it is proposed to dispose of the drilling waste by bringing its consistency to a condition that allows the mixture to be compacted in the usual way. The complexity of the process is the amount of soil it needs. It is also possible to reduce the moisture of the drilling waste by its physical dehydration or by the introduction of certain chemical additives.

It is planned to build a life-size prototype of such storage for long-term observations of its operational parameters: leakage, radioactivity, etc., with the help of observation wells and research of surrounding soil samples, water, air, and various radiation.

Advantages of the sludge storage construction with the device of a covering from a soil-cement which is put on the drilling waste condensed to a rigid plastic consistency with addition of soil of a construction site is low cost of production due to use of a water-resistant soil layer as the bottom of a structure. It is also possible to solve the problem of utilization of the soil sludge extracted during construction.

The positive thing is that the sludge tank walls have a high water resistance W8–W12; significant compressive strength—2 MPa; environmental safety, and durability.

It is suggested to dry the soil of the construction site in the atmospheric air, with stirring in the warm season and introduction of sand, dry little bonded soil, slag, and inactive ashes, which are laid in the form of drainage layers or water-absorbing layers.

References

1. Timofeeva, K. (2012). Application of soil cement as an anti-filtration screen of barns sludge pits for waste drilling and oil and gas wells. *Collection of Scientific Papers (Industry Engineering, Construction), 4*(34), 67–70.
2. Can oil and gas be mined without harming the environment. http://kolo.news/category/biznes/11737. Last accessed 25 Aug 2019.
3. Senenko, N., Storozhenko, D., & Senenko, A. (2017). Analysis of technogenic impact (gas-industrial complex) on the state of soil and groundwater. *Physics, Mathematics and Chemistry*, 258–260.
4. Timofeeva, K. A. (2013). Patent for utility model 87868. Kyiv: State Patent Office of Ukraine.
5. Zotsenko, M. L., & Timofeeva, K. A. (2015). Patent for utility model 101183. Kyiv: State Patent Office of Ukraine.
6. SNIP 3.06.03-85 Highways. Accepted. 08/20/1985 Introduction Date 01/01/1986—M. FSUE CPP, 2006, p. 131.
7. Zotsenko, M. L., Vinnikov, Y. L., & Zotsenko, V. M. (2016). *Soil-cement piles by drilling-mixing technology: Monograph*. Kharkiv: Madrid Printing House.
8. Goryk, O. V., Pavlikov, A. M., & Kyrychenko, V. A. (2009) Calculation of statically indeterminate composite beam elements by using refined boundary conditions and with account of their state diagrams. *Mechanics of Composite Materials, 45*(1), 53–58. https://doi.org/10.1007/s11029-009-9058-9.
9. Zotsenko, M., Vynnykov, Y., Doubrovsky, M., (…), Krysan, V., Meshcheryakov, G. (2013). Innovative solutions in the field of geotechnical construction and coastal geotechnical

engineering under difficult engineering-geological conditions of Ukraine. In *18th International Conference on Soil Mechanics and Geotechnical Engineering: Challenges and Innovations in Geotechnics, ICSMGE.*

10. Zotsenko, N. L., & Vinnikov, Y. (2016). Long-term settlement of buildings erected on driven cast-in-situ piles in loess soil. *Soil Mechanics and Foundation Engineering, 53*(3), 189–195. https://doi.org/10.1007/s11204-016-9384-6.
11. DBN B.2.1-10. (2018) *Foundations and Foundations of Buildings and Structures Main Provisions*. Adopted: 09/22/2009 Date of introduction: 01/10/2010, p. 32.
12. Yermakova, I. A. (2005). Features of dynamic compaction of soil mixtures using mining waste—"tails". *Canadian Technology Sciences*, 05.23.02. Poltava.
13. DSTU B B.2.1-17. (2009). Foundations and foundations of buildings and structures. Soils. *Methods of Laboratory Determination of Physical Properties*. Adopted: 02/08/2018 Date: 01/01/2019, p. 36.
14. GSTU 41-00 032 626-00-007-97. (1998). Industry standard of Ukraine. Environment protection. *Construction of exploration and exploitation wells for oil and gas on land.* Date of introduction-01.01.1998, p. 80.
15. Zotsenko, M. L., & Timofeeva, K. A. (2011). Patent for utility model 71256. Kyiv: State Patent Office of Ukraine.
16. Zotsenko, M., Vynnykov, Y., Lartseva, I., & Sivitska, S. (2018). Ground base deformation by circular plate peculiarities. *Paper Presented at the MATEC Web of Conferences*, 230. http://doi.org/10.1051/matecconf/201823002040.

Analysis of Emergency Management Methods in Oil and Oil-Product Reservoirs

Oleksandr Zyma, Roman Pahomov, and Evgen Dyachenko

Abstract The article is devoted to the analysis of emergency management methods in storage tanks of crude oil and petroleum products in the territory of Ukraine. The peculiarities of reservoir operation, possible emergencies in reservoir parks, the nature of the destruction of reservoirs during the explosion of the air–vapor mixture, and the process of petroleum products combustion in them are considered. Methods of fire localization and extinguishing in oil and oil-product reservoirs are investigated. Their advantages and disadvantages are analyzed. The feasibility of using a "sublayer" method of fire extinguishing in liquid reservoirs has been determined. This method has significant advantages over traditional methods of ceasing combustion. The modernized installation of "sublayer" fire extinguishing with the use of self-powered foam pipeline for the submission of foam under the product layer, which, in comparison with other methods of fire extinguishing, allows to guarantee a high fire safety level and personnel security of fire and rescue units involved in the fire elimination. The practical significance of the analytical studies is formulated.

Keywords Emergency · Tank · Oil product · Methods of extinguishing

1 Introduction

Every year, energy production and consumption increase, as well as the production of chemical enterprises and, consequently, the number of people who use petroleum products in their lifetime. Therefore, the issue of safe handling of petroleum products is of particular importance. Reservoirs are the main storage place for crude oil and petroleum products at refineries and chemical plants, transshipment and distribution oil depots, road, rail, water, and air transport enterprises. Accumulation of flammable and combustible liquids in a relatively small area of the tank park leads to increased fire risk of such industries. A tank fire is one of the most dangerous emergencies that cause not only significant material damage, but also human casualties. Fires occurring

O. Zyma (✉) · R. Pahomov · E. Dyachenko
Poltava National Technical Yuri Kondratyuk University, Poltava, Ukraine
e-mail: zymaae@gmail.com

V. Onyshchenko et al. (eds.), *Proceedings of the 2nd International Conference on Building Innovations*, Lecture Notes in Civil Engineering 73,
https://doi.org/10.1007/978-3-030-42939-3_34

at oil and petroleum product storage and processing facilities, in production facilities where flammable liquids are used, are accompanied by rapid propagation, high temperature, and dense thermal exposure and smoke. The explosion of the vapor–air mixture results in partial or complete destruction of tanks, process equipment, and structures [1]. Damage from fires consists of losses from destruction and damage to process equipment by excess pressure during an explosion, loss of mechanical properties under the action of high temperature in the combustion zone, or overheating from thermal radiation of the flame. Significant component of the set losses are the losses on the extinguishing agents and the cost of restoring the functioning of the facility after the fire.

The situation is complicated by the economically driven tendency to move to larger tanks, which further increases the volume of flammable liquids per unit area of the tank. This, in turn, increases the risk of fire spreading to adjacent tanks in the absence of timely localization and elimination of the combustion center.

2 Formulation of the Problem

According to the statistics of the occurrence of fires in the reservoirs of all oil producing countries, there is a clear tendency to increase the fire danger with the increase of the reservoir park scale, which indicates the need for further improvement of fire hazard measures during their design and operation.

3 Analysis of Recent Studies

Many publications have been devoted to the safe operation of oil and oil-product reservoirs, to prevent emergencies and to eliminate fires, which provide methods for the safe and efficient use of machinery, equipment, and extinguishing agents. In article [2], the issues of prevention and elimination of emergencies in oil-tanked reservoirs were considered, but the use of water jets was intended only for cooling the walls of the burning tank and the adjacent tanks, and not for extinguishing flammable liquids.

In article [3], the quenching of combustible substances by the cooling method was considered. The data obtained indicate that the consumption of flammable and combustible liquids consumes carbon dioxide in the solid state 3.7 times and finely sprayed water 5.2 times, respectively, less than the consumption of the foaming agent.

Methods and technical means of extinguishing petroleum products in tanks based on solid carbon dioxide granules are given in works [4, 5]. However, it should be noted that the use of systems that provide carbon dioxide from balloons or isothermal tanks as well as liquid nitrogen can cause the tank structure to collapse. A large amount of gaseous substance with low temperature, although under certain circumstances, may properly cool the combustible liquid and allow it to be extinguished, but will necessarily come in contact with the metal structures of the tank heated to high temperatures under the influence of the flame.

In works [6–10], a method of stopping combustion in a tank by extinguishing aerosol is considered. With the help of it in the combustion zone, the effect of inhibition of the combustion zone, the effect of phlegmatization of the combustible medium with inert gases, and the effect of mixing the burning fluid are achieved.

In addition to these sources, the works of OM Volkov, GA Proskuryakov, MG Topolsky, AF Fedorova, OE Basmanov, and others are devoted to the issues of fire safety of reservoir parks.

Due to the complexity of the fires' elimination in oil and oil products in reservoirs, analysis as well as classification of methods of fire elimination is an urgent task and requires constant attention.

4 The Purpose of the Article

The question of the possibility of using "sublayer" extinguishing fires in tanks is urgent. It is necessary to determine the possibility of extinguishing fires in reservoirs with liquids by the "sublayer" method in general and to establish the effectiveness of such method.

5 Method

The analytical method of research is used in the work. The methodological basis of the study is a systematic approach based on principles such as integrity, structure, interconnection of the system, and the environment.

6 Results

To date, oil and petroleum product depots are one of the most important elements of Ukraine's oil supply system. According to the State Fiscal Service, in the territory of Ukraine, at the beginning of 2018, 5.9 thousand payers of fuel excise taxes were registered. They bring together 20,000 fuel sales points, of which 9000 filling stations hold tanks for the storage of fuel and lubricants up to 100 m^3 [11]. Currently, more than 11,500 metal tanks with a total capacity of more than 5 million m^3 have been operated at the oil depots. The oil pipelines have reservoirs with a total capacity of more than 1.35 million m^3 in the pipeline system. But the available oil storage capacities will be sufficient only to form less than 45% of Ukraine's required reserves. Therefore, it is necessary to modernize some of the tanks that have been in operation for a long time (620 thousand cubic meters) and to build new ones, designed for at least 1 million [12] cubes.

Oil and oil-product fires in reservoirs are usually complex and large scale, eliminated with great difficulty, cause great harm, and often lead to death. According to statistics, in the USA in the reservoir parks, annually there are 20 fires; in Japan, each year there is one fire. In average, there are two fires in Ukraine over 3 years [13]. Every fourth fire results in the complete combustion of oil. 93.4% of fires occurred in land tanks, with 32.1% of them occurring in crude oil tanks, 53.9% in gasoline tanks, and 14.0% in tanks with other oil products (oil-fuel, kerosene, etc.). The fires occurred mainly in vertical steel tanks (227 cases), of which 198 cases (87.2%) were in gasoline and crude oil tanks [14–16].

Fires in tanks usually begin with an explosion of the vapor mixture in the gas space of the tank without breaking the roof, however, with some damage in some of its places. The explosion force will usually be higher in those tanks where the volume of gas space filled with a mixture of oil and air–vapor (low liquid level) is greatest. Depending on the force of the explosion in a vertical metal tank, the following may occur: The roof of the tank is completely demolished, thrown away by a distance of 20–30 m, and the surface of the liquid in the tank burns over the entire area; the explosion shifts the roof slightly (the tank partially opens), after which the roof is stuck and half immersed in the burning liquid; the roof of the tank is deformed, and cracks appear in the places of its attachment to the walls of the tank, as well as in the welds of the roof. In cylindrical tanks, the bottom is most often destroyed, with oil spreading over a large area.

As a tragic example, on June 8, 2015, an explosion and fire occurred on the territory of the BRSM-Nafta oil depot near the village of Kryachki in Kyiv region. After rescuers arrived, it was reported that three gasoline tanks were burning. At night, the fire spread and the fire engulfed several more tanks. On June 9, the SES of Ukraine reported that 16 tanks were burning. There are 17 reservoirs at the oil depot (Fig. 1).

Gasoline in the tanks burned until June 16, 2015. Almost from the start of the fire, rescuers pumped it into the tanks of a nearby oil depot, which lasted until June 22.

The fire caused both physical and material losses: Five people were killed in the liquidation process, 12 were injured: six SES of Ukraine employees and six railroad workers, and the state suffered losses of UAH 1.2 billion.

The accident caused significant pollution of the environment with oil and its products of combustion. According to experts of the Institute of Sorption and Problems of Endoecology of the NAS of Ukraine, oil spills amounted to about 240 tons (90 tons—in soil, 150 tons—in reservoirs). The soil in the territory of the oil depot was impregnated with oil to a depth of 20–45 cm.

Together with the flue gases, benzopyrene, sulfuric anhydrides, carbon monoxide (IV), nitrogen oxides, gaseous and solid incomplete combustion products, vanadium compounds, sodium salts, and others were found in atmospheric air.

Any fire is relatively easy to fight at its initial stage, taking all measures to localize the fire, preventing the spread of the combustion zone over large areas and masses of combustible substances. The effectiveness of extinguishing a fire at this stage

Fig. 1 Fire at the territory of the "BRSM-Naphtha" oil depot

depends on the correct choice of extinguishing agents and means, their skillful use by all employees, as well as on the promptness of communication, signaling, and activation of extinguishing installations and systems.

The basis of fire fighting is the forced cessation of the combustion process. There are several methods of stopping combustion in practice (Fig. 2), and all these methods have both advantages and a number of disadvantages.

The method of cooling is based on the fact that the combustion of a substance is possible only when the temperature of its upper layer is higher than its flash point for flammable liquids Γ'_{comb}—28 °C. If you remove the accumulated heat from the surface of the combustible substance, i.e., cool it below the flash point, the combustion stops.

The method of dilution is based on the ability of the substance to burn when the oxygen content in the atmosphere is more than 14–16% by volume. With the decrease in oxygen in the air below the specified value, the flame combustion ceases and then the smoldering ceases due to the decrease in the rate of oxidation. Reduction of oxygen concentration is achieved by the introduction of inert gases and vapors

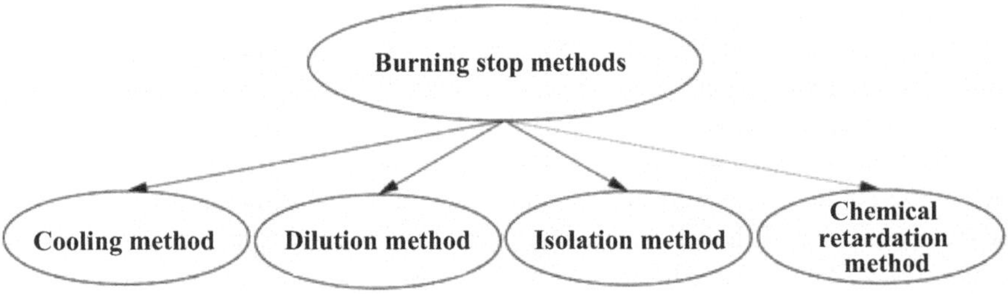

Fig. 2 Classification of combustion cessation methods

into the air and with the exterior or the dilution of oxygen by combustion products (in isolated rooms).

The insulation method is based on stopping the supply of oxygen to the burning substance. Different insulating fire extinguishers (chemical air-mechanical foam, powder, etc.) are used for this purpose. Features of fire extinguishing action of air-mechanical foam consist of isolation of a surface of a burning liquid from a flame torch, reduction as a result of this speed of evaporation of a liquid and reduction of partial pressure of a pair of combustible liquid entering the combustion zone, and also in cooling of the burning liquid. The role of each of these factors in the process of extinguishing varies depending on the properties of the burning liquid, the quality of the foam, and the method of its supply. When the foam is fed, the foam from the flame torch and the heated fuel surface is destroyed at the same time. The accumulated foam layer shields part of the surface of the combustion from the radiant heat flow of the flame, reduces the amount of vapor entering the combustion zone, and reduces the intensity of combustion. The foaming solution, which is simultaneously released from the foam, cools the fuel. In addition, during the quenching process, a convection heat–mass exchange occurs in the volume of the fuel, which results in the liquid temperature being equilibrated throughout the volume, except for the "pockets," in which the heat–mass exchange occurs regardless of the bulk of the liquid.

A method of chemical inhibition of combustion reactions is the introduction into the combustion zone of halogen-derived substances (methyl bromide and ethyl, freon and the like), which, when they get into the flame, break up and connect with the active centers, stopping the exothermic reaction, i.e., the release of heat. As a result, the combustion process stops.

In the world practice, as an improvement of extinguishing agents, a promising trend is to combine existing methods to enhance the physical–chemical properties of components that are intended to extinguish flammable liquids. The use of certain extinguishing agents for extinguishing fires of combustible liquids, in particular petroleum products, is limited by the conditions of technical implementation of the method of supply and the influence of thermophysical indicators of burning fluids (open flames, convective flows, high temperature). Therefore, the process of extinguishing fires in tanks with combustible liquids is extremely complex, long-lasting, and quite costly, which necessitates the development of newest means and methods of extinguishing.

Oil and petroleum products are mainly stored in vertical steel tanks. Due to their prevalence, they account for about 90% of tank fires. Among them, more than 50% are fires of gasoline, about 33%—crude oil, about 15%—oil-fuel, kerosene, diesel, and other petroleum products.

In 30–40% of cases, fires are caused by tank damage due to heat or explosion. In 60% of the cases, fire in the tanks is accompanied by deformation of the dry wall and the appearance of closed spaces—"pockets," which significantly complicates the extinguishing of the fire and in about half of the cases led to the complete combustion of oil in the tank or the spread of fire to the adjacent tanks. The formation of such "pockets" indicates an insufficient amount of workforces and means involved in cooling the tank or inefficient use of them.

Therefore, the primary task of the SES units is to prevent the destruction of the burning tank and to prevent the spread of fire to the adjacent tanks by cooling them with water. The main methods and means of cooling tanks are stationary automatic installations (cooling rings) and mobile equipment (carriage trunks, mobile hydromonitors). But, as practice shows, in 50% of cases, stationary installations are destroyed by the explosion that precedes the fire, and the upper circuit of the tank is destroyed. Therefore, the main task of the tank cooling rests on mobile machinery. Recommendations for its use indicate the intensity of water flow without taking into account the type of oil, the height of the dry wall of the tank, the type of fire barrels, and the water pressure.

According to the regulatory documents, all tanks with oil and petroleum products with a volume of 1000 m^3 or more must be equipped with a fixed fire protection system.

The most common fire extinguishing systems are the use of foam generators and foam cameras in the upper belt of tanks. Each foam generator is connected to a dry pipe that serves as a foam solution from fire trucks.

But the reliability of such systems is too low. After all, it is known that an explosion of gas–vapor mixture first occurs in the tank, after which the combustion process begins. The explosion disrupts or deforms the roof of the tank, destroys the net-splitters and cases of foam generators, and often, they simply fly away. Even if the foam generators survive from the destructive force of the blast wave, then in the first minutes of the fire at 80–100% of the tank filling due to intense thermal impact, they will fail—both net-splitters and cases—even before the foam solution is applied. If the tank is filled with fuel from 10 to 50%, the jet of foam falling from a height of 6–10 m may not reach the surface of the product due to the powerful convection flows generated during the combustion process.

Recently, systems with facilities for "sublayer" fire extinguishing of tanks began to appear on the objects of the oil and gas complex of Ukraine. The fire is extinguished by supplying low foam from a high-pressure foam generator to the fuel layer. Foam generators are installed by the bund wall, and a dry pipe with a foam nozzle approaches the tank.

The foam rises through the fuel layer directly above the inlet to the burning surface to form a film that spreads over the surface. In this case, foam in the pockets formed by fragments of roof metal and walls of vertical tanks is possible, which complicates the extinguishing process. To quickly cover the entire surface mirror, a special foam inlet system (with multiple inlet points or foam collector) is required.

Sublayer fire extinguishing systems are considered efficient, but the equipment is too expensive.

At the same time, since 1972, in the system of the State Reserve used for "sublayer" extinguishing of tanks RVS 2000, RVS 5000 with a sleeve, which is brought to the surface. Bottom Foam Installations (BFI) are equipped with over 2000 tanks in Ukraine. This method of fire extinguishment was only reflected in regulations in 1994.

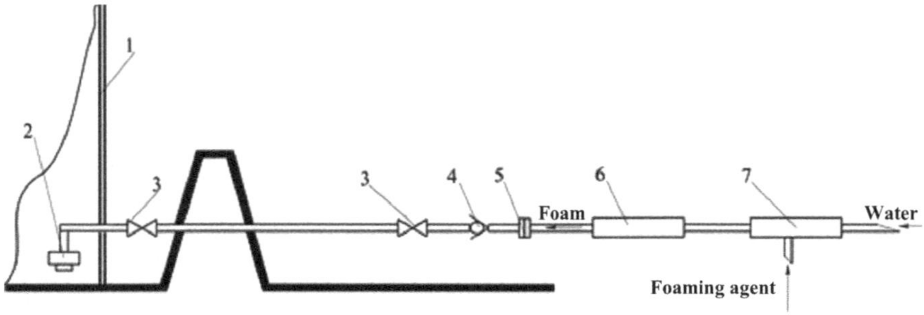

Fig. 3 Schematic diagram of a "sublayer" fire extinguishing system using an independent foam pipe to apply foam to the product layer: 1—tank; 2—diffuser; 3—locking valve; 4—back pressure

However, since then, these settings have been upgraded. In November 2006, a seminar on "Fire safety in tank parks. Fire Extinguishing Systems" was held, where they tested a refined installation. In 2008, low-cost units for the supply of foam under a layer of flammable liquid USFL-U began to be installed at the facilities of the Ukrainian Oil and Gas Complex. The schematic diagram of which is shown in Fig. 3.

The foaming solution is fed through the tubes to the installation USFL-U, installed in the lower belt of the tank, from a self-powered source (fire pump or fire truck). They use a common foaming agent.

Due to the modernization, the dimensions of the installation were significantly reduced and the reliability was increased.

7 Scientific Novelty

The scientific task of developing an effective system for the elimination of the consequences of emergencies in oil and petroleum reservoirs was further developed. A modernized installation of a "sublayer" fire extinguishing with the use of an self-powered foam pipeline for applying foam to the product layer is considered. Compared to other methods of extinguishing fires in tanks, it allows to increase the level of safety at the minimum expense of materials and human resources

8 Practical Importance

The advantage of the "sublayer" method over the traditional, where the foam is served from above, is the protection of the foam generators and foam chutes from the explosion of the air–vapor mixture. It is important that in the implementation of the "sublayer" method, the personnel of the fire departments and equipment are located behind the bund walls and are less exposed to the immediate danger of the ejection or boiling of burning oil.

Also, the "sublayer" way of fire extinguishment in tanks has the following advantages:

- The installation of "sublayer" fire extinguishment makes it possible to eliminate combustion in the tank despite the destruction of the upper belt and the presence of sections closed at the top;
- The effectiveness of the installation of a "sublayer" extinguishing little depends on the duration of the fire and the temperature that can be acquired by oil or oil products due to combustion;
- In the case of "sublayer" extinguishing manner, all the foam enters the tank, which allows to reduce the consumption of the foaming agent;
- The installation of "sublayer" extinguishing goes well with gas trapping technology;
- Installation of "sublayer" extinguishing allows to increase the effective volume of the tank;
- During a fire in the tank, virtually eliminates damage to the equipment of the "sublayer" installation of extinguishing from explosions and heat flows;
- The effectiveness of the "sublayer" way of extinguishing is almost independent of atmospheric phenomena (wind, precipitation);

When foam is supplied, the installation works automatically (from the foam pressure) and closes automatically after extinguishing, so the USFL-U can be used as part of automatic fire extinguishing systems (without electric and pneumatic drive).

9 Conclusions

"Sublayer" fire extinguishing systems are suitable for extinguishing liquid hydrocarbons (primarily oil and petroleum products) stored in vertical steel tanks with fixed roof (without pontoons or with pontoons). In the case of equipment with "sublayer" fire extinguishing of pontoon tanks, it is necessary to ensure that all the foam enters the annular gap between the tank wall and the pontoon. During the calculations, it is necessary to apply foam to the entire surface of the liquid mirror in the tank, provided that it is extinguished by mobile extinguishing installations.

The technological features of the "sublayer" fire extinguishing installations make it possible to guarantee a high level of fire safety and safety of the personnel of the fire and rescue units involved in the elimination of fires at the objects of the oil and gas complex of Ukraine.

References

1. Abramov, U. A., & Basmanov, A. E. (2006). *Prevention and liquidation of emergency situations in tank farms with oil products* (p. 251). Kharkiv: AGZU.
2. Linchevsky, E. A., & Sirovy, V. V. (1999). Development of tactical support for pulse extinguishers. *Fire Safety. Scientific Collection*, 3, 21–23. Cherkasy: CIPB.
3. Lisniak, A. A. (2011). Extinguishing flammable liquids by cooling Problems of fire safety. *Scientific Collection*, (29), 110–115.
4. Koroliov, R. A., Kovalishyn, V. V., & Shtain, B. V. (2017). Analysis of the method of canceling fire in the tanks with oil products in a combined way. *Scientific Journal Science Rise Technical Sciences, 6*(35), 41–50.
5. Vasiutiak, A. O., & Shtain, B. V. (2015). Experimental dossing of easily borrowed ridins with granules of dry ice. *Fire Safety. Scientific Collection* (pp. 18–25). Lviv: LDU.
6. Morgun, Y. O., Balaniuk, V. M., & Mikhalychko, B. M. (2009). Prospects for the cancellation of class "B" in reservoirs with extinguishing aerosols. *Fire Safety, 15*, 84–89.
7. Balaniuk, V. M., Mikhalychko, B. M., & Morgun, Y. O. (2012). Gas blanking with aerozole burned combustible ridins in reservoirs by method. *Fire Safety,* 21, 19–22.
8. Zhao, H., & Liu, J. S. (2016). The feasibility study of extinguishing oil tank fire by using compressed air foam system. *Procedia Engineering, 135*, 61–66. https://doi.org/10.1016/j.proeng.2016.01.080.
9. Oguike, R. (2013). Study of fire fighting foam agent from palm oil for extinguishing of petrol fires. *Science Postprint 1, 1*. https://doi.org/10.14340/spp.2013.12a0002. https://doi.org/10.14340/spp.2013.12a0002.
10. Hutzler, S. (2005). *The Physics of Foam Drainage Proceedings of MIT European Detergents Conference* (pp. 191–206). Wurzburg. https://doi.org/10.1063/1.1366070.
11. In 2017, the number of gas stations in Ukraine increased by 10%. Retrieved from https://delo.ua/business/v-2017-godu-na-10-vyroslo-kolichestvo-azs-v-ukraine-338904/.
12. How Ukraine was preparing to reduce its exports of petroleum products from the Russian Federation. Retrieved from https://www.ukrinform.ua/rubric-economy/2714323-ak-ukraina-pidgotuvalas-do-skorocenna-eksportu-naftoproduktiv-z-rf.html.
13. Ulinets, E. M. (2009). Localization of oil spill fires in reservoir parks (abstract of Ph.D. dissertation). Kharkiv, p. 18.
14. Chernetsky, V. V. (2015) Influence of thermal fire factors on the integrity of vertical steel tanks with oil products (Ph.D. dissertation). Lviv, p. 121.
15. Pichugin, S., Zyma, O., & Vynnykov, P. (2016). Reliability Level of the Buried Main Pipelines Linear Part Recent Progress in Steel and Composite Structures. In *Proceedings of the 13th International Conference on Metal Structures, ICMS 2016* (pp. 551–558). https://doi.org/10.1201/b21417-76.
16. Zyma, O. E., Dyachenko, E. V., Pahomov, R. I., & Zhyhylii, S. M. (2018). Works execution organization at reconstruction and renovation of buildings after the fire with usage of slabs lifting method. *International Journal of Engineering and Technology(UAE), 7*(2), 242–246. https://doi.org/10.14419/ijet.v7i2.23.11951.

Planning of Cities, Buildings and Engineering Networks

Features of Formation of Branding of Historical Settlements in the Context of Development of Religious Tourism (on the Example of Small Settlements of Poltava Region)

Larysa Borodych, Oleksandr Savchenko, Pavlo Vasyliev, and Maryna Borodych

Abstract Branding approaches of the territories, which are the driving force for the development of the recreational and tourism system, in particular, the religious tourism system of the Poltava region, are analyzed in the article. Religious tourism, based on a well-developed network of religious objects, most of which are architectural monuments, is one of the main and most important factors for the formation and development of small towns in Poltava region nowadays. On the example of Poltava region centers of religious tourism analysis, it is proved expediency of its usage for small settlements development.

Keywords Territory branding · Recreational tourism · Religious tourism · Brand of territory · Tourist branding · Orthodox cult objects · Natural resources · Health-improvement tourism

The dynamic development of tourism in the world encourages public authorities to use all possible means to improve the management of this process in order to attract more tourists. In developed countries, branding is widely used to promote the benefits of territories to tourists.

One way to claim competitive advantage in the market and make the area attractive to potential tourists is to create an attractive brand for the area. Ukraine is seeking European integration that is why the problem of competitiveness in the world market can be solved by combining the efforts of the state and business.

Analysis of Recent Research and Results The research of problems that arise in the study of tourism branding, is devoted to numerous scientific works. In particular, in the works of Havrylyuk [1] he proved the innovative importance of national tourism branding; the British expert in this field Anholt [2] proposed to use national branding as an element of economic marketing; Nagornyak [3] explored country branding;

L. Borodych · O. Savchenko (✉) · P. Vasyliev · M. Borodych
Poltava National Technical Yuri Kondratyuk University, Poltava, Ukraine
e-mail: al_sav@ukr.net

L. Borodych
e-mail: larborodich@ukr.net

V. Onyshchenko et al. (eds.), *Proceedings of the 2nd International Conference on Building Innovations*, Lecture Notes in Civil Engineering 73,
https://doi.org/10.1007/978-3-030-42939-3_35

Vizgalov in his work [4] and also Sanker and Brown in the thesis of the report [5] have clarified the essence of the brand category of the city, etc. Issues of religious tourism are covered in the works Jackowaski and Valene [6], Smith [7], Rinschede [8], Cohen [9], Alderman [10].

But the works of these scientists do not sufficiently reveal the problems of tourism branding and especially in the context of religious tourism.

Territory branding is therefore a strategy to increase the competitiveness of territories in order to gain foreign markets, attract investors, tourists, new inhabitants and skilled migrants. Interest in area branding indicates that there is finally a perceived benefit in implementing a consistent resource management, reputation, and image strategy for any city.

According to Muzychenko-Kozlovskaya [11], tourism branding, as a separate category, involves the formation of advantages (tourist attractiveness) of tourist destinations in the process of managing the image and promoting tourist attractiveness of a city or country with the help of a tourist brand consisting of a visual image and slogan.

Tourism branding goals can be:

- attractive image of the country, city, territory promotion;
- culture, peculiarities of history, national mentality of the country or city promotion where the tourist products or services are produced;
- creation and consolidation in the minds of consumers of the image of the tourist product or service of the country or city where they are produced in order to maintain the planned sales volume;
- profit increase in the results of the spreading information about the unique qualities of tourist products or services that are introduced through an attractive image of the place of their production or provision.

Historical forms of territorial promotion.

Since ancient times, people have been intuitively engaged in what we would now call territorial marketing and branding. In an effort to create favorable conditions for attracting resources, they have intuitively used various promotion strategies that can be divided thematically into religious and secular ones. The promotion of the city through the establishment of a religious center made it possible to attract pilgrims, and the increase in the quantity of church officials was accompanied by appropriate infrastructure creation.

As an example of an approach where settlements had a boost in their development as a religious center thanks to cult objects which are such famous cities of Ukraine as Pochaev and Svyatogorsk. On the territory of Poltava region, there are also some settlements with cult monuments of history and architecture and considerable historical and cultural potential for development of religious and pilgrim tourism, in particular the settlement of Kozelshchyna, towns of Hadyach and Lubny. Their historical development is linked to the placement of orthodox religious objects, which played a significant role in the existence of these settlements. The most famous places of worship such as the Mgar Holy Spaso-Preobrazhenskiy and Kozelshchyna

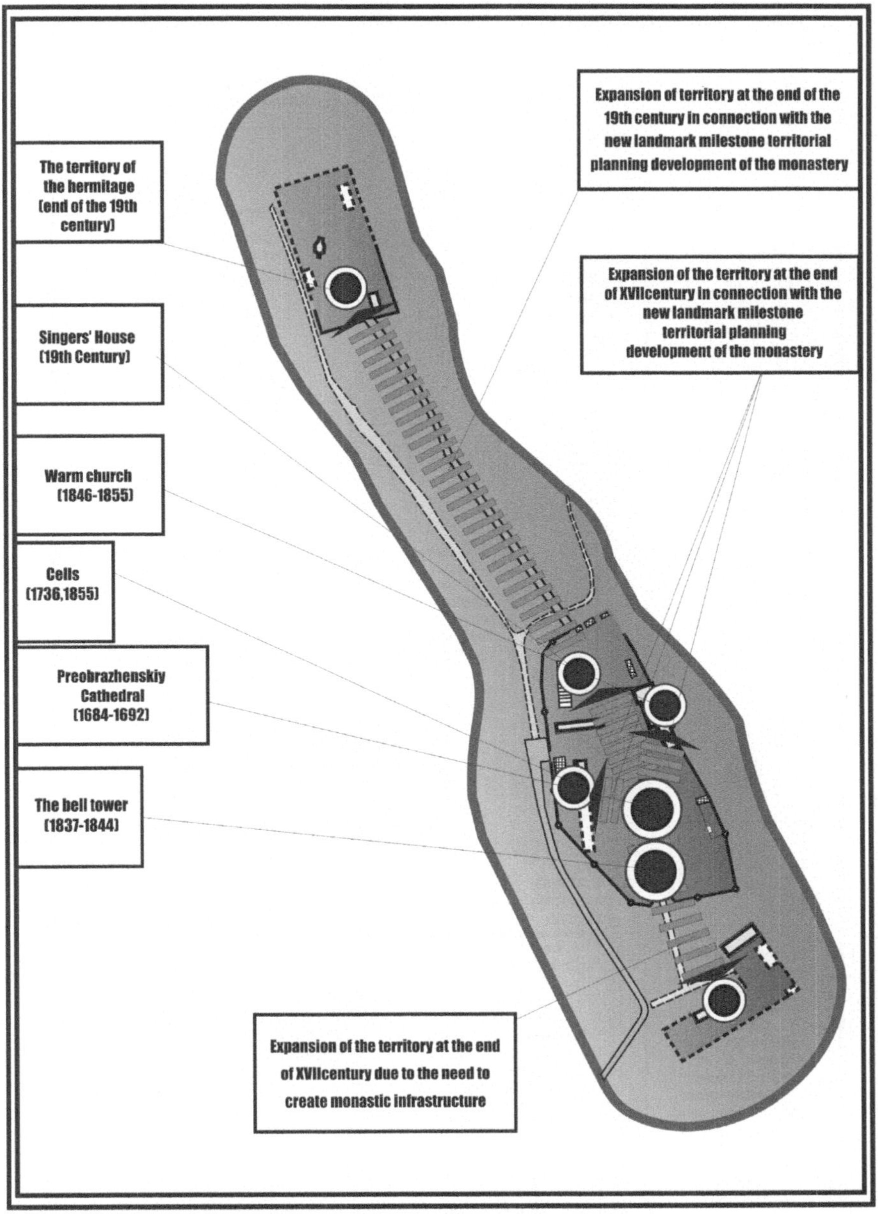

Fig. 1 Stages of formation of the territory of the Mgar Spaso-Preobrazhenskiy monastery

Rizdvyano-Bogorodychniy monasteries, whose chronological description of development stages is given in Figs. 1 and 2, have traditionally become the cultural centers of the settlements where they are located.

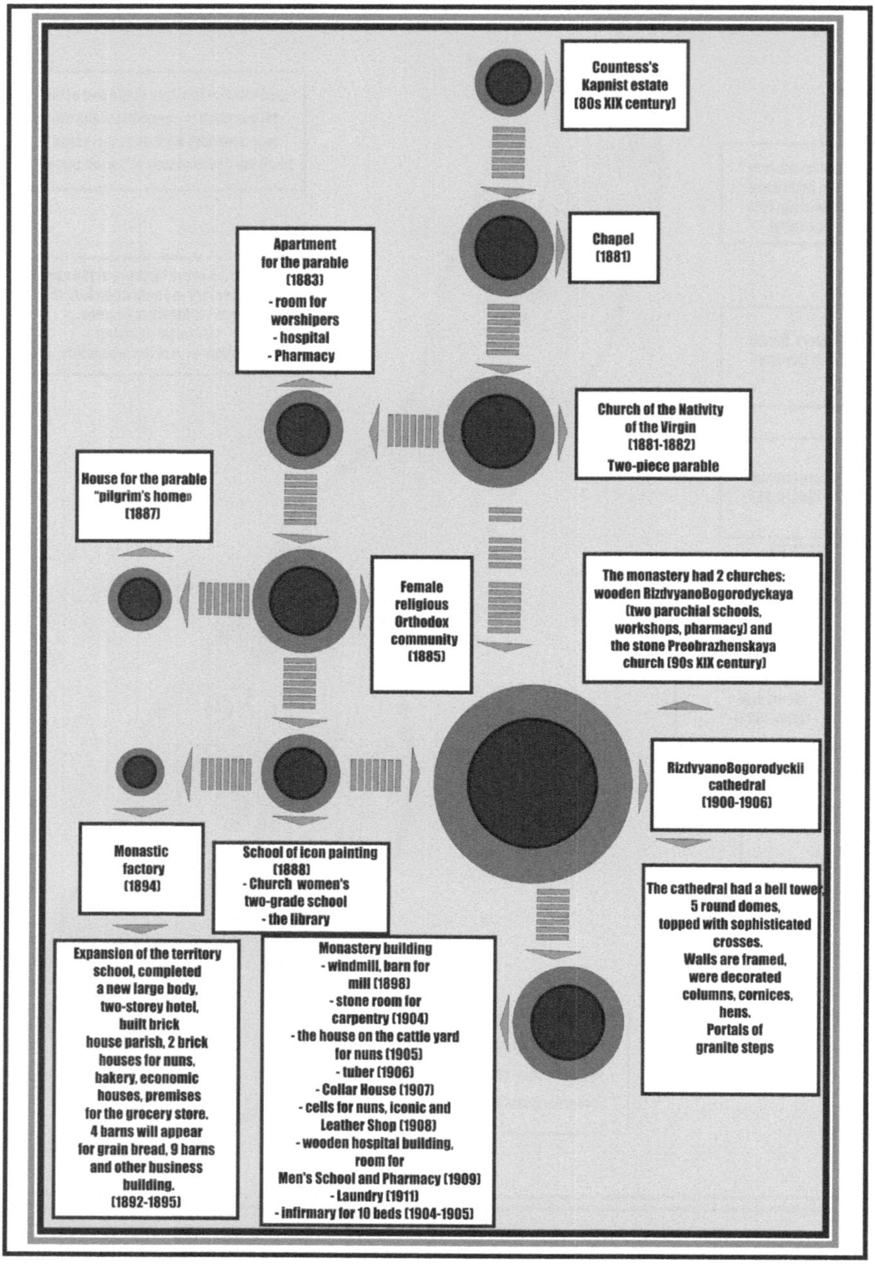

Fig. 2 Stages of formation of the Kozelshchyna Rizdvyano-Bogorodychniy monastery

Along with other objects in the urban environment, they create a reputation and brand image for the city—a determining criterion that influences investment attraction.

For the successful socio-economic development of settlements, it is necessary to consider the following factors:

- Economic: innovation (state of economic development, investment attractiveness and investment climate, availability of municipal economic development programs;
- Social: quality of life, social structure of the population, development of the network of objects of social infrastructure;
- Infrastructure: quality of the territory (availability of certain resources), diversity and quality of the environment, accessibility of service facilities, safety of the urban environment;
- Historical and cultural factors: architectural and urban planning, historical and cultural, etc.

These factors combine to provide economic investment, and the brand of the city must embrace the essence of its identity.

But what can be revealed by the identity of any city—in the first place is the presence of historical and cultural potential inherent in this locality. The strategic plan and the brand development plan determine which activities will be supported and whether the development of historical and cultural heritage objects should be taken as the basis for urban development as a basic element of branding of the settlement in the context of the development of tourist and recreational activities.

In terms of trends in the development of modern tourism, the division into religious and pilgrimage tourism is not fundamental to the further economic development of places of worship. The main denominator is attracting more tourists by creating favorable conditions for staying in the above territories and promoting their availability. The area of information should differ depending on the potential of a cult building and be determined by appropriate calculations.

There are three levels of attractiveness of cult sites: within the micro-zone of the tourist region (several administrative regions), within the tourist region (several regions), and within the macro-tourist region (countries, in some cases, the continent).

For the Poltava region, the largest area of attraction is the Mgar Holy Spaso-Preobrazhenskiy Monastery, which has at least ten shrines, among which are icons, a tomb, and an ark. In addition to religious values, the monastery has great architectural and urban value. This is a monument of architecture of national importance, made in the style of Mazepa Baroque. It stands on a hill on the northeast outskirts of Lubny. It was founded in 1619 by Princess Raina Mohylyanka Vyshnevetskaya and Metropolitan Isaiah Kopynsky. Pilgrims and organized groups of tourists from different parts of Ukraine will come to it. As the monastery stands on a hill near the Kyiv-Kharkiv-Dovzhansky highway of international importance, forming an unsurpassed silhouette of the suburbs, and closer to the panorama, it creates a business card for travelers. The monastery's infrastructure is constantly expanding to meet the various needs of visitors: spiritual, cultural, and material.

Another cult object that has entered the international tourist level is the Hador Center for Admiral Azaken in Gadjach, where the tomb of the great tzadik Alter Rebe is located. The center is mainly focused on pilgrims, but in the context of the large number of cult sites in the city and its surroundings, it has considerable potential for tourists.

The town of Gadjach takes small and somewhat chaotic steps to shape the brand of this settlement. This town needs outsourcing of strategic planners to systematically enhance its tourist attractiveness.

The cult buildings of the township have a slightly smaller area of attraction, Kozelshchyna, Kotelva, Dykanka, and others. Tourists visiting these places are mostly residents of this tourist region. According to the scheme of formation of the tourist system of Poltava region, all religious tourist routes are linear in nature with the tourist center in Poltava.

Taking into account, the branding of historic settlements should be conducted in the context of potential consumers of a religious tourism product, gradually expanding them.

As a result, benefits of branding in the tourism industry are [12]:

- opportunity to earn additional income;
- the ability to protect the manufacturer in the process of working with partners;
- facilitating the selection of the tourist product by the consumer;
- the ability to identify the travel company and its services among competitors' tourism products;
- facilitating the entry of new products to the neighboring markets;
- the opportunity to invest in the future.

Recently, the process of updating and developing of settlement master plans has been intensified, so it is advisable to synchronize these two processes—development of the master plan and brand strategy of the city. An example of this approach is the concept of development of settlements in Novi Sanzhary and Myrhorod in the Poltava region, which is based on the idea of developing them as tourist resort cities, where the idea of developing historical and therapeutic potential is based on the brand strategy and master plan. An example of these shows that settlement data is promising for the development of the recreational industry, which can lead to an increase in the total number of residents in the summer season more than twice.

Conclusion

Tourism in Ukraine is developing at a rapid pace, but it occupies a modest place in the world tourism services market, especially in the religious tourism segment. Ukraine has great potential for the development of religious tourism. Due to the fact that religion and tourism use the same cultural heritage, it is necessary to ensure their close cooperation and integrated approach to tourism and culture.

Despite the existence of many approaches and strategies in developing the brand of the city, its implementation must take into account the historical features of the development of the city. In turn, the development strategy underlying the development of the concept and master plan should also be based on a thorough economic analysis and the calculation of the feasibility of a particular direction of development.

Such approaches can serve as a basis for the parallel formation of both the settlement brand and the concept of town-planning solutions for it, and, after some quantitative and qualitative refinement, can be used in the solving of the same problems for other settlements.

Territorial marketing activities in tourism branding can be: development of attractive positioning and image for the territory; supplying products and services in an efficient and accessible manner; promotion of attractive and useful qualities of the territory in order to inform the users about its advantages widely.

References

1. Havryliuk, A. (2011). Derzhavne rehuliuvannia komunikatsiinoho zabezpechennia turystychnoi industrii v Ukraini (avtoref. dys. na zdobuttia nauk. stupenia dokt. ek. Nauk). Kyiv.
2. Anholt, S. (2007). *Competitive identity: The new brand management for nations, cities and regions*. Basingstoke: Palgrave Macmillan.
3. Nahorniak, T. (2008). Kraina yak brend. Natsionalnyi brend "Ukraina". *Stratehichni priorytety: naukovo-analitychni shchokvartalnyi zbirnyk, 4*(9), 220–228.
4. Gabdrakhmanov, N., Rubtzov, V., & Mustafin, M. (2016). Role of historical cities in the tourist breading: (Case study of Republic of Tatarstan). *International Business Management, 10*(22), 5267–5272.
5. Nolan, M., & Sidney, N. (1992). Religious sites as tourism attractions in Europe. *Annals of Tourism Research, 19*, 68–78.
6. Jackowaski, A., & Valene, S. (1992). Polish Pilgrim-Tourists. *Annals of Tourism Research, 19,* 92–106.
7. Smith, V. (1992). Introduction: The Quest in Guest. *Annals of Tourism Research, 19,* 1–17.
8. Rinschede, G. (1992). Forms of religious tourism. *Annals of Tourism Research, 19,* 51–67.
9. Cohen, E. (1998). Tourism and religion: A comparative perspective. *Pacific Tourism Review, 2,* 1–10.
10. Alderman, D. (2002). Writing on the Graceland wall: On the importance of authorship in pilgrimage landscapes. *Tourism Recreation Research, 27*(2), 27–33.
11. Muzychenko-Kozlovska, O. (2014). Turystychnyi brendynh: sutnist, skladovi ta perevahy. *Visnyk Natsionalnoho universytetu "Lvivska politekhnika",* 797, 396–402.
12. Vyzghalov, D. (2011). Brendynh horoda. M., Fond "Ynstytut əkonomyky horoda".
13. Zenker, S. (2010). The place brand centre—a conceptual approach for the brand management of places. In *39th European Marketing Academy Conference*. Copenhagen.
14. Eidelman, B., Fakhrutdinova, L., Gabdrakhmanov, N., & Nayda, A. (2016). Ways of formation of regional brands in modern conditions. *Academy of Marketing StudiesVol., 20,* 39–44.

Scientific and Technical Activities Management Automation of the Department of Structures from Metal, Wood, and Plastics

T. A. Dmytrenko, A. O. Dmytrenko, T. M. Derkach, and L. A. Klochko

Abstract The problem of introducing the latest information technologies into the management of higher educational institutions training specialists in the construction industry departments is considered. The theoretical and methodological aspects analysis of improving reporting in educational institutions and a special attention are paid to the current state of ERP—systems development. Creating an intellectual information management system necessity for scientific and technical activities of the teaching staff at higher educational institution technical direction departments is substantiated. The main functions should provide this developed software module are given. Using the Unified Modeling Language for more detailed module designing of the scientific and technical activities department management is provided. The possibility of using use-case diagrams that visually reflect various interaction scenarios between users and use-cases and describe the functional system aspects is presented. Particular attention is drawn to the detailed developed database description of the module for the scientific and technical activities management of the department of Structures from Metal, Wood, and Plastics. The developed interface is presented; the main requirement for this was accessibility and understandability for all users. The information management system of the department of Structures from Metal, Wood, and Plastics provides for checking the text for uniqueness by the teacher or user. Requirements for testing a software product are given. Based on the study, a software-based module for managing the department scientific and technical activities is proposed, which general information system part is "Portal—Department."

T. A. Dmytrenko (✉) · A. O. Dmytrenko · T. M. Derkach · L. A. Klochko
Poltava National Technical Yuri Kondratyuk University, Poltava, Ukraine
e-mail: dmitr_tat@ukr.net

A. O. Dmytrenko
e-mail: andmyt@ukr.net

T. M. Derkach
e-mail: tanider@ukr.net

L. A. Klochko
e-mail: lina.dmitrenko@gmail.com

V. Onyshchenko et al. (eds.), *Proceedings of the 2nd International Conference on Building Innovations*, Lecture Notes in Civil Engineering 73,
https://doi.org/10.1007/978-3-030-42939-3_36

Keywords Intellectual system · Semantic text analysis · Department scientific activity

1 Introduction

Currently, the task of developing a mechanism for implementing strategies by creating a strategic management system has been set at the state level. In the context of the innovative economy development, shifts of interest toward the strategic planning system have become a necessity for educational institutions as well. Many higher educational institutions introduce strategic management into their activities, which is one of the methods that determine the competitive ability and investment of organization attractiveness in the long term [1, 2].

One of the important elements of strategic planning is the implementation of modern information systems for managing various objects. The problem of building an automated control system for scientific and technical activities of the departments of technical direction has involved universities in Ukraine and abroad. Currently, software products that meet the requirements of full accounting for monitoring the scientific and pedagogical staff activities of higher educational institutions departments in the technical field could not be found. Therefore, this issue becomes even more relevant.

To build information systems to support the higher education institutions scientific activities management and improve the education quality, the need arises for the ERP—systems implementation. When developing the technical task for the problem study, "Standards and recommendations for ensuring the quality of higher education in the European space" was used, developed by the European Association for the Assurance of Higher Education Quality ENQA [3, 4].

2 Main Body

2.1 The Analysis of Recent Research Sources and Publications

To create key characteristics of the department scientific and technical activities management module that is being developed, an existing services analysis was carried out and their features and disadvantages were determined. Improving reporting methodological and theoretical directions studies in educational institutions using automated information systems has been carried out by many domestic and foreign scientists. The information systems nature and significance analysis are reflected in the number writings of authors [5–8]. Scientists pay special attention to the current state of development of ERP—systems and their classification [9, 10]. At the same time, there is a need to develop and put into practice a specific information system to automate the department scientific work accounting of training specialists in the construction industry [11, 12].

2.2 The Selection of Previously Unsolved Parts of the General Problem

The need to create an intellectual information management system for the scientific and technical activity of the higher educational institution departments a technical direction arises to optimize the reporting process, reducing organizational costs to ensure the scientific activities of the department and increasing staff productivity.

2.3 Problem Definition

The main objective of this study was the design and software implementation of the module for the management of scientific and technical activities of the department of Structures from Metal, Wood and Plastics (SMWP), which is one of the information system elements "Portal—Department" [13].

2.4 Scientific Novelty

For the first time, an applied intellectual content management system with an expanded toolkit of Web portals was implemented, which contains software for managing the scientific and technical activities of the technical field departments scientific and pedagogical staff. The system will enable optimizing the reporting procedure, expanding the possibilities for analyzing the results, and ensuring transparency and objectivity in the procedure for evaluating the teachers' activities.

2.5 The Main Material and Results

The software module for managing the department scientific and technical activities for the intellectual information system "Portal—Department," should ensure the following functions:

(1) distinction between the user and the administrative part of the software module (administrator/editor/user);

(2) registered user should be able to freely enter the module for managing scientific, scientific and technical activity using individual unique login and password. Upon entering the system, he gets access to scientific publications (where he can add a new publication, with the ability to add a co-author, select the appropriate categories and other details);
(3) reports generation on scientific, technical, and innovative activities automatically;
(4) system should be able to verify the scientific publication uniqueness;
(5) ensure the development or addition (replenishment) of new publications and their categories;
(6) include a system for planning the date and time of scientific material publication in advance, which will make it possible to automatically publish an article at a specified date and time;
(7) ability to change the published article status:

   - private status (the publication will not be displayed in the information system, even if it was published earlier);
   - public status (the publication will be displayed in the information system, even if it has not been published previously);

(8) administrator ability to clear the database of already irrelevant data;
(9) ability to save all the stages of creating a scientific publication on the possibility of returning it to the selected version;
(10) department information presence on the Internet, the availability of reports on scientific, technical, and innovative activities, and scientific materials;
(11) simplicity and ease of navigation, that is, an intuitive interface for both student and teacher. We have an administrator/editor/user.

At the information system design stage, the main technical task sections were drawn up:

- architecture description of the information intellectual system and database structures;
- scenario description for automating system processes;
- scenario development for integrating the system with external software products;
- sources of initial data and options for the initial system content;
- development of the access rights concept and user and administration authority.

For more detailed module design for managing the department scientific and technical activities, UML models were used.

UML is a broad profile language; it is an open standard that uses graphic notation, with the help of which an abstract of this module system model was created.

A precedent diagram is developed that visually reflects various scenarios of interaction between actors (user) and precedents (use-cases) and describes the functional aspects of the system (business logic).

In the simplest case, the use-case chart is determined in the discussing process with those user's functions that he wanted to implement. These diagrams are the basis for achieving mutual understanding between the programmer, the project developer, and the project customers [14].

One of the important module elements for the scientific and technical duality management of the department of Structures from Metal, Wood and Plastics (SMWP) is given in the following main table's database:

(1) wp_users (user table), it stores all the data about users:

- user_login (user login field)
- user_pass (user password field)
- user_nicename (user nickname field)
- user_email (user email field)
- user_registered (field of student registration date in the system);
- user_activation_key (user activation key (sent to the mail to confirm registration))
- display_name (name for display)

(2) wp_quiz_aditional (table for saving scientific publications):

- user_ id (user id)
- post_id (post id, i.e., publication)
- date (publication date)

(3) wp_posts (post table):

- post_autor (post author, in this case administrator or editor)
- post_date (post creation date)
- post_ content (filling the post, for example, the text of a scientific publication)
- post_title (post title, for example, article title);
- post_modified (date of last post changes)
- post_status (post status);
- post_type (post type, for example, article, slider, main page content)

(4) wp_postmeta (table for additional post data):

- post_id (post id, i.e., publication)
- meta_id (field for writing id of custom fields)
- meta_key (field for writing the name of custom fields)
- meta_walue (field for writing the contents of custom fields)

(5) wp_cimy_uef_data (user category table):

- user_ id (user id)
- field_id (category confirmation);

Most tables are linked to one or more others using a single field. This field will be a unique identifier for each record (example post_id).

This model of database organization is most suitable for this system. This organization enables to make some changes to the user system profiles. If the administrator changes the editor scientific degree to the teacher or the scientific article publication date, he may indicate other data. Therefore, in some tables, separate fields are duplicated, providing the ability to change information, rather than static attachment to registration data.

The main encoding of the database is utf8, the utf8_general_ci collection. The choice of such encoding was due to the need to use Cyrillic characters to display most text information, which enables this encoding. The password length field is 32 characters, and this length was chosen because of the need to store the password encrypted by the md5 system in it, encoding it up to a size of 32 characters.

During the study, an interface was designed and developed. The most effective approach to designing an interface is to develop it by modeling user-defined functions. At the beginning of the prototyping process, paper interface layouts were created; then screen forms were developed that simulate user interaction. The interface design principles bases are human factors.

The main requirement in the interface development is accessibility and understandability for all users. Therefore, a popular model for building an information system was chosen.

Since designing useful and comprehensive information for the user is a serious matter, it should be evaluated at the same level as the system architecture or program code. Designing messages require considerable time and/or little effort.

Processing error messages is one of the teaching means the user how to use the system. Having accidentally performed a certain action with an error, the user should receive a message about this, in an intuitive form, what exactly the user did wrong and what needs to be done to correct this error.

Information search is the providing means the user with the opportunity to receive only the information that he wants to see from a large amount of other information. Therefore, this function should be provided in the system interface in the place necessary for it. An example of its use in the program module is the login form; entering invalid data, the user receives a message and a link by which you can solve the problem.

In the process, Ajax technology was used—Web application development technology that uses code on the client machine to modify data on the Web server. As a result, Web-pages are dynamically updated without reloading a full page, interrupts data exchange. Using Ajax, it was implemented to obtain real-time information about the h—index and the number of scientific citations and pedagogical workers, graduate students and doctoral students on Google—Scholar Citations.

One of the problems in developing the module was that when updating the contents of the page, the Web application changes to a new state, but constantly reloads the page while the user waits for data on the h—index and the citations number of scientific and pedagogical workers, graduate students and doctoral students on Google—Scholar Citations are not advisable, so the choice of this technology is quite reasonable.

A content management system has been developed that provides the administrator in the information system with the ability to perform the following actions [15]:

(1) adding and deleting text;
(2) text editing;
(3) adding and removing graphic design;
(4) change of supporting materials and applications;
(5) adding and deleting new and scientific works (articles, publications)
(6) registration of new users;
(7) management of the display of scientific and scientific-technical and informational publications;
(8) adding and removing new categories for publications;

The presented system of the department of Structures from Metal, Wood and Plastics is aimed at teachers and researchers, students and ordinary users. The purpose of its creation is to ensure the management of the department scientific and technical activities, which will greatly facilitate the work of the above-listed users.

When developing a software product, modern Web-technologies were selected and justified, which allow creating interactive Web-pages (see Fig. 1).

One of the main information system parts is the page for checking the text for uniqueness by the teacher or user.

After or while writing a scientific or scientific—technical work, the teacher can check the text for uniqueness. On this page, you can specify a link to a site which contents the algorithm will ignore (if this publication has been already posted by the teacher on another Internet resource).

After analyzing the text, the user receives a percentage of the text and links uniqueness to the pages from which the text was taken if the publication is not completely unique. The system also provides data on which specific words coincided.

To verify the system operation and its implementation compliance with the requirements, testing was used, which consisted of three parts:

(1) regression testing;
(2) functional testing;
(3) stress testing.

3 Conclusion

Based on the research results, an applied intelligent system for managing the scientific and technical activities of the scientific and pedagogical staff of the departments of training specialists in the construction industry of higher educational institutions was designed and implemented.

The system enables optimizing the reporting procedure, expanding the possibilities for analyzing the results, and ensuring transparency and objectivity in the procedures for evaluating the teachers' activities.

A description of the system architecture and database structures and the process automation script is provided.

Fig. 1 Main information system page

A scenario has been developed for integrating the system with external software products and source data sources and options for the initial content of the system.

References

1. Ashimova, D. E., Amirov, A. Z., Sultanova, B. K., & Kabylova, D. A. (2016). Information system of the results of scientific activities at the University. *Scientific Almanac, 2*(11), 22–25.
2. Dmytriv, K. I., & Shpak, Y. N. (2017). Research information systems in enterprises management: Experience and perspective. Economic bulletin of NTUU KPI , 231–239.
3. The Standards and guidelines for quality assurance in the European Higher Education Area (ESG). http://www.enqa.eu. Last accessed 15 May 2019.

4. Dmytrenko, T., Semko, O., Dmytrenko, A., Derkach, T., & Voskobiinyk, O. (2018). Development and implementation of algorithms of building structure engineering calculations for shear fraction under pressing-through. Paper presented at the MATEC Web of Conferences, 230. https://doi.org/10.1051/matecconf/201823002004.
5. Faisal, A. M. (2016). Information and Communication Technology (ICT). Social Changes and Transformation, 60960. https://doi.org/10.13140/RG.2.2.22371. Last accessed 11 March 2019.
6. Fedyakova, N. N. (2016). Improving university management information systems. *Integration of education, 2*(83), 198–210.
7. Prorok, V. Y., Zykov, A. M., & Karytko, A. A. (2013). Method of calculating the required performance of computing elements in high-loaded multiprocessor computing systems. *High technology in space research of the Earth, 5*(1), 46–51.
8. Sviridov, S. (2017). Information and analytical accounting system for the scientific activity of academic institution. In *First International Conference on Ocean Thermo hydromechanics-2017*. https://doi.org/10.29006/978-5-9901449-3-4-2017-1-144-147. Last accessed 03 May 2019.
9. Mouna, J., Latifa, B. A., & Anis, B. A. (2014). Classification of security threats in information systems. *Procedia Computer Science, 32,* 489–496.
10. Russell, S. J., & Norvig, P. (2010). *Artificial Intelligence: A Modern Approach*. Prentice Hall: Upper Saddle River. Publisher.
11. Bellatreche, L., Valduriez, P., & Morzy, T. (2017). Advances in databases and information systems. *Information systems, 70,* 1–2.
12. Castro, J., Kolp, M., & Mylopoulos, J. (2002). Towards requirements-driven information systems engineering: the Tropos project. *Information systems, 27,* 365–389.
13. Ivas'ko, I. A., Dmitrenko, T. A., Derkach, T. N., & Dmitrenko, A. O. (2018). Development of the department of management of scientific and scientific and technical activities of the chair for informational intellectual system "Portal-kafedra". *Systems of Control, Navigation and Communication, 6*(52), 104–112.
14. Ivanova, T. V., & Baranov, V. V. (2016). The current state of the development of information systems. *Scientific notes, 10,* 224–226.
15. Bazargan, K. (2014). Review: A new approach to peer review, using WordPress. In *Septentrio Conference Series*. https://doi.org/10.7557/5.3041. Last accessed 20 Apr 2019.

Big Cities Industrial Territories Revitalization Problems and Ways of Their Solution

Mykola Dyomin, Andrii Dmytrenko, Denys Chernyshev, and Oleksandr Ivashko

Abstract The purpose of the article is to identify characteristic for Ukraine organizational, functional planning and design features of art formations creating on the revitalized industrial enterprises' basis (on condition that the main part of the existing building is preserved). The method of comparative analysis was used. Existing art formations examples were analyzed based on revitalized industrial enterprises in major cities of Ukraine, in post-Soviet countries and in the European Union. The classification of industrial enterprises and individual buildings according to their suitability for revitalization was first developed. The techniques of multi-story industrial buildings' reconstruction for other functions, including art function, have been improved. The hierarchical model of art formations was developed for the first time, the art cluster definition was specified, and the classification of art clusters by functional feature was developed. The specific functional and planning art clusters features in the post-Soviet space were revealed. The study results can be used in the development of promising cities concepts, detailed plans of industrial areas to be revitalized and reconstruction projects of individual industrial enterprises and buildings.

Keywords Construction · Revitalization · Industrial territories · Evaluation criterion · Art formation

1 Introduction

The problem of industrial territories redevelopment, which today appeared in the central areas of large cities and in residential areas, is today an international problem, taking into account the latest analytical studies in the cities of Lodz (Poland), Kyiv, Kharkiv, Lviv (Ukraine), Zurich (Switzerland), Moscow, St. Petersburg (Russia).

M. Dyomin · D. Chernyshev · O. Ivashko
Kyiv National University of Construction and Architecture, 37 Povitroflotskyi Avenue, Kyiv, Ukraine

A. Dmytrenko (✉)
Poltava National Technical Yuri Kondratyuk University, 24 Pershotravnevyi Avenue, Poltava, Ukraine
e-mail: metr5555@ukr.net

V. Onyshchenko et al. (eds.), *Proceedings of the 2nd International Conference on Building Innovations*, Lecture Notes in Civil Engineering 73,
https://doi.org/10.1007/978-3-030-42939-3_37

There are two ways of using the former industrial territories, which usually occupy a vast area and consist of a whole complex of structures. The first option involves the complete clearing of the site for new construction. The second option involves the partial preservation of an existing industrial building, especially if it is listed architectural object, with its redevelopment into a business function, trade function or art function.

The analysis of recent domestic and foreign studies has proved that the problem is still relevant, different ways of its solution are offered, and one of the modern options for industrial territories redevelopment is a possibility of their transformation into art formation. In particular, the problem of industrial territories revitalization, the analysis of urban formation factors, technologies of architectural monuments restoration and modernization with the use of modern materials and technologies are covered in the scientific works of Antonova [1], Gyurkovich [2], Ivashko [3–6], Ivashko [7], Kobylarczyk et al. [8], Makarova [9], Orlenko [6, 10], Pleshkanovska and Biriuk [11], Fedotova [12].

The influence of creative industry and art formations as its variety on urban development has been investigated by Boix et al. [13], Karachalis and Deffner [14], Działek and Murzyn-Kupisz [15].

Concepts of Ukrainian cities development (Lviv, Chernivtsi, Vinnytsia, Zhytomyr, Poltava, Podilskyi district of Kyiv) developed in such international projects as Integrated Urban Development in Ukraine (with the governments of the Federal Republic of Germany and the Swiss Confederation financial support) [16] also consider the development of the creative industry (including art formations) as one of the industrial territories revitalizing means.

However, organizational, functional, planning and design aspects of art formations created on the basis of revitalized industrial enterprises in Ukraine require further investigation.

2 Main Part

2.1 Purpose of the Article

The purpose of the article is to identify specifics for Ukraine organizational, functional planning and design features of art formations' creating on the revitalized industrial enterprises basis (on condition that the main part of the existing building is preserved).

2.2 Research Methodology

The method of comparative analysis was used. Existing art formations examples were analyzed on the basis of revitalized industrial enterprises in major cities of Ukraine, in post-Soviet countries, in the European Union. Such factors as the population of the city, its role and place in the settlement system, placement in the structure of the city, space planning and structural features of the revitalized industrial buildings, the degree of their historical and cultural value, etc., were taken into account.

2.3 Results

Today, industrial territories built up by capital industrial buildings have lost their original function and purpose that are in the public eye. These areas also have become the focus of environmental protection and historical and cultural heritage protection institutions. The subject of public discussion is these territories use with maximum preservation of separate buildings and entire streets appearance that define the face of the area, and changes in their functional use directions, while simultaneously solving the problems of urban environment gentrification and increasing its aesthetization level. At the same time, attention is paid to the latest technologies and building materials.

There are two main types of industrial territories and individual enterprises revitalization—with a complete change of function or with partial preservation of production functions. Both options can be implemented using only the territory when the existing buildings are completely demolished (shopping mall "Manufactory" in Sumy at the site of light industry enterprise or shopping mall "Kyiv" in Poltava on the part of the still functioning turbo-mechanical plant's territory), or using existing buildings ("Mystetskyi Arsenal" in Kyiv, "Equator" shopping mall in Poltava, located in the former workshops of the "Poltavamash" plant). If the existing function of the industrial enterprise is partially preserved, for other functions, industrial buildings (Vorskla garment factory in Poltava city) or office and amenity buildings (fitness club creation on the basis of plant "Ltava" office and amenity buildings in Poltava) may be used either in part or in whole.

In general, the revitalization process of an industrial building with renovation consists of several stages, depending on the degree of complexity. This can be the restoration of load-bearing structures reliability and the improvement of performance, partial redevelopment and restoration of operational properties without changing the typical design decision, increasing the building volume due to the extension and additional stories of the building. When an industrial building is an architectural monument, it is necessary to define clearly what should be protected and authentically preserved.

In order to evaluate the effectiveness of the re-profiling of industrial architecture objects for other functions, the possibility of revitalizing enterprises in different industries was analyzed, not all of them are subject for structural replacement and redevelopment due to the specifics of functioning and structural schemes. As a rule, enterprises are not subjects for similar reconstruction if they belong to those industries which, by the functioning peculiarities, envisage large areas of territories, are highly harmful, or if the enterprise is located in a long building, and there is a specific concomitant construction and transport network oriented around the production process, if the enclosures contain large equipment that cannot be removed.

The Soviet era enterprises (they make up the majority in the post-Soviet space) of consumer industry and food processing, consumer services and commercial and warehouse buildings could be revitalized in easiest way. Taking account of their unification, the use of metal and reinforced concrete frame, revitalization is mainly about changes in external appearance, redevelopment and completion. In the case of food industry enterprises, they are characterized by the use of variable columns and a frame scheme (precast reinforced concrete frame). In addition, in many cases, the design scheme does not provide for the possibility of its replacement full or partial.

In developed countries and in many projects in Ukraine, the process of industrial facilities revitalization is based on thermal modernization by improving the thermal insulation properties of building enclosures (due to the insulation of external walls, roofs, windows replacement for energy-saving ones) and the reconstruction of engineering systems. It also provides for the replacement of the roof and insulation of the attic space, additional facade insulation, replacement of windows, basement ceiling insulation, updating of the heating system, as well as cold and hot water risers, updating of the ventilation system, integration of heat regeneration systems and updating of entrance doors and stairways.

Internal redevelopment is also possible, including arrangement of internal space additional division both horizontally and vertically (with arrangement of additional floors).

If the height of the existing building floor is 6 m or more, it may be necessary to reconstruct it vertically, with additional floor slabs, for more efficient use of the building volume. However, the use of cranes for the installation of large construction products is very limited or impossible.

In one-story industrial buildings, this problem is usually solved by the arrangement of a separate steel frame (the so-called shelves) and monolithic reinforced concrete floor slabs. It is more difficult to solve this problem in multi-story industrial buildings. Due to the difference in the design load on the floor slabs for industrial and public functions, it is possible to arrange additional floor slabs using the existing frame and foundations (possibly with a slight reinforcement). An example of such solution is shown in Fig. 1.

However, such an option imposes significant restrictions on the slab's weight. In this case, it is advisable to use steel beams and monolithic reinforced concrete ribbed overlap with a reduced thickness of about 70 mm (Fig. 2). For forming of such floor slab can be used, in particular, block-modular formwork, developed by experts of PoltNTU [17].

There are many options for industrial territories redevelopment, one of which is reprofiling for an art function, which is a fairly new but promising area of revitalization that is spreading rapidly around the world. In recent years, revitalized objects called

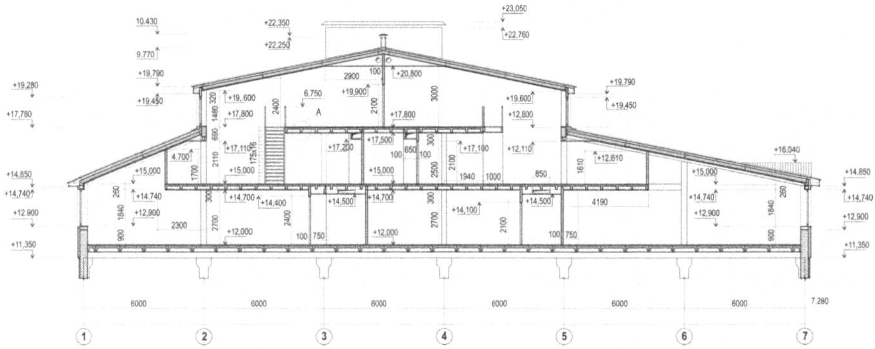

Fig. 1 Reconstruction of the industrial building attic space for office premises, Kyiv. Cross section (fragment). Project proposition developed by PoltNTU experts (Riabiko et al.)

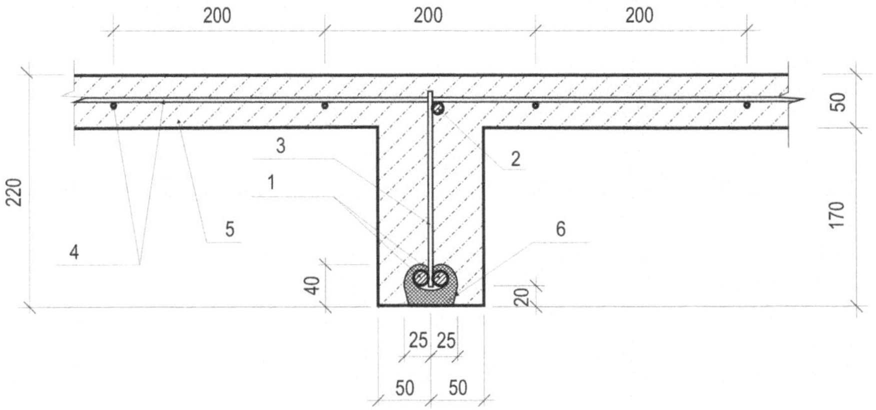

Fig. 2 Detail A (fragment of monolithic floor slab): 1—steel armature $d = 14$; 2—steel armature $d = 10$; 3—steel armature $d = 4$; 4—steel armature $d = 4$; 5—concrete class C 20/25; 6—plastic liner for attaching the false ceiling frame

"art cluster" have appeared in many cities (Berlin, Zurich, Paris, Lodz, Krakow, Katowice, Kyiv, Kharkiv, Moscow, St. Petersburg, Tallinn). The numbers are growing so fast that it gives the impression that a new type of institution or complex is associated with the arts. As stated in the scientific publications of Ivashko [3, 4], despite the fact that the term "art cluster" is widely used in print and electronic sources and is the name of many revitalized industrial objects, it was proved and created by Dyomin and Ivashko structural logical models based on that not all such objects fall under the defining features of the art cluster and argue for the existence of contradictions in the definition of meaningful filling of the term "art cluster."

The number of objects that function under the name of "art cluster," if not massive, is quite widespread, which allows and requires to analyze it as an object of the scientific study of the essence, functioning and typological features. In this regard, the social and urban role of such objects is assessed.

As the analysis of the Ukrainian experience of industrial territories revitalization shows, the functional filling of the revitalized industrial development depends largely on the size of city. In big cities (Poltava, Sumy) with a population of 250–500 thousand people, viable projects are mainly viable, which involve the adaptation of the facility to the shopping (shopping and entertainment) function of the "Equator" shopping mall in Poltava in the buildings of the former plant "Poltavamash," tax inspection in the buildings of the former Elektromotor plant, etc.). In such cities, even promising development concepts [18, pp. 104–107] suggest the creation of art objects only within technoparks (innovative parks, "Creative industries parks," etc.). The creation of viable self-sufficient art objects is mostly possible only in major cities with a population of more than 800,000 people (cultural platform "Izolyatsia" in Donetsk in 2010–2014 on the basis of the former plant of insulation materials, since 2014—in Kyiv, "Mystetskyi Arsenal" in Kyiv). The smaller the size of the city, the closer to its central part art formation should be.

According to the applied study of Dyomin and Ivashko system-structural analysis, art formation can be represented as a hierarchical model, where the primary unit is an art object. An art center is made up of several art objects united by a common infrastructure. An art cluster is a higher level of art formation where art objects are interconnected and aimed at achieving a common result. Based on the combination of several art schemes, it is possible to create a megacluster with a combination of several simple clusters and additional art objects related to common infrastructure in its structure.

Dyomin and Ivashko analyzed examples of revitalized industrial enterprises in different countries of the world where the art function is present. Three variants of art formations functioning were identified on the basis of the analysis. The first direction is when the art formation is created by the building owner, who selects the tenants of the premises according to the holistic art concept. In such case, the participants are connected by production processes and an art center is created. In case when their activities are aimed at a common result, an art cluster is created. The second direction—when the operation is created from below by the participants cooperative agreement, with a common economic basis, possibly accounting, the participants themselves choose a place depending on the activity and formulate space requirements. The third direction—when the art formation is created spontaneously, the set of participants is random.

According to the functional scheme, there are three types of art clusters: with the combination of many equivalent functions, none of which is the principal; with the combination of one principal and minor functions; with the presence of several basic functions, and additionally the so-called microclusters, with only a few functions present.

The comparative analysis of the functional filling of art clusters in the post-Soviet space and in the cities of Western Europe allows to distinguish the main difference between them: for the post-Soviet countries there is a clear functional zoning of the premises and integration of the premises similar in function into separate units, and

European art formations tend to be petty-bitty and allow a free mixture of rooms with the ability to change their functions.

2.4 Scientific Novelty

Classification of industrial enterprises and individual buildings according to their suitability for revitalization was first developed. The techniques of multi-story industrial buildings reconstruction for other functions, including art function, have been improved. The hierarchical model of art formations was developed for the first time, the art cluster definition was specified, and the classification of art clusters by functional feature was developed. The characteristic functional and planning features of art clusters in the post-Soviet space are revealed.

2.5 Practical Importance

Results of the study can be used in the development of cities promising concepts, detailed plans of industrial revitalized areas and reconstruction projects of industrial enterprises and buildings.

3 Conclusions

The problem of industrial territories redevelopment in major cities is an urgent one and requires the search for new revitalization ways, one of which is the art formations' creation.

Possibility of industrial enterprises reconfiguration with the existing buildings preservation depends on the industry, volume planning and structural features of the building, as well as its technical condition.

The problem of the construction volume rational use in revitalized industrial buildings can be solved by arranging additional floors. In multi-story industrial buildings, this type of reconstruction requires a minimum weight of floor slabs, which can be provided by the use of steel beams and monolithic reinforced concrete slabs with a reduced thickness of not more than 70 mm, for forming which it is advisable to use block-modular formwork.

Art formation can be represented as a hierarchical model with following levels: art object, art center, art cluster and megacluster. There are three main organizational models for the art formations functioning, with different centralization level.

Industrial territories (enterprises) revitalization with creation of a certain hierarchical level art formations largely depends on the city size, its place and role in the settlement system and the revitalized territory location in the city structure.

Depending on the number and combination of basic and auxiliary functions, art clusters are divided into three main types. In addition, the so-called microclusters with only a few features are distinguished.

There is a significant difference in the functional and planning organization of art formations in the post-Soviet space (clear functional zoning with the integration of premises into large blocks) and in Western European cities (mainly small-scale structure with flexible function of separate rooms).

References

1. Antonova, A. (2015). Art-klastery kak prostranstvo dlya razvitiya kulturnogo potentsiala goroda. In *Proceedings of the XLVII Intern. Research and Practical Conf. V mire nauki i iskusstva: voprosy filologii, iskusstvovedeniya i kulturologii, 4*(47), pp. 56–59. SibAK, Novosibirsk.
2. Gyurkovich, M. (2019). Wybrane przykłady transformacji zespołów poprzemysłowych. *Wiadomości Konserwatorskie, 57,* 142–157.
3. Iwaszko, O. (2018). Społeczna Przyroda Nowych Artystycznych kierunków w Miejskim Środowisku. *Przestrzeń/Urbanistyka/Architektura, 2*, 167–176.
4. Ivashko, O. (2018). Comparative analysis of Art-clusters—a new type of multi-functional buildings and their location in the city's structure. *Środowisko Mieszkaniowe (Housing environment), 25,* 13–17.
5. Ivashko, O. (2019). The issues of conservation and revitalization of the monuments of industrial architecture. *Wiadomości Konserwatorskie, 58,* 143–147.
6. Orlenko, M., & Ivashko, O. (2017). Art-clusters as a new type of buildings: The specificity of the spatial solution and the features of restoration during the redevelopment process (the experience of Poland). *Środowisko Mieszkaniowe (Housing environment), 21,* 109–115.
7. Ivashko, Y. (2013). *The fundamentals of formation of the art nouveau style in Ukraine (the end of the nineteenth and the beginning of the twentieth century) (doctoral dissertation).* Kyiv: Kyiv National University of Construction and Architecture.
8. Furtak, M., Kobylarczyk, J., & Kuśnierz-Krupa, D. (2019). Beton w adaptacjach i rozbudowach obiektów zabytkowych (na wybranych przykładach z Porto). *Wiadomości Konserwatorskie, 58,* 15–22.
9. Makarova, K. (2010). Postindustrializm, dzhentrifikatsiya i transformatsiya gorodskogo prostranstva v sovremennoy Moskve. *Neprikosnovennyi zapas, 2,* 150–152.
10. Orlenko, M. (2018). *Issues and methods of restoration of architectural monuments in Ukraine (11th century—the early 20th century) (Doctoral dissertation).* Kyiv: Kyiv National University of Construction and Architecture.
11. Pleshkanovska, A., & Biriuk, S. (2014). Mistse promyslovykh terytorii v planuvalnii strukturi krupnoho mista ta napriamy ikh transformatsii (na prykladi m. Kyieva). *Suchasni problemy arkhitektury ta mistobuduvannia, 37*, 263–267.
12. Fedotova, N. (2013). Kreativnyi klaster v kontekste sotsiokulturnykh problem regiona. In *Proceedings of All-Russian Scientific and Practical Conference "Socio-Cultural Space of Modern Russia: Challenges of the XXI Century"* (pp. 59–68). Moscow: Knizhnyi Dom.
13. Boix, R., Capone, F., De Propris, L., Lazzeretti, L., & Sanchez, D. (2016). Comparing creative industries in Europe. *European Urban and Regional Studies, 23*(4), 935–940.
14. Karachalis, N., & Deffner, A. (2012). Rethinking the connection between creative clusters and city branding: The cultural axis of Piraeus Street in Athens. *Quaestiones Geographicae, 31*(4), 87–97.
15. Działek, J., & Murzyn-Kupisz, M. (2014). Young artists and the development of artistic quarters in Polish cities. Belgeo, 3. http://journals.openedition.org/belgeo/13012. Last accessed 23 Aug 2019.

16. Integrated urban development in Ukraine—GIZ. https://www.giz.de/en/worldwide/39427.html.
17. Riabiko, H., Liakh, V., & Dmytrenko, A. (2011). Blochno-modulna opalubka dlia monolitnoho perekryttia. Patent of Ukraine for utility model No. 59927 dated 10 June 2011.
18. Kontseptsiya intehrovanoho rozvytku mista "Poltava 2030". http://www.2030.poltava.ua/ua/kontseptsiya-intehrovanoho-rozvytku-mista-2030.

Environmental Areas of Poltava Planning Development

Yuri Golik, Oksana Illiash, Yuliia Chuhlib, and Nataliia Maksiuta

Abstract The ecological potential of the city has a significant impact on all strategic goals of city development. To successfully use the existing environmental potential within the framework of the integrated city development strategy, detailed expert analysis and forecasting of its future changes have been carried out. These analytical developments databases take into account the important results of expert assessments of the ecological state of the city made in previous years and the results of environmental and public opinion polls.

Keywords Environmental aspects of planning · Environmental potential · Integrated urban development

Urban sustainability indicators are tools that enable urban planners, city managers and policymakers to assess the socio-economic and environmental impact. For example, these indicators can be used for contemporary urban projects, infrastructure, policies, waste disposal systems, pollution, and access to public services. They allow to diagnose problems and pressures and, therefore, to identify areas that may arise from problem solving through proper management and evidence-based responses. They also enable cities to monitor the success and impact of sustainability measures [1]. One of these indicators (directions) is environmental sustainability which is shown in Fig. 1 [2].

Within development planning, it is important to consider the following [3]:

- Without useful monitoring-based data, it is not possible to develop the right development direction.
- Measures should be put in place to ensure that the goals are met.
- National, geographic, cultural, financial properties, etc., must be taken into account during planning.
- Sets of development indicators change through time.
- Changing indicator (indicators) also changes the hike to planning, so it should be revised from time to time.

Y. Golik · O. Illiash · Y. Chuhlib · N. Maksiuta (✉)
Poltava National Technical Yuri Kondratyuk University, Poltava, Ukraine
e-mail: mns7000@yahoo.com

V. Onyshchenko et al. (eds.), *Proceedings of the 2nd International Conference on Building Innovations*, Lecture Notes in Civil Engineering 73,
https://doi.org/10.1007/978-3-030-42939-3_38

Fig. 1 Venn diagram representing the standard dimensions of sustainable development and referencing concepts proposed in WCED, 1987

1 Purpose

Modern Poltava is a city with significant natural resource potential, advantage economic and geographic location, rational urban planning of the territory and well-known historical and cultural heritage. These all characteristics in the complex determine the significant ecological potential of the city.

Accordingly, such potential has a significant impact on all areas of its development. Today, the city's ecological potential is conducive to introducing the principles of environmentally friendly production and consumption, improvement, development of recreational and recreational areas and facilities, improvement of green areas and streamlining of public spaces accessible to all residents, improving mobility and environmental situation.

2 Methodology, Practical and Scientific Significance

To successfully use the existing environmental potential of the city, detailed expert analysis and forecasting of its proposed changes were carried out based on the SWOT analysis methodology. These analytical developments databases take into account the important results of expert assessments of the ecological state of the city made in previous years [4] and the results of environmental and public opinion polls. Environmental analysis is an important part of the strategic management planning process. SWOT means the complex of »strengths« , »weaknesses«, »opportunities« and »threats« (Fig. 2) [5]. Many scientists offer this analysis as an analytical tool that should be used to classify significant environmental factors, both internal and external.

Fig. 2 SWOT components

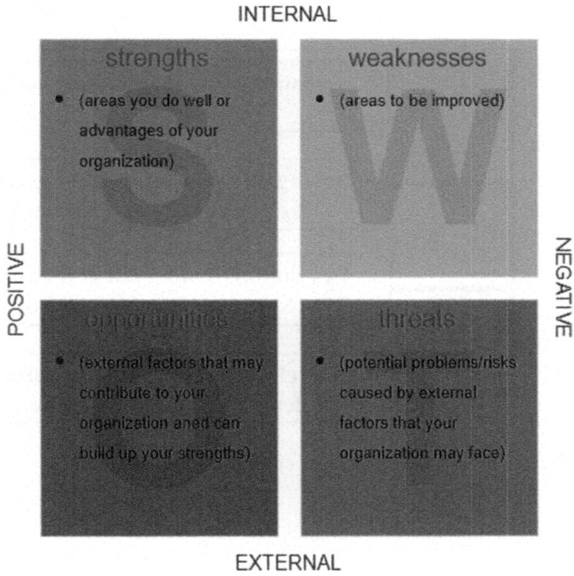

SWOT should not be considered as a static analytical tool with a focus on its results only. It should be used as an active part of the management and development process [6]. In other words, it is one of the bases for further planning.

The main practical guideline of these analytical studies is to identify the main directions of the environmental development of the city. It will ensure Poltava in the future to preserve the environment, available natural potential and improve the ecological situation. The directions of city development planning cited in the article are scientifically grounded and interrelated, which explains the reasons for setting the above goal and the need to achieve it.

3 Results

There are four main directions (Fig. 3), which according to experts and the results of environmental and public opinion polls should be considered in the strategic planning of cities [7–9].

Direction 1 Establishment of an information-analytical system of environmental monitoring and management in the city.

At present, monitoring of the environmental state of the city is not regular and is conducted with a limited list of individual indicators. It does not allow for an objective assessment of the current ecological state and to predict the impact of environmental factors on the development of the city. Accordingly, the information system for controlling parameters and objects of the environment is almost absent, which requires creation from scratch and a fundamental change in the approaches to its creation.

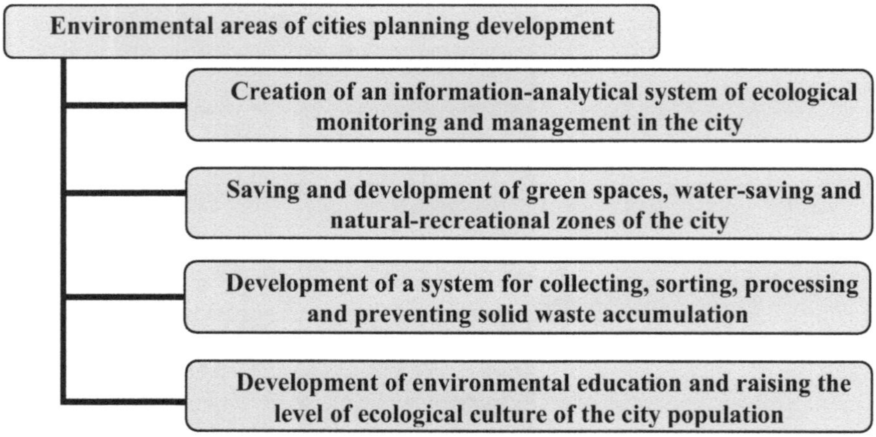

Fig. 3 Environmental areas of cities planning development

The main objectives (tasks) are the following:

1. Development of the program and organization of an independent system of information support, monitoring and laboratory control of parameters and objects of the environment: air quality and greenhouse gases level; ecological status of surface water bodies; hydrogeological situation and quality of drinking water; quantitative and qualitative parameters of the water supply and sewerage system, in particular the efficiency of wastewater treatment; targeted use and level of soil contamination; climatic parameters; noise impact on the environment.
2. Organization of an early warning system and prompt response to changes in the state of objects and environmental parameters.
3. Creating an effective environmental management system.

The information-analytical system of environmental monitoring and management is planned to be organized throughout the city and the surrounding suburban areas concentrating on the most important objects and zones in terms of their impact on the general ecological state and living conditions of the residents [10].

The control of air quality and the level of greenhouse gases will be carried out in the first place:

(1) along the main busy highways of the city;
(2) on the border of sanitary protection zones of existing industrial enterprises of the city;
(3) in areas of dense residential buildings;
(4) in all park areas of the city.

The monitoring of the ecological status of surface water bodies will cover all water bodies in the city and surrounding suburban areas, in particular, the Vorskla River within the city territory and the pond in the land of the Dendropark, the culture and recreation park »Peremoga«, »Pushkarivska Balka« [11].

The control of the hydrogeological situation and the quality of drinking water will be carried out:

(1) in the zones of sanitary protection of five water intakes of the city;
(2) five water intakes straight before the city water supply system;
(3) directly in the places of public water supply (»residential taps«).

Controlling the quantitative and qualitative parameters of the city's water supply and sewerage system, in particular, the degree of wastewater treatment will include monitoring the composition of the return water after treatment at the Zaturinsky and Suprunivsk treatment plants and the composition of the city's storm sewage.

The control of the targeted usage and level of soil contamination will be carried out: in residential areas, in park zones, at the border of sanitary-protective zones of industrial objects and enterprises, in water protection zones, within nature-protected areas.

Climate control will be phased in throughout the city, but first and foremost that it will be set up in residential areas and recreation areas.

Control of noise impact on the environment of the city will be primarily carried out along the main highways and in park areas surrounded by highways.

Direction 2 Saving and development of green spaces, water-saving and natural–recreational zones of the city and their components.

In order to ensure the sustainable development of the territory and the ecologically balanced principle of environmental management, the spatial structure of the natural ecological framework, including areas and sites of the nature memorable parks (NMP)—207.36 ha, water bodies—about 94 ha, coastal protective strips—about 355 has been partially created, DLF forests—61.13 ha, green spaces for public use, recreational areas, etc. However, the "green frame" of the city requires constant saving and renewal, which will significantly improve living conditions, improve the urban environment and increase its environmental resistance to human-made loads.

The critical tasks for the implementation of this direction development are the following: inventory and removal of the boundaries of the nature green areas, water conservation and nature recreational zones; scientific research on the resilience of green spaces to climate change; development of the program of maintenance and development of green spaces, objects of the nature reserve fund, natural and recreational potential and its realization; development of the Dendropark—monuments of landscape art of general importance "Poltava City Park"; landscaping and organization of landscaping in the territories of former and existing industrial enterprises; creation of resistant to environmental influences of city zones; increasing the level of utilization of the potential and the role of KP "Decorative Cultures" of NMP.

The basic approaches aimed at the development of green spaces, water conservation and natural–recreational zones of the city are the following [12]:

- optimization of green spaces in the city under the etiology of their habitat and resistance to biogenic and anthropogenic factors, in particular climatic factors, and establishment of appropriate ecological links;

- zoning and mapping of the city territory by the level of sustainability of green spaces;
- mapping of biotopes as biodiversity habitats in the city and surrounding areas;
- zoning, ordering and arrangement of Poltava City Park zones under the requirements of the current legislation and bringing its boundaries into nature;
- optimization of the collection block of dendroflora of Poltava City Park;
- creation of new, streamlining and expansion of existing lawn zones in the city from drought-resistant and grass mixtures;
- formation of further water-saving and natural–recreational zones necessary for the rehabilitation of the atmospheric air, consolidation of soil from erosion on steep slopes and organization of complete rest of the city population;
- landscaping of gardens and parks in the territories of former and existing industrial enterprises;
- organization of regular wetting of the city's green space during the hot season;
- organization of vertical landscaping of the city;
- organization of measures for the prevention and rehabilitation of green spaces, that are subjected to orito- and mycoinvasions;
- coordination of the areas mentioned above of work by the Information and Monitoring Ecological Center (IMEC).

Direction 3 Development of collecting, sorting, recycling and preventing the accumulation of solid household waste.

In the Poltava, solid household waste (SHW) is one of the ambiguous and challenging environmental and socioeconomic solutions. Approximately 600,000 m^3 of the landfill are generated annually and dumped over 3 million tons, which has a significant negative impact on all ecological components and impairs the living conditions of people in nearby villages. Therefore, the municipal landfill for its ecological status should not accept waste but requires professional closure and technical and biological reclamation.

SHW sorting is not centralized but is done manually at the municipal landfill site with the involvement of third-party contracting entities. The process of separation of hazardous wastes from the mixed mass and their future separate storage and processing remains uncontrolled. An important problem in the city remains oversized waste, which causes the formation of unauthorized landfills.

Therefore, establishing an effective waste management system requires concerted action on the part of the government, business, community and every citizen [13].

The following tasks will be key to the development of a waste management system in the future: developing a solid waste management strategy and program for the Poltava for the period up to 2030 and establishing a monitoring system for its implementation; creation of an administrative organizational system of waste management based on municipal cooperation, including hazardous, medical, construction waste and introduction of separate collection of resource fractions of waste; definition of new territory for solid waste landfill; professional closure and rehabilitation of the existing landfill [14].

It is planned to develop the effective waste management system in Poltava by next steps:

- organization of municipal cooperation in the field of waste management;
- creating a public–private partnership to address waste management issues effectively;
- setting up a municipal utility to address waste management issues;
- optimization analysis of decisions regarding the choice of sorting and processing technologies;
- construction of solid waste sorting and processing complex;
- organization of a monitoring system for soil, groundwater and atmospheric air in the area of the closed waste landfill;
- organizing educational and preparatory work among representatives of municipal enterprises and official local self-government bodies, public activists, etc., on the issues of solid waste management [13].

Direction 4 Development of environmental education and raising the city population ecological culture level.

Within the last 10 years, ecological and educational activities belong to the priorities of the regional environmental policy. This priority has led to the formation of effective public organizations and the formation of an active youth ecological movement, the creation of better conditions in the higher educational institutions of the region for the training of specialists–ecologists, active involvement of scientists, academic community and using the potential of ecological students in the development of regional programs on environmental areas [8, 9].

Increasing the level of population culture is possible within the conditions of a systematic approach [15]. Accordingly, it is relevant to create a multi-level system of environmental information and education for all segments of the population and visitors of the city. It can be done by organizing an expert assessment of the ecological situation; organization of the system for informing people about the environmental situation in the city; organizing a network of raising the level of qualification of officials in matters of ecological management; developing a program of environmental education for children and teenagers.

The development of environmental education and raising the level of ecological culture are planned to be implemented among the entire population of the city with the coverage of suburb residents. The policymakers' groups are student and school youth, educators, representatives of public organizations of different directions, representatives of local self-government bodies and state authorities.

The development of environmental education and raising the levels of the ecological culture of the city population will be implemented through:

- establishment of an independent information-monitoring activity structure and an expert environmental council;
- establishment an Emergency Environmental Assistance Service;

- organization of cooperation between scientific institutions, higher educational establishments, public organizations, state ecological institutions for the formation of the system of professional training on issues of environmental management;
- organization of voluntary movement of social and environmental orientation;
- implementation of the environmental education program for children and teenagers;
- organizing community outreach and awareness campaigns on solid household waste management;
- creation of a "Visiting Center" in the territory of Poltava City Park for centralized environmental events in the city.

4 Conclusions

The presented analytical finding is a component of the latest comprehensive studies of the current state and prospects Poltava development, based on the SWOT analysis data and takes into account the results of conducted ecological and sociological surveys and professional expert assessments. The selected directions (goals) of environmental development are the basis of the concept of integrated city development. These findings are a science-based basis for the development of investment projects.

References

1. European Union. (2015). In-depth report: indicators for sustainable cities.
2. Wced our common future: report of the world commission on environment and development Oxford university press, Oxford. (1987). Especially point, particularly Chapter 2, III, pp. 1–300. Online via http://www.un-documents.net/wced-ocf.htm.
3. Food and agriculture organization of the united nations. (2002). Pressure-state-response framework and environmental indicators. Available from: http://www.fao.org/ag/againfo/programmes/en/lead/toolbox/refer/envindi.htm.
4. Golik, Y., & Illiash, O. (2014). *Environment of Poltava region*. Kopi center, Poltava, p. 265.
5. SWOT analysis tutorial. Available from: https://online.visualparadigm.com/tutorials/swot-analysis-tutorial/.
6. Pickton, D. W., & Wright, S. (1998). What's SWOT in strategic analysis? *Strategic Change, 7*(2), 101–109.
7. Poltava 2030. Conception of integrated development of the city. Available from: https://drive.google.com/file/d/1Rzf_AaUJ29PNKfNoDts2Ar4BF7Z7Nt1D/view.
8. Onyshenko, V. O., Golik, Y. S., Illiash, O. E., Stepova, O. V., Smoliar, N. O., Bredun, V. I., & others. (2017). *Regional program for environmental protection, rational use of natural resources and ensuring ecological security, taking into account regional priorities of Poltava region for 2017–2021 (Environment–2021)*. Poltava, p. 131.
9. Holovko, V. A., Onyshenko, V. O., Golik, Y. S., Illiash, O. E., Pruglo, O. E., & Reva, E. V. (2017). *Comprehensive program for solid waste management in Poltava region for 2017–2021*. Poltava, pp. 134.

10. Golik, Y., Illyash, O., & Maksiuta, N. (2018). Urban "heat-island effect" and its connection with architectural and climatic features on the example of Poltava. *International Journal of Engineering and Technology, 7*(3.2), 597–601. https://doi.org/10.14419/ijet.v7i3.2.14598.
11. Stepova, O. V. (2016). Analysis of phosphate pollution of surface water bodies of Poltava region. *Bulletin of the Kharkiv National Karazin University Ecology series*, (14), 78–82.
12. Smoliar, N. O. (2018). Urban green zones planning concept and ecological functionality (through the example of Poltava, Ukraine). In N. O. Smoliar, V. I. Bredun, & O. M. Toronchenko (Eds.) *International Journal of Engineering and Technology, 7*(3.2), 522–527.
13. Tolkovanova, V. V., & Illyash, O. (2018). Solid waste management in the context of local self-government development and inter-municipal cooperation. *Training manual*. Kyiv, p. 367.
14. Golik, Y., Illyash, O., & Bilous, M. (2017). Subregional solid waste management strategy for Poltava region. *Journal Environmental Safety, 23*(1/2017), 20–25.
15. Golik, Y., & Illyash, O. (2002). Formation of professionals training system in the field of waste management/*The collective monograph The development of the natural sciences in the countries of the European Union in view of the challenges of the 21st century* (pp. 15–19). Latvia Izdevnieciba Baltija Publishing.

Ways to Improve the School Buildings Capital Fund

Vadym Kutsevych, Halyna Osychenko, Volodymyr Rusin, and Olga Tyshkevych

Abstract The article investigates effective directions of improving existing rural school buildings capital fund (on the example of Ukraine). The study was performed using an algorithm consisting of four steps. The study considered the prerequisites for improving the network and types of rural school buildings. Based on the analysis of modern socio-pedagogical concepts, the requirements for the material and technical base of schools in rural areas have been identified. The authors considered the existing fund of school buildings in rural areas of Ukraine regarding its compliance with modern social and pedagogical requirements. The study found that the main areas for improving the capital stock of school buildings in rural areas are modernization and reconstruction. The methodology of reconstruction works has been expanded; the method of calculating the effectiveness of reconstructive measures has been improved. It is revealed that the cost of reconstruction depends on the technical condition, building type and the number of premises that are insufficient (unnecessary).

Keywords Rural school building · Network · Reconstruction · Modernization

V. Kutsevych
Kyiv National University of Construction and Architecture, 31, Povitroflotsky Avenue, Kiev 03037, Ukraine

H. Osychenko
O. M. Beketov National University of Urban Economyin, 17, Marshal Bazhanov Street, Kharkiv 61002, Ukraine

V. Rusin
State Enterprise “UkrStateBuildExpertise”, 1, Zygin Street, Poltava 36014, Ukraine

O. Tyshkevych (✉)
Poltava National Technical Yuri Kondratyuk University, 24, Pershotravnevyi Avenue, Poltava 36011, Ukraine
e-mail: olgatyshkevych3639@gmail.com

V. Onyshchenko et al. (eds.), *Proceedings of the 2nd International Conference on Building Innovations*, Lecture Notes in Civil Engineering 73,
https://doi.org/10.1007/978-3-030-42939-3_39

1 Introduction

In the conditions of reforming the educational sector of Ukraine, there is a need to improve the network and buildings of educational institutions [1–3]. In the country, the vast majority of educational institutions are rural general secondary education institutions. In the total number of institutions, such schools make up two-thirds (11,300), with about 1 million students enrolled [4]. The problem of reorganization of the material and technical base of rural educational establishments in the conditions of limited economic resources cannot be solved without improvement, rational use of school buildings capital fund, which were erected in large numbers in the past years [5].

The existing school network in rural Ukraine has significant drawbacks. Mostly, rural general secondary education institutions are housed in buildings that are morally obsolete (Fig. 1). These buildings do not meet the modern socio-pedagogical requirements for the capacity of buildings, warehouse and floor space. This leads to a deterioration in the learning environment and quality of education of children living in rural areas [6].

All this was caused by measures to consolidate general secondary education institutions, the elimination of schools, the use of a limited number of school types and approaches to the school network formation that did not take into account local characteristics [7]. In the Soviet period leading research and design institutes, a significant number of scientists such as Kovalsky L.M., Sarkisov S.K., Stepanov V.I., Svitko V.A., Antoshkin V.F., Reshetnikova N.V. investigated the issues of school

Fig. 1 Existing school buildings in Ukraine. Left: a former zemstvo school in the village of Baranivka, Shishatsky district, Poltava region (the end of the 19th century was built); right: a school in the village of Velyka Ryblivka, Poltava region, construction of the 1980s. Source: Olga Tyshkevych's photo

network organization and calculation, the school buildings formation, the school buildings capital fund rational use. Some improvements have not lost their relevance even nowadays.

However, much of the past years research and recommendations cannot be used in modern conditions for rural construction in Ukraine due to significant changes in social, pedagogical and economic conditions. At the same time, the need to research the issues related to the use of the school buildings capital fund, the formation of a school network in rural areas in the new socio-economic conditions is extremely acute [7].

2 Purpose

The purpose of the article is to identify effective directions of improvement existing buildings of general secondary education institutions in rural areas on the example of Ukraine.

3 Research Methodology

In the course of the study, the school network and buildings of general secondary education institutions in rural areas was systematically considered in the light of current requirements. In developing proposals for the use of the capital fund of rural school buildings, modern social—pedagogical, sanitary and hygienic, socio—demographic requirements, specific conditions of rural settlement and cultural and community services to the population were taken into account [6].

The methodological algorithm of scientific research consists of the following steps:

- Identification of prerequisites for improvement of the network and buildings of general secondary education institutions in rural areas, namely social and pedagogical requirements for material and technical base, current tendencies of rural school buildings and network formation, analysis of the status of existing school buildings in rural areas;
- identification of modern ways of improving existing school buildings;
- development of scientifically substantiated proposals to determine the economic feasibility of undertaking measures to improve existing school buildings.

The methods of qualitative, quantitative, comparative analysis of modern pedagogical research, experience of innovative pedagogical activity were used to determine the socio-pedagogical requirements for the material and technical base of rural schools. Methods of statistical and factual analysis of literary sources, project documentation on the subject of research, field surveys and photos of existing school buildings were used to determine the status and characteristics of the existing material and technical base of general secondary education in rural areas.

4 Preconditions and Proposals for Improvement of the Capital Fund of Rural School Buildings

The solution to the problem of improving the capital fund of rural secondary education institutions buildings is connected with a thorough analysis of socio-pedagogical requirements in close relation with the socio-economic and logistical conditions of rural settlements of Ukraine.

The authors analyzed the existing capital fund of school buildings. It is revealed that it was formed mainly from buildings that were built in the 30–90 years of the twentieth century.

There are three groups in the existing school building fund:

- buildings built before 1965;
- buildings built in 1966–1975;
- buildings built in 1976–1990.

The first group consist of zemstvo (Local government) schools and church-parish schools which were built at the end of the 19th century, as well as the Soviet period school buildings of the late 50s and 60s on typical projects. These typical designs met the standard design and construction requirements of 1958. The most widespread were the buildings, erected on typical projects of schools for 80 students—№ 2-02-40, for 160 students—№ 2-02-41, for 192 students—№ 2-10 M-5, for 320 students—№2-10 M-6, for 400 students—№260 and №357-2 (Fig. 2).

The second group is characterized by the massive construction of school on typical projects that were designed in accordance with the rules of designing schools in 1962. In Ukraine, the most widespread are the buildings erected on a typical project for 192 students—№224-1-160. In the third period, the school buildings construction was carried out on typical projects that met the design standards of 1973. During the same period, the reconstruction of building that were built the first and second periods was started. The largest number of school buildings in this period was built on typical projects for 192 students—№224-1-160/75, for 320 students—№224-1-182, for 464 students—№224-1-135, for 624 students—№224-1-134, for 784 students—№ 224-1-221, for 1176 students—№224-1-88 (Fig. 3).

Due to the fact that typical projects were designed in accordance with the normative requirements of previous years, the composition and space of the premises of these rural school buildings does not meet the modern requirements and needs improvement.

Improvement of morally (functionally) outdated buildings, improvement of their composition and space of premises in accordance with modern requirements should be carried out in compliance with the requirements of insulation, aeration, sound

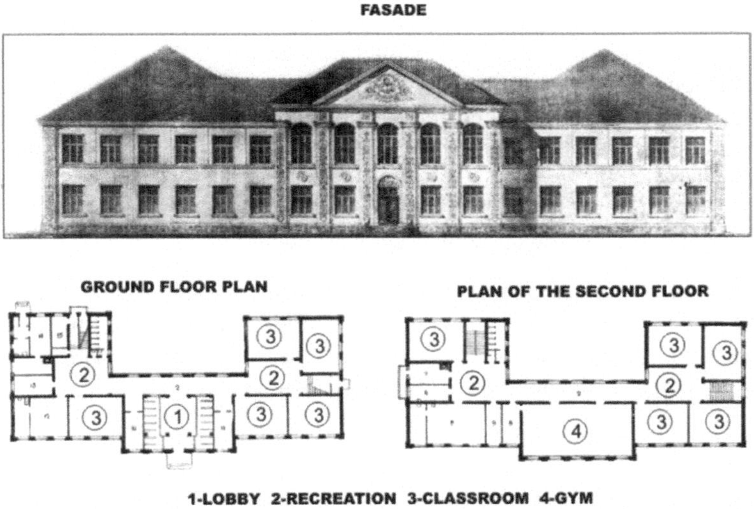

Fig. 2 The typical project of school building for 400 students—№260. Source: PAT "KYIVZNDIEP" Archive, Part Number 1023/1953

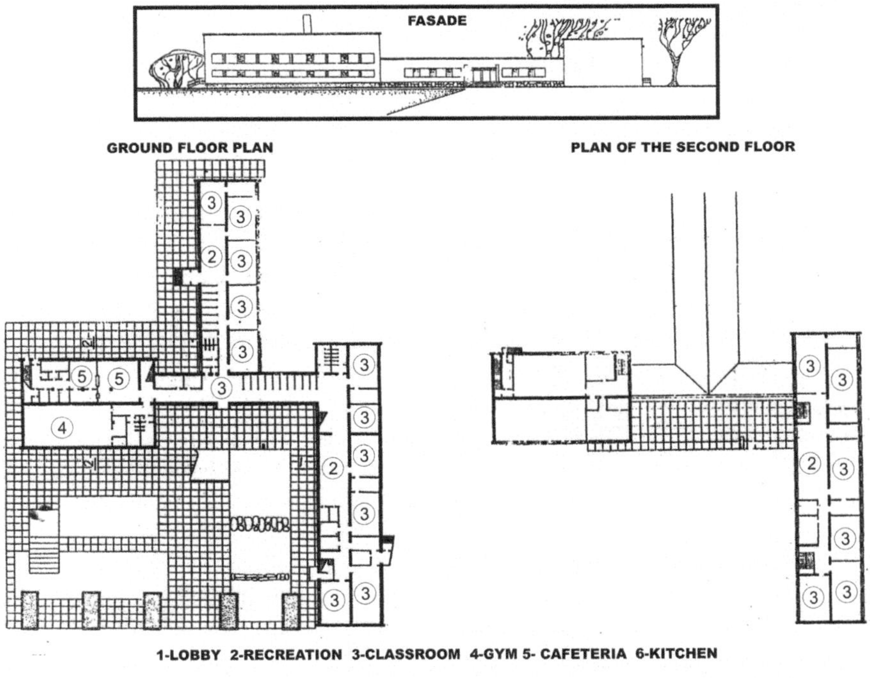

Fig. 3 The typical project of school building for 624 students—№244-1-134. Source: author's own resources

insulation and fire safety [8, 9]. For this purpose, it is possible to use two directions of rural school buidings capital fund improvement, i.e., modernization and reconstruction [7, 10].

The modernization involves the preservation of the building functional purpose in the existing building dimensions, the redevelopment and relocation of non-supporting structures. It is recommended that this direction of improvement be used in the event that the school building is full.

Reconstruction can be done in two types:

- with partial or complete alteration of the building functional purpose in the existing building dimensions, redevelopment and relocation of its non-bearing elements;
- preservation or alteration of the building functional purpose with the use of building blocks and simultaneous redevelopment [10].

The feasibility of the school buildings reconstruction and modernization is determined on the basis of a comparison of previously developed options, their feasibility study. Determination of expediency of carrying out works on modernization or reconstruction should be performed in the following sequence:

- determination of the general secondary education institution type, number and filling of classes, total number of students;
- determination of the list and areas of premises that are insufficient (extra);
- expert evaluation of the building technical condition, intended for modernization or reconstruction;
- development and analysis of options for school building modernization or reconstruction, calculation of economic efficiency [8, 10].

Type of general secondary education institution, number and occupancy of classes, total number of students, type of institution with which integration of general secondary education institution is possible is determined by calculating the school network [11]. It is necessary to take into account the characteristics of the existing general secondary education and cultural institutions network, factors that influence its formation: prospective changes in the settlement system, road transport infrastructure of the area, the number and population of rural settlements, the intensity of economic and transport links [11, 12]. The calculation of the network is performed on the basis of real indicators of the existing and perspective children's contingent, which will allow to determine the capacity of school buildings and occupancy of student groups [6].

Determination of the composition and area of the premises is carried out in accordance with the types of establishments that will be placed in the building and modern architectural and planning requirements for their design [8].

The building technical condition depends on the building capitality and the degree of structural elements deterioration. Capacity of the building depends mainly on the durability and degree of building fire resistance. Physical (technical) deterioration of a building occurs as a result of partial or complete loss of its primary technical qualities, deterioration of the working capacity of individual structural elements and so on. The degree of wear and tear depends on the nature and geometric parameters of the structures, the physical qualities of the materials, the quality of the construction work, the features of the location of the building on the ground, the conditions and terms of operation [10].

Visual, mechanical methods, laboratory testing methods of individual samples, field tests of structures, physical and complex methods are used to determine the degree of building structural elements deterioration. The most accessible and versatile is the visual method. It allows to determine the condition, degree of wear of structures and individual units without special devices and equipment. Determination of the building deterioration degree begins with the main supporting elements—foundations and walls. Their condition can correspond to three stages. The first stage is characterized by a 20% wear of the masonry. In this case, its solidity remains. In the second stage (masonry wear is 21–40%), there is a loss of adhesion of the stone with the solution; the masonry is divided into several elements, but the strength of the solution remains. The third stage (masonry wear is 41%) is characterized by the presence of hair cracks in the stone, loss of strength of the solution. Buildings with a deterioration degree of the masonry and walls more than 60% are not subject to reconstruction [10].

To determine the economic feasibility of school building reconstruction, the cost of reconstruction and new building construction are compared with the terms of their next operation:

$$\frac{Vr}{Tr} \leq \frac{Vn}{Tn} \tag{1}$$

where, Vr – the cost of the building reconstruction, Tr – the period of subsequent after reconstruction, Vn – the cost of new similar building construction, Tn – the term of subsequent operation after construction [10].

The economic efficiency of the school buildings reconstruction in general is defined as a comparison of the obtained useful result with the costs necessary to achieve the goals:

$$E = (Kr - Kn) \geq 0 \tag{2}$$

where, Kr – the total useful results obtained the conduct reconstruction, Kv – the total capital investment required to achieve this result [10].

According to the analysis of design documentation of the school building reconstruction projects performed by PAT "KYIVZNDIEP," the amount of costs for the dismantling of the structural elements is generally small in total and is about 2.7% of the total cost of the total construction work. The cost of materials suitable for later use is 40% of the total cost of construction works.

Choosing the best option for improving existing school buildings is based on comparing the results of economic calculations [10, 13]. Preference is given to a solution that achieves greater economic impact or prioritizes an indicator that is recognized as the main one. As the analysis of the project documentation shows, the solutions that provide the least changes are the most effective. This is especially true of rooms and units that are saturated with engineering equipment (physical and chemical laboratories, sanitary facilities, etc.).

5 Scientific Novelty

The scientific novelty is that:

- promising areas for improving existing school buildings in modern socio-economic conditions have been identified;
- the method of reconstruction works has been improved;
- the method of determination has been improved.

6 Practical Importance

The study found that the capital fund of rural school buildings in Ukraine was formed mainly in the second half of the twentieth century. It consists of buildings designed according to the normative requirements of the past; their composition and the area of the premises do not meet modern requirements. The recommendations developed will help to improve the rural school buildings capital fund.

7 Conclusions

Modernization and reconstruction are promising areas for improving the capital fund of existing school buildings in rural areas. The feasibility of conducting these activities is established based on project proposals analysis and economic justification. For this purpose, the composition and area of the premises, which depend on the general secondary education institution type and its role in the school network structure are consistently determined, the building technical condition is evaluated, and project proposals for the building improvement are developed.

The economic calculation of the reconstruction or modernization feasibility is performed on the basis of determining the construction work cost, as provided by the project. The most effective are the project proposals, which imply a small number of engineering-saturated units changes, significant completions of insufficient premises and planning decisions changes.

References

1. Law of Ukraine. (2002). On the National Doctrine of Education Development. № 347/2002.
2. Law of Ukraine. (2017). About education. № 2145-VII.
3. Law of Ukraine. (1999). About general secondary education. № 651-XIV.
4. Socio-economic situation of rural settlements of Ukraine: Stat. Collection. (2016). Kyiv, UA: State Statistics Committee of Ukraine.
5. Boarin, P., & Davoli, P. (2015). Deep renovation of the school building stock: the European opportunity and the Italian strategy. *Journal of Technology for Architecture and Environment, 0*(9), 96–105. https://doi.org/10.13128/techne-16110.
6. Tyshkevych, O. P. (2010). *Architectural-planning organization of rural small schools (Extended abstract of candidate's thesis).* Available from Vernadsky National Library of Ukraine. (RA373035).
7. Stepanov, V. I. (1975). *School buildings.* Moscow, USSR: Stroyizdat.
8. Methodical recommendations for the design of institutions of general secondary education. (2017). Kyiv, UA: PAT "KYIVZNDIEP".
9. Perkins, B. (2001). *Elementary and secondary school.* New York, NY: Wiley.
10. Guidelines for the Improvement and Use of the Rural Secondary Schools Building Fund. (1991). Kyiv, UA: UkrNIIPgrazhdanselstroy.
11. Tyshkevych, O., & Obidniy, A. (2018). The main approaches for increasing the efficiency of the rural school network. *International Journal of Engineering and Technology(UAE), 7*(3.2), Special Issue 2, 686–691. https://doi.org/10.14419/ijet.v7i3.2.14614.
12. Guryanova, M. P. (2003). Reserves for the modernization of the rural small-scale school in Russia. Moscow, RU: IPSR RAO.
13. Congedo, P. M., D'Agostino, D., Baglivo, C., Tornese, G., & Zacà, I. (2016). Efficient solutions and cost-optimal analysis for existing school buildings. *Energies 2016, 9*(10), 851. https://doi.org/10.3390/en9100851.

The Spatial Arrangement and Structural Solutions Concept of a Small Rural Public Building in Ukraine

Tetiana Kuzmenko, Vasyl Liakh, and Andrii Dmytrenko

Abstract Purpose of the article is to develop the small rural public building spatial arrangement and structural solution concept that meets following requirements: compactness; planning decisions flexibility; inclusivity; construction scheme integrity and rigidity; construction and operational building costs reduction; materials consumption reducing with their energy efficiency increasing; construction time reducing; durability; labor costs reducing and heavy machinery use reduction. The research methodology involves a comprehensive approach in formulating a concept, based on literary sources analysis and domestic and foreign experience, identifying the criteria that it should meet. Actual rural public buildings types are identified: cooperative public buildings; libraries; administrative service centers; clinical primary care centers; public safety centers. The small rural public building spatial arrangement solution is determined: one-story, maximum with one inner bearing wall, low plinth and pitched roof with large overhangs. The basic structural elements characteristics are determined: bases of bored and cast-in-place piles, connected by monolithic reinforced concrete grill; insulated floor on soil, which upper coupler structure allows partitions direct bracing; single-layer walls made of light concrete blocks; prefabricated reinforced concrete floor slabs; wooden rafter system; simple, double-slope roof, covered with steel profiled sheets with a protective coating. The scientific novelty is to formulate a requirements list for small rural public building spatial arrangement and structural solution, which corresponds to rural united territorial communities' modern needs and capabilities. The developed concept introduction into design and construction practice throughout Ukraine will enable territorial communities to solve a large range of tasks independently according to available resources.

Keywords Current nomenclature of rural public buildings · Spatial arrangement and structural solution concept · Inclusivity of buildings · Flexibility of planning solutions · Unification

---

T. Kuzmenko · V. Liakh · A. Dmytrenko (✉)
Poltava National Technical Yuri Kondratyuk University, 24 Pershotravnevyi Avenue, Poltava, Ukraine
e-mail: metr5555@ukr.net

V. Onyshchenko et al. (eds.), *Proceedings of the 2nd International Conference on Building Innovations*, Lecture Notes in Civil Engineering 73,
https://doi.org/10.1007/978-3-030-42939-3_40

1 Introduction

The state policy of Ukraine in the local self-government field envisages power decentralization. That means the transfer of a large part of the powers, resources and responsibilities to the united territorial communities (UTC) and rests on the interests of the rural settlements inhabitants. This policy is based on the provisions of the European Charter of Local Self-Government and world standards for public relations. Within the scope of their mandate, and in accordance with the resources available, UTCs today address a number of economic, environmental and social problems, including, inter alia, the development of community infrastructure and the provision of quality services to the public over the full public services range. The infrastructure development and the expansion of services, including administrative ones, necessitate the forming of a rural public buildings new topical types nomenclature, which would be economically and energy efficient and would meet the new, more stringent norms of inclusivity. Therefore, the issue of developing the concept of a spatial arrangement planning and constructive solution of a small public building for a village that meets the resource and economic capabilities of the united territorial communities and is in line with the Energy Strategy of Ukraine for the period until 2035 [1].

Researches in the field of effective economic structures of low-rise buildings were carried out by domestic and foreign authors: Dudykevych [2], Melnyk et al. [3], Sanytskyi et al. [4], Tymchenko et al. [5], Lobodenko et al. [6], Date et al. [7], Cheng et al. [8], Craig and Grinham [9].

Modern foreign practice of design and construction of small rural public buildings is covered in sources [10–15].

Features of actual public rural buildings architectural and planning organization were investigated in the works of Lytvynenko [16], Kuzmenko et al. [17–19], Shulyk et al. [20].

However, the most cost-effective, energy-efficient, and inclusive spatial arrangement and structure solutions for small public buildings, most relevant to rural settlements today, have received insufficient attention in existing research and design practice.

2 Main Part

2.1 Purpose of the Article

The purpose of the article is to develop the small rural public building spatial arrangement and structural solution concept that meets following requirements: compactness; planning decisions flexibility; inclusivity; construction scheme integrity and rigidity; construction and operational building costs reduction; materials consumption reducing with their energy efficiency increasing; construction time reducing; durability; labor costs reducing and heavy machinery use reduction.

These requirements meet the capabilities of the united territorial communities and allow in the shortest possible time to complete the construction and commissioning of a public building on their own.

2.2 *Research Methodology*

The research methodology involves a comprehensive approach in formulating a concept, based on analysis of literary sources and domestic and foreign experience, identifying the criteria that it should meet, and the development of this complex architectural and structural a solution that covers all stages of the construction of the building and to a certain extent operational costs and meets the capabilities of the united territorial communities in providing infrastructure tours of rural settlements and satisfaction of services to the population.

2.3 *Results*

The Main Types of Buildings The following types of public buildings relevant to the modern Ukrainian countryside can be distinguished: a cooperative building of a public center, a public safety center, an administrative services center (ASC), a clinical primary care center. If the first type usually includes large hall-type premises and the second—large garages for storing both firefighters and police special vehicles (and sometimes school buses), the last two types can be classified as small public buildings.

Basic Spatial Arrangements of Buildings Given the current requirements of inclusivity and the need to exclude the use of elevators/lifts for reasons of economy, such buildings should be exclusively one-story. On the second floor, it is advisable to have staff apartments. However, given the small size of staff apartments, the need to maximize its isolation from the public part of the building, the desirability of providing communication with the land plot, as well as sufficient, as a rule, the area of land, even when designing public buildings with office accommodation for staff, it is advisable to provide such housing also one-story, locking it horizontally to the public part. The locking can be done according to one of three schemes: linear, offset and angular (Fig. 1). Based on the requirements of compactness, it is necessary to design such buildings rectangular in plan, approximate in proportion to the square.

It is advisable to use a frameless structural system. Maximum design flexibility can be ensured by using single or double span designs (preferably with longitudinal load-bearing walls). The distance between the load-bearing walls is determined by the most common dimensions of reinforced concrete circle hollow floor slabs: 7200 mm for single spans and 6000 mm for two spans (Fig. 2a, b). In this case, a one-span

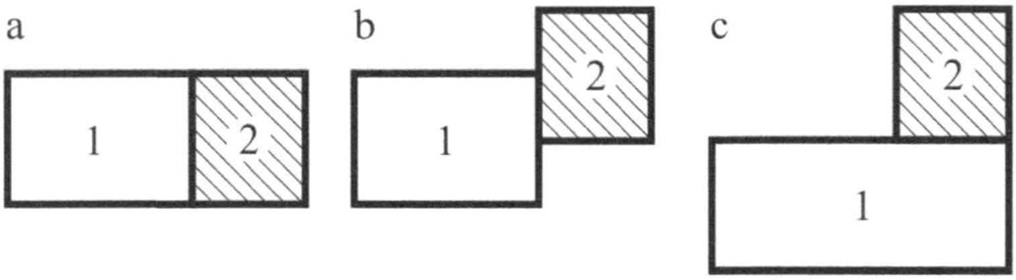

Fig. 1 Block diagrams of public (1) and residential (2) parts of rural public buildings with housing for staff: **a**—linear; **b**—offset; **c**—angular

scheme can be used for the smallest buildings (for example, ASCs), and a two-span scheme—for buildings of a larger size.

The height of the premises is assumed to be a minimum normatively permissible—3.0 m.

It should also be noted that in accordance with the new rules on the inclusivity of buildings [21, p. 13], with new construction it is necessary to provide all entrances and exits flush with the ground without arranging a porch, which certainly requires a deepening of the entrance node inside the building and leads to an increase in the transit area and, thus increasing the volume and making the building more expensive. Such a decision is relatively unacceptable for relatively small buildings. Planning the floor at the level of the land marking requires a complex of constructive (arrangement of storm sewage system with possible heating of the site before entering) and organizational (regular cleaning of the paving around the perimeter of the building from the snow. In rural areas, such measures may be too costly. Therefore, it is advisable to place the floor 0.45 m above the perimeter paving level, and to connect the ground before the entrance, limited by the retaining walls, to a 5% gradient (1:20) with a steep descent. It is also necessary to arrange a deep overhang above the entrance, which will create appropriate comfort for visitors and provide protection of the building from atmospheric precipitation and other adverse conditions.

Foundations and Floor Arrangement It is proposed to use bored and cast-in-place piles with a monolithic reinforced concrete grill, which provides rigidity and integrity of the structure, reduces the building mass and, accordingly, reduces the cost of materials and construction. The grill forms the basement of the walls, which is proposed to be insulated with extruded expanded polystyrene. The floor is offered to arrange the combined on soil. In this case, a monolithic reinforced floor slab (100–150 mm thick) is proposed to be placed above the insulation of extruded expanded polystyrene. The plate should allow the internal partitions of brick (up to 120 mm thick), aerated concrete blocks (up to 200 mm thick), as well as of gypsum boards on a metal frame to support it.

Walls Modern requirements for the level of resistance of heat transfer of external walls ($R_{q\,min} \geq 3.3\ m^2$ K/W for most of the territory of Ukraine, except for the southern regions) significantly narrow the range of choice of materials for single-layer walls. In our opinion, light concrete blocks will be appropriate for rural areas,

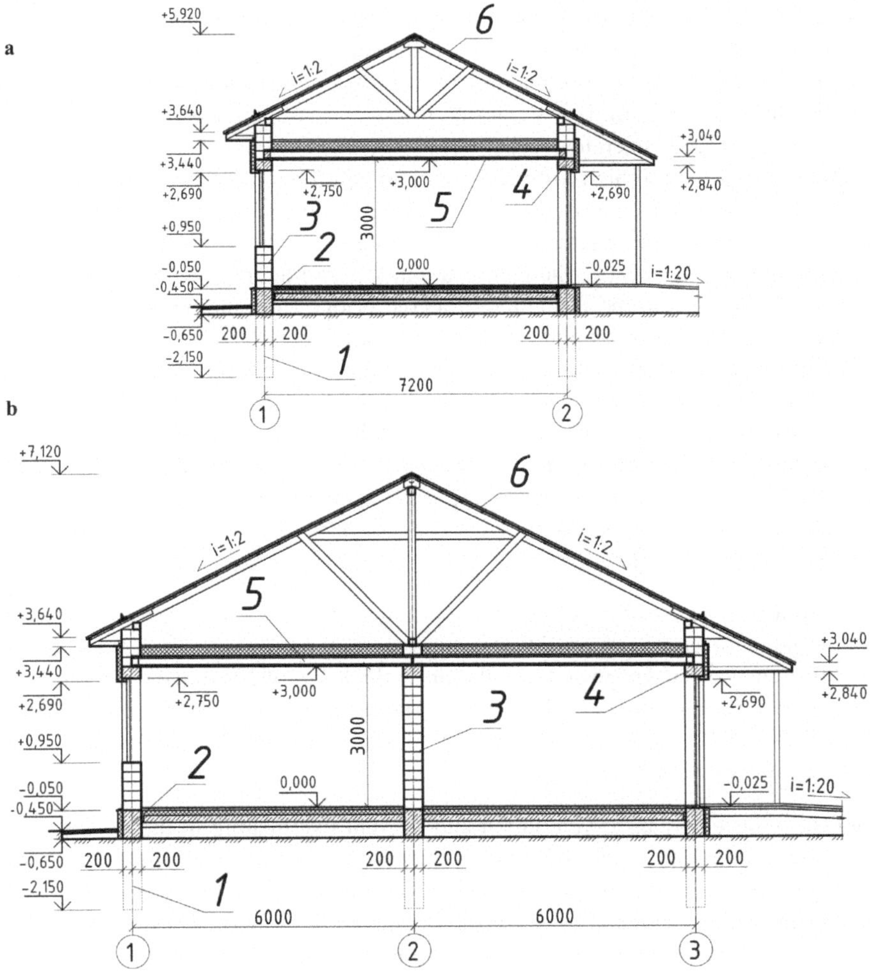

Fig. 2 Cross-sectional diagram of a small rural public building (**a**—one-span scheme, **b**—two-span scheme): 1—bored and cast-in-place piles; 2—monolithic reinforced concrete grill; 3—walls of aerated concrete blocks 400 mm thick; 4—monolithic reinforced concrete belt under the floor slabs; 5—concrete hollow floor slabs insulated with mineral wool slabs; 6—pitched roof of metal profiled sheets on wooden patches and rafters

as the most common material in Ukraine. They are quite lightweight, so they do not require heavy handling, and generally lead to savings in transport costs, which is very important for limited-capacity UTCs.

It should be noted that aerated concrete blocks are a fairly good insulation material. The thermal conductivity of D500 gas concrete in dry form is 0.12 W/m K, which is 4 times less than in solid brick (0.45–0.55 W/m K) and slightly below the thermal conductivity of wood (0.15 W/m K). Despite its excellent thermal properties, aerated concrete is a sufficiently strong material for erection of load-bearing walls. The most common aerated concrete blocks have a B2.5 compression strength class and can have a density of D350–D600. Such blocks can be erected supporting walls with a total height of up to 20 m. It is suggested to use 1–2 structural spans to reduce the weight of reinforced concrete overlap.

The use of a monolithic reinforced concrete belt to support the reinforced concrete hollow floor slabs allows not only to increase the strength and spatial rigidity of the building (including resistance to deformation of the foundations under the foundations), but also to use the belt as a window and door lintel in the outer and inner walls. To prevent the formation of a cold bridge from the outside, the belt is insulated with mineral wool plates at least 100 mm thick.

The use of aerated concrete blocks with a thickness of 400 mm allows to achieve resistance of heat transfer of walls not less than 4.16 m^2 K/W (without taking into account external and internal equipment). In addition, if necessary, the wall can be additionally insulated with mineral wool panels up to 150 mm thick, which will increase the heat transfer resistance of the wall to at least 7.9 m^2 K/W. Alternatively, heat-efficient multilayer polystyrene concrete wall blocks developed by PoltNTU scientists can be used [22].

Overlapping The use of proven standard products (6 and 7.2 m reinforced concrete hollow slabs) for such a responsible structural element ensures proper fire resistance, durability and ease of design and installation. A thickness of 200 mm of mineral wool slabs is proposed to provide a reduced overlap heat transfer resistance of at least 5.0 m^2 K/W (at $R_{q\ min} \geq 5.0$ m^2 K/W). The design of the overlap allows increasing the layer of insulation if necessary.

Roof In order to organize ventilation and prevent overheating in the winter, it is advisable to provide a roof attic with an increase in roof removal (up to 700 mm) in winter protection. It will allow to manage without means of drainage of rain and melt water, will create conditions for the organization of thermal insulation in the future. It also does not require the construction of an anti-ice system, which is very energy-intensive and expensive, especially in rural settlements and will be almost unrealistic to implement within the UTC. The most rational is a double-slope roof. The roof is made of profiled zinc-coated steel sheets, painted. Roofing system and patches can be made of local wood in the form of truss trusses (for one-span scheme) or in the form of sling rafters with support on the inner wall (two-span scheme). A 1:2 roof slope ensures good precipitation in all climatic zones of the country.

2.4 Scientific Novelty

The scientific novelty is to formulate a list of requirements for a large-scale planning and constructive solution of a small public building for rural settlements, which not only corresponds to the current normative documents of Ukraine in the field of construction, but also to the modern needs and opportunities of rural united territorial communities.

2.5 Practical Importance

The concept of a large-scale planning and constructive solution of a small community building for rural settlements can be used in the design of such buildings to solve the important social problem of developing the infrastructure of the village and expanding the full range of services for the population. The introduction of the developed concept into design and construction practice throughout Ukraine will allow the territorial communities to independently solve a large range of tasks within their powers and according to available resources.

3 Conclusions

Thus, the concept of small rural public buildings spatial arrangement and constructive solution has been developed, enabling the development of integrated territorial communities' infrastructure and extension of services to the rural population at low cost. This is of great socio-economic importance for the sustainable development of settlements.

References

1. Enerhetychna stratehiia Ukrainy na period do 2035 r. (2017). Bezpeka, enerhoefektyvnist, konkurentospromozhnist. https://zakon.rada.gov.ua/laws/show/605-2017-%D1%80.
2. Dudykevych, Y. (2013). *Enerhooshchadni kotedzhi: metodyky proektuvannia budynkiv bez hazu*. Lviv: Spolom.
3. Melnyk, A., Mavdiuk, A., & Pidlutskyi, V. (2017). Analiz zarubizhnoho dosvidu v perekhodi na masove zvedennia pasyvnykh budivel zarady enerhetychno nezalezhnoho maibutnoho. *Mistobuduvannia ta terytorialne planuvannia, 65*, 360–369.
4. Sanytskyi, M., Pozniak, O., & Marushchak, U. (2013). *Enerhozberihaiuchi tekhnolohii v budivnytstvi*. Lviv: Lvivska politekhnika.
5. Timchenko, R., Krishko., D., & Pluzhnik, A. (2017). Energeticheskaia nezavisimost zdanii. *Mistobuduvannia ta terytorialne planuvannia, 63*, 373–378.
6. Lobodenko, E., Mikhailova, E., & Gusev, K. (2018). Prospects for reinforced autoclaved cellular concrete production technology. *Vestnik MGSU, 13*(117), 740–748.
7. Data, A., Angeli, D., & Kalyanova-Larsen, O. (2017). Naturally ventilated double-skin fasade in modeling and experiments. *Energy and Buildings, 144*, 17–29.
8. Cheng, C. Y., Cheung Ken, K. S., & Chu, L. M. (2010). Thermal performance of a vegetaled cladding system on facade walls. *Building and Environment, 45*(8), 1779–1787.
9. Craig, S., & Grinham, J. (2017). Breathing walls: the design of porous materials for heat exchange and decentralized ventilation. *Energy and Buildings, 149*, 246–259.

10. The Rural Library Project: Building Libraries, Building Community. https://tandfonline.com/doi/abs/10.1080/01616846.2014.910721?src=recsys&journalCode=wplq20.
11. Capital Funding for Rural healthcare. https://www.ruralhealthinfo.org/topics/capital-funding.
12. What is good rural design? https://www.pps.org/article/what-is-good-rural-design.
13. Building the rural hospital of the future. https://www.healthcaredesignmagazine.com/architecture/building-rural-hospital-future.
14. Australia. A public building bringing together a rural community. https://www.domusweb.it/en/architecture/2018/10/03/australia-a-public-building-bringing-together-a-rural-community.html.
15. Service provision in rural areas concepts from vital rural area. http://www.vitalruralarea.eu/scientific-articles/238-service-provision-in-rural-areas-concepts-from-vital-rural-area.
16. Lytvynenko, T. (1999). *Formuvannia silskykh maloobiemnykh kooperovanykh budivel ta ikh konstruktyvni osoblyvosti (PhD dissertation)*. Poltava: Poltava State Technical University.
17. Kuzmenko, T., & Dmytrenko, A. (2019). Aktualni typy silskykh hromadskykh budynkiv v konteksti terytorialnoi reformy. In *Proceedings of the 10th International conference "Science and society" (March 15, 2019)* (pp. 313–321). Hamilton, Canada: Accent Graphics Communications and Publishing.
18. Kuzmenko, T., Dmytrenko, A., & Milikova, N. (2019). Arkhitekturno-planuvalna orhanizatsia klinichnykh tsentriv pervynnoi medychnoi dopomohy v silskii mistsevosti Ukrainy. *Arkhitekturnyi visnyk KNUBA, 17–18,* 502–508.
19. Kuzmenko, T., & Dmytrenko, A. (2019). Tsentr hromadskoi bezpeky yak aktualna hromadska budivlia dlia silskykh poselen. Arkhitekturnyi visnyk KNUBA, 17–18, 509–513.
20. Shulyk, V., Obidnyi, O., & Halchenko, O. (2016). Pro formuvannia tsentriv hromadskoi bezpeky, peredumovy I dosvid proektuvannia. Mistobuduvannia ta terytorialne planuvannia, *62*(part 1), 555–562.
21. DBN. (2018). V.2.2-40:2018. Inkliuzyvnist budivel i sporud. Minrehion Ukrainy, Kyiv.
22. Onyshchenko, O., Riabiko, H., Liakh, V., Dmytrenko, A., & Bondarenko, R. (2009). Teploefektyvnyi stinovyi blok. Patent of Ukraine for utility model No. 42123 dated June 25, 2009.

Application of the Modern Finishing Materials in Interiors of the Preschool Educational Institutions

N. E. Novoselchuk

Abstract The article is aimed to systematize the list of finishing materials and requirements which can be used for interiors of preschool institutions. In order to study the declared subject, factual, empirical, and systemic methods have been used in the article. The scientific novelty is in systematization and supplement of a list of finishing materials which can be used in interiors of the preschool institutions. The criteria of environmental friendliness, operational safety, compliance with fire safety requirements, and several other requirements are taken into account, depending on the functional purpose of a premise. The practical value is concentrated in obtaining the results that can help to make the microclimate of preschool institutions better and reduce the risks for children's health. Main issues associated with the use of hazardous finishing materials in children's preschool institutions, in particular, the problem of the formaldehyde migration level are highlighted in the article. The basic requirements are defined in the article. The list of finishing materials which can be used in these institutions without harming children's health based on the study of regulatory documents and scientific sources is systematized.

Keywords Finishing materials · Construction materials · Children's preschool institutions · Interior · Environmental friendliness

1 Introduction

The major development trends of global economy and trade comprise the extended range of all type consumable products differentiation, increment in wares transfer, growth of manufacturers' quantity using various technologies, augmentation synthetics, and non-natural materials in the production of commodities [1]. And today, the construction market offers a wide variety of options for finishing materials. Proper choice of such materials for children's institutions is a major task.

N. E. Novoselchuk (✉)
Poltava National Technical Yuri Kondratyuk University, Poltava, Ukraine
e-mail: NovoselchukNE@gmail.com

V. Onyshchenko et al. (eds.), *Proceedings of the 2nd International Conference on Building Innovations*, Lecture Notes in Civil Engineering 73,
https://doi.org/10.1007/978-3-030-42939-3_41

It is known that children spend much of their time in kindergarten. And it is essential that the ambient should be ecological and high quality. Various factors influence on that as follows: air quality, electromagnetic radiation, household chemicals and hygiene accommodations, substances defined by the emitted material substance existence, etc. The wrong choice of modern finishing materials is also dangerous. Some researchers [2–5] established that ammonia, acrylonitrile, phosphoric anhydride, butyl and vinyl acetates, hydrogen cyanide, hexamethylene diamine, complex of organic phthalates, aromatic hydrocarbons, acrylates, methyl, butyl, and isopropyl alcohols, formaldehyde, phenols, and a number of other impurities can be emitted from finishing materials into the human environment in significant concentrations [1].

The issue of finishing materials choice for the internal environment of preschool institutions is directly related to the indoor environment condition and is becoming increasingly relevant. It should be mentioned that in the context of the increase in the quantitative diversity of materials, the quality level significantly decreases in course of co-option for the kindergarten rooms finishing. Mainly, it is resulted because non-professionals are tolerated in the purchase process, and the choice of any material often depends on the taste preferences and financial capacity of the parents whose children attend the associated kindergarten. Meanwhile, the fundamental requirements for the finishing materials used for such type of buildings are ignored, notwithstanding that they are recorded in the regulatory documents of Ukraine and must be strictly adhered. The basic requirement hereunder is safety, which ensures both the operational characteristics of material depending on the type of the room, and its environmental friendliness, which is directly related to children's health. Failure to comply with this requirement may lead to the formation of undesirable risk factors that adversely affect children's health and cause various allergic reactions, as well as affect the emotional and mental state. A significant tool for creating a healthy environment in kindergartens is observance of the concept such as "ecological environment—children's health" and involvement of professionals in this process.

Scholars regularly report on the results of their research in numerous scientific publications. Builders, architects, designers, physicians, ecologists, and teachers who deal with the outlined problems publish the results of their research in order to improve the situation at the level of constructional, architectural design, and maintenance work in childcare facilities.

Some scientists at the empirical and theoretical levels investigate the environmental friendliness of the used finishing materials, declare the relevance, and demonstrate the importance of this problem. These are works of V. P. Knyazeva, B. V. Gusev, V. M. Dementiev, I. I. Mirotoretsev, K. L. Antonov, V. V. Ovchinnikov, N. V. Zaitseva, V. A. Bardonov, Yu. D. Gubernski, A. Hodgson, L. Yu. Caranastas, M. J. Mendell, K. B. Rumchev, and K. I. Stankevich et al. Issues on material science for architects are discussed in the following works: D. P. Ayrapetov, V. G. Mikulsky, V. E. Bayer, P. V. Krivenko, M. P. Burak, Yu. M. Tikhonov, Yu. P. Panibratov, Yu. G. Meshcheryakov, T. V. Sheina, etc. Works of such scientists are focused on the studies of the internal environment of preschool institutions formation aspects, including the competent use of the finishing materials: V. O. Ryzhikov, E. N. Dautov, D. M. Kharitonov, A. N. Stasyuk, E. G. Yakushev, I. S. Katunin, P. S. Barkov, N. V. Dryukov, A. O. Kadurina, N. B. Blokhin, L. T. Vikhrov, G. M. Davydov. Scientific works of S.M. Novikov, L.Yu. Karanastas, A.A. Gladkovskaya, K.B. Rumchev, M.J. Mendell, C. G. Bornehag, and others are focused on the studies of negative health consequences of consumers under the influence of substances that migrate from products to the environment.

Thus, available accessible research activities analysis, constructional and architectural sources, and architectural and engineering materials enabled the allocation of factual material for research. The field and visual inspection of the interiors at a number of Ukrainian preschool institutions field inspection carried out by the author enable to confirm the validity of the applied theoretical and empirical methods and the reliability of the results of scientific research.

1.1 The Objective of the Study

The objective of the study is to determine the finishing materials and requirements. These materials can be used in the interiors of preschool institutions consistent with fundamental requirement.

1.2 The Methodology of the Study

Based on the existing methods of scientific research experience, the following methods were used in the paper: factual—for study of systematization of the available constructional, architectural, literature, and scientific sources; empirical—basing on analysis and comparison of the studied objects, with due regard to the current requirements; systemic—enabling to study this issue comprehensively, taking into account all its components.

1.3 Scientific Novelty and Practical Importance

Scientific novelty and practical importance consist in the systematization and addition, a list of finishing materials that can be used for the interiors of preschool institutions. The criteria of environmental friendliness, operational safety, compliance with fire safety requirements, and a number of other requirements depending on the functional purpose of the facilities are taken into account. The obtained results will help to make the internal environment and the microclimate of kindergartens more healthy, which will help reduce the risks of children's health.

2 Results

Currently, the quality of raw materials for the production of building and finishing materials and the materials themselves is determined by such regulatory documents as the state building norms (DBN), the state standards of Ukraine (DST), SNiP, and specifications and is evaluated by technological and technical characteristics. The sanitary safety of finishing materials is determined by a set of sanitary-hygienic characteristics. They determine the potential hazard of the material to human health and compliance with sanitary requirements. The danger of finishing material may be manifested in air pollution of premise, or in direct human contact with such material. Deleterious health effect is the result of interactions combination between the material, environment, and individual [6].

The most dangerous material hazard factor for children's health is formaldehyde migration levels: in general, for all types of products, up to 16–25% for certain groups of goods (plywood, fiberboard, particleboard). The formaldehyde levels of 2.5 were registered in children's blood, which is higher than the comparison level ($p < 0.05$). Also among children being under conditions of the prolonged exposure, the presence of immunodependent inflammation is registered [1]. These scientific data are the source of serious concern for children's health and cause to select the finishing materials with all responsibility.

In order to prevent air pollution in children's educational institutions, it is extremely important to develop the industry of maxi safe finishing materials designed specially for kindergartens [1].

The master regulatory documents that regulate the use of finishing materials for kindergartens in Ukraine are the state building norms "Preschool Educational Institutions" B.2.2-4: 2018 and No. 563/28693 (Sanitary Regulations for Preschool Educational Institutions) approved on March 24, 2016. These documents indicate the main operational requirements and types of materials for various premises of these types of buildings [6]. In addition, the materials must meet fire safety requirements and several special requirements, depending on the functional purpose of the premise.

Requirements and materials for the interior decoration of the main premises of kindergartens are represented in Table 1. However, it should be pointed out that in the regulatory documents ignore data as for finishing ceilings. Also, finishing materials for the floor and walls are not indicated for all types of rooms, which are displayed in the table as empty cells (Table 1).

In this regard, based on interiors objects under investigation analysis, scientific, regulatory sources, and the existing market for finishing materials, the author has a supplemented list of the recommended materials for the interiors of preschool institutions.

Thus, the use of combustible materials is prohibited for walls and ceilings in the rooms which are used as evacuation routes (corridors, foyers, recreation areas, vestibules, stairwells). For that occasion, the best solution would be painting the walls with paint materials which form non-combustible film on the surface [8].

Table 1 Finishing materials and their requirements used in preschool institutions under the regulatory documents of Ukraine [7, 8]

Room designation	Floor finishing material	Requirements for the floor finishing material	Wall finishing material	Requirements for the wall finishing material
Common premises • Playroom • Changeroom • Bedroom	Homogeneous linoleum	• Humidity resistance • Base with glass wool • Low heat conductivity • Resistance to detergents and disinfectants • Non-slip surface • Safe heating systems are allowed		• Smooth surface • Humidity resistance • Possibility of wet cleaning • Light reflectance 50–70%
Tiring room	Ceramic tiles	Slip-resistant surface	Ceramic glazed tile	Minimal height of tile paving 1.5 м
Halls for PE and music	• Parquet board • Homogeneous linoleum		Non-combustible materials	• lMoisture resistance • Possibility of wet cleaning
Swimming pools Showers Toilets	• Ceramic tile • Terrazzo concrete	Non-slip surface	Ceramic glazed tile	Minimal height of tile paving 1.8 м
Medical facilities		Possibility of wet cleaning		
Administration premises Educational premises	• Homogeneous linoleum	• Moisture resistance • Possibility of wet cleaning • Light reflectance 50–70%		
Food department	Ceramic tile	• Non-slip surface • Minimal slope to the sewage system 0.03%		• Moisture resistance • Possibility of wet cleaning • Minimal height of tile paving 1.8 м

Latex and silicone paints are recommended materials for the walls for large premises, halls for athletics and music-making, and office premises. They are made using water-dispersed components, and therefore, do not form combustible films, toxic substances, and do not thermally decompose in case of fire. These paints are abrasion-resistant and enable wet cleaning with non-aggressive non-abrasive detergents [9].

It is also possible to use the fiberglass wallpaper, which has a glass base and consists of fibers of quartz sand, soda, lime, and dolomite—environmentally friendly minerals. Such fiberglass does not enable breeding ground for microorganisms; it is characterized by strength, water tightness, environmental friendliness, and fire resistance. This coating, "breathes," does not accumulate the static charge, is not susceptible to infection by fungi and mold, and helps to maintain microclimate in the room. Ceilings in the high air humidity rooms (food processing department, showers, laundry rooms, washrooms, toilets, etc.) should be painted with oil paint [10], Fig. 1.

All materials used for the interior decoration of the preschool educational institutions should have a positive conclusion by the state sanitary-epidemiological expertise.

In addition, it is necessary to turn attention to the specialized and innovative materials and technologies which can be used in the preschool institutions. These are the bioactive ceramics (innovative self-cleaning ceramic material Bios), developed by specialists from Casalgrande Padana in Italy. This material is able to reduce the number of four major bacterial strains by 99.9%, as evidenced by the Department of Microbiology of the University of Modena, Italy [11]. The base of the material is silver with antibacterial properties, and Bios does not contain health chemicals harmful to human.

The HYDROCERA technology merits attention, it breaks down bacteria that cause unpleasant odors and yellow spots by means of hybrid photocatalyst technology (a combination of photocatalyst and antibacterial metal) [12], Figs. 2 and 3.

It is also possible to use the antibacterial disinfectant paint with the addition of silver ions, in which composition cleans and disinfects the air, and can be effectively used in the preschool institutions. The effect is based on the principle of photocatalysis, which sets in motion under the influence of light [13].

3 Conclusions

The adverse effect of finishing materials on the human body is resulted from the release of harmful substances into the environment. In practice, this can be eliminated only by removing such material from the room. Therefore, at the design stage, it is necessary to predefine the right choice and introduce into the project such materials that are safe for humans only and reject materials that contain at least a microdose of the dangerous substances. This will encourage manufacturers. They shall be focused on the production of environmentally friendly materials only. Its implementation into the construction market should be predetermined by the consumer's deliberate

Fig. 1 Fiberglass wallpaper development phases [10]

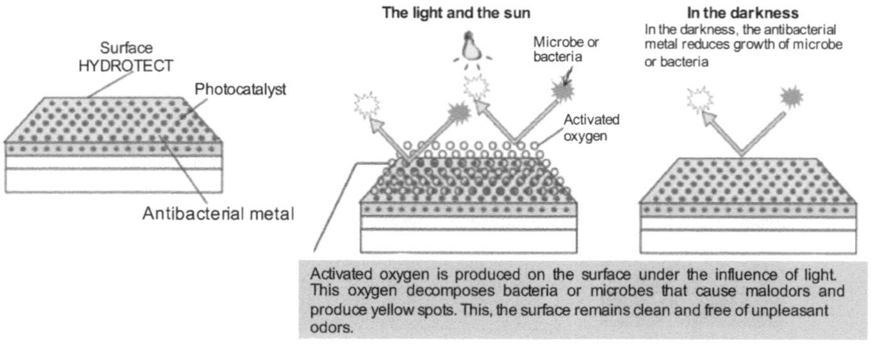

Fig. 2 Technology HYDROCERA [12]

Fig. 3 Field of HYDROTECT application [12]

choice—his refusal to purchase hazardous materials containing harmful substances [1, 14, 15]. It is also required that all materials, regardless of their range of application, shall be subjected to general requirements—they should not emit harmful substances into the environment. The use of materials containing substances detrimental to humans should always be avoided. In addition, strict control over the quality of finishing materials that are used in preschool institutions is required.

References

1. Nikiforova, N. V. (2017). *Sanitary assessment according to health-risk criteria of the construction, finishing materials and furniture products safety as objects of technical regulation.* Thesis for a Candidate Degree in Medicine Sciences/Perm (179 p).
2. Karanastas, L. Y. (1990). *Study of the influence of environmental polymer materials (homes, schools) on the health status of adolescents and high school students.* Dissertation thesis, Candidate of Medecine,14.00.07/Karanastas Larisa Yurievna - Rostov-on-Don (24 p.
3. Mendell, M. J. (2007). Indoor residential chemical emission as risk factors for respiratory and allergic effects in children: review. *Indoor Air 17*(4), 259–277. https://www.ncbi.nlm.nih.gov/pubmed/17661923.
4. Zaitseva, N. V., Dolghikh, O. V., & Dianova, D. G. (2012). Features of the cellular element of children's immunity in conditions of environmental exposure of toluene, formaldehyde, phenol. *Bulletin of the Samara Scientific Center of the Russian Academy of Sciences, 14*(5–2), 341–343.
5. Mitra, S., Noor Ezlin, A. B., Bexzad, N., & Othman, J. (2019). Selection of impactful environmental parameters using an integrated community feedback system for environmental assessment. *International Journal of Engineering & Technology, 8*(1.2), 89–95. http://dx.doi.org/10.14419/ijet.v8i1.2.24878.
6. Sautov, I. (2005). *Environmental assessment of building materials.* Retrieved from http://art-con.ru.u7984.argon.vps-private.net/node/1017.
7. State Building Standards, 2.2–4 (2018). Institutions of preschool education—introduction 2018-10-01—state publishing house 'Ukrarchbudinform" (40 p).
8. Sanitary Regulations for Preschool Institutions № 563/28693 (2016), 16 p.
9. Wall Requirements in Kindergarten. (2018). Retrieved from https://yokvadro.ru/stati/trebovanija-k-stenam-v-detskom-sadu.html.
10. Fiberglass Wallpaper. (2019). Retrieved from https://ru.wikipedia.org/wikifiberglasswallpaper.
11. Bioactive Ceramics Casalgrande Padana. (2013). Retrieved from https://keram.ru/articles/bioaktivnaya-keramika-casalgrande-padana. Last accessed 2019/09/11.
12. Rizhikov, V. (2013). *Interior design of the subject-developing environment of educational institutions implementing the educational program of preschool education*, Moscow (254 p).
13. Antibacterial Disinfectant Paint. (2018). Retrieved from https://isalain.net/antibakterial_naya-dezinfitciruyushaya-kraska-s-dobavleniem–p1923202.htm.
14. Chernyshev, D., Ivakhnenko, I., Ryzhakova, G., & Predun, K. (2018). Implementation of principles of biospheris compatibility in the practice of ecological construction in Ukraine, *7*(4.8), 424–427. http://dx.doi.org/10.14419/ijet.v7i4.8.27283.
15. Khairul, S. B., Mohamad, B. A., Ali Al-Shami, A. M., Ahmad, R., & Ahmad, Z. (2019). Safety assessment method for prefabricated timber roof truss, *8*(1.12), 68–74. http://dx.doi.org/10.14419/ijet.v8i1.12.28849.

Basics of Forming a Network of School Objects Network in Rural Administrative Area

Oleksandr Obidniy

Abstract The purpose of the article is to identify the basics of forming a network of school facilities of rural administrative district, to analyze the existing and most advanced means of forming a network of alternative teaching methods. Based on the existing network of school facilities, it is proposed to supplement the traditional network with an alternative system of education based on their interaction. In the course of the study, the current state of the education system, traditional and alternative teaching methods, and their influence on the formation of the functional space of DDZ and ZNZ were investigated. The current state of the network of preschool and school services of NovoSanjar district of Poltava region is analyzed. The study considered the factors and principles of forming a school education network. Methods of optimization of the network of preschool and school services of the district are formed and defined. The feasibility of using mobile-transforming structures in the formation of educational buildings and school buildings using mobile-transforming tools was analyzed during the study.

Keywords Network of schools · Secondary education · Alternative education systems · Demographics · Mobile facilities · Collapsible systems · Optimization

1 Introduction

The Law of Ukraine "On Education" sets a strategic task: to bring Ukraine to the level of developed countries by reorganizing existing and creating fundamentally new types of educational institutions based on multivariance, flexibility of educational and educational technologies, forms and methods of education improvement, and transformation centers of cultural and educational work of residential complexes (neighborhoods or quarters), taking into account national traditions and world achievements in educational culture [1]. New concepts of education in Ukraine determine the core of the educational process of the child's personality: his/her tendencies, abilities, needs, interests, social experience, self-awareness, and character.

O. Obidniy (✉)
Poltava National Technical Yuri Kondratyuk University, Poltava, Ukraine
e-mail: obidniy.alex@gmail.com

V. Onyshchenko et al. (eds.), *Proceedings of the 2nd International Conference on Building Innovations*, Lecture Notes in Civil Engineering 73,
https://doi.org/10.1007/978-3-030-42939-3_42

But the existing network of preschool and general kindergartens does not correspond to resettlement existing system. The educational institutions were built according to the projects and regulations of the 1990s, at that time their design capacity was different, which did not correspond to the present state (the school, which has a capacity of 990 students, cannot accommodate 660).

According to statistics, the quantity of preschool children in Poltava region has increased significantly over the past five years. The existing quantity of preschool institutions is not able to provide the required number of places. In some cases, kindergartens are redeveloped for other purposes, but not always successfully, as in conditions of high-density construction, it is difficult to find conditions that will meet all needs.

The introduction of a new-type educational process and new technologies in the education system also requires a reorganization of the internal structure of educational institutions. The architectural and planning concept of a comprehensive school in its traditional sense has lost its relevance. The problem of creating a harmonious learning environment is closely related to the environmental aspects of school building. In view of the above, alternative education systems, in particular, Rudolf Steiner's anthroposophical teachings and Waldorf pedagogy, are becoming more relevant. This educational system combines the doctrine of anthropological features of the child with the architectural concept of functional and three-dimensional formation of school buildings.

However, the system of education in practice faces a number of difficulties, including organizational order. For example, urban schools are characterized by a multi-system where students are involved in several shifts, while in rural areas, classes are conducted mostly in unfilled classes. The low occupancy rate of rural schools is explained by the extensiveness of rural settlements and the high level of migration of the population, especially young people. Hence, the irrational cost of education and the extra cost of a large number of teachers. The level of education in a rural small school is much lower than in a large urban school.

All that testifies to the urgent need to improve education system organization, its rationalization, and in the future-optimization. Modern computer technology usage is a real basis for the practical solution of optimization problems in the field of public education.

Today, the architecture of most educational institutions does not contribute to the spiritual and aesthetic development of students, does not take into account the peculiar stereotypes of life, Ukrainian mentality, ethnic features of regions of Ukraine, and new educational technologies.

The dynamism inherent in modern civilization, the growth of the social role of the individual, the humanization and democratization of society, the intellectualization of labor, and the rapid change of technology and technology around the world puts new demands on preparing a young person for life and professional activity, and thus the formation of a qualitatively new system and institutions using the latest information and architectural technologies.

The socioeconomic and spiritual development of Ukraine today requires a radical reform of the education system, including the architectural and planning concept of the school in order to create comfortable conditions for learning and rest for students.

The analysis of the state of the traditional education system outlines several interrelated negative trends that create unfavorable learning, work, and leisure conditions for students and cause many diseases in children. The architectural and planning concept of a comprehensive school in its traditional sense has lost its relevance and has become ideologically exhausted. Efforts to "diversify" the experimental construction of school buildings did not gain popularity, because the single curriculum for all did not require too variable space-planning decisions of the school. Unified teaching methods, standard work programs, and textbooks were entirely consistent with unified schools. Today, the architecture of most educational institutions does not contribute to the spiritual and aesthetic development of students, does not take into account the peculiar stereotypes of life, Ukrainian mentality, ethnic features of regions of Ukraine, new educational technologies.

As evidenced by pedagogical studies of scientists-psychologists, sociologists, physiologists and other specialists involved in the organization of education, the crisis of the traditional education system, the contradiction between the requirements of society to provide a single level of education and the variability of personality development in the multicolored range of its slopes and abilities can be altitudes and abilities educational institutions that build on new concepts.

Pedagogical and architectural science has contributed a great deal of scientific research into the field of school building in the development of those state strategic acts. New types of school buildings, alternative schools, have been developed and tested in experimental construction.

2 Research Methodology

The work is carried out using complex and mathematical research methods, which includes the historical–theoretical method used in the analysis of literary sources and design materials, the method of comparative analysis of domestic and foreign design experience, the systematization of the results of scientific research, literary, and information sources; analysis of statistics; modeling of educational objects.

Problems of formation of educational institutions in different years were devoted to the research of Sarkisov S.K., Stepanova V.I., Kovalsky L.M., Urenova V.P., Svitka V.A., Antoshkin V.F., and Reshetnikova N.V. et al.

3 Prerequisites and conceptual principles of forming a network of rural schools

General secondary education in Ukraine has a well-developed institutional network, which is generally able to meet the educational needs of children and adolescents of school age. New-type general education institutions are being created, reducing the number of schools that work in two or three shifts; private schools are expanding. At the same time, the state of secondary education in rural areas is close to critical; the network of evening schools has been shrinking while the number of students has increased over the last years.

Comprehensive educational institutions are divided into: institutions of the first degree (elementary school; four years of study); second degree (primary school; five years of study); and third degree (high school; three years of study). The senior school should function mainly as a vocational one, in which no one is fully educated.

The network of general secondary education consists of comprehensive educational institutions of all types and forms of ownership, incl. for children and adolescents in need of social assistance and social rehabilitation.

The network of general secondary education also includes extracurricular educational establishments, inter-school educational and industrial factories and those vocational–technical educational establishments (VET) and higher educational establishments of I–II levels of accreditation, which provide a complete general secondary education.

The contingent of pupils of general educational institutions is formed, as a rule, from children from 6 to 18 years, with the exception of students of evening (variable) educational institutions.al.

3.1 Distribution of Educational Establishments by Type

Full-time schools account for the vast majority (about 98.9%) of comprehensive schools: as of 2017/2018, there were 21,965 of them.

There is a steady tendency to decrease the share of full-time educational institutions working in two or three shifts: in 1990/1991, it was 20%; the proportion of students who studied two or three shifts dropped from 15% in 1990/1991 up to 8% in 2000/2001; 87% of them are in one shift; 13% in two or three shifts.

There are 572,000 teachers working in day schools. The ratio of teachers to students is 1: 11.6.

3.2 Establishments of a New Type

New-type institutions include high schools, lyceums, colleges, and educational complexes. As of the beginning of 2017/2018, there were a total of 1832 units, accounting for 8.3% of the total number of day-care institutions. Since 1995, the network of new-type establishments has shown quite high growth rates: yes, in 2017/2018, the quantity of gymnasiums increased almost twice (from 150 to 296), compared to 1995/1996, and the number of lyceums more than doubled (from 138 to 283).

Comprehensive educational institutions are for children with special needs. The network of schools for children with special needs has 402 institutions, which is nine units less than in 1990/1991. The number of children enrolled in these institutions during the period 1995/2005 remains almost constant—at the level of 68–69 thousand.

3.3 Distribution of Educational Institutions by Ownership

The vast majority (99%) of comprehensive secondary schools is state-owned; 6743.7 thousand students (99.7% of the total) enroll in them; 571.5 thousand (99.1%) teachers work. The ratio of teachers to students is 1: 11.8.

Most—99.7%—of the state general educational institutions are subordinated to the Ministry of Education and Science of Ukraine. Others are run by various ministries, agencies, and organizations.

The private general education sector accounts for only 1% of the total network. All private establishments belong to day care; the vast majority—to new type of establishments.

At the beginning of 2004/2005, there were 229 private schools with 20.3 thousand children and adolescents (0.3% of the total number of students); 5125 teachers (0.9% of the teaching staff of secondary schools) worked.

The ratio of teachers to pupils was 1: 4, which is almost three times higher than the level of full-time public schools.

The largest number of private educational institutions is in Kharkiv (31), Odessa (29) regions; ARC (25); Kyiv (24); Dnipropetrovsk (23); and Donetsk (19) regions. At the same time, there are no private secondary schools in the Volyn and Ternopil regions.

3.4 Distribution of General Secondary Schools by Location. Rural Schools

More than two-thirds (67.1%; 14.9 thousand units) of the total number of Ukrainian secondary schools are located in rural areas. Almost one-third (32.4%; 2.2 million) of the country's students are enrolled in them.

Rural schools. Among educational institutions operating in rural areas, the share of new types of institutions is very small, in addition, there is a tendency to reduce them.

There are virtually no private schools in the countryside. Due to the worsening demographic situation, the number of small-scale schools is increasing, and the content of these schools is becoming too low for local budgets. For example, in the Chernihiv region (second in the territory, 18th in population) in 2004/2005. 147 small-scale schools operate, there are districts (Ripkinsky, Novgorod-Siverskyi), where there are about 40% of such schools.

At the beginning of 2017/2018, the number of teaching staff vacancies in rural schools was about 3000 (most in Kherson, Mykolaiv oblasts).

Due to the lack of teachers in some schools, only an invariant part of the curricula is implemented, which reduces not only the opportunities for students to receive

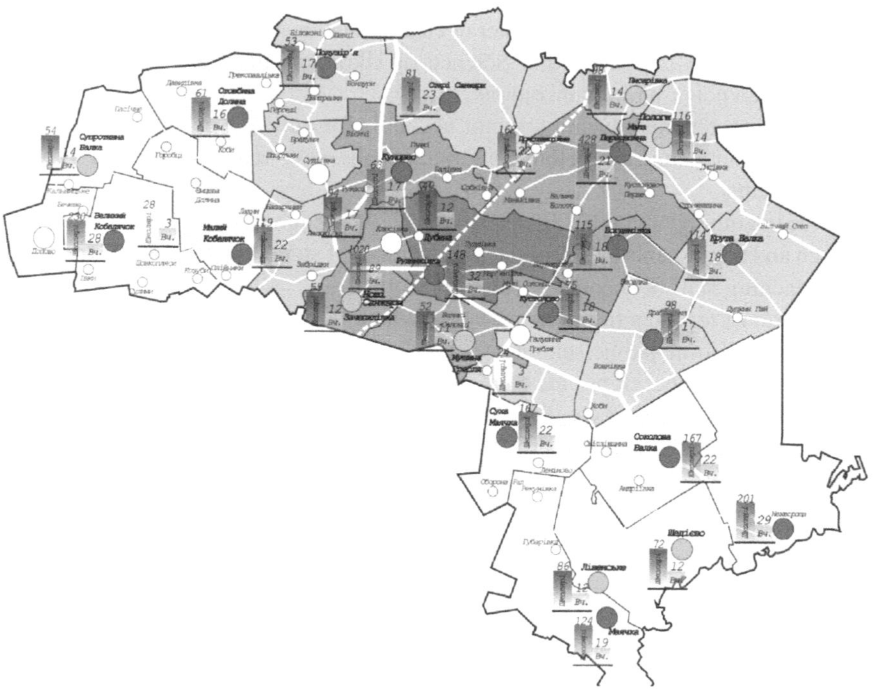

Fig. 1 Scheme of the existing network of school facilities in Novosanjarskiy district

quality general education, but also their chances of continuing their education in higher education.

Novosanjarsky district is located in the southern part of Poltava region. Its area is 1.3 thousand km^2 (ninth place in the region), with a population of 36.4 thousand people. In the area, there are settlement and 28 village councils, 77 settlements.

Novosanjarsky district is an agricultural district. There are 19 farms operating on 69 farms.

The educational sector includes 30 general education institutions with about 3200 students (see Fig. 1).

4 Analysis of Alternative Teaching Methods

A historical analysis of the centuries-old development of the Ukrainian national school and the study of foreign experience have made it possible to identify alternative educational technologies in which the architecture of the school is included as an educational and educational function. Of all the models of non-traditional teaching technologies, the Waldorf School is distinguished, which is based on the ideas of anthroposophy, whose teaching is close to the new system of education in Ukraine, with a sharp increase in attention to the problems of the spiritual life of people.

4.1 Distance Education

Globalization and computerization trends in the world require education, especially given the important role of computer and information technology, a radical change in the educational paradigm in the modern learning culture. Modern learning environments based on multimedia systems and telecommunication networks are capable of storing large amounts of information, providing the opportunity to work with its various types: textual, graphic (static and dynamic images), and audio information.

The study and analysis of automated distance education systems (ASDO) show that for the implementation of most of the functions assigned, it must contain information (knowledge) of three types:

1. About the subject area (domain model—IPY). The domain model should reflect the structure of the domain. It can be used to determine the sequence of topics studied and the control exercises.
2. About the person being taught (student model). The learner model includes a dynamically updated set of parameters that reflect the general characteristics of the learner and the projection of their knowledge into system knowledge (on a domain model).

3. About teaching methods. The teaching methodology determines the sequence of the study of topics (based on the model of the subject area) and forms tasks for the control of human knowledge (based on the model of the student).

The semantic model is based on the semantic network. The semantic network is a framework for representing knowledge as nodes connected by arcs.

Features of the structure of semantic networks:

1. Nodes of semantic networks are concepts of objects, events, and states;
2. Arbitrary nodes of the same concept refer to different values unless they are marked as belonging to the same concept;
3. Arcs of semantic networks create relationships between concept nodes (notes above the arcs will indicate the type of relationship).

Public policy is to create the conditions for compulsory comprehensive secondary education; the achievement of pupils of a certain educational and cultural level, which is ensured, in particular, by a developed network of comprehensive educational institutions of different types and forms of ownership, as well as the existence of alternative forms of obtaining a general secondary education. The state policy in the field of educational content is not only aimed at acquiring students with the knowledge and skills, but also the formation of key competences based on them, which will facilitate the ability through communication to find understanding with other people, to equip their own lives and to be useful to family, community, and society [2].

4.2 Pedagogy A.S. Makarenko

Makarenko concluded on the basis of his pedagogical activity that age is not one of the decisive beginnings in education. Makarenko considered one of the most serious prejudices to hinder the correct resolution of the problems of education, that in the educational work, it is possible to be guided by intuition, that the talent of the teacher is crucial. Through all his activities, he argued that decisive in education is our "accurate opinion," knowledge of the basics and techniques of educational affairs.

Makarenko openly stated that pedagogical science is needed, but not detached from life, but connected with it and assisting the tutor in practical work.

The initial stage of the whole educational process is pedagogical design. Makarenko divides the projected personality qualities into two categories:

- typical, common to all;
- individual, arising from the characteristics of each pet, taking into account his inclination, ability.

At the core, the core of Makarenko's pedagogical theory is his teaching about the collective. The team goes through its development in three stages.

1. There is no team yet, and the teacher at this time is playing the role of dictator, speaking with the requirements for the pupils.
2. There is an asset group of the most active pupils, wishing to participate in various types of work, supporting the teacher's endeavors and his requirements for the pupils.
3. Self-governing bodies are formed, the team becomes able to independently solve a variety of educational, economic, cultural, and other issues, and the requirements go to a separate pet from the whole team.

A school year, a school day, a schedule of lessons, study vacations, changes or, more precisely, breaks between lessons are also signs of a class-defined system. Noting the merits, one cannot fail to see in this system a number of significant shortcomings, namely: The classroom-oriented system is focused mainly on the average student, creates intractable difficulties for the weak, and delays the development of abilities in the more powerful; creates difficulties for the teacher in accounting for the individual characteristics of students in organizationally individual work with them both in content and in pace and methods of learning; does not provide organized communication between senior and junior students, etc.

The purpose of Makarenko's pedagogy is to bring up fighting, active, vital character. The pupil must have a sense of duty and a notion of honor, feel his obligations to society, be able to obey a comrade and punish him, be polite, harsh, and kind depending on living conditions. He must be an active organizer; be persistent and hardened; and be able to control yourself and influence others. He must be fun, cheerful, fit, able to fight and build, able to live and love life, and he must be happy. And so, it must be not only in the future, but also in each of its present day [3, 4].

4.3 Pedagogy of "Free Education"

"Pedagogy of Free Education" is one of the most general concepts of reform pedagogy. Elements of this concept penetrate practically all other pedagogical directions.

Its ideological source and its representatives consider the ideas of Rousseau and Disterweg about the nature of education. The motto of "pedagogy of free education" was the German motto "Everything for the child around the child." The American teacher and philosopher D. Dewey explained the credo in this direction: "There is a shift of centers of gravity. The child becomes the sun, around which the upbringing revolves."

The guiding principles of free education:

- the teacher's belief in the creativity of the child coupled with the conviction that any external (even the most charitable) influence on the child's creative potential has inhibitory effects;
- focusing the caregiver's efforts on acquiring the child's own experience, on the basis of which the full development of personality takes place;

- stimulating active attitude to life, culture, educational and cognitive activity, and the need for systematic self-education and self-education;
- the interpretation of the school as a living organism that is constantly evolving in accordance with the childish nature;
- understanding of the role of the teacher as a senior companion of his or her pupils, who organizes the educational and educational environment for the children to display their creative opportunities freely;
- organizing the life of the school community on the basis of self-government (by community type).

Bright representatives of "free education" are G. Charrelman and M. Montessori. The founder of this direction in pedagogy was the Swedish writer and teacher E. Kay.

Representatives of "free education" advocated the breaking of pedagogical traditions that restricted the freedom and initiative of the child. Instead, they suggested creating conditions for children that would provide them with self-nourishment. The main disadvantage of this area is excessive pedocentrism [5].

For citations of references, we prefer the use of square brackets and consecutive numbers. Citations using labels or the author/year convention are also acceptable. The following bibliography provides a sample reference list with entries for journal articles [1], an LNCS chapter [2], a book [3], proceedings without editors [4], as well as a URL [5].

4.4 Montessori Pedagogy

Montessori pedagogy refers to the system of teaching children developed by Dr. Maria Montessori (beginning in practice—1906), as well as the philosophy of teaching in schools and kindergartens. Montessori's pedagogy is based on open teaching as opposed to closed teaching, such as frontal teaching. This pedagogy, in contrast to ideological pedagogy, can be called experimental in the sense that the child's own observations lead the teacher to choose appropriate didactic techniques for the child and the situation. The essence of Montessori pedagogy is the motto: "Help me do it myself."

Principles of Montessori pedagogy

According to Maria Montessori, the process of developing a child's personality is divided into three phases:

1. the first stage of childhood (0–6 years);
2. the second stage of childhood (6–12 years);
3. youth (12–18 years).

Each of these stages represents a distinct independent segment of development.

The phase of the first stage of childhood (0–6 years) is the most important time of life, because it is during this time that the personality and abilities of the child are formed. Montessori understands the first six years of life as the second embryonic growth phase in which the child's spirit and soul develop. While the adult filters his or her perceptions, the child absorbs into the environment and becomes part of his personality.

Montessori defines the second stage of childhood as a phase of lability. During its development, the child goes through so-called sensitive or sensitive periods. During such periods, the child is particularly sensitive to certain environmental disorders, such as movements, language, or social aspects.

Crucial to the development of Montessori pedagogy and related materials is the following observation: One of the most important phases of sensitivity for each child is the phase of "improving the senses."

The Montessori method is based on observing the child in natural conditions and accepting her as she is. The main idea of the Montessori method is to encourage the child to self-development. In Montessori groups, the child learns mostly independently through a specially designed Montessori environment. The environment prepared for the child should be appropriate to the growth and proportions of the child of the appropriate age. Each student can decide at his own discretion what to do today. Children are engaged in age groups [6].

4.5 Waldorf Pedagogy

Waldorf pedagogy is a concept of parenting. The main purpose of Waldorf pedagogy is to assist the individual in his own formation and to discover talents in harmony with the laws and tendencies of the development of nature, man, and the world. The author of the method of Waldorf pedagogy is the Austrian scientist and philosopher Rudolf Steiner, and is based on his philosophy of anthroposophy. The first Waldorf school was opened in Stuttgart on September 7, 1919, for the children of workers at the Waldorf Astoria factory (hence the name).

The main method of teaching—the method of "heartfelt economy"—the basic method of Waldorf pedagogy. The method is that in the process of learning, children develop activities that the child can master at this stage of development without internal resistance of the body. Thus, from the period of change of teeth to puberty, necessarily develop memory, work with the imaginative thinking of the child, and appeal to the feeling, not to the intellect. After puberty, educational material includes concepts that work with the child's abstract thinking.

Waldorf pedagogy is used by about 900 educational institutions in 60 countries. Since 1919, they have been developing in Europe, America, Africa, Australia, Asia, post-Soviet countries, and since 1990—in Ukraine [6].

5 Means of Transformation of Buildings in the Formation of School Buildings

Dynamic architecture is a new trend in modern architecture. The very word "dynamic" indicates movement in architecture. To make the building move, designers use the so-called bottom-up floor of independent floors on a solid foundation, as pieces of meat on a skewer, between them, have blades through which the wind passes, and this drives the floors in motion and still produces energy. Thus, the facade of the building does not have a clear architecture, and it constantly takes different forms.

With such a "moving" design, it is very important to use heavy-duty materials, using metal columns and a monolithic overlay for the frame. Not only are they very durable and comfortable to build, but they are also not very expensive, which is important for a project even in 60 floors. Apartments in such a dynamic building have from two to four rooms, an approximate area of 126–166 m^2. Meters, storerooms, wardrobes built-in, and the number of bathrooms are made based on how many rooms in the apartment. Thanks to the frame structures, apartment owners will be able to easily remodel their housing, based on their advantages in interior design projects. Using the principle of dynamic houses with appropriate structures, not only residential apartments are developed, but also used in the design of office space.

Are there any advantages to such dynamic projects? Yes, there are many positives. First, reducing construction time. When building floors on a conventional system, it takes about three weeks, the same construction was moving, according to experts in the field of dynamic architecture and can be collected in only three days, and the construction of the entire structure will be reduced to 18 months (construction under the standard program takes about 30 months). Secondly, this project can reduce the number of workers up to 90 people! Finally, you can save on construction equipment. It will be necessary at the initial stage of construction during the erection of the trunk of the building, when stringing floors no crane or any other equipment is required.

One of the known disadvantages to date is the issue of shaking of residents while rotating floors. So, before you buy an apartment in one of the dynamic buildings, it is worth checking if you have any diseases related to the violation of the vestibular apparatus.

5.1 *Advantages of Using Mobile Designs*

In modern day-to-day life, we are increasingly using the term "mobility." With the development of new technologies and human needs, it is time to take a new look at our familiar architecture.

Differently interpret the concept of "mobility": in one case—it is a mobile home on wheels, in another it is a collapsible structure, and in the third it is a reinforced concrete house with a small building area. In the years 1950–1960, the twentieth century saw the first manifestations of mobile architecture in the world. This concept boiled down to the function of mobile housing for temporary residence in one place, but gradually it grew into more individual. Lightweight homes, mobile or hotel rooms do not require significant material costs and time for installation work, so they are cost-effective.

"Internal mobility" is characteristic of mobile architecture. This means adapting the object to new conditions: social or economic status, changing family composition, changing generations, or simply changing the place of residence "according to the mood", without changing the general three-dimensional parameters by transforming the internal space. And it is these parameters that characterize the mobile architecture.

Japan is a prominent representative of mobile architecture development. One of the draws is Atsushi and Mayumi Kawamoto's architects from the mA-style architectural bureau. Their example of mobile architecture is called the Riverbank House, which has the shape of an isosceles triangle and stands on a tiny stretch near the river.

Usually, a mobile home is economical, easy to install, practical, and occupies a small area. These homes are fully functional, where everything has a purpose. They have their own source of electricity (solar panels), some collect rainwater, but mostly a mandatory connection to the sewer and water supply. Mobile homes can be designed and built anywhere. For example, they can be used as hotel rooms, which are placed on trees and even on water. People who constantly want to change their environment or travel, use mobile architecture on wheels (these are trailers, buses, and other objects that are specially designed for a comfortable living).

Mobile architecture is a prime example of modern housing. There are already many mobile home projects available today. Young architects are particularly interested in this area. The most unusual projects are the houses of the Dutch architect Kaas Osterruis, "portable" houses, houses of embryos. Most of these projects are still concepts, but some have begun to be implemented in glass, steel, concrete, and other materials. It is worth mentioning Tokyo's Origami House by architect Yasuhiro Yamashita's.

This building exhibits great lines and shapes that are a fantastic solution for densely populated cities.

For many architects today, the main priority is cost-effectiveness and environmental friendliness, so now more and more projects are being created that allow a person to live in harmony with nature without harming it. And despite modern technologies and innovative building materials, more and more people are returning to nature, in harmony with it.

6 Proposals for the Organization and Calculation of the Rural School Network

The optimal formation of the architectural environment of modern kindergartens should be carried out taking into account the influence of both "external" factors:

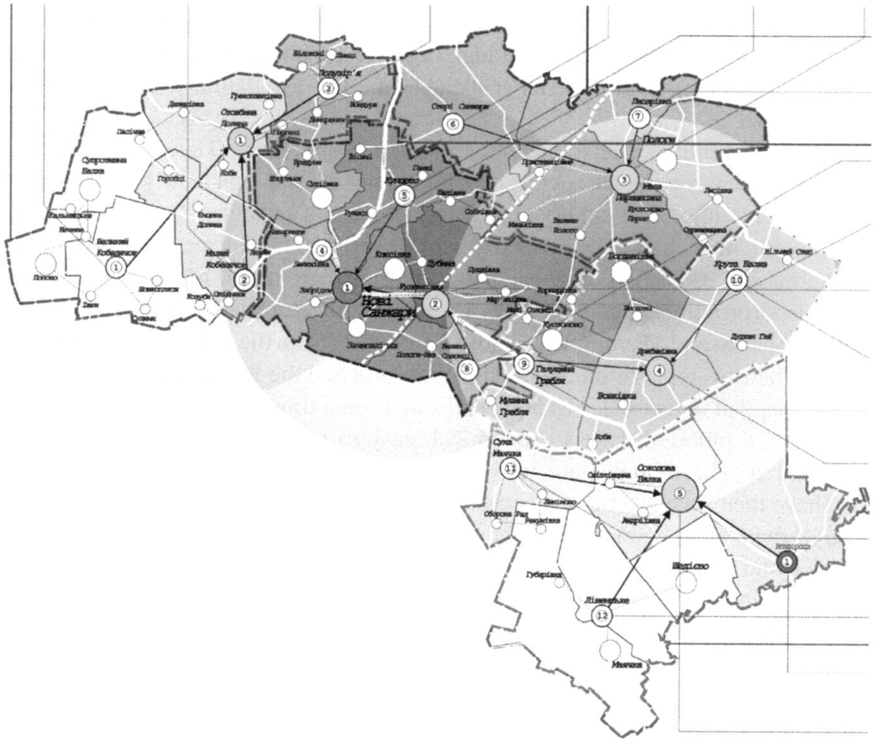

Fig. 2 Scheme of network of school facilities of Novosanjarskiy district with alternative methods of education

landscape-urban, socioeconomic, cultural–historical, natural-climatic, factor of constructive opportunities, and “internal”: sanitation-physiological, functional–organizational, scale-spatial, imaginative–psychological, sensory, aesthetic, as well as environmental factors.

When forming a network of alternative educational institutions, it is important to understand that the education system, in turn, is one of the elements of a more complex architectural and territorial system of the city as a whole and is in constant close interaction with other systems of the city: communicative, industrial, residential, environmental, etc. That is, it is a subsystem of a more complex system that has a significant impact on its functioning and development. An important aspect of the formation of a network of schools is to take into account the main factors that influence the development of the network.

To address the formation of a network of school facilities, the adoption of differentiated network optimization consists of dividing the existing network of educational institutions in the city into two parts—traditional and alternative—and sequential optimization of each network according to the needs of the population [7]. The network is divided by a percentage of 80:20 (traditional: alternative). Considering that

Fig. 3 Scheme of the network of school facilities of Novosanjarskiy district with traditional teaching methods

alternative education techniques are a new under-tested practice in Poltava, their percentage at the initial stage of formation is only 20%.

In optimizing the mathematical model of a network of school buildings, the Sarkisov SK method was taken as a basis [8]. The task of locating school network objects and selecting the optimum capacity, depending on the settlement system and the size of the school contingent, can be solved separately by modifying known mathematical methods. The following groups of variables occur in solving these problems: the number of students living in districts "j" and enrolling in the schools located in points "i," the variants of the capacity of schools and the corresponding consolidated costs per place, as well as travel expenses related to moving students.

Taking into account the degree of probability of the factual initial information that is now implicit, it can be considered sufficient at this stage of the study of transformation by replacing a nonlinear function with a linear one. In this form, the task is to determine the optimal capacity and the optimal transportation plan and can be solved by modifying linear programming methods, which is important in view of the widespread practical implementation of the model. It is likely that the number of places in schools in any district will not meet the need for them. To prevent

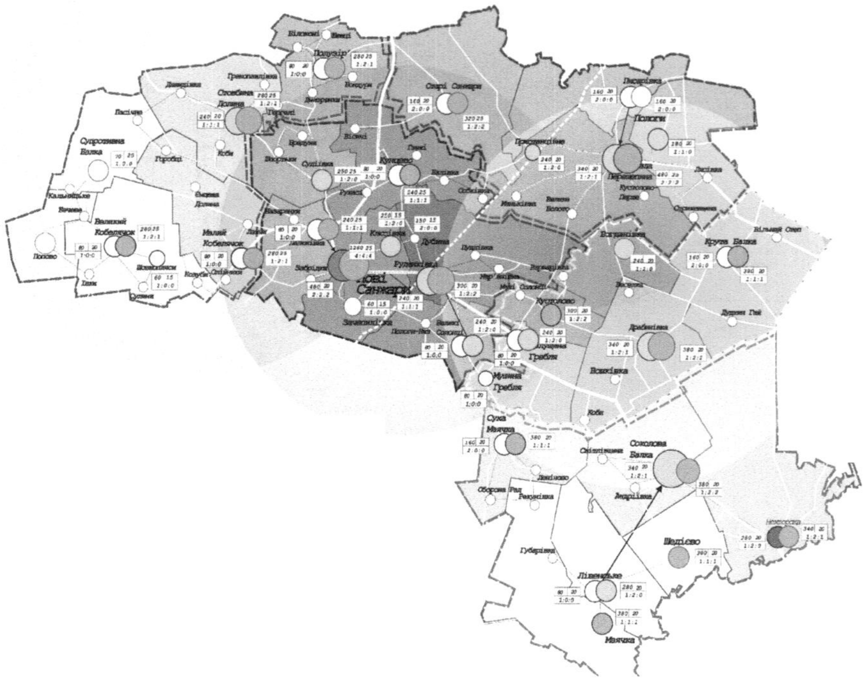

Fig. 4 General outline of a perspective network of school facilities in the Novosanjarskiy district

overcrowding, it is necessary to exceed the total capacity of schools beyond the need for student places. So, in general, we use an open model of a transport problem. However, for the sake of convenience, by introducing a dummy clause, we reduce the problem to a "closed" form, that is, when the total number of students in the district corresponds to the total capacity of all schools in the area in question.

Scheme of a network of school facilities in Novosanjar district with alternative methods of education.

The formation of a promising network of secondary education institutions is based on the existing network of schools and kindergartens, identified factors, and nomenclature of secondary schools and based on the model built. The alternative network of educational institutions is designed for 20% of children of preschool and school age.

The main structural elements of the alternative network, given the current developments in reforming the structure of the education system, selected two types of objects:

Alternative school of the full cycle of secondary education (includes kindergarten, combined with the elementary school—I degree of education, and high school (II–III degree of education), which allows to organize a full cycle of secondary education in rural areas.

Primary school, combined with kindergarten (primary educational complex).

The placement of primary schools is limited by the condition of the radii of reach, which, according to the standards, is accepted at 500 m.

Thus, the alternative network covers the entire city. Alternative schools are predominantly located in quiet areas, in areas of estate development or near parks or squares.

The provision of these areas with traditional education is not lost due to the reach of other traditional schools.

The transportation of middle and senior students to primary school is carried out by school buses. The scheme of routes of school buses is offered. Transportation of students to school is carried out within the established alternative district. The transport service is extended to pupils living at a distance greater primary education. The bus stops are designed so that the maximum pedestrian access of the students to the gathering point at the stop does not exceed 500 m.

The number of bus routes is equal to the number of primary schools to which students are transported.

As a result of the calculations, a prospective number of preschool children in the districts revealed the need to enlarge some schools.

The improvement of the network of traditional preschool and secondary educational institutions was carried out according to the adjustment of the district master plan. Based on the master plan and taking into account the new network of alternative educational institutions, which is 20% complementary to the traditional one, 20 establishments providing full secondary education (I, II, PI degrees) and 20 preschool institutions were located in the city.

With regard to preschool institutions, according to three analyzes, their number is insufficient already. The project proposes to supplement the NES network with alternative primary education facilities in the existing districts and new kindergartens in the areas of prospective development.

According to the analysis, the number of preschool institutions is insufficient already. The project proposes to supplement the NES network with alternative primary education facilities in the existing districts and new kindergartens in the areas of prospective development.

The general network of school facilities of Novosarzharsky district is a combination of two networks—the traditional (see Fig. 3) perspective and alternative (see Fig. 2) network of schools of Novosarzharsky district.

According to the analyses, there was a need for the construction of new and reconstruction of existing schools, provision of transportation for middle and high school students to school.

Thus, the promising network of schools serving the Novosarzharsky district is 80%, and the other 20% is occupied by the sector of schools with alternative teaching methods. In total, 12 schools were reconstructed in the area, three new general secondary schools were built, and two primary schools were built. Three junior high schools were redeployed into primary schools.

In general, the network of traditional institutions (see Fig. 4) was built on perspective calculations of the population and the accessibility of students to educational institutions.

7 Conclusions

Alternative educational institutions should be considered as part of the overall network.

The tendencies of development of a network of educational institutions in some regions of Poltava region are investigated. It is revealed that improvement of the network of the general education system is possible provided the expansion of the development of the alternative system of education and its separation into a separate network of secondary schools.

The main factors that influence the construction of a network of educational institutions in the district are identified.

The basic technique of differentiated linear programming (network optimization) for forming networks of educational institutions is determined.

Networks of traditional and alternative educational institutions have been developed, and an existing network of traditional educational institutions has been supplemented, taking into account the development of the district.

References

1. Law of Ukraine.(2019). *On general secondary education*, №651-XIV, as of 06.06.2019.
2. Krivul, G. F., Krivul, A. F., Barbaruk, V.N. (2002). *Computer technologists in education. Book 1. The department's secretary's arm* (175 p). Lugansk: Publishing House Vostochnoukr. nat. them. V. Dal.
3. Pavlov, M. P. (1956). *A.S. Makarenko's pedagogical views*. Moscow: Trudrezervizdat.
4. *A.S. Makarenko Collection of Works* (Vols. 1, 2, 5).
5. Ex. Kornetov, G. B., Boguslavsky, M. V., & Kornetov, G. B. (1995). *Anthology of pedagogy of free education*.
6. Dyachok, O. M. (2000). *Principles of formation of school architecture with non-traditional teaching methods*: *Author* (20 p). Candidate of Architecture (18.00.02) KNUBA - K.
7. Obidniy, A. (2018). The main approaches for increasing the efficiency of the rural school network/Olga Tyshkevych, Aleksandr Obidniy// *International Journal of Engineering & Technology, 7*(3.2): Special Issue 2., S. 686–691. Rezhym dostupu. http://dx.doi.org/10.14419/ijet.v7i3.2.14614.
8. S.K. Sarkisov Principles of designing and optimization of a network of school buildings. Kiev NMK VO 1990.

Problems of Construction of Industrial Buildings in Ukraine

Viktor Rudenko, Taras Rudenko, and Mariia Rudenko

Abstract The work is devoted to the improvement of the design and construction of industrial buildings on the basis of a systematic approach, using the principles of functional structure, which ensure the unification of architectural and construction and technological solutions that allow inspecting the processes of forming objects of factory architecture and the construction of these facilities. Construction of Ukrainian industrial architecture in the period of the twentieth and early twenty-first centuries has three periods, different in degree of industrialization. The beginning of the twentieth century is characterized by the individual design and construction of production facilities for the needs of the customer. There were no national requirements (rules and regulations) that had imposed certain limitations on the customer and the builder.

Keywords Traditional solutions · Modular (adaptive) type of solution · Enterprise · Module of enterprise · Functional units of modules · Blocks of aggregated equipment

1 Introduction

1.1 The Early Development of the Ukrainian Industrial Architecture

From the 1930s onwards, in the context of a planned economy, the question of finding an ideology of industry formation that would meet new planning approaches arose. What should be the plants and factories? The experience of the USA and Europe was analyzed. In the USA, large enterprises prevailed, in Europe—smaller ones. For

V. Rudenko (✉) · T. Rudenko · M. Rudenko
Poltava National Technical Yuri Kondratyuk University, Poltava, Ukraine
e-mail: rudenko.v.arch@gmail.com

T. Rudenko
e-mail: rudenkoforwork@gmail.com

M. Rudenko
e-mail: rudenko.formail@gmail.com

V. Onyshchenko et al. (eds.), *Proceedings of the 2nd International Conference on Building Innovations*, Lecture Notes in Civil Engineering 73,
https://doi.org/10.1007/978-3-030-42939-3_43

the planned system of the economy, large enterprises were advisable. The decisive indicator was the number of products per square meter of the enterprise area. The next task that arose after 1945 was the rebuilding of the destroyed businesses and the massive construction of new ones in the shortest possible time. Until 1956, old approaches were still used as individual objects. It has become evident that to meet the five-year targets, new approaches are needed to ensure the intensification of the architectural and construction industry of the national economy.

1.2 The Intensive Development

Since the late 1950s, a system of rigid unification and typing has been introduced in all areas of design and construction. The vast majority of industrial facilities built in the 60s were built on typical projects, the bulk of which was developed outside Ukraine. An important contribution to the theory of industrial approaches to the formation of factories and factories was made by Sc. D. Kim, proposing to use sectional-type production facilities [1]. At the same time, there was a creation of a system of industrial production of unified building structures, which provided the implementation of typical projects. The main purpose of this approach was the general reduction of the terms of design and construction of all types of architectural objects, including industrial, which generally corresponds to the current conditions of the Ukrainian economy. After all, this is one of the main tasks of the economy—improving productivity. The use of a unification and typing system in construction not only gave positive results in the construction of Ukraine in the second half of the twentieth century but also created some problems in the creation of rational factory facilities. So the basis of design and manufacture of columns of frames of industrial objects was laid height module 1200 mm, that is, regardless of the peculiarities of the spatial parameters of production technological processes, the height of shops was always a multiple of 1200 mm. But, as the practice of design and construction has shown, such large-scale modular coordination of the size of industrial buildings does not correspond to the spatial coordination of technological equipment, in which modular coordination is absent at all. That is, in a simplified form, if the spatial organization of technological equipment is not "fit" into the building parameters by at least a few millimeters, the next uniform size of the internal space of the building in height is plus 1200 mm. In this case, the volume of the building being heated significantly increases; the costs of materials and energy are increased unnecessarily. As a result—the mismatch of the building part of the enterprise to its technological part, which contradicts the requirements of the system-wide principle of compactness, formulated for architectural systems by Dr. Lavryk G. I. [2], according his works, the spatial organization of architectural objects should correspond to their functional features in the conditions of external factor influence—natural-climatic, socio-economic and scientific–technical.

Nowadays, we feel lack of economic research of dynamics of changes in the enterprise's economic efficiency in terms of the variability of their functional and spatial structure, although in the late 80s of the last century in the research of specialized research institutes, for example, at the Central Research and Design-Experimental Institute of Industrial Buildings located in Moscow, work was carried out, with the participation of one of the authors of this article as a responsible executor, on the formation of a new type of enterprise based on compact structure of process equipment and its spatial coordination in the buildings. Considerable results have been obtained in the study of the methods of spatial formation of technological processes and in the experimental design of coal-mining enterprises of modular-block type. The project was based on the use of unified indicators of the spatial organization of the building. Namely the beam size—24 m, the bay of columns—12 m, the height of columns—19.2 m. The building is a full-frame. The frame is metal. Bridge lifts were used to ensure the movement of process equipment for repair and replacement. The equipment is located on its own supports, not on floors of a multi-story building. Thus, the problem of "dictatorship" of unified parameters of the building over the technological layout of the equipment and vice versa was solved.

It was found that, unlike the traditional multi-story factories, the main buildings of coal-mining factories can have a single-story spatial structure, which can be expanded if necessary. A unified building and technological module were adopted per unit of extension. The enterprise module is a part of the whole factory and performs the functions of it. The hierarchical structure of the modular-block enterprise is presented in Fig. 1.

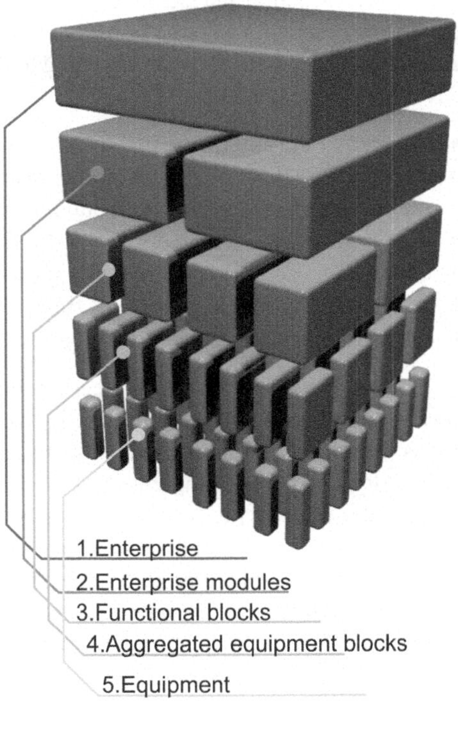

Fig. 1 Hierarchical structure of the modular-block enterprise

2 Results of Research

This formation of industrial enterprises is consistent with the systematic approach as a scientific method. The above structure defines features of their spatial formation implements the "hierarchical" principle of construction. The "hierarchical" principle of construction implies that from several spatially and functionally completed elements of the lower level, a larger element of the next level is created and so on to the highest one [3]. For example, units of aggregated equipment form functional units, functional units form modules, and modules form a factory. The elements of each level are formed according to their specific laws. At the same time, the requirement of spatial coordination of elements acts as a universal regularity at all levels. In the experimental design of coal-processing factories, it was found that such enterprises can have unified construction and technological solutions in 24 options. These options are created for six coal enrichment schemes with four production capacities (250 tons per hour, 500 tons, 750 tons, and 1000 tons per hour). To create technological layouts of all factory variants, 20 functional unit variants are required. Thus, there are all prerequisites for the transition from individual design of coal-processing plants to the creation of typical projects of these enterprises. The consequence is the unification and typing not only of technological layouts but also of the construction part (general plans, volume-planning decisions of buildings, structural elements of production buildings with maximum consideration of the features of the location of technological equipment).

New construction and technological solutions for the two-module coal-processing plant were compared with the project of a similar enterprise [4]. At the same time, the second enterprise was formed according to the traditional approach. The comparison gave the following results:

- the area of the enterprise site decreased 2.15 times;
- the length of the main highways decreased 1.45 times;
- the expanded area of the main building was reduced 4.13 times;
- the construction volume decreased 1.4 times;
- concrete consumption decreased 1.4 times;
- steel consumption decreased 2.1 times;
- the estimated cost decreased 1.5 times.

Since 1990, the Poltava Engineering and Construction Institute (now—Poltava National Technical Yuri Kondratyuk University) within the framework of diploma and course design works at the Faculty of Architecture has been developing projects of foundry, food processing enterprises, car service station, ceramic factories based on modular-block type. In 2013, in Makiivka, a Ph.D. thesis was defended on the topic: "Principles of architectural—planning organization of food industry enterprises (on the example of production buildings of dairies)" Author—Taras Rudenko. The thesis analyzes the evolution of industrial design methods (Fig. 2). The principles of dairies forming in modular-block design were also determined. It is obvious that the difference between traditional enterprises, whose ideology was created in

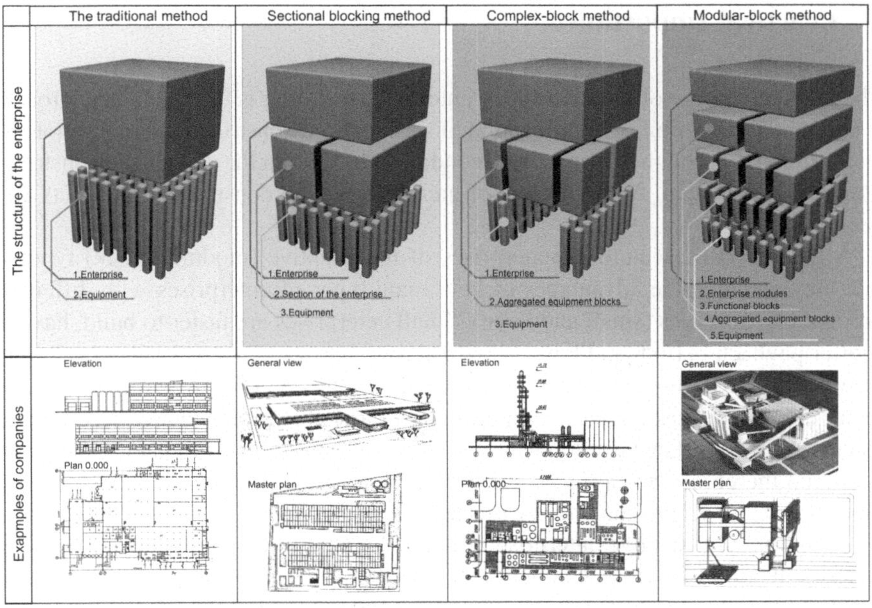

Fig. 2 Design methods for adaptive production buildings

the 1930s, and enterprises of adaptive (modular-block) [5] type, lies in the number of hierarchical levels of the functional and spatial structure of these objects.

3 Scientific Novelty

Traditional solutions were created to get the maximum final capacity, which was consistent with the planned system of economy. The commissioning of production facilities in turn allowed early production before the end of enterprise formation.

But the quality of early-stage products was much lower than those produced by a fully-formed enterprise. This is due to the fact that equipment with excess capacity was used in the production. That means that for some period, the enterprise had a mismatch of the building space with the technological layout of the production processes. This problem can be solved by the use of modular-block (adaptive) type enterprises. Modular increase of production capacity allows to observe the functional and spatial correspondence of technological lines of their production capacity and the volume of the module, which corresponds to the action of the system-wide principle of compactness [6] and to market conditions of economy.

4 Practical Importance

The main advantage of modular-block enterprises building is that such an approach is based on the possibility of unification of architectural construction and technological solutions. Through this approach, conditions to shorten the design and construction time are created. Also, the cost of architectural and construction products are decreased.

A characteristic feature of enterprises of the adaptive (modular-block) type is that they combine the advantages of both main types of enterprises with different production capacities (small and large). Small enterprises are faster to build, have a shorter payback period, and have faster reaction to external factors changes. Large enterprises have better economic indicators of building area construction per square meter cost, as well as a greater proportion of marketable products produced on the square meter. But they have longer creating period, which leads to later production start, and therefore have a longer payback period [7]. It leads to the later to profit time. Below are graphs that clearly illustrate this (Fig. 3).

A multimodal adaptive enterprise differs from a traditional large one. The main differ in that the first module is built and operates as a small enterprise, that means it pays off in the shortest possible time. The following modules are being added to an existing working enterprise. In addition, partial stopping of production by disconnecting one or more modules may be possible if necessary.

Architectural compositional solutions of adaptive type objects are directly related to the modular-block method of forming as the basis for dynamic solutions. The

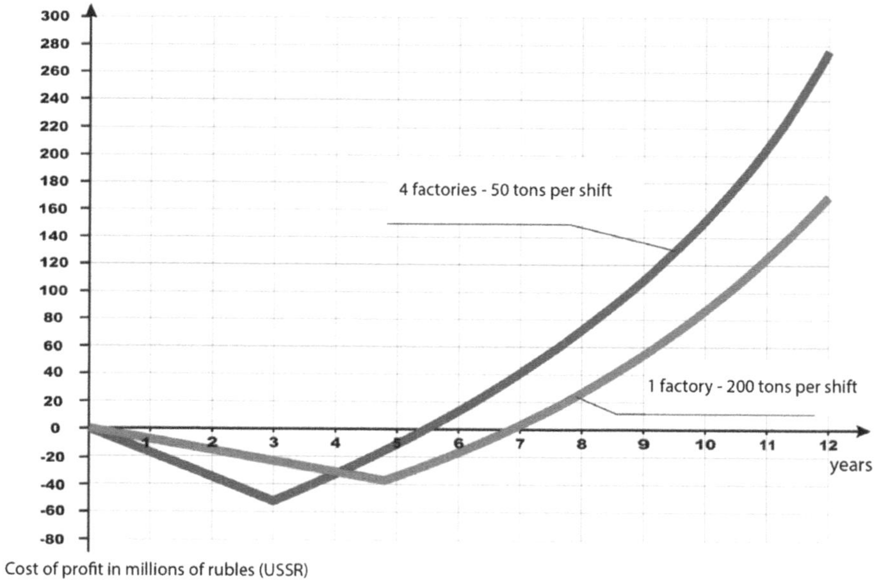

Fig. 3 Costs of profit comparison for four 50 tons factories and 1200 tons factories [8]

unification of spatial-planning and design solutions complicates the process of forming an architectural composition for all types of objects. But industrial enterprises are the basis of a functional environment that provides the production of goods, and therefore affect the quality of the economy system functioning in general. Therefore, architectural and compositional solutions of production buildings should be consistent with the external influence of socio-economic factors. Because of that reason, the enterprise may be built in the shortest time and have laconic architectural and compositional solutions of so-called traditional (rational) type, or may have complicated composition solutions of image type [9]. It is certain that rational solutions, having predominantly unified design and space-planning solutions, will be put into operation faster than objects of the image type. Since "time is not a factor influencing the design decision but is the main criterion for evaluating the design decision" [10] and the project implementation time is decisive in obtaining positive economic indicators, the vast majority of manufacturing enterprises will be formed as rational.

5 Conclusions

The history of industrial architecture development in Ukraine testifies to the expediency of the use of unification in the architectural and construction business and the absence of the need for the traditional formation of factories and factories in new socio-economic conditions.

In the practice of research, design, and construction of industrial enterprises, there are developments that create the prerequisites for the intensification of research and design work and industrialization of construction based on unification not only the buildings of enterprises but also their technological assemblies.

References

1. Kim, N. N. (1988). *Industrial architecture* (2nd edn.), Revised. and add. Stroyizdat, Moscow.
2. Lavrik, G. I. (2007). Methods for assessing the quality of housing. *Research, design, examination. BSTU named after V.G. Shukhov*. Belgorod.
3. Blauberh, Y. V., Sadovskyi, Y. N., & Yudyn, Э. H. (1969). *Systemnyi podkhod: predposylky, problemy, trudnosty*. Moscow: Nauka.
4. Rudenko, T. (2016). Evoliutsiia naukovoi paradyhmy u formuvanni promyslovoi arkhitektury - Suchasni problemy arkhitektury ta mistobuduvannia.
5. Rudenko, T. (2012). Poniattia modul v promyslovii arkhitekturi - Suchasni problemy arkhitektury ta mistobuduvannia (2012).
6. Sidorenko, V. D. (head, chief editor), Puchkov, A. O., Sitkarev, O. V., et al. (2010). Forgotten architectural tectology is a small treatise by A.V. Rosenberg "The philosophy of architecture" of 1923 and a great contemporary architecture study. In *Zb. sciences Proceedings of Art Studies, Architecture and Cultural Studies* (pp. 284–317).
7. Rudenko, T. V. (2012). Chas yak osnovnij kriterij ocinki yakosti proektnih rishen v promislovij arhitekturi. Tradiciyi ta novaciyi u vishij arhitekturno-hudozhnij osviti, pp. 109–110.

8. Rudenko, T. (2013). Vzaiemozviazok systemy vyrobnychykh protsesiv z prostorovymy obiemamy promyslovykh budivel kharchovoi haluzi - Suchasni problemy arkhitektury ta mistobuduvannia.
9. Rudenko, V., Rudenko, T., Rudenko, M. (2018). Functional structure of industrial objects. *International Journal of Engineering & Technology, 7*(4.8.), 631–635.
10. Rudenko, V., Rudenko, T., Rudenko, M. (2018). Industrial architecture as a system. *International Journal of Engineering & Technology, 7*(3.2.), 661–666.

The Use of Bricks in the Facade Decoration of Architectural Structures of Poltava of the Late Nineteenth–Early Twentieth Centuries

Tetiana Savchenko

Abstract On the basis of field surveys and comparative analysis, the principles and techniques of the use of decorative brick decoration of Poltava buildings of the late nineteenth–early twentieth centuries were revealed. Features of application of a brick decor depending on style of a building are established.

Keywords Brick · Decoration · "Brick style" neo-gothic · Neo-renaissance

1 Introduction

The architecture of the late nineteenth–early twentieth centuries is an important stage in the formation of the historical development of the central part of Poltava city, as it is a significant part of it. This period is characterized by the widespread use of decorative qualities of brick for facade decoration. Construction of new objects in the historic center of the city in unity with the existing buildings, reconstruction and possible adaptation of the old objects to new functional needs requires a deep and comprehensive study of the architecture of the late nineteenth–early twentieth centuries, including its decorative.

Decorative trimming of brick facades of the late nineteenth–early twentieth centuries is considered in the works of I. Abramyuk, Y. Aseev, I. Bartnev, O. Borisova, Y. Biryuleva, V. Goryunova, Y. Ivashko, A. Ikonnikov [1], T. Kazantseva, S. Linda, S. Ponkalo, V. Timofienko [2], M. Tubli, V. Yasievich and others. Decoration of Poltava buildings of the studied period is mentioned in the works of K. Hladysh, Y. Ivashko, T. Skibitskaya. Separate systematic studies of the brick decor of Poltava of the late nineteenth–early twentieth centuries are currently absent. That is why the purpose of the study is to identify, describe and classify decorative brick decoration of Poltava buildings of this period and methods of its application.

Empirical and theoretical methods of scientific research were applied in the work. Empirical studies include field surveys of objects and their photo fixation, study of

T. Savchenko (✉)
Poltava National Technical Yuri Kondratyuk University, Poltava, Ukraine
e-mail: stv-26@ukr.net

V. Onyshchenko et al. (eds.), *Proceedings of the 2nd International Conference on Building Innovations*, Lecture Notes in Civil Engineering 73,
https://doi.org/10.1007/978-3-030-42939-3_44

literature on the topic of research. To theoretical-generalization of literature, comparative analysis of decorative details, systematization of types of decorative trimming and brick form.

2 Outline of the Basic Material

2.1 New Technologies for the Production of Bricks in the Late of the Nineteenth–Early Twentieth Centuries

The development of Poltava architecture in the late nineteenth–early twentieth centuries was in line with the nation-wide process of urban environment formation and stylistic development of the era of capitalism and was similar to European trends. For this period, Russia and Ukraine (which was under its control) were characterized by urban growth, which was accompanied by intensive economic development of the city and significant growth of construction. Style innovations were introduced into the architecture of Poltava with a significant delay, which was characteristic of the provincial cities of the late nineteenth–early twentieth century. In the second half of the nineteenth century, a rationalist trend of historicism, the so-called brick style, developed in Ukraine [3]. The artistic and technical qualities of the brick were widely used to achieve the artistic significance of the buildings. The walls of buildings of different style directions were left unstained; decorative elements were made of bricks or used stucco details.

The origins of the rationalist stream are traced in the work of the representative of the late classicism K.-F. Schinkel (Germany) [4]. The building of the Berlin Building Academy, according to his project, initiated the "brick style." The development of this trend is related to the work of Dutch architect Berlage, Hendrik Petrus (1856–1934). He and his followers, representatives of the Amsterdam School of Rationalism, used the aesthetic properties of the brick. In the future, the "brick style" expanded all over the Europe and became widespread in Russia, which at that time included most of modern Ukraine (including Poltava). It was originally used for the construction of barracks and industrial buildings. From the late nineteenth century, it was applied to residential and public buildings. The use of bricks for the decorative trimming of the facades made it possible to create a clear three-dimensional solution for buildings. The tsar's permission for the construction of non-plastered residential buildings [5] contributed to the spread of the use of bricks.

In the nineteenth century, new technologies for the production of bricks were introduced, which contributed to the growth of production and improvement of its quality. The brick of the second half of the nineteenth–beginning of the twentieth century had a mark on the bed (the working face of the brick) in the form of the surname or initials of the owner of the plant (Fig. 1) [6]. The sizes of bricks in the middle of the nineteenth century were regulated by the tsarist rule within the limits of $26.5 \times 13.3 \times 6.7$ cm, although they differed somewhat depending on the

Fig. 1 The stamped brick of the late nineteenth–early twentieth centuries of Poltava province (from V. Smirnov's collection): pottery and brick workshops in Oposhny: **a** VO Bordyug; **b–d** brothers Philip, Nicephorus and Silas (Silant) Flesh; **e** Ivan Onachko; **f** Jacob Stanko (J.S.); **g** YP Gresko; **h** Ivan Pruglo; **i** Sergey Shabliy; **j** the factory of the merchant Peysah Mordkovich Corfu, Poltava, 1910–1916; **k** Plant Railway Invalidity Committee employees of the house them. Emperor Alexander II", Kobelyatsky County in the 1880s; **l** factory Morduk (Mordukha) Borohovich Radunsky, Poltava, 1894–beginning of the 20th century; **m** the plantV.P.Taranushenko, Poltava, founded in 1888; **n** the plant of Elizabeth Ostrogradskaya, with. Pogrebenskoe, Kremenchuk county, 1840–1914; **o** Company B and Co. (true unknown); **p** the workshop of Princess Maria Kantakuzina (KMK), **p** Big Buromka, Zolotonosha district, **q** brick from Deymanovka village, Piryatinsky county (belonged to the landowner Zakrevsky); **r** the plant of Prince Kochubey, comm. twentieth century, Dikanka

manufacturer. Until the end of the nineteenth century, the main type of decoration was brick or gothic, characterized by the consistent alternation of dots and spoons in each row. In the late nineteenth and early twentieth centuries, chain dressings, which alternate dotted and spoon rows, with the placement of spoons one above the other became widespread (Fig. 2). Stitching, with open masonry, was performed by different types of stitching—cutting, roller, corner.

In the nineteenth century, several brick factories appeared in Poltava to establish a wide range of bricks, including curly ones (Fig. 3). The serial production of brick parts led to the unification of the architectural decoration of the facades. It was used to create various architectural details—cornices, drafts, plinths, pilasters, etc. [7]. The use of brick for the three-dimensional solution of buildings has given new opportunities to create expressive compositions of facades of different style directions.

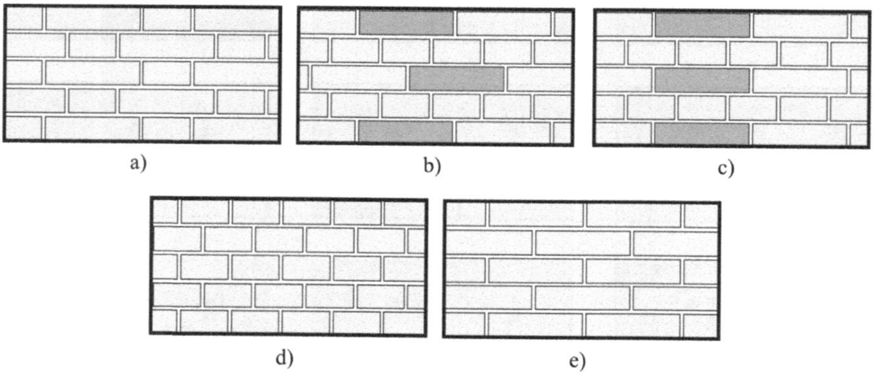

Fig. 2 Types of bricks: **a** verstovy (Old Russian, Gothic, Polish, pole-spoons); **b** cross; **c** chain; **d** dull; **e** false

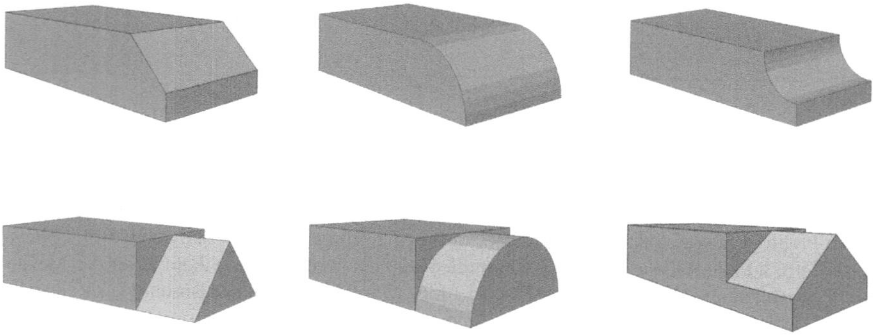

Fig. 3 Types of figured brick

2.2 Principles and Techniques of Applying Brick Decor in the Buildings of Poltava in the Late Nineteenth–Early Twentieth Centuries

On the basis of field surveys and comparative analysis of the brick buildings of Poltava in the late nineteenth–early twentieth centuries, the following principles of the use of bricks in decoration are distinguished:

- *The principle of historicism.* In the architecture of the late nineteenth–early twentieth centuries, there is a phenomenon of historicism, which involves the use of historical prototypes—neostyle [8]. In the architecture of Poltava in the late nineteenth–early twentieth centuries, the neo-renaissance trend is widely represented. Some of the buildings in this area have no stuccoed brick facades: the house of the camomile society (16 Pushkin Str., 1901 by architect O. Shirshov); house of Poltava regional archive, ul. Pushkin, 18/24 (former Zemsky Library building, 1901, architect A. Zinoviev) residential building with a shop, ul. Sobornosti, 24/15 (former excise company building, 1900), residential building on the street. Stritenskaya, 19 (Fig. 4) and others. For the decorative decoration of these buildings, the details of the renaissance—rusting of the facade planes, triangular and arcuate sands, brackets, pilasters and columns were used.

Several buildings are represented by the Gothic. Residential building on Pushkin Street, 40 (former PA Pertsovich apartment building, 1890), administrative building, ul. Sobornosti, 3 (former mansion of doctor Akim Glazer, beginning of twentieth century) (Fig. 5). Gothic techniques are used in the decorative decoration—asymmetry of the composition, arrowheads, biphorias, triphoria spiers, loopholes, etc.

- *The principle of tectonic arrangement of architectural decoration* [9] is most characteristic of the rationalistic direction—"brick style." These objects include residential buildings on the street. Pushkin, 1/12 (1901) (Fig. 6); st. Shevchenko, 13; Gogol St., 15/24 (1900); 41 Shevchenko St. (1900) and others. A characteristic feature of these buildings is the creation of the volumetric composition of the facade only due to the decorative brickwork using figured—curved brick [10]. The decorative elements are located on the eaves, friezes, horizontal drafts, above the windows, in the form of castle stones and sandracks and under the windows, on the vertical articulations of pilasters and niches.
- *The principle of innovation* is implemented in the buildings of decorative and rationalist modernity: a dwelling house on Kotliarevsky Street, 30 (formerly profitable house of merchant S. Kogan, 1905, architect V. Wessel); an apartment building on European Street, 10 (the former apartment building of Samoilovich, 1913, architect P. Klein); hostel of polytechnic technical school, ul. Stritenskaya, 49 (former apartment building of the pharmacist Zaslavsky), residential buildings on the street. Gagarin, 1 (Fig. 7); street Stritenskaya, 27, city mansion on the street Korolenko, 2 and others. These buildings are characterized by the use of forms of facade decoration that do not have historical prototypes.

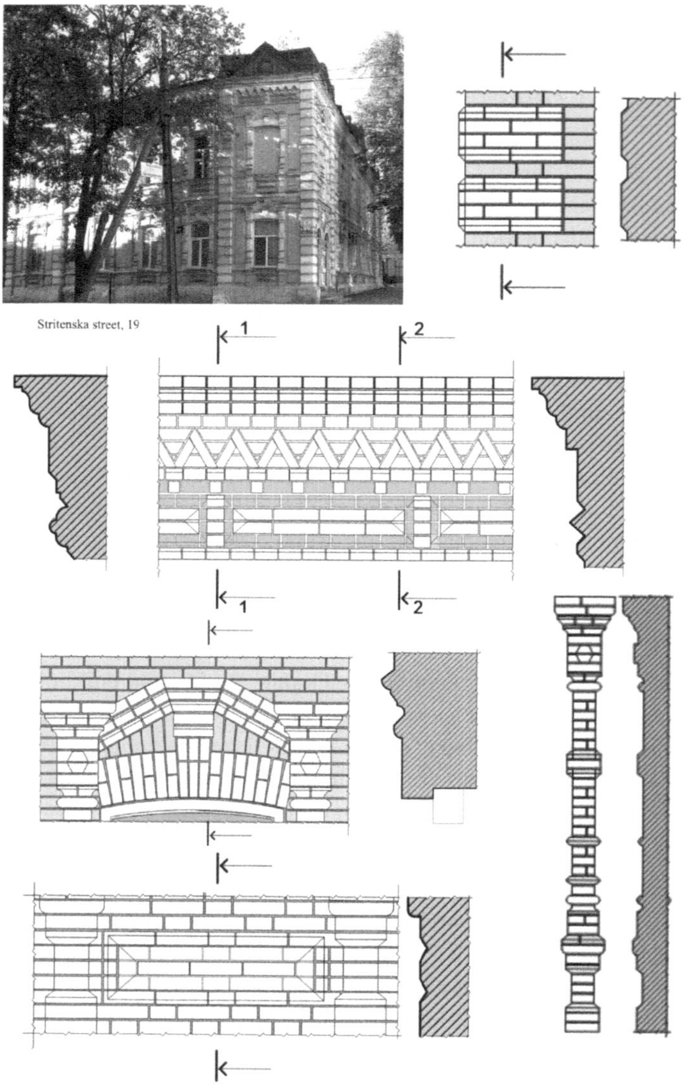

Fig. 4 The principle of historicism (neo-renaissance)

The main options for decor placement in the objects of decorative modernity are:

- decor in the plane of the wall (under the eaves, over-, under- or over- and window sills, inter-storey zones);
- decor on the speakers (bay windows, entrances) [11].

The modernity is characterized by the dynamic linearity of decorative elements based on asymmetric and symmetrical compositions of complex curvilinear forms.

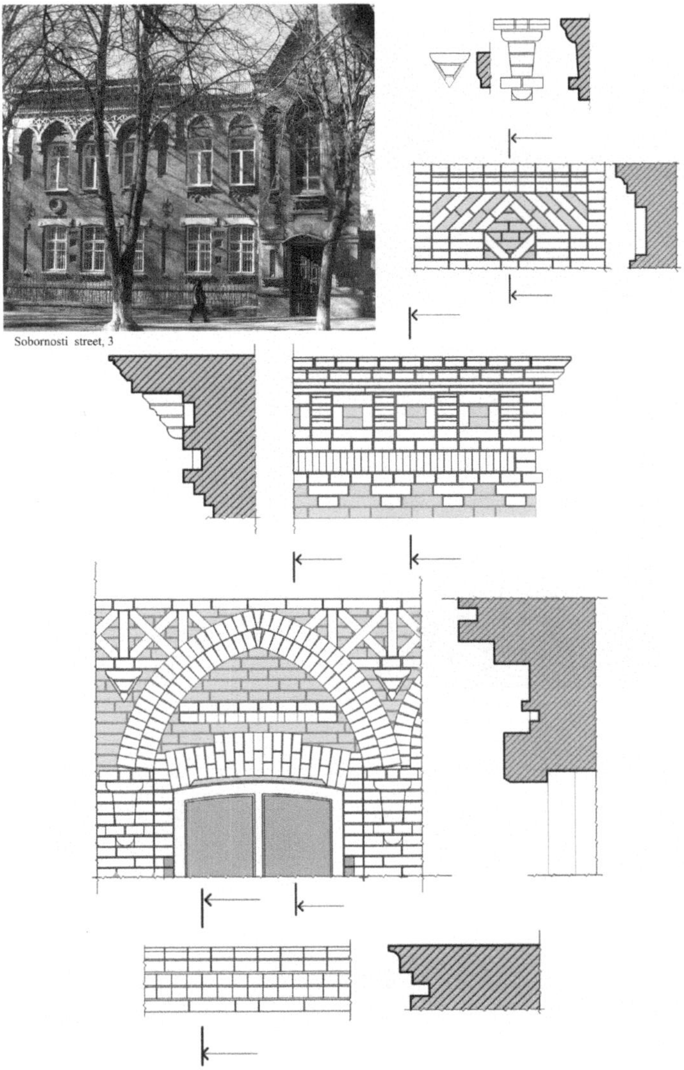

Fig. 5 The principle of historicism (neo-Gothic)

The decoration of modern buildings is individually tailored to each object. The attitude of secession artists made this division for decoration—structural elements were given in a special, unusual configuration, which was also the decoration, and in decorative elements non-bearing, decorative features were emphasized [12]. A rationalist modernist is characterized by the combination of decoration and design, by separating the forms of structural elements.

For brick non-plastered buildings of Poltava of the late nineteenth–early twentieth centuries, the following techniques of decorative decoration are typical:

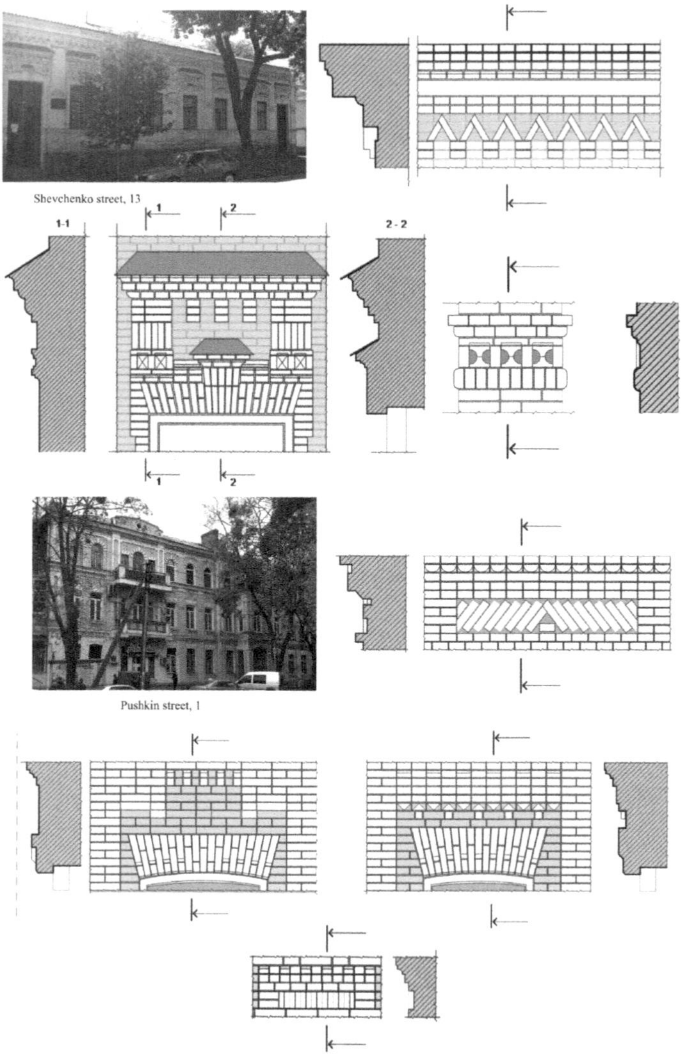

Fig. 6 The principle of tectonic arrangement of architectural decoration

- *The principle of the combination of exposed brickwork with stucco or stucco work*—applied to architectural objects of different styles. In this case, the main plane of the wall was left unpaved, only painted. Decorative elements were brick stucco or stucco.
- *Brick—the main formative element*—the volumetric composition of the facade of the building is created only due to the decorative qualities of the brick and masonry, all decorative elements are made of brick.

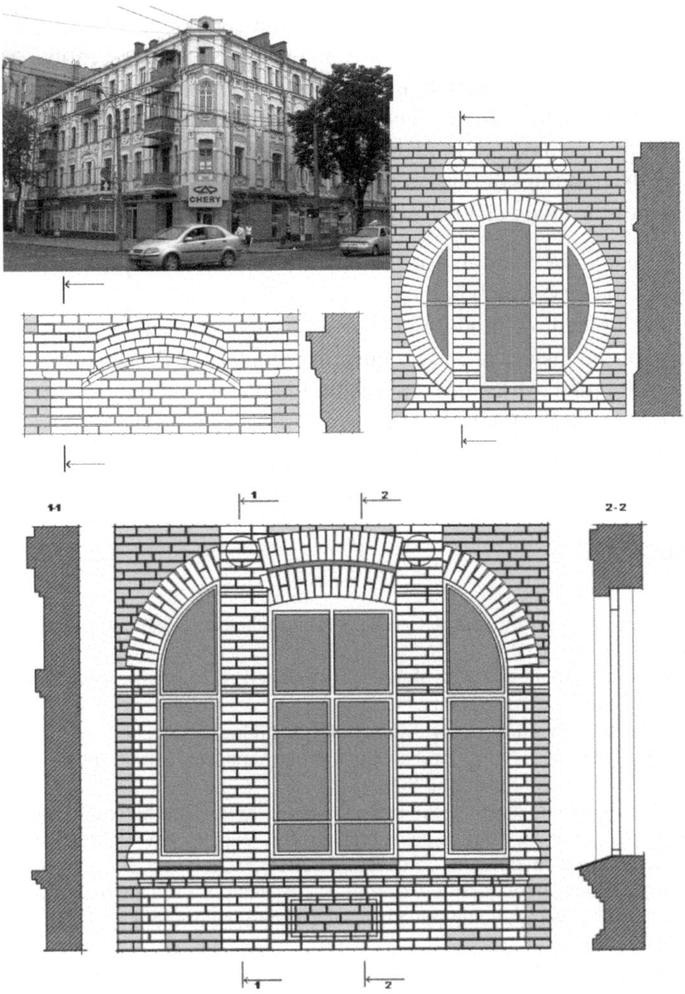

Fig. 7 The principle of innovation. Yevropeiska str., 1

- *The principle of stylistic subordination*—decorative elements made of brick ordinary masonry, which is "adjusted" in the form of architectural details.

3 Conclusion

Unplastered brick buildings of the late nineteenth–early twentieth centuries make up a large part of the construction of the central part of the city of Poltava. Their study is relevant in terms of reconstruction and new construction in the historic city center. During the research, the following principles of decorative decoration of brick buildings were highlighted: historicism, tectonics and innovations. Among the techniques of decoration revealed: a combination of brick masonry with stucco or stucco parts, brick—a formative element, style subordination.

The results of the study are to identify the principles and techniques of applying brick decor in the buildings of Poltava in the late nineteenth–early twentieth centuries.

The results of the study can be used in practical field:

- to develop recommendations for the restoration and appropriate use of historic buildings of the city, which belongs to this period;
- for the development of modern design solutions, made taking into account the peculiarity of the historical and architectural context of Poltava development;
- for the development of lecture courses on the history of architecture of Ukraine and regional features of architecture.

References

1. Ikonnikov, A. (2001). Arhitekrura HH veka. Utopii i realnost. Izdanie v dvuh tomah. T.1. M., Progress – Traditsiya.
2. Istoriia ukrainskoi arkhitektury/Yu. Asieiev, V. Vecherskyi, O. Hodovaniuk ta in.; Za red. V. Tymofiienka. Kuiv, Tekhnika (2003).
3. Aseev, Y. (1989). *Styly v arkhytekture Ukraynu*. Budyvelnuk: Kuiv.
4. Horiunov, V., Tubly M. (1994). Arkhytektura epokhy moderna. Kontseptsyy. Napravlenyia. Mastera P.– S.-Peterburh: Stroiyzdat SPb.
5. Abramiuk I. Zasoby vyraznosti «tsehlianoho styliu» v zhytlovii arkhitekturi Lutska. Arkhitekturna spadshchyna Volyni. Zbirnyk nauk. prats. Rivne, Vyp. 2., 188–193 (2010).
6. http://www.v-smirnov.ru/coll.htm. Last accessed 2019/09/21.
7. Arkhitektory i misto: [monohrafiia]/ Kariuk M., Bieliavska O., Dmytrenko A.., Nehai H., uklad. ta red. Bieliavska O., Poltavska orhanizatsiia Natsionalnoi spilky arkhitektoriv Ukrainy, Poltava, TOV «Asmi» (2018).
8. Linda S. Historicism in the architecture of Lviv: tendency across centuries//Architectus. – # 2 (38), 45–52 (2010).
9. Ponkalo, S. I. (2016). Pryntsypy ta pryiomy zastosuvannia arkhitekturnoho dekoru na budivliakh Lvova kintsia XIX st. – pochatku XX st., Suchasni problemy arkhitektury ta mistobuduvannia, K: KNUBA, Vyp. 45, 98–108.
10. Dekoratyvne ozdoblennia budivel Poltavy 1900 – 1910-kh rr./T.Skibitska, I.Shuleshko// Studii mystetstvoznavchi, №2, 65–72.
11. Ivashko, Y. (2013). Osnovy styleutvorennia modernu v arkhitekturi Ukrainy (kinets XIX - pochatok XX stolittia): avtoref. dys. d-ra arkhit.: 18.00.01/Yu. V. Ivashko; Kyiv. nats. un-t bud-va i arkhit., K.
12. Kazantseva, T., Ponkalo, S. (2014). Classification of facade decoration of secession buildings built by I. *Levynsky company in Lviv, Przestrzeń i Forma*, №21, 171–182.

Second Life of the Residential Building Area of the Middle of the 50s—Early 80s of the Twentieth Century in Ukraine: Opportunities and Perspectives

L. S. Shevchenko

Abstract The object of scientific research is the residential buildings of the mid-1950s and early 1980s. The residential environment of that period is predominant in the central historical districts of most Ukrainian cities. The purpose of this publication is to analyze the domestic housing environment of the mid-50s and early 80s of the twentieth century, to find out its status and to find the solution about their future existence. The novelty of this work is to identify such methods of modernization of residential areas of the outlined period, which could be adapted during the experimental design into the existing living environment of the modern city, in particular—Poltava (Ukraine). The study provided on two levels—theoretical and empirical. The author's proposals for one of the residential quarters of the central part of Poltava (Ukraine) are novelty for both inhabitants and authorities. Their practical importance lies in the possibility of introducing real positive changes in the quality of the living environment, physical, and moral conditions of its buildings and territories.

Keywords Residential building · Living environment · Housing stock · Modernization · Structure · Constructive elements

1 Introduction

Second half of the 50s of the twentieth century is characterized by massive construction of typical and fast-moving residential buildings and quarters. This was typical not only of Ukrainian cities, but also of the majority of cities of the post-Soviet countries. At that time, it was one of the ways to quickly resolve the post-war housing crisis. The majority of scientists consider that it was a breakthrough in housing construction, because since over 224,545 m^2 of housing were put into operation in only 15 years (from 1951 to 1965). This is convincingly evidenced by the schedule

L. S. Shevchenko (✉)
Poltava National Technical Yuri Kondratyuk University, 24, Pershotravneviy Ave, Poltava, Ukraine
e-mail: Ls.shevchenko@ukr.net

V. Onyshchenko et al. (eds.), *Proceedings of the 2nd International Conference on Building Innovations*, Lecture Notes in Civil Engineering 73,
https://doi.org/10.1007/978-3-030-42939-3_45

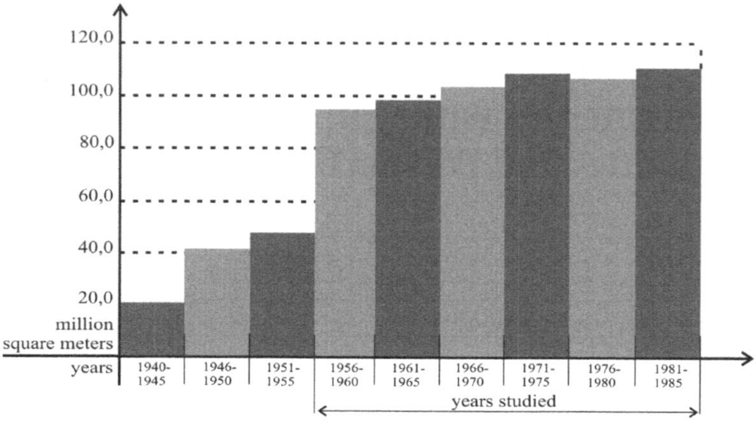

Fig. 1 Housing volumes in the USSR (based on data [1])

of housing growth in the Soviet Union during the study period compared with the previous years (Fig. 1).

Residential development was the result of engineering research and a solution to the actual problem of the lack of own housing in the cities of that time Soviet Union. In particular, in Ukraine, this process took place using industrial house building technologies, which are characterized by strong typing and unification of structural elements, planning schemes, and space-spatial solutions. "The standardization and industrialization of apartment buildings and the construction of large housing estates or new towns were global phenomena that seemed and efficient solution to the post-war housing shortage" [2].

Such buildings were not only typical of the cities of the post-Soviet countries. Similar construction was carried out in other European countries—Germany, France, Sweden, Denmark, and Finland. New residential areas have appeared here with typical four-story and five-story residential buildings. On the one hand, the housing environment of the cities has acquired a common and rather modest panel face. On the other—it has received the most effective way of mass-housing construction with new technologies and constructive solutions. Such housing was cheap, comfortable enough and agrees with the social standards at the time. In addition, this was a new higher level of comfort for the owners of such dwellings despite the modest appearance of the houses and the simplicity of their functional organization. "There is an utter uniformity in detail and construction technique used throughout all the houses. All the parts were standardized mass-produced components that could be arranged in various compositions, which in the end proved to create a diverse and provocative system. Use of repetitive, standardized elements in varied combinations emphasizes the fundamental dichotomy between industrial production and a poetic architecture" [3]. The inhabitants of the five-story residential building had the opportunity to use territories around the house and internal courtyards. It was where the Soviet life was concentrated. They had a good planning with a track system, beautiful green spaces, playgrounds, and children's pools. There was formed a unique microclimate with a developed social infrastructure: domestic services, kindergartens, nurseries, schools, clinics, and shops.

2 Directions, Scientific Methods, and Source Base of the Research

The object of the scientific research is the residential buildings of the mid-1950s and early 1980s. The residential environment of that period is predominant in the central historical regions of most cities of Ukraine. Buildings of this type of the first generation of complete housing are designed for a 125-year life span. They have stood for 40–50 years. Some of them, least durable and not promising will be demolished. Other must be overhauled or upgraded according to the needs of today. A number of previous studies have shown that the life span of such housing can be extended up to 150 years. But it will be with timely overhaul. So, the relevance of this research stems from the agenda—what to do with the housing stock of cities in the post-Soviet territories? How to strengthen the structural system of such buildings in case of their modernization? How to change the gray panel face of residential quarters in the central parts of the cities? Is it worth modernizing such buildings at all?

Therefore, the purpose of this publication is to analyze the domestic housing environment of the mid-50s and early 80s of the twentieth century, to find out its status and to find possible answers to the above questions. The novelty of this work is to identify such methods of modernization of residential areas of the outlined period, which could be adapted during the experimental design into the existing living environment of the modern city, in particular—Poltava (Ukraine).

The author of the paper studies the housing on two levels—theoretical and empirical. The first step, in this case, is the accumulation of necessary information and its systematization. Libraries, archives, and special funds of project research institutes came to the author's help. This made it possible to identify the main typical design decisions of the study period. There are 2–3-story brick residential buildings of 1955–1960 series 1-255; 4-5-story brick houses of 1954–1961, series 1-411 and 1-433; 5-story panel houses of 1958-1967, series 1-464, 1-335, 1-434, 1-434C, 1-464A, 1-335A, and MK-5; 5-story brick houses 1966–1970, series 1-OPB for small families; 9-storied panel houses 1968–1976, series M-464; 12-story panel houses of 1973–1985, series of M-464 and M-335; 9-story prefabricated panel houses 1976–1985, series M-464, M-335-BC; large panel 1-480 series.

The analysis of historiographic materials, scientific works of previous researchers, architectural and design materials, and typical design decisions are systematically classified according to the main components. The Ukrainian scientific community has come together to find solutions to this problem. Noteworthy are the scientific publications of N. Demin, Ye. Klyushnichenko, G. Lavryk, I. Gnes, M. Byvalina,

M. Gabrel, and others [4–6]. Scientists not only outline the problem of outdated residential buildings in major cities of Ukraine (Kiev, Odessa, Kharkiv, Dnipro, Lviv), analyze its condition, but also express their own considerations regarding the humanization of such living environment in accordance with current requirements and needs. In this case, the reconstruction of residential buildings is considered in all aspects—spatial, economic, social, technical, and aesthetic [6]. Some of the scientists are trying to solve the issues outlined by modernizing the housing stock of the study period. Among them are A. Schreiber [7], S. Bulgakov [8], K. Kartashova [9], Ye. Poletayev [10] and others. Interesting for us is the practical experience of the European countries of the former socialist camp on the modernization of housing environment, which became the object of their own research by scientists Kostetsky [11], Lukmanova, and Slobodeniuk [12] and their results.

As we can see, these problems are also on time in the post-Soviet countries. This is evidenced by publications of foreign colleagues, including K. Snopek, A. Martin, D. Erickson, R. Lorens, M. Branczik, M. Benko, J. Muliuolyte, and others [13–15]. Solution of these problems in post-Soviet areas is also interesting for foreign specialists far beyond Europe [16]. The European experience in the conservation and renewal of the housing stock under study shows that its reconstruction and modernization are achieving positive social, technical, and economic results.

Housing building in the mid-50s–early 80s of the last century is the object of study not only for architects, but also construction practitioners, designers, and engineers. They are looking for real practical possibilities of modernization of buildings in accordance with modern requirements of heat engineering, acoustics, and strength of structural elements. Effective practical results of researches are covered in works of Ukrainian scientists and engineers V. Balytskyi, V. Andrukhov, L. Martynov, O. Semko, O. Filonenko, and others [17–20]. An important aspect of the implementation of the proposed solutions is the legislative database. The Minregionstroy and Ukrainian State Scientific and Research Institute of Design Cities "Dipromisto" named after Yu. Bilokon (Ukraine) have developed a new bill "On complex reconstruction of quarters (neighborhoods) of outdated housing stock" [21], which outline favorable conditions for the practical implementation of reconstruction measures in Ukraine.

The empirical level of the study is based on a visual survey of existing (realized) objects of residential buildings of the outlined period. It makes possible to confirm the authenticity of the scientific search results and the correctness of the applied empirical and theoretical methods. In addition, it is an integral part of the process of accumulating factual material. A visual survey of the objects under study gives them a real visual response, including an assessment of the technical condition of the residential buildings. In order to fix the current state of the objects, photographs are taken. They are the most accurate documentary images of the objects.

Further processing of materials is based on historical factual, monographic, and topographical methods, in which the objects of study are the living environment of different levels of urbanization and its landscape (yard) organization. The historical and factual method has made it possible to investigate the historical basis of the formation of residential buildings of the study period, its structural elements. Using a monographic method, it was possible to investigate in detail specific residential buildings in an urban structure. In substantiating the specifics of architectural planning and structural design of residential buildings, a comparative-historical method of research is used. It is based on the comparison of domestic objects with their analogues, not only in the post-Soviet space, but in Europe as a whole.

3 The State of Residential Buildings in the Mid-50s–Early 80s of the Twentieth Century

The presentation of the material should begin with the results of a thorough analysis of residential buildings in the mid-1950s and early 1980s. The planning organization of the housing areas, the characterization of each type of buildings by architectural planning and design decisions, their functional organization are found out. Some scientists believe that the value of five-story residential buildings is not so much in the buildings as in the planning structure of the neighborhoods they form. They had good planning with a system of tracks, green areas, and playgrounds for children and swimming pools. For the first time in the territory of these residential yards, complex landscaping with perennial plants was applied in the Soviet Union, minimizing the effects of noise and wind.

With all the modesty of exterior, the dwelling houses of the studied period were a breakthrough—both in the level of comfort received by their inhabitants at that time and in the technology of construction. According to M. Sutavičius, "mass-housing program had to be based on the construction of the most cost-effective, standardized buildings. On Khrushchov's initiative, experts were sent to France to study the precast systems and to acquire the Camus system patented by the French engineer R. Camus in 1948. Later new buildings methods from prefabricated concrete panels were developed" [22]. After the All-Union Assembly of Builders in 1954 and the subsequent decisions of the Central Committee of the CPSU and the Meeting of Ministers of the USSR on capital construction, a new material and technical base for the construction of housing by industrial methods was created at a rapid pace. These were the buildings from large elements of industrial production. The main role in the mass construction of the 50–60s of the twentieth century went to technology. In the first phase of the transition to new methods of industrialization, an experimental construction was of great importance. During this process, various planning designs of a dwelling house, methods of building residential groups and residential districts, landscaping, basic structural solutions, elements, components and assemblies were elaborated and tested (Fig. 2).

For the whole period of existence of the residential building under study, scientists of different generations made its thorough and comprehensive analysis [4–6, 8, 14, 17–19, 22]. The author made a comprehensive analysis of such living environment in the central districts of Poltava (Ukraine). The main purpose of the analysis was to find out the possibility of modernizing the living environment in accordance with the

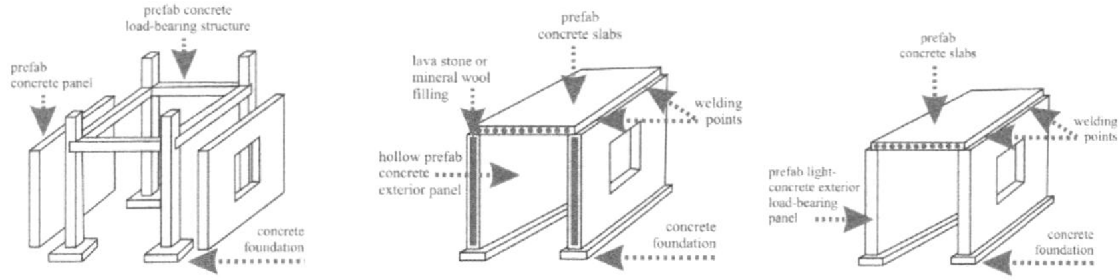

a - precast load-bearing concrete structure to which solid panels of precast concrete were attached

b - load-bearing large concrete panels that were joined together by welding reinforced bars and pre-installed metal plates on several points of the perimeter

c - load-bearing large concrete panels that were welded together, but exterior panels as well as interior ones were not hollow

Fig. 2 Basic structure typologies for residential buildings (by Sutavičius [22])

contemporary needs, opportunities, and challenges of the society. By systematizing the results obtained, we have the opportunity to summarize the moral and physical deterioration of residential buildings in the mid-1950s and early 1980s:
moral and aesthetic aspect:

- the uniqueness and lack of clarity of the architectural design, the similarity of the appearance of buildings and quarters as a whole in different cities and countries. This led to the "degradation of the exterior of cities" [18], the loss of features of architectural, historical, and cultural traditions of different regions of the country;
- uncomfortable though in some places homely planning of quarters, outdated equipment of courtyards, neglected economic territories. This situation complicates the functioning of the living space with the current transport load and changes in the recreational filling of the yard;
- moral aging of interior planning designs, that is—minimal room areas, the presence of adjacent rooms, and unusable small kitchen area (inability to equip modern household appliances). It does not correspondent to the ergonomic needs of modern housing and its architectural and planning standards.

physical and constructive aspect:

- neglected state of internal power grids that cannot withstand the load of modern household appliances;
- wearing-out of all engineering networks;
- low thermal conductivity of external walls (3–5 times lower than the minimum permissible in accordance with modern requirements [18, 20]). Consequently, high energy supply in the cold season and significant overheating of the premises during the warm season;
- emergency condition of joinery;
- nonconformity of internal walls and partitions with modern standards on sound insulation;
- emergency condition of architectural and structural elements—balconies, joints between slabs, parapets on roofs, staircases;
- deterioration of protective structures of buildings.

socio and economic aspect:

- "chaotic settled" in the first floors of residential buildings;
- financial inability of residents of the investigated dwelling houses to carry out their modernization by their own efforts;
- absence of an "effective financial and investment mechanism" [18];
- limited budget funds for reconstruction works.

The results obtained indicate the novelty and timeliness of this study. In general, all of these problems are related to two main points. The first is the real physical wearing of the basic engineering and structural components of buildings (Fig. 3). The second is moral wear and non-conformity with modern construction standards and comfort criteria.

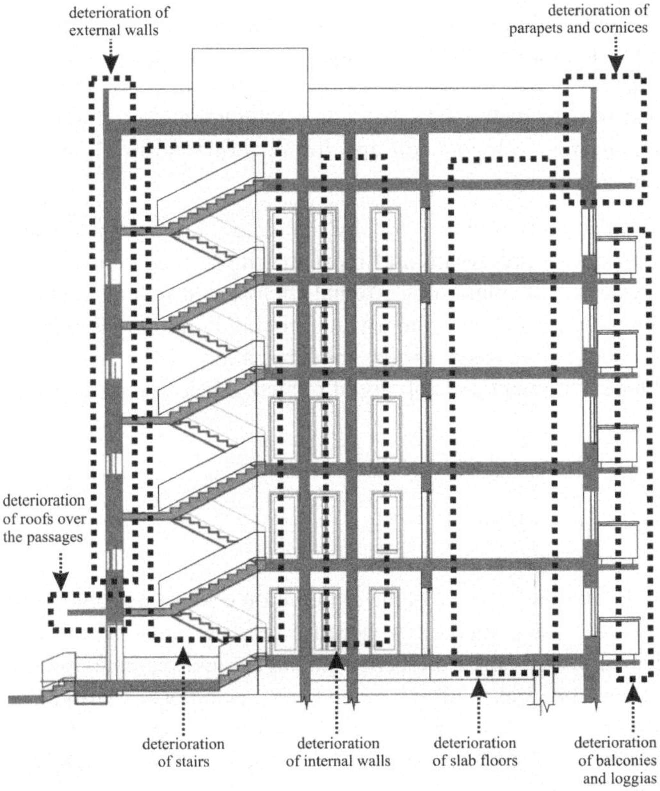

Fig. 3 Main physical and structural damage to the elements of residential buildings (according to [6, 18, 19, 22])

4 Perspective Approaches to the Modernization of Residential Buildings in the Mid-1950s–Early 1980s

The scientific community has formed two fundamentally opposite approaches to solving the questions raised previously. The first is aimed at demolition of this residential inheritance and construction in the vacated places modern multi-story comfortable housing in the central areas of the city. The second is aimed at modernization of these residential buildings. The results of the analysis of predicting of a more favorable approach show the benefit of modernization. The basis—the problems of recycling a significant amount of construction waste during the demolition of these houses, the economic justification for the elimination of satisfactory technical condition of the housing, which has a real market value [18]. In addition, the basis for supporting the second approach is a positive experience of housing reconstruction in the foreign countries of the post-Soviet space, increasing the strength of the main load-bearing structures of residential buildings from the design and analysis of the degree of residual operational resource for individual structural parts of residential buildings in our country (load-bearing walls, slabs, balconies, foundations and stairs [19]). A number of professionals are developing concepts for modernizing such housing stock. There are two main approaches:

1. *thrifty*, working to eliminate the physical deterioration of residential buildings. It is based on the preservation of most structures, their reinforcement in places where necessary, insulation of external enclosure structures and so on;
2. *cardinal*, which is aimed primarily at eliminating the moral deterioration of the housing stock. It is based on the modernization of the apartments with their radical redevelopment and solving of related physical and structural problems.

4.1 Thrifty Direction

Supporters of this trend argue that the elimination of design problems interfere with the structural features of buildings, because they are made of durable material—concrete. Partitions are often the communications and elements of stiffness of the wall. They cannot be weakened and even more they cannot be rearranged. Specialists, among whom are designers, architects, builders, offer to keep planning, but to occupy apartments with a few families waiting. To compensate for the lack of utility rooms, allocate a little more floor space to them than provided for by housing standards. In this case, the redevelopment is recommended to be limited to the minor improvements—an extension of the kitchen-niche with a gas hob to electric; an increase in the area of the hallway at the expense of the common room; an increase of the size of a room by incorporating a loggia into its structure. A large number of specialists are inclined to reconstruct the local character, in particular, buildings to attach volumes of kitchen-bay windows in the form of rigid three-dimensional structures. It will

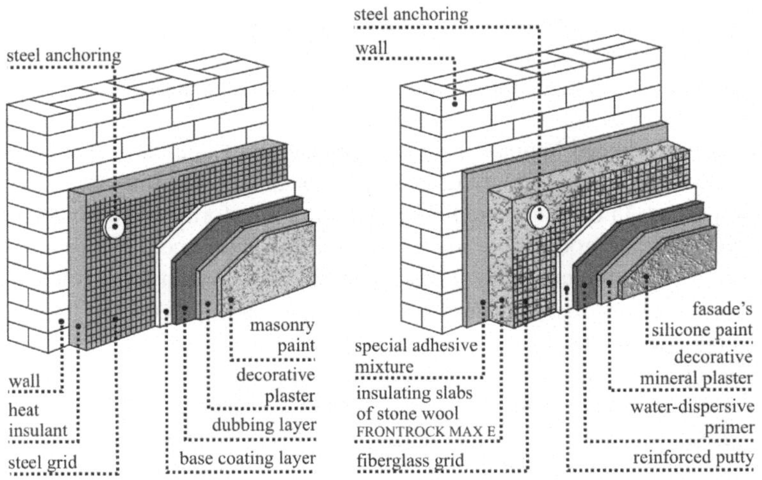

a - sample of fasade thin-layer plaster system, proposed by [20]

b - thermal insulation of the external walls using insulating slabs of stone wool FRONTROCK MAX E (ROCKWOOL company proposal [23])

Fig. 4 Options for thermal modernization of exterior wall structures for large-block and brick residential buildings

provide additional spatial stability of the whole building, protecting it from potential destruction.

Resolving physical problems of residential buildings is aimed at a significant improvement in thermal and sound insulation indicators. Variants of insulation of external walls and slabs of floors and coverings are developed and offered by both scientists and companies-manufacturers of thermal insulation materials, which is shown in Fig. 4 [20, 23]. According to estimates by specialists [19], such actions contribute to saving heat costs for the existing building by 30–40% and improving the thermal regime in the apartments (temperature increase from 18–20 °C to 22–23 °C).

4.2 Cardinal Direction

This direction is based on the interior redevelopment of apartments, aimed at bringing them to the current building standards and modern comfort requirements. Such modernization consists of the completion of several floors, an attic floor, additional areas in the form of bay windows and insulated loggias. In addition, work is done to strengthen the main load-bearing structures of the house, to improve thermal and sound insulation characteristics.

Architectural and planning reorganization of the internal living environment can occur with a minimal or significant cost. At minimal cost, the room remains in the same internal bearing structural elements. Increasing the space of rooms and improving the functional organization of the apartment are due to the completion of bay windows, insulated loggias, and the organization of open terraces. Bathrooms are separated by a partition, and the area of rooms and kitchens is increased, which is very important. In Ukrainian traditions, there is a transformation of kitchens into a room (kitchen-dining room) for family members and even reception of guests. With such solution, the structures are not broken. Reconstruction may consist of sanitary-engineering works on the transfer of risers and connection of devices. The status of the dwelling does not change.

Significant actions include the involvement during modernization additional space from adjacent apartments with changing status of the apartment. Such reorganization techniques allow to improve the functional processes (organization of utility rooms), to increase the parameters and the number of rooms in the apartment, to organize a large outdoor terrace, and to greatly increase comfort. But, at the same time, such modernization leads to a reduction in the number of apartments within one floor, and consequently to a house in general. Each of these directions is able to solve "related problems" connected with the elimination of transit rooms, insulation of buildings, and the adaptation of the first floors for people with disabilities, improving communication links horizontally and vertically of buildings, providing the required number of parking places, functional areas in yard spaces, and so on.

The result of the scientific searches of the author was the proposal for the modernization of the quarter with residential buildings of the studied period in Poltava (Ukraine), bounded by Sobornosty, Nechuy Levytsky, Zygin streets, and Klubny lane. The analysis of the area of the quarter and the plans of residential buildings showed the non-conformity of the architectural and planning design to the current normative requirements and the modern level of comfort of living, physical, and moral demolition of the buildings of the quarter. Among the most relevant:
architectural and planning direction:

- uniformity of construction;
- narrow and right-of-way;
- the absence of turning and cutting grounds;
- lack of temporary places for cars (as a result of placing cars near children's playgrounds);
- outdated improvement;
- lack of modern equipment elements (both game and physical culture);
- narrow corridors and stairwells, transit rooms.

structural and functional direction:

- physical and moral demolition of buildings;
- low thermal and sound insulations;
- the presence of cracks at the joints of reinforced concrete slabs;
- lack of technical floor (or it is small);

- lack of elevators, waste tray;
- critical deterioration of communication ladders;
- low energy efficiency of buildings;
- low level of comfort of living in apartments;
- minimum area of apartments;
- bash, combined with WC.

Using a method of experimental design, a project proposal was developed aimed at eliminating the identified defects and bringing this living environment to modern needs. Figure 5 presents a fragment of the author's project proposal for the modernization of the living environment on the example of one-room apartment in the building of the late 50s–early 80s of the twentieth century.

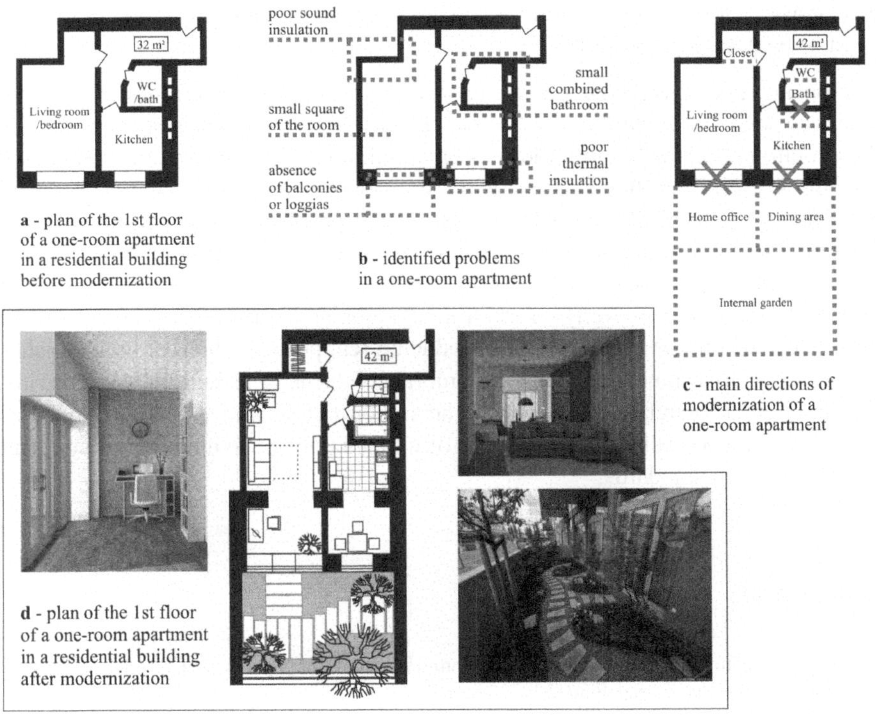

Fig. 5 Main physical and structural damage to the elements of residential buildings (according to [6, 18, 19, 22])

5 Conclusions

The problems outlined in the article on the continued existence of residential buildings in the mid-1950s and early 1980s are relevant not only for post-Soviet countries. The research activity is related to the urgent need to modernize such living environment. The worldwide practical experience of solving the problem of five-story dwellings has revealed the most used methods of their modernization and, consequently, transformations into a modern comfortable living environment:

- reconstruction with replacement of communications, insulation, and unique architectural exterior;
- reconstruction with replacement of communications, insulation, unique architectural exterior, and completion of the attic floor;
- extension of the building, reinforcement of structural elements, and completion of several additional floors.

Housing renovation shows that its reconstruction and modernization are achieving positive social and economic results. This is quite possible in Ukraine as well, with the adoption of a new bill that foresees the possibility of reconstruction of outdated dwellings of different floors, expands the sources of financing for such works (and not only by authorities), and helps to resolve the problem of compensation to the owners of such dwellings and other problems related to the process of modernization.

The author's proposals for one of the residential quarters of the central part of Poltava (Ukraine) are novelty for both inhabitants and authorities. Their practical importance lies in the possibility of introducing real positive changes in the quality of the living environment, physical, and moral condition of its buildings and territories. With her research and experience, the author tried to join a group of scientists obsessed in a good sense with the ideas of humanizing the living environment and raising its social standards.

References

1. Housing in the USSR and the RSFSR from 1918 to 1990. (2016). Retrieved from https://burckina-faso.livejournal.com/1527935.html.
2. Plessein, B., & Tsaplev, N. (1981). Developments in large-panel construction. Batiment International. *Building Research and Practice, 9,* 348.
3. Demchak, G. (2000). *Towards of post-industrial architecture: Design and construction of houses for the information age*. Massachusetts: Massachusetts Institute of Technology.
4. Demin, N., & Byvalina, M. (2005). Town-planning and socio-economic issues of reconstruction of the territory of the five-storey large-panel building. *Urban planning and territorial planning, 20,* 90–94.
5. Byvalina, M. (2016). Problems of modernization and reconstruction of urban territories built by first series large panel buildings. *Modern problems of architecture and urban planning, 45,* 146–151.

6. Gabrel, M. (2016). Problems and principles of humanization of the residential environment of residential buildings of the 70's of the XX century. *Modern problems of architecture and urban planning, 45,* 160–169.
7. Schreiber, A. (1993). Technical and economic assessment of organizational and technological solutions when designing the reconstruction of residential buildings. *Construction Economics, 3,* 25–27.
8. Bulgakov, S. (2001). *Reconstruction of residential buildings of the first mass series and low-rise residential settings.* Moscow: Globus.
9. Kartashova, K. (2003). Reconstruction of urban housing taking into account modern social needs. *University News. Stroitelstvo, 7,* 125–131.
10. Poletayev, Ye. (2003). On the issue of reproduction, conservation and modernization of the housing stock. *Construction Economics, 10,* 62–64.
11. Kostetsky, N. (2003). Foreign experience in the reproduction of the housing stock, its conservation and modernization. *Construction Economics, 5,* 33–45.
12. Lukmanova, I., & Slobodeniuk, S. (1997). Experience in the reconstruction of panel buildings in Germany. *Construction Economics, 3,* 48–54.
13. Branczil, V. (2012). Planning of standardized housing types in Hungary in 1948–1960. *Architecture & Urbanizmus, 46*(3–4), 180–194.
14. Benko, M. (2015). The lifespan of large prefabricated housing estates in post-communist cities: An international comparison. *Architecture & Urbanizmus, XLIX*(3–4), 180–197.
15. Muliuolyte, J. (2013). Rediscovering large scale housing estate in post socialist cities. *Journal of Architecture and Urbanism, 37*(1), 51–58.
16. Molnar, V. (2013). *Building the state: Architecture, politics, and state formation in postwar Central Europe.* New York: Routledge.
17. Balytsky, V., Franivsky, A., & Skrypka, D. (2006). "Khrushchevka"—Reconstruction without resettlement of residents. *Budivnitstvo Ukraine, 7,* 11–17.
18. Andrukhov, V., Kolesnyk, A., Martynova, L., & Matviychuk, V. (2010). Assessment of the technical condition of residential buildings of the first mass series of industrial erection and options for their future prospects. *Budivnitstvo Ukraine, Том, 8*(1), 103–111.
19. Andrukhov, V., Kolesnyk, A., Martynova, L., & Otamanenko, M. (2011). Ways to restore physical deterioration in buildings of the first mass series of buildings. *Budivnitstvo Ukraine, Том, 10*(1), 118–125.
20. Filonenko, O., Yurin, O., & Kodak, O. (2018). Energy performance of residential buildings. *Series: Industrial Machine Building, Civil Engineering 1(50),* 189–196. https://doi.org/10.26906/znp.2018.50.1075.
21. Law of Ukraine of December 22, 2006 No 525 "On complex reconstruction of quarters (neighborhoods) of outdated housing stock". (2012). Retrieved from http://zakon3.rada.gov.ua/laws/show/525-16.
22. Sutavičius, M. (2014). *Mass-housing: Tendencies and modernization* (Research paper). Retrieved from https://www.google.comsearchsource=hp&ei=vUFIXb_RIq6OrwSttKuwBg&q=MODERNIZATION+OF+RESIDENTIAL+HOUSES+OF+THE+MIDDLE+OF+THE+50TH+-THE+BEGINNING+OF+THE+80TH+YEARS+OF+XXTH+CEN.
23. Catalogue ROCKWOOL®. (2018). Retrieved from https://www.rockwool.ua/frontrock-max-e-ua/.

Residential Marinas and Marina Villages on Inland Waterways

Matvey Shkurupiy, Volodimir Nikolaenko, Yulia Kuznetsova, and Tamara Kutiak

Abstract The purpose of this article is to systematize and specify the existing scattered material which deals with the design and modern operation of residential marinas, as well as to determine their main architectural and planning features. In the article, the author considers the world experience of the residential marinas architectural and planning organization, such as marina villages, with allowance for functioning on inland waterways. An attempt was made to explain the terms "residential marina" and "marina village," general provisions for the architectural–planning organization of marina villages are given, and a brief description of the statistically average marina village development and its features are defined.

Keywords Residential marina · Marina village · Inland waterways · Housing · Complex · Development

1 Introduction

1.1 Problem Definition

The study relevance of "residential marina" and "marina village" architectural and planning structure, their terminological definition and the features of functioning on inland waterways has been determined:

M. Shkurupiy (✉) · V. Nikolaenko · Y. Kuznetsova · T. Kutiak
National University «Yuri Kondratyuk Poltava Polytechnic», Pershotravnevyj Ave 24, Poltava 36011, Ukraine
e-mail: shkurupiy1990@gmail.com

V. Nikolaenko
e-mail: Dastam@i.ua

Y. Kuznetsova
e-mail: yulia_france@ukr.net

T. Kutiak
e-mail: abricoskcomka@gmail.com

V. Onyshchenko et al. (eds.), *Proceedings of the 2nd International Conference on Building Innovations*, Lecture Notes in Civil Engineering 73,
https://doi.org/10.1007/978-3-030-42939-3_46

- firstly, by the steady growth of their number which is related to the development of a living sustainable trend inside the marina with a certain contingent of shipowners and visitors;
- secondly, by their direct urban development impact on the development of nearby communities;
- thirdly, their significant urban development potential, since they can act as tourist, recreational and public urban centers, accumulating a certain amount of labor, material and technical resources.

1.2 The Analysis of Recent Research Sources and Publications

At the moment in professional literature, there is not enough material on this subject.

Nevertheless, among the existing works on the problem of designing, arranging, and developing marinas, including residential ones, the following specialists should be noted: A. Bogomolov (Ph.D.), B. Tobiasson, C. Thoresen, D. Carevic, D. Natchez, G. Tsinker, L. Gucma [8], M. Pechenik (Ph.D.), Michael J. Oldham [12], N. Grishin [7], P. Novoselov, S. Roff, T. Luković [9, 10] and Z. Nagayeva (Grand Ph.D.). Also, the article used materials from a number of specialized organizations: Association of Inland Navigation Authorities [1], British Waterways Association [3, 4], Global Marina Institute [6] and American Society of Civil Engineers [14].

1.3 Scientific Novelty

"Residential marina" and "marina village" are fairly new urban development objects. Therefore, the article attempts to define terms "residential marina" and "marina village", classify them according to determining features, analyze world experience of the architectural and planning organization of marina villages, highlight their main characteristics and give examples of the largest modern marina villages.

1.4 Practical Significance

The results of this study can serve as a theoretical basis for the development of practical recommendations and regulatory documents regarding the placement of "residential marinas" and "marina villages" in the development of yacht tourism on inland waterways.

2 The Main Material and Results

2.1 Terms Definition: «Residential Marina», «Marina Village» and «Residential Mooring»

For the moment, the yachting industry and tourism existing conceptual apparatus do not provide a sufficiently adequate idea of the «residential marina» and «marina village» concepts. The absence of officially fixed explanations led to a large number of interpretations. This situation is also complicated by the lack of a single comprehensive official statistics on yacht infrastructure in Europe. «It is hard to give specific data on the size, type and capacities of the marina industry in Europe because there is no professional association that would collect, process, research, unify the data and encourage the development of the industry» [10, p. 405].

Nevertheless, at this stage of the yachting industry development, the terms "residential marina" and "marina village" have the following meanings.

«Residential marina» is a generic name for the marina group, including marina villages, which are characterized by the availability of residential areas, public service facilities and infrastructure in their planning structure, together providing all conditions for comfortable living inside the marina.

«Residential marina provides freehold or strata title permanent accommodation adjoining the marina» [6, p. 18]. «A shoreline classification involves the non-project use of project lands and waters for facilities where watercraft can be launched, retrieved, or moored for the purpose of providing access to the lake for certain residential property owners (e.g., off-water and project-front lots, non-transient campgrounds, multi-family dwellings). Residential properties associated with this classification include townhouses, condominiums, apartments, some campgrounds (long-term campgrounds >14 days) and subdivision access lots» [5, p. J-89].

However, residential marinas in Europe, especially in Great Britain, where a whole culture of living on the water has developed over several centuries, which are usually called separate special marinas (narrowboat marinas) or yacht marinas in the structure of which «residential moorings»—specially equipped and intended for the parking of narrowboats (English version of the houseboat, *author's note*) used as a permanent home are envisaged. The structure of such sites includes obtaining a special license for their construction and equipment, tax payment and registration of a fixed residence place. One of the features of such sites is the ability to leave it on the watercraft for a certain period with the obligatory condition of returning to the registration (parking) place [3, 13].

Marina village is a fundamental concept of the architectural and planning association of residential development, public service facilities and marina [11, p. 129].

«At the beginning of its appearance, this term meant the type of residential development directly on the seashore. Today, they are called harbors with marinas, residential buildings with a marina and a zone with individual houses, having a place to moor the yacht» [2, p. 245]. «The most promising yachting objects are marinas with a developed infrastructure and residential buildings directly on the berths (with certain assumptions, an analogy with motels can be drawn). They are the most attractive for a wide range of tourist contingent» [7].

2.2 Architectural and Planning Organization of «Residential Marinas» and «Marina Villages» on Inland Waterways. General Provisions

The rapid development of the yachting industry and tourism from the middle of the twentieth century, along with the emergence of new infrastructure facilities, a global increase in the size of the recreational small fleet and market changes in the yacht services market, led to the formation of **residential marinas** and **marina villages**.

Today, marina villages represent a specially designed multifunctional architectural environment that combines an attractive residential, yacht, and recreational component, as well as a wide range of various services for the maintenance of watercraft, its owners, and visitors.

One of the main tasks of residential marinas and marina villages is to attract a certain contingent of shipowners, by placing living space for the purpose of its selling or renting, as well as the possibility of its placing in their structure of residential moorings for houseboat dwellers living on houseboats or floating homes. The desire to increase the range of sizes and types of accepted vessels, which corresponds to one of the latest consumer trends in the yachting industry in recent years, complicates the architectural and planning structure of marina villages.

> The development of marinas on the seas, rivers, lakes and canals of Europe occurs in two different forms: as a private investment project, and as an urban municipal investment [9, p. 461].

As a rule, the exploitation of «residential marinas» is typical for calmer inland waterways; meanwhile, «marina villages», due to their greater visual and functional attractiveness, form prestigious coastal buildings and are equally represented both on the coast and inland waterways.

An analysis of existing examples of residential marinas and marina villages has shown that they can be formed in two different ways.

Precisely like settlements (Fig. 1) imitating the planning structure and a certain architectural atmosphere of a small settlement (village). This concept of the complex is typical for most newly built marina villages and is used to enhance the marina prestige by creating a unique architectural environment, increasing the effect of seclusion and visual appeal. One of the peculiarities of this concept is the predominance of residential buildings based on natural and alluvial foundations, as well as the possibility of living in the water landscape background and yacht entourage. In this case, the total built-up area is larger than the water area [7].

Fig. 1 Architectural–planning imitation of a small settlement structure: **a** Port Solent, Portsmouth, UK; **b** Burton Waters (marina village), Lincoln, UK; **c** Hythe Marina Village, Southampton, UK [8]

> This concept for the waterfront locates development around the existing marina to create a focused activity center. A combination of living and work spaces permit the wide range of activities associated with actual small town life. Additionally, water-oriented uses are highlighted in this plan. A project of this size would be able to realize successfully the "village" atmosphere and avoid overwhelming the visitor with the magnitude of the development [12].

It is worth noting a rather large building density—the trend is the location of the largest possible number of buildings and infrastructure in a limited area.

Directly marinas. This architectural and planning solution for residential marinas and marina villages envisages the predominance of the functions related to the direct maintenance of recreational boats (parking, storage, refueling, repairs, etc.) and personal watercrafts.

The main representatives of this architectural–planning model are marinas for houseboats (narrowboats), floating houses, etc. The building area is less than the water area, for example: Canoe Pass Village, Freiser river, Vancouver, Canada; Hartford Marina, Great Uz river, UK; Westbay Marine Village, Saanich, Canada.

2.3 «Residential Marinas» and «Marina Villages» Development on Inland Waterways

Inside modern residential marinas the development provides the possibility of temporary (seasonal), long-term (in the case of rental housing) and permanent residence.

Depending on the marina the quantitative ratio between the living space for temporary, long and permanent residence may be different, and this is one of the examples of housing development for both temporary, and permanent residence.

Residential marina can specialize in providing a certain type of housing (e.g., for only permanent residence) and several other types in various proportions.

As a result, today, there are two functional types of residential marinas (marina villages).

Purpose-build marina villages are designed to serve permanent residents, usually those who own certain property, use the services of marina constantly, etc. Such marina villages are often alternative residential formations, such as houseboat (floating homes) communities and narrowboat marinas (with residential moorings): Westbay Marina Village in Saanich (Canada), Canoe Pass Village on Freiser river in Vancouver (Canada).

Mixed-use marina villages are intended to serve the general public: from shipowners and residents living directly in the marina to random residents living directly in the marina to casual visitors, tourists, and vacationers who are not attached in any way to the marina.

> The importance of the interdependence and integration of commercial, residential and recreational elements of such schemes should not be underestimated. Apart from giving vitality to a scheme, the commercial elements of a working marina are necessary to properly service the needs of the yachtsman or boat owner. Other elements such as pubs, restaurants and shops grouped together should also be viable, depending on the size of the project. The important consideration is that the design should aim to produce an environment which satisfies the needs of the yachtsman, the resident and the visitor [12, p. 187, 15].

The most common planning techniques for the construction of residential marines and marina villages are in many ways similar to the layouts of ordinary marinas:

- location of the main block of buildings along the perimeter around the water area (pools) of the marina (Fig. 1a, b);
- development depthward the coast with a wide or narrow front (from the water's edge to the coast along one side).

Like other types of marinas, buildings and structures of marina villages can be located both on land and on water.

There are two main options for the development of marina villages as follows:

- development is located both on land and on water (Fig. 1a, c)
- living space is mainly located on the water surface inside the water area (these are marinas villages, where more than 70% of the buildings are located on the water);
- development is entirely located on land (on natural soils).

> In case of the Hythe marina village (Fig. 1c) «the land which had been designated for development was extended at an early stage to provide more space for bigger houses and to allow a more imaginative development [12, p. 186].

The most frequently used formula is coastal building on natural soils + availability of artificial alluvial foundations. They are formed as island and peninsular plots connected with the coast and with each other by small bridges (intended for cars and pedestrians (Fig. 1a, c).

> The creation of the islands, apart from making the best use of the land, created visual interest as well as allowing the development of distinct areas with a strong sense of place and identity. However, as with car parking, there are strict dimensional constraints in providing manoeuverability and moorings for boats [12, p. 186].

The peculiarity of the marina villages that are built up with capital residential buildings is the presence of their own parking space, attached to the property.

Today, the following methods of placing a parking space for boats regarding housing in marinas villages are common:

- **separate parking space** (private berth), known as a **pen**, is common in marina villages, completely built up by individual and blocked residential houses, where every house or apartment has an equipped individual mooring space in the form of a mooring wall, pier, shed, etc. (Fig. 2);
- **personal parking space** as a part of the public berth for a certain number of places (this technique is typical for marina villages built up with middle-rise and few high-rise buildings (apartment buildings) (Fig. 3).

The chief development of most residential marinas and marina villages located within inland waters is currently represented by individual (single detached), blocked (terraced and semi-detached) residential houses, and multi-family (low-rise apartment buildings) residential houses of small and medium number of floors. Development including multi-story residential buildings is represented by single objects.

> The development of the marina village concept is therefore one which offers a well thought-out solution to the problem, particularly if it can be situated reasonably close to urban centers. The main opportunities for the development of such projects have therefore largely been related either to the redevelopment of dockland areas where patterns of employment and land use have changed, or have arisen in a few suitable locations in conjunction with the development of reclaimed land or derelict land. «Dockland areas have one principal advantage in that they are generally very near centers of large urban populations and have

Fig. 2 Examples of the private pier location relatively to the dwelling: **a** Hythe Marina Village, Southampton, UK; **b** Burton Waters Marina Village, Lincoln, UK [5]; **c** Texoma Harbor Village, Big Mineral Arm river, Texas, USA; **d** Marina Uitgeest, Uitgeest, Netherlands; **e** Rheinsberg Marina, Rheinsberg Harbor Village, Germany

Fig. 3 Examples of the location of a private parking place as a part of the general mooring: **a** Kip Marina Apartments, Scotland; **b** Caswell Cove Marina Season Apartments, Housatonic river, USA; **c** Port Solent Apartments, UK; **d** Marina Village Apartments, Florida, USA

> tended therefore to be developed fairly intensively, with high-rise apartments or apartments in converted warehouse buildings. St. Katherine's Dock in London is a typical and early example. Land reclamation sites have been developed in a variety of ways with houses perhaps being more dominant than flats [12, p. 185].

3 Conclusions

As a result of this work, it was found that the notion of "residential marina" is an extremely generalized definition of a group of marinas, including marina villages, which are characterized by the living space availability (for seasonal, long-term, and permanent residence), in different ratios, catering enterprises and infrastructure that provide together all conditions for comfortable living inside the marina. It was also clarified that individual marinas for narrowboats or marinas, in the structure of which "residential moorings" are envisaged, are called residential marinas in certain countries (Great Britain, Netherlands, USA).

Marina villages are the principal concept of an architectural–planning association of capital residential buildings or any other living space, public service facilities, and marina, designed in the small settlement form (a village).

As a result of the analysis of existing examples, it becomes clear that the residential marina often represents more utilitarian objects with a free territory planning (we mean the location of buildings and structures) which allow for the placement of household and repair and maintenance areas in the main development structure and their interaction.

Marina villages, in their turn, have densely composed development, the most visually appealing environment (in some cases they have a unique general stylistic solution of buildings and structures), a full range of catering enterprises, and full or partial removal of household and repair and maintenance areas outside of common development.

References

1. Association of Inland Navigation Authorities. (2010). *Residential use of inland waterways (advisory document)*. Fearns Wharf, GB: Association of Inland Navigation Authorities.
2. Bohomolov, A. E. (2007). K voprosu o terminologicheskih oboznacheniyah i klassifikacii obyektov yahtinga. *Vestnik Odesskoj gosudarstvennoj akademii stroitelstva i arhitektury, 9*, 244–251.
3. British Waterways. (2010). *Moorings along the banks of BW waterways: explanatory notes and background to policies*. Hillmorton, GB: Waterway Conservation & Regeneration group.
4. British Waterways. (2003). *Waterways & Development plans*. Hillmorton, GB: Waterway Conservation & Regeneration group.
5. Catawba-Wateree Project, Shoreline management guidelines (SMG), Catawba. (2017). Available online: https://trishgreer.com/wp-content/uploads/2017/01/LNMC-Lake-Norman-SMP.pdf.

6. Global Marina Institute (GMI). (2014). *Glossary of Marina terms*. Retrieved from: https://issuu.com/bmfweb/docs/global_marina_institute_gmi_-_marin.
7. Grishin, N. A. (2011). *Sovremennye tendencii razvitiya marin. Gradostroitelnye aspekty*. Retrieved from: https://www.blackseanews.net/read/9741.
8. Gucma, L., Drwięga, K., & Butrymowicz, R. (2018). Statistical analysis of parameters of selected worldwide yacht ports and marinas in terms of design guidelines. *Annual of Navigation, 25*(1), 205–217. https://doi.org/10.1515/aon-2018-0014.
9. Kizielewicz, J., & Luković, T. (2013). The phenomenon of the Marina development to support the European model of economic development. *The International Journal on Marine Navigation and Safety of Sea Transportation, 7*(3), 461–466. https://doi.org/10.12716/1001.07.03.19.
10. Luković, T. (2012). Nautical tourism and its function in the economic development of Europe. *Visions for Global Tourism Industry—Creating and Sustaining Competitive Strategies, 399–430*. https://doi.org/10.5772/38058.
11. Lukyanova, L. G., & Cybukh, V. I. (2004). *Rekreacionnye kompleksy*. Kiev, UA: Visha shkola.
12. Oldham, M. J. (1993). Residential Marinas: The case of Hythe Marina village. *Ekistics, 60*(360/361), 183–189. JSTOR. Retrieved from: https://www.jstor.org/stable/43623154?&seq=1#page_scan_tab_contents.
13. Smith, P. (2011). *What is the difference between residential and leisure moorings?* Retrieved from: https://livingonanarrowboat.co.uk/what-is-the-difference-between-residential-and-leisure-moorings.
14. The Task Committee on Marinas 2020 of the Coasts, Oceans, Ports, and Rivers Institute of the American Society of Civil Engineers. (2012). *Planning and design guidelines for small craft harbors* (3rd edn.). Virginia, USA: American Society of Civil Engineers.
15. Zinenko, T. N., Chepenko, N. S., & Litvinenko, T. E. (2000). Congress of ukrainian ceramic society. *Ogneupory i Tekhnicheskaya Keramika, 12*, 51.

Residual Life Cycle of the Motorway Bridge

G. O. Tatarchenko, N. I. Biloshytska, M. V. Biloshytskyy, and P. Ye. Uvarov

Abstract The study is devoted to the problem of the transportation constructions' residual life cycle management. The condition of the motorway bridge is analysed, the construction's life cycle predictive model is developed, and its main problem areas are identified. A predictive model of the bridge construction life cycle, priority troubleshooting measures as well as urban planning measures to improve the environment and comprehensive landscaping are suggested. The research objective can be formulated as a search for an analytic function that integrally reflects the process of degradation of a reinforced concrete element as a function of time and serves as a fundamental function for the life cycle model in a form.

Keywords The bridge reinforced concrete construction's life cycle predictive model · Motorway bridge · Physical deterioration · Landscaping · Engineering and transport system

1 Introduction

The problem of the residual life cycle assessment and service life prediction for Ukraine is particularly important due to the number of negative factors. The main ones are: difficult economic and financial situation of the state; low road infrastructure development, almost the last in Europe; and short service life of the reinforced concrete elements of bridge constructions, lack of appropriate operation systems, as well as the consequences of the hostilities in the area of the Joint Forces Operation.

In the context of extremely limited financing of the transport construction industry, it is evident that in order to optimize the costs for repair and reconstruction works of the engineering infrastructure, the systematic technical information on the condition of bridges and a scientifically sound system for predicting the residual resource are required. It is the systematization of the information on the technical condition and prediction of the residual resource of the "UKRAVTODOR" motorway bridges that is the subject of this study.

G. O. Tatarchenko (✉) · N. I. Biloshytska · M. V. Biloshytskyy · P. Ye. Uvarov
Volodymyr Dahl East Ukrainian National University, V. Dahl EUNU Pr. Central 59-a, Severodonetsk 93400, Ukraine
e-mail: tatarchenkogalina@gmail.com

V. Onyshchenko et al. (eds.), *Proceedings of the 2nd International Conference on Building Innovations*, Lecture Notes in Civil Engineering 73,
https://doi.org/10.1007/978-3-030-42939-3_47

The bridge structures of Ukraine are the most valuable part of the national transport infrastructure. The term of their operation is estimated in tens of years, been maintained by the specialists of several generations. Therefore, the task of predicting the residual resource of the life cycle of the bridge structures is urgent and has significant economic and social weight. A bridge structure is a complex set of elements; therefore, we see the formation of a predictive model only as a set of interrelated models for forecasting the life cycle of both the structure as a whole and the totality of its structural elements.

2 Research Analysis and Problem Statement

Today, a significant deterioration of the road network in Ukraine as a whole, particularly bridges, is stated. The number of bridges requiring repair and reconstruction works is increasing at an alarming rate (Fig. 1) [1]. An attempt of a linear regression forecast shows a significant increase in the number of bridges that need urgent repair from 2010 to 2018 from 1622 to 2743 bridges.

Transport experts note a significant gap between the level of scientific support for the design of the elements of the bridge structures and their operation systems. There was a clear need for an up-to-date methodology for scientific support of the issue of the residual life cycle management of transport structures. Of particular importance in this methodology is the task of assessing the operational status and forecasting the service life of the reinforced concrete elements of structures, which is collectively called the "life cycle of the construction object". A much broader sense is invested in this term than before. It is understood as a model of a structure theory that enables to predict and manage the residual life cycle of the transport structures.

The problem of engineering life cycle management is becoming a priority in many countries of the world. Despite the unconditional relevance of this issue, there are too few publications on this subject in Ukraine. Theoretical studies on the problems

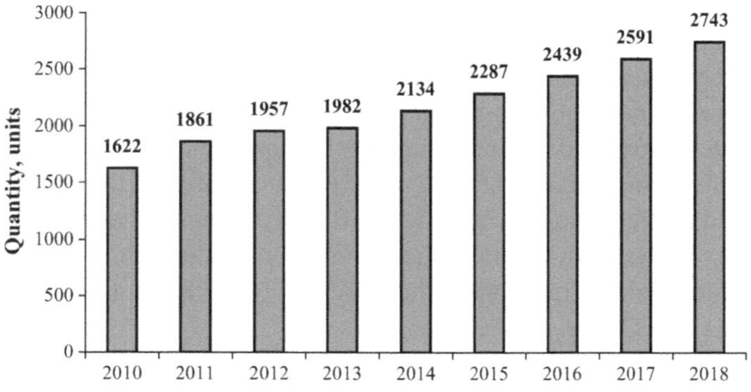

Fig. 1 Dynamics of growth of the bridges requiring immediate repair and reconstruction

of reliability, durability and prediction of the residual service life of the road bridges are found in the materials of famous Ukrainian [2, 3] and world scientists [4–6].

In the USA and Europe, experts consider this problem as urgent in developing the theoretical foundations of the facility operation system, so there is enough research material. Among foreign publications on this specific topic, we name a few of the most famous. The experts Zhongjie Zhang, Louisiana Transportation Research Center, and Xiaoduan Sun and Xiang Ying Chen, University of Louisiana at Lafayette, suggest using simplified three-element bridge preservation model in terms of bridge deck, substructure and superstructure components [4]. The model is made with the help of the deterioration matrix. It can be obtained by transforming the probability matrix of state change using empirical formulas based on the analysis of the performance history of each structure separately.

In the works of Finnish scientists Asko Sarja and Erkki Vesikari, Technical Research Center of Finland, Building Technology [5], the process of physical changes in the structure is described by the matrices of uncertainty and degradation impacts.

Researchers' particular attention is attracted to the American professor Dan M. Frangopol from Lehigh University works on bridge management and designing the residual service life predicting models based on its management statistic data [6].

Existing models for predicting the residual life of the bridge structure life cycle are a central scientific problem of the bridge operation system. In addition, there is also the problem of assessing and forecasting the preservation of the residual resource of bridges, which has become especially relevant in the last 20 years [7]. Until 2002, the building codes of Ukraine provided neither a methodology for residual service life assessment nor criteria to assess the technical condition of the bridges in use. Today, for the first time, we have a package of regulations for the bridge technical condition assessment and prediction [8].

The main document in the operation system is the State Standards of Ukraine (ДСТУ-Н Б В.2.3-23:2012) "Guidance on the Assessment and Prediction of the Motorway Bridges Technical Condition" [9], State Building Regulations (ДБН Б.1.1-5:2007) "The Composition, Content, Designing Procedure, Coordination and Approval of the Section of Engineering and Technical Measures of Civil Protection (Civil Defence) in Urban Planning Documentation" [10] and State Building Regulations (ДБН В.2.3-22.2009) Bridges and Pipes, and Basic Design Requirements [11].

3 Research Objects

Research objects analyse the preservation of the motorway bridge across the River Borova, Luhansk Oblast, to design a predictive model of the reinforced concrete structure life and a complex of the prioritized measures on conducting repair and rehabilitation operations and on improvement of the engineering–transport system of the region.

4 Research Results

The Sievierodonetsk–Lysychansk–Rubizhne triangle is one of the most densely populated areas in the region. The basis of the economic agglomeration base is the chemical industry and instrument engineering. The agglomeration's external communications are provided by the road transport that runs along the bridges over the Seversky Donets and Borova rivers. Both bridges were extremely physically deteriorated, and after one of them was damaged due to the hostilities in 2014, they became unusable. Therefore, the solution to the problematic situation that has arisen due to the damage to the engineering and transport infrastructure in the area of the existing motorway bridge is relevant, the timely resolution of which will lead to improved traffic on the regional highway P 66 and entry to the city of Sievierodonetsk.

The bridge was built in 1962 according to the project developed by Kharkiv branch of DPI "PromtransNDIproekt" in 1956. It is a reinforced concrete bridge structure with dimensions of (11x11.36 m) with beam span elements. (A detailed plan of the bridge structure is shown in Fig. 2). The full length of the bridge is 125.36 m. The dimensions of the bridge are $7.2 + 2 \times 0.75$ m, width—8.4 m, which does not comply with existing standards, motorway road—III category, and number of lanes—2. Absolute elevations along the road vary from 55.00 to 59.00 m above sea level. Overall length of approaches that need to be reconfigured is about 420 m. The length of the exits from the P 66 motorway towards Objizdna Street, which must be rebuilt, is about 100 m.

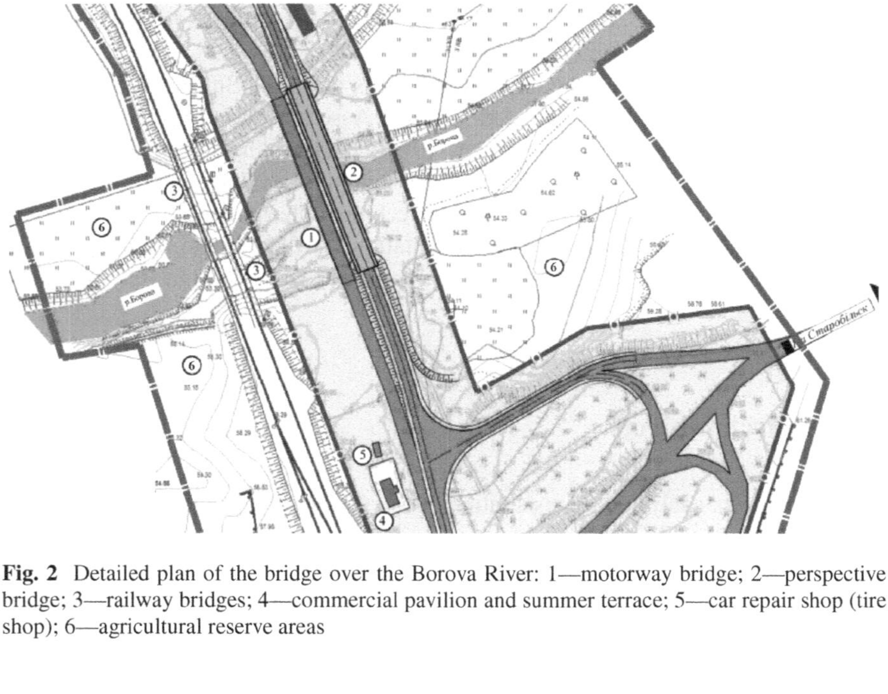

Fig. 2 Detailed plan of the bridge over the Borova River: 1—motorway bridge; 2—perspective bridge; 3—railway bridges; 4—commercial pavilion and summer terrace; 5—car repair shop (tire shop); 6—agricultural reserve areas

Vertical dynamic load on the bridge is accepted H-13 and НГ-60. The bridge substructure transfer the load to the foundation with the help of reinforced concrete piers-pillars. The footpaths are made of assembled concrete slab. On top of them is a layer of monolithic concrete 170 mm thick. Railings are made of precast concrete sections; railings are made of monolithic reinforced concrete. Engineering utilities were laid along the extreme beams, on both sides of the bridge: communication cables, heat pipes, water pipes, 0.4 kW power cable.

From the social welfare facilities between the railway line and the road, there is an existing shopping pavilion with a summer terrace and a car repair shop (tire shop). The territory is subjected to restoration, located in the north-western part of Sievierodonetsk, and is about 24 ha.

The territories adjacent from the east to the bridge are not built up and are reserve territories of the city of Sievierodonetsk for agricultural purposes. On the west side of the road is the land allocation boundary of the railway. To the south of the junction area are the lands of industrial enterprises. The roads have a hard surface—bitumen concrete. The footpaths, motorway and railway bridge have electric lighting. In other nearby territories, any improvement and organized landscaping are absent. Melt and rainwater are discharged from existing parts of the road outside the driveway to the adjacent territory.

The main exit from the bridge links the grade-separated intersection from Bogdana Leshchyny Street and two railway access roads. This intersection directs the P66 road main traffic flow towards Promyslova Street and further towards Lysychansk.

Within the framework of the study, the classification of existing models to predict the technical condition of bridge structures is proposed (Fig. 3).

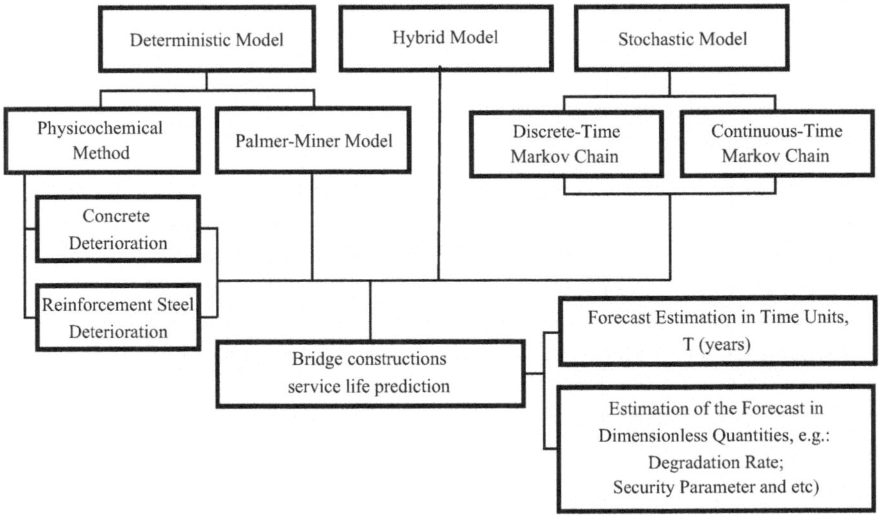

Fig. 3 Classification of the predictive models of the road bridges reinforced concrete element durability

Based on the authors' work [1, 12, 13], a comprehensive analysis was conducted and a predictive model of life cycle of the reinforced concrete elements of the bridge structure was developed.

The life cycle model of the transport structure is formed by means of an analytical procedure that enables to determine the safe service time of elements (or the structure as a whole) at given loads under certain operating conditions. It is a function of time that describes the process of deterioration of the technical, physical, mechanical, chemical and aesthetic characteristics of an element during the service life of the element/structure. In general, this function looks like:

$$T_l = R(t)L(P_1, P_2, \ldots, P_n), \tag{1}$$

where

T	service life term of the predicted element;
$L(P_1, P_2, \ldots, P_n)$	a random function;
$P_1, P_2, \ldots, P_n$	parameters characterized by: material properties, stress–strain state, type of construction, environment, level of maintenance;
$R(t)$	generalized factor, the degradation of which is described by the model.

This can be, for example, the calculated resistance of the material, the bearing capacity of the element, the width of the cracks, the reliability parameter, etc. The degradation model is an analytical dependence that gives the average value of the degradation factor as a function of time:

$$R(t) = R_l \cdot f(t), \tag{2}$$

where

R_l	the initial value of the degradation factor;
$f(t)$	degradation function, t—time.

The research objective can be formulated as a search for an analytic function of type (2), which integrally reflects the degradation process of a reinforced concrete element as a function of time and serves as a fundamental function for the life cycle model in the form (1). The concept of "integrally" here implies the meaning of generalization in the fact that the degradation model is expressed through one degradation factor. In this case, the crack formation parameter is taken as such a factor.

The life cycle model uses a cracking model. The function of the indicator of degradation of reinforced concrete element is supplemented by operational coefficients:

$$f(t) = n_1 \cdot n_2 \cdot n_3 \cdot s \cdot t^2, \tag{3}$$

where

n_1 operating condition coefficient;
n_2 design type coefficient;
n_3 coefficient of environmental function.

$$n = n(p_1,\ p_2,\ p_3),$$

where

p_t environmental parameters (range of average max and min temperatures, relative humidity of the environment, the presence of chlorides);
s scale factor which has a dimension $1/t^2$ (unit of time – year).

Taking into account (3), the life cycle model of a reinforced concrete element has the form

$$A(t) = a \cdot \left(1 + n_1 \cdot n_2 \cdot n_3 \cdot s \cdot t^2\right). \tag{4}$$

From Eq. (4), we obtain the ultimate degradation, equating $A(t) = \alpha_\tau$, where α_τ is the crack width limit value. In other words, we obtain a life cycle forecast model in the form (1):

$$T_l = \left(\frac{a_\tau - a}{a \cdot n_1 \cdot n_2 \cdot n_3 \cdot s}\right)^{0.5}, \tag{5}$$

where

T service life term of the predicted element.

Priority measures are proposed to eliminate the problems of the site: firstly, the expansion of the structure is 12 m, including two sidewalks of 1 m each, the entrances to the new bridge should have two lanes, the width of the lane is 3.75 m, the shoulder is 2.5 m, and type of coverings is asphalt concrete (Fig. 4). The design loads fully take into account the required design situations and limit states according to the project. Energy-saving electric lighting is provided on one side of the bridge.

In order to improve the environment, a number of planning and engineering activities may be envisaged that include:

- Measures that affect all components of the environment and generally improve sanitary conditions;
- Construction management according to the designed functional zoning;
- Establishment and organization of sanitary protection zones for residential development (within the sanitary protection zones, new housing construction, overhauls of existing residential buildings with completion and reconstruction are prohibited) and landscaping with the help of green spaces is supposed;

Fig. 4 View of the bridge structure across the Borova River, Luhansk Oblast

- Engineering preparation of the territory and vertical layout.

The upgrading and partial landscaping are expected to be within 50 m on the north side of the existing lane of the road and green lawn in the area between the existing and the project road. Green spaces are designed along the road from the north side of the existing bridge to the exit to Rubizhne, and part of the area for the lawn located between the roads—in the area between the Borova River and the existing interchange. To protect against noise and gas contamination along a public road, green spaces are required with a strip width of at least 10 m.

5 Conclusions

The problems of managing the residual resources of the life cycle of transport facilities are analysed. Based on the analysis of the condition of the road bridge and the identified problems, priority measures are proposed to eliminate them, urban development measures to improve the environment and comprehensive improvement of the surrounding area.

Based on the studies, a detailed plan of the bridge construction across the Borova River was designed and the engineering and transport system of this territory were adjusted, which organically fits into the existing structure at the appropriate hierarchical level and significantly improves traffic safety and the amount of traffic flow.

References

1. Lantuh-Lyaschenko A.I. (2012). *Nadiynist i dovgovichnist avtodorozhnih mostiv. Naukovi rozrobki z normativnogo regulyuvannya.* Zb. Transportna akademiya Ukrayini: 20 rokiv (1992–2012) – K.: NTU.
2. Davidenko O. O. (2013). *Analiz dovgovichnosti avtodorozhnih mostiv Ukrayini.* Mizhvid. nauk.-tehn. zb. Naukovo-tehnichni problemi suchasnogo zalizobetonu.
3. Lantuh-Lyaschenko, A. I. (2009). *Markovski modeli degradatsiyi zalizobetonnih elementiv mostiv.* Promislove budivnitstvo ta inzhenerni sporudi. – Kiyv.
4. Zhang, Z., Sun, X., & Wang, X. (2003). *9th International Bridge Management Conference*, Orlando Airport Marriott Orlando, Florida.
5. Vesikari, E. (2003). Life cycle management of concrete infrastructures for improved sustainability. LIFECON. *Project funded by the European Community under the Competitive and Sustainable Growth Programmer.*
6. Frangopol, D. M., Kong, J. S., & Emhaidy, S. (2001). Reliability-based life-cycle management of highway bridges. *Journal of Computing in Civil Engineering.*
7. Li Chun, Q. (2005). Time dependent reliability analysis of the corrosion affected concrete structures. *International Journal of Material & Structurale Reliability.*
8. Yatsko, F. V. (2010). Prognoz dovgovichnosti zalizobetonnih elementiv mostiv. Statistichniy pidhid. *VIsnik nats. un-tu «Lvivska politehnika» – Lviv.*
9. DSTU-N B V.2.3–23:2012 Nastanova z otsinyuvannya i prognozuvannya tehnichnogo stanu avtodorozhnih mostiv. –K.: Minregion Ukrayini (2013).
10. DBN B.1.1–5:2007 «Sklad, zmist, poryadok rozroblennya, pogodzhennya ta zatverdzhennya rozdilu inzhenerno-tehnichnih zahodiv tsivilnogo zahistu (tsivilnoyi oboroni) u mistobudivniy dokumentatsiyi»–K.: Minregionbud Ukrayini (2008).
11. DBN V.2.3-22.2009. Mosti ta trubi. Osnovni vimogi proektuvannya: - K.: Minregionbud Ukrayini (2009).
12. Uvarov, P. E., Kravchunovskaya, T., Shparber, M. E. (2010). Innovatsionnyie tehnologii modelirovaniya organizatsii zhiznennogo tsikla proektov-ob'ektov stroitelstva (kontseptualno-metodologicheskiy aspekt). *Zb. nauk. prats Pridniprovskoyi derzhavnoyi akademiyi budivnitstva ta arhitekturi. «Stroitelstvo, materialovedenie, mashinostroenie»–Dnipropetrovsk: PDABtaA.*
13. Yanchuk, L. L. (2010). Obgruntuvannya modeli prognozu zhittevogo tsiklu zalizobetonnih elementiv mostovogo perehodu. *VisnikNatsIonalnogo unIversitetu «Lvivska politehnIka». Lviv.*

Modern Information Technologies in System Architecture—Urban Planning—Building Constructions

G. O. Tatarchenko, O. A. Chernih, V. M. Sokolenko, and Z. S. Tatarchenko

Abstract The toolkit of modern versions of SAPFIR-3D and LIRA-SAPR allows to carry out the idea of an architect through a detailed analysis of a tense deformed state and verification of the bearing capacity of building structures in the software complex LIRA-SAPR to produce design documentation with the requirements of the necessary regulatory documents in the SAPFIR-3D program.

Keywords Information technology · Geometry · Form · Architecture · Urban planning · Building constructions · Designing

1 Introduction

One of the most important areas of implementation of modern information technologies is construction. Computer-integrated design leads to enhanced quality and speed of work, financial control, communication and access to shared data and productivity. Unfortunately, many firms in the third world countries are yet to understand this essential value and its importance to the development of their construction sector [1].

Design of products which satisfy the consumer not only in operational, but also esthetic qualities which are broadcast in requirements to creation of certain forms of objects of design is the purpose of the modern designer of any direction. The form, in turn, first of all, is inseparable from its geometrical characteristics. Achievements in the field of geometry, construction mechanics, physics and other sciences enable to create optimum constructive forms which meet obviously set requirements.

Understanding these geometrical images, different behind plastic character, bear in themselves harmonious coherence, special behind tectonics, which in synthesis with composition and graphic means gives the opportunity to receive samples with high esthetic rates [2], has to be the basic component in the course of computer design of objects of graphic design.

G. O. Tatarchenko (✉) · O. A. Chernih · V. M. Sokolenko · Z. S. Tatarchenko
Volodymyr Dahl East Ukrainian National University, V. Dahl EUNU Pr. Central 59-a, Severodonetsk 93400, Ukraine
e-mail: tatarchenkogalina@gmail.com

V. Onyshchenko et al. (eds.), *Proceedings of the 2nd International Conference on Building Innovations*, Lecture Notes in Civil Engineering 73,
https://doi.org/10.1007/978-3-030-42939-3_48

2 Analysis of Publications and Problem Statement

The architecture, by Patrik Schumacher definition, is coded by the double code: usefulness (functionality) and beauty (the formal decision), and one-sided aspiration only to formal aspects is valid anomaly [3]. The social functionality of the built-in environment keeps in its communicative potential. Development of space complexes as systems of importance is a key to modernization of the main competence of architecture.

The Council on Tall Buildings and Urban Habitat (CTBUH) UK Chapter hosted presentations and panel discussion at the RIBA London, supported by Zaha Hadid Architects and ABB Group [4].

Reflecting on the lessons learned from the past 50 years, the presentations and panel discussion explored the future of our built environment in cities designed to be sustainable and smart, socially responsive and self-learning: the main discoveries of the past 50 years that have influenced our built environment; how artificial intelligence and big data inform the city of the future; how smart technologies can improve social, environmental and economic balance of our cities; improving the integration of research into smart tall buildings and smart cities.

The panel of speakers presented their latest research with a focus on how greater collaboration can be implemented across the disciplines shaping our urban environment: Peter Murray of New London Architecture, Simon Giles of Tyrens London, Carolyn Dwyer of City of London Corporation, David Nicholl of ABB, Patrik Schumacher of Zaha Hadid Architects. Organized by Viviana Muscettola and Katrin Förster.

Striking example of parametrical design of the city residential district is the project for developing a part of Admiral Serebryakov Embankment by Novorossiysk consortium led by Zaha Hadid Architects (Fig. 1). The team has suggested to arrange on the embankment dynamic composition from nine buildings which at the different levels are connected among themselves by passes, the areas and podiums and also have continuation in the form of the moorings going to the sea. The configuration which fits into the available urban development and, at the same time, can become modern architectural symbol of the city has as a result turned out. The project integrates business, ecological, cultural and recreational functions, creating unique consecutive agglomeration and creatively programming the territory [5].

Currently, one of recognized and available in Ukraine and neighboring countries is the LIRA-SAPR program complex [6]. It includes such modules as SAPFIR-3D, MONOMAKH-SAPR and ESPRI (Fig. 2).

The SAPFIR-3D module enables to carry out: design of various construction objects—a lot of floor residential and public buildings, constructions of any assignment of superstructures, small architectural forms; preparation of analytical models of construction objects for the subsequent strength calculation and the analysis of design in the LIRA-SAPR program complex; search of versions of volume decisions and rational constructive schemes on the basis of parametrical modeling and interactive space shaping.

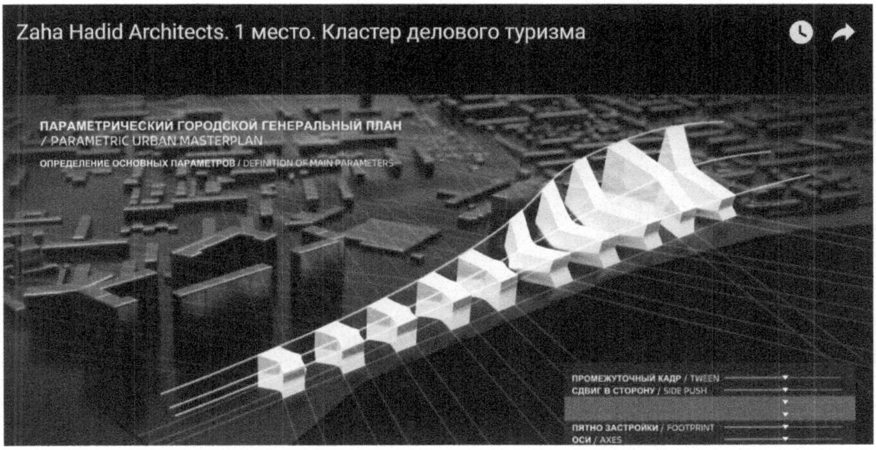

Fig. 1 Concept of developing a part of Admiral Serebryakov Embankment by Novorossiysk consortium led by Zaha Hadid Architects

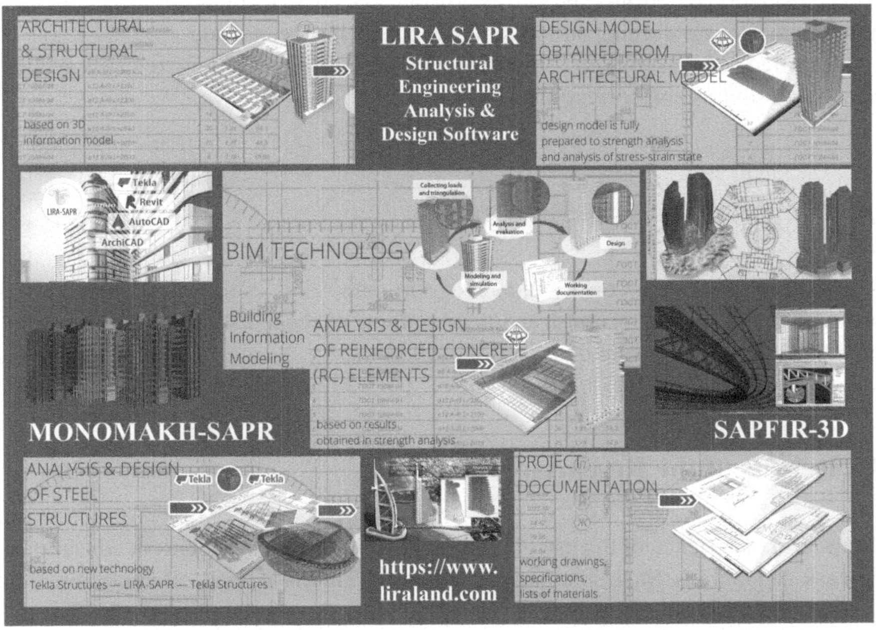

Fig. 2 LIRA-SAPR program complex

Fig. 3 New version of software LIRA-SAPR

The MONOMAKH-SAPR module is intended for calculation and structural design of buildings from monolithic steel concrete, with brick walls. In the course of work calculation of the building and its separate parts with forming working drawings and schemes of reinforcing structural components has been made.

The LIRA-SAPR program complex is constantly updated (Fig. 3): releases include new opportunities and corrections [7].

3 The Purpose of the Research

Application of modern construction program complexes in system architecture—urban planning—building constructions at training of specialists of the higher school with the solution of practical tasks of the enterprises.

4 Results of the Research

Long-term experience of the use of modern information technologies in educational process at Donbass State Technical University (DonSTU) and effective research cooperation with the LIRA-SAPR company under the direction of the D. Sc., the Professor Gorodetsky O.S. [8] have provided an opportunity to receive in the fall of 2018 for Volodymyr Dahl East Ukrainian National University (V. Dahl EUNU) the license sets of training programs of ACADEMIC set 2018 and to begin the process of implementation in educational process at the departments "Architecture and Urban Planning" and "Constructions, Urbanistic and Space Planning" this modern construction program complexes LIRA-SAPR FULL 2018, MONOMAKH-SAPR PRO 2016 and ESPRI 2018.

The use of scientifically capacious tools of modern versions of the SAPFIR-3D and LIRA-SAPR programs allows:

- To fulfill the original idea of the architect on the basis of geometrical premises of art shaping by means of parametrical shaping when developing the concept of the residential district of the city (Fig. 4) and project of the building (Fig. 5);
- To carry out the detailed analysis of tensely state of strain and check of bearing capacity of building constructions in the LIRA-SAPR program complex (Fig. 6);

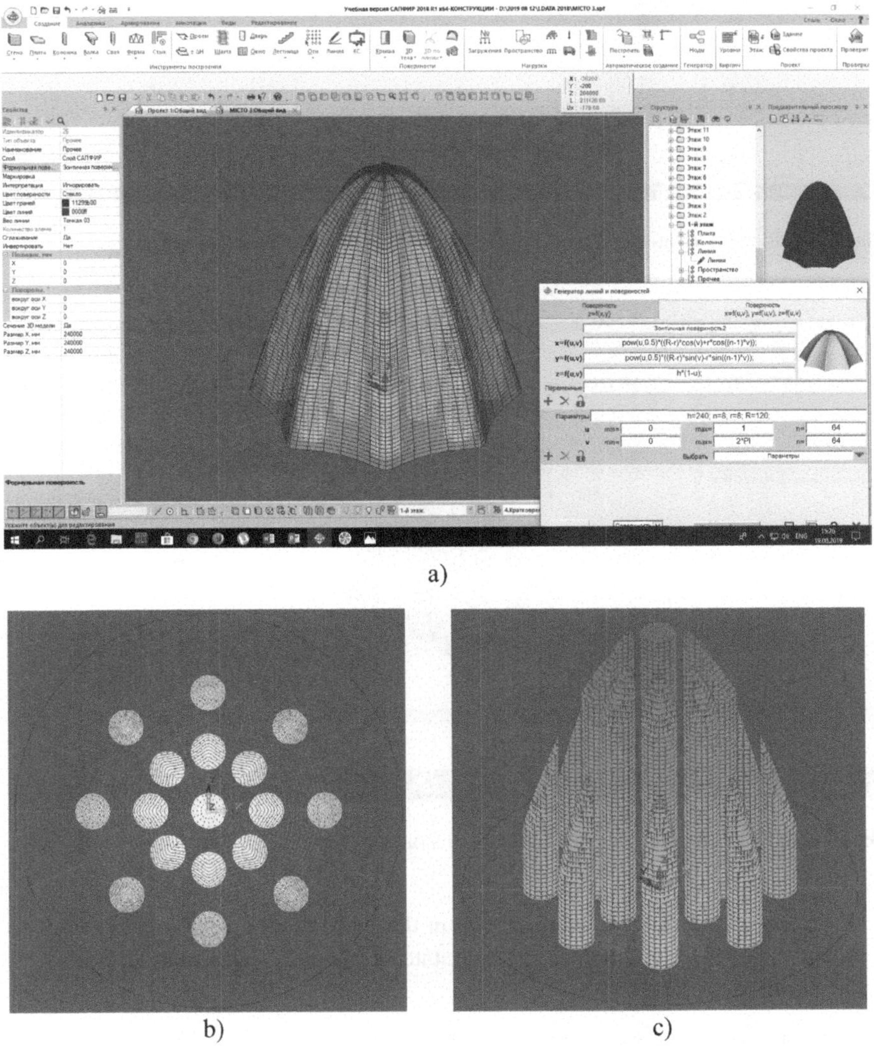

Fig. 4 Creation of conceptual architectural model of the residential district of the city in the SAPFIR-3D program: **a** parametrical shaping, **b** bottom view, **c** look in isometry

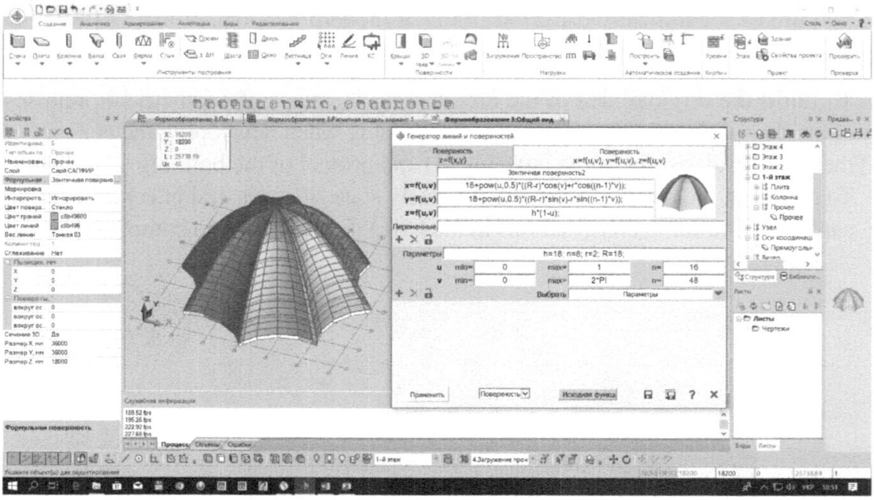

Fig. 5 Creation of architectural model of the building in the SAPFIR-3D program

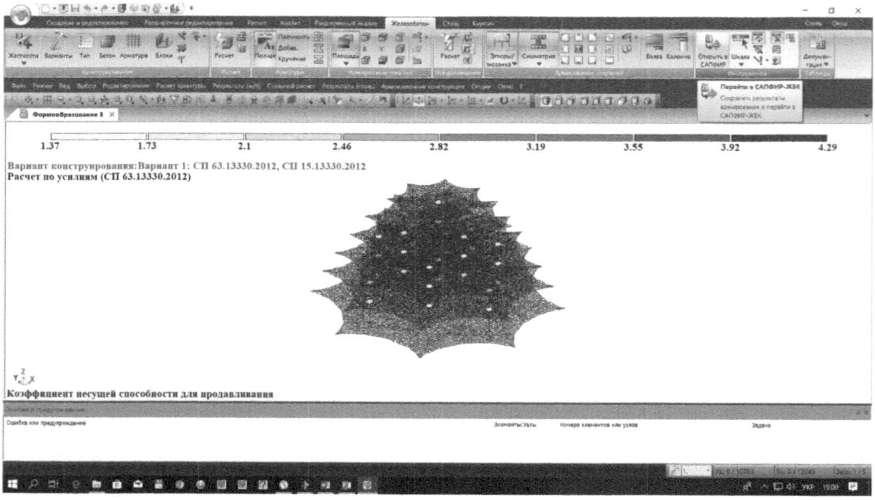

Fig. 6 Check of bearing capacity of building constructions in the LIRA-SAPR program complex

- To prepare the project documentation in the SAPFIR-3D program (Fig. 7) with observance of requirements, relevant state normative documents depending on the country of construction.

It is especially necessary to emphasize the role of the LIRA-SAPR company in assistance on implementation of modern programs in educational process in institutions of higher education of the construction direction for training specialists. On the official site of the LIRA-SAPR company [9], there is provided different information

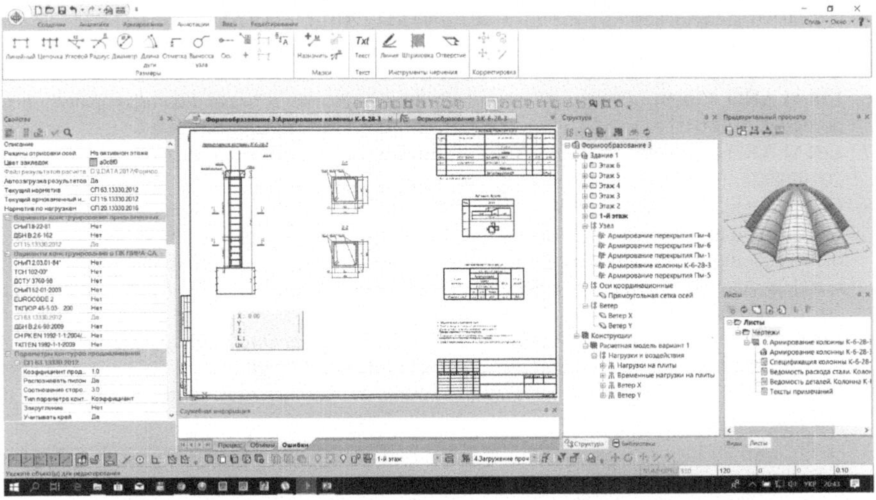

Fig. 7 Creation of the operating documentation in the SAPFIR-3D program

on modern software products. Students have opportunity free of charge to load educational versions of programs and to obtain exhaustive information for its use [10] both in educational process and during the work in construction companies.

5 Conclusions

Attraction of the accumulated international experience in the field of architecture, use of modern construction information technologies, effective research cooperation with the LIRA-SAPR company, design organizations and industrial enterprises give all opportunities for effective implementation in educational process of dual technique of training future specialists capable to solve complex problems of creating projects of reliable buildings and constructions with high esthetic rates.

References

1. Onyegiri, I., Nwachukwu, C. C., & Jamike, O. (2011). Information and communication technology in the construction industry. *American Journal of Scientific and Industrial Research*, 2153–649X https://doi.org/10.5251/ajsir.2011.2.3.461.468.
2. Mihaylenko, V. E., & Yakovlev, M. I. (2004). Osnovi kompozitsiyi (geometrichni aspekti hudozhnogo formotvorennya): *Navch. posib. dlya stud. vischih navch. zakladiv.* – K.: Karavela.
3. Schumacher, P. (2010). *The autopoiesis of architecture. A new framework for architecture* (Vol. 1). New York: Wiley.
4. Hadid, Z. (2019). *Architects homepage, 'Smart Cities. Smart Buildings' seminar*. https://www.youtube.com/watch?v=5fI92t8zViw, last accessed 2019/08/21.

5. Agentstvo strategicheskogo razvitiya «CZENTR». (2019). *Domashnyaya stranicza, «Zaha Hadid Architects. 1 mesto. Klaster delovogo turizma»*. https://www.youtube.com/watch?v=-4Jo1_pe4_A&t=13s, last accessed 2019/08/21.
6. LIRALAND (2019). *Group: Company Profile*. https://www.liraland.com/company/, last accessed 2019/08/21.
7. LIRALAND. (2019). *Group: News*. https://www.liraland.com/news/. Last accessed 2019/08/21.
8. Chernih, O., & Sokolenko, V. (2017). Dosvid zastosuvannya suchasnih informatsiynih tehnologiy u navchalnomu protsesi v DonDTU Suchasni metodi i problemno-oritntovani kompleksi rozrahunku konstruktsiy i yih zastosuvannya u proektuvanni i navchalnomu protsesi: tezi dopovidey *Mizhnarodnoyi naukovo-praktichnoyi konferentsiyi, m. Kiyiv*, 25–26 zhovtnya 2017. – K.: Talkom.
9. LIRALAND. (2019). *Group: LIRA-SAPR for students and teaching/training purposes*. https://www.liraland.com/services/forstudents.php, last accessed 2019/08/21.
10. LIRALAND. (2019). *Group: Knowledge base*. https://help.liraland.ru/. Last accessed 2019/08/21.

Light Facilities Complex in Architectural Design

Aleksandr Vasilenko and Andrii Koniuk

Abstract The study determines methodological bases of forming an integrated light facilities complex of modern residential architecture. Light reveals the form of the architectural object and creates an image. Architecture becomes a source of artificial illumination using light technology. One of the main priorities of the research is scientific approach in the creation of bioclimatic and ecological architecture. The problems of the effective use of the aesthetic potential of natural and artificial light have been considered. The analysis of scientific works has brought focus on the following issue—in the process of developing facade systems of housing buildings, the necessary value of the role of functional formation of form by light is not provided. Today, in the architectural planning, it is necessary: to introduce European norms in Ukraine, to conform the national normative base to the norms of the European standards.

Keywords Architectural composition · Compositional thinking · Light means · Residential architecture · Methodological principles · Natural and artificial lighting

1 Introduction

The problems of psychological comfort for life activity of people, the problem of providing high-quality architectural environment, as well as those of effective aesthetic potential of natural and artificial light and innovative light technologies are topical nowadays. Determining the power parameters of the environment, necessary to optimize the form of housing buildings needs to the improved, with European norms being introduced in Ukraine and of national normative base being conformed to the norms of the EU.

A. Vasilenko
Odessa State Academy of Civil Engineering and Architecture, Odessa, Ukraine
e-mail: abvasilenko10@gmail.com

A. Koniuk (✉)
Poltava National Technical Yuri Kondratyuk University, Poltava, Ukraine
e-mail: konyk.a.e@gmail.com

V. Onyshchenko et al. (eds.), *Proceedings of the 2nd International Conference on Building Innovations*, Lecture Notes in Civil Engineering 73,
https://doi.org/10.1007/978-3-030-42939-3_49

The analysis of scientific publications enables to distinguish open questions that lie in the fact that within the development of the housing building facade systems not enough attention is given to the role of shaping function of the light. Topicality of the issue is predetermined by the necessity to reinforce scientifically reasonable practice of forming modern light facilities, which has become the pre-condition for the formulation of the purpose and the tasks of this research.

In a methodological aspect, the existing studies of the predecessors do not summarize all the complex of tasks that must be solved within the framework of forming the complex of light facilities in the architecture of low-rise accommodation. The systematic vision of the architecture of these buildings in the context of such an important point as ethno cultural identity and the life activity environment has not been provided yet.

Among basic research papers in the field of theory of solar radiation, calculations of insolation in architecture, relief treatment of facades of housing building there can be highlighted the works of such scientists as: O. Sergejchuk, V. Belikova, S. Vetoshkin, N. Gusev, L. Dashkevich, V. Drozdov, B. Dunaev, D. Lazarev, I. Ckrill, J. Hraska, R. Jonson, R. Kittler, F. Erisman, R. Hopkinson, J. Koso, A. Olgaj, T. Rodgers.

Thus, the practices of forming light facilities complexes in architecture of housing buildings are in the center of attention for ecology and architecture.

The purpose of the research is to determine methodological principles of forming the complex of light facilities in architecture of housing buildings.

The tasks of the research consist in the necessity: (1) to formulate methodological principles of forming the complex of light facilities in modern residential architecture during the educational process at the academic and research institute of architecture; (2) to define progress of typology of housing building trends in the aspect of using the complex of light facilities; (3) to define and systematize the traditional architectural-compositional shaping means; (4) to analyze chiaroscuro correlations that are the composition means of architectural shaping; (5) to show the role of the light as a shaping basis of architectural composition.

2 Materials and Methods

Methods of the research on forming the complex of light facilities in the architectural housing systems are based on ecosystem deco approach and the analysis of the light environment that forms interaction of light with architecture.

The general research method of the "light component" of architectural systems is proposed. The methods for assessing the factors influencing the formation of the complex of light facilities are determined. A research methodology and criteria for assessing the form snapping are elaborated.

The study contains an analytical review of the given well-known research methods used in the theory of architecture, research methods and criteria for evaluating the "light component" of architectural systems, as well as the method of research formative and "comfortable function" of light in architecture. The research methodology focuses on aesthetic aspects.

Architectural activity is regarded as an ecological system "man–environment." The ecological system has four components:

1. Production sphere (production of technical products);
2. Non-production sphere (service);
3. Recreational sphere (special type of activity);
4. Communication sphere.

The role of solar energy for each of these components is specific and requires separate case studies. There have been improved the methods for the experimental study of sunlight in the exterior and interior of modern residential buildings, including:

- The method of visual observation of the influence of natural light on architectural form and its emotional-aesthetic perception of a person;
- The method of expert estimations of the conditions of the light environment for productive implementation of home activities, to create optimum mental attitude and rest.

The methods suggested take into account the trends of development of modern functional and spatial architectural structures, layouts, typological and compositional characteristics.

3 Light Complex

In this research, the following definition is taken as basic for the complex of light facilities: the totality of sources of natural and combined lamplight that provides insolation, sanitary-hygienic norms, comfort and aesthetic light environment, integrated in the systems of engineering constructions with the aim of solving modern architectonically composition tasks.

As already noted above, with the change of natural illumination, the character of illumination of architectural forms changes for a day, but always in such a way that these forms continue to be perceived with the personal touches peculiar to them. Unlike natural illumination, lamplight creates visual illusions (curvature of form, unclearness, fabulousness) [1].

The stream of natural light that has the only direction puts as if to it by volume oriented forms in the equal terms of illumination. Natural light is usually sent to the surface that is illuminated by the stream of parallel rays, thus, every curvilinear surface in every point meets with it at another corner, in turn, and determines different luminosity in these points (see Fig. 1).

Every plane in all points will meet with the homogeneous stream of rays sent to it strictly at the same corner and, thus, will get the same luminosity all throughout [2].

Fig. 1 Kogor Courtyard. Architect—Norman Foster

Therefore, the idea of a plane is bound in our consciousness with the idea of practically unchanging luminosity. An idea about curvilinear surfaces is related to the idea about luminosity appropriately variable of this surface. Significant in size after size forms (planes), the sources of artificial light are illuminated by the divergent pinches of light rays sent to them. Thus, a plane cannot get even luminosity in all points, because in its every point, it is at the different distance from the source of light and other corner with the ray sent to its forms by a plane in this point (see Fig. 2).

Fig. 2 Luvr Abu Dhabi. Architect—Jean Nouvel

A curvilinear surface at the sources of illumination, located on its geometrical wasp or in a geometrical center, appears to be the friendliest, particularly, to even distribution of luminosity on it [3].

4 Compositional Means of Architectural Light Forming

Studying the issue of detecting architectural form by light, A.S. Sheepanov deems it wise to divide the considered material into three parts (see Fig. 3): detecting general, basic architectural form (general forms of ceiling) by light; detecting the plastic, of the details [4].

The visuognosis of the form is determined by distribution of brightness. An ordinary plane can at certain illumination angle concave, protuberant or wave. The illusion of undulating curvature of the even surface is by the rhythmic change of its luminosity is called "bright by waviness." The sources of light influence the degree of "bright waviness" of architectural surfaces. Not eliminated is the probability of achieving such an effect also by means of metrical rows of light-pervious architectural constructions [5].

Light forms an artistic image. Light can create different tints of mood. Light can be sullen and mysterious, joyful and soothing, intimate and relaxing, etc. Herein, the intuition of the designer has a primary value as playing the chiaroscuro on the emotional coloring of interior (see Fig. 4).

Fig. 3 Shopping center, New York. Architect—Santiago Calatrava

Fig. 4 Station Liege. Architect—Santiago Calatrava

5 Light-Form Shaping Basis of the Architectural Composition

Among the tasks related to the improvement of the quality of light environment, visibility and perception of objects and surfaces are of great importance. Quality of visibility and perception to a great extent is determined not only by the parameters of the illumination of building and development of territories but also by the properties of a human eye. Human eye is well adjusted to the sunbeams as the light energy sources. Observing the colored surfaces at the change of brightness level within the limits of the corresponding area of photopia is accompanied by the change of the color feeling (phenomenon of Bezold-Brückk), especially noted at sunny illumination of surfaces of facades and details that got among architects an estimation "sunlight destroys color" (see Fig. 5).

New concepts "bioarchitecture" and "eco-building" come from the architectural culture of the old times, when people lived exceptionally in eco-buildings. Environmental materials were used for the construction. Building was adapted to the local climate conditions. Bioarchitecture and sunlight mutually influenced each other and coexisted in unity. In modern conditions, the main building principle remains launching the building in a surrounding landscape and carrying out the intention of the customer to take into account local traditions in the use of materials. Since ancient times, the basic building materials have been natural materials—soil, clay, stone, timber, grass and reed. In Central Europe, the most widespread materials were an adobe brick, rammed soil and wooden ceiling with the reed roof. A roof protruding from every quarter is protected from summer heat and scalding sun. The basic requirement that is applied to all types of eco-building is that after the completion of the term of life the house is to be laid out in a natural environment or get fit for a new application (see Fig. 6).

Fig. 5 Kogor Courtyard. Architect—Norman Foster

Fig. 6 Pinakothek der Moderne. Architect—Stephan Braunfels

While designing and constructing eco-friendly houses, architectural, planning and constructive techniques for improving energy efficiency are combined. With the use of environmentally friendly building materials and structures, such as wall panels of pressed rye straw and stuccoes with clay, we get an increase in the level of environmental friendliness. Moreover, the environmental friendliness is reinforced in three directions: (1) the environmental friendliness and renewability of materials of basic structures (walls and floors); (2) the reduction of energy for wall production and installation (compared to ordinary brick—less than 300 times); (3) the cost of the construction of 1 m^2 of a building made of straw panels (blocks) can be three times less compared to traditional structural systems (brick, reinforced concrete flooring) [6–8].

On the territories with different climatic conditions, forms of different types of eco-building are constructed. On flat, cool territories the most widespread "houses with the form of the hump are popular." On territories with temperate climate, buildings can be more open, on while large glass surfaces that serve as traps for heat. The orientation of the house is also important.

Technological developments in architecture enable to use energy of the Sun multilaterally. Application of active and passive facilities of sunny architecture allows to decrease the part of fossil power media in exploitation of houses. The advantageous difference of "sunny" architecture is contained in the fact that the requirement diminished to the possible minimum in energy was satisfied by means of energy of the Sun. The situation of eco-building and environments must be created consciously. When choosing a construction method, you should always consider solar radiation as a viable source of the Sun. If a house expresses the inner world of people who live in it, then it becomes the physical "projection" of the spiritual and moral state of maintaining the ties between the man and the house.

For Europeans the period of time, when it is necessary to ward off the Sun by shading, lasts approximately from May to September. On the south side of the house for shading, there are cultivated leafy plants that will cool it [9]. On the eastern, western and northern parts of the house the best idea is to plant evergreens. In summer, period plants give shade and coolness. Going back to a Sun, leaves create shade.

In Europe standards and norms of planning the house have obligatory character both for an architect and for those who will use the building. Some positions that are operated in this period in Central Europe in relation to the standards of natural illumination and insolation are as follows: general norms in relation to a garden, general norms in relation to minimum distances between buildings and general norms in relation to ventilation of apartments, sizes of apartments, their insolation and natural illumination [10].

Besides these norms of the estimation of the house quality, the basic factor will be natural light, and the quality factor will be natural insolation. If they provide a wonderful panorama from the window, an accommodation can already be subsumed « luxury » [11].

6 Conclusions

1. Methodological principles of forming the complex of light facilities of modern architecture forms are "subordination of the light constituent"; architecture and plan decisions; "symbiosis of climatological and architectural forms";

"introduction of new building technologies"; "the use of innovative means of illumination."

2. Light-shadow correlations are the important composition means of architectural shaping that carries out the influence on other composition facilities. By means of light, it is possible to define the relief of flat surface (frontal), strengthen then or weaken the feeling of weight or lightness by volume form (long shadows give the impressions of heavy form, short—vice versa).
3. The role of light is shown as shaping basis of architectural composition. Quality and comfort of architecture depend on the ability of the master to use and to combine these elements, both in internal and external space.

References

1. Ajzenberg, Yu. B. (2010). Energoeffektivnoe osveshhenie. Problemy i resheniya. In Yu. B. Ajzenberg & O. V. Maloxova (Eds.). *Energosovet*, (Vol. 6(11)S, pp. 20–26).
2. Shhipanov, A. S. (1960). Osveshhenie v arxitekture interera. A. S. Shhipanov (p. 116s). M.: Gosstrojizdat.
3. Seppanen, O. (2010). Trebovaniya k energoeffektivnosti zdanij v stranax ES. O. Seppanen. *Energosberezhenie*, *7S*, 42–51.
4. Koso, J. (2008). Solnechnyj dom. Estestvennoe osveshhenie v planirovke i stroitelstve. In J. Koso (Ed.), *Per. s vengerskogo A. I. Guseva* (p. 174s). M.: ZAO « Izdat. gruppa « Kontent » ».
5. Hraska, J. (1988). Doba insolacie okien tienenych zastavbou. In J. Hraska (Ed.), *Zbornik vedeckych prac Stavebnej fakulty SVST 1985*. Bratislava, Alfa.
6. Koniuk, A. (2018). Problemy arkhitekturnoi orhanizatsii ekolohichnoho ta enerhoefektyvnoho zhytla na prykladi ekolohichnoho blokovanoho zhytlovoho budynku v m. Poltava. Andrii Koniuk, Kateryna Danko, Enerhoefektyvnist v budivnytstvi ta arkhitekturi (Vol. 11S, pp. 112–119). Rezhym dostupu. https://doi.org/10.32347/2310-0516.2018.11.112-119.
7. Budivnytstvo z solomianykh blokiv. Osnovni etapy zvedennia Ekobudynku z solomy [Construction of straw blocks. The main stages of erection of EcoBuilt from straw]. http://www.gidproekt.com/stroitelstvo-iz-solomennyx-blokov-osnovnye-etapy-vozvedeniya-ekodomov-iz-solomy.html. (in Ukrainian).
8. Vyrobnytstvo i budivnytstvo karkasnykh budynkiv z solomianykh panelei [Production and construction of frame houses with straw panels]. http://www.eco-bud.com. (in Ukrainian).
9. Kittler, R. (1985). Luminace distribution characteristics of homogeneuns skies. R. Kittler, Light. *Research and Technology*, *17*(4).
10. Vasylenko, O. (2019). Visual optical effects in architecture. In O. Vasylenko, N. Polshchikova, M. Stashenko, & V. Zajarko (Eds.), *Problems of theory and history of Ukrainian architecture: Collection of scientific papers: output.19* (Vol. 389s. S, pp. 16–24). Odessa: Astroprint. Rezhym dostupu. https://doi.org/10.31650/2519-4208-2019-19-16-23.
11. Sergejchuk, O. V. (2015). Vimogi norm ES—osnova rozroblennya kompleksu normativnix dokumentiv z prirodnogo ta shtuchnogo osvitlennya. In O. V. Sergejchuk (Ed.), *Suchasni problemi texnichnogo regulyuvannya u budivnictvi: Zbirnik naukovix prac/*. K.: Knuba.

Modeling of Shell-Type Spatial Structural Forms by Superpositions of Support Nodes Coordinates

Oleg Vorontsov, Larissa Tulupova, and Iryna Vorontsova

Abstract In the design of modern building structures and architectural coating forms, while at the sketch stage basic geometric forms with all their advantages and disadvantages are already determined, geometrical design takes an important role. Modeling of shell-type spatial structural forms and coatings and their calculations, taking into account their physical properties, features of manufacturing and installation technology, requires presentation of information about the object in a discrete form. So it is advisable to carry out the process of their formation immediately in this discrete form. The current state of curvilinear architecture objects and constructions modeling requires taking into account as many as possible data and requirements to ensure the accuracy of the model. In geometric modeling, geometric characteristics are initial data—coordinates or parameter values. They are also geometric conditions. In most cases, these initial data are represented in a numerical form. Their arrays can be quite large. In these circumstances, methods of global continuous modeling, when only a single solution must be found, are ineffective. It happens because they require to use sufficiently complex mathematical algorithms and cannot provide necessary adequacy of models. Methods of discrete geometric modeling do not have these disadvantages. The method, proposed in this research, allows to form construction coverings in the form of balanced discrete frames, including surfaces of parallel transfer without compilation and solving bulky systems of linear equations. This is impossible to be obtained using the known finite differences method and static-geometric method.

Keywords Static-geometric method · Geometrical apparatus of superpositions · Value of recurrent dependence · Superposition coefficients

---

O. Vorontsov (✉) · L. Tulupova
Poltava National Technical Yuri Kondratyuk University, Poltava, Ukraine
e-mail: uaag.poltava2012@gmail.com

I. Vorontsova
Poltava Oil and Gas College of Poltava National Technical Yuri Kondratyuk University, Poltava, Ukraine

V. Onyshchenko et al. (eds.), *Proceedings of the 2nd International Conference on Building Innovations*, Lecture Notes in Civil Engineering 73,
https://doi.org/10.1007/978-3-030-42939-3_50

1 Introduction

Today, discrete geometrical modeling is the most promising direction of the development of applied geometry. There are two branches of this modeling—sampling of continuous geometric images and forming, using initial dates [1–4].

One of the most common methods of discrete surface modeling is the finite elements method. The essence of this method is a discrete representation of surfaces as a set of separate elements. These elements interact with each other in a certain number of nodal points.

The method of finite differences varies from the method of finite elements by its simplicity. But the first of these methods loses in versatility and some accuracy of results.

Using static interpretation of the method of finite differences, Professor S. Kovalev created a static-geometric method to form discrete geometric images. All these images are to possess some specific properties. This method is the simplest one in terms of understanding and the most obvious method in discrete modeling of continuous images. Beside this, in many cases it takes into account static features of different objects [5].

Article [6] by Professor S. Pustyulga dwells on the extension and development of the static-geometric method to shape properties. In this work, for forming geometric images he proposed the usage of a mathematical apparatus and geometric interpretation of numerical sequences in combination with the classical finite difference method and the static-geometric method.

This method enables passing on simply and effectively to continuous analogs of discrete models, which were already formed, solving a number of problems of discrete geometric modeling of balanced images without solving cumbersome systems of linear equations.

There are some advantages and disadvantages in solving practical problems for each of these methods. Because of this, their studying, enrichment by new effective methods, research of a possibility of compilation, and using all this, expansion of an initial data set are relevant. The further development and improvement of those methods in general are also significant.

Using geometric apparatus of superpositions together with the classical method of finite differences, static-geometric method and mathematical apparatus of numerical sequences for discrete surface modeling makes it possible to enrich them by new effective algorithms, allows improving their modeling capabilities and extending the range of practical problems [7–17].

2 Purpose

The purpose of this article is to develop a method to form modeling shell-type spatial structures and coatings in a form of balanced discrete frames of parallel transfer surfaces without compilation and solving cumbersome systems of linear equations.

3 Methods

Development of geometric models was carried out on the basis of applied geometry, in particular, discrete geometric modeling methods (static-geometric method, geometric interpretation of numerical finite differences method and mathematical apparatus of numerical sequences), elements of topology, analytical geometry and numerical analysis. In the process of practical implementation of the created models and algorithms, MathCAD 14.0 software environment was used.

4 Results

Formula

$$z_{i,j} = k_1 z_{i-1,j} + k_2 z_{i+1,j} + k_3 z_{i,j-1} + k_4 z_{i,j+1}, \tag{1}$$

where k_1, k_2, k_3, k_4 are coefficients of the superposition of applicates z given adjacent nodes, is identical to the finite difference five-point dependence

$$4z_{i,j} = z_{i-1,j} + z_{i+1,j} + z_{i,j-1} + z_{i,j+1}. \tag{2}$$

Therefore, the recurrent dependence value, which is a prototype of an external forming load, for the formation of a discrete surface frame based on the superposition of given four nodal points can be written in the form:

$$P_{i,j} = z_{i,j} - k_1 z_{i-1,j} - k_2 z_{i+1,j} - k_3 z_{i,j-1} - k_4 z_{i,j+1} \tag{3}$$

where P_i is a discrete value of recurrent dependence.

Provided that $\sum_{i=1}^{4} k_i = 1$,

$$P_{i,j} = z_{i,j} - k_1 z_{i-1,j} - k_2 z_{i+1,j} - k_3 z_{i,j-1} - (1 - k_1 - k_2 - k_3) z_{i,j+1}. \tag{4}$$

Consider an option of organizing a chain of superpositions to determine applicates of eighty-one (taking into account the symmetry of the support circuit—only fifteen) internal nodes as superpositions of given applicates of the central node and all contour ones, a segment of the discrete surface frame, which is presented, respectively, in Fig. 1.

Taking into account the results from [11], we can write

$$z_{i+4} = 0,5z_{i+3} + 0,5z_{i+5} - P;$$
$$z_{i+5} = 2z_{i+4} - z_{i+3} + 2P;\ z_{i+5} = 2z_i + 32P - z_i - 9P + 2P;$$
$$z_{i+5} = z_i + 25P;\ ;\ P = \frac{z_{i+5} - z_i}{25}; \tag{5}$$

$$z_i = z_{i+5} - 25P;\ z_{i+1} = z_{i+5} - 24P;\ z_{i+2} = z_{i+5} - 21P;$$
$$z_{i+3} = z_{i+5} - 16P;\ z_{i+4} = z_{i+5} - 9P;$$
$$z_j = z_{j+5} - 25P;\ z_{j+1} = z_{j+5} - 24P;\ z_{j+2} = z_{j+5} - 21P;$$
$$z_{j+3} = z_{j+4} - 16P;\ z_{j+4} = z_{j+5} - 9P; \tag{6}$$

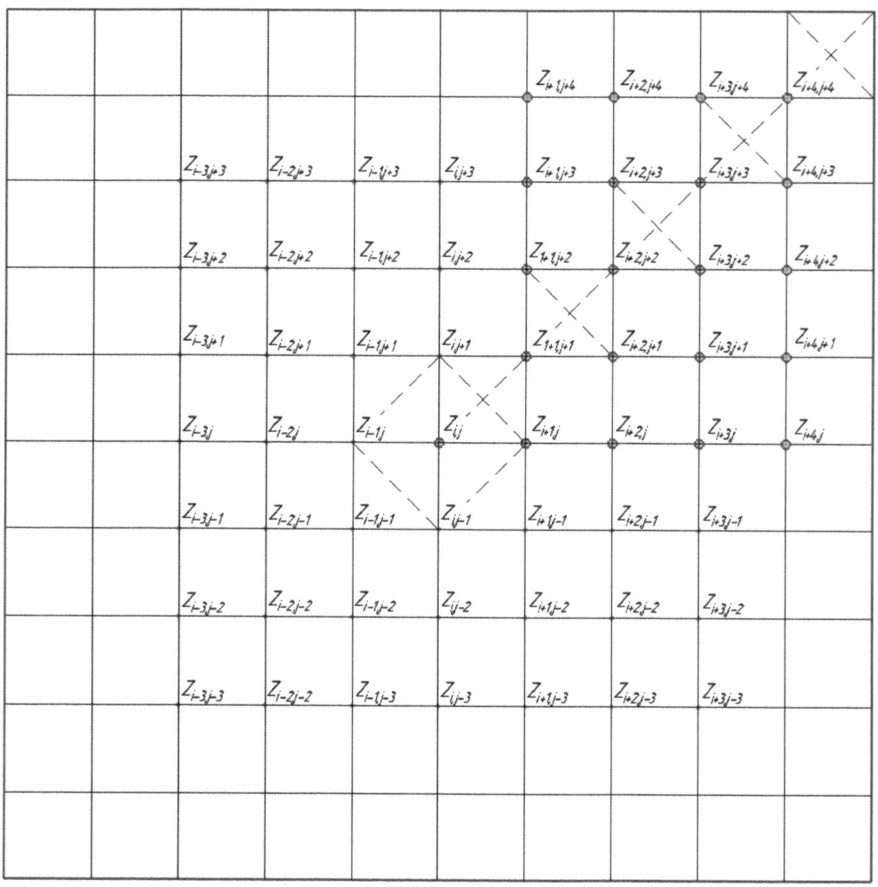

Fig. 1 Plan of the surface segment frame for the 81st inner node

Adding (6) to (5), we obtain formulas for determining applicates of internal nodes:

$$2z_{i,j} = z_{i+5,j} - 25P + z_{i,j+5} - 25P;\;\; 2z_{i,j} = z_{i+5,j} + z_{i,j+5} - 50P;$$

or:

$$z_{i,j} = 0,5z_{i+5,j} + 0,5z_{i,j+5} - 25P;\;\; P = \frac{0,5z_{i+5,j} + 0,5z_{i,j+5} - z_{i,j}}{25}; \tag{7}$$

$$2z_{i+1,j} = z_{i+5,j} - 24P + z_{i+1,j+5} - 25P;\;\; 2z_{i+1,j} = z_{i+5,j} + z_{i+1,j+5} - 49P;$$

or:

$$z_{i+1,j} = 0,5z_{i+5,j} + 0,5z_{i+1,j+5} - 24,5P;\;\; P = \frac{0,5z_{i+5,j} + 0,5z_{i+1,j+5} - z_{i+1,j}}{24,5}; \tag{8}$$

$$2z_{i+2,j} = z_{i+5,j} - 21P + z_{i+2,j+5} - 25P;\;\; 2z_{i+2,j} = z_{i+5,j} + z_{i+2,j+5} - 46P;$$

or:

$$z_{i+2,j} = 0,5z_{i+5,j} + 0,5z_{i+2,j+5} - 23P;\;\; P = \frac{0,5z_{i+5,j} + 0,5z_{i+2,j+5} - z_{i+2,j}}{23}; \tag{9}$$

$$2z_{i+3,j} = z_{i+5,j} - 16P + z_{i+3,j+5} - 25P;\;\; 2z_{i+3,j} = z_{i+5,j} + z_{i+3,j+5} - 41P;$$

or:

$$z_{i+3,j} = 0,5z_{i+5,j} + 0,5z_{i+3,j+5} - 20,5P;\;\; P = \frac{0,5z_{i+5,j} + 0,5z_{i+3,j+5} - z_{i+3,j}}{20,5}; \tag{10}$$

$$2z_{i+4,j} = z_{i+5,j} - 9P + z_{i+4,j+5} - 25P;\;\; 2z_{i+4,j} = z_{i+5,j} + z_{i+4,j+5} - 34P;$$

or:

$$z_{i+4,j} = 0,5z_{i+5,j} + 0,5z_{i+4,j+5} - 17P;\;\; P = \frac{0,5z_{i+5,j} + 0,5z_{i+4,j+5} - z_{i+4,j}}{17}; \tag{11}$$

$$2z_{i+1,j+1} = z_{i+5,j+1} - 24P + z_{i+1,j+5} - 24P;$$
$$2z_{i+1,j+1} = z_{i+5,j+1} + z_{i+1,j+5} - 48P;$$

or:

$$z_{i+1,j+1} = 0,5z_{i+5,j+1} + 0,5z_{i+1,j+5} - 24P;$$

$$P = \frac{0,5z_{i+5,j+1} + 0,5z_{i+1,j+5} - z_{i+1,j+1}}{24}; \tag{12}$$

$$2z_{i+2,j+1} = z_{i+5,j+1} - 21P + z_{i+2,j+5} - 24P;$$
$$2z_{i+2,j+1} = z_{i+5,j+1} + z_{i+2,j+5} - 45P;$$

or:

$$z_{i+2,j+1} = 0,5z_{i+5,j+1} + 0,5z_{i+2,j+5} - 22,5P;$$
$$P = \frac{0,5z_{i+5,j+1} + 0,5z_{i+2,j+5} - z_{i+2,j+1}}{22,5}; \tag{13}$$

$$2z_{i+3,j+1} = z_{i+5,j+1} - 16P + z_{i+3,j+5} - 24P;$$
$$2z_{i+3,j+1} = z_{i+5,j+1} + z_{i+3,j+5} - 40P;$$

or:

$$z_{i+3,j+1} = 0,5z_{i+5,j+1} + 0,5z_{i+3,j+5} - 20P;$$
$$P = \frac{0,5z_{i+5,j+1} + 0,5z_{i+3,j+5} - z_{i+3,j+1}}{20}; \tag{14}$$

$$2z_{i+4,j+1} = z_{i+5,j+1} - 9P + z_{i+4,j+5} - 24P;$$
$$2z_{i+4,j+1} = z_{i+5,j+1} + z_{i+4,j+5} - 33P;$$

or:

$$z_{i+4,j+1} = 0,5z_{i+5,j+1} + 0,5z_{i+4,j+5} - 16,5P;$$
$$P = \frac{0,5z_{i+5,j+1} + 0,5z_{i+4,j+5} - z_{i+4,j+1}}{16,5}; \tag{15}$$

$$2z_{i+2,j+2} = z_{i+5,j+2} - 21P + z_{i+2,j+5} - 21P;$$
$$2z_{i+2,j+2} = z_{i+5,j+2} + z_{i+2,j+5} - 42P;$$

or:

$$z_{i+2,j+2} = 0,5z_{i+5,j+2} + 0,5z_{i+2,j+5} - 21P;$$
$$P = \frac{0,5z_{i+5,j+2} + 0,5z_{i+2,j+5} - z_{i+2,j+2}}{21}; \tag{16}$$

$$2z_{i+3,j+2} = z_{i+5,j+2} - 16P + z_{i+3,j+5} - 21P;$$
$$2z_{i+3,j+2} = z_{i+5,j+2} + z_{i+3,j+5} - 37P;$$

or:

$$z_{i+3,j+2} = 0,5z_{i+5,j+2} + 0,5z_{i+3,j+5} - 18,5P;$$
$$P = \frac{0,5z_{i+5,j+2} + 0,5z_{i+3,j+5} - z_{i+3,j+2}}{18,5}; \quad (17)$$

$$2z_{i+4,j+2} = z_{i+5,j+2} - 9P + z_{i+4,j+5} - 21P;$$
$$2z_{i+4,j+2} = z_{i+5,j+2} + z_{i+4,j+5} - 30P;$$

or:

$$z_{i+4,j+2} = 0,5z_{i+5,j+2} + 0,5z_{i+4,j+5} - 15P;$$
$$P = \frac{0,5z_{i+5,j+2} + 0,5z_{i+4,j+5} - z_{i+4,j+2}}{15}; \quad (18)$$

$$2z_{i+3,j+3} = z_{i+5,j+3} - 16P + z_{i+3,j+5} - 16P;$$
$$2z_{i+3,j+3} = z_{i+5,j+3} + z_{i+3,j+5} - 32P;$$

or:

$$z_i + 3, j + 3 = 0,5z_{i+5,j+3} + 0,5z_{i+3,j+5} - 16P;$$
$$P = \frac{0,5z_{i+5,j+3} + 0,5z_{i+3,j+5} - z_{i+3,j+3}}{16}; \quad (19)$$

$$2z_{i+4,j+3} = z_{i+5,j+3} - 9P + z_{i+4,j+5} - 16P;$$
$$2z_{i+4,j+3} = z_{i+5,j+3} + z_{i+4,j+5} - 25P;$$

or:

$$z_{i+4,j+3} = 0,5z_{i+5,j+3} + 0,5z_{i+4,j+5} - 12,5P;$$
$$P = \frac{0,5z_{i+5,j+3} + 0,5z_{i+4,j+5} - z_{i+4,j+3}}{12,5}; \quad (20)$$

$$2z_{i+4,j+4} = z_{i+5,j+4} - 9P + z_{i+4,j+5} - 9P;$$
$$2z_{i+4,j+4} = z_{i+5,j+4} + z_{i+4,j+5} - 18P;$$

or:

$$z_{i+4,j+4} = 0,5z_{i+5,j+4} + 0,5z_{i+4,j+5} - 9P;$$
$$P = \frac{0,5z_{i+5,j+4} + 0,5z_{i+4,j+5} - z_{i+4,j+4}}{9}; \quad (21)$$

Thus, taking into account above formulas (5–21) and the results of [3], to determine applicates of an arbitrary node, using the known applicate of a given contour node in the direction of i axis, we can write

$$z_{i+n} = z_i + n^2 P;\ \ P = \frac{z_{i+n} - z_i}{n^2};$$
$$z_i = z_{i+n} - n^2 P; z_{i+1} = z_{i+n} - (n^2 - 1^2)P; z_{i+2} = z_{i+n} - (n^2 - 2^2)P;$$
$$z_{i+3} = z_{i+n} - (n^2 - 3^2)P; \ldots; z_{i+k} = z_{i+n} - (n^2 - k^2)P,$$

or:

$$z_{i+k} = z_{i+n} + (k^2 - n^2)P, \tag{22}$$

where k is a desired node number; n is a number of the given contour node.

To determine applicate of an arbitrary node by the known applicate of a given contour node in the direction of j axis, we can also write

$$z_{j+m} = z_j + m^2 P;\ \ P = \frac{z_{j+n} - z_j}{m^2};$$
$$z_j = z_{j+m} - m^2 P; z_{j+1} = z_{j+m} - (m^2 - 1^2)P; z_{j+2} = z_{j+m} - (m^2 - 2^2)P;$$
$$z_{j+3} = z_{i+m} - (m^2 - 3^2)P; \ldots; z_{j+l} = z_{j+m} - (m^2 - l^2)P,$$

or:

$$z_{j+l} = z_{j+m} + (l^2 - m^2)P, \tag{23}$$

where

l is a desired node number; m is a number of the given contour node.

By adding (23) to (22), we obtain formulas for determining applicates of internal nodes, as well as values of recurrence dependence:

$$2z_{i+k,j+l} = z_{i+n,j+l} + (k^2 - n^2)P + z_{i+k,j+m} + (l^2 - m^2)P;$$
$$2z_{i+k,j+l} = z_{i+n,j+l} + z_{i+k,j+m} + (k^2 + l^2 - n^2 - m^2)P;$$

or:

$$z_{i+k,j+l} = 0,5 \cdot z_{i+n,j+l} + 0,5 \cdot z_{i+k,j+m} + (k^2 + l^2 - n^2 - m^2) \cdot 0,5 \cdot P; \tag{24}$$

$$P = \frac{z_{i+k,j+l} - 0,5 \cdot z_{i+n,j+l} - 0,5 \cdot z_{i+k,j+m}}{(k^2 + l^2 - n^2 - m^2) \cdot 0,5} \tag{25}$$

where

k is a desired node number, n is a number of the given contour node,
$z_{i+n,j+l}$ is the given applicate of a contour node in the direction of axis i;
l is a desired node number, m is a number of the given contour node,
$z_{i+k,j+m}$ is the given applicate of a contour node in the direction of axis j;

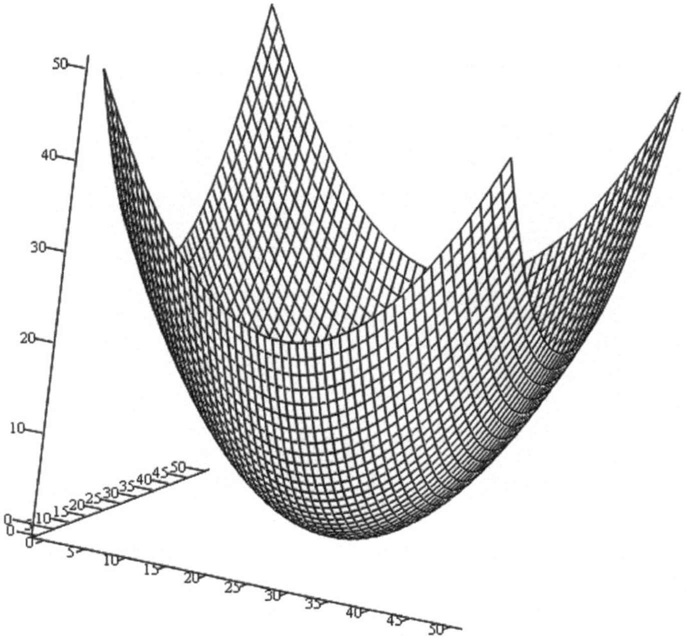

Fig. 2 Discrete frame of two-dimensional numerical sequence $z_{i,j} = i^2 + j^2$

P a value of recurrence dependence, which is equal to 0.25 of a value of an external formative load of the static-geometric method:

$$P = 0,25 \cdot KP.$$

We shall verify the validity of formulas (24, 25) on a test example of determining applicates and a value of recurrence dependence of the discrete frame of the two-dimensional numerical sequence shown in Fig. 2 (26):

$$z_{i,j} = i^2 + j^2. \tag{26}$$

For verification, we take the discrete values of the applicates from Table 1.

$$z_{i,j} = z_{3,2} = 13,\ z_{i-1,j} = z_{2,2} = 8,\ z_{i+1,j} = z_{4,2} = 20,$$
$$z_{i,j-1} = z_{3,1} = 10,\ z_{i,j+1} = z_{3.3} = 18.$$

1. By static-geometric method:

$$4z_{i,j} = z_{i-1,j} + z_{i+1,j} + z_{i,j-1} + z_{i,j+1} - KP$$
$$\Rightarrow 4z_{3,2} = z_{2,2} + z_{4,2} + z_{3,1} + z_{3/3} - KP$$

Table 1 Values of coordinates $z_{i,j}$ of the sequence points $z_{i,j} = i^2 + j^2$

j	*i*										
	−5	−4	−3	−2	−1	0	1	2	3	4	5
5	50	41	34	29	26	25	26	29	34	41	50
4	41	32	25	20	17	16	17	20	25	32	41
3	34	25	18	13	10	9	10	13	18	25	34
2	29	20	13	8	5	4	5	8	13	20	29
1	26	17	10	5	2	1	2	5	10	17	26
0	25	16	9	4	1	0	1	4	9	16	25
−1	26	17	10	5	2	1	2	5	10	17	26
−2	29	20	13	8	5	4	5	8	13	20	29
−3	34	25	18	13	10	9	10	13	18	25	34
−4	41	32	25	20	17	16	17	20	25	32	41
−5	50	41	34	29	26	25	26	29	34	41	50

$$\Rightarrow 4 \cdot 13 = 8 + 20 + 10 + 18 - KP$$
$$\Rightarrow 52 = 56 - KP \Rightarrow KP = 4.$$

2. By formulas (24, 25):

$$n = 5,\ m = 5,\ k = 3,\ l = 2,\ P = 0,25 \cdot KP = 0,25 \cdot 4 = 1;$$
$$z_{i+k,j+l} = z_{3,2},\ z_{i+n,j+l} = z_{5,2} = 29,\ z_{i+k,j+m} = z_{3,5} = 34.$$
$$z_{i+k,j+l} = 0,5 \cdot z_{i+n,j+l} + 0,5 \cdot z_{i+k,j+m} + (k^2 + l^2 - n^2 - m^2) \cdot 0,5 \cdot P$$
$$\Rightarrow z_{3,2} = 0,5 \cdot z_{5,2} + 0,5 \cdot z_{3,5} + (3^2 + 2^2 - 5^2 - 5^2) \cdot 0,5 \cdot 1$$
$$\Rightarrow z_{3,2} = 0,5 \cdot 29 + 0,5 \cdot 34 + (3^2 + 2^2 - 5^2 - 5^2) \cdot 0,5 \cdot 1$$
$$\Rightarrow z_{3,2} = 14,5 + 17 + (13 - 50) \cdot 0,5 \cdot 1 \Rightarrow z_{3,2} = 13;$$
$$P = \frac{(z_{i+k,j+l} - 0,5 \cdot z_{i+n,j+l} - 0,5 \cdot z_{i+k,j+m})}{(k^2 + l^2 - n^2 - m^2) \cdot 0,5} \Rightarrow$$
$$P = \frac{(z_{3,2} - 0,5 \cdot z_{5,2} - 0,5 \cdot z_{3,5})}{(3^2 + 2^2 - 5^2 - 5^2) \cdot 0,5} \Rightarrow$$
$$P = \frac{(13 - 0,5 \cdot 29 - 0,5 \cdot 34)}{(13 - 50) \cdot 0,5} \Rightarrow P = \frac{-18,5}{-18,5} \Rightarrow P = 1.$$

The results of this test example confirm the correctness of the obtained formulas (24, 25).

5 Scientific Novelty

On the basis of the geometric apparatus of superpositions, a new method of discrete modeling has been developed. This allows formatting balanced discrete frames of parallel transfer surfaces, the second-order curves of which are components of the frame. The value of recurrence dependence P is a forming one.

6 Practical Significance

The usage of the geometric apparatus of superpositions together with the static-geometric method and the classical method of finite differences for discrete modeling of surfaces enabled us to expand the range of practical problems and optimize the models created for their implementation.

7 Conclusions

The method, which is proposed in this work, allows forming spatial structures and shell-type coatings of building structures in the form of balanced discrete frames, including parallel transfer surfaces without compilation and solving cumbersome systems of linear equations. This is impossible to be obtained, using the known finite difference method and static-geometric ones.

References

1. Guoliang, Xu, Pan, Oing, & Bajaj, Chandrajit L. (2006). Discrete surface modelling using partial differential equations. *Computer Aided Geometric Design, 23*(2), 125–145. https://doi.org/10.1016/j.cagd.2005.05.004.
2. Meek, D., Ong, B., & Walton, D. (2003). Constrained interpolation with rational cubics. *Computer Aided Geometric Design, 20*(5), 253–275. https://doi.org/10.1016/S0167-8396(03)00044-X.
3. Mykhaylenko, V. YE., & Koval'ov, S. M. (2004). *Dyskretne heometrychne modelyuvannya: Vynyk-nennya, rezul'taty, pidsumky. Prykladna heometriya ta inzhenerna hrafika* (Vyp. 74, S. 3–8). K.: Knuba.
4. Naydysh, V. M., & Naydysh, A. V. (2004). *Novyy pohlyad na problemy dyskretnoho heometrych-noho modelyuvannya. Prykladna heometriya ta inzhenerna hrafika* (Vyp. 74, S. 14–19). K.: Knuba.
5. Kovalev, S. N. (1986). *Formirovaniye diskretnykh modeley poverkhnostey prostranstvennykh arkhitekturnykh konstruktsiy: Dis.... doktora tekhn nauk*: *05.01.01* (p. 348s). S.N. Kovalev, M.: MAI.
6. Pustyul'ha, S. I. (2004). *Pryntsypy uzahal'nennya statyko-heometrychnoho pidkhodu do modelyuvannya dyskretnykh struktur. Prykladna heometriya ta inzhenerna hrafika* (Vyp. 74, S. 114–121). K.: Knuba.
7. Vorobkevich, R. I. (1989). *Konstruirovaniye liniy i poverkhnostey na osnove spetsial'-nykh operatsiy nad funktsiyami i trigonometricheskimi splaynami. Dis. kand. tekhn. nauk. 05.01.01*. K.: KISI.
8. Samchuk, P. V. (1991). *Upravleniye formoy diskretno zadannykh poverkhnostey v zadachakh proyektirovaniya obolochek. Dis. kand. tekhn. nauk.05.01.01*. K.: Kisi.
9. Khay, C. K. (1994). *Upravleniye formoy rastyanutykh sistem na osnove funktsional'-nogo slozheniya. Dis. kand. tekhn. nauk. 05.01.01*. K.: Kisi.
10. Vorontsov, O., Tulupova, L., & Vorontsova, O. (2018). Geometric and Computer Modeling of Building Structures Forms [Special Issue 8]. *International Journal of Engineering and Technology. 7*(4.8), 560–565. ISSN: 2227-524X. https://doi.org/10.14419/ijet.v7i4.8.27306 Published on: October 13, 2018.
11. Kovalev, S. N. Superpozitsiyah, O., & S.N. Kovalev (2010). Prikladna geometrIya ta Inzhenerna grafIka.: Zb. nauk. prats (Vip. 84. S. 38–42). K.: Knuba, ISSN 0131-579X.
12. Vyazankin, V. A. (2010). Zamena superpozitsii konechnogo chisla tochek tsepyu posledovatelnyih superpozitsiy par tochek. In V. A. Vyazankin, & A. V. Mostovenko (Eds.), *Prikladna geometrIya ta Inzhenerna grafIka.: Zb. nauk. prats* (Vip. 84, S. 296–300) K.: Knuba. ISSN 0131-579X.
13. Vorontsov, O. V. (2018). Vyznachennia koordynat vnutrishnikh vuzliv, yak superpozytsii zadanykh koordynat tsentralnoho ta dvokh konturnykh vuzliv dyskretno predstavlenoi kryvoi. In O. V. Vorontsov, L. O. Tulupova, & I. V. Vorontsova (Eds.), *Visnyk Khersonskoho natsionalnoho tekhnichnoho universytetu*. (Vyp. 3(66), S. 120–124). TOM 2. Kherson: KhNTU.

14. Cherniha, R., & Serov, M. (2006). Symmetries, ansätze and exact solutions of nonlinear second-order evolution equations with convection terms. *II. European Journal of Applied Mathematics, 17*(5), 597–605. https://doi.org/10.1017/S0956792506006681.
15. Yakovlev, S. V., & Valuiskaya, O. A. (2001). Optimization of linear functions at the vertices of a permutation polyhedron with additional linear constraints. *Ukrainian Mathematical Journal, 53*(9), 1535–1545. https://doi.org/10.1023/A:1014374926840.
16. Cherniha, R., & Pliukhin, O. (2013). New conditional symmetries and exact solutions of reaction-diffusion-convection equations with exponential nonlinearities. *Journal of Mathematical Analysis and Applications, 403*(1), 23–37. https://doi.org/10.1016/j.jmaa.2013.02.010.
17. Pustyul′ha, S. I., & Samostyan, V. R. (2008). Analiz suchasnykh pidkhodiv do dyskretnoho heomet-rychnoho modelyuvannya ob″yektiv. Prykladna heometriya ta inzhenerna hrafika (Vyp. 80, S. 60–66). K.: Knuba.

Energy Efficient Economy in Ukraine, Azerbaijan, and the EU Problems of Present and Future

English Compound Construction Economic Terminology: Current Aspects of Professional Text Cohesiveness

Anna Ageicheva, Alla Bolotnikova, Yuliia Hunchenko, and Iryna Perederii

Abstract The current paper refers to peculiar features of scientific and scientific and popular texts on construction economics. Among those being regarded here, there are professional text coherence and cohesion. Constructio economic terminology is an example of several scientific field integrations and interrelations, made up of construction and economic term aggregates, correlated with these science notion systems. Term is a word or word combination used in professional language, aimed at detonating special notions or objects characterized by normativity, systematicity, definitivity and precision. National language terminology enrichment is predominantly made up due to terminologization of word combinations, forming in turn compound terms. Professional lexis plays a special role among the lexical means of cohesion connection in professional texts. It is an important tool of actualization of identity, opposition, implication, particular and common, class, variety, etc. The term's important feature is its ability to be the key cohesion component of professional texts. Term's cohesion function provides discursive cohesiveness of professional construction economic text organization. Discursive importunateness of cohesive lexis consists in its ability to connect terms into nominative construction.

Keywords Construction economics · Terminology · Compound terms · Coherence · Cohesion

A. Ageicheva · A. Bolotnikova · Y. Hunchenko (✉) · I. Perederii
Poltava National Technical Yuri Kondratyuk University, Poltava, Ukraine
e-mail: yuliyagunchenko@gmail.com

A. Ageicheva
e-mail: ageicheva@ukr.net

A. Bolotnikova
e-mail: a.p.bolotnikova@gmail.com

I. Perederii
e-mail: iryna.perederii@gmail.com

V. Onyshchenko et al. (eds.), *Proceedings of the 2nd International Conference on Building Innovations*, Lecture Notes in Civil Engineering 73,
https://doi.org/10.1007/978-3-030-42939-3_51

1 Introduction

Construction economics is a set of social and industrial relations in construction, a science that studies the laws of construction development, factors that determine efficiency of labor and use of means of production in this important branch of economics, as well as forms and methods of economic work in construction production. As an independent branch of science, construction economics was formed after the Second World War; construction economics explores specific forms and manifestations of economic laws in construction industry, examines ways of construction industry development and links with other sectors of economics, methods and means of improving its efficiency. At the same time, it is intended to solve specific economic problems that arise in the processes of development of construction production management organizational forms (including specializations, cooperatives, concentrations, etc.), development of capital investments, use of available resources, etc.

Construction economics develops a system of methods for planning capital investments and construction production, examines its efficiency (factors, criteria, indicators, standards), conditions for increasing labor productivity, increasing capital efficiency, improving use of material and other resources, accelerating development of the existing and new production facilities, achievement of technical and economic project indicators, etc. Its objectives also include development of economic bases of construction design, construction typing, industrialization, economic conditions and urban development by means of improving social and economic benefits necessities of estimated cost reduction. These are essential for ensuring technical and economic regulation of construction production at all stages of the investment process and controlling it. Construction economics ought to form and improve relevant regulatory framework, which is a system of interrelated cost and natural standards (planning, production, estimates, accounting, etc.). It also studies capital construction resources and efficiency of their use, focusing on issues of material and technical base development, fixed assets, their reproduction, formation and regulation of construction industry working capital, training and use of personnel, organization and improvement of labor efficiency, material and technical supply. Construction economics should provide economic justification of methods applied in management and stimulation in capital construction, economic calculation, financing and crediting of construction, system of settlements between participants of the construction process, based on the need to strengthen the role of economic incentives to accelerate scientific and technological progress. Economic and mathematical apparatus of calculations, modern tools of modeling and forecasting bring construction economics to the exact sciences. Its entire arsenal serves as scientific base reliable foundation for selecting and implementing optimal solutions for planning, designing the organization and providing processes for construction production. As a science, construction economics develops in close collaboration with economic theory and with many specialized disciplines (finance and credit, accounting statistics and business analysis, etc.), as well as with sciences that study production and technical basis of construction.

Considering the above mentioned, we can note that English terminology of construction economics is an example of several scientific field integrations and interrelations, made up of construction and economic term aggregates, correlated with these science notion systems, regulated and controlled. The term's functioning can be best traced in the text, thereby making the text an object of numerous scientific researches. Although texts have already been the objects of investigation, the most of attention has been paid to literary or journalese texts. Investigation of scientific-popular texts seems to be of no less importance; especially nowadays, when constant, still rapid development of scientific knowledge makes them a verbal means of preserving and transferring human cognitive experience.

Scientific-popular economic and construction texts have already been an object of linguistic research. Thus, the problem of English economic texts is reflected in the works by K. Tarasova (study of economic discourse); K. Petushinskaya (English journalese texts as an example of economic discourse); I. Humovska (English juridical terminology in economic texts); V. Artiuh (terminological verbal word-combinations in English economic texts); I. Kryvoruchko (text as a unit of business communication); S. Rybachok (means of cohesion of English economic texts) and others. Constructing terminology aspects of functioning are revealed in the works by I. Kuznietsova (architectural terminology), A. Pysmychenko (construction terminology), N. Knyshenko (compounding construction terminology) and others. Yet, peculiarities of functioning compound nouns—construction economic terms—in the scientific-popular texts were not investigated. Hence, determining the aim of our research, this is the following: defining the main functions of the terms being investigated in the stated texts and studying their usage. It should also be noted here that diversified study in the field of modern English construction economic terminology contributes to identification of its formation regularity, functioning, systematization and helps to predict the main tendencies of its further development. This is what determines the field of our research.

2 Results

Scientific-popular texts in periodical publications are considered to be secondary sphere of term's functioning. Using terms in the texts of this type is aimed at performing informative and educational objectives, as well as keeping intellectual contact with readers.

Term is a word or word combination used in professional language, aimed at detonating special notions or objects characterized by normativity, systematicity, definitivity and precision. National language terminology enrichment is predominantly made up due to terminologization of word combinations, forming in turn compound terms. Used to define various spheres of human professional activity, compound terms together with other terms make up a certain system within the national language. It has to be noted that compound terms are currently dominating within modern terminological systems. This can be explained by the fact that it is compound term, which is able to most completely reflect the necessary characteristic features of the notion being named. Dominance of such terms could be caused by rapid occurrence of new notions to be named, which is a result of technology development, including those in construction sphere. Terms of this kind are characterized by high ability to be differently combined, and the number of their components can be enlarged.

The current research conducted at the material of English compound nouns—construction economic terms—suggests taking into account both terminological and world-building aspects. As far as compounding is considered to be one of the main areas in English lexis development, its study appears as an important aspect in world-building investigation process.

Terms in professional language perform a number of functions, i.e., the role they play as a means of special notion denotation. As far as term is a polyfunctional unit, it is necessary to consider its functions in their connections. Term's functioning system is much more complicated than corresponding non-term's system, which is due to the fact that term is used to express the complex general notion, e.g., scientific, technical, construction, economic, political, etc. Among the most significant functions of the terms, the following ones are usually named: *nominative, definitive, informative, communicative, text-forming, cohesion, thematic, pragmatic* and *heuristic*. One of the main term's functions is its nominative or representative function that is denomination of a definite meaning. Special feature of a term as a nominative unit lies in special knowledge fixation. Thus, term is a key element connecting scientific-popular text.

Henry Widdowson defines text as an "actual use of language, distinct from a sentence which is an abstract unit of linguistic analysis." [8]. Though the text has already been an object of numerous linguistic studies, the scientists' primary attention was turned toward literary texts or publicistic texts. Investigations in the field of scientific-popular texts, although of the less importance, especially nowadays, when constant development of newest knowledge is being reflected in dictionaries, workbooks, study guides, etc., are still less typical. In modern times of scientific-technical progress development, scientific text becomes a significant verbal means of preservation and communication of human experience and as a means of scientific knowledge further development. Thus, it deserves the same attention as the other types of the texts mentioned.

Among the main features of scientific text, the most significant include its *informativity*, *integrity*, *cohesion* and *ability to be articulated*. Cohesion, which is a text category enabling information transfer through the system of interrelated and integrated elements in the complex of their structural, semantic, cognitive and functional characteristics, is one of the most important features of scientific text [5]. Coherence and cohesion are the two concepts, which are closely connected and even sometimes regarded as synonymous. Still, coherence is a cognitive phenomenon. Its recognition involves text and refers to logical flow of interrelated topics in a text, thus establishing a mental textual world. Cohesion can be regarded as an explicit indicator of relations between topics in the text. It refers to text-internal relationship of linguistic elements linked by lexical and grammatical devices across sentence boundaries.

Cohesion can be of several types, which are phonetic, morphemic, morphological, lexical or semantical. It also can be distant or contact one. Such division is based on the placement of repeated elements. The main types of cohesion generally specified are the following: coreference, ellipsis, conjunction and lexical cohesion.

Specificity of English scientific construction economical text, much alike to scientific economical text, consists mainly in comparatively small amount of key concepts being used in it, which are in close connection with each other. Terms in such a text, together with terminological meaning, can supply the elements of the common language meaning, while the non-terminological lexis can supply scientific information. It ought to be noted that professional text enables the most compete disclosing of the term's meaning. Peculiar feature of scientific-popular texts is the inhomogenuity and inequality of term's placement in the text structure, graduality of term's representation, usage of complete lexical repetitions, doublets, non-terminological elements, analogues and tropes [4].

Transformational changes in society determine linguists' interest to terminology, including those used in scientific-popular construction economic texts. Informational processes in turn contribute to researchers' interest to construction economics as a professional subject. Accessibility of such kind of texts, including those in English, not to the specialists only, but to common mass readers, provokes necessity in including construction economic terminology into future specialists in translation professional training, including simultaneous translation. Skills and abilities of making translational transformations constitute a significant component of a translator's competence; therefore, considerable attention should be paid to their training and correct application [1, 559].

Professional lexis plays a special role among the lexical means of cohesion connection in professional texts. It is an important tool of actualization of *identity, opposition, implication,* particular and common, class, variety, etc. [4]. Terminological lexis is a system factor in the text and therefore contributes to its cohesion [3, 21].

The term's important feature is its ability to be the key cohesion component of professional texts. Term's cohesion function provides discursive cohesiveness of professional construction economic text organization. Discursive importunateness of cohesive lexis consists in its ability to connect terms into nominative construction. The means of the term's cohesive potential maintenance on semantic level include *reiteration, general scientific notions and antonyms.*

Reiterations demonstrate a high cohesive ability. They contribute to easier understanding of professional information in the text and keep reader's attention to the key data read. Cohesive nature of a text can be also realized through such forms of text composition as paragraphs [4, 11–12].

Terms occupy a special place among lexical means of the professional text cohesion coherence, being an important instrument of maintenance of identity, opposition, involvement, a part and a whole, a class and plurality, etc. Terms in the text serve as a system factor, i.e., serve cohesion.

According to the results of our research, compound nouns—construction economic terms—mainly serve the actualization of such means of cohesion as lexical reiteration, general scientific words and antonymic means. Lexical reiteration cohesion ability contributes to accentuation and keeping the addressee's attention and consequently to better text understanding. Examples of lexical reiteration in scientific-popular economic texts are the following:

All of the consumption ***outputs*** *of trees are in one way or another fed back to support production, replenishing the resources required to create more trees. Very little of the* ***outputs*** *from consumption involved with the life cycle of a building are kept "in the loop."*

It should be noted that text cohesion maintenance is more often met through complete lexical reiteration of compound terms rather than synonymic reiterations. This is due to the higher information value of a compound term comparing to a simple term.

Antonymy as a means of cohesion can also be actualized by compound terms:

Construction markets are characterized by supply schedules where the ***short-term price elasticity*** *of supply is lower than the* ***long-run elasticity****.*

One more example of antonymy usage to serve cohesion is the following:

The implication of this assumption is that construction prices (i.e., the cost of a construction project to the client) will rise and fall, with lags, in relation to changes in demand. Suppliers of increasingly scarce inputs are able to raise prices during ***upturns*** *in demand but face falling prices during* ***downturns****.*

We contrasted ***private costs*** *and* ***external costs****—a distinction that helps to explain a broad set of environmental problems. The related analysis represents an important tradition in welfare economics, stretching back to the beginning of the twentieth century.*

Usage of antonymic compound terms in the scientific-popular economic text also contributes to text cohesion ability actualization as far as it emphasizes reader's attention with the help of opposition.

The research conducted revealed examples of synonymic lexical reiteration of compound economic terms in scientific-popular construction economic texts:

In direct contrast, ***environmental economics*** *does not accept that the ecosystem, or nature, is merely another sector of the economy that can be dealt with by market forces.* ***Environmental economists*** *proceed from the basic premise that there is an extensive level of interdependence between the economy and the environment, and there is no guarantee that either will prosper in the long term unless governments enforce measures that make firms acknowledge the complete life cycle costs arising from their economic activity.*

As for the contiguous and distant peculiarities of compound terms' positioning in the text, we should note that both types of positioning can be seen in the investigated texts, though contiguous position is more frequent. Contiguous and distant terms' positioning in the scientific-popular economic text enables gradual actualization with systematic widening of a term semantic meaning.

Text compression function is among the main text-forming functions of compound nouns economic terms in the scientific-popular construction economic texts. It is because a compound term main stylistic-structure feature is its ability to combine two full meaning stems. Semantic meaning of such compound term is usually bigger than simply a sum of two terms' meanings included into a compound; therefore, a compound or parasyntheton term is more informative compared to a simplex, enabling it to perform a text compression function.

3 Conclusions

In the given paper, the authors revealed characteristic terminological features of a specific branch of science, which is construction economics. Specificity of English scientific construction economical text consists mainly in comparatively small amount of key concepts being used in it, which are in close connection with each other. Terms in such a text, together with terminological meaning, can supply the elements of the common language meaning, while the non-terminological lexis can supply scientific information. The authors' concern includes coherence and cohesion means of specialized text that are grounded on using compound terms. Terms occupy a special place among lexical means of the professional text cohesion coherence, being an important instrument of maintenance of identity, opposition, involvement, a part and a whole, a class and plurality, etc. Terms in the text serve as a system factor, i.e., serve cohesion. The investigation conducted has shown that compound terms contribute to text cohesion actualization mainly by the means of lexical reiteration, synonymic reiterations, antonymy, contiguous and distant positioning. Further research in the noted field will contribute to identification of its formation regularity, functioning, systematization and will help to predict the main tendencies of its further development.

References

1. Ageicheva, A., & Hunchenko, Y. (2018). Grammar peculiarities of scientific and technical translation in construction sphere. *7*(3.2) (Special Issue 2).
2. Danny, M. *Construction economics: A new approach*.
3. Drozdova, T. V. (2003). Scientific text and problems of its understanding (at the material of English economic texts).
4. Fiorentino, R., & Poeppel, D. (2007). Compound words and structure in the lexicon. *Language and Cognitive Processes*.
5. Menzel, K., Lapshynova-Koltunski, E., & Kunz, K.: New perspectives on cohesion and coherence. Implications for translation.
6. Rybachok S. M. (2005). Terminological lexis as a means of cohesion in English economic text. Zaporizha Mational University Press.
7. Smith, C. S. (2003). *Modes of discourse. The logical structure of texts*. Cambridge: University Press.
8. Widdowson, H. G. (2007). *Discourse analysis*. Oxford: University Press.

Application of the Modern Motivational Theories for Increasing Competitive Advantages of Construction Companies

Konul Asaf Aghayeva

Abstract The purpose of the research was to determine the general performance principles of motivation mechanism and its elements based on the modern motivational theories for further increasing competitive advantages of construction companies. The results of the research allowed substantiating basic performance principles of motivation mechanism and the general system elements of motivational approaches, applicable to human resources that would be useful when managing construction projects and enterprises in general.

Keywords Motivation · Encouragement · Construction organization · Employees · Human resource management · Motivational theory · Building sector

1 Introduction

In modern market, one of the most important challenges facing any organizational leadership (including building) is determination and future efficient use of organizational resources [1]. The changes of environmental conditions require the organization to have corporate management flexibility to remain competitive [2]. During this process, a great significance is given to effective use of such organizational resources as human resources [3]. To maintain and increase the effectiveness of staff ability to deliver good results, there is a need in constant motivation—the use of a system of encouragements that encourage corporate employee to effective, highly productive labor [4]. The efficiency of the current system of motivation and encouragement in building organization largely affects the labor, social and creative activity of each employee, resulting in the final business performance and competitiveness of building organization [5].

Independent assessment of building organization capacity in carrying out certain types of work, investing advisability and partnership depends on the extent, to which building organization has the ability to use motivated employees' effective work. Employees' skills and background, sociopsychological climate within the group and

K. A. Aghayeva (✉)
Azerbaijan University of Architecture and Construction, 11 Ayna Sultanova, Baku, Azerbaijan
e-mail: konul.aghayeva@gmail.com

V. Onyshchenko et al. (eds.), *Proceedings of the 2nd International Conference on Building Innovations*, Lecture Notes in Civil Engineering 73,
https://doi.org/10.1007/978-3-030-42939-3_52

labor efficiency of certain employees affect the labor quality, as well as the effectiveness of management decisions and therefore the ultimate business performance of a building organization.

2 Main Part

2.1 Purpose of the Article

The purpose of this study is to identify the optimum system of motivation and encouragement in human resource management in modern building organizations based on modern motivational theories in order to achieve high productivity.

The study draws on the works of world scientists dedicated to the problems of forming and developing a system of human resource management of construction companies in a competitive environment, the management and reproduction of their labor potential and principles of management problems comprehensive solutions.

2.2 Research Methodology

Fulfillment of the tasks set by the research was carried out by applying a systems approach (while conducting examination of motivation potential, the specifics of the activity of construction companies and their influence on the process of human resource management in these companies) and abstract-logical examination (when analyzing terminology).

2.3 Results

The paper is aimed at perfecting research and methodological foundations of human resource management in construction companies by means of motivation and incentives.

Attainment of the set goal presupposes fulfillment of the following tasks:

- studying the specifics of construction companies' staff management;
- developing conceptual guidelines for setting up a system of construction companies' staff motivation, taking into consideration the specificity of the industry;
- perfecting the mechanism of incentives for the personnel of construction companies.

Construction is one of the largest sectors of economy, which provides an increase in production capacity and assets [6]. It has a number of specific features, each of which is characterized by specific human resource management issue. Today, the features of employees' activity are often ignored in features' assessment of motivation and encouragement in building organizations.

The construction industry, like all the other industries, has its own characteristics in terms of its management. For example, mobility of workplaces and construction machines, equipment and technological labor infrastructure are crucial. These characteristics are constantly tied to the price situation in labor and land resources market. At the same time, regulatory and legislative acts are tied to construction projects' location. Strong correlation with other economy sectors makes the industry sensitive to the effects of negative direct factors. It should also be noted that most projects in the construction business are long-term projects, and employees' involvement in such projects for a long time requires special mechanisms of work motivation and approaches in the personnel management system. Building organizations must operate with regard of environmental friendliness principles while choosing materials and do not pose a threat to the lives and health of potential buyers; this fact significantly affects the market value of the project. Specific institutional instability of the most post-Soviet regions and volatile external environment of construction enterprise operation contribute to the necessary adaptation of enterprises and the development of new modern economic methods of analyzing and forecasting the construction projects. This significantly affects the priority resources in the overall system of constraints—human resources.

Modern approach to human resource management demonstrates the need in combining rational employment with motivation and employee recognition. The disadvantage of building organizations in market is the lack of guidance note in motivation and employee recognition and quality assessment of encouragement methods. To increase efficiency, it is necessary to improve the approaches in appropriate matching of workforce policies and development business strategy under management.

The research of encouragement necessity in building organizations under modern conditions, taking into account the specifics of this sector, makes it possible for management subjects to form real practical encouragements that will encourage employees to work effectively. The system of motivations, in contrast to individual encouragements, takes into account different situations and necessary encouragements that determine employees' behavior.

The efficacy of motivational process is determined by a degree of personal goals' satisfaction and task solution that are faced by the company. It is advisable to focus on encouragement methods of labor activity [3]. Classification of encouragement methods is common to all sectors of economy. In addition, it is appropriate to consider individual, group, internal and external encouragements provided by labor (sense of labor importance, self-esteem, etc.). The mentor role of organizational leadership in relation to employees is in positive effect of maintaining their initiative, employees' identification with the organization in which they work. It is advisable to conduct employee recognition in terms of commitment [7], facilitating employees' perception of themselves as a single team [8], their interaction with not only organizational leadership, but also with each other, knowledge and experience flow [9]. In this case, the entrants will not be afraid to start an unfamiliar job, and skilled workforce will be even more needed.

Identifying a leader, we need to know who will coordinate and develop the motivation of employees. They point to the need in both intrinsic (satisfaction from activity, elimination of guilt or shame, work performance as a particularly valuable job) and external motivation (encouragement, avoidance of punishment and career development) [10].

According to Dillon and Manz, the lead man has to create the way of emotional development to conduct emotional motivation of employees (pride, guilt, shame). The importance of understanding a full sense of purpose, promoting the formation of deep intrinsic value of job, understanding by employees of their labor, a sense of its value. The importance of interaction between the leader and employee, for which he will play an inspiring, intellectually stimulating role, will be an example of an idealized behavior and will contribute to their social identity and career promotion.

Motivation of production capacity increases as a set of techniques and methods of influence on the employee by means of organizational leadership. They encourage employees to certain behavior in labor to achieve the goals of the organization based on necessity to meet personal needs.

A labor motivation as a purposeful encouragement to highly efficient labor is reached through the effect on personal needs, interests and goals. The idea is that employees' motivation to work performance involves motives, influences and encouragements—conscious and unconscious ones, resulting in employee desire to achieve a certain goal. The theory of motivation is experimental and does not offer clear conclusions.

In conclusion, it should be noted that the main tasks of motivation are the following: human resource training and management training in terms of psychological foundations of corporate communication; liberal approach formation to human resource management with the use of modern methods of motivation (organizational leadership); understanding the formation of nature and importance of motivation in labor (employees).

Majority of authors define motivational mechanism as a set of organizational, economic, logistical and sociopsychological tools and methods of effective labor encouragement to ensure the achievement of motivational policy objectives.

The motivational mechanism converts a set of factors, principles, encouragements, motives, values, expectations and behavioral reactions out of linear discrete state into constantly iterative process. Structurally, motivational mechanism includes financial and economic methods and tools, a specific staff composition, rules and regulations of employee recognition, frequency of measures assessment that were put into effect, etc. (Fig. 1).

Motivational mechanism should be built by taking into account the features of employee, which determine their needs, interests, attitudes and values, and by taking into account the current human resource management structure, factors influencing management internally and externally, as well as historical experience.

Motivation takes place at three interrelated levels, each of which has its own distinctive features.

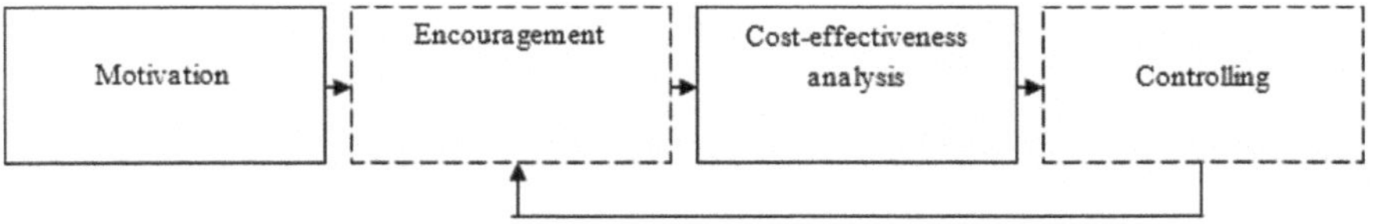

Fig. 1 Process of motivation

Managerial level:

long-term and medium-term motivation;
advertising and staff retention;
coordination of personal, group and managerial levels of motivation.

Group level—mandatory account of group (relations within group) impact on personal motivation to labor.

Personal level—plays a key role, as a group of unmotivated employees cannot have a total positive motivation.

Accordingly, human resource development can be considered as organizational leadership management in building sector aimed at identification, motivation of personal fulfillment of employees in terms of creativity, encouragement of its manifestation in strategic goals' implementation of the organization (Fig. 2).

2.4 *Scientific Novelty*

The study provides an opportunity to present its results as a theoretical generalization of knowledge settled in modern world scientific community. It suggests the need to consider motivation and employee recognition of labor activity in the building company as the most important factor in work performance improvement. It is important to take into account the features of building sector, which cause the specifics of human resource management in modern building companies that are considered in the article.

2.5 *Practical Importance*

However, the research in the field of human resource management in building organizations in the majority of cases is descriptive in nature; some issues of motivation and encouragement are considered poorly or not considered at all. The lack of science-based approach that takes into account specific features of building sector is one of the reasons for this kind of research. Based on this approach, there can be made a motivational mechanism and human resource value management in building companies.

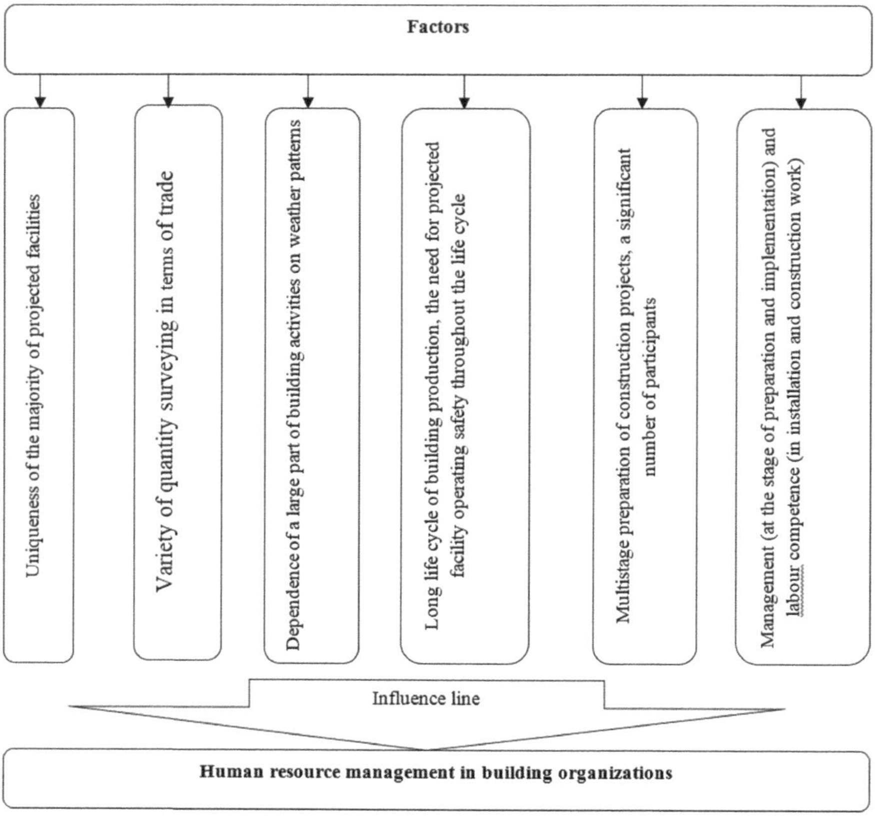

Fig. 2 Specific factors affecting human resource management in building organization

The guidance notes for practical application of theoretical leadership developments in building companies also need improvement. Modern leadership methods do not meet the requirements of practical activity. With this aim, the results of the study can be used in the managing of construction organizations by making motivation more systematic and theoretically based.

3 Conclusion

Management plan is the key point in human resource management, as well as encouragement for achieved planned performance.

Business performance of building organization is expressed in different forms: industrial, social, economic, educational and psychological at the same time. Business performance appropriate for organization in general is advisable to put in one group—first-order business performance. Business performance of organizations, the achievement of which is necessary for employees—second-order business performance.

Since the term "employee" means not an abstraction, but concrete people, whose development is beneficial for organization, the best result in this area is achieved only by high motivation. Almost all negative aspects of organization can potentially affect employees' satisfaction, which, in turn, affects motivation.

In a market economy, building companies must pay particular attention to socioeconomic factors of employee performance, as it affects team performance and competitiveness of a building company in the market. Motivation plays an important role here as a basis of any human behavior and his or her efforts to achieve the goals of the organization.

References

1. Becker, K., Hartmann, B. L., & Miller, J. M. (2014). Fostering successful career paths in construction: motivation, evaluation. *Feedback Practice Periodical on Structural Design and Construction, 19,* 159–167.
2. Peklar, J., & Bostjancic, E. (2012). Motivation and life satisfaction of employees in the public and private sectors. *Uprava.*
3. Armstrong M. (2004). *Practice of human resource management,* 8th ed. St. Petersburg: Publishing House "Peter".
4. Jaiswal, N. K., & Dhar, R. L. (2015). Transformational leadership, innovation climate, creative self-efficacy and employee creativity: A multilevel study. *International Journal of Hospitality Management, 51,* 30–41.
5. Prokhorova, M. V. (2009). *Human resource rendering a profit.* Moscow: Eksmo Publishing House.
6. Wang, X.-H., Kim, T.-Y., Lee, D.-R. (2016). Cognitive diversity and team creativity: Effects of team intrinsic motivation and transformational leadership. *Journal of Business Research. 69,* 3231–3239.
7. Moriano, J. A., Molero, F., Topa, G., & Lévy Mangin, G. P. (2014). The influence of transformational leadership and organizational identification on intrapreneurship. *International Entrepreneurship and Management Journal, 10,* 103–119.
8. Barrick, M. R., Thurgood, G. R., Smith, T. A., & Courtright, S. H. (2015). Collective organizational engagement: Linking motivational antecedents, strategic implementation, and firm performance. *Academy of Management Journal, 58,* 111–135.
9. Mittal, S., & Dhar, R. L. (2015). Transformational leadership and employee creativity. *Management Decision, 53,* 894–910.
10. Yan, L., Chuan-Hoo, T., & Hock-Hai, T. (2012). Leadership characteristics and developers' motivation in open source software development. *Information & Management, 49,* 257–267.

Promising Directions for the Development of BIM Technologies in Ukraine on Its Way to European Integration

A. I. Bielova, N. Y. Zhuravska, and A. Y. Kochedikova

Abstract The current state of the construction market in Ukraine is characterized by a small, but still booming, signaling shifts in economic development, application of new approaches and mechanisms both in the management of construction projects and works. The article considers a fundamentally new approach for Ukraine—the use of BIM technologies—which goes beyond the design and is inseparably used for the production, operation and diagnostics of buildings, and serves as cluster of information in interaction between building systems, models of their degradation in real conditions, ergonomics data, ecology in exploitation and recycling, i.e., the creation of the Digital City.

Keywords BIM technologies · Design · Construction · Construction market · Innovative development · Buildings · Structures

1 Introduction

Ukraine's accession to the European Community requires domestic enterprises and industries to change the deadlocked approaches to production and to a new level of development in accordance with the requirements of the European Community and market requirements.

It is worth noting that the current state of the construction market in Ukraine is characterized by a small, still noticeable booming, compared to previous periods, and Europe is one of the most solvent innovative markets in the world (Table 1), which successful countries have tried to master. This is due to the fact that innovation of

A. I. Bielova
"Institute of Innovation Education" of Kyiv, National University of Construction and Architecture, Kiev, Ukraine

N. Y. Zhuravska (✉)
Kyiv National University of Construction and Architecture, Kiev, Ukraine
e-mail: nzhur@ua.fm

A. Y. Kochedikova
Boryspil Institute of Municipal Management at the Interregional Academy of Personnel Management, Kyiv Region, Ukraine

V. Onyshchenko et al. (eds.), *Proceedings of the 2nd International Conference on Building Innovations*, Lecture Notes in Civil Engineering 73,
https://doi.org/10.1007/978-3-030-42939-3_53

Table 1 Countries with the highest Global Innovation Index 2018 rankings by country group income [22]

High-income countries		Countries with income higher than average		Countries with income below average		Low-income countries	
Rating	Country	Rating	Country	Rating	Country	Rating	Country
1	Switzerland	17	China	43	Ukraine	117	Bolivia
2	Holland	34	Latvia	45	Vietnam	118	Nigeria
3	Sweden	35	Malaysia	48	Moldova	119	Guinea
4	United Kingdom	36	Slovakia	53	Mongolia	120	Zambia
5	Singapore	37	Bulgaria	57	India	121	Benin
6	USA	40	Lithuania	59	Georgia	122	Niger
7	Finland	47	Chile	68	Armenia	123	Ivory Coast
8	Denmark	49	Romania	71	Peru	124	Burkina Faso
9	Germany	52	Montenegro	85	Indonesia	125	The Republic of Togo
10	Ireland	54	Costa Rica	102	Guatemala	126	Yemen

production in the EU exceeds 75%. The EU is a territory of a highly efficient innovation economy [1], where the average GDP per capita is almost eight times higher than in Ukraine (2008 level [2, p. 19]) and much higher than the level of economic development. On this background, it can be predicted that the future path of innovative development for both sectors and national economies will be closely linked to the model of innovative development of the European Union, which involves not only close cooperation in the field of R&D but also integration of the European countries' economies [3, p. 170]; these will inevitably lead to harmonization of national legislation in the innovation field with international agreements and conventions on the contractual legal field of international scientific and technological cooperation with strategically important partners.

At present, the European Union's innovative development is being pursued within the framework of the Europe 2020 Development Strategy, agreed in March 2010 and formally endorsed by the Brussels European Council, where the priorities of this strategy are: intellectual growth—the development of a knowledge-based economy; sustainable growth—moving forward toward a more resource-efficient, environmentally friendly and competitive economy; socially integrated growth—promoting a high-employment economy that provides social and territorial cohesion [4, p. 118].

2 Strength Problem Solving

To solve these problems in the framework of the "Europe 2020" strategy, seven major initiatives are launched. The initiative, which is responsible for the scientific and technical sphere of the European Union up to 2020 are the following: innovation Union, aimed at reorienting policy research and innovation the main challenges (climate change, energy and resource efficiency, health and demographic change); strengthening all the links in the innovation chain; improving framework conditions and access to Finance for R&D with the aim of turning innovative ideas into products and services that provide growth and employment [4, p. 119].

In the information overview of the "Horizon 2020" new framework program of the European Union on scientific-technological and innovation development [5, Sects. 63–65], it was noted that priority will be given to high-performance technologies: eco-, nano-, bio- and info-technologies that focus on solving social and global problems. It is also expected to overcome the obstacles of cooperation between countries by creating multinational consortia involving researchers from all over the world; different types of organizations, namely universities, research centers, commercial and private enterprises, different in size and fields of activity, the national funding that will contribute to the further development of information, knowledge and technology.

As of 2016, the share of expenditure on research and development of new ideas and technologies exceeded the limit of 3% of GDP only in two countries of Europe–Sweden (3.25%) and Austria (3.09%); in three countries, it is approaching this level (Germany—2.94%, Denmark—2.87% and Finland—2.75%), and ten countries (Bulgaria, Croatia, Cyprus, Latvia, Lithuania, Malta, Poland, Romania, Slovakia and Serbia) are critically low—less than 1% of the GDP. The average for European Union countries' share of expenditure on research and development is approximately 2.08% of GDP [6]. As we have shown so far, Ukraine not being included into this rating is due the fact that introduction of new technologies is not a priority as separate sectors and government policy in our country.

It should also be noted that choosing the way of integration with the European Union for the Ukrainian society should be relevant to study not only the strategic initiatives as prospects of the construction market development, as the leading economic sector of Ukraine, but the experience of developed countries. Because each of the members of the European Union has its own history of providing innovative development of the construction industry, that is, different initial conditions, programs and tools to stimulate and, as a consequence, different results.

Analysis of national programs for promoting innovation development in EU countries showed that in most cases, these programs are aimed at promoting the creation of research and innovative projects of industries with the possibility of using state financial support instruments. Therefore, it is necessary to distinguish three types of state programs of innovation development:

- state technological or research programs;
- programs aimed at commercialization of results;

- national cluster programs.

The most common are state technology or research programs. They "take into account life cycle of innovation, allocating time and resources between basic research, scientific training, applied sciences, technological development and stages of commercialization" [7, pp. 25–26]. Some of these programs are aimed at attracting public funds to specialized technological areas, such as nano-technologies, biomaterials, which are determined by the priorities of socioeconomic development of the state. As a rule, these programs of the European Union countries are aimed at supporting innovative enterprises in research institutions, as well as creation of new business lines or other means for the commercial application of research results [7]. Just as with state technology or research programs, the focus of innovation commercialization programs depends on the socioeconomic development priorities of the country, which, in particular, can be attributed to the revival of certain industries, health, safety, climate change, etc.

With regard to national cluster programs, they are widespread in the countries of the European Union, where in most countries one or two programs are state funded [8, pp. 25–26].

Consequently, each country chooses its own way of ensuring innovative development of economic sectors and the country as a whole, based on the needs and priorities of socioeconomic development.

Note, however, that the purpose and objectives stated in the program were effectively achieved, and the government must use special tools to ensure innovative development and adapt them for their own needs. The role of the analyzed tools attention will be focused on information technologies, because they are the most widely used in modern economic–social space of the most developed countries of the world. It is the BIM technology that is rapidly gaining wide popularity in the construction industry and is already implemented by international companies. It should also be noted that Informatization of the construction industry currently occupies one of last places in the world, so building information modeling is only a matter of time.

The limit of twentieth–twenty-first centuries, associated with the rapid development of information technology, saw the emergence of a fundamentally new approach in architectural and structural design, creating a computer model of the new building and carries all the information about the future of the site. It is a natural human reaction to updated information richness of life. Moreover, the flow of this information does not stop even after the building is designed and constructed as a new object enters into the stage of operation, in its interaction with other objects and the environment, that is, in modern parlance, starts the active phase of the life cycle of the building [9].

As a result of the above mentioned, the concept of building information modeling has emerged, which is of far greater, broader and deeper value in design than just a new method. This is a fundamentally different approach to the erection, equipping, maintenance, operation and repairing of the building. Moreover, even decommissioning of the building and/or its dismantling (liquidation) can also be foreseen by

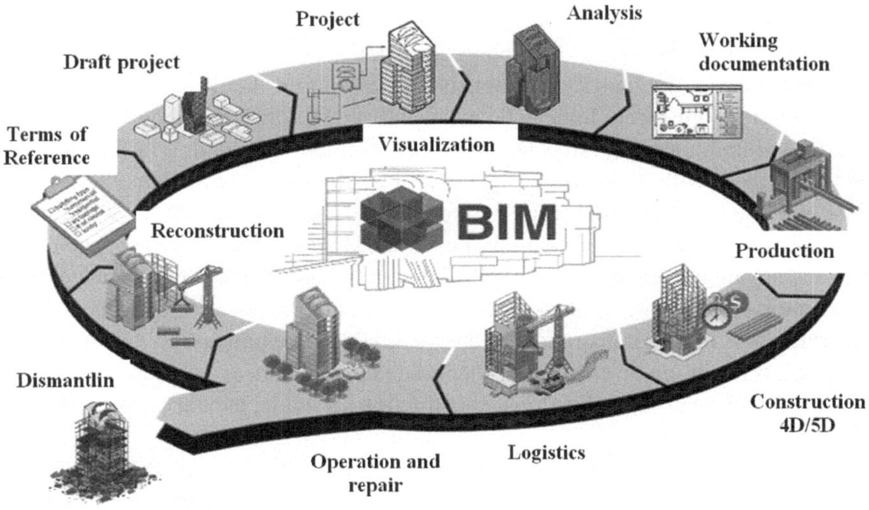

Fig. 1 Life cycle of BIM technologies

BIM technology (Fig. 1). BIM is our new perspective on the world and rethinking ways of human impact on this world [10].

Currently, the Ministry of Regional Development is analyzing the use of BIM technologies in design and construction, carefully examining foreign experience, and since this year, BIM technologies in Ukraine must be implemented at the state level, and it must be regulated by the law.

European Union countries have already officially launched BIM technologies in 2013, and in July 2018, the BIM Working Group has issued a Guide to Technology Implementation for European Public Customers.

Experts estimate that the design, construction and operation of BIM facilities are more efficient. Thus, in the EU countries, the expected annual savings from the use of BIM technologies at the design and construction stage is more than 20%. In the UK, BIM plans to reduce project implementation time by 50% by 2025 [11].

Advantages of using BIM technologies in modern design and construction are as follows:

- reduction of terms of project documentation preparation;
- reduction of the errors in design probability;
- control of key indicators and observance of terms of the works performance;
- prompt submission of information on research and test results, project documentation and reports in electronic form;
- prompt adjustment of construction costs;
- reduction of cash costs;
- reducing the time of the building commissioning.

Features of BIM are as follows: Information modeling of buildings is an integrated approach to the construction, equipping, operation and repair of the building, which involves the collection and comprehensive processing in the design process of all architectural design, technological, financial and other information about the building with all its interactions and dependencies. In information modeling in the building and all that goes with it, this is considered as a single object. Each elementary module, a building is a spatial information model that is associated with the knowledge base, and in which each element can be assigned additional attributes. Such features and advantages of organically stem from global differences of knowledge from information—their compositions, hierarchy, procedurals and descriptiveness [12]. Construction site since then, in fact, is designed as a single unit and any change in its setting leads to automatic change of other related parameters and objects, changes in drawings, renderings, specifications, construction schedule, etc., at all stages of the life cycle.

Numerical information regarding existing or planned facility in BIM can be used for: making specific design decisions; creating high quality project documentation; predicting performance of the facility; development of cost estimates and construction plans; ordering and fabrication of materials, structures and equipment; managing construction of the building and its operation and the means of technical equipment during the whole life cycle; building control as a commercial activity; design and reconstruction or repair of the building, its demolition and recycling, etc. (Fig. 2).

Applying a building information model makes it much easier to handle the object and has a number of advantages over classic design methods. First of all, BIM allows to virtually develop, link and reconcile components, systems and future systems created by different specialists and organizations, to test their viability, functionality and performance in advance. BIM makes it possible to create a model where architects, designers, engineers and other professionals involved in the project can work in parallel. The BIM environment supports collaboration functions throughout the life of a building without the risk of inconsistency or loss of data, and makes it impossible to transmit and transform them. Making informed decisions in the early stages of an

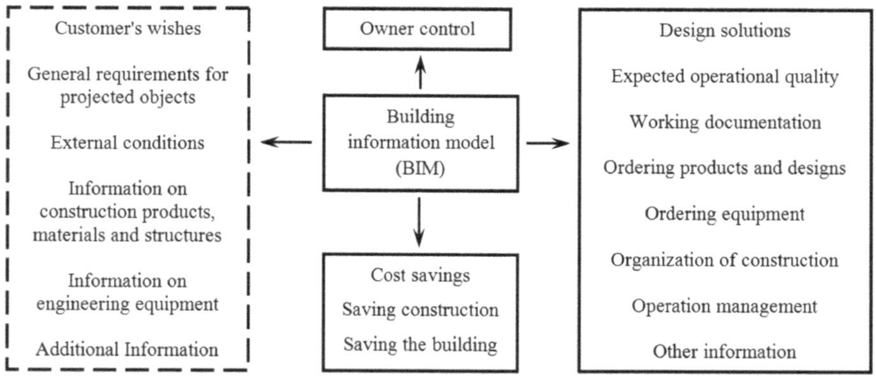

Fig. 2 An integrated BIM information link scheme [21]

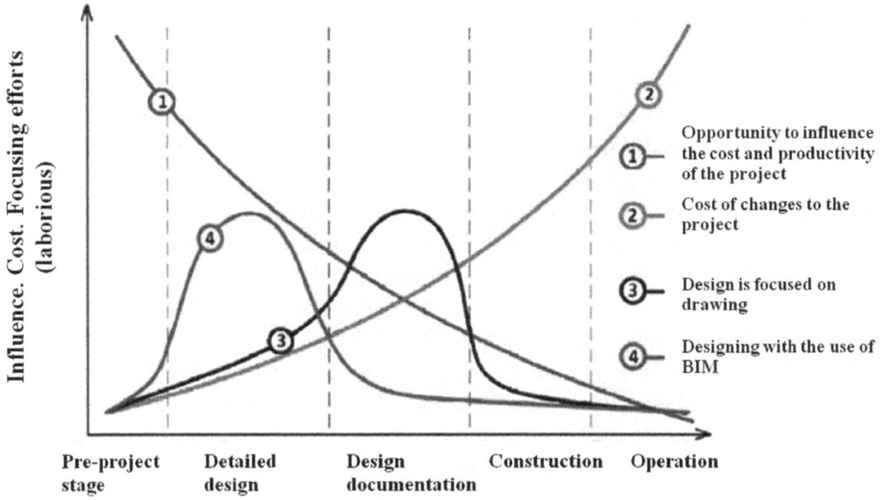

Fig. 3 Changes in the price and the ability to make changes to the project over time from the beginning of the project work in the ordinary design and with the use of BIM

object's existence allows to save money in advance, as it is known that the cost of making changes to a project increases exponentially over time (Fig. 3).

Modern Information Modeling—Building Information Modeling—is inextricably linked with Building Performance Management and Building Life cycle Management. BIM does not only make it easy to manufacture, accelerate the erection of structures, but also to track the efficiency of investments, to accumulate qualitative and quantitative data used in various fields under the scheme Product—Processes—Resources [13].

Modern direction of the construction industry development is moving toward a combined paradigm of architectural and structural form of algorithmic architecture. High-precision BIM models to meet the technological requirements of manufacture allow to obtain a new constructive and architectural forms. Now most of the leading developers of construction CAD—Autodesk, Nemetschek, Graph iSOFT and others—support the BIM technology in their products, where compatibility of the various programs has developed specific data exchange format IFC. Originally introduced for Autodesk Revit and Tekla, IFC has gradually become a clipboard full of data without losing valuable information content.

BIM is the sum technologies, the result of evolution of the simulation systems, the kind of response to the increasing complexity of the functions and subsystems maintenance to the demands of modernity to form structures from both architectural and structural point of view. Modern BIM enlarged subsystems of a building is one super-object that has already been implemented in some complexes [14]. Obviously, the consolidation and mutual settlement BIM cannot remain within the building. Now the BIM system of each construction seamlessly implementing onto the level of integration in the urban environment determines the transition of BIM technologies in 4D and 5D systems. 4D is widely used in the local BIM, allowing simulating the mounting of the frame elements and fencing. At the same time, 5D systems imply the accumulation of qualitative data of BIM and distribution unity with GIS technology. Thus, the present BIM systems are part of information systems (I-Model), which accumulate and carry information regarding the phenomena of nature, with which we interact and the social and economic history of people's lives [13]. As for the global perspective that is already becoming reality, in 2015, the Government of Britain announced a program of total introduction of BIM in the country: Digital Built Britain [15, 16]. This program systemizes the existing development of BIM at four levels (from "0" to "3") and envisages the transition of the UK to the third BIM level (Fig. 4). Level 1 engagement was achieved during the design and construction of the fifth Heathrow airport terminal, but participants' expectations were not met. The increase in productivity compared to the unorganized forms of work was only 10%. An example of a Level 2 environment software implementation is Autodesk Navisworks, Solibri Model Checker, Bentley Navigator. At this level, organized engagement can provide up to 50% reduction in project non-production costs and accessible visual planning and construction management—4D as well as project cost management—5D.

In this form, BIM is a technology that symbolizes the advent of "digital construction." In 2011, the UK government decided that from April 2016, all public procurement in the construction industry would be implemented only for projects

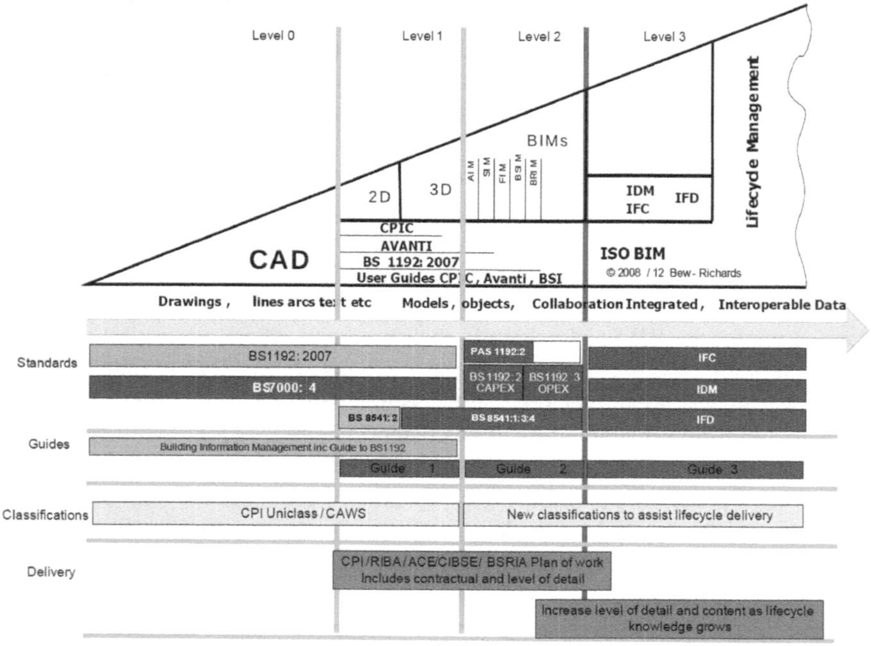

Fig. 4 British systematization (levels) of BIM

implemented in BIM Layer 2 technology. Thus, the industry received strong incentive for progress. A special working group was formed (BIM Task Group), which was tasked with developing the necessary standards and protocols for working with free access (since the lion's share of the market is represented by small and medium-sized companies).

The introduction of BIM technologies in the world is growing at a rapid pace, and often with government support. Ukraine is also experiencing a revival of interest in information modeling of construction systems, but this process is unique to some integrated enterprises or companies with foreign investments. BIM is actively used in the construction industry of Ukraine, where its effectiveness is obvious: the construction of large shopping centers and multifunctional facilities with complex internal infrastructure. At the same time, it is advisable to highlight the main barriers to the introduction of BIM in Ukraine, namely:

- high cost of BIM software complexes compared to the cost of project services;
- profitability only for large, typical or foreign projects;
- non-regulation of the regulatory framework regarding the status of information modeling and its implementation in the construction process at all stages;
- imperfect legislation allowing the production of structures by unqualified participants;
- uncertainty in the distribution of responsibilities and intellectual property rights;
- unwillingness of investors to invest further in information models that can be used not only in the construction, but also in the operation of facilities;
- inertia and tradition of the construction industry, lack of understanding of BIM benefits;
- compatibility between different software products, development of uniform standards for data transmission;
- infertility of the construction industry with regard to the introduction of BIM, unpreparedness of design contractors; asymmetry of risks and rewards in construction; the lack of standardized business and contract models in construction that could be linked through the BIM process [17].

The stimulating factors for the introduction of BIM in Ukraine can be identified as follows:

- orientation of designing to foreign markets for which BIM is natural;
- implementation of European building standards, which are organic for BIM complexes;
- the rising cost of energy, forcing developers and owners to switch to information technology for design, construction and operation with a high level of forecasting and control;
- implementation of energy-saving programs and reforms, which prompts the state to become an effective savings owner;
- expectations of foreign investments and programs and the need for effective control over their implementation.

World experience shows that the transition of companies to BIM requires step-by-step changes, according to the concept (primarily the execution of small, typical objects), in a separate part of the staff (the so-called BIM team), which, if homogeneous and gradual, should lead to increased productivity works [18]. A total transition to BIM in the future is inevitable, but it should be understood that it is possible only if the technology and the design process are changed. For the active use of BIM technologies in Ukraine, first of all, it is necessary to carry out explanatory work and change the approach of customers and designers of construction sites, and the effective customer must be the state.

BIM technology goes beyond design and is inextricably used for the production, operation, diagnostics of buildings, serves as an information cluster of information on the interaction between building systems, models of their degradation in real conditions, data on ergonomics, ecology during operation and utilization—this is how Digital City is formed SMART City. The filling sources are automated monitoring systems with fixed sensors in real time, as well as people who are end-users of the building and have sensors in mobile devices. The imminent integration of BIM with other global information tools, such as social networks, GPS, systems for monitoring data on loads and impacts on the building, its interaction with the environment [19]. In this way, the BIM building with all the subsystems allows managing and adjusting its condition as a whole object, accumulating qualitative and quantitative data that forms the knowledge base for decision making for the following structures.

The detailed information model of the building allows to optimize its parameters, reveals sensitivity to changes of conditions and parameters and reveals all their interdependence with each other. During the construction and operation of the building, information model in real time accumulates the history of states of system elements deviations occurrence, and their eliminations. The use of intelligent work execution tools and integration with augmented reality systems minimizes the difference between virtual and actual models, identifying unplanned situations in a timely manner and offering ways to respond. The invaluable experience can be used to plan maintenance and repair programs, build models of degradation of system elements, both for a particular building and for analogues. BIM allows shaping the economy of sustainable development, recording and creating the history of our civilization [20–24].

3 Conclusion

Currently, BIM technology is a modern approach to design, construction and operation. It can be said that BIM is a complete numerical description and properly organized information about the object, which is used both at the stage of design and construction of the building, during its operation and even demolition.

An important component of this technology is a single information space, a database containing all the information about the technical, legal, property, operational, energy, environmental, commercial and other characteristics of the building.

Due to the very accurate and detailed processing of the model, this technology makes it possible to perform different calculations, analyses and simulations (in classical design for each calculation, you need to do separate additional work). Obviously, using BIM technology, it is much easier to simulate the entire life cycle of a building and use the results to adjust the project, resulting in a better solution.

More and more architects and engineers in the world are taking steps in the direction of BIM technology, and more and more construction organizations are pushing for BIM. This technology saves money at all stages of a building's life cycle, but it delivers the most efficiency when it comes to an integrated approach to the facility, since the more accurate the information model is initially, the more useful it will be then where it shrinks, the number of errors and downtime in the construction process improves the understanding between the customer, the designer and the contractor.

References

1. Sidenko V. R. (2011). An innovative model for EU development—from the Lisbon strategy to Europe 2020. In *Problems of innovation and investment development* (No. 1, pp. 113–126).
2. Commission of Independent Experts and Leading Experts of EU Ministries. (2017). *Final report of an independent European audit of Ukraine's national research and innovation system*. Horizon 2020 Political Support Tool. EU.
3. Eurostat. https://ec.europa.eu/eurostat/data/database.
4. Innovation in Ukraine. (2011). In *European Experience and Recommendations for Ukraine, Vol. 3. Innovation in Ukraine: Proposals for policy measures* (76 p.). The final version (draft from 10/19/2011). Phoenix.
5. Heitz, V. M. (Ed.). (2015). *Innovative Ukraine 2020: National report/for the head* (336 p). NAS of Ukraine (Table and Figure).
6. Benavente, J. M. (2002). *The role of research and innovation in promoting productivity in Chile* (29 p). Department of Economics, University of Chile, WP No. 200.
7. Shlepakov, L. N. (1995). Systems with databases for solving the problems of recognition and classification of information messages. In *Intellectualization of systems. Processing* (pp. 11–38). NASU.
8. Azhar, S., & Justin Brown, Farooqui, R. *BIM-based sustainability analysis: An evaluation of building performance analysis software*. http://ascpro.ascweb.org/chair/paper/CPRT125002009.pdf.
9. Levchenko, O. V. (2013). *BIM in the course "Information Technologies of Modern Architectural Design". Modern problems of architecture and urban planning* (Issue 34). Department of Information Technologies in Architecture, KNUBA, UDC 004, 721.021, 721.024.
10. Khoroshov, A. N. (1999). *Introduction to mechanical systems design: a tutorial* (372 p). Belgorod.
11. Minregin Official Website. (2019) *2019 should be the year of the beginning of the introduction of BIM technologies in Ukraine,—Partskhaladze*. Minregin Official Website. http://www.minregion.gov.ua/press/news/2019-rik-maye-stati-rokom-pochatku-vprovadzhennya-bim-tehnologiy-v-ukrayini-partshaladze/.
12. Bew, M., & Richards, M. BIM Task Group. http://digital-built-britain.com.
13. Onikienko, V. V. (2006). *Innovative policy of EU and CIS countries: Problems and practice of implementation* (No. 3, pp. 170–183). Ukrainian Society.
14. King, M. The British told the world what a Level 3 BIM is: This—Digital Built Britain.

15. Ashcraft, H., & Shelden, D. R. (2007). *BIM implementation strategies*. Gehry Technologies.
16. Friedrich, M. (2011). Using a four-dimensional building information model to identify temporal spatial threats to construction safety. In *Geodesy, architecture & construction 2011 (GAC-2011)*, 24–26 November 2011, Lviv, Ukraine (pp. 78–79).
17. Varnaliy, Z., Onishchenko, S., & Masliy, A. (2016). Threat prevention mechanisms of Ukraine's economic security. *Economic Annals-XXI, 159*(5–6), 20–24. https://doi.org/10.21003/ea.V159-04.
18. Kozlov, I. M. (2010). Evaluation of the economic efficiency of the implementation of building information modeling. *Architecture and Modern Information Technologies (AMIT): Electronic Journal*.
19. Cherniha, R., & Pliukhin, O. (2013). New conditional symmetries and exact solutions of reaction-diffusion-convection equations with exponential nonlinearities. *Journal of Mathematical Analysis and Applications, 403*(1), 23–37. https://doi.org/10.1016/j.jmaa.2013.02.010.
20. Bilyk, A. S., & Belyaev, M. A. (2015). HIV modeling. Overview of opportunities and prospects in Ukraine. *Journal of Industrial Construction and Engineering, 2*.
21. Talapov, V. V. (2011). *BIM basics: An introduction to building information modeling* (392 p). DMK Press.
22. Gaidutsky, P. Problems of integration. Ukraine: EU. http://gazeta.zn.ua/international/ukraina-es-problemy-integracii-_.html.
23. Cherniha, R., & Serov, M. (2006). Symmetries, ansätze and exact solutions of nonlinear second-order evolution equations with convection terms, II. *European Journal of Applied Mathematics, 17*(5), 597–605. https://doi.org/10.1017/S0956792506006681.
24. Yakovlev, S. V., & Valuiskaya, O. A. (2001). Optimization of linear functions at the vertices of a permutation polyhedron with additional linear constraints. *Ukrainian Mathematical Journal, 53*(9), 1535–1545. https://doi.org/10.1023/A:1014374926840.

Management of Production Processes in the Construction of Logistics Complexes

L. M. Boldyrieva, K. M. Kraus, and O. V. Stanislavyk

Abstract It is proved that a mandatory infrastructure element for international transport corridors, as well as for any large enterprise, is the availability of logistics complexes (warehouse). The expediency construction of logistics complexes with the help of modern metal structures and sandwich panels has been proved. The concepts of "logistics complex," "distribution center" and "logistics center" are analyzed. Classification of logistics centers by functional purpose and types of the most popular fast-growing logistics complexes by classes are given. Types of production processes in construction (design with the latest standards; professional approach combined with years of experience and availability of advanced construction technologies, modern equipment and advanced equipment; warranty obligations; free consultations) with management functions are characterized. Project management system for the construction of logistics complexes is substantiated, which takes into account the management functions (planning, organization, motivation, control) and includes such estimation indicators as construction costs and optimal location of the complex (by the criterion of minimum logistics costs). The logistics complex is characterized by the following criteria: main purpose; localization factor; warehouse processing conditions of inventories; stored goods; degree of ownership. It is suggested to calculate the optimal localization of logistics complex using the coordinate method.

Keywords Construction · Logistics · Logistic systems · Design · Processes · Technologies · Management · Functions

L. M. Boldyrieva (✉)
Poltava National Technical Yuri Kondratyuk University, Poltava, Ukraine
e-mail: Boldyrewaljuda@ukr.net

K. M. Kraus
Borys Grinchenko Kyiv University, Kyiv, Ukraine

O. V. Stanislavyk
Odessa National Polytechnic University, Odessa, Ukraine

V. Onyshchenko et al. (eds.), *Proceedings of the 2nd International Conference on Building Innovations*, Lecture Notes in Civil Engineering 73,
https://doi.org/10.1007/978-3-030-42939-3_54

1 Introduction

A mandatory infrastructure element for international transport corridors, as well as for any large enterprise, is the availability of logistics complexes (warehouse). Therefore, at the stage of production optimization and commercial activity, it is necessary to effectively resolve the issue of design and construction of modern logistics complexes. Due to the systematic approach to the construction of logistics complex, while observing all the necessary design standards, it is possible to minimize the costs of logistics.

The realities of time require from construction contractors to build high-speed industrial, material and technical facilities in the shortest possible time with metal structures and sandwich panels that have high quality and low metal consumption. The construction of logistic complexes is to be performed with the help of modern metal structures and sandwich panels. The construction of logistic complexes with the help of modern metal structures and sandwich panels is an effective solution both technically and economically. Such modern technologies satisfy needs in time and space, in particular of following criteria: speed, quality, reliability, availability and low cost (cost depends on the parameters: type and dimensions of logistics complex, design, volume of additional works). Many domestic companies attract additional external sources of financing for the construction of their own logistics complexes, usually foreign capital or its share.

The analysis displays that logistics complexes for different industries in Ukraine have not yet been formed. Active construction of logistics centers will contribute to improvement of national, regional and transnational transport and logistics systems, as well as to the development of the economy of both individual regions and the country as a whole.

Transport and logistics systems, in turn, will provide effective research, production, supply and transportation communications between economic agents. However, such infrastructure projects as construction of logistics centers are accompanied by a large amount of investment. For example, in developed countries, such investments account for 3–3.5% of annual GDP growth. For Ukraine, it is necessary to implement projects on the organization of logistics complexes at the intersection of transport routes of ethnic importance, connecting the north to the south and the west to the east.

2 An Overview of the Latest Sources of Research and Publications

The following foreign scholars have considered the issues of construction logistics:

- Susanna Hedborg Bengtsson, who studied coordinated construction logistics. This topic is extremely popular and is widely discussed in the Swedish construction industry. To improve transport efficiency, Swiss are proposing to reduce the amount of resources used and to ensure workplace health and safety coordination [1].

- C. Leifgen and S. Kujajewskias are convinced that the construction logistics system should be focused on the planning, control and monitoring functions, as this will ensure a high level of flow of resources and people at the construction site.

Logistics system in construction should also cover repository management and facility planning, enabling it to be more systematic and complete. The stages of a construction project, where its planning and implementation takes place, differ significantly in specific conditions: depending on the complexity of the project, stakeholder involvement or the volume of the construction project [2].

American Society of Civil Engineers Stakeholders (2014) can positively assist or hinder attempts at innovation. Much depends on the nature of their engagement. The stakeholder engagement process can be complex and unpredictable, more so, if no strategic plan is put in place or if no systematic thinking is invested in the innovation. Stakeholder engagement in innovation process in general and innovation diffusion in particular is examined in the context of construction.

From a theoretical perspective, analysis of the effect of stakeholders could be expected to help in refining innovation process so that it produces decisions and outcomes more likely to lead to successful innovation and diffusion. To cover diverse applications from product development to the drafting and implementation of national standards, 19 innovation projects were studied. Accordingly, an explicit plan for communication and engagement with identified key stakeholders is necessary as a condition for successful innovation and diffusion [3].

The purpose of the study should be to identify and classify supply chain management (SCM) problems, and to clarify their causes during the construction, supply chain or at the intersection between these processes. This allows for identification of how on-site problems affect SCM in construction projects and how they can be mitigated [4].

The result of the research should be, first, the accumulation of theoretical knowledge (based on the theory of industrial networks), and secondly, the improvement of logistics in practice directly by a specialist [5]. However, this requires adjusting logistics off-site.

Customer satisfaction and continuous improvement are the main goals of construction logistics. Many researchers focus on relationships between contractors and ultimate customers in order to improve the understanding of the importance of customer satisfaction. However, the purpose of the study is to extend the logistics framework for building materials to satisfy the customer from owner to project manager level [6].

Problems of logistic complexes construction in the territory of Ukraine are hardly considered in literary sources. They are mostly considered in scientific and literary sources: first, problems faced by logistics at regional, national and international levels are highlighted; second, the issues of warehousing logistics process are analyzed; third, the classification of logistics centers by functional purpose, warehouses and warehouse networks, the typology of warehouse buildings is carried out. In particular, the “Program for the Development of National Network of International Transport

Corridors in Ukraine" is listed as one of the ways of solving problems of creation of transport and warehouse logistics centers [7].

Some literature references point to the existence of logistics complexes in Ukraine that fulfill certain logistic functions. Most of them are concentrated in Kyiv and Odessa regions, in the East of the country. The purpose of research is to study the management of production processes in the construction of logistics complexes and to develop practical recommendations for its improvement.

The theoretical and methodological basis of scientific research is the dialectical method of scientific cognition, systematic approach to the study of economic phenomena, the provisions of modern economic theory, scientific works of scientists and economists who research of problematic issues of design activities in construction and logistics, recommendations of research institutions problems.

Scientific research is based on the use of general scientific methods of research: theoretical generalization, system analysis and synthesis; terminological analysis, abstraction and formalization.

3 Main Body

Historically, Ukraine has geographical borders with seven countries (Belarus, Moldova, Poland, Russia, Romania, Slovakia and Hungary). Therefore, the expediency of forming international logistics complexes is a kind of platform in the trade between the CIS countries, Europe, Asia and the Caucasus. Thus, there are three pan-European transport corridors (ITC No. 3, 5, 9) and five OGS corridors (No. 3, 5, 7, 8, 9) operating in Ukraine. Transportation on the MTK TRACECA (Europe—Caucasus—Asia) is developing [8]. According to research by the English Rendall Institute, Ukraine's transit potential is the highest in Europe and is 3.11 points [9]. In economically developed countries, companies form their own logistics facilities or use the services they provide.

It is advisable to locate logistics complexes in such cities of Ukraine as: Izmail, Kyiv, Lviv, Odessa, Uzhgorod/Chop, Kharkiv, Chernivtsi, Sarny (Rivne region), which are located near customs posts, railways and highways of regional and international importance, which are part of the MTC system, in particular, in places with the highest density of turnover or transit.

Despite the large amount of scientific research in the field of logistics, logistics complexes are not separated into a separate category. The problem with identifying the logistics complex among other logistics objects arises from the interpretation of the concept of "logistics complex." In the scientific literature, there are three concepts: "logistics complex," "distribution center" and "logistics center."

According to scientists, logistics centers are multimodal hubs aimed at improving efficient transportation, provide local concentration and implementation of various logistics tasks and functions (distribution, long-term storage and sale of goods and services) and integration of logistics systems, view macros infrastructure unit of region's economy logistics. For example, Polish scientists interpret the concept of "logistics center" as a spatial object of defined functioning with a defined infrastructure and organization of activities, through which logistics services related to the reception, warehousing, distribution and dispatch of goods, as well as support services are implemented which may be provided independently by business entity to the sender or recipient" [10].

Domestic scientists in their works emphasize that logistics center is a hub object of logistics networks, in which processes of distribution of cargo flows, change of vehicles, storage and inventory management. Also, scientists provide the classification of logistics centers by functional purpose: international logistic distribution centers (economic area: 100–150 ha; radius of action: 500–800 km); regional logistics distribution centers (economic area: 20–50 ha; radius: 50–80 km); local logistic distribution centers (main task: completion of the formation of modern distribution network system); branch logistics distribution centers (intended to serve a particular industry or entrepreneurs); logistics service centers [10].

Some scientists identify logistics centers with distribution centers and warehouses. In turn, the term "distribution center" is a warehouse complex that receives goods from manufacturing or wholesale companies located in other regions of the country or abroad and distributes them in smaller batches to customers (small and wholesale enterprises and retail) through them or their distribution network.

Logistics complexes are large-scale professional warehouses with auxiliary facilities (office buildings, service premises, security point, etc.). Unlike conventional warehouses, logistics complexes include: super-easy storage of goods, simplified and as easy as possible moving them, reducing the time of their search, loading, sorting and shipment of goods, availability of modern hoisting and transport equipment, logistic service.

We believe that it is wrong to identify a logistics center and distribution or distribution centers. Characteristics of the logistics complex are given in Table 1.

The most popular high-speed logistics complexes are distinguished by following classes:

Table 1 Logistics complex characteristics

Characteristic	Logistics complex	
	Logistics center	Distribution center
Main purpose	Consolidation/deconsolidation	Stockpiling
Localization factor	Intersection of highways, integration of different modes of transport	Sales market
Conditions of warehouse processing of inventories	Transit transfer type	Distribution type (distribution centers)
Stored goods	Universal	Special
Degree of ownership	General purpose	Individual purpose (corporate, contractual)

1. Warehouses for responsible storage of any type of products by modern technologies (maximum permissible ceiling height is 10 m, operation is calculated for a rather long period of time). For example, Roshen Corporation Logistics Center in Yagotyn [11] meets all the requirements for a Class A logistics facility;
2. An analogue of previous variant, but in a reduced proportion (most often it is a reconstruction or simplified model with a ceiling height of 6–8 m);
3. The building was reconstructed (a warehouse with limited premises);
4. Underground and other unsuitable premises (for short-term storage of goods and archive).

Management of production processes in the construction of logistics complexes, first of all, provides for designing according to the most up-to-date standards, i.e., at all stages of designing, the development of complex projects (design, working design, working documentation); professional approach combined with years of experience and availability of advanced construction technologies, modern machinery and advanced equipment; guarantee obligations (guarantee of high quality of completed construction works, complete safety of structures); free consultations.

Managerial activity is important and involves combining different functions or activities. Management functions in construction can be called with all the specialized types of various construction works and services. Term "function" is translated from Latin language and means "exercise, fulfillment." Main management functions include: planning and forecasting, organization, motivation and control.

Thus, each management function in construction is aimed at solving specific, diverse and complex problems of interaction between individual units, which require the implementation of large set of specific measures, in particular, in the construction of logistics complexes.

When designing engineering networks (power and water supply systems, heating, ventilation and sewerage, special systems of video surveillance, access control, security and fire alarms, fire extinguishing, etc.), it is necessary to take into account the specificity and type of logistics complex. For example, designing:

- Refrigeration warehouses to ensure uninterrupted power supply in the system of warehouse equipment to ensure uninterrupted operation of refrigeration units;
- The composition of fuel and lubricants goods of chemical industry must take into account the safety requirements for storage of the relevant goods and materials.
- The stage of preparation of a set of documents includes:
- Design of premises for storage and distribution of products for retail chains, shops and supermarkets;
- Design of warehouse complexes for storage of goods or raw materials of industrial, economic and industrial enterprises;
- Design of wholesale food warehouses, commodity bases, vegetables and granaries;
- Design of the rolling stock as part of large logistics complex, as well as a separate solution;

- Design of warehouses for goods with special requirements for storage conditions (pharmaceutical warehouses, warehouses of fuel and lubricants, warehouses of goods of the chemical industry, etc.).
- Structurally and technologically designed logistics complex should include:
- Effectively located access areas and parking (taking into account the specific nature of the existing access roads by factors: the intensity of traffic, depending on the time of day and seasonality, etc.);
- Modern systems of automation of the accounting of goods, engineering systems and warehouse equipment for uninterrupted operation of logistic (warehouse) premises in the conditions of peak load;
- Rationally designed and placed areas of reception and unloading of warehouse products;
- Clear and clear scheme of placement of goods on the shelves inside the logistic (warehouse) premises;
- Reasonably planned and designed schemes for moving conveyors and loaders inside logistics (warehouse) premises.

The value of production cost of construction of distribution complex should be determined by formula (1):

$$\mathrm{PC} = \sum \frac{(\mathrm{DMC} + \mathrm{DLC} + \mathrm{TC} + \mathrm{VMC} + \mathrm{FMC} + \mathrm{TCI})}{\mathrm{PP}}, \tag{1}$$

where PC—production cost of construction, thousand UAH;
DMC—direct material costs, thousand UAH;
DLC—direct labor costs, thousand UAH;
TC—transport costs, thousand UAH;
VMC—variable material costs, thousand UAH;
FMC—fixed material costs, thousand UAH;
TCI—total capital investment in the construction of distribution complex, given by the factor of time, thousand UAH;
PP—payback period of construction (number of years).

It is possible to construct a graphical interpretation of the optimal localization of logistics complex using the coordinate method. To do this, we need to find the coordinates of points of X_{complex}, Y_{complex}. Thus, the coordinates of the center of gravity of cargo flows ($\mathrm{X_{complex}}$, $\mathrm{Y_{complex}}$), i.e., the point at which logistics complex can be located are given by formulas 2–3:

$$X_{\mathrm{complex}} = \frac{M_1 \times X_1 + M_2 \times X_2 + \cdots + M_n \times X_n}{M_1 + M_2 + \cdots + M_n}, \tag{2}$$

$$Y_{\mathrm{complex}} = \frac{M_1 \times Y_1 + M_2 \times Y_2 + \cdots + M_n \times Y_n}{M_1 + M_2 + \cdots + M_n}, \tag{3}$$

where M_i—material flow distributed to the i-consumer ($i = 1, 2, \ldots n$);
X_i, Y_i—coordinates of i-consumer ($i = 1, 2, \ldots n$).

Therefore, in time, international corporations are in great demand for the individual design of logistics complexes of any size and complexity. Domestic builders, taking into account the wishes of customers, make architectural and construction sections of the project documentation strictly adhering to the requirements of current rules of design of logistics centers.

Types of production processes in construction, taking into account management functions are described in Table 2.

Scientific novelty is to justify project management system for the construction of logistics complexes, which takes into account management functions (planning, organization, motivation, control) and includes such estimated indicators as construction costs and the optimal location of complex (by the criterion of minimum logistics costs). Practical significance of scientific research results is to determine the relationship between the elements of production process management system in the logistics complexes construction; methodological approaches analysis and development an improved system of production processes management in the logistics complexes construction.

Table 2 Types of production processes in construction characteristics, taking into account management functions

Management functions	Types of production processes in construction
Planning, forecasting	Designing (development of complex projects: project, working project, working documentation)
Organization of construction	• Preparation of land for construction • Erection of foundation • Erection of the frame of building • Wall construction and erection • Installation of the facade of building • Roofing works • Connection of communications and engineering systems • Internal facing works • Installation of equipment and production systems • Arrangement of the adjacent territory (if necessary)
Customer motivation	• Guarantee of high quality of completed construction works • Full responsibility for the quality of work performed • Safety of structures • Guarantee of timely implementation and commissioning • Use of certified materials for construction • Free consultations • Implementation of projects in the shortest possible time
Control	Quality control at every stage of the project [12]

4 Conclusion

Thus, in Ukraine, the main reasons hampering the development of logistics complexes construction are the following: insufficient demand for these objects, centralization of commodity, financial, management resources, lack of large companies in the regions, existence of outdated goods transportation scheme, significant financial costs, lack of investors, high capital intensity and high risks for long-term projects, lack of government policy and support in this area and lack of qualified specialists as problems with the purchase (lease) of land.

References

1. Hedborg Bengtsson, S. (2019). Coordinated construction logistics: an innovation perspective. *Construction Management and Economics, 37*(5). https://doi.org/10.1080/01446193.2018.1528372.
2. Leifgen, Ch., & Kujajewski, S. (2018). Integrated digital and model-based construction logistics management based on lean thinking approaches (pp. 428–435). The International Association for Automation and Robotics in Construction. https://doi.org/10.22260/ISARC2018/0060.
3. Widén, K., Olander, S., & Atkin, B. (2014). Links between successful innovation diffusion and stakeholder engagement. *Journal of Management in Engineering, 30*(5).
4. Thunberg, M., Rudberg, M., & Karrbom Gustavsson, T. (2017). Categorising on-site problems: A supply chain management perspective on construction projects. *Construction Innovation, 17*(1), 90–111. https://doi.org/10.1108/CI-10-2015-0059.
5. Sundquist, V., Gadde, L.-E., & Hulthénk, K. (2018). Reorganizing construction logistics for improved performance. *Journal of Management in Engineering, 36*(1), 49–65. https://doi.org/10.1080/01446193.2017.1356931.
6. Jang, H., Russell, J. S., & Yi, J. S. (2003). A project manager's level of satisfaction in construction logistics. *Canadian Journal of Civil Engineering, 30,* 1133–1142. https://doi.org/10.1139/L03-068.
7. Programa rozvitkunats Ionalno Yimerezh Im Izhnarodnih transport nihkoridor Iv v UkraYinI. (2006). Postanova KabInetu Min IstrIv Ukra Yiniv Id 12 kvItnya 2006 r. № 496. https://zakon.rada.gov.ua/laws/show/496-2006-%D0%BF.
8. OfIts Iyniyveb-sayt Ukrzal IznitsI. http://www.uz.gov.ua.
9. Sidorchuk, O. G. (2019). Strateg IyasotsIalnoYibezpeki Ukra Yini: poslIdovn Istformuvannya ta napryamireallzatsIYi. *ProblemiekonomIki, 1,* 176–183. https://doi.org/10.32983/2222-0712-2019-1-176-183.
10. Krikavskiy, E. V. (2008). Logisticheskiytsentr – etouzlovoyobektlogisticheskiysetey. *Logistika: problemyiiresheniya, 5*(18), 38–39.
11. OfIts Iyniyveb-saytkorporatsIyi «ROSHEN». https://roshen.com/ua/uk/pro-roshen/logistychnyy-tsentr.
12. Varnaliy, Z., Onishchenko, S., & Masliy, A. (2016). Threat prevention mechanisms of Ukraine's economic security. *Economic Annals-XXI, 159*(5–6), 20–24. https://doi.org/10.21003/ea.V159-04.

Infographic Modeling of Heat Exchange of Energy-Efficient Building

Natalia Bolharova, Mykola Ruchynskyi, Volodymyr Skochko, and Vitalii Lesko

Abstract A method of visualizing the relationship between the architectural and design parameters of the projected object and the volumes of energy consumption and energy consumption was developed and implemented based on the construction of special diagrams of heat receipts and heat losses. It is proposed to use a new type of diagrams, which is a symbiosis of parallel sets and Sankey diagrams, to visualize the energy flows within the projected object. The diagram gives the designer an opportunity to visually analyze heat flows and heat losses in separate rooms, individual elements of enclosure structures, and ventilation systems. This type of diagram shows the visually relationship between the model variation parameters and the resulting energy and energy costs of the individual building components; i.e., the model is dynamic. Informative and convenient for the designer from a technical point of view, graphical presentation of information about heat consumption, and heat losses in the building will allow him to quickly and accurately make an energy assessment of the model. With infographics, a large amount of data can be presented in an accessible and compact form. This will allow you to comprehensively evaluate your decisions and work with a large amount of data at the same time. In addition, this type of diagram will allow the designer to solve the inverse problem—when it is possible to set limits on the amount of heat losses due to certain structures in the diagram itself, moving the corresponding points on it. Taking into account the required amount of insulation of interior spaces and other constraints, the system would propose design options, in particular the dimensions, position, and required resistance to heat transfer of structures.

Keywords Energy efficiency · Heat balance · Infographics · Sankey diagrams · Parallel sets' diagrams · Energy flows

N. Bolharova (✉) · M. Ruchynskyi · V. Skochko · V. Lesko
Kyiv National University of Construction and Architecture, Povitroflotsky ave., 31, Kiev 03037, Ukraine
e-mail: ruchinsky@ukr.net

V. Onyshchenko et al. (eds.), *Proceedings of the 2nd International Conference on Building Innovations*, Lecture Notes in Civil Engineering 73,
https://doi.org/10.1007/978-3-030-42939-3_55

1 Introduction

1.1 Problem Statement

Issues of energy and resource conservation are relevant today, so finding the best ways to systematize and streamline the energy resources used is also an important task. When performing technical and economic justifications, designers generally compare possible technical and economic decisions that can be made to further develop the final project documentation. Many factors are analyzed, and the types of energy consumed are considered. For maximum clarity, the data being compared is presented in the form of certain graphs, charts, or other means of visual display of information. These graphical tools help accelerate decision making, so informativeness and simplicity are the keys.

However, analyzing energy flows in a projected object using tables is not always convenient and informative. Therefore, finding an algorithm for visually displaying the relationships between the model variation parameters and the resulting energy and energy losses of individual building components is a pressing task.

For such important parameters as heat consumption or heat loss, visualization tools should be not only attractive and visual, but also technically informative. Infographics should reflect not only the level of the studied indicators, but also their hierarchy. The designer must comply with the conditions of the balance of consumed energy in the building. Therefore, the task of visualizing the distribution of this energy or its routing becomes a search for template solutions and systematization of all categories, by which each corresponding distribution will be analyzed.

An approach has been developed that allows informative visualizing the thermal balance of energy-efficient buildings and structures based on special functional diagrams. It is proposed to make diagrams of heat loss and heat gain, based on their percentage distribution by separate zones or premises, elements of enclosure structures, and ventilation systems. Such diagrams will give information about the distribution of energy flows in the building in the form of heat balance most accurately and taking into account the architectural features of the object.

1.2 Analysis of Recent Research

Methodological basis of graphical data visualization tools based on the idea of graphics functions was created in [1]. Three hierarchical groups of graphics functions were distinguished: image functions, active functions, and creative functions.

Bolharova [2] analyzed infographic tools that may be suitable for describing heat exchange processes and have different informative, pictorial, or modeling capabilities, and, in general, different operating capacity indicators. In the work, a qualitative analysis of all possible infographic means from the point of view of functionality is carried out.

One of the most flexible and appropriate tools for solving this problem is the synthesis of a model of parallel sets with the Sankey diagrams. The use of Sankey diagrams to analyze the distribution, transformation, and energy losses that will be used in the work for infographic modeling of heat transfer in a building was considered in [3]. Also, the possibility of using Sankey diagrams to describe heat fluxes is presented in [4]. The ability to use different visualization tools [5] presents a system that allows users to interactively explore complex flow scenarios using Sankey diagrams, allows users to zoom in and learn details on demand. The work demonstrates the ability to support quantitative flow tracking, as well as the ability to present data at different levels of detail, making it easier to understand complex flow scenarios. Visualization of heat inflows and heat losses in a building using Sankey diagrams is discussed in detail in [6]. The work [7] presents the calculations of all the energy consumption of a small factory, the results of which are shown in the Sankey diagram. This work [8] deals with network visualizations, including Sankey diagrams, which help to describe the structural changes in world car sales at the system level. This paper [9] focuses on the search visualization of electronic networks. The proposed visualizations, which are similar to the Sankey diagrams, use the force or weight between the links, as well as a hierarchy of nodes to help investigate large network nets. The paper discusses how principles of interaction between nodes help the user in research.

However, the Sankey charts do not have the integrity of the graphs of parallel sets, which complicates the analysis and comparison of volumes of heat revenues and heat losses. Therefore, using only the Sankey diagram in its pure form, as proposed in these papers, will not solve the problem of analyzing the model variation parameters and the magnitudes of energy and energy losses. Lupton and Allwood [10] presented a systematic method for generating various hybrid Sankey diagrams that approximate the proposed infographic.

For maximum approximation to practical application, the paper also used the method of complex calculation of temperature indicators, which was demonstrated in [11–15]. With this technique, it is possible to calculate the temperature indices in the typical places of the house, as well as to determine the impact of changes in energy consumption on these indices, as well as this allows to solve the inverse design problems. This allows us to investigate the effect of architectural, structural, and engineering parameters of variation on energy consumption.

2 The Purpose of the Research

The purpose of this work is a graphical representation of the relationship between the thermal balance characteristics of the building and its geometric parameters, with the possibility of further computer implementation of this technique.

To achieve this goal, the following tasks were set:

1. To analyze infographic modeling tools, which are suitable for visual analysis of heat exchange processes in a building.
2. Calculate heat losses through individual types of enclosure structures.
3. Develop a rational graphical model of presentation of the results of calculations on the basis of graphical–technological approach.

3 Research Methods

In conducting this research, the basic principles of construction physics and the laws of heat transfer were used. Graphic methods of visual representation of discrete data were applied in the construction of geometric models and their interpretations.

4 Infographic Representation of Heat Exchange Processes of Energy-Efficient Building

4.1 Basic Principles for the Construction of Geometrically Interpreted Physico-Mathematical Models of Heat Exchange Processes for Energy-Efficient Buildings

The paper proposes to construct a heat balance diagram for a specific projected building. As an example, we used the results of calculations of the problem, demonstrated in paper [12]. In paper [11], the basic principles of construction of geometrically interpreted physico-mathematical models of heat exchange processes of energy-efficient buildings were revealed.

A general approach and mathematical tools were demonstrated to reflect the temperature in the air and on the walls of building walls. The approach is based on the method of thermoelectric analogy, which implies the image of a discrete calculated building understudy in the form of a non-planar graph, the nodes (vertices) of which represent points in the air and on the surfaces of the outer and inner walls, and the links (edges) are the supports of the heat transfer between these points. Modeling using such a mathematical tool will be represented by the summation of the heat balance equations for each of these nodes, with subsequent resolution of the obtained system with respect to their temperatures. The equation of heat balance for each point can be written in the following form [Eq. (1)]:

$$\sum_{j=1}^{n} K_{i,j} \cdot (t_j - t_i) \pm Q_i = 0 \quad (1)$$

t_i and t_j these are the temperatures at the *i*th and *j*th points;
$K_{i,j}$ is the heat transfer coefficient between the *i* and *j* points of the discrete calculation model;
Q_i is the sum of all heat and heat losses in a room or on a wall surface, including energy of sources or leaks.

For different parts of the building environment, the heat transfer coefficients are defined differently. It depends on how the energy is transmitted between the nodes: transmission (conductive), convective, beam or mixed. These coefficients are calculated by Eq. (2).

$$K_{i,j} = \left(1 \bigg/ \sum_{p=1}^{m} R_p\right)\bigg|_{i,j} \tag{2}$$

R_p is the resistance of the heat transfer of the pth layer of a structure or air medium, which lies at the intersection of the trajectory of the transmission (conductive), convective or radiative transfer of thermal energy between the *i*th and *j*th points

In paper [11], these coefficients were calculated for different cases.

It is proposed to explore a one-story flat-roofed building with three rooms, which are provided with air using forced exhaust ventilation (see Fig. 1). The plan was applied points in the air (in the centers of the rooms), on the ceilings (as shown in the conditional fragments of pok-digging), and on the inner surfaces of all walls (or wall fragments). The applied nodes are interconnected so that the drawn connections reflect all possible ways of heat exchange between the walls of the house, the air masses inside and outside the house. We obtained a discrete design model of a house with different types of connections between vertices. This model is not the final model of the house, because to simplify the calculations there are some types of heat loss (through the floor, etc.)

Following a series of simplifications described in paper [12], the equilibrium heat balance for each individual study point will look like as Eq. (3). Example equation created for point No. 1.

$$\begin{aligned} &-(K_{1,4} + K_{1,5} + K_{1,6} + K_{1,7} + K_{1,8} + K_{1,9} + K_{1,21} + K_{1,22} + K_{1,2}) \cdot t_1 \\ &+ K_{1,4} \cdot \tau_4 + K_{1,5} \cdot \tau_5 + K_{1,6} \cdot \tau_6 + K_{1,7} \cdot \tau_7 + K_{1,8} \cdot \tau_8 + K_{1,9} \cdot \tau_9 \\ &+ K_{1,21} \cdot t_{21} + K_{1,22} \cdot t_{22} + K_{1,2} \cdot t_2 + Q_1 = 0 \end{aligned} \tag{3}$$

Substituting to the system of equations the magnitudes of heat in rooms Q_1, Q_2, and Q_3, the system is solved for relatively unknown temperatures. In the example, it was assumed that all the thermal energy enters the premises convectively (directly through the air). That is, there are no other sources of energy (except nodes 1, 2, and 3) in the model.

This approach allows to vary Q_i volumes and to predict expected temperatures in the air, as well as on the surfaces of wall structures in the building. In addition,

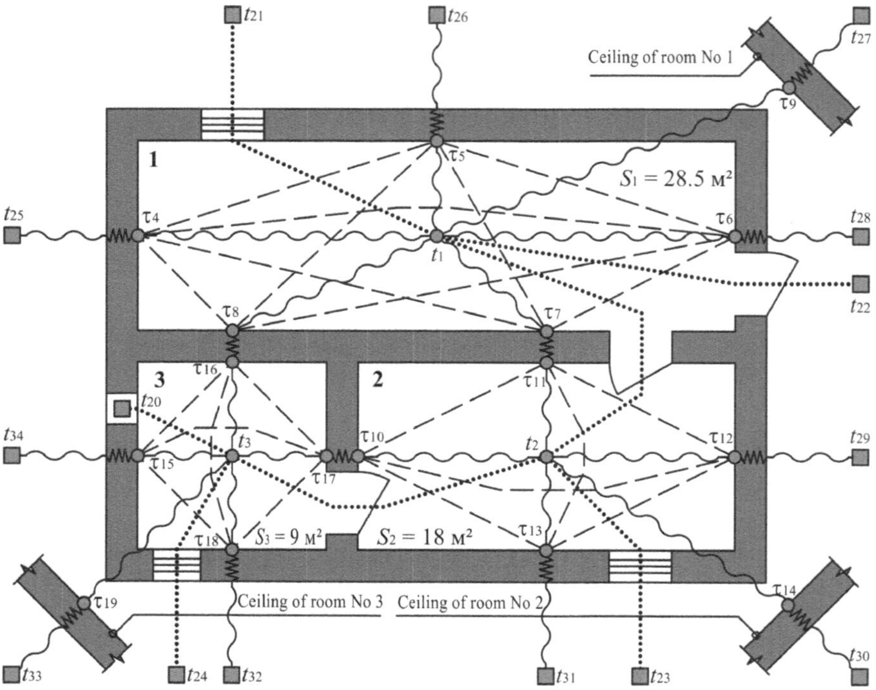

Fig. 1 House plan and discrete heat transfer model in it
Notations:
t_i—air temperature;
τ_i—temperature of internal surfaces of walls;
••••••••••••••••••••—resistance to heat transfer for the movement of air masses entering the house through windows and exterior doors, which are removed by the forced ventilation system;
WWWWWWWWWWWWWW—resistance to transmission heat transfer through internal walls and external enclosures;
∽∽∽∽∽—resistance to convective heat transfer;
— — — — — — — —resistance to radiation heat transfer

a different formulation of the problem is possible if it is necessary to determine the volumes of heat inflow to the air of the premises at which the air temperature corresponds to the set parameters. An example of the results of such an inverse simulation task is calculated and presented in Table 1.

4.2 Analysis of Infographic Modeling Tools for Visual Analysis of Heat Exchange Processes in a Building

In order to analyze the interaction between volumes in a building, such infographic diagrams are interesting, which show lines that combine the following characteristic

Table 1 Results of calculations of the design task to determine the volumes of heat inputs Q, W at given room temperatures

Type of data			Task number		
			No. 1	No. 2	No. 3
The parameters are set	Indoor air temperature	t_1, °C	21	22	25
		t_2, °C	20	21	25
		t_3, °C	19	20	25
	The specific support of heat transfer of translucent and door structures is accepted	$R'' = R''_{5.26} = R''_{6.28} = R''_{13.31} = R''_{18.32}$, m^2 K/W	1.3		
	Expected air temperature in exhaust ventilation	t_{20}, °C	18	19	22
	The outdoor temperature is set	$t_{EXT} = t_{21} = t_{22} = t_{23} = t_{24} = t_{25} = t_{26} = t_{27} = t_{28} = t_{29} = t_{30} = t_{31} = t_{32} = t_{33} = t_{34}$, °C	0	– 5	– 22
Values are calculated	Heat input	Q_1, W	711.983	905.616	1516.794
		Q_2, W	424.628	549.297	976.572
		Q_3, W	170.573	227.453	532.239
	Temperature of internal surfaces of walls	τ_4, °C	20.605	21.493	24.123
		τ_5, °C	20.638	21.537	24.207
		…	…	…	…
		τ_{13}, °C	19.666	20.565	24.213
		τ_{14}, °C	19.612	20.495	24.088
		τ_{15}, °C	18.678	19.573	24.178
		τ_{16}, °C	18.932	19.882	24.61
		τ_{17}, °C	18.902	19.852	24.613
		τ_{18}, °C	18.67	19.563	24.159
		τ_{19}, °C	18.591	19.462	23.989

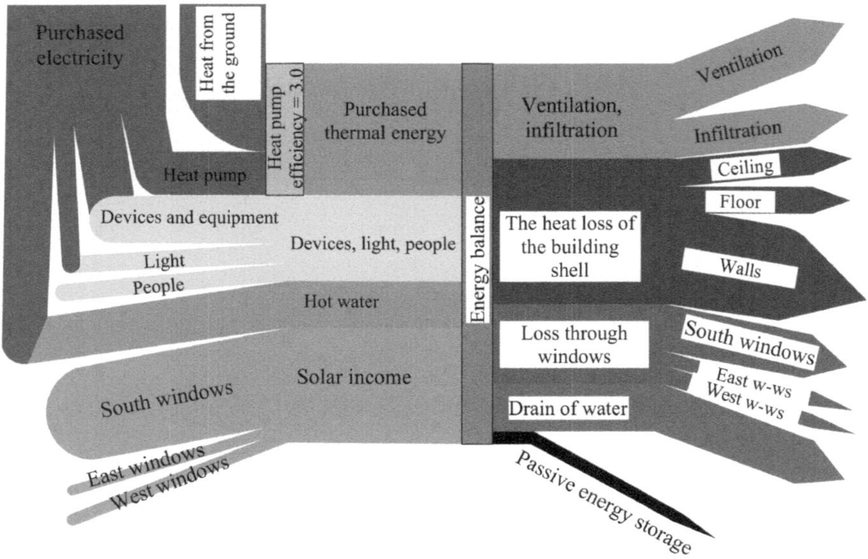

Fig. 2 An example of a general view of the Sankey diagram with the heat flows in the building

features: The line itself shows the relationship of the objects, and the width of the line—k_i—quantitative indicator of this connection. Arrows and streams at different stages can be split or combined.

This type of chart has the specific "Sankey" name. Sources consider Matthew Sankey to be the first to use this type of visualization.

Most often, such diagrams are used to visually display the movement of specific energy, resources, or materials, but they can also be used to represent the flow of any process.

The following is a Sankey diagram (see Fig. 2), showing the energy balance of a particular home at some point in time. A certain amount of energy enters the house, and this amount is equal to the energy lost. The total amount of heat energy flows is measured in kWh. In such Sankey diagrams, as a rule, the energy flows are distributed as follows: The left receives heat into the house (heat from the ground, added electricity, hot water, solar revenue, etc.), and on the right are energy losses (ventilation, heat loss by the shell buildings, losses through windows, etc.), and in the middle of the scheme there is the energy balance.

Such a diagram makes it possible to trace the amount of energy that flows through a building from its entry into the building and ending with its loss.

Also interesting for displaying heat fluxes in a building are graphs of parallel sets that are similar to the Sankey diagrams—they also operate with flows and proportions. But the difference from the Sankey diagrams is that the path of a certain flow is divided into sets that are quantified and move on to the next level of the object hierarchy (see Fig. 3).

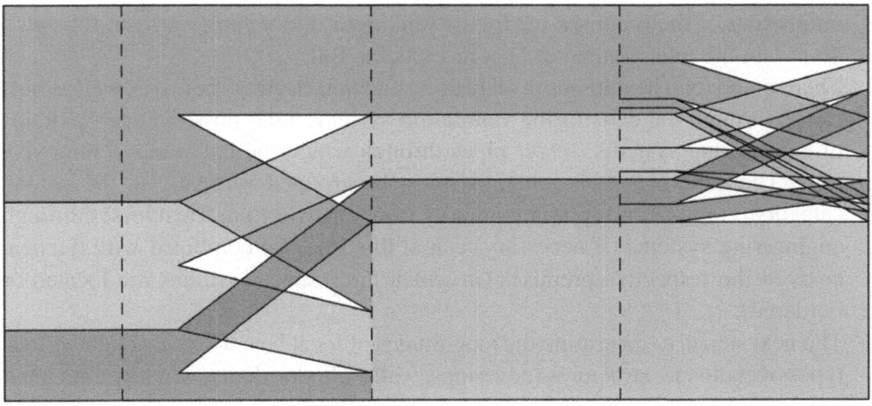

Fig. 3 General view of the diagram as a graph of parallel sets

The line in the diagram of parallel sets corresponds to a particular data set, where the values are represented by the thickness of the line divided by subsets. The thickness of each line and its flow path is represented by the proportional coefficient of the category as a whole.

4.3 Visualization of Heat and Heat Losses in the Modeling of Heat Exchange of Buildings

To build a diagram, first we need to determine the main categories into which it will be broken down. The house can be analyzed as a complex system containing "inputs" as well as "apparatus of distribution of input resources" (energy resources) and "outputs." We will conclude that scaled energy revenues should be placed at the entrance to this system, and energy consumption indicators will be placed at the exit. To simplify the perception of the results of all calculations and to understand that the total amount of energy revenues and energy consumption must be equal, we will reflect all energy flows in percentages. Also, we will consider the ratio of heat losses of the building based on design practice. It should also be remembered that Sankey charts are most often constructed on a plane and allow categories and distribution parameters (resources) to be structured both horizontally and vertically. With this in mind, as well as guided by the principles of providing energy to buildings, it is possible to set the following algorithm for visualization of heat balance (placement of distribution categories is offered from left to right):

1. It is necessary to set the values of all heat flows to each of the heating rooms or groups of rooms. The quantities to be obtained should be combined into the first category (column) vertically and split individually from top to bottom by the

numbering of the premises (or by the ranking of the volumes of heat receipts). To accept the total volume of heat receipts for 100%.

2. Then set the total heat losses in all heating and non-heating rooms, as well as heat losses through the exits of the ventilation system, other possible energy drains of engineering systems (sewer pipes through which the hot water is removed, etc.). The quantities to be obtained must be grouped vertically in the second category (column) and split individually from top to bottom. Heat losses through engineering systems (if necessary) can at this stage be combined with the heat costs of the respective premises (in which the respective drains are located or bordered).
3. The next step is to determine the magnitudes of local heat losses through various types of enclosure structures (coverings, walls, plinths, doors, windows, etc.) and through the ventilation systems and other heat sinks of engineering systems in these premises. The quantities to be obtained must be grouped vertically in the third category (column) so that all vertical connections with the second category are retained.
4. It is necessary to calculate the overall rates of heat loss through the same types of fencing structures (coverings, walls, plinths, windows, doors, etc.), as well as through the same types of engineering systems throughout the building.
5. Finally, check the error that can occur between the total heat inflow and the total heat loss and estimate its magnitude. If the magnitude of this error is too large, determine the cause of its occurrence.

Calculation of heat losses through individual types of enclosure structures is performed using formulas that have the following form:

1. $G'_{i,j}$—loss of thermal energy through opaque wall structures:

$$G'_{i,j} = (F_j - F''_i) \cdot (t_i - t_j)/R'_{i,j} \tag{4}$$

where $R'_{i,j}$—is the resistance of heat transfer of a wall structure:

$$R'_{i,j} = \left(1/\alpha_{Ki} + \sum_{p=1}^{n} (l_p/\lambda_p) + 1/\alpha_{K_j}\right) \cdot \eta'_{i,j} \tag{5}$$

$$\eta'_{i,j} = (F_i - F''_i)/(F_j - F''_i) \tag{6}$$

t_i and t_j	are the temperature at the ith point of the interior space of the study room or the jth point in the outside air;
l_p and λ_p	are the thickness of the pth layer of the structure and the coefficient of thermal conductivity of the material of the respective layer;
F_i	is the area of the inner surface of the wall.

$$F_i = F_i' + F_i'' \tag{7}$$

F_i' is the area of the opaque part of the wall;
F_i'' is the area of the glazing or door leaf;
α_{K_i} and α_{Kj} these are the coefficients of the convective heat exchange of the inner and outer surfaces of the enclosure with air.

2. $G''_{1,j}$—heat loss of window (and other translucent) and door structures:

$$G''_{1,j} = F_i'' \cdot (t_i - t_j)/R''_{i,j} \tag{8}$$

$R''_{i,j}$ this is the resistance to heat transfer of translucent or door structures.
$G_{L_{i,j}}$ heat loss due to the drainage of engineering systems (ventilation ducts, etc.) with the volumetric flow rates $\Delta L_{i,j}$:

$$G_{L_{i,j}} = (t_i - t_j)/R_{L_{i,j}} \tag{9}$$

$R_{L_{i,j}}$ is the resistance of heat transfer at heat losses to the total mass of air (or water) removed from the premises of the building:

$$R_{L_{i,j}} = 1/\left(\Delta L_{i,j} \cdot c \cdot \rho\right) \tag{10}$$

where

c an indicator of the heat capacity of the air or liquid being removed from the premises;
ρ the density of air or liquid being removed from the premises

An example for constructing a heat balance diagram of a building is calculated for problem No. 3 (Table 1). The total heat consumption of the proposed three-room forced ventilation building should be $\Sigma Q = 3025.605$ W. The heat consumption of the first room is $Q_1 = 1516.794$ W; second room is $Q_2 = 976.572$ W; third room is $Q_3 = 532.239$ W. All kinds of heat losses according to Eqs. (4)–(10) have been calculated, and their total value throughout the house is $\Sigma G = 3036.167$ W. The relative error between the values of heat supply and heat loss is: $\xi = 0.35\%$. This indicates a sufficiently high accuracy of the calculations. Therefore, it is suggested that the chart be constructed in percentage terms, unchanged and with high accuracy.

The calculated percentage distribution of heat and heat losses of this house is presented in Table 2.

The final diagram, which is based on the principles set out in the paper and according to Table 2, is shown in Fig. 4.

Table 2 Distribution of heat receipts and home heat losses

Categories (types of structures and design values) ↓	Heat losses of rooms (%)			Total metrics by category (%) (W)
	Room No. 1	Room No. 2	Room No. 3	
Outside walls	13.66	8.07	6.31	28.04 (851)
Coating	10.17	6.65	4.11	20.93 (635)
Windows	1.79	1.79	0.36	3.93 (119)
Doors	2.38	–	–	2.38 (73)
Ventilation	–	–	44.72	44.72 (1358)
Total heat losses on premises	27.99	16.51	55.50	100 (3036)
Heat	50.13	32.28	17.59	100 (3026)

Building a diagram of a home's heat balance illustrates the informativeness of the results of the calculations and can also serve as a basis for analyzing its overall energy balance. Charts can easily be supplemented by other types of energy or categories of their losses.

Such diagrams can be similarly arranged to visualize the processes of heat exchange in a building during the warm season, when the premises require cooling. Due to the high accuracy of the calculation method, it is possible to represent the data in percentages so that the height of each column of data (broken down by category) will be the same and always equal to 100%. Therefore, such a diagram can be used as a visual calculator and to calculate the combined percentage data (indicators) of selected categories, to use a predetermined scale, to build interactive relationships between diagrams and models of buildings in order to manage their heat losses (see Fig. 4).

5 Scientific Novelty

For the first time, a geometric model of the interrelation of architectural, structural, and engineering parameters of a building with energy balance parameters has been created, and a rational graphical model of presentation of computer calculations of heat receipts and heat losses of a house is proposed.

6 Practical Importance

The advantage of the results of this study is the clarity, convenience, and informative nature of the proposed method of data presentation, which greatly simplifies the

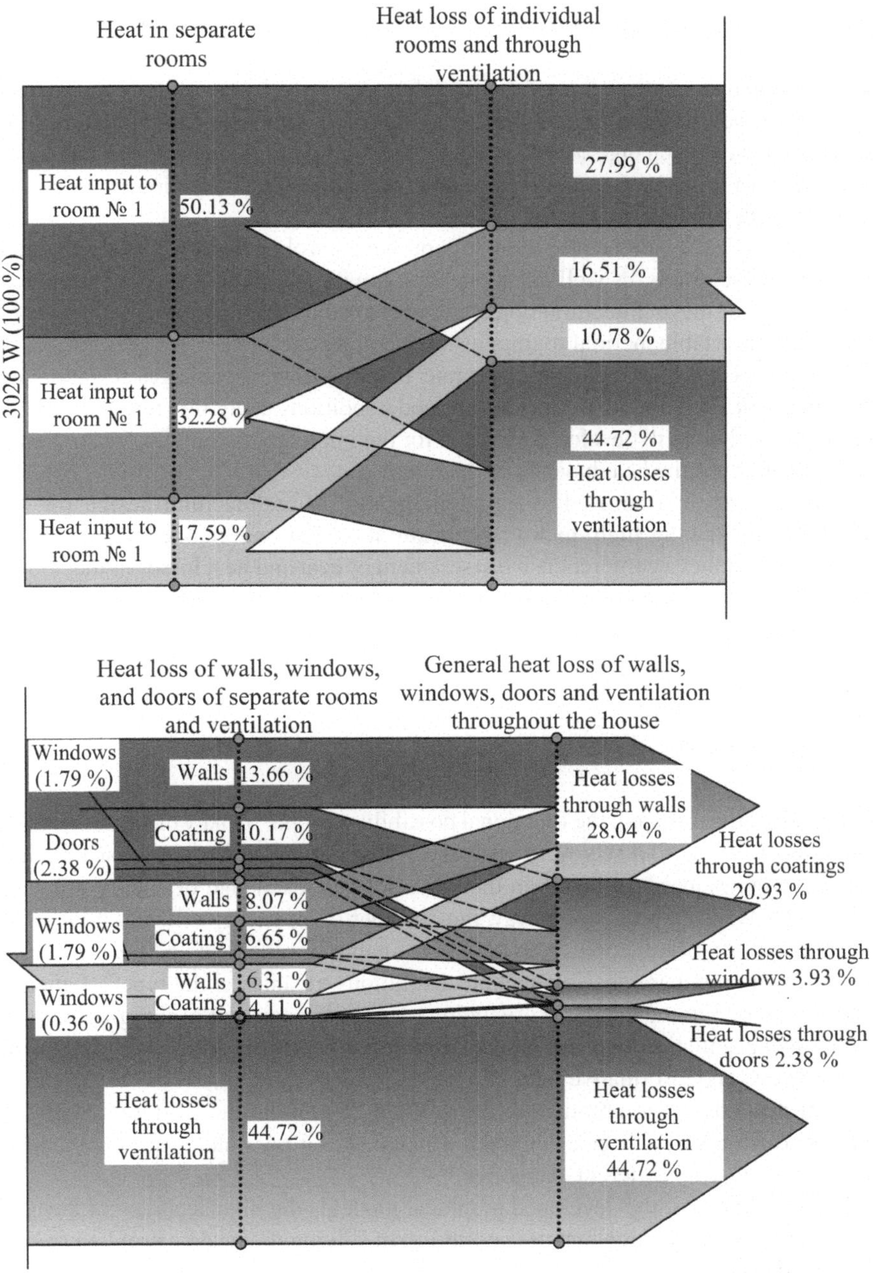

Fig. 4 Home heat balance diagram

analysis of large amounts of data in the design of heat exchange energy-efficient building.

This work is a continuation of the modeling proposed by the authors of the heat exchange of buildings described in papers [11, 12], in which an experimental simulation of the heat exchange of a building on the example of a one-room cottage was performed. In this paper, a graphical model of heat loss visualization in modeling the heat exchange of such a house is developed.

The paper deals with the possibility of optimal control of the model of the internal structure of the object, as well as its interaction with the model of the environment in the design of the architectural object in order to adjust the energy requirements to ensure a comfortable thermal regime inside the object.

It is promising to continue to integrate it into software packages to automate the process of sketching under predetermined conditions using this research. In the future, the authors will develop a proposal for the implementation of software based on the diagram of the heat balance of the house.

This research can be used as an auxiliary tool of architectural design for the possibility of making high-quality architectural design and engineering decision, and also to conduct a comprehensive assessment of heat and heat losses in the house, to be able to trace the connection between architectural and design solutions and energy balances.

7 Conclusions

As a result of the research, the functional possibilities of the Sankey diagrams and the diagrams of the parallel sets were analyzed, and a graphical model of visualization of heat revenues and heat losses in the model of heat exchange of energy-efficient buildings was developed. In this way, the designer of an energy-efficient home will have the opportunity to conduct a qualitative analysis of the heat and heat losses of the facility, to compare possible technical solutions, to analyze the types of energy used. Compared to the tabular presentation of data, this method of infographic presentation of calculation results will accelerate decision making due to its ergonomics, compactness, and informative nature.

Total heat losses on premises and local heat losses through different types of fencing structures have been calculated, which makes it possible to analyze the heat losses of a house by different categories.

The peculiarity of the developed graphical model is the high accuracy of the calculation results calculation and the possibility of solving the inverse problem (when embedded in computer-aided design)—varying the parameters of the model, you can get energy consumption. This demonstrates the prospect of developing computer aids that will allow the architectural model to be controlled in the design process to minimize energy requirements to provide a comfortable thermal environment within the facility.

References

1. Ploskyi, V. (2016). Funktsionalnist hrafiky yak metodolohichna osnova hrafichnykh tekhnolohii. *Prykladna heometriia ta inzhenerna hrafika, 92,* 93–99.
2. Bolharova, N. (2018). Heometrychna model formuvannia ratsionalnoi struktury arkhitekturnoho obiektu za parametramy enerhoefektyvnosti. Dys. kand. tekhn. nauk: 05.01.01. 167.
3. *Sankey diagrams for energy balance*. Eurostat Website. (2018). Retrieved from https://ec.europa.eu/eurostat/statistics-explained/index.php/Sankey_diagrams_for_energy_balance#What_you_can_do_with_the_Sankey_tool.3F.
4. Subramanyam, V., Paramshivan, D., & Kumar, A. (2015). Using Sankey diagrams to map energy flow from primary fuel to end use. *Energy Conversion and Management, 91,* 342–352. https://doi.org/10.1016/j.enconman.2014.12.024.
5. Riemann, P., & Hanfler, M., & Froehlich, B. (2005). Interactive Sankey diagrams. In *IEEE Symposium on Information Visualization. IEEE Symposium on Information Visualization*. https://doi.org/10.1109/INFVIS.2005.1532152.
6. Abdelalim, I., O'Brien, W., & Shi, Z. (2017). Data visualization and analysis of energy flow on a multi-zone building scale. *Automation in Construction, 84,* 258–273. https://doi.org/10.1016/j.autcon.2017.09.012.
7. Osborn P. (1985). *Handbook of energy data and calculations including directory of products and service* (260 p). Butterworth & Co (Publishers). https://doi.org/10.1016/c2013-0-04062-5.
8. Krempel, L., & Plümpe,r T. (1999). International division of labor and global economic processes: An analysis of the international trade in automobiles. *Journal of World-Systems Research, 3,* 487–498.
9. Eick, S. G., & Willis G. J. (1993). Navigating large networks with hierarchies. In *IEEE Proceedings Information Visualization* (pp. 204–210).
10. Lupton, R., & Allwood, J. (2017). Hybrid Sankey diagrams: Visual analysis of multidimensional data for understanding resource use. *Resources, Conservation and Recycling, 124,* 141–151. https://doi.org/10.1016/j.resconrec.2017.05.002.
11. Bolharova, N., Ploskyi, V., & Skochko, V. (2017). Praktychni aspekty pobudovy fizychnoi dyskretnoi modeli teploobminu enerhoefektyvnoi budivli. *Tekhnichna estetyka ta dyzain. Naukovo-tekhnichnyi zbirnyk, 13,* 9–20.
12. Bolharova, N. (2018). Modeliuvannia protsesiv teploobminu pry proektuvanni enerhoefektyvnykh budynkiv. Enerhointehratsiia-2018: materialy vosmoi Mizhnarodnoi naukovo-praktychnoi konferentsii, 56.
13. Pavlenko, A., & Koshlak, H. (2015). Design of processes of thermal bloating of silicates. *Metallurgical and Mining Industry, 7*(1), 118–122.
14. Yurin, O., Azizova A., & Galinska, T. (2018). *Study of heat shielding qualities of a brick wall corner with additional insulation on the brick*. Paper presented at the MATEC Web of Conferences (Vol. 230).
15. Yurin, O., & Galinska, T. (2017). *Study of heat shielding qualities of brick wall angle with additional insulation located on the outside fences*. Paper presented at the MATEC Web of Conferences (Vol. 116). https://doi.org/10.1051/matecconf/201711602039.

Analysis of the Current State of Construction of High-Rise Monolithic Reinforced Concrete Buildings

S. A. Farzaliyev, S. R. Quluzadeh, and T. F. Mehtiyeva

Abstract The high-rise construction production in our republic has a high development rate and good potential for future development. Among a number of socioeconomic issues, this is of great importance in the organization of a cheap housing market. The scarcity and high cost of the construction site in the republic and major cities of the world make the construction of high-rise monolithic reinforced concrete buildings an actual one. There are special requirements for the construction of high-rise buildings. Among these requirements is improvement of normative technical documents, design issues, application opportunities of modern techniques and technologies, professionalism of engineering and technical staff, quality control, security work, etc. The article analyzes the current state of high-rise monolithic reinforced concrete buildings in the world and in our republic, and the results are determined. The analysis shows that construction of high-rise buildings in the world contributes to the creation of a cheap housing fund and a comfortable living space. Meanwhile, existing organizational and technological problems must be solved.

Keywords High-rise · Monolithic reinforced concrete · Construction production · Mold · Concrete · Fittings · Organizational–technological problems · Quality control

1 Introduction

Analysis of the current state of the construction of monolithic reinforced concrete buildings in foreign countries and our republic shows that the construction of monolithic reinforced concrete carcass buildings in the construction industry for many years will retain its dominance. The scale of the construction of high-rise monolithic reinforced concrete buildings in the world and in our country and the pace of development require the solution of a number of problems existing among them:

S. A. Farzaliyev (✉) · S. R. Quluzadeh
Azerbaijan University of Architecture and Construction, Baku, Azerbaijan
e-mail: sferzeli@gmail.com

T. F. Mehtiyeva
Azerbaijan State Pedagogical University, Baku, Azerbaijan

V. Onyshchenko et al. (eds.), *Proceedings of the 2nd International Conference on Building Innovations*, Lecture Notes in Civil Engineering 73,
https://doi.org/10.1007/978-3-030-42939-3_56

urban planning, architecture, constructive, technological–organizational solutions. To carry out effective construction in the background of rising prices in our country and in the world day by day will always remain urgent.

High-rise buildings are a modern form of efficient development of large-scale construction production without alternatives and with certain regularities [1]. This is mainly due to the increase in the number of population in large cities and capitals due to the shortage of construction sites and the steady increase in the economic growth of the city. At the same time, these economically very efficient buildings are provided with floors of various purposes and favorable conditions are created. The placement of shops, cafes, restaurants, offices, guest houses and residential areas in such high-rise buildings creates favorable conditions for the development of business environment and economy and for the living. At the same time, the density is prevented by the construction of this type of buildings separately [2–4].

2 Main Part

2.1 *Purpose of the Article*

The main purpose of the article is to improve the efficiency of construction production by analyzing the current state of construction of high-rise monolithic reinforced concrete carcass buildings in the world and our republic and identifying organizational and technological problems for development and their solutions.

2.2 *Research Methodology*

Comparative analysis and graphic–analytical methods were used in the article. At this time, the current situation of the construction of high-rise monolithic reinforced concrete buildings in major cities of the republic, former Soviet countries and world countries has been analyzed. The analysis shows that construction of high-rise buildings creates conditions for the creation of a cheap housing fund and a comfortable living space. At this time, a number of design and organizational and technological problems must be solved.

2.3 *Result*

The pace of development of high-rise buildings in our republic is compared to the period of Soviet Union. After our republic gained its independence, the construction of high-rise buildings with monolithic iron concrete frame began to develop more rapidly. The development rate of the high-rise residential house built in the last 5 years is given in Table 1 and its graphical analysis in Fig. 1 [5].

High-rise monolithic reinforced concrete constructions are mainly carried out in major cities of our republic. Information on the number of population in major cities of the Republic and the area of the built residential houses is given in Table 2 and their comparative analysis graph in Fig. 2.

75% of these buildings are built by private construction companies and 25% by state construction companies. If 9–12-story buildings were built in the country until the 1990s, the construction of 354,554-story buildings was carried out in recent years. During the construction of such buildings, the power of foreign specialists was used both on the design stage and at the stage of construction work. The main reason for this is the low experience of local specialists in the design of high-rise monolithic reinforced concrete buildings, low ability to organize and manage high-rise work, poor knowledge of construction technology, low ability to use modern technology machines, lack of highly qualified personnel, low labor costs and other issues [6].

For one reason or another, the period of implementation of monolithic reinforced concrete works in our republic is 2–3 times more than in foreign countries. Thus, although the existing monolithic reinforced concrete works are carried out in our republic for 15–20 days, they cover 5–7 days in foreign countries [7, 8].

Table 1 Development rate of the high-rise residential house

Type of construction	Construction area				
	2013	2014	2015	2016	2017
Total construction area (thousand sq. m)	2403	2197	1932	2121	2017
Including High-rise	1442	1358	1197	1315	1250
Low-rise	961	839	735	806	767

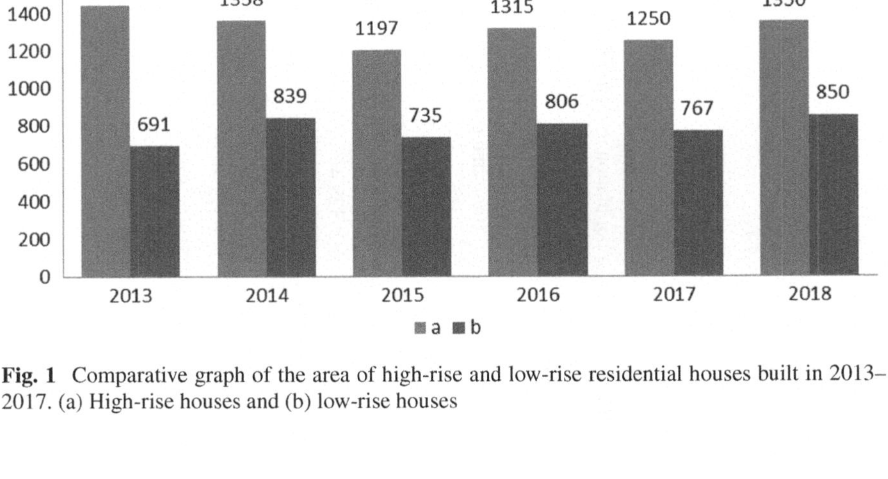

Fig. 1 Comparative graph of the area of high-rise and low-rise residential houses built in 2013–2017. (a) High-rise houses and (b) low-rise houses

Table 2 Number of population in major cities of the republic and the area of the built residential houses

No.	Cities	2015		2016		2017	
		Number of population (thousand people)	Construction area (thousand.m2)	Number of population (thousand people)	Construction area (thousand.m2)	Number of population (thousand people)	Construction area (thousand.m2)
1	Bakı	2204	333.1	2225.8	724.3	2245.8	659.9
2	Sumgaıt	332.9	28.9	336.2	33.0	339.0	44.4
3	Absheron	202.8	144.8	205.2	94.4	207.5	58.4
4	Gandja	328.4	36.6	330.1	58.9	331.4	19.6
5	Nakhchivan	90.3	380.1	91.1	380.3	92.1	385.9

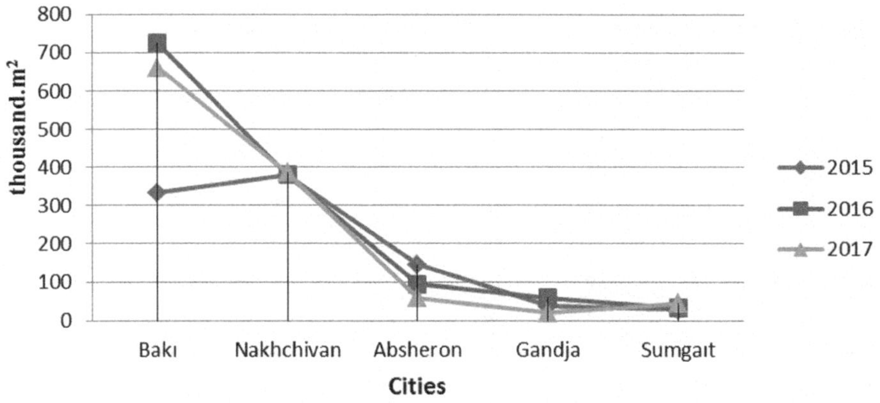

Fig. 2 Comparative graph of the areas of residential houses built in large cities

If we look at the experience of foreign countries, we can see that the weight of monolithic reinforced concrete buildings is 50–60% on average. This is 60% in Russia, 67% in Germany, 68% in Italy, 70.7% in France, 71.9% in Japan, 72.6% in the USA, 89% in Turkey, 95% in Israel and 87.5% in Azerbaijan [9].

If we look at the statistics of residential houses built by foreign countries during the year, we can see that in Belgium with a population of about 10.4 million people, 42–45 thousand apartments are built annually, in Norway with a population of about 5 million, 26–30 thousand apartments are built annually, in Spain with a population of 40 million. The number of apartments built by a number of foreign countries during the year is listed in Table 3.

Table 3 Number of apartments built by a number of foreign countries

Countries	Over the years (thousand apartments)							
	2008	2009	2010	2011	2012	2013	2015	2016
Azerbaijan	17	14	18	16	20	21	17	15
Germany	176	159	160	183	201	215	248	278
USA	1128	792	648	588	648	768	969	1061
Belgium	50	45	48	42	45	43	43	44
Bulgaria	21	22	16	14	10	9	8	8
UK	188	158	137	142	144	138	171	151
Spain	615	388	257	168	115	65	46	76
Norway	29	22	18	20	26	29	28	29
Poland	165	160	136	131	153	145	148	142
Japan	1 094	788	813	834	883	895	909	886
France	400	334	346	421	347	332	338	379

Table 4 Number of apartments for every 10,000 people in the CIS countries

Countries	Over the years (thousand apartments)						
	2008	2011	2012	2013	2014	2015	2016
Azerbaijan	17	16	20	21	19	17	15
Armenia	6	11	11	7	5	7	4
Belarus	62	74	60	67	73	59	52
Kazakhstan	37	33	35	34	37	45	50
Kyrgyzstan	15	16	16	16	17	21	20
Moldova	19	14	14	16	15	21	17
Russia	54	55	59	65	748	56	52
Tajikistan	15	23	16	17	18	17	15
Uzbekistan	23	25	27	28	24	27	29
Ukraine	20	18	20	23	25	28	26

The number of apartments for every 10,000 people in the CIS countries for 2008–2016 is given in Table 4.

Grounded world experience shows that large cities with very scarce and expensive land should have 30–50 floors for the construction of economically efficient buildings.

Concrete is the most widely used in the construction of high-rise buildings. The volume of monolithic reinforced concrete use in the world is 1.5 billion m^3. The volume of monolithic iron concrete per 1 person is 0.75 m^3 in the USA, 1.2 m^3 in Japan, 0.8 m^3 in Germany, 0.5 m^3 in France, 1.7 m^3 in Belgium, 2.0 m^3 in Israel, 0.4 m^3 in Turkey, 1.1 m^3 in Norway and 0.15–0.2 m^3 in the CIS countries [10].

Analysis of the current state of high-rise monolithic reinforced concrete buildings shows that the desire of people around the world to live in high-rise buildings is one of the perspectives of development of this type of buildings. The main factors characterizing this are the placement of all necessary infrastructures in high-rise buildings. World experience shows that the locations of hotels, stores, coffee bars, business centers, offices, training centers, medical institutions, kindergartens, personal vehicles, entertainment centers, etc., in a high-rise building create comfortable conditions.

Analysis of the current state of multi-story monolithic reinforced concrete buildings shows that a number of serious issues for perspective development must be resolved:

- the lack of perfect normative bases for the design and construction of high-rise buildings;
- to use molds requiring less labor (quick and easy disassembly and assembly of modular molds, polymer light molds, molds made of composite materials, molds not disassembled from concrete, self-rising molds);
- to use armature requiring low labor costs (ready-made products in the form of armature carcasses, using automated tools of junction units of on-site carcasses);
- using high-speed concrete pumps for concrete works;

- to use high-strength and quick-drying concrete.
- compaction, drying and maintenance of concrete should be carried out using special equipment:
- serious attention should be paid to the technology of implementation with special control over concrete works;
- highly skilled workers and professional engineers should be involved in the correction of monolithic reinforced concrete structures;
- ensuring the seriousness of monitoring and quality control of the implementation of technological processes of high-rise buildings.

One of the problems of perspective development of high-rise monolithic reinforced concrete buildings is to ensure reliable durability of these buildings. One of the main issues in the design of such buildings is ensuring its earthquake resistance.

A number of scientific–practical studies have been conducted in this direction in the world and former CIS countries [11]. The results of these studies show that the design of high-rise buildings depends on the chosen constructive scheme of ensuring earthquake resistance to ensure the durability of the building.

The application of the latest achievements of modern technologies in the future development of high-rise buildings is also of great importance. In the world, the construction of "smart homes" is already set as a goal. Introduction of modern technologies and automated systems in high-rise buildings that envisage factors such as people's comfort, safety, time savings, etc., is a new stage in the prospective development of buildings.

The future development prospective of the construction of high-rise monolithic reinforced concrete buildings in our republic is very large. Thus, in the average developed countries, the housing area per 1 person is 30–35 m^2, while in our country it is 12.6 m^2. It takes about 60 years for the population to meet the required today's resources. Therefore, the construction of high-rise residential houses in our country will still be needed for many years.

With the improvement of the normative documents of monolithic reinforced concrete construction, the issue of designing the construction of monolithic reinforced concrete buildings taking into account the interaction of different technologies of production should be effectively resolved. It is also an important condition for the training and professional development of construction personnel, as well as the contribution of construction organizations carrying out the construction of high-rise monolithic buildings with modern equipment. At the same time, reducing the cost of construction and reducing labor costs should be solved by improving the production technologies of the works that make up these processes. Thus, the share of manual labor in the performance of monolithic reinforced concrete works is high. In order to reduce labor costs and ensure that work is carried out with high quality, it is necessary to organize the proper management of construction processes and the automation of manual work. However, in order to shorten the construction period, it is important to protect the concrete. Intensification of concrete hardening reduces the duration of its stay in the mold, accelerates the removal of mold and speeds up the circulation of mold.

Improvement of quality control systems is also one of the prospective directions of development of monolithic construction production. This control system includes material control, technological operations control and construction control functions. This control can be carried out mainly by persons with special permit licenses and authorized representatives of state institutions.

Analysis of the current state of the construction of monolithic reinforced concrete buildings in foreign countries and our republic shows that the construction of monolithic reinforced concrete carcass buildings in the construction industry for many years will retain its dominance. The scale of the construction of high-rise monolithic reinforced concrete buildings in the world and in our country and the pace of development require the solution of a number of problems existing among them: urban planning, architecture, constructive and technological–organizational solutions. Against the backdrop of rising prices day by day in our country and in the world, high-rise monolithic reinforced concrete will permanently keep the construction of buildings actual.

2.4 Scientific Novelty

For the first time, the analysis of design, technical, technological, organizational and economic condition of construction of the high-rise monolithic reinforced concrete buildings constructed in our republic was carried out. It is determined that to improve the efficiency of construction of high-rise monolithic reinforced concrete buildings, new organizational and technological methods of high-level design, mechanization and automation of technological processes of construction production should be developed, the choice of effective options for the organization of construction production and the use of special purpose machinery and equipment and collaboration of highly qualified professionals should be implemented.

2.5 Practical Importance

Research results can be used in the development of concepts of perspective urban planning, organizational and technological solutions for the construction of high-rise monolithic reinforced concrete buildings in the creation of cheap housing market.

3 Conclusions

Analysis of the construction of high-rise buildings shows that there are a number of organization–technological problems such as low experience of local specialists in the management and organization of high-rise work, low skills in the use of modern technology machines, poor knowledge of construction technology, lack of highly qualified personnel, low labor costs, etc.

1. Analysis of the situation of the construction of high-rise monolithic reinforced concrete buildings in foreign countries shows that construction of high-rise buildings in the world creates conditions for the creation of a cheap housing fund and comfortable housing.
2. Analysis of the current state of the construction of high-rise monolithic reinforced concrete buildings in foreign countries and our republic shows that mechanization and automation of the implementation of the perfect normative technical base, high-level design solutions, technological processes of construction production in order to increase the efficiency of the construction of high-rise monolithic reinforced concrete buildings, the selection, new organizational and technological methods of high quality control and joint work of professional highly qualified specialists should be developed.

References

1. Shchukina, M. N. (2007). *Modern high-rise construction* (440 p). Monograph. M.: SOP "ETC Moscomarchitecture".
2. Bukhtiyarova, A. S., Kolchunov, V. I., Rybakov, D. A., & Filatova, S. A. Study of the survivability of residential and public buildings with new constructive system of industrial panel and frame elements. *Construction and Reconstruction, 6*(56), 18–24.
3. Mordich, A. I., Vigdorchic, P. I., & Belevich, V. N. (1999). New universal frame system of high-rise buildings. *Concrete and Reinforced Concrete, 1*, 2–4.
4. Maklakova, T. G. (2008). *High-rise buildings. Urban planning and architectural and structural design problems* (p. 160). Publishing House ASV.
5. The Website of the State Statistical Committee of the Republic of Azerbaijan. (2018). *The main socio-economic indicators of construction.* https://www.Stat.gov.az/source/construction/.
6. Farzaliyev, S. A. (2018). İmproving the efficiency of using monolithic reinforced concrete structures in high-rise buildings. In *III International Scientific-Practical Conference "Research in the Field of Technical Sciences*" (p. 30–34). Saratov.
7. Minh, N. S. (2002). *Rational planning of housing construction programs* (Dissertations Ph.D.) (p 149).
8. Tien, N. T., Van Binh, P., & Dung, L. X. (1996). Cong nghe xay dung nha cao tang. Bao cao tong ket de tai. Ha noi, 134tr.
9. Telechenko, V. I. (2009). *Organization and technology of construction* (520 p). Publishing House ACB.
10. Markovsky, M. F. (2011). High-rise housing. Drove right to the error. *Architecture and Construction, 1,* 75–79.
11. Costanzo, S., D'Aniello M., & Landolfo, R. (2017). Seismic design criteria for Chevron CBFs European versus North American codes (part 1). *Journal of Constructional Steel Research,* 83–96.

Scientific Approaches for Planning the Architecture for Urban Economic Space

Petro Gudz, Maryna Gudz, Olga Vdovichena, and Oksana Tkalenko

Abstract The basic scientific approaches to the planning of urban economic space architecture based on the methods of structural and logical analysis, generalization, and comparative analysis are investigated as the purpose of the work. An analysis of urban planning methods and interests of residents of territorial communities has been carried out, and successful practices of urban and space revitalization have been investigated. The novelty of the work is formulated—the methodical approach to the choice of methods of planning the architecture of urban economic space has been further developed, which allows to integrate landscape, architectural-construction, financial-economic, marketing-logistical, digital tools of their realization in the triangle of urban entities "power—business—the public." In urban planning, the methods of solving the problems of planning the architecture of urban economic space are distinguished: rational planning of development, limiting the growth of large cities, segregative distribution in urban areas into functional zones (residential, industrial, communal-warehouse, external transport, suburban), and digitalization of urban space management. Cities compete for development resources, specialists, etc., and are forced to move to master plans for urban development, development strategies, plans for the revitalization of industrial territories, harmonization of urban planning, and the architectural environment of settlements. It is established that effective methods of planning the urban economic space are such methods as: strategy, modernization, reconstruction, revitalization, digitization, social activities, technological innovations, green economy, and logistics.

Keywords Urban planning · Master plan of the city · Economic space · City · Territorial community · Revitalization

P. Gudz (✉) · M. Gudz
Zaporizhzhia Polytechnic National University, Zaporizhzhia, Ukraine
e-mail: pitgudz@gmail.com

O. Vdovichena
Chernivtsi Trade and Economic Institute of Kyiv National Trade and Economic University, Chernivtsi, Ukraine

O. Tkalenko
State University of Telecommunications, Kiev, Ukraine

V. Onyshchenko et al. (eds.), *Proceedings of the 2nd International Conference on Building Innovations*, Lecture Notes in Civil Engineering 73,
https://doi.org/10.1007/978-3-030-42939-3_57

1 Introduction

Contemporary urban environment is evolving along with its subsystems: construction, housing, communal, economic, social, digital, management, etc. Ukrainian cities lack state-of-the-art construction technologies and new machines; on the other side, they fail local government leaders, innovations, and activity. Only under these conditions, the prospects of investment development within urban spatial systems and subsystems are open.

Specialists in urban science and architecture, design and management, economics and marketing, sociology and ecology, as well as regional economics define the city as their subject matter. Currently, urban planners state the accumulation of problems of architectural and construction character, which include unplanned, accumulated construction, poor sanitary provision, noise, and air pollution.

Experts on regional economics highlight the need to address the issues of quality planning of urban space architecture, such as defining planning criteria and choosing urban development strategies, and developing mechanisms for their implementation. There is a large number of additional problems in the development of urban space such as high level of entropy in digitalization of construction. That is manifested in lack of structured information on engineering and construction communications in industrial zones and housing construction, terms of their exploitation, modernization, low level of digitization of economy, especially in old industrial territories, use of big industrial fields provision of social services, and administration of sustainable development of territories.

There are various approaches for planning the architecture of the urban economic space among foreign and domestic scientists. Thus, diverse tools are used in the planning of urban economic space: strategizing [1], modernization [2], reconstruction [3], revitalization [4], digitization [5, 6], social activities [7], technological innovations [8], green economy [9, 10], and logistics [11, 12]. Scientists, applying sustainable approaches, cannot still offer effective tools for providing an innovative leap for the development of urban communities. Therefore, further interdisciplinary research is needed for the development of an ecological-based SMART-cities planning methodology and the revitalization of industrial sites to socialize urban space.

The purpose of the work is to determine the basic scientific approaches for planning the architecture for urban economic space.

Method. The methodological basis of the study lies in understanding of competitive advantages of urban space, which in the process of regulatory influences should maintain and increase attractiveness for residents, investors, and tourists. Methods of structural and logical analysis, generalization, and comparative analysis are used as the basis of the research methodology.

Results 1 Harmonization of planning methods and interests of residents of territorial communities.

Paul R. Krugman defines "second nature factors" in his classification. They are benefits that are created by activities of people and society, i.e., agglomeration effect (high population density in cities, which saves economies of scale); human capital (education, health, work motivation, mobility, and adaptability of the population); institutional capital, which contributes to improving the investment climate, mobility of the population, diffusion of innovations, etc. He also outlines development of infrastructure as a factor that reduces economic distances [13].

Urban planning is a complex multifaceted activity of society aimed at creating material and spatial environment of human activity in urban locality and settlement areas. A specific branch of science and technology devoted to research on engineering, social, and economic as well as ecological problems of living environment formation. It includes construction of settlement areas infrastructure, their planning, and construction in tandem with combination of hierarchies of objects and levels. Urban planning activity covers research, design, and management of the processes of implementation of measures that determine the formation and development of functional and architectural and planning structure of settlements and districts in accordance with demographic, social, economic requirements, and natural and environmental conditions; development of engineering and transport infrastructure; preservation and enrichment of the environment. The resources for solving urban planning problems are rational and systematic organization of the territory planning and coordinated location of residential areas, industrial complexes, recreational zones, community centers, etc.

The results of urban development should help to ensure the managerial processes of settlements and territories development, construction planning, reconstruction and operation of settlements, and regions in accordance with the needs of the population and production.

Planning methods in urban planning include:

scientific definition of territorial and urban development objects and systems of urban economy, their functional, planning parameters and evaluation criteria, developing their typology basics;

theory, methodology, techniques of engineering-planning and space-spatial formation, and reconstruction of urban-planning objects of different types;

engineering and technological, social and economic, environmental, and technological factors that influence the formation of the environment;

optimization methods of architectural and engineering-planning decisions of settlements and regions taking into account peculiarities of social and demographic, economic, ecological processes, and natural conditions on the basis of modern information technologies;

technology of complex design and planning works technology, process management of functioning and development of regions, cities and villages with the use of methods and tools of applied information technology, and heuristic methods of creativity in urban planning;

methods of multifactor assessment of the qualities of urban planning decisions at different phases and stages of design;

methods of creating and maintaining urban register of settlements, urban planning databases, and other territorial information systems (TIS);

regularities and tendencies of settlement, organization of production activity, functioning of urban economy objects, social sphere, urban transport systems, street-track network and their elements, systems of engineering equipment and engineering preparation of the territory, provision of urban amenities, and landscape architecture;
urban ecology and resource conservation;
economics of urban planning and evaluation of the territory.

Taking into consideration existing types of landscaping for the residential area, the best one in real estate development is a residential neighborhood. This form of organization helps to organize water supply, sewerage, a network of institutions of medical, consumer, and sport services for the population in the most appropriate way. It is necessary to take into account the distances to the place of work, residence, and ways of transportation.

Among the objects of the urban hierarchy, the master plan is the leading and most important document for the development of the city for the next 15–20 years: "Master plan of a settlement - urban planning documentation, which defines the basic decisions of development, planning, construction and other use of the territory of a settlement." [14]. The legal and organizational principles of the development, approval, registration, and application of construction codes are regulated in detail by a special Law of Ukraine "On Building Rules" [15].

The analytical part of the master plan contains a comprehensive assessment of the current state of the territory of the settlement and the actual problems of its urban development. Among the tasks of the justifications and proposals of the master plan, the following ones are specified: specification of principled decisions of district planning projects in accordance with local conditions and state and public interests; ensuring the health and epidemic well-being of the population; forecasting the need for housing, public services, manufacturing, recreational and wellness facilities, transport and communications, engineering equipment, landscaping and public works, utilities, environmental protection, cultural heritage and identifying the means to meet those needs; determination of priority and permissible types of use and development of the territory and their location; setting restrictions on certain uses of the territory in accordance with the requirements of legislation, construction, sanitary, environmental, and other state standards.

Thus, the master plan in its content is a document that contains information of public interest and is important for the society. It divulges such aspects as determination of permissible types of territorial development, environmental protection insurance, cultural heritage preservation, selection of territories for housing, etc. Above-mentioned issues are priorities for the life of the community of any settlement.

Master plan should be transparent for governing bodies and public representatives to determine the effectiveness of relevant authorities and/or local governments in catering to the needs of local community. Any planning decision is information about the disposal of land owned by the territorial community. Therefore, the master plans should be open and cannot be considered as confidential information that is owned by the state, because it is the property of the territorial community, and therefore, so it must be relevant information for the public.

Every citizen has the right to know whether his or her home is in the vicinity of radiation hazard, chemical, or noise pollution. The cost of housing and the convenience of living are directly related to the master plan of the settlement. In order to find out where to buy a new apartment or house, you should get acquainted with the general plan of the city. It uncovers neighborhood infrastructure, as well as construction plans for the next 15–20 years. In 10 years time, maybe, there will be an airport or train station, or, a multi-stored office buildings will grow right in front of the windows.

Expropriation of real estate for public use. Often people living primarily in the old part of the city do not suspect that their private property may be subjected to expropriation. The master plan may involve reconstruction of certain areas of the city, which is usually accompanied by the resettlement of residents of the reconstruction areas. In such cases, the owners of the expropriated property must get amends. The master plan section "Phase One of the Construction Project" reveals the reconstruction plans and conditions. Expropriation of real estate can affect not only the residents of the old districts but also the owners of suburban areas and towns, farmers who own land in close proximity to the city limits.

In the theory and practice of urban planning, this issue is connected to spatial planning that is an integral part of urban planning. The competitiveness of a modern city, as noted in the research of R. Giffinger and Mariana Stallbohm, is directly related to its ability to fulfill the specific functions of the metropolis, and includes the following:

- spatial "expansion" of the city, which involves the creation of social and economic sub-centers, intensive creation of workplaces on the basis of polycentric model;
- intensive development of a knowledge-based economy both in the manufacturing sector and in the services sector, in the center and in the periphery of agglomeration;
- decision-making centers concentration, such as international and interregional production and service business structures, political, civic, and cultural organizations [1].

2 Urban Economics: Successful Practices in Urban and Space Revitalization

Modern city is an adaptive, open for interactions, organized, and structured economic system, which consists of interdependent and interacting participants. Common purpose and economic interests unite them. It is designed to optimize the use of resources in economic flows within subsystems, where residential space subsystem takes first place in architectural and construction subsystems. Subsystem reflects its system-forming influence in the city [16, pp. 36–37]. Scientists also define the central office space subsystem; industrial-production spatial subsystem; transport spatial subsystem; as well as the landscape and recreational spatial subsystem.

Urban scientific research has been actively aimed at energy-saving technologies in construction and science of materials. Therefore, 75% of residential areas need to be reconstructed in order to reduce energy costs. EU countries arrange ambitious plans, for example, by 2020, Denmark will have reduced energy costs by 75% in comparison with old buildings. Norway, the Netherlands, and Germany rely on internal-heating of buildings. The UK and Hungary are based on home systems of non-carbon dioxide gas. France relies on energy-generating structures [5].

The implementation of logistical approaches to urban planning has allowed overcoming a number of functional traffic problems to local authorities. The authorities have taken the following measures in order to overcome these problems in cities [17, pp. 163–164]:

Copenhagen (Sweden) has a clear list of zones and points where commercial freight transport is allowed to be unloaded;
Stockholm (Sweden) established city-distribution centers and resource points, located outside the city. Being transported by heavy transport, they are redirected on trucks with a capacity of up to 3.5 tons across the city. The routes of movement, based on the orders of the final recipients, are considered in such a way, that the truck will be overloaded from the center of the city;
In Stockholm, Gothenburg, Malmö, and Lundi (Sweden), the movement of trucks of more than 8-year exploitation is restricted in certain areas of the city;

Barcelona (Spain) proposed to transport material resources to cities at night. In particular, to operate two runs from 23.00 p.m. to 5.00 a.m., which will be equivalent to seven runs, performed during rush hours.

Rotterdam (the Netherlands) and Osaka (Japan) are encouraging transport companies to expand the practice of using hybrid and electric vehicles in truck designs, allowing them to operate in areas where it is forbidden to work with internal combustion engines.

Zurich (Switzerland) started operation of existing electric transport networks (tram) for garbage disposal transport.

Analysis of foreign and domestic scientific sources shows that one of the most effective forms of general implementation and strategic plans for urban development is the planning of a complex of measures for the transformation of individual urban territories, localities, and industrial infrastructure out of decline state into transformed renewed life cycle activities. It is so according to the theory of the life cycle of organizations. The purpose of revitalization is the removal of urban areas out of crisis and the modernization or provision of new functional purpose for them, e.g., for recreation, industrial tourism, cultural space [18, pp. 71–76, 19–21].

The peculiarity of urban planning is its participatory nature. The guiding principle of participatory planning is to enable stakeholders to participate actively in city planning. Social technologies of participation are of great importance. They are the following: the user group identification, social studies, workshops, aimed at strengthening the idea of mutual decision-making, and building overall responsibility for its adoption, because the city is being built and reconstructed for people. Local government's initiative and citizen activity can lead to comfortable and rational use of urban spaces.

According to foreign thought-leaders, 1% of the active community population is enough to initiate and launch change projects, including revitalization of social orientation [7, p. 131]. Thus, in Lodz the "bottom-up" initiative is being implemented by more than 80 social cooperatives and more than 120 non-governmental organizations. Bicycle infrastructure is modernized; regional cultural palaces and centers appear; projects for the protection of architectural monuments are launched; and discussions about the problems of the city are held thanks to their efforts. This is an important ally in revitalization projects, though it is often undervalued.

Successful practices of revitalization in Polish industrial cities have been effectively implemented in Silesia. It has gradually become clear that these are useful objects. Thousands of private owners of small objects have also changed their thoughts. They began to repair, adapt, and preserve them from destruction. It became even fashionable to live in a loft of the former factory. Such lofts appeared in Bytom, Gliwice, Gyrardow, Krakow, Lodz.

Polish city Lodz became the place of a huge urban experiment such as revitalization and reconstruction of entire neighborhoods. In the historical district of Old Polissia, reconstructions are subject to the streets. There are "Wuerfers" some pedestrian slow-motion zones, so-called (Dutch. *Woonerf* means "street for life") green islands. The project also includes a program to support local residents. It includes the organization of cultural and leisure centers for pedestrian accessibility, integration centers that will help to cope with addictions, and educational programs. The peculiarity of revitalization is their socialization; otherwise, the revitalization will be just next repair. For example, such transformation of the territories as the former power plant into the "New Business Center of Lodz" [19, p. 152].

In the Czech Republic during the industrial era, the city of Ostrava and its suburbs Bartowice were distinguished for the developed industrial complex. They were called "steel heart of the republic" with their mines, metallurgical, and coke plants. In order to avoid the collapse of the city's development and social tension, the concept of a cultural city was proposed as a concept of revitalization. Vitkovice Metallurgical Works, being the largest and the only full-cycle enterprise in Czechoslovakia, had mine in the territory, coke production, workshops for the production of iron, steel, rolled ferrous and non-ferrous metals, pipes, equipment for the metallurgical and chemical industries, employed 40 thousand employees. Further, it changed the concept of development from an industrial age monument into a space of tourist and investment development [20].

Thus, examples of urban architecture planning of foreign settlements indicate the adaptability to changes, the process of information and digitization, and the segregate nature of different functional zones planning.

The novelty of the work is the further development of a methodical approach to the planning methods way of the architecture of urban economic space, which allows integrating landscape, architectural and construction, financial and economic, marketing and logistics, and digital tools for their implementation.

3 Conclusions

In urban planning, the methods of solving the problems of architecture planning of urban economic space are the following: rational planning of development, limiting the growth of large cities, segregate distribution of urban areas into functional zones (residential, industrial, communal-warehouse, external transport, and suburban), and digitalization of urban space management.

References

1. Giffinger, R., & Stallbohm, M. (2014). Changes of metropolitan development: strategic efforts in comparison of Barcelona and Vienna. Retrieved from https://www.researchgate.net/publication/228600700_.
2. Monastyrs'kyj, H. L. (2010). Modernization paradigm of managing the economic development of the territorial base of the base: [monohrafiia] (p. 464). Ternopil', Ekonomichna dumka, TNEU.
3. Pleshkanovs'ka, A. M. (2009). *Demographic change your mind reconstruction* (pp. 345–355). Urban Construction and Area Planning, KNUBA, *33*.
4. Klekhovskyj, D. (2016). *Urban metamorphoses: New life of old industrial zones*. Retrieved from https://www.novayagazeta.ru/articles/2016/11/30/70726-uroki-polskogo.
5. Urban Agenda. (2016). Moskovskyj urbanystycheskyj forum. *6*, 227. Retrieved from http://blog.mosurbanforum.ru/futurecities.
6. Hrudynyn, M. Y. (2016). *Come to plan new places*. Retrieved from https://drive.google.com/file/d/0B3q2U4GLSfdddGZDZEZpN2Z6RU0/view
7. Lendry, Ch. (2005). *Creative city*. Per. s anhl (p. 399). Yzdatel'skyj dom «Klassyka-KhKh1».
8. New York City Department of Technology and Innovation. Retrieved from http://www1.nyc.gov/site/forward/index.page
9. Labour Market Research Study (2010). *Defining the green economy*. Labour Market Research Study. ECO Canada. Retrieved from http://www.eco.ca/pdf/Defining-the-Green-Economy-2010.pdf.
10. London Gardens Online. (2012). Retrieved from http://www.londongardensonline.org.uk/gardens-online-record.asp?ID=KAC055.
11. The Barcelona Agenda 21 indicators: Rogers, D., Tibben-Lembke, R. (2001). An examination of reverse logistics practices. Journal of Business Logistics, *22 (2)*. 129–145.
12. Vdovichena, O., Vdovichen, A., & Chychun, V. (2018). Managing the advertising activities in the system of integrated brand promotion of an enterprise. *European Research Studies Journal, XXI*(2), 124–136.
13. Krugman, P., Obstfeld, M. (2014). *MauriceInternational economics: Theory and policy*, Global Edition (Inglese) Copertina flessibile. 3 lug.
14. Law of Ukraine. (2011). About the regulation of public servicing. *Vidomosti Verkhovnoi Rady Ukrainy, 34,* 14.
15. Warehouse and Utility Plan for the Settlement (2012). Ministerstvo rehional'noho rozvytku, budivnytstva ta zhytlovo-komunal'noho hospodarstva Ukrainy. Pro zatverdzhennia DBN

B.1.1-15:2012 Nakaz 13.07.2012 № 358. Retrieved from https://zakon.rada.gov.ua/rada/show/v0358858-12?lang=ru
16. Herasymchuk, Z. V., & Nischyk, T. O. (2011). *Spatial development of the city: [monohrafiia]* (p. 212). LNTU: Luts'k.
17. Averkyna, M. F., & Herasymchuk, Z. V. (2012). Institutional protection of Green Logistics in the city. *Aktual'ni problemy ekonomiky, 11*(137), 161–169.
18. Gudz, P., Dawydenko, I., & Shykina, O. (2019). Support system of solutions for planning sales actavaties in the tourizm industry. *International Journal of Engineering and Advanced Technology (IJEAT), 8*(6), 3979–3983. https://doi.org/10.35940/ijeat.f9082.088619.
19. Pancewicz, Ł., & Zbieranek, P. (2013). *Pomoran cities—How to transform them for a general good* (p. 268). Wydawnictwo Uniwersytetu Gdańskiego: Gdańsk.
20. Baraniuk, A. (2014). *How the Czech Republic is trying to solve the problems of decaying industrial regions* (p. 55). Hazeta: «Prav. Da».
21. Varnaliy, Z., Onishchenko, S., & Masliy, A. (2016). Threat prevention mechanisms of ukraine's economic security. *Economic Annals-XXI, 159*(5–6), 20–24. https://doi.org/10.21003/ea.V159-04.

The Influence of Oxygen Regime on Aerotank-Displacer with Fixed Biocenosis Operation

Andriy Karahiaur, Tamara Airapetian, Valeriy Novokhatniy, and Oleksandr Matyash

Abstract The influence of oxygen regime on aerotank-displacer with fixed biocenosis operation using mathematical modeling is given. **Methods.** The mathematical model has been developed. It includes equations: for transfer concentrations of organic pollutants, oxygen, and activated sludge along the length of the aeration tank; for transfer concentrations of organic pollutants and oxygen over the biofilm's thickness; and kinetic equations (dependencies for determining substrate utilization rate by suspended and fixed biocenosis and the oxygen consumption rate in aeration tank and in biofilm). This model is implemented numerically. **Results**. Using the mathematical model, the influence of oxygen regime on the oxidation of substrate in free volume of aeration tank and in the biofilm was studied. The effect of aeration intensity, as well as the location and distribution density of elements with fixed biocenosis on the efficiency of aeration tank was also researched. **Scientific novelty**. A two-level mathematical model has been developed that takes into account simultaneous purification by activated sludge and biofilm, as well as the effect of oxygen content on the process. **Practical significance**. The research results allow us to calculate the rational parameters of the aerotank-displacer with fixed biocenosis.

Keywords Aerotank-displacer · Active sludge · Biofilm · Kinetics · Transfer · Modeling

1 Introduction

Biological removal organic contaminants (OC) in aerotank-displecers are common technique in domestic wastewater treatment. The discharge of insufficiently treated

A. Karahiaur
Kharkov National University of Civil Engineering and Architecture, Kharkiv, Ukraine
e-mail: vkg.knuca@ukr.net

T. Airapetian · O. Matyash (✉)
O.M. Beketov National University of Urban Economy in Kharkiv, Kharkiv, Ukraine
e-mail: 19831010@gmail.com

V. Novokhatniy
Poltava National Technical Yuri Kondratyuk University, Poltava, Ukraine

V. Onyshchenko et al. (eds.), *Proceedings of the 2nd International Conference on Building Innovations*, Lecture Notes in Civil Engineering 73,
https://doi.org/10.1007/978-3-030-42939-3_58

effluents into water bodies is one of the main reasons for their environmental degradation. This leads to stricter standards for purification quality. Often, functioning facilities for biological treatment of OC under aerobic conditions cannot provide the necessary effect. In this regard, there are new solutions and approaches to intensify the aeration tank. Among them are multi-stage biological treatment scheme [1], organization of a rational scheme of hydrodynamic flows [2, 3], creation areas with aerobic and anaerobic oxidation conditions [4], and using reagents [5]. Also, along with the removal of OC by activated sludge, purification by an immobilized biocenosis can be used [6–8]. In this case, the biofilm can be formed either on fixed additional elements, or on floating carriers [9]. Fixed biocenosis in comparison to activated sludge due to a higher concentration of microorganisms and a lower value of the half-saturation constant has a better extracting ability.

Mathematical modeling allows you to take into account many factors, and justify the rational parameters of the biological treatment process and necessary improvements [2, 9]. In addition to the kinetics of the process, in modeling, it is necessary to take into account the transfer of substrate and oxygen along the length of the aeration tank and thickness of biofilm. The oxygen content has a significant effect on the OC oxidation, not only in the biofilm [10] but also in aeration tank [11], which must be taken into account in mathematical modeling.

2 The Aim of Research

The aim of the research is to study, using mathematical modeling, the effect of oxygen regime (OR) on effectiveness using an additional charge with fixed biocenosis in aerotank-displacer for removing organic contaminants.

3 Methodology of Theoretical Studies for Removing OC in Aeration Tank with Fixed Biocenosis

In the process of mathematical modeling, which describes the biological treatment process in aerotank-displacer with fixed biocenosis, we made the following assumptions:

1. velocity of wastewater along the height and length in aeration tank is constant, and concentration of transfer substances along the height of the structure is neglected;
2. diffusion coefficients of the substrate, oxygen, and suspended biocenosis in the free volume of aeration tank have constant values;
3. concentration of microorganisms in a biofilm has a constant value in its thickness and in time; the thickness of biofilm also has a constant value in time;

4. biofilm is saturated by oxygen as a result of dissolved oxygen flow from the waste fluid; the interphase transfer of oxygen directly from air bubbles to biofilm is neglected.

Based on these assumptions, the basis mathematical model of aerotank-displacer with a suspended and fixed biocenosis operation and the consideration of OR in the aeration tank and biofilm are the following equations:

1. Organic contaminants transfer equation along the length of the aeration tank:

$$\frac{\partial L_{\mathrm{a}}}{\partial t} + V\frac{\partial L_{\mathrm{a}}}{\partial x} = D_{L_a}\frac{\partial^2 L_{\mathrm{a}}}{\partial x^2} - R_{\delta_L} - \varepsilon R_{L_{\mathrm{a}}}, \quad (1)$$

where x—horizontal coordinate, m, which varies from 0 to S (aerotank length); t—times, s; L_{a}—concentration of OC in aerotank, mg/l; $V = Q_{\mathrm{a}}/F$—average flow velocity in aerotank, m/s; F—cross-sectional area of aeration tank, m^2; Q_{a}—purified water consumption, m^3/s; $D_{L_{\mathrm{a}}}$—diffusion coefficient of OC in the free volume of aeration tank, m^2/s; $R_{\delta_L} = \lambda_\delta N_L$—OC utilization rate by fixed biocenosis in the aerotank volume, mg/(l s); $R_{L_{\mathrm{a}}}$—OC utilization rate by suspended biocenosis (activated sludge), mg/(l s); N_{Lbf} —OC flow through the biofilm surface for disposal by fixed biocenosis (biofilm), (m/s)/(mg/l); $\lambda_\delta = F_{\delta 1}/F$—design parameter, m^{-1}; $F_{\delta 1}$—specific surface area of the biofilm (per unit length of the aeration tank), m; $\varepsilon = 1 - W_\delta/W_{\mathrm{a}} = W_{\text{ж}}/W_{\mathrm{a}}$—coefficient that takes into account that the decrease in the free volume of the aeration tank is caused by the placement of an additional load (nozzles) with fixed biocenosis; W_{a}—aeration tank displacement, m^3; $W_{\text{ж}}$—fluid volume in aeration tank, m^3; W_δ—loading volume (nozzles) with fixed biocenosis, m^3.

Equation (1) is solved under the following boundary conditions:

- initial condition: $t = 0\ L_{\mathrm{a}} = L_{\mathrm{a}0}$;
- border conditions: $x = 0\ \ D_{L_{\mathrm{a}}}\frac{\partial L_{\mathrm{a}}}{\partial x} = V(L_{\mathrm{a}} - L_{\mathrm{a}0})\ \mathrm{x} = \mathrm{S}\ \frac{\partial L_{\mathrm{a}}}{\partial x} = 0$,

where $L_{\mathrm{a}0}$—the concentration of OC in the source water, mg/l.

2. Equation of the transfer organic contaminants by the thickness of biofilm (necessary for calculating the parameter N_L)

$$\frac{\partial L_{\mathrm{bf}}}{\partial t} = D_{L_{\mathrm{bf}}}\frac{\partial^2 L_{\mathrm{bf}}}{\partial y^2} - R_{L_{\mathrm{bf}}}, \quad (2)$$

where y—coordinate, m, which varies from 0 to δ (thickness of biofilm); L_{bf}—concentration of OC in biofilm, mg/l; $D_{L_{\mathrm{bf}}}$—coefficient of molecular diffusion of OC in biofilm, m^2/s; $R_{L_{\mathrm{bf}}}$—OC utilization rate by fixed biocenosis in biofilm, mg/(l s).

Equation (2) is solved under the following boundary conditions:

- initial condition: $t = 0\ L_{\mathrm{bf}} = 0$;

- border conditions: $y = \delta\ \frac{\partial L_{\text{bf}}}{\partial y} = 0$;

$$y = 0\, N_L = -D_{L_{\text{bf}}} \frac{\partial L_{\text{bf}}}{\partial y} = K_L (L_{\text{a}} - L_{\text{bf}}|_{y=0}),\ L_{\text{bf}}|_{y=0} = L_\delta \tag{3}$$

where K_L—the mass transfer coefficient of OC in liquid film, m/s; L_δ—OC concentration on biofilm surface, mg/l.

3. Equation of oxygen transfer along the length of aeration tank:

$$\frac{\partial C_{\text{a}}}{\partial t} + V \frac{\partial C_{\text{a}}}{\partial x} = D_{C_{\text{a}}} \frac{\partial^2 C_{\text{a}}}{\partial x^2} + \varepsilon \alpha K_{C_{\text{a}}} (\beta C_{\text{p}} - C_{\text{a}}) - R_{\delta_C} - \varepsilon R_{C_{\text{a}}}, \tag{4}$$

where C_{a}—oxygen concentration in free volume of aeration tank, mg/l; $D_{C_{\text{a}}}$—diffusion coefficient of oxygen in free volume of aeration tank, m^2/s; $R_{\delta_C} = \lambda_\delta N_C$—oxygen consumption rate of fixed biocenosis in aerotank volume, mg/(l s); N_C—oxygen flow through biofilm surface for oxidation of OC by fixed biocenosis (biofilm), (m/s)/(mg/l); $R_{C_{\text{a}}}$—oxygen consumption rate by suspended biocenosis (activated sludge), mg/(l s); C_{p}—equilibrium oxygen concentration, mg/l; α, β—coefficients that take into account the effect of impurities on the dissolution of oxygen in wastewater; $K_{C_{\text{a}}}$—volumetric coefficient of mass transfer oxygen from a bubble into water, s^{-1}.

Equation (4) is solved under the following boundary conditions:

- initial condition: $t = 0\ C_{\text{a}} = 0$;
- border conditions: $x = 0\ \ D_{C_{\text{a}}} \frac{\partial C_{\text{a}}}{\partial x} = 0 = V(C_{\text{a}} - C_{\text{a0}})$; $x = S\ \frac{\partial C_{\text{a}}}{\partial x} = 0$,

  where C_{a0}—concentration of active sludge in source water, mg/l.

4. Oxygen transfer equation for thickness of biofilm (necessary for calculating the parameter N_C):

$$\frac{\partial C_{\text{bf}}}{\partial t} = D_{C_{\text{bf}}} \frac{\partial^2 C_{\text{bf}}}{\partial y^2} - R_{C_{\text{bf}}} \tag{5}$$

where C_{bf}—oxygen concentration in biofilm, mg/l; $D_{C_{\text{bf}}}$—coefficient of molecular diffusion of oxygen in biofilm, m^2/s; $R_{C_{\text{bf}}}$—oxygen consumption rate of fixed biocenosis in biofilm, mg/(l s).

Equation (5) is solved under the following boundary conditions:

- initial condition: $t = 0\ C_{\text{bf}} = 0$;
- border conditions: $y = \delta\ \frac{\partial C_{\text{bf}}}{\partial y} = 0$;

$$y = 0\ N_C = -D_{C_{\text{bf}}} \frac{\partial C_{\text{bf}}}{\partial y} = K_C (C_{\text{a}} - C_{\text{bf}}|_{y=0}),\ C_{\text{bf}}|_{y=0} = C_\delta, \tag{6}$$

where K_C—mass transfer coefficient of oxygen in liquid film, m/s; C_δ—oxygen concentration on the biofilm surface, mg/l.

5. Kinetics equations (dependencies for determining the rate of utilization of OC by suspended and fixed biocenosis $R_{L_{\text{a}}}$ and $R_{L_{\text{bf}}}$, as well as oxygen consumption rates in the aeration tank $R_{C_{\text{a}}}$, and in the biofilm $R_{C_{\text{bf}}}$):

$$R_{L_{\text{a}}} = \frac{\mu_{\text{a}} X_{\text{a}}}{Y_{\text{a}}} \frac{L_{\text{a}}}{K_{\text{m}_{L_{\text{a}}}} + L_a} \frac{C_{\text{a}}}{K_{\text{m}_{C_{\text{a}}}} + C_{\text{a}}}, \tag{7}$$

$$R_{L_{\text{bf}}} = \frac{\mu_{\text{bf}} X_{\text{bf}}}{Y_{\text{bf}}} \frac{L_{\text{bf}}}{K_{\text{m}_{L_{\text{bf}}}} + L_{\text{bf}}} \frac{C_{\text{bf}}}{K_{\text{m}_{C_{\text{bf}}}} + C_{\text{bf}}}, \tag{8}$$

$$R_{C_{\text{a}}} = \alpha_{1\text{a}} R_{L_{\text{a}}} + \alpha_{2\text{a}} b_{\text{a}} \frac{C_{\text{a}}}{K_{\text{m}_{C_{\text{a}}}} + C_{\text{a}}} X_{\text{a}}, \tag{9}$$

$$R_{C_{\text{bf}}} = \alpha_{1\text{bf}} R_{L_{\text{bf}}} + \alpha_{2\text{bf}} b_{\text{bf}} \frac{C_{\text{bf}}}{K_{\text{m}_{C_{\text{bf}}}} + C_{\text{bf}}} X_{\text{bf}}, \tag{10}$$

where μ_{a}, μ_{bf}—maximum specific growth rates of microorganisms biomass, s^{-1}; $K_{\text{m}_{C_{\text{bf}}}}$, $K_{\text{m}_{L_{\text{bf}}}}$, $K_{\text{m}_{C_{\text{a}}}}$, $K_{\text{m}_{C_{\text{bf}}}}$half-saturation constants, mg/l; X_{a}, X_{bf}—microorganism concentrations, mg/l; Y_{a}, Y_{bf}—transformation ratio of substrate into biomass; $\alpha_{1\text{a}}$, $\alpha_{2\text{a}}$, $\alpha_{1\text{bf}}$, $\alpha_{2\text{bf}}$,—stoichiometric oxygen consumption coefficients during oxidation of oxygen unit and for self-oxidation products of death microorganisms; b_{a}, b_{bf}—microorganism dying constants, s^{-1}; a, bf—indices in the above notation related, respectively, to the free volume of aeration tank and biofilm.

6. Equation of activated sludge transfer along the length of aeration tank:

$$\frac{\partial X_{\text{a}}}{\partial t} + V \frac{\partial X_{\text{a}}}{\partial x} = D_x \frac{\partial^2 X_{\text{a}}}{\partial x^2} + \frac{\mu_{\text{a}} X_{\text{a}} L_{\text{a}}}{K_{\text{m}_{L_{\text{a}}}} + L_a} \frac{C_{\text{a}}}{K_{\text{m}_{C_{\text{a}}}} + C_{\text{a}}}, \tag{11}$$

where D_x—diffusion coefficient of activated sludge in free volume of the aeration tank, m^2/s.

Equation (11) is solved under the following boundary conditions:

- initial condition: $t = 0\ X_{\text{a}} = 0$;
- border conditions: $x = 0\ \ D_x \frac{\partial X_{\text{a}}}{\partial x} = V(X_{\text{a}} - X_{\text{a0}})$; $x = S\ \frac{\partial X_{\text{a}}}{\partial x} = 0$,

  where X_{a0}—concentration of active sludge in source water, mg/l.

7. The dependence for calculating the coefficient ε when the nozzles are located not along the entire length of the structure, but only in its part. In this case,

preservation of their volume and, accordingly, increase distribution density are taken into account.

$$\varepsilon = 1 - (1 - \varepsilon_0)\frac{S}{S_{\mathrm{bf}}}, \tag{12}$$

where ε_0—coefficient of reduction free volume of aeration tank with even distribution of nozzles along the entire length facility; S_{bf}—length of aeration tank section in which the nozzles are located, m.

The system of equations and dependencies (1)–(12), which forms basis of mathematical model, is solved numerically by finite difference method.

4 Results of Theoretical Studies for Removing OC in Aerotank with Fixed Biocenosis

We performed a partial check of adequacy of the developed mathematical model by comparing the numerical results with the analytical solution of Eq. (1). A particular option was considered when biological treatment is carried out in aeration tank of traditional design (only with suspended biocenosis) with sufficient oxygen supply to the process. In this case, Eq. (7) takes the form:

$$R_{L_{\mathrm{a}}} = \frac{\mu_{\mathrm{a}} X_{\mathrm{a}}}{Y_{\mathrm{a}}} \frac{L_{\mathrm{a}}}{K_{\mathrm{m}_{L_{\mathrm{a}}}} + L_{\mathrm{a}}},$$

If we neglect diffusion transfer, then Eq. (1) has the following solution:

$$L_{\mathrm{a}0} - L_{\mathrm{a}} - K_{\mathrm{m}_{L_{\mathrm{a}}}} \ln \frac{L_{\mathrm{a}}}{L_{\mathrm{a}0}} = \frac{\mu_{\mathrm{a}} X_{\mathrm{a}}}{Y_{\mathrm{a}} V} x. \tag{13}$$

Figure 1 presents the results of comparison.

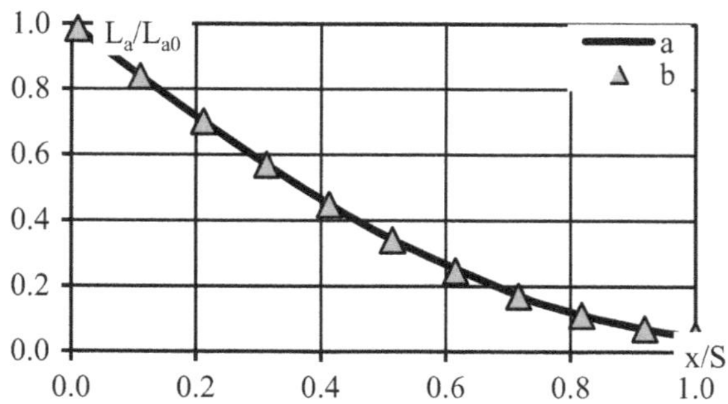

Fig. 1 Comparison of calculation results by Eq. (13) (**a**) with the results of numerical experiment (**b**)

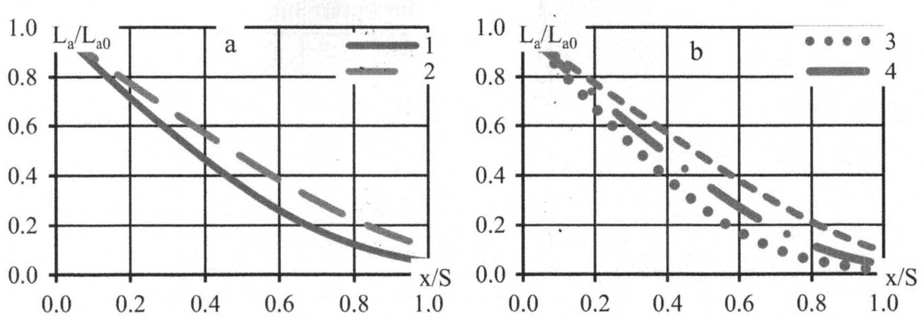

Fig. 2 Distribution of concentration of OC along the length of aeration tank: **a** traditional aeration tank; **b** with fixed biocenosis; 1—OR is not taken into account; 2—OR is taken into account; 3—OR is not taken into account either in the aeration tank or in the biofilm; 4—OR in the aeration tank is not taken into account, but in the biofilm it is taken into account; 5—OR is taken into account both in aeration tank and in biofilm

Using the developed mathematical model, the effect of OR on operation of both traditional aerotank-displacer and aerotank-displacer with fixed biocenosis was studied (Fig. 2).

Our studies have shown that OR has a significant impact on the quality of purification when using both biocenosis (suspended and fixed). As a result of calculations, distribution of oxygen concentration along the aerotank's length was obtained. These concentrations correspond to oxygen content in a real facility. When calculating the values of constants from equations, systems (1)–(12) were taken according to the recommendations

It is known that the volumetric coefficient of mass transfer oxygen from a bubble into water K_{C_a} depends on aeration intensity I ($m^3/(m^2 s)$ [12].

$$K_{C_a} = \frac{6}{d_{\text{bub}}} \frac{1}{V_f},$$

where d_{bub}—bubble diameter, m; V_f—bubble rise velocity, m/s.

In this regard, the effect of aeration intensity, as well as the reduced length of area with fixed S_{bf}/S biocenosis on the efficiency of structure was studied (Fig. 3). The option is considered when additional elements are located either along the entire length of aeration tank, or in its second part along the fluid flow. As criterion of efficiency accepted parameter E.

$$E = \frac{[L_a(S)]_a}{[L_a(S)]_{bf}},$$

where $[L_a(S)]_{bf}$—concentration of OC at the end of aerotank-displacer with fixed biocenosis, mg/l; $[L_a(S)]_a$—concentration of OC at the end of traditional aerotank-displacer, mg/l.

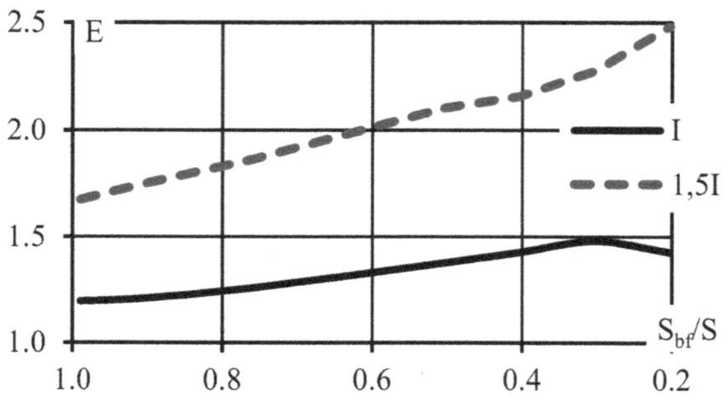

Fig. 3 Effect of aeration intensity, location and distribution density of elements with fixed biocenosis on the efficiency of aeration tank

The efficiency of using nozzles with fixed biocenosis increases with their denser arrangement, but up to a certain point. Then the efficiency decreases. This can be explained by the fact that oxygen saturation of biofilm decreases, the influence of its content becomes more and more limiting. With increasing aeration intensity, the extreme value of the parameter E shifts to the right, and the efficiency of the fixed biocenosis increases.

5 Scientific Novelty

Scientific novelty of the presented studies is two-level mathematical model of process removing organic contaminants in aerotank-displacer, which contains additional elements (fixed biocenosis). The model takes into account simultaneous purification with activated sludge and biofilm, as well as the effect of oxygen content on the process.

6 Practical Significance

The practical significance of the presented studies lies in the fact that the developed mathematical model allows calculating the rational parameters of aerotank-displacer with fixed biocenosis, at which the efficiency of removing organic containments by biological treatment will be maximum.

7 Conclusion

The two-level mathematical model has been developed. The model takes into account the simultaneous wastewater treatment from organic contaminants by activated sludge and biofilm. This allowed us to study the influence oxygen regime of aeration intensity, as well as location of additional elements on the efficiency of aeration tank with fixed biocenosis.

References

1. Shamsutdinova, Z. R., & Khafizoy, I. I. (2016). Analysis of the aerotanks efficiency in wastewater treatment system. In *Proceedings of the Voronezh State University of Engineering Technologies* (Vol. 4, pp. 245–249). https://doi.org/10.20914/2310-1202-2016-4-245-249.
2. Biliaev, M. M., & Lemesh, M. V. (2018). Modeling of biological wastewater treatment on the basis of quick-computing numerical model. *Science and Transport Progress. Bulletin of Dnipropetrovsk National University of Railway Transport, 1*(73), 15–23. https://doi.org/10.15802/stp2018/124882.
3. Gornostal', S. A. (2013). *Improving the efficiency of biological wastewater treatment plants* (Ph.D. dissertation).
4. Grycyna, O. O. (2013). *Improvement of calculation method and constructions of aerotanks with anoxic and aerobic zones* (Ph.D. dissertation).
5. Kobeleva, J. V. (2017). *Biological treatment of municipal wastewater with the use of reagents* (Ph.D. dissertation).
6. Gebara, F. (1999). Activated sludge biofilm waste water treatment system. *Water Research, 13*(1), 230–238.
7. Marzec, M. (2017). Reliability of removal of selected pollutants in different technological solutions of household wastewater treatment plants. *Journal of Water and Land Development., 35*(1), 141–148. https://doi.org/10.1515/jwld-2017-0078.
8. Revilla, M., Galán, B., & Viduri, J. R. (2018). Analysis of simulation tools and optimization of the operational conditions for biofilm activated sludge industrial process. *International Journal of Environmental Science and Technology, 15*(12), 2499–2510. https://doi.org/10.1007/s13762-017-1626-2.
9. Shreve, M. J., & Brennan, R. A. (2019). Trace organic contaminant removal in six full-scale integrated fixed-film activated sludge (IFAS) systems treating municipal wastewater. *Water Research, 151,* 318–331. https://doi.org/10.1016/j.watres.2018.12.042.
10. Airapetian, T. S. (2017). Determination of rational parameters of aerotanks-displacers with fixed biocenosis and taking into account the oxygen regime. *Problems of Water Supply, Sewerage and Hydraulics, 28,* 12–18.
11. Henze, M. M., Van Loosdrecht, M. C., Ekama, G. A., & Brdjanovic, D. (2008). *Biological wastewater treatment.* London: IWA Publishing.
12. Oleynik, A., & Airapetian, T. (2019). Practical recommendations to oxygen calculation modes for biological strain water treatment in aerothenes with closed and referring bioecenosis. *Journal Municipal Economy of Cities, 1*(147), 175–180. https://doi.org/10.33042/2522-1809-2019-1-147-175-180.

The Management of Organizational Processes of the Transport Use in Construction

O. V. Komelina, Iu. V. Samoilyk, L. M. Boldyrieva, and V. V. Krapkina

Abstract Classification of vehicles in construction has been proposed with such classification features as future of goods transportation, direction of movement loads, specialized vehicles and intensity of work freight transport. Scientific approaches to the management of transport processes in construction have been further developed. Factors affecting the management efficiency of transport processes in construction have been identified. These factors include delivery time; frequency of departure of cargo; safety; reliability of compliance with the delivery schedule; conductive ability; admissibility in geographical terms; cost of transportation (cost) and resource efficiency. It has been proved that the total cost of goods delivery from sender to recipient includes the cost for goods delivery (cost per 1 ton/km); costs of goods loading and unloading and costs of goods safety on the road. It has been proved that automation of construction is carried out by means of specialized machines on loading and unloading and transport-warehouse works. Road transport efficiency is greatly enhanced when using semi-trailers and trailers.

Keywords Construction · Construction and installation work · Organizational processes · Transport · Management

1 Introduction

Construction and installation works involves moving a large amount of construction materials, semi-finished products and finished products, varying in size and weight. Therefore, execution of construction and installation works is inextricably linked to use of different types of vehicles, which are also the link between construction sites

O. V. Komelina (✉) · L. M. Boldyrieva
Poltava National Technical Yuri Kondratyuk University, Poltava, Ukraine
e-mail: Komelinaolha@gmail.com

Iu. V. Samoilyk
Poltava State Agrarian Academy, Poltava, Ukraine

V. V. Krapkina
Kyiv-Mohyla Academy Poltava, Kiev, Ukraine

V. Onyshchenko et al. (eds.), *Proceedings of the 2nd International Conference on Building Innovations*, Lecture Notes in Civil Engineering 73,
https://doi.org/10.1007/978-3-030-42939-3_59

and suppliers of materials, structures, parts and equipment. Complex mechanization of transport process (loading—transportation—unloading) requires special attention.

Choice of transport means is made taking into account a number of factors. There are types of cargo, conditions of loading and unloading works, distance of transportation, road-transport conditions and volume of transport services performed. In the system of organizational processes management of transport in construction, it has been possible to distinguish problems of inefficient use of vehicles. The first problem is imperfects organizational forms (idle workers due to late delivery of materials make up about 8% of the work shift). The second problem is outdated specialized rolling stock (technical support).

Effectiveness of managing organizational processes of transport using is directly related to specifics of material flows inherent in each project of a construction company.

2 An Overview of the Latest Sources of Research and Publications

The purpose of scientific research is summarizing scientific approaches to managing organizational processes of transport using in construction and developing practical recommendations for its effective use. Theoretical and methodological bases of scientific research have been general scientific methods of research. There is theoretical generalization, system analysis and synthesis, abstraction and formalization.

Questions of transport processes management in construction have been widely researched by scientists. The research by Andersson and Nilsson [1] deserve considerable attention. These scholars focus their research on the issues of planning logistics processes in construction. The authors consider the issues of the construction industry development and identification of vectors for improving logistics in this field relevant. Also, the scientists explore the chain of goods movement in construction, formulating the main tasks of logistics at this stage.

Topicality of planning logistics in construction is discussed in the works by Kalsaas et al. [2], Dubois et al. [3] and Kenny [4]. Scientists emphasize the need to improve methods of transport logistics management in construction.

Sobotka and Czarnigowska [5] unite elements of construction logistics planning system in five groups. The first group is its initiating (preparation of logistical guidelines for the construction site, strategic planning for the project management). The second group is its design (preparation of logistical guidelines for design; preparation of requirements specification; logistical guidelines for tender preparation; quality system of logistical service). The third group is planning (preparations of schedules and charts of labor and equipment; utilization, sub-contractors work and material consumption; preparations of logistics concept of the construction site; design of the installations on site and how to disassemble these; preparations of guidelines for lease or purchase of machines; preparations of suppliers selection; assessment of logistical service efficiency and impact on the environment; preparation of planning and placing orders, scheduling and deliveries; waste management planning). The fourth group is execution (work progress monitoring; schedules and plans updating; adjusting orders to current demand for resources, workforce, materials and sub-contractors; planning and coordinating deliveries, loading, un-loading and warehousing, distributing; implementing logistic service quality standards; managing waste). The fifth group is commissioning (dismantling of site installation; managing information flows and documentation) [5].

The relevance of research into the development of transport logistics in the cities construction can be seen in the works by Connekt [6], Balm et al. [7], Ekeskär and Rudberg [8], Fang and Ng [9]. In the works by Janné et al. [10] and Connekt [6], thesis about the relevance of transport logistics development in smart cities can be seen. Scientists consider that "the growth in urban population and economic upturn is leading to higher demand for construction, repair and renovation works in cities. Houses, public utilities, retail spaces, offices and infrastructure need to be adapted to cope with increasing number of residents and visitors, urban functions and changing standards. Construction projects contribute to more attractive, sustainable and economically viable urban areas once they are finished. However, transport activities related to construction works have negative impact on surrounding community if not handled appropriately. It is estimated that 15–20% of heavy goods vehicles in cities are related to construction, and 30–40% of light commercial vans [6, 10]. Scientists [10] point out that "smarter, cleaner and safer construction logistics solutions in urban areas are needed for environmental, societal and economic reasons. However, in many European cities and metropolitan areas the sense of urgency is not evident or a lack of knowledge is creating passivity" [10].

Despite considerable interest of scientists to the problem of transport logistics in construction, the issues of managing transportation processes in construction need further research.

3 Main Body

Choosing a type of transport is one of the construction logistics main tasks. Results of construction and installation works depend on its effective implementation. Efficiency of transport operations in construction depends to a large extent on the method of their implementation and the factors that influence choice of transport kind: delivery time, frequency of consignments, reliability of adherence to the delivery schedule, ability to carry different goods, ability to deliver cargo to any what is the point of the territory, cost of transportation.

Decision to transport building materials and equipment is made taking into account a number of factors of objective and subjective nature. These factors include economic and legal factors, natural-geographical conditions, versatility or specialization of vehicles, development of economic relations with partners involved in the formation of the supply chain, level of environmental impact, ecological security, etc.

Economic factors included level of transportation costs in construction, time of delivery and the safety of the goods transported. Natural and geographical conditions make it possible to carry out any type of communication, taking into account the location of suppliers and consumers. Versatility or specialization of vehicles used to deliver goods is traced through the following indicators:

- for rail transport: mass transportation of goods over long distances (more than 600 km); continuity and uniformity of transportation under any suitable conditions; speed of cargoes delivery (speed of movement);
- for road transport: high maneuverability; direct delivery of the goods (freight from the shipper to the consignee is delivered without overloading to other modes of transport); low cost of cargo transportation for short distances; a greater variety of rolling stock types;
- for air transport: the highest speed of delivery.

So, for example, if it is necessary to quickly deliver construction materials, it is necessary choose air or road transport. The choice of a particular type of transport for construction is interrelated with other logistics tasks (creation and maintenance of an optimal stocks level).

The types of transportation are also taken into account:

- unimodal (delivery of goods by only one mode of transport);
- intermodal (the system of cargo delivery according to an agreed scheme involving several modes of transport, as a rule, under a single contract of carriage for the entire route of passage).

Minimization of total costs is the main criterion for evaluating effectiveness of managing the organizational processes of transport using. The value of this indicator is result of material flows specifics between the customer (construction organization), supplier (suppliers), the contractor (carriers) during implementation of certain projects at the construction company. Achievement of the optimum value of the management efficiency criterion of organizational processes of transport using in construction depends on a number of other indicators and limitations.

For example, a lot of criteria affect the choice of transport types, the main ones being:

- delivery time;
- frequency of departure of cargo;
- security;
- reliability of compliance with the delivery schedule;
- capacity;
- admissibility in geographical terms;
- cost of transportation, resource savings.

Transportation process of a truck consists of the following elements: preparation of cargo for transportation, loading on rolling stock, removing of rolling stock with cargo from the point of departure to the unloading point, unloading and delivery of cargo, moving of rolling stock under next loading.

Evaluation and analysis of the rolling stock, separately of each of its units and the park as a whole is carried out by means of a technical and operational indicators system characterizing the quantity and quality of the work.

The main task of transport logistics in construction is profits increasing. Comprehensive indicator of the freight wagons using (the distance of the cargo movement) is its productivity. Modern wagon park in Ukraine consists mainly of 4-axle wagons. The average carrying capacity of such a cargo train is 62.4 tones. To ensure transportation of certain cargo types, eight-axle wagons with increased capacity (up to 180 tones) are used.

With reduction of the freight transportation distance, coefficient of empty movement of wagons (before loading) increases because of the probability of using unloaded wagons for loading decreases.

It is necessary to use formula 1 to determine the cost of transporting cargo:

$$\mathrm{TTC} = C_{\mathrm{delivery}} + C_{\mathrm{loading+unloading}} + C_{\mathrm{goods\ safety}} \tag{1}$$

where TTC is the total transport costs of cargo delivery from the sender to the recipient;

C_{delivery} is the costs for goods delivery (cost per 1 ton/km);

$C_{\mathrm{loading+unloading}}$ is the costs of the goods loading and unloading;

$C_{\mathrm{goods\ safety}}$ is the costs of the goods safety on the road.

Thus, calculation of the transport costs per 1 ton (Formula 1) takes into account all the main cost elements of the shipping cargo from producer to consumer.

Classification of vehicles in construction is given in Table 1.

Generalization and systematization of classification features on the use of vehicles in construction (Table 1) can serve as a basis for solving organizational and economic optimization problems of transport logistics in construction in order to increase the effectiveness of specific project implementation in a construction company.

Automation of construction is carried out by means of specialized machines on loading and unloading and transport-warehouse works. Road transport efficiency is greatly enhanced when using semi-trailers and trailers. Special heavy-duty semi-trailers are used for transportation of oversized heavy-duty heavy goods and heavy vehicles on solid-surface roads and improved dirt roads. Traffic support team helps to organize transportation in difficult pass areas.

The effectiveness of managing the organizational processes of transport using in the construction of enterprises can be evaluated by the method of value approach with the allocation of strategic and tactical value of the logistics chain, taking into account the alternative of choosing transport organizations for transportation [11]. These organizations can in turn be evaluated on such criteria as "price", "quality", "delivery time", "company reputation" and their respective combinations.

Table 1 Classification of vehicles in construction

Classification features	Types
Futures of goods transportation	• External • Internal
Direction of movement loads	• Vertical and sometimes inclined, intended for lifting loads (cranes, lifts) • Horizontal, intended for movement of building materials, structures, parts of equipment from places of extraction or preparation to places of use: rail transport [normal track (1524 mm)] and narrow (600 and 750 mm); railless transport (road, tractor); water and air transport (steamers, helicopters); special transport (transporters, cableways, cable cranes, etc.); pipeline transport
Specialized vehicles	• Autoconcrete trucks (transportation of concrete mixtures in the body of the mold-like form, provided with a lid and heating) • Auto for solute transportation provides transportation of mortars with mechanical impulse inside and mechanism of portion delivery of a solution • Concrete mixers are used to transport concrete over long distances and to prepare the mixture on the road • Auto for cement transportation equipped with a device for loading and unloading of cement, gypsum, lime, dry ash, ground powder and consist of a tractor with a tanker-semi-trailer, which is located with a slope of 69° in the direction of unloading • Auto for plate transportation and auto are used for beam transportation used for transportation of plates, beams, columns, piles, etc. • Panel transporters are used for transportation of wall panels and have special adaptations for fixing of structures in an upright position • Farm trucks are used for farms transporting and other products that require transportation in working position • Trailers for transportation of heavy loads and cars • Auto for bitumen transportation are used for transportation of bitumen at temperatures up to 200 °C from equipment for the production of bitumen to warehouses or bases of consumption and consist of a car tractor and semi-trailer with a system of heating and pumping of bitumen • Auto for container transportation are used for delivery of small pieces and tare packing goods
The intensity of work is characterized by the load flow (the number of loads that move through a certain section of the transport network over a specified period of time (days, months, quarter, years)	• External goods flows are carried out centralized according to contracts, by all types of transport from the storage of suppliers to warehouses of construction and assembly enterprises • Internally goods flows are constructed of intermediate warehouses

The implementation of this approach to achieve flexibility in project management and to harmonize the stakeholder relationships that arise during its implementation is an important result [12, 13].

4 Conclusion

The complication of managing organizational processes of transport used in construction requires systematic approach, taking into account the totality of organizational, production and economic aspects has been investigated. It has been substantiated that organizational and economic optimization tasks of transport logistics in construction are multi-criteria in content, and the proposed approaches to solving them allow to evaluate the effectiveness of managing organizational processes of transport using in the construction enterprises. With this in mind, classification features and criteria for evaluating the efficiency of vehicles used in construction have been developed, as well as the expediency of widespread use of a value approach with highlighting the strategic and tactical value of the logistics chain when selecting transport organizations for transportation. Practical application of these developments ensures the optimum result from the point of view of ultimate consumer of construction products, enables to reduce inventories, reduce risks in construction, to ensure optimality of material, financial and information flows, and thereby to ensure high efficiency of the entire investment and construction cycle.

References

1. Andersson, O., Nilsson, A. (2018) *Planning for Construction Logistics an evaluation and development of a construction logistics plan at Serneke* (Master's thesis in the Master's Programme Design and Construction Project Management). Retrieved from https://hdl.handle.net/20.500.12380/255901.
2. Kalsaas, B. T., Skaar, J., & Thorstensen, R. T. (2015). Pull versus push in construction work informed by last planner.
3. Dubois, A., Hulthén, K., & Sundquist, V. (2019). Organising logistics and transport activities in construction. *International Journal of Logistics Management, The, 30*(2), 620–640. https://doi.org/10.1108/IJLM-12-2017-0325.
4. Kenny, C. (2009). Transport construction, corruption and developing countries. *Transport Reviews, 29*(1), 21–41. https://doi.org/10.1080/01441640802075760.
5. Sobotka, A., & Czarnigowska, A. (2005). Analysis of supply system models for planning construction project logistics. *Journal of Civil Engineering and Management, 11*(1), 73–82. https://doi.org/10.1080/13923730.2005.9636335.
6. Connekt. (2017). *Outlook city logistics.* Topsector Logistiek.
7. Balm, S., Spoelstra, J., & Quak, H. (2015). Applying a behavioural change model to the adoption of freight electric vehicles: Lessons for effective instruments. In *URBE Conference*, Rome, Italy.
8. Ekeskär, A., & Rudberg, M. (2016). Third-party logistics in construction: the case of a large hospital project. *Construction Management and Economics, 34,* 174–191.
9. Fang, Y., & Ng, S. T. (2011). Applying activity based costing approach for construction logistics cost analysis. *Construction Innovation, 11,* 259–281.
10. Janné, M., Fredriksson, A., Berden, M., van Amstel, W. P (2018). *Smart construction logistics.* Retrieved from https://www.researchgate.net/publication/327281275.
11. Popovichenko. (2011). Logistics as means of survival of building enterprise in current economic conditions. *Economic Journal-XXI, 3–4,* 55–57.

12. Komelina, O. V., Hrynko, O. V., Khrystenko, O. V. (2018). Stakeholder management in the development of building organizations. *International Journal of Engineering & Technology, 7*(3.2), 191–194.
13. Varnaliy, Z., Onishchenko, S., & Masliy, A. (2016). Threat prevention mechanisms of Ukraine's economic security. *Economic Annals-XXI, 159*(5–6), 20–24. https://doi.org/10.21003/ea.V159-04.

Methodical Approach to Optimization of Housing Cost in the Housing Market of Ukraine

O. V. Komelina, L. H. Shcherbinin, S. A. Shcherbinina, and B. M. Ivanyuk

Abstract In this article, for the first time, a system of econometric models is developed. The system reflects the dependence of the mediate cost for 1 m^2 of housing in the Ukraine housing market on the demand (population income) and supply (volumes of commissioned housing and manufactured construction products), as well as the dependence of demand and suggestions from factors of influence. On the ground of the developed econometric models system, an optimization function of the minimum mediate cost for 1 m^2 of housing was obtained, as well as the estimated values of supply and demand. According to the results of the calculations, the obtained predicted value is compared with the minimum, which indicates a reserve for improving the efficiency of the budget funds usage, which is directed to the construction or purchase of housing for some privileged categories of citizens. This research methodology can be used to determine the minimum mediate cost for 1 m^2 of affordable housing at the level of Ukrainian regions.

Keywords Housing market · Mediation cost for 1 m^2 of housing · Econometric models · Optimization function · Decomposition

1 Introduction

The housing construction market reflects the problems of Ukraine's current economic situation. The pace of new housing construction is declining, but demand for it is rising, and so are the prices for new buildings. The housing market is a promising area for investing. It is important not only to regulate the demand and supply in the housing market, but also to solve the problem of providing affordable housing for citizens. As the developments in the housing market are quite dynamic, taking into account the factors affecting the cost of housing, there is a need to develop a methodological approach that allows to determine the optimal mediate cost for 1 m^2 of housing in order to attract and use public funds as a major investor, and at the

O. V. Komelina · L. H. Shcherbinin · S. A. Shcherbinina (✉) · B. M. Ivanyuk
Poltava National Technical Yuri Kondratyuk University, Poltava, Ukraine
e-mail: scherbininasveta@gmail.com

V. Onyshchenko et al. (eds.), *Proceedings of the 2nd International Conference on Building Innovations*, Lecture Notes in Civil Engineering 73,
https://doi.org/10.1007/978-3-030-42939-3_60

same time it strengthens the financial system and helps to solve a number of social problems.

2 An Overview of Recent Research Sources and Publications

Many foreign and domestic scientists have considered problems of mathematical modeling methods' application for the analysis of supply and demand in the housing market. The behavior of buyers, sellers and the pricing mechanism of the housing market with the usage of the simulation modeling (ABM) method was investigated by Pangallo et al. [1]. The correlation between housing costs rising and rising of per capita income was discussed in researches of Kulkarni et al. [2]. Zheng et al. [3] studied factors which influence the real demand of homebuyers (households) in the housing market and the cost of housing. An important factor explaining the dynamics of housing prices, in their view, is the speculative behavior of investors. Napoli et al. [4] treated the issues of housing affordability, the particularities of housing problems, the analysis of wealth distribution both in terms of income and market value of residential buildings and the calculation of threshold income (based on the HAI housing affordability index) that filter access to the housing market [4]. Modeling the price change for 1 m^2 of housing in the housing market of Ukraine is presented in the research of Galchinsky et al. [5]. Yu. V. Kalynichenko, V. V. Dobrovolskaya, A. R. Abramchuk, researched principles of tendencies modeling methodology in the real estate market [6]. V. V. Lagovsky and T. M. Soldering used econometric methods for modeling the dynamics of the construction industry development [7]. The method of construction for economic and mathematical models of price dynamics in the Kyiv housing market is given in the research of Shaposhnikov [8].

3 Purpose of the Article

The main purpose of the work is developing and implementation of methodological approach for optimization of the housing cost, in particular, forecasting and finding the optimal mediate cost for 1 m^2 of housing, taking into account the demand and supply in the Ukraine housing market.

4 Methodical Approach to Optimization of House Building Cost

4.1 Theoretical Aspects

First of all, the factors affecting the productive feature (mediate cost for 1 m^2 of housing) were identified. The tightness and nature of the various factors on the result and their correlation were detected using the mathematical methods. Multiple regression procedures were used in the research. A scientific approach, such as decomposition, was used. The approach allows considering any investigated system as complex, including separate interconnected subsystems, within which lower-level subsystems can be distinguished [9]. A system of econometric models has been constructed to obtain the optimization function of the cost for housing building in Ukraine (Fig. 1).

4.2 Modeling Outcomes

To calculate the optimal mediate cost for 1 m^2 of housing in the Ukraine housing market, data from the State Statistics Service of Ukraine for the period from 2010 to 2018 were used [10]. STATISTICA 8.0 software application package was used to develop and implement econometric models. It is a multiple-purpose integrated system intended for statistical analysis and data processing [11]. The modeling outcomes are presented in Fig. 2. The constructed system of econometric models establishes the correlation between the cost for 1 m^2 of housing, household income, the volumes of putting in operation of housing and the volume of produced construction products.

Using the coefficient of multiple correlation, *F*-statistics and student's *t*-test, the parameters of the constructed models of multiple linear regression were estimated as shown in Table 1. The degree of values scattering observed relatively to the regression line, that is, the standard error of estimation is determined. This value is used to construct a border set of projected values.

According to the results of the evaluation, it is possible to make a conclusion about the adequacy of the constructed models with usage of the experimental data and possibility of its usage for analysis and forecasting. The optimization function of the housing cost in the Ukrainian housing market was calculated. According to the results of calculations, the minimum mediate cost for 1 m^2 of housing is 10,200 UAH/m^2 as shown in Fig. 3.

5 Scientific Novelty

A system of econometric models has been developed. The system reflects the dependence of the mediate cost for 1 m^2 of housing on demand (population income) and

The first stage

Formation of dependence of mediate cost model for 1m^2 of housing from demand and supply in the Ukraine's housing construction market $Y_1 = f(Y_2, Y_3, Y_4)$

$$Y_1 = \delta_0 + \delta_1 Y_2 + \delta_2 Y_3 + \delta_3 Y_4$$

Y_2 – household income, billion UAH.

Y_3 – putting in operation of housing, ths. m^2

Y_4 – volume of produced construction products, million UAH

The second stage

Formation, parameterization and models verification of the household income, dependence level, supply and demand in the housing market in the region on the indicators: x_1, x_2, x_3, x_4; s_1, s_2, s_3; z_1, z_2, z_3.

$$\begin{cases} Y_2 = b_0 + b_1 x_1 + b_2 x_2 + b_3 x_3 + b_4 x_4; \\ Y_3 = c_0 + c_1 s_1 + c_2 s_2 + c_3 s_3; \\ Y_4 = a_0 + a_1 z_1 + a_2 z_2 + a_3 z_3. \end{cases}$$

x_1 – average monthly nominal wage, UAH.

x_2 – Household deposits, million UAH.

x_3 – total resources per household in a month on average, UAH.

x_4 – loans to households, millions UAH.

s_1 – funds of population for housing construction, million UAH.

s_2 – current assets of construction industry (at year end), million UAH.

s_3 – other taxes related to construction, million UAH.

z_1 – investments in fixed capital of construction, million UAH.

z_2 – commissioning of new major construction methods, million UAH

z_3 – remuneration of employees in construction, million UAH

The third stage

An optimization function formation of mediate cost for 1m^2 of housing in the Ukraine's housing market.

$$Y_1 = \delta_0 + \delta_1 [b_0 + b_1 x_1 + b_2 x_2 + b_3 x_3 + b_4 x_4] + \delta_2 [c_0 + c_1 s_1 + c_2 s_2 + c_3 s_3] + \\ + \delta_3 [a_0 + a_1 z_1 + a_2 z_2 + a_3 z_3] \rightarrow \min$$

Limitation: $xi \in [x_i^1; x_i^2]$, $s_i \in [s_i^1; s_i^2]$, $z_i \in [z_i^1; z_i^2]$,

$x_i^1 = \min\{x_i\}, x_i^2 = \max\{x_i\}; s_i^1 = \min\{s_i\}, s_i^2 = \max\{s_i\}, z_i^1 = \min\{z_i\}, z_i^2 = \max\{z_i\}$.

Fig. 1 Methodical approach to optimization of mediate cost for 1 m^2 of housing in the Ukraine's housing market

The first stage

Modeling results of dependence of mediate cost for 1m^2 of housing from demand and supply in the Ukraine's housing construction market: $Y_1 = \delta_0 + \delta_1 Y_2 + \delta_2 Y_3 + \delta_3 Y_4$

Multiple linear regression equation is obtained:
$Y_1 = 3758{,}46 - 0{,}999 \cdot Y_2 - 0{,}137 \cdot Y_3 + 0{,}441 \cdot Y_4$
The parameters of the multiple linear regression equation were estimated: the multiple correlation coefficient R =0,987; By F-test F_t (3,5) < F_r (64,97). According to the Student's t-test, the obtained coefficient estimates are statistically significant. The standard error of estimation is equal to 555,88. Therefore, obtained mathematical model is adequate to the experimental data.

The prediction of the productive feature was obtained:
$Y_1 = 3758{,}46 - 0{,}999 \cdot 3300 - 0{,}137 \cdot 9000 + 0{,}441 \cdot 29400 = 12188{,}4$
The mediate cost for 1m^2 of housing is 12188,4 UAH/m^2 with a 95% confidence interval (10739.30; 13637.51).

The second stage

Economic-mathematical models of dependence of Y_2, Y_3, Y_4 on the factors of influence are constructed:

$$\begin{cases} Y_2 = 72{,}11 + 0{,}042 \cdot x_1 + 0{,}001 \cdot x_2 + 0{,}24 \cdot x_3 - 0{,}003 \cdot x_4; \\ Y_3 = 518{,}77 + 0{,}091 \cdot s_1 + 0{,}039 \cdot s_2 - 3{,}246 \cdot s_3; \\ Y_4 = -16340{,}0 + 0{,}5 \cdot z_1 - 0{,}9 \cdot z_2 + 0{,}6 \cdot z_3. \end{cases}$$

The parameters of multiple linear regression equations are estimated, all models are adequate to the experimental data.
Projected values of the successful indications were obtained:

$$\begin{cases} Y_2 = 72{,}11 + 0{,}042 \cdot 11000 + 0{,}001 \cdot 519700 + 0{,}24 \cdot 10000 - 0{,}003 \cdot 1101900 = 3264{,}29; \\ Y_3 = 518{,}77 + 0{,}091 \cdot 34500 + 0{,}039 \cdot 270750{,}2 - 3{,}246 \cdot 1100 = 10705{,}44; \\ Y_4 = -16340{,}0 + 0{,}5 \cdot 56000 - 0{,}9 \cdot 3500 + 0{,}6 \cdot 33000 = 28376{,}8. \end{cases}$$

The level of the population income is UAH 3264,29 billion with 95% of confidence interval (3143.36; 3385.23).
The volume of housing commissioning is 10705,44 thousand m^2 with 95% of confidence interval (9461,69; 11949,20).
The volume of produced construction products is UAH 28376,8 million with 95% of confidence interval (23935,7,69; 32818,0).

The third stage

Optimization functions of mediate cost for 1m^2 of housing in the Ukraine's housing market are constructed, taking into account the limitations:

$$Y_1 = 3758{,}46 - 0{,}999 \cdot [72{,}11 + 0{,}042 \cdot 2239 + 0{,}001 \cdot 275093 + 0{,}24 \cdot 3481 - 0{,}003 \cdot 732823] -$$
$$- 0{,}137 \cdot [518{,}77 + 0{,}091 \cdot 17589{,}2 + 0{,}039 \cdot 94487{,}1 - 3{,}246 \cdot 507] +$$
$$+ 0{,}441 \cdot [-16340{,}0 + 0{,}5 \cdot 29767 - 0{,}9 \cdot 99634{,}3 + 0{,}6 \cdot 18028] = 10200 UAH / m^2$$

Fig. 2 Results of the optimization of mediate cost for 1 m^2 of housing in the Ukraine's housing market

Table 1 Summary statistics

Mathematical model	R	F_r	F_t	t	Std. error
$Y_1 = \delta_0 + \delta_1 Y_2 + \delta_2 Y_3 + \delta_3 Y_4$	0.987	64.97	3.5	1.464	555.88
$Y_2 = b_0 + b_1 x_1 + b_2 x_2 + b_3 x_3 + b_4 x$	0.999	751.76	4.4	0.49	35.626
$Y_3 = c_0 + c_1 s_1 + c_2 s_2 + c_3 s_3$	0.975	31.5	3.5	5.457	787.92
$Y_4 = a_0 + a_1 z_1 + a_2 z_2 + a_3 z_3$	0.974	30.7	3.5	−3.567	2231.3

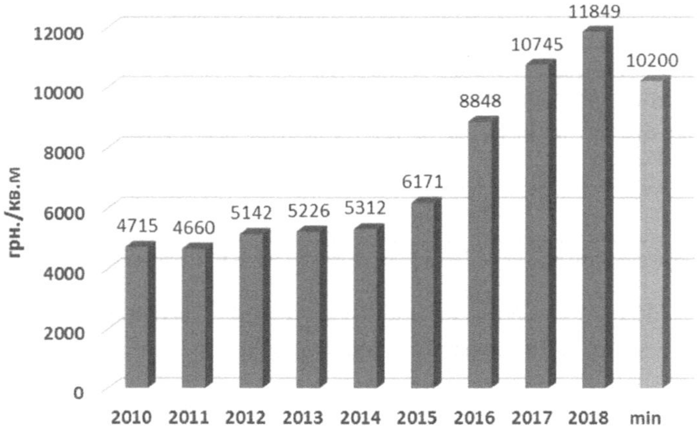

Fig. 3 Dynamics of indicators of indirect cost for 1 m^2 of housing in the Ukrainian housing construction market

supply (volumes of commissioned housing and manufactured construction products), as well as the dependence of demand and supply on the factors of influence. On the basis of the developed econometric models system, an optimization function of the minimal mediate cost for 1 m^2 of housing was obtained.

6 Practical Importance

Indicator of the mediate cost for 1 m^2 of housing is used:

- in determining the amount of public investment aimed at housing construction for citizens who need to improve housing conditions and state support in accordance with the legislation of Ukraine;
- in determining the size of the public share of national investment in the construction of housing for such citizens;
- in determining preferential loans for certain categories of citizens for the specified purpose [12, 13].

Consequently, determining the optimal value of the mediate cost for 1 m^2 of housing contributes to improving the efficiency of budget funds usage that are directed to the construction or purchase of housing for certain privileged categories of citizens.

7 Conclusion

In the framework of this research, a methodological approach to optimizing the cost of housing in the Ukrainian housing market has been developed. Based on the developed by the authors methodological approach, using the data of the State Statistics Service were obtained predicted and optimal indirect costs of 1 m^2 of housing. The predicted value of the mediate cost for 1 m^2 is at the level of 12,188.4 UAH/m^2, and the minimum value is at the level of 10,200 UAH/m^2, which is 16.3% below the predicted.

Thus, improving the efficiency of budget funds usage is possible if the use of state regulatory instruments and management decisions, in which it will be economically advantageous for the construction organization to set the lowest value for 1 m^2 of affordable housing, taking into account the conditions set by the internal and external environment housing market in Ukraine. This research methodology can be used to determine the minimum mediate cost for 1 m^2 of affordable housing at the level of Ukrainian regions.

References

1. Pangallo, M., Nadal, J.-P., & Vignes, A. (2019). Residential income segregation: A behavioral model of the housing market. *Journal of Economic Behavior & Organization, 159*, 15–35. https://doi.org/10.1016/j.jebo.2019.01.010.
2. Kulkarni, O., Jakhar, S., & Hudnurkar, M. (2014). A comparative study of relation between the national housing & building material cost and economic gap in India. *Procedia Economics and Finance, 11*, 695–709. https://doi.org/10.1016/S2212-5671(14)00234-2.
3. Zheng, M., Wang, H., Wang, C., & Wang, S. (2017). Speculative behavior in a housing market: Boom and bust. Economic Modelling, vol. 61, 50–64 (2017). https://doi.org/10.1016/j.econmod.2016.11.021
4. Napoli, G., Trovato, M. R., & Giuffrida, S. (2016). Housing affordability and incomethreshold in social housing policy. *Procedia-Social and Behavioral Sciences, 233*, 181–186. https://doi.org/10.1016/j.sbspro.2016.05.345.
5. Halchynskyi, L. Y., & Stanislavchuk, Y. S. (2010). Ekonomiko-matematychne modeliuvannia dynamiky zminy stanu rynku budivnytstva v Ukraini. *Ekonomichnyi visnyk NTUU "KPI", 7*, 246–249.
6. Kalynichenko, Y. V., Dobrovolska, V. V., Abramchuk, A. R. (2016). Modeliuvannia tendentsii rynku nerukhomosti. Ekonomichnyi analiz: zb. nauk. prats. *Ternopilskyi natsionalnyi ekonomichnyi universytet, 23*(1), 28–36.
7. Lahovskyi, V. V., & Paianok, T. M. (2018). Modeliuvannia dynamiky rozvytku budivelnoi haluzi Ukrainu. *Hlobalni ta natsionalni problemy ekonomiky, 23*, 710–716.
8. Shaposhnikova, I. O. (2018). Analiz chasovykh riadiv pervynnoho rynku zhytlovoi nerukhomosti m. Kyieva. *Ekonomichnyi visnyk universytetu, 36*(1), 139–147.

9. Komelina, O. V., & Shcherbinina, S. A. (2018). Methodology of estimation of energy reserves and energy efficiency of the housing fund of Ukraine. *Marketing and Management of Innovations, 1,* 382–390.
10. Derzhavna sluzhba statystyky Ukrainy. http://www.ukrstat.gov.ua/. Last accessed September 10, 2019.
11. Khalafian A. A. (2007). STATISTICA 6. Statisticheskii analiz dannykh: uchebnik. 3-e izd. OOO "Binom-Press".
12. UNIAN. https://www.unian.ua/economics/realestate/10372926-u-minregioni-pidrahuvali-skilki-koshtuye-budivnictvo-kvadratnogo-metra-zhitla-v-ukrajini.html. Last accessed September 12, 2019.
13. Varnaliy, Z., Onishchenko, S., & Masliy, A. (2016). Threat prevention mechanisms of Ukraine's economic security. *Economic Annals-XXI, 159*(5–6), 20–24. https://doi.org/10.21003/ea.V159-04.

Construction Enterprises Innovating Activities on the Basis of Industry 4.0 and "Deep" Digital Transformations

Nataliia Kraus, Olena Zerniuk, and Alina Chaikina

Abstract Characteristic features of the research and innovation impact on construction enterprises digitization were investigated and generalized in the article. Creation of digital industrial platforms of construction industry in different countries of the world and their adaptation to market realities was analyzed. It was established by authors that in order to transform construction enterprises management system in Ukraine, and the implementation of industry digitalization must be primarily aimed at developing mechanisms for adapting country's economic system to European space and realizing projects that require collective efforts with involvement of public and private stakeholders on regional and national levels. It was argued that the main factors slowing down dynamics of construction complex development in Poltava region include: significant currency fluctuations, liquidity crisis of banking institutions, and reduction of the population solvent demand. Introduction of new forms of relationships between enterprises, government, public authorities, universities, in particular, by forming a cluster on the basis of innovative hub makes possible ensuring exit of construction complex from economic crisis and perspective development of construction complex in Poltava region. It was confirmed by researches that there should be effective interaction in the "technical education—business of the construction sphere—innovative activity" chain and professional consultations by establishing a clear organization of "technical universities—scientific researches—production—construction" should be provided. Measures to combine potential of research, innovative development directly with industry and practical construction were substantiated and revealed. Authors developed a real model of construction cluster in Poltava region created on the basis of an innovative hub. It was also concluded that the main advantage of proposed clustering approach on the basis of

N. Kraus
Borys Grinchenko Kyiv University, Kiev, Ukraine
e-mail: k2205n@ukr.net

O. Zerniuk
Poltava National Technical Yuri Kondratyuk University, Poltava, Ukraine

A. Chaikina (✉)
Poltava National Technical Yuri Kondratyuk University, Golovko Str. 14/11, Poltava 36004, Ukraine
e-mail: alinachaikina@ukr.net

V. Onyshchenko et al. (eds.), *Proceedings of the 2nd International Conference on Building Innovations*, Lecture Notes in Civil Engineering 73,
https://doi.org/10.1007/978-3-030-42939-3_61

innovative hub is that due to such development of cluster systems, higher school can effectively integrate results of higher, academic and sectoral science of Ukraine, as well as advanced results of the world community science in development and implementation of innovative projects in the field of construction.

Keywords Innovation · Construction sector · Construction enterprises · Innovation hub · Industry 4.0 · Digital transformation

1 Introduction

In Ukraine, the innovation activity of industrial enterprises remains at a very low level for many years, thus in 2016, only 18.9% of enterprises were engaged in innovative activity. In 2015, only 723 enterprises were innovating: 400 of which implemented new technological processes and 570 realized innovative products. Since 2010, amount of sold innovative products has decreased by almost 32%, from 33 697.6 to 23 050.1 million UAH, and amount of innovative products supplied for export was only 21%. Total innovation expenditures in the industry during this period increased by almost 72% from 8045.5 million UAH to 13,813.7 million UAH, while expenditures of foreign countries decreased from 2411.4 million UAH to 58.6 million UAH or 97%. The main source of innovation financing costs is institutions and enterprises own funds, which account for 85% of the total funding. State has practically shied away from financial impact on these processes. Domestic industrial complex is based on industries focused on low-tech production and export of raw materials, and factors that ensure the competitiveness of the national economy: availability of labor, natural resources, and capital.

Deterioration of technological structure of production testifies consolidation of export-raw model of economic development in Ukraine. Thus, if in 2002 share of high-tech and medium-tech industries accounted for 22.2% and 43.1%, in 2017 it decreased to 17% and 41.7%, share of low-tech industries increased from 34.7% to 41.3% instead. In 2017, Cabinet of Ministers of Ukraine established the National Committee for Industrial Development, which should lay the foundations for institutional renewal of domestic production. The issue is about formation of value chains, which entails changing approaches to innovation and investment, as well as moving closer to private business. The main task is development of domestic production, creation of new jobs and new added value, rather than increasing imports [1, p. 21].

1.1 An Analysis of Recent Research and Publications and Selection of Previously Unsolved Parts of the General Problem

Construction enterprises innovation activity on the basis of Industry 4.0 and "deep" digital transformations were investigated by such famous scientists and inventors as W. Isaacson, S. Brand, J. Wells, E. Williams, B. Gates, G. Jacobs, B. Elbrecht, D. Engelbart, J. von Neumann, E. Reinert, S. Huntington, and J. Shapiro. Economists from Ukraine are also actively involved in analyzing positive and negative consequences that Industry 4.0 brings to economy and production, revealing product content and technological innovations generated by the Fourth Industrial Revolution and determining emergence of "Industry X.0" in the world. Among them are V. Heitz, V. Galasyuk, A. Hrytsenko, S. Kubiv, A. Medvedev, D. Oliynyk, R. Pustovit, and O. Yaremenko. But, at the same time, there is a number of actual issues such as revealing the content of risks that construction companies face as they grow; development of innovative hub model in the Poltava region construction cluster, remains still relevant and needs to be studied.

1.2 Purpose of the Article

Purpose of the article is to study features of creating digital industrial platforms of construction industry in different countries of the world, presentation of preventive blitz-diagnosis of problems and risks of construction enterprises at the stage of their growth. As well as to find out options for correlating the life stages of a "construction company—innovation market—new product" with development of preliminary recommendations for all options, analysis of the innovative hub developed model for construction cluster in the Poltava region, substantiation, and disclosure of measures to combine potential of scientific research, innovative development directly with industry and practical construction.

2 Main Material of Research Work with Justification of the Obtained Scientific Results

Modern innovative technologies such as: cloud technologies, modern ways of collecting and analyzing big data, crowdsourcing, cryptocurrency, blockchain technologies, unmanned vehicles, etc., are radically changing entire sectors of the economy. Based on these technologies, digital revolution becomes the Fourth Industrial Revolution (Industry 4.0), essence of which is mass introduction of cyber-physical systems into production, which blurs boundaries between physical, digital, and biological spheres, and emergence of an entirely new type of industrial production based on processing of a large array of data to achieve complete automation of production and implementation of the latest scientific and technological advances in technological processes. It is anticipated that these cyber-physical systems will integrate into one self-regulating network, connect to each other in real-time, and promote radically new ways of interacting in the value-adding process [2, p. 7].

In Ukraine, the development of construction enterprises is based on various investment funds. For these reasons, there is a need to develop a complete system of support for innovative projects, which should include the creation of innovation-venture funds and hubs of construction. For these purposes, it is proposed to bring the number of small construction enterprises in Ukraine up to 90% of their total. However, it should be noted that it is difficult for Ukrainian construction companies to compete with foreign enterprises that promote modern innovative projects, because they use material and technical base that they have inherited since the Soviet era.

The activity of a construction enterprise is the equilibrium of two vectors: [owner, will, freedom, innovation] + [state, obligation, necessity, coercion, subjugation, dependence]. The complexity of analysis lies in the fact that essential factors of first and second groups condition each other [3, p. 63]. Preventive blitz-diagnosis of problems and risks of construction industry at the stage of their growth is presented in Table 1.

Creating digital industrial platforms for construction industry and adapting them to market realities are essential processes to ensure scale and coverage of national and regional initiatives to digitize construction industry and economic entities initiatives [2, p. 13]. Digitizing European Industry (DEI) seeks to bring together common interests on the Platform Economy and secure future global standards for connecting smart construction enterprises and involves investing in digital innovation opportunities based on information and communication technologies (ICT) standards and adaptation of workforce by training human capital to acquire necessary digital transformation skills [2, p. 14].

Innovation leads to a closer interdependence between the advancement of digital technologies and their use in different industries and requires creation of highly innovative digital sectors and renewal of digital innovation capacity of all industries. In this purpose, several national and regional initiatives have been launched to take advantage of the opportunities offered by digital innovation in Europe, including: Industrie 4.0 (DE), Smart Industry (NL), Catapults (UK), and Industrie du Futur (FR) [6]. For example, for Industrie 4.0 in Germany was created Reference Architectural Model Industrie 4.0 (RAMI 4.0) based on the Standard IEC62264 for integrating management systems in construction industry, which promotes an understanding of what standards are required to implement Industry 4.0. Standard IEC62264 details models of production operations objects and attributes, integration and management of production operations, messaging and business services and more. In addition, in Germany associations, ProSTEP and ViP have developed a directory of compatibility criteria for infrastructure, interfaces, standards, architectures, and more in the form of Product Lifecycle Management (PLM-Code for PLM Openness, CPO) [7].

Examples of national and regional programs that improve industrial production digitization, namely construction companies, are initiatives launched by different

Table 1 Preventive blitz-diagnosis of problems and risks of construction enterprises at the stage of their growth (compiled by the authors on the basis of sources [4, p. 102]; [5, p. 425–446])

Possible options for the life cycle ratio: "construction enterprise (CE)—innovation market (IM)—new products (NP)"	Preliminary diagnosis of major problems and risks of CE	Previous recommendations
1. Growth CE—embryonic state IM—technical idea for NP creation	Problem is related to the need for rapid implementation of the technical idea and NP entering on IM. The most significant risk is the mismatch of the expected and actual consumer response to the NP	Develop a rigorous timetable for activities that provide a fast and efficient transition from a technical idea to a serial NP release. Clearly motivate managers to implement the plan in terms of time and quality
2. CE Growth—IM Crystallization—Technical Idea of Creating NPs	Problem is the same, but time is limited. The main risks are events that can slow down the output process IM (counterparties, internal inconsistencies)	
3. CE Growth—IM Growth—Technical Idea of Creating NPs	Problem is the same, but the time resource is almost exhausted. The main risks are the same	Mobilize all resources for organizing NP output activities on IM
4. CE Growth—IM Saturation –Technical Idea of Creating NPs	Technical idea was "late"	Use technical idea for a second product that addresses the needs of a more promising IM (second market niche) sector
5. CE Growth—IM maturity—Technical Idea of Creating NPs		
6. CE Growth—IM decline—Technical Idea of Creating NPs		
7. CE Growth—Embryonic state IM—NP development	There are no visible problems. There is a good chance for a successful CE development in the future. The main risk is the mismatch of expected and actual consumer response to the NP	Perform continuous IM status analysis. Develop the program for the fastest and most effective product promotion
8. CE Growth—IM Crystallization—NP development	There are no visible problems. But the time available for IM output is limited. The main risk is the same + events that can slow down output on IM (counterparties, internal mismatches)	Develop a timetable for activities that enable rapid and efficient transition from product development to serial production. Clearly motivate managers to implement the plan

(continued)

Table 1 (continued)

Possible options for the life cycle ratio: "construction enterprise (CE)—innovation market (IM)—new products (NP)"	Preliminary diagnosis of major problems and risks of CE	Previous recommendations
9. CE Growth—IM Growth—NP development	The main problem is the limited time resource for NP output to EM. The main risks are the same	Mobilize all resources to accelerate NP exit measures on IM
10. CE Growth—IM Saturation—NP development	Development is "late." It is obvious that there are problems in management and marketing	Use development for a second NP that is more demand-oriented
11. CE Growth—IM maturity—NP development	Organization of the main processes is "limping." The main risk is deterioration of financial condition	Reorganize the main processes
12. CE Growth—IM decline—NP development		
13. CE Growth—Embryonic state IM—NP output to IM	Has a chance for successful development of CE in the future. The main risk is mismatch between the expected and actual consumer response to the NP	Continuously monitor consumer response to NP and IM status. Develop the program for fastest and most effective promotion of NP
14. CE Growth—IM Crystallization—NP output to IM		
15. CE Growth—IM Growth—NP output to IM	Unbeatable/wonderful! There are no problems. An internal nature is risks of paramount importance: will the company "drive" such rapid/rapid growth?	Analyze and refine basic processes. Pay more attention to discipline, responsibilities and powers
16. CE Growth—IM Saturation—NP output to IM	Exit to IM NP "with some delay." Underestimation of management. Difficulties with CE financial condition	
17. CE Growth—IM maturity—NP output to IM	Same, but financial problems can be more serious	Same + modify NP for more promising IM (other market niche) sector
18. CE Growth—IM decline—NP output to IM	Exit NP on IM came too late. Obvious problems in management and marketing. Financial losses can be critical for CE	Reorganize management system and major CE processes. Modify NP for more promising sector of IM (other market niche)
19. CE Growth—IM crystallization—growth in sales of innovation	Situation is quite promising for CE. The main problem is to maintain a position in the emerging market. The main risks are actions by competitors	Constantly analyze the status of IM, and especially actions of competitors. Ensure that CE (in organization of key processes) is prepared for its potential growth

(continued)

Table 1 (continued)

Possible options for the life cycle ratio: "construction enterprise (CE)—innovation market (IM)—new products (NP)"	Preliminary diagnosis of major problems and risks of CE	Previous recommendations
20. CE Growth—IM growth—NP sales growth	Fantastic situation! The problem is to save this situation. Internal risks are most clearly seen: in terms of analyzing trends in the development of IM, as well as the organization of major CE processes	Constant IM analysis. Ensure CE readiness to move to another product line or to another (niche) IM sector
21. CE Growth—IM Saturation—stable sale of NP	There are no obvious problems. The main risk is the financial loss that can occur if CE "hangs" on IM with an "old" product	Develop a production upgrade program: move to another product range or to different sectors (niches) of IM
22. CE Growth—IM maturity—NP sales stability	Problem is in the limited time available for "production upgrades." The most important are marketing risks as well as risks related to reorganization of internal processes	Implement the production upgrade program

European countries: in Sweden, it is "Produktion 2030"; in Spain—"Industry 4.0"; in France—Industrial Alliance "Industrie du Futur"; in Italy—Italy's National Industrial Plan and others. China's manufacturing companies also show interest in digitizing construction companies, but take a different approach, relying more on direct investments in European companies, such as Krauss-Maffei, Stoll, Manz Group, Kuka, which are important to them. China's level of investment in relevant technologies exceeds the EU level. Common programs in China are "Made in China 2025", which is considered as Chinese equivalent of Industry 4.0, and Internet Plus (IP).

In order to transform construction enterprise management systems in Ukraine, implementation and realization of industry digitization should be primarily aimed at developing mechanisms for adapting country's economic system to European space and for Projects of Common Interest (PCIs) implementation which is a matter of common interest and require collective efforts involving both public and private stakeholders at regional and national levels. Due to this, international practice has established a framework for identification, planning, and implementation of PCIs based on the provision of a single coherent regulatory framework [2, p. 15].

Aspects that can determine success of an innovative "young" construction company in Ukraine are the following: speed of development, which corresponds to the speed of development of construction market (if the company has an idea, it is necessary to act because there is no time for its "bearing"—it will be necessary to supplement and “grind” it during implementation); ability for partnership and synergy (no construction company in innovation economy will “survive” alone—only forming alliances, hubs, and partnerships is inevitable); ability for innovation and continuous innovation development; cultivation of talents; globality in innovative thinking (projects should target both local construction markets as well as to the world).

Within enterprise of construction industry, functions that performed by ideology at institutional level nowadays move to organizational culture. In any institute, enterprise, trade union or political party there is a common cognitive component—organizational knowledge [8, p. 195]. Organizational culture of a construction company materializes in rules and traditions on which relationship between “principal” and “agent” are based. Interpretive function of organizational culture of construction enterprise is based on the use it by “agents” to evaluate validity of “principal” decisions in case of any unforeseen circumstances. In our case, this could be a risk during realization an innovative project by a construction company.

Organizational cultural of a construction company, in our opinion, should be formed taking into account level of innovative behavior of its employees. By innovative behavior, we mean a set of actions while creating innovations and reaction to conditions of innovative construction activity. Complexes of organizational routines create an organizational innovation culture of entrepreneurial structure of construction industry.

Construction companies are a set of principal-agent relationships as a mechanism of limited divergence between economic entities and an internal organization, taking into account decisive role of transactional construction costs. Hence, it can be argued that intercompany relations is a contract whereby one or more persons (“principals”) hire other persons (“agents”) to carry out innovative construction projects or to transact based on delegated authority and principal decisions [9, p. 14–15], on innovative activity in the field of construction [10].

Effective interaction in the chain of “technical education—construction business—innovation activity” should be worked out and professional consultations should be provided by establishing a clear organization “technical universities—scientific research—production—construction.” It should continue to follow the concept of “innovation triangle,” which is based on rapid and successful implementation of breakthrough innovations in the industry and requires cooperation of three different subjects: consumer, developer, and inventor. The main purpose of such collaborations is to activate innovative activities of technical university graduates. Universities will train young professionals and research staff, capable of initiating and implementing innovative projects, as well as releasing independent entrepreneurs of high-tech construction businesses. Working on real innovative projects will allow students to acquire new knowledge and transfer technology in the course of communication with the staff of construction enterprises. The transfer will also take place through migration of scientists between construction companies and technical university, which will allow them to acquire necessary knowledge without expense of intellectual property rights.

In the context of the study, we give an example of an innovative construction cluster within the Poltava region as a spatially localized association of participants in the investment-building process, innovative enterprises of building materials and other related industries, research institutes, universities, banking institutions, public authorities, local authorities in order to increase level of construction products competitiveness and increase economy of the region as a whole.

The main factors slowing down dynamics of construction complex development in the Poltava region include: significant currency fluctuations, liquidity crisis of banking institutions, and reduction of solvent demand of population. Ensuring the building complex recover from the economic crisis and prospective development is possible with introduction of new forms of relationships between enterprises of the industry, government, public, universities, in particular, by forming a cluster on the basis of innovative hub. The real model of a construction cluster in the Poltava region created on the basis of an innovative hub is presented in Fig. 1.

As can be seen from Fig. 1 that to construction cluster of the Poltava region should be included all participants of investment and construction complex.

Among investment and construction companies that we believe should join this cluster firstly are large construction and assembly organizations that have experience, reputation, and consistently high profits.

We consider including to the cluster of 3–4 investment and construction companies focused on implementation of individual investment projects to be effective. It should be noted that this innovative hub of the construction cluster is a node to support high intensity of innovation, through which all logistics of innovative development of the region goes. This type innovative hub will allow to form regional innovation markets, to ensure a dominant presence in them, to organize capital inflows (primarily venture capital), to set "rules of the game," to define norms of relations between the entities, and to form a creative and innovation-oriented human resources reserve. The innovation hub will facilitate rapid and high-quality integrated "processing" of innovations (as ideas) into innovative products that will be successfully implemented in the high-tech markets of the region.

The main advantage of our innovative hub-based clustering approach is that due to this cluster systems development, higher education can effectively integrate results of higher, academic and sectoral science of Ukraine, as well as advanced results of science of the world community in development and implementation of innovative projects.

Thus, the formation of an institutional environment based on clustering provides a powerful impetus for building an effective innovation economy in the context of comprehensive modernization. Today's the stage of modernization should foresee, first and foremost, a large-scale upgrade of production under influence of formation of advanced VI technological structure.

Due to instability, the extreme variability of the basic elements of Ukrainian society and presence of residual phenomena of previous development, Government of the country is facing fundamentally new tasks that it must solve in the course of institutional changes.

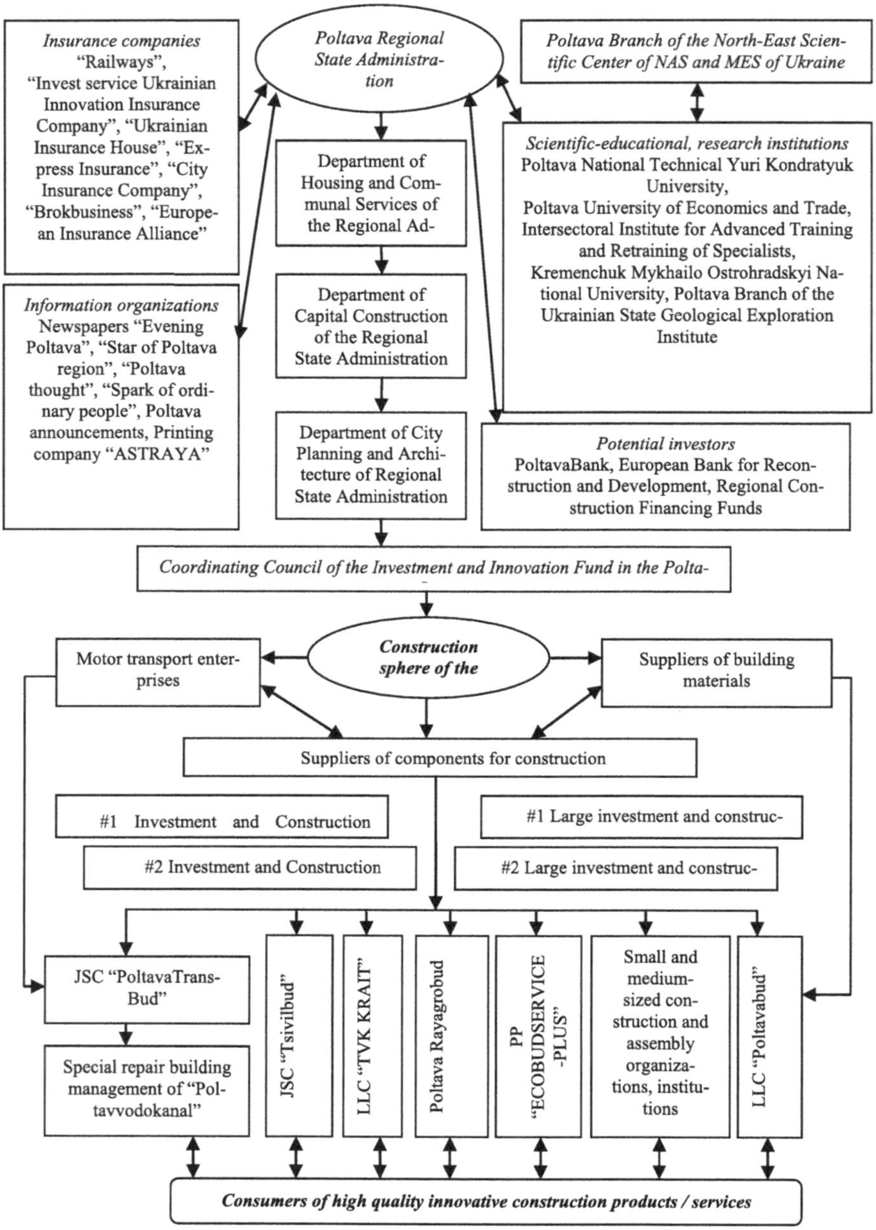

Fig. 1 Model of construction cluster innovation hub in the Poltava region (development by authors)

We believe that the creation of clusters based on innovative hubs is one of the key factors of the country economic growth, which provides a comprehensive system of support for economic and technological interdependence. Therefore, we propose to consider the interaction between participants of innovation process in innovation cluster as a basis for increasing competence, competitiveness and level of cooperation between firms and other institutes, institutions in the sphere of innovation.

From an institutional point of view, a necessary condition for construction an innovative type of national economy is conformity of the relations of innovation development main participants with principles of the triple helix. Verticalization of Ukrainian economy is a "stumbling block" for innovative development of construction sector. The "technical university—government—construction business" model contrasts with vertical mechanisms of managing innovative development of establishing and improving horizontal links between participants in innovation process of construction industry.

Institutions of consolidation, consistency, quality are the basic parameters of this model. To ensure right momentum, collegial bodies, industry alliances of innovators should be created, and systematically decentralized decision-making processes, turning them into joint initiatives.

We are convinced that today a number of measures need to be taken to combine potential of research and innovative development directly with industry and practical construction, including:

1. Define nature, legal forms, and scope of activities (including financing and revenue structure) of innovation support and construction business infrastructure (techno parks, business incubators, technology transfer structures, innovation hubs). It is needed to focus efforts on creating conditions for control of innovative and commercial activities and benefits for construction techno parks. The number of techno parks should be clearly limited (up to 5) and they should operate within control and incentive mechanism, which would prevent inappropriate use of benefits for innovative projects by other (non-innovative) economic operators.
2. On the basis of the leading national technical universities of Ukraine activity of university construction hubs as "platforms" for acceleration of ideas should be created/started. Technical universities that are characterized by efficient operation of innovative construction hubs should be given full financial autonomy to enable them to carry out certain financial transactions, bypassing in most cases lengthy and tedious procedures for applying to the Treasury of Ukraine. Allow technical universities engaged in innovative construction business to use funds provided as assistance, grants, gifts, sponsorship, without following public procurement procedures provided for by Ukrainian law.
3. Encourage technical universities and research institutes to innovation construction activities, allowing them not to tax all the revenues from licensing and spin-off companies and directs them to modernization of universities' laboratories and equipment.
4. Determine the amount and procedure of paying for intellectual property to the creators. In case of acquisition of inventions rights and other objects of intellectual

property right to public authorities, to introduce procedure of licensing and access to information about such objects.

5. Through government programs, promote involvement of scientists to work in construction industry. Provide joint funding for employment of such scientists in specific innovative construction projects. Establish a doctoral and postgraduate institute with joint funding from industry and state. State should stop practice of national publications domination in the course of scientific evaluation and move to an internationally recognized system of scientific research evaluation.
6. It is necessary to develop a new state program for forecasting scientific, technological, and innovation development in search of innovative breakthrough directions and ensuring competitiveness of the Ukrainian economy in the field of construction. It is advisable to implement a "single window" principle during providing participants of industrial park such services as: approval of project documentation, approval of documents for maintenance in the territory of land plot, issuance specifications for reinforcement objects to the engineering networks and structures.
7. While ensuring development of industrial parks, more attention should be paid to application of public–private partnership principles and mechanisms for granting certain preferences as an example is deferred payment mechanism "Tax Increment Finance," harmonization of industrial park activities with the needs of technological modernization and sectoral priorities of innovative development.

3 Conclusions

Participants and leaders of developed countries at the 46th World Economic Forum in Davos, held in January 2016, emphasized beginning of the Fourth Industrial Revolution. It was noted at the Forum that this revolution would lead to elimination of barriers between machine and person and promote their integration, and consequently would cause dramatic technological changes in the world economy and way of people life. In order to avoid possible socio-economic, technological upheavals and institutional chaos that this revolution is carrying, it is already necessary to calculate today possible negative consequences and to consider additional problems and aspects that need to be addressed. In the context of innovative development of a knowledge-based society, it is necessary to avoid two "end points" on the innovation spectrum. First one is to support exclusively science and research in the construction industry. Second one is to focus on construction of large structures that house high-tech companies.

References

1. Draft Parliamentary Hearing Recommendations on the theme: "National Innovation System: State and Legislative Support for Development". Homepage, http://ua-ekonomist.com/archive/2018/3/rekom.pdf, Accessed 18 Jan 2019.
2. Oliynyk, D. I. (2018). *Innovative development of territorial communities in the conditions of the fourth technological revolution: priorities and prospects (analytical. report)*. Kiev: NISD.
3. Yaremenko, O. L., & Pankratova, O. M. (2007). Institutions and economic freedom of economic entities. *Economic Theory, 3,* 63.
4. Medvedeva, A. M. (2011). Blitz-analysis of a growing company and risk management. *Modernization Innovation Development, 3*(7), 102–104.
5. Kraus, N. (2019). *An innovative economy in the globalized world: an institutional basis for formation and a trajectory of development*. Kiev: Agrarian Media Group.
6. Communication from the Commission to the European Parliament, the Council, the European Economic and Social Committee and the Committee of the Regions Digitising European Industry Reaping the full benefits of a Digital Single Market {SWD (2016) 110 final} Homepage, http://eurlex.europa.eu/legalcontent/EN/TXT/?uri=CELEX:52016DC0180, Accessed 15 Jan 2019.
7. Code of PLM openness homepage, https://www.techniatranscat.com/abouttechniatranscat/abouttechniatranscat/codeofplmopenness, Accessed 14 Jan 2019.
8. Sapir, J. (1995). Culture economique, culture technologique, culture organisationnelle elements pour une interpretation de l'histoire economique russe et sovietique. *Cahiers du Monde Russe, XXXVI*(1–2), 195.
9. Pustovit, R. F. (2005). Institutional specificity of internal-firm relations "principal-agent" researches. *Scientific works of DonNTU, 89–2,* 14–15.
10. Varnaliy, Z., Onishchenko, S., & Masliy, A. (2016). Threat prevention mechanisms of Ukraine's economic security. *Economic Annals-XXI, 159*(5–6), 20–24.

Determination of Critical Depth of Cutting Soil by Cutters with Building Excavators

S. V. Kravets, O. P. Lukianchuk, O. V. Kosiak, and O. O. Gaponov

Abstract In civil and reclamation construction, when laying the foundations and other elements of structures, work is carried out, which is accompanied by deep soil development. The energy intensity of the working process increases significantly at subcritical values of the chips due to the blocking of the part of the soil. Let us determine the critical depth of the asymmetric blocked one side. The calculated dependences show that as the cutting angle of the cutter decreases from 50° to 20°, the relative critical depth of asymmetric blocked cutting ($h_{\kappa p1}/b_p$) increases: for refractory and semi-solid clay, from 1.71 to 3.12; for semi-hard loam from 1.76 to 3.27; for hard sand from 1.78 to 3.75. For semi-blocked cutting, this depth increases: for refractory clay, from 2.92 to 5.21; for semi-solid clay from 2.78 to 5.03; for semi-hard loam from 2.77 to 5.14; for solid sandwiches from 2.65 to 5.45. The mathematical models for determining the critical depth of cutting for the extreme lateral incisors of the multi-slip chains of trench excavators, operating in conditions of asymmetric lateral cutting and semi-block cutting, are obtained. Based on the obtained approximated mathematical models in each particular case, based on the data on the thawing soils recommended by BNiP, it was possible, depending on the cutting angle of the cutter, within the cutting angle of 20°…50°, to determine the rational depth of cutting for the extreme lateral incisors of the multi-slip chains of trench excavators.

Keywords Construction · Laying the foundations · Excavator · Cutting of the soil · Critical depth

---

S. V. Kravets · O. P. Lukianchuk (✉) · O. V. Kosiak
Department of Building, Road, Melioration, Agricultural Machinery and Equipment (BRMAME), National University of Water and Environmental Engineering, 11, St. Soborna, Rivne, Ukraine
e-mail: o.p.lukyanchuk@nuwm.edu.ua

O. O. Gaponov
Kharkiv National Automobile and Highway University, Kharkiv, Ukraine

V. Onyshchenko et al. (eds.), *Proceedings of the 2nd International Conference on Building Innovations*, Lecture Notes in Civil Engineering 73,
https://doi.org/10.1007/978-3-030-42939-3_62

1 Formulation of the Problem

In civil and reclamation construction, when laying the foundations and other elements of structures, work is carried out, which is accompanied by deep soil development. Trenching with single-bucket and multi-bucket excavators is carried out under conditions of limited working bodies by lateral vertical walls. The destruction of the soil by the extreme lateral incisors (teeth) installed on the working equipment of excavators is carried out with the restriction of the working process by these vertical walls on one side of the incisor. On the other hand, the cutter interacts with an array of soils, forming a break with a one-way collapse, or does not interact, as with free cutting. In the first case, the cutter performs blocked asymmetric cutting, and in the second case—half blocked cutting.

It is known that an increase in the cutting depth of a knife to a critical depth $h \leq h_{кр}$ leads to a decrease in the energy intensity of the working process, because the growth intensity of the resistance to cutting is smaller than the growth of the cross-sectional area of the slit [1, 10, 11].

The energy intensity of the working process increases significantly at subcritical values of the chips due to the blocking of the part of the soil. Therefore, the calculation of parameters of single-bucket and multi-lorry trench excavators must be carried out on the basis of critical depth-cutting of soils.

Average cutters of chain-scraper trench excavators and bucket teeth of single-hull excavators work in conditions of symmetric blocked, semi-blocked, free and combined (under binary scheme of cutting) cutting, and extreme lateral incisors—in conditions of asymmetric blocked or semi-blocked cutting.

Let us determine the critical depth of the asymmetric blocked cutting of the soil in a trench at the vertical wall, limiting the space of destruction of the tool on one side. For this purpose, the designed scheme of interaction of a straight-line incisor with soil (Fig. 1) is developed. The problem is solved with known assumptions [3, 7].

2 Presenting Main Material

On the elemental volume of a chip element in height dh before the chipping of the ground, the forces acting on the vertical plane of the OAD (see Fig. 1a). The system of equations of equilibrium of all forces on the normal (n) and the tangent (τ) of the axis to the frontal plane of the clamping have the following form (1). From the first equation of system (1), we obtain (2),

$$dN_c = -\frac{dN}{\cos\varphi}\cos(\alpha_p + \varphi + \psi) = -\frac{qb_p}{\cos\varphi\cdot\sin\alpha_p}\cos(\alpha_p + \varphi + \psi)dh \quad (1)$$

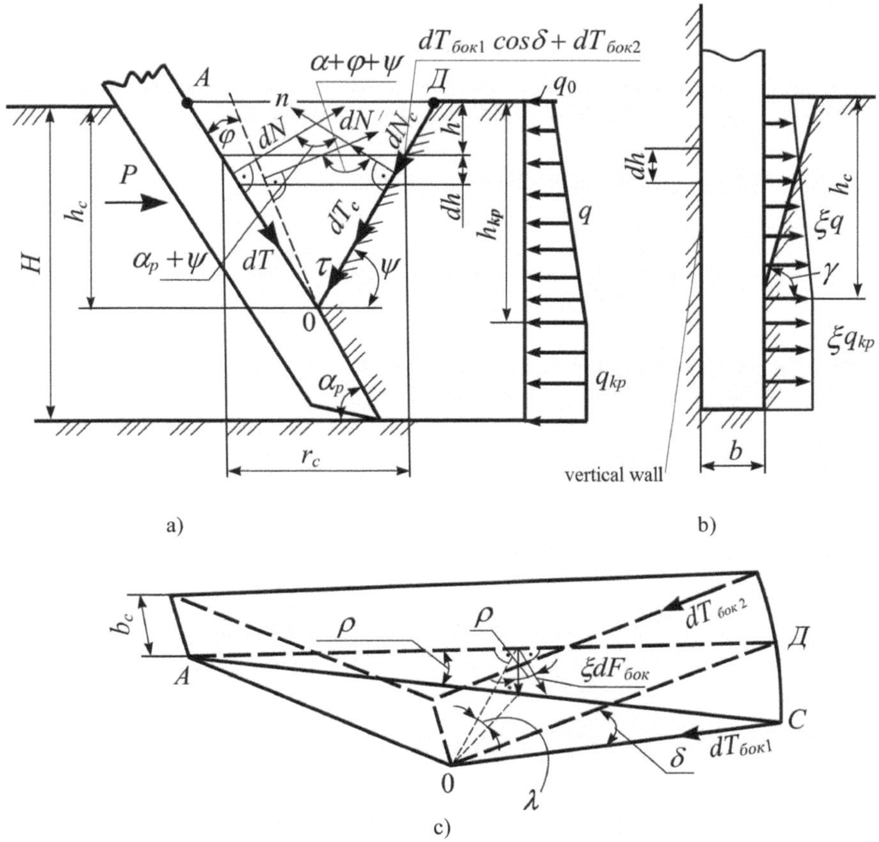

Fig. 1 Scheme of interaction of an extreme lateral incisor with soil under asymmetric blocked cutting: **a** in the longitudinal plane, **b** in the transverse plane, **c** the shape of the chips element in the recess process

$$\begin{cases}\sum P_n = dN_c + dN'\cos\left(\alpha_p + \varphi + \psi\right) = 0 \\ \sum P_\tau = dT_c + dT_{бок1}\cos\delta + dT_{бок2} - dN'\sin\left(\alpha_p + \varphi + \psi\right)\end{cases} \tag{2}$$

where: q—the law of the distribution of normal pressure on the forehead of the cutter in depth; b_p—is the width of the cutter; φ—angle of external friction of the soil; α_p—cutting angle of the cutter; ψ—angle of the soil shift with a cutter in the longitudinal plane; dh—elementary depth of cutting.

The law of the distribution of normal pressure per knife in depth is substantiated in the literature [3, 7] and is presented in the form (3),

$$q = q_0 + \frac{q_{кр} - q_0}{h_c} k_{пер} h \tag{3}$$

where: q_0, $q_{кр}$—respectively, the minimum pressure on the surface of the day and the maximum possible pressure on the bearing capacity of the soil (critical pressure) is determined by known dependencies [3, 7]; $k_{пер}$—the ratio of the guaranteed gravel depth h_c to the critical depth of cutting $h_{кр}$ ($k_{пер} = 0.9 \cdots 0.95$); h—current depth value.

Given (3) the expression (2) will be rewritten as (4).

$$dN_c = \frac{\cos(\alpha_p + \varphi + \psi)}{\cos\varphi \sin\alpha_p} b_p \left(q_0 + \frac{q_{кр} - q_0}{h_c} k_{пер} h \right) dh \tag{4}$$

Elementary tangential forces acting on the frontal (dT_c) and lateral ($dT_{бок1}$, $dT_{бок2}$) chip planes are determined by the law of the Coulomb for soils (5), (6), (7),

$$dT_c = \operatorname{tg}\varphi_0 dN_c + c dF_c \tag{5}$$

$$dT_{бок1} = (\xi q \cos\rho \cos\lambda \operatorname{tg}\varphi_0 + c) dF_{бок1} \tag{6}$$

$$dT_{бок2} = (\xi q \operatorname{tg}\varphi_0 + c) dF_{бок2} \tag{7}$$

where: φ_0—angle of internal friction of the soil; c—coefficient of clutch in the soil; ξ—is the coefficient of the side pressure of the soil chip element; dF_c, $dF_{бок1}$, $dF_{бок2}$—elementary areas corresponding to the frontal and lateral surfaces of the clapping; ρ, δ, λ—are the angles formed by the lateral surface of the skidding with the parallel vertical side wall of the trench plane (see Fig. 1c).

In this case, the values of elementary areas, respectively, are: (8), (9), (10),

$$dF_c = (b_p + \rho r_c) \frac{dh}{\sin\psi} = [b_p + \rho(\operatorname{ctg}\alpha_p + \operatorname{ctg}\psi)(h_c - h)] \frac{dh}{\sin\psi}, \tag{8}$$

$$dF_{бок1} = r_c \frac{dh}{\cos\lambda} = (\operatorname{ctg}\alpha_p + \operatorname{ctg}\psi)(h_c - h) \frac{dh}{\cos\alpha}, \tag{9}$$

$$dF_{бок2} = r_c \cdot dh = (\operatorname{ctg}\alpha_p + \operatorname{ctg}\psi)(h_c - h) dh, \tag{10}$$

where r_c—current value of the radius of rock placement of soil. If we substitute the expressions (5–7) into the second equation of system (1), taking into account the dependences (4), (8–10) and integrate it in the range from 0 to h_c, then after the intermediate transformations we obtain Eq. (12).

$$\frac{\operatorname{ctg}\alpha_p+\operatorname{ctg}\psi}{2}\left\{\left[\frac{c\rho}{\sin\psi}+c\left(1+\frac{\cos\delta}{\cos\lambda}\right)\right]+\xi\left(q_0+\frac{q_{\kappa p}-q_0}{3}k_{nep}\right)\cos\rho\cos\delta\cdot\operatorname{tg}\varphi_0+\xi\operatorname{tg}\varphi_0\right\}h_c=$$
$$=\left[\left(q_0+\frac{q_{\kappa p}-q_0}{2}k_{nep}\right)\frac{\sin(\alpha_p+\varphi+\varphi_0+\psi)}{\cos\varphi\cos\varphi_0\sin\alpha_p}-\frac{c}{\sin\psi}\right]b_p \tag{11}$$

$$h_{\kappa p}=\frac{h_c}{k_{nep}}=\frac{\left(q_0+\frac{q_{\kappa p}-q_0}{2}k_{nep}\right)\frac{\sin(\alpha_p+\varphi+\varphi_0+\psi)}{\cos\varphi\cos\varphi_0\sin\alpha_p}-\frac{c}{\sin\psi}}{\frac{\operatorname{ctg}\alpha_p+\operatorname{ctg}\psi}{2}k_{nep}\left[\begin{array}{l}c\left(1+\frac{\rho}{\sin\psi}+\frac{\cos\delta}{\cos\lambda}\right)+\\+\xi\left(q_0+\frac{q_{\kappa p}-q_0}{3}k_{nep}\right)\operatorname{tg}\varphi_0(1+\cos\rho\cos\delta)\end{array}\right]} \tag{12}$$

$$\cos\rho=\sqrt{1-\left(\frac{\operatorname{ctg}\gamma}{\operatorname{ctg}\alpha_p+\operatorname{ctg}\psi}\right)^2} \tag{13}$$

$$\cos\lambda=\frac{1}{\sqrt{1+\left(\frac{\operatorname{ctg}\alpha_p\operatorname{ctg}\gamma}{\operatorname{ctg}\alpha_p+\operatorname{ctg}\psi}\right)^2}} \tag{14}$$

$$\cos\delta=1-2(\operatorname{ctg}\alpha_p+\operatorname{ctg}\psi)^2\sin^2\psi\ \sin^2\frac{\rho}{2} \tag{15}$$

$$\gamma=\arccos\left(\frac{c\cdot\cos\varphi_0}{\left(1-0{,}74\operatorname{tg}\varphi_0-\frac{1{,}52\cdot c}{q_{\kappa p}}\right)q_{cep}}\right)-\varphi_0 \tag{16}$$

$$q_{cep}=\frac{q_0+q_{\kappa p}}{2} \tag{17}$$

The mathematical model (12) can be used to determine the critical depth of cutting when layer cutting of the soil. The conditions for the destruction of the soil in the upper and lower tiers are different. The upper soil-forming organ (cutter), due to the influence of the day-surface, operates in conditions of asymmetric blocked cutting: on one side the cutter is limited by the vertical side wall of the trench, on the other side—an array of soil that collapses and goes out to the surface. From the side of the lateral collapse, the soil is destroyed mainly due to the deformation of the separation [10, 11], and therefore, there is no lateral pressure and frictional forces on the lateral plane of the shaving element of the soil with the collapse. From the side of the vertical wall, the soil is destroyed due to the deformation of the slice and the shear, and therefore the normal pressure and frictional forces appear on the lateral vertical plane of the soil shaving element. In this case, the upper cutter creates a free space with the size bp on the daytime surface for the release of soil from the lower tier. As a result, the lower soil-forming organs form a rectangular crack. The formation of the slit in the lower tier is also due to the deformation of the slice and the displacement.

Taking into account these features, the dependence (12) for determining the critical depths of asymmetric blocked cutting in the upper and lower tiers is rewritten as follows (18), (19).

$$h_{\kappa p1} = \frac{\left(q_0 + \frac{q_{\kappa p} - q_0}{2} k_{nep}\right) \frac{\sin\left(\alpha_p + \varphi + \varphi_0 + \psi\right)}{\cos\varphi \cos\varphi_0 \sin\alpha_p} - \frac{c}{\sin\psi}}{\frac{\operatorname{ctg}\alpha_p + \operatorname{ctg}\psi}{2} k_{nep} \left[c\left(1 + \frac{\rho}{\sin\psi} + \frac{\cos\delta}{\cos\lambda}\right) + \xi\left(q_0 + \frac{q_{\kappa p} - q_0}{3} k_{nep}\right) \operatorname{tg}\varphi_0 \right]} b_p \tag{18}$$

$$h_{\kappa p2} = \frac{\left(q_0 + \frac{q_{\kappa p} - q_0}{2} k_{nep}\right) \frac{\sin\left(\alpha_p + \varphi + \varphi_0 + \psi\right)}{\cos\varphi \cos\varphi_0 \sin\alpha_p} - \frac{c}{\sin\psi}}{\left(\operatorname{ctg}\alpha_p + \operatorname{ctg}\psi\right) k_{nep} \left[c + \xi\left(q_0 + \frac{q_{\kappa p} - q_0}{3} k_{nep}\right) \operatorname{tg}\varphi_0 \right]} b_p \tag{19}$$

Dependence (19) is identical to the dependence for determining the critical depth of cutting in the lower tiers, which is given in the literature [3, 7]. This confirms the reliability of the mathematical model (12).

If the extreme lateral cutters make half blocked cutting (blocking one side edge of the cutter with an upright wall), $dT_{бок1} = 0$, then we obtain (20).

Comparing the dependences (19) and (20), we can conclude that the critical depth of half blocked cutting by the extreme lateral incisors in the upper tier is 2 times greater than the critical depth of blocked cutting in the lower tier if $\gamma = \pi/2$.

$$h_{кр1нбл.} = \frac{\left(q_0 + \frac{q_{кр} - q_0}{2} k_{пер}\right) \frac{\sin(\alpha_p + \varphi + \varphi_0 + \psi)}{\cos\varphi \cos\varphi_0 \sin\alpha_p} - \frac{c}{\sin\psi}}{\frac{\operatorname{ctg}\alpha_p + \operatorname{ctg}\psi}{2} k_{пер} \left[c + \xi\left(q_0 + \frac{q_{кр} - q_0}{3} k_{пер}\right) \operatorname{tg}\varphi_0\right]} b_p \quad (20)$$

In the dependences (18), (19) and (20), the angle ψ is unknown. It is determined on the basis of the assumption that the resistance to cutting the knife should be minimal [3, 7]. The angle of displacement of the soil is determined by the maximum value of the depth of the cleavage [2–9, 12].

The dependencies of the relative critical depth of cutting in the upper tier from the angle of soil shift are given in Fig. 2, 3, and the cutting angle of the cutter in Fig. 4.

Calculated dependencies show that with decreasing the cutting angle of the incisor from 50° to 20°, the relative critical depth of asymmetric blocked cutting ($h_{кр1}/b_p$) increases: for tight plastic clay and semi-hard clay from 1.71 to 3.12; for semi-solid loam from 1.76 to 3.27; for a hard clay sand from 1.78 to 3.75.

For semi-block cutting, this depth increases: for tight plastics clay from 2.92 to 5.21; for semi-hard clay from 2.78 to 5.03; for semi-solid loam from 2.77 to 5.14; for a hard clay sand from 2.65 to 5.45.

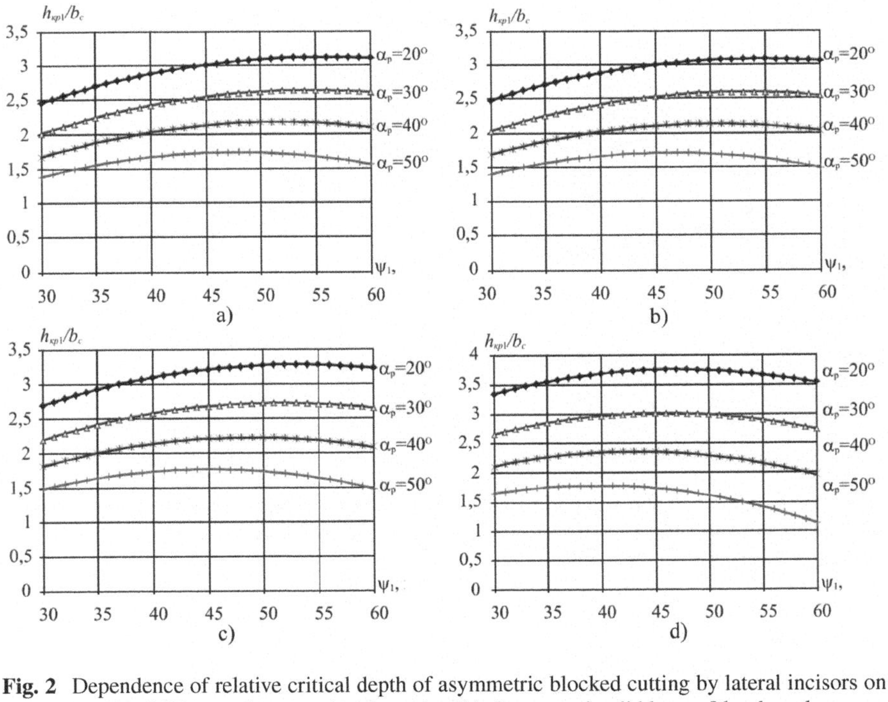

Fig. 2 Dependence of relative critical depth of asymmetric blocked cutting by lateral incisors on the angle of landslide: **a** refractory clay, **b** semi-solid clay, **c** semi-solid loam, **d** hard sand

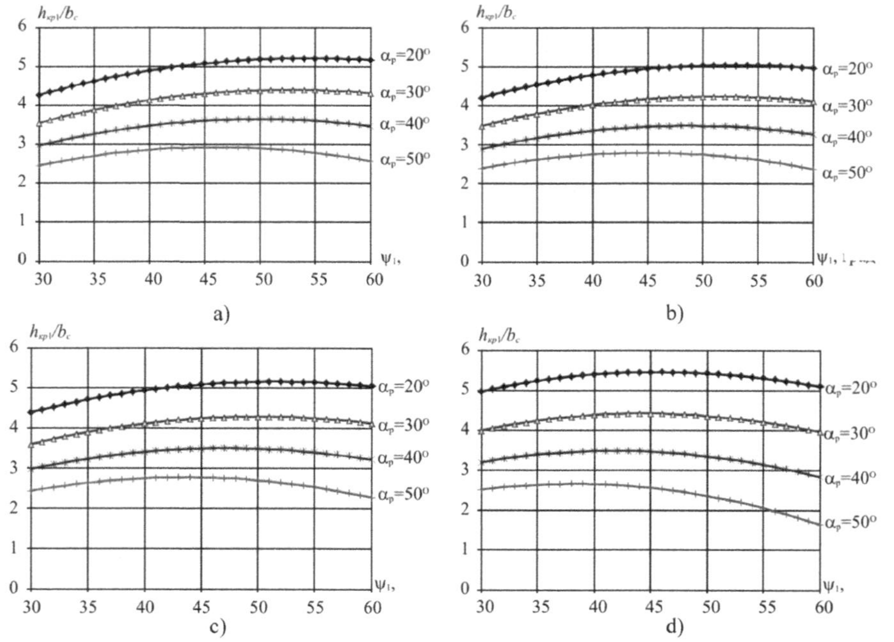

Fig. 3 Dependence of relative critical depth of half blocked cutting with lateral incisors from the angle of landslide: **a** refractory clay, **b** semi-solid clay, **c** semi-solid loam, **d** hard sand

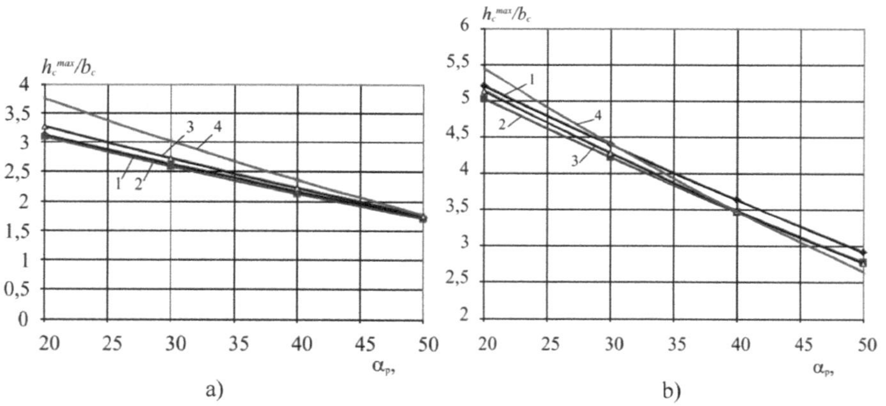

Fig. 4 Dependences of relative critical depth of cutting of soils from the angle of cutting of lateral incisors: **a** asymmetric blocking, **b** semi-blocked; **(i)** refractory clay, **(ii)** semi-solid clay, **(iv)** semi-solid loam, **(v)** hard sand

Table 1 Value of approximation coefficients for lateral incisors

Type of cutting Type of soil	Asymmetric blocking		Semi-Blocked	
	a'	*n'*	*a'*	*n'*
Refractory clay	4.02	0.046	6.70	0.076
Semi-solid clay	3.98	0.046	6.50	0.075
Semi-solid loam	4.26	0.050	6.69	0.079
Hard sand	5.02	0.066	7.26	0.093

Mathematical models (18) and (20) are cumbersome and inconvenient to calculate the critical depth of cutting. In order to simplify it, an approximation was made within the angle of cutting of the cutter $\alpha_p = 20\ldots50°$. Approximation error does not exceed 3.4%. As a result, a linear decreasing dependence was obtained (21),

$$\frac{h_{\kappa p1}}{b'_p} = a' - n' \cdot \alpha_p \quad (21)$$

where *a'* and *n'* are coefficients of approximation, depending on the type of soil and cutting; α_p—cutting angle of the cutter in degrees.

The values of the approximation coefficients for lateral incisors are given in the table (Table 1).

Based on the obtained approximated mathematical models in each particular case, based on the data on the thawing soils recommended by BNiP, it was possible, depending on the cutting angle of the cutter, within the cutting angle of 20°…50°, to determine the rational depth of cutting for the extreme lateral incisors of the multi-slip chains of trench excavators.

3 Conclusions

The mathematical models for determining the critical depth of cutting for the extreme lateral incisors of the multi-slip chains of trench excavators, operating in conditions of asymmetric lateral cutting and semi-block cutting are obtained.

The calculated dependences show that as the cutting angle of the cutter decreases from 50° to 20°, the relative critical depth of asymmetric blocked cutting ($h_{\kappa p1}/b_p$) increases: for refractory and semi-solid clay, from 1.71 to 3.12; for semi-hard loam from 1.76 to 3.27; for hard sand from 1.78 to 3.75. For semi-blocked cutting, this depth increases: for refractory clay, from 2.92 to 5.21; for semi-solid clay from 2.78 to 5.03; for semi-hard loam from 2.77 to 5.14; for solid sandwiches from 2.65 to 5.45.

Mathematical models of the relative critical depth of asymmetric blocked and semi-block cutting for the extreme lateral incisors of the multi-slip chains of trench excavators are approximated (error of approximation does not exceed 3.4%) linear dependence on the angle of cutting.

References

1. Alekseeva, T. V., Artemev, K. A. & Bromberh, A. A. (1972). *Dorozhnyie mashyni. ch. I. Mashyni dlia zemlianikh rabot* [Road machines. Machines for earthmovings]. (p. 504) Moscow: Mashynostroenye.
2. Andresen, L. & Jostad, H. P. (2002). A constitutive model for anisotropic and strain-softening clay. Proceedings Numerical Modern in Geomechanics (pp. 79–84). NUMOG VIII, Rome, Italy.
3. Artem'yev, K. A. (1989). *Teoriya rezaniya gruntov zemleroyno-transportnymi mashinami* [Theory of soil cutting by earth-moving machines]. (p. 80). Omsk: OmPI.
4. Hashash, Y. M. A., & Whittle, A. J. (2002). Mechanism of load transfer and arching for brand excavations in clay. *J. of Geotechnical and Geoenvironmental Engineering, 128*(3), 187–197.
5. Karlsrud, K. & Andresen, L. (2008). Design and performance of deep excavations in sof clays. *International Conference on Case Histories in Geotechnical Engineering*. 9.
6. Koval′, A. B., Musiyko, V. D., & Leychenko, YU. B. (2015). *Osnovy syntezu komponuval′noyi skhemy universal′nykh zemle-ryynykh mashyn bezperervnoyi diyi* [Fundamentals of the synthesis of the layout scheme of universal ground-breaking machines of continuous action]. Systemy i srodki transportu samochodowego. Wubrane zagadnienie. Monografia Nr 4.38 Seria: Transport pod redakcja naukowa Kazemierza Lejdy (pp. 263–268). Rzeszów (Polska): Politechnika Rzeszowska.
7. Kravets, S. V., Kovanko, V. V. & Lukyanchuk, O. P. (2015). *Naukovi osnovy stvorennia zemleryino-yarusnykh mashyn i pidzem-norukhomykh prystroiv: monohrafiia* [Scientific basis for the creation of earth-tiered cars and underground moving devices: a monograph]. (p. 322) Rivne: NUVHP.
8. *Krytychnohlybynni dvoyarusni hruntorozpushuvachi*: *monohrafiya* [Critical deep two tier soil rippers]/S.V.Kravets′ ta in.; za zah. red. S.V. Kravtsya. (p. 235) Rivne: NUVHP. ISBN 978-966-327-384-6 (2018).
9. Musiyko, V. D. (2015). *Masshtabni efekty pry fizychnomu modelyuvanni protsesiv rizannya gruntiv* [Large-scale effects in the physical modeling of soil-cutting processes]. Visnyk Natsional′noho universytetu vodnoho hospodarstva ta pryrodokorystuvannya. Zbirnyk naukovykh prats′. Tekhnichni nauky (Vol. 2(70), pp. 112–119). Rivne: NUVHP, Vyp.
10. Stanevskyi, V. P. (1984) *Sovershenstvovanye robocheho protsessa zemleroinikh mashyn* [Perfection of the working process of digging machines] (p. 128). K.: Vyshcha shkola. Yzd-vo pry KHU.
11. Vetrov, A. Yu. (1972). *Rezaniye gruntov zemleroynymi mashinami* [Soil cutting by earth moving machines]. (p. 359). Moscow: Mashinostroyeniye.
12. Zav'yalov, A. M. (2012). *Matematicheskoye modelirovaniye rabochikh protsessov dorozhnykh i stroitel'nykh mashin: imitatsionnyye i adaptivnyye modeli* [Mathematical modeling of work processes of road and construction machines: imitation and adaptive models]: monografiya (p. 411). Omsk: SibADI.

Calculation Method of Safe Operation Resource Evaluation of Metal Constructions for Oil and Gas Purpose

Valerii Makarenko, Andrii Manhura, and Iryna Makarenko

Abstract The safe life of pipeline structures depends largely on the corrosion-mechanical properties of the pipe steels. But the scientific-technical and technological developments to improve the operational reliability and durability of the pipelines still reveal contradictions and uncertainties both among researchers and practitioners; the lack of a clear idea of the reasons and factors that cause failure and destruction of metal structures, as well as scientifically substantiated practical recommendations regarding the optimal choice of pipe steels, operated under conditions of chemical-aggressive environments under variable temperature-barrier modes of production, delays the correct choice of organizational and technical measures to eliminate corrosion-mechanical damage and reduce the bearing capacity of metal structures. This, in turn, can lead to severe technological and economic and environmental consequences. This problem is exacerbated by the lack of scientifically sound concepts of technical diagnostics and methods of determining the working (accident-free) resource and the lack of efficiency of traditional methods and means of nondestructive testing. In this regard, at the present time are important computational methods using modern criteria for assessing reliability of complex structures based on fracture mechanics using mathematical tools involving probability theory. The paper presents a new approach in the estimation of the safe (working) resource of pipeline structures of oil and gas purpose.

Keywords Pipeline · Corrosion · Damage · Reliability · Resource · Fracture · Bearing capacity

1 Introduction

The fuel and energy complex of Ukraine is the basis of economic development and significantly influences the growth of scientific and technological progress, the intensification of industrial processes, the increase and improvement of their technological level, as well as in many cases determines the rate of growth of national well-being of

V. Makarenko · A. Manhura (✉) · I. Makarenko
Poltava National Technical Yuri Kondratyuk University, Poltava, Ukraine
e-mail: mangura2000@gmail.com

V. Onyshchenko et al. (eds.), *Proceedings of the 2nd International Conference on Building Innovations*, Lecture Notes in Civil Engineering 73,
https://doi.org/10.1007/978-3-030-42939-3_63

the country. Despite the decline in oil production, caused by the economic difficulties of the transition period in our country and fluctuations in prices on the world market, the major part of the transport systems for pumping and storage of oil and petroleum products continues to be actively exploited.

The problem of ensuring the required operational reliability of oilfield equipment and engineering steel structures is becoming more and more urgent every year due to the prevailing tendency of the equipment aging in comparison with the rates of technical re-equipment of the oil and gas industry of Ukraine. Therefore, one of the priority tasks in difficult circumstances, when updates are physically and morally obsolete fixed assets with financial difficulties occur in limited quantities, is the storage and increase the oilfield service systems through the use of effective methods and ways to increase overhaul cycles.

The safe life of piping structures depends largely on the corrosion-mechanical properties of the pipe steels. However, the scientific and technical and technological developments that exist so far to improve the operational reliability and durability of the pipelines reveal contradictions and uncertainties both among researchers and practitioners. Lack of a clear idea of the causes and factors that cause failure and destruction of metal structures, as well as scientifically substantiated practical recommendations regarding the optimal choice of pipe steels, which are operated under conditions of chemical-aggressive environments at variable temperature and baric conditions of production [1–23]. This, in turn, does not make it possible to develop effective organizational and technical measures for the prevention of failures and accidental destruction of pipelines, which can lead to severe technological and economic-environmental consequences. This problem is exacerbated by the lack of scientifically sound concepts of technical diagnostics and methods of determining the working (accident-free) resource and the lack of effectiveness of traditional methods and means of nondestructive testing. In this regard, calculation methods using modern criteria for estimating the reliability of elements of complex structures based on fracture mechanics using a mathematical apparatus with the involvement of probability theory are of great importance nowadays [1, 6, 8].

Therefore, the purpose of this work was to research and develop a method of engineering assessment of the safe operational resource of oil pipeline structures.

The method of collecting and processing information about the technical condition of the research object. It is known [4, 6, 8–10, 13–15] that the technical condition and operational safety-free residual life of pipelines is determined by mechanical stresses in the pipe wall. During operation, the pipeline is exposed to the following factors:

- The internal pressure exerted by the transported product;
- Elastic bending (pipeline distortion) in the vertical and horizontal planes;
- Temperature influence.

Internal pressure causes annular (circular) and longitudinal stresses in the pipe wall, which are determined by the following Laplace formulas:

Circular stresses

$\sigma^{P}_{\text{кц}} = [\, n \cdot P\,(D - 2\delta_{\text{т}})] / 2\delta_{\text{т}}$, where P—is the internal pressure in the pipeline during repair, PA; δ_t—is the pipe wall thickness (actual), m; D—is the outer diameter of the pipe, m; n—reliability coefficient for load—internal working pressure in the pipeline;

Longitudinal stresses

$\sigma^{p}_{пр} = \mu\sigma^{p}_{кц} - \alpha\, E\, (T_г - T_р) = [\mu \cdot n \cdot P \cdot (D - 2\delta_т) / 2\delta_т] - \alpha \cdot E \cdot (T_г - T_р)$,

where $\alpha = 12\ 10^{-6}$ 1/deg—coefficient of thermal expansion of steel; $E = 206\ 10^9$ PA—modulus of longitudinal elasticity of steel; T_g—soil temperature during laying (during construction); T_r—soil temperature during repair work, $\mu = 0.3$—coefficient of transverse deformation.

During the normal operation of the pipeline (as well as in the course of repair work), the total stresses must not exceed the permissible stresses $\sigma_{\text{dop}} = [\sigma]$, determined taking into account the parameters of the actual technical condition of the pipeline, including mechanical properties of steel, service life, level of defect, safety category, and coefficients reliability.

The permissible stresses of the pipe wall are determined in accordance with the requirements set out in SNiP 2.05.06-85 * and other normative documents [5, 11, 18, 19].

The allowable stress calculations use the calculated metal tensile (compression) resistance R_1 and R_2:

$$R_1 = (\sigma_B m)/(k_1 k_H),\ R_2 = (\sigma_n m)/(k_2 k_H),$$

where $\sigma_в$ is the minimum value of the tensile strength; σ_T is the minimum value of the yield strength; m—coefficient of working conditions; k_1 and k_2 are material reliability coefficients; k_n—coefficient of reliability for the purpose of the pipeline.

The values of σ_t and σ_b for new pipes are determined from the certificates for the long-term pipelines—experimentally or calculated taking into account the effect of the aging of the pipe metal and welds.

To perform the calculations take the values of the listed coefficients, which correspond to the largest safety margin.

The value of the coefficient m depends on the category of the site and is determined by Table 1 SNiP 2.05.06-85 *. The smallest value of $m = 0.75$ for category I and II pipelines under review.

The coefficient k_1 depends on the peculiarities of the technology of pipe production and is determined by Table 9 SNiP 2.05.06-85 *. The largest value is $k_1 = 1.55$.

The coefficient k_2 depends on the features of the technology of pipe production and the ratio σ_t/σ_v, determined by Table 10 SNiP 2.05.06-85*. The largest value $k_2 = 1.15$.

The coefficient k_n depends on the diameter of the pipeline, determined according to Table 11 SNiP 2.05.06-85 *. For pipelines with pressure up to 5.5 MPa, the highest value is $k_n = 1.0$.

The permissible stresses are determined by the following conditions.

The first condition: The circular tensile σ_{kc} must be no more than the permissible value $[\sigma_{kc}]_1$:

Table 1 Results of calculations of allowable internal working pressure during operation of OJSC Ukrnafta pipelines

Mark steel	Diameter, mm	Wall thickness, mm	The yield stress, MPa	Tensile strength, MPa	The rate of attenuation from aging	Permissible internal pressure, MPa
17HS	530	7.2	372.4	509.6	0.787	4.97
ST10	426	8.3	205.8	392	0.787	5.48
ST20	426	8.3	345	455.7	0.787	6.46
17H1SU	508	9.8	310.7	519.4	0.781	7.28
14HGS	529	9	338.1	509.6	0.781	6.22
10H2S1	529	9	372.4	519.4	0.781	6.34
09H2S	529	12	294	346.9	0.781	6.54
Old oil pipeline (>40 years of operation)						
ST10	508	8.1	310.6	519.4	0.641	4.87

$\sigma_{\text{кс}} \leq [\sigma_{\text{кс}}]_1 = R_1\ \psi_1 / n_p$.
n_p—is the coefficient of reliability for the internal pressure in the pipeline, the highest value being $n_p = 1.15$; ψ_1—is the coefficient taking into account the two-axis stress state of the pipeline wall, which is determined by the formula:

$$\Psi 1 = \sqrt{1 - 0.75^2} - 0.5\xi,$$

where ξ—is the coefficient of biaxiality of the stress state, defined as follows: $\xi = 0$ for tensile longitudinal stresses, $\xi = R_1\ [\sigma_{pr}]_1$—for the compressive longitudinal stresses.

The second condition: The longitudinal stress σ_{pr} in absolute value must not exceed the permissible value $[\sigma_{pr}]_2$, determined by the formula:

$$\sigma_{\Pi p} \leq \left[\sigma_{\Pi p 2}\right] = R_1 \cdot \psi_2.$$

Here, ψ_2 is the coefficient taking into account the biaxial stress state of the pipeline wall determined by the formula:

$$\Psi 2 = \sqrt{1 - 0.75\eta^2} - 0.5\eta,$$

where η—is the coefficient of biaxiality of the stress state, defined as follows: $\eta = 0$—for tensile longitudinal stresses, $\eta = 1.15[\sigma_{\text{кц}}]/R_1$ – for compressive longitudinal stresses.

The third condition: The total longitudinal stress σpr in absolute value must not exceed the permissible value $[\sigma]_3$, determined by the formula:

$$\sigma_{\Pi p} \leq \left[\sigma_{\Pi p}\right]_3 = \psi_3 \cdot S.$$

$S = \sigma_t \cdot m/0.9\ k_n$; the coefficients m and k_n defined above; ψ_3—coefficient taking into account the biaxial stress state of the pipeline wall, determined by the formula:

$$\Psi 3 = \sqrt{1 - 0.75\omega^2} - 0.5\omega,$$

where ω is the coefficient of biaxiality of the stress state, defined as follows: $\omega = 0$—for tensile total longitudinal stresses σ_{pr}; $\omega = \sigma_{kc}/S$—for compressive longitudinal stresses σ_{pr}.

The fourth condition: The circular tension σ_{kc} shall not exceed the permissible value σ_{kc} determined by the formula:

$$\sigma_{kc} \leq [\sigma_{kc}]_4 = S.$$

The S value is defined above.

Fifth condition: The total longitudinal stress σ_{pr} must not exceed the absolute value of 0.7 from the yield point of the metal:

$$\sigma_{\Pi p} \leq [\sigma_{\Pi p}]_5 = 0.7\sigma_T.$$

1. Defectiveness and aging of metal. When defects are found on the wall of the pipeline, it is necessary to determine the coefficients of attenuation of the pipe wall in the annular and longitudinal directions σ_{kc} and σ_{pr}, by special methods [2, 4, 8].

   The aging of the pipe metal in the long-term operation of the pipeline is taken into account by the input coefficient of deformation aging K_S, determined by the formula

   $K_S = 1 + 0.025\ C_{экв} \cdot t$,

   where t—is the life of the pipeline (years); $C_{ЭКВ}$—is the carbon equivalent of a metal, expressed as a percentage.
2. The limit values of the annular and longitudinal stresses shall be reduced by taking into account the coefficients $\alpha_{кц}$, $\alpha_{пр}$, K_S as follows:

   $$\sigma_{кц} \leq \frac{\alpha_{кц}}{k_S}[\sigma_{кц}]_1 = \frac{\alpha_{кц}}{K_S} \cdot \frac{R_1}{n_P} \cdot \psi_1;\quad \sigma_{пр} \leq \frac{\alpha_{пр}}{K_S}[\sigma_{пр}]_2 = \frac{\alpha_{пр}}{k_S} \cdot R_1 \cdot \psi_2;$$

   $$\sigma_{пр} \leq \frac{\alpha_{пр}}{K_S}[\sigma_{пр}]_3 = \frac{\alpha_{пр}}{k_S} \cdot \psi_3 \cdot S;\quad \sigma_{кц} \leq \frac{\alpha_{кц}}{k_S}[\sigma_{кц}]_4 = \frac{\alpha_{кц}}{k_S} \cdot S;$$

   $$\sigma_{пр} \leq \frac{\alpha_{пр}}{k_S}[\sigma_{пр}]_5 = \frac{\alpha_{пр}}{k_S} \cdot 0.7\ \sigma_Т.$$
3. The algorithm of calculation: The calculation of the predicted engineering assessment of the residual (working) resource, taking into account the deformation aging of the metal is carried out in the following sequence:

a) The condition of the pipeline is evaluated, for which the following operations are performed:

- The defects and thickness of the wall of the pipeline are diagnosed;
- Technical changes of the pipeline VAT are measured.

b) A calculation model is created;
c) The residual resource is calculated.

A special computer program is used to determine the allowable internal pressure. The solution algorithm consists of the following steps:

1. Enter the input data:

   D—pipe diameter, m; δ_t—is the thickness of the pipe wall, m; σ_t—is the yield point of the metal, PA;
   σ_v—Limit of strength of metal, PA; $E = 206\ 10^9$ PA—is the modulus of elasticity of the metal pipe;
   $\mu = 0.3$—Poisson's ratio; $\alpha = 12\ 10^{-6}$ 1/deg—coefficient of thermal expansion of metal; T_g—pipeline laying temperature, deg;
   T_r—Temperature of operation or repair, deg; m—coefficient of working conditions; k_1 and k_2 are the material reliability coefficient;
   k_n—Coefficient of reliability for the purpose of the pipeline; P_0 and ΔP—the initial internal pressure and the set accuracy of pressure determination, Pa;
   n—Reliability factor for load—internal working pressure in the pipeline;
   α_{kc}—Coefficients of pipe attenuation from defect in annular and longitudinal directions;
   t—The service life of the pipeline.

2. Determine the calculated metal tensile resistance (compression) of R_1 and R_2 (see formulas above).
3. Calculate the annular and longitudinal stresses from the action of the first group of forces—internal pressure and temperature influence
   $\sigma_{\text{кс}(1)} = [nP\,(D - 2\delta_T)] / 2\delta_T$,
   $\sigma_{\text{пр}(1)} = \mu\sigma_{\text{кц}(1)} - E\alpha\,(T_\text{г} - T_\text{р})$.
4. Calculate the total circular tension $\Sigma\sigma_{\text{кц}} = \sigma_{\text{кц}(1)}$.
5. Calculate $\sigma_{pr(2)}$ and $\sigma_{pr(3)}$ – flexural stresses in the vertical and horizontal directions.
6. Determine the minimum and maximum values of longitudinal stresses by formulas with respect to signs:

$$\min\sigma_{\Pi p} = \sigma_{\Pi p(1)} - \sqrt{\left(\sigma_{\Pi p(2)}\right)^2 + \left(\sigma_{\Pi p(3)}\right)^2};$$

$$\max\sigma_{\Pi p} = \sigma_{\Pi p(1)} - \sqrt{\left(\sigma_{\Pi p(2)}\right)^2 + \left(\sigma_{\Pi p(3)}\right)^2}.$$

7. Determine the auxiliary quantities ξ and ψ by the formulas:

$$\xi = 0 \text{ when } \sigma_{pr} \geq 0;\ \xi = -\left(\min \sigma_{pr}\right)/\mathrm{R}_1 \text{when } \min \sigma_{pr} < 0$$

$$\Psi 1 = \sqrt{1 - 0.75\xi^2} - 0.5\xi$$

8. Determine the auxiliary quantities η and ψ_2 by the formulas, with $\eta = 0$ when $\sigma_{pr} \geq 0$:
   $\xi = 1.15\, \sigma_{\text{кц}}/R_1$, when $\sigma_{\text{пр}} < 0$; $\Psi 2 = \sqrt{1 - 0.75\eta^2} - 0.5\eta$.
9. Determine the auxiliary quantities S, ω and ψ_3 by the formulas:
   $S = (\sigma_T \cdot m)/0.9\, k_н$; $\omega = 0$, when $\sigma_{\text{пр}} \geq 0$;
   $\omega = \sigma_{\text{кц}}/S$, when $\sigma_{\text{пр}} < 0$; $\Psi 3 = \sqrt{1 - 0.75\omega^2} - 0.5\omega$.
10. Determine the attenuation coefficients of defects and the coefficient of aging of the metal k_c.
11. Check strength and durability conditions (five conditions above).
12. When all strength and durability conditions are met, the test pressure should be increased by ΔP and all calculations repeated (3–10).
13. When at least one condition is not satisfied, the pressure in the previous cycle of the algorithm is accepted.

The scheme of calculation of the predicted engineering estimation of the residual (working) resource taking into account the deformation state of metal of pipeline structures is shown in Fig. 1.

The systematic analysis of statistics on the failures of pipelines over the long term of operation shows that, as a whole, metal structures are in a state that meets the requirements of design and technical and technological documentation for such

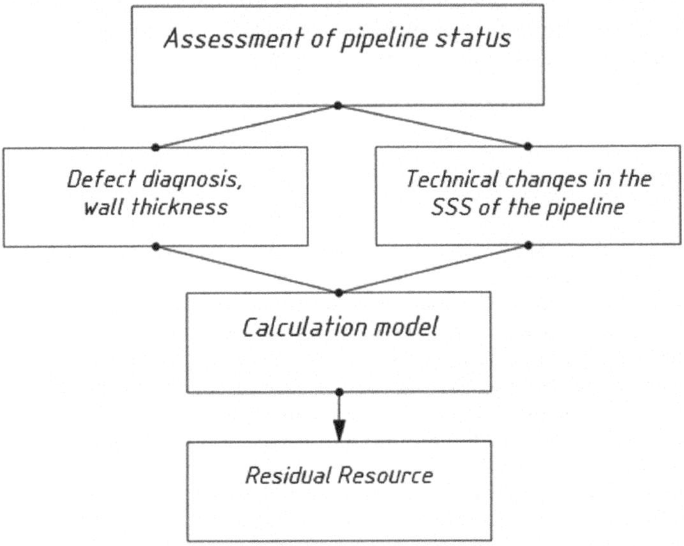

Fig. 1 Scheme of pipeline calculation residual resource

objects. However, many of them have a mode of intensive corrosion-mechanical wear of individual sections of metal structures, which requires timely technical diagnostics and routine preventive control and repair.

The table shows, as an example, the results of calculations of the allowable internal pressure of the sections of the three surveyed pipelines, taking into account their characteristics. The calculations are made at the coefficients of attenuation and aging $\alpha_{\text{пр}}/K_S$, which correspond to the terms of operation of the pipelines and their defects.

The calculations show that the old pipeline, which is operated at the facilities of NJSC "Nadvirna Nafta" of OJSC "Ukrnafta," with a permissible internal pressure 4.87 MPa.

When assessing the possibility of extending the life of the surveyed pipelines with a pressure of medium to 4 MPa, the comparison was made with the data of the technical requirements laid down in SNiP 2.05.06-85 "Industrial pipelines" and other regulatory and scientific and technical documents.

2 Conclusions

A method for calculating the safe resource of pipeline systems has been developed, which can be used to evaluate the feasibility of carrying out both major repairs and development of design and technological solutions with the aim of extending the operational life of pipelines and other equipment for oil and gas purposes.

An algorithm for calculating the residual (working) life of industrial pipelines was developed, taking into account the fracture toughness of metal, which is based on modern criteria for fracture mechanics and programs for mathematical modeling of corrosion-fatigue fractures of steel pipelines working under pressure and long-lasting medium conditions temperatures (+ 40... −30 °C).

References

1. Makarenko, V. D. (2006). *Reliability of oil and gas production systems*. Chelyabinsk Central Scientific Research Institute.
2. Borodavkin, P. P. (1992). *Underground pipelines Design and construction*. Moscow: Nedra.
3. Lutak, V. P., & Boychuk, I. Ya. (2002). Operational reliability of oil pipelines under the conditions of the Nadvirnaftogaz NGDU. *Oil and gas industry., 2,* 38–40.
4. Gumerov, A. G., Yamaleev, K. M. & Zhuravlev, G. V. et al. (2001). *Fracture resistance of metal pipes of oil pipelines*. Moscow: LLC "Nedra-Business Center".
5. Makarenko, V. D. (1985). SNiP 2.05.06-85. *Calculation of pipelines for strength and stability* (p. 25–36). Moscow: Stroyizdat.
6. Makarenko, V. D., Paliy, R. V. & Galichenko, E. N. et al. (2002). *Physicomechanical fundamentals of hydrogen sulfide corrosion destruction of field pipelines*. Chelyabinsk Central Scientific Research Institute.
7. Paliy, R. V. (2002). *System-technological methods for managing the safety of field pipelines*. TSNTI: Chelyabinsk.

8. Pokhmursky, V. I. (1982). *Corrosion fatigue of metals*. Kiev: Naukova Dumka.
9. Radkevich, O. I., Chumalo, G. V. (2003). Metal damage to industrial pipelines in a hydrogen sulfide environment. Physics Chemistry Mechanics of Materials. *4*, 112–114.
10. Makarenko, V. D., Paliy, R. V., Mukhin, M. Yu. (2001). *Technological basis for ensuring the operational reliability of pipelines*. Moscow: LLC "Nedra-Business Center".
11. (1981). SNiP II-23-81*. *Steel structures*. Moscow: FSUE CPP.
12. Honeycomb, R. (1968). *The plastic deformation of metals*. Мир: Edward Arnold. Москва.
13. Maas, W. B. (1981). *Metalloberfache*. Schw. u.Schn. *35*(5), 382–391.
14. Little, B., Wanner, P., & Mansfeld, F. (1991). Intern. *Materials, 36*(6), 253–262.
15. Migel, R., & Ruge, V. (2011). Hydrogen as alloy element. *Schw. u. schn, 7,* 250–252.
16. Marvin, C. W. (2012). Determining the strength of Corroded Pipe. *Materials propection and Perfolmance, 11,* 38–40.
17. Vasilkovsky, O., & Rivard, A. (2013). The effect of hydrogen sulfide in gruide oil on fatigue ckack grouth in pipe line steel. *Corrosion, 38*(1), 19–22.
18. (2006). DBN B.1.2-2.2006. *System reliability and safety of construction projects.*
19. NACE Standard TM-01-77(90). (2009). *Standard Test Method NASE.* Houston. P.O.BOX 218340.
20. Makarenko, V. D., Muravjev, K. A., & Kalyanov, A. I. (2006). Special features of manual ars weldinq of root joints in nonrotatinq welds in pipelines in Westem Siberia. *Weldinq International., 10*(5), 64–71.
21. Makarenko, V. D., Shatilo, S. P., & Astafev, V. I. (1998). Methods of increasing the corrosion resistance of oil pipelines. *Welding International, 12,* 34–39.
22. Makarenko, V. D., & Shatilo, S. P. (1999). Increasing desulphurisation of the metal of welded joints in oil pipelines. *Welding International, 12,* 56–61.
23. Makarenko, V. D., Beljaev, V. A., Protasov, V. N., & Shatilo, S. P. (2000). Mathematical model of the mechanism of resistance of welded joints in oil and gas pipelines to static hydrogen fatigue. *Welding International, 4,* 83–88.

Hydrate Formations Modeling for the Oil and Gas Facilities Reconstruction

Valeriy Makarenko, Yuriy Vynnykov, Anna Liashenko, and Oleksandr Petrash

Abstract Absence of full-scale tests results on real models that adequately reflect the physical and thermodynamic processes occurring in the intertubular space during oil production does not permit to develop effective technological methods as well as organizational and technical measures for the prevention and reduction of the probability of hydrate formations, corrosion-active against the surface of casing, and tubing in the annular space of oil wells. To extend the lifetime of oil and gas facilities, the information-measuring system and equipment have been developed, which permit to fully approximate the conditions and modes of hydration of iron and its oxides to real down-hole processes and characteristics, which make it possible to investigate hydrate formations with a wide probability in a wide variation range of temperature and pressure. Obtained through information and measuring system, the results of the statistical experiments gave the opportunity to determine the conditions necessary for hydrate formations of iron (Fe) and its oxides in the temperature range from −15 to +60 °C and a pressure of 0–60 MPa, which confirmed thermodynamic calculations of phase equilibria in the annular space of wells known from the literature.

Keywords Hydrates · Corrosion · Activity · Well · Paraffin · Bactericides

1 The Actuality of the Topic and the Formation of Problems

Oil production and transportation are inevitably accompanied by the sedimentation and accumulation of oil sludge in wells and the field pipelines, which lead to a decrease in the effective diameter of the production string and complicates the operation of oil pipelines and reservoirs, resulting in a decreased oil production (by 20%) and increased power consumption when oil is pumped out (by 10–15%). According to [1, 2, 5], the deposited conglomerate (oil sludge) consists of paraffin (up to 60–70%), resins, asphaltenes, water, and mechanical impurities (iron hydrates, hydrates of iron oxides, and hydrate formations of other metals, iron oxides, clay, sand, carbonates, and other salts). According to the literature [7–11], asphalt, resin,

V. Makarenko · Y. Vynnykov · A. Liashenko (✉) · O. Petrash
National University "Yuri Kondratyuk Poltava Polytechnic", Poltava, Ukraine
e-mail: anliashenko14@gmail.com

V. Onyshchenko et al. (eds.), *Proceedings of the 2nd International Conference on Building Innovations*, Lecture Notes in Civil Engineering 73,
https://doi.org/10.1007/978-3-030-42939-3_64

and paraffin deposits are difficult to dissolve in petroleum solvents like gasoline, kerosene, diesel, etc., due to the high content of refractory paraffin. The solubility of the petroleum conglomerate increases sharply at temperatures close to the melting point of paraffin (50–60 °C).

Oil production at the deposits of the Poltava region is accompanied by intense paraffin hydrate formation in production strings and annular space (between the outer surface of the tubing and the inner surface of the casing). To extend the lifetime of oil and gas facilities, there are significant labor, material, and financial resources used by oil companies.

In order to prevent the plug formation and preventive removal of paraffin hydrates, the following methods are generally used in the world practice [3–18] according to literature: (1) wells cleaning with inhibitors or chemical reagents and hot water (hydrochemical method); (2) the use of scraper tools (mechanical method); (3) the use of lined tubing by applying a granular glass or epoxy resin to its inner surface; (4) application of bactericidal protection; (5) the use of magnetic methods of protection; (6) the use of special electric heating devices, which include, for example, a tubular barrel with heating elements placed on a spiral on its surface; the heater is lowered into the column and fed with power, and the thermal energy of the heater is transmitted through the tube column, which leads to the destruction of hydrates and frozen paraffin deposits in the tubing, melting them; (7) application of direct electric heating of wells with the help of special electrical installations using the heating element of the string and casing.

To select the best way to combat hydration, further studies are needed to study the conditions, regimes, zones, and patterns of paraffin hydrate formation in the production string and annular space in oil wells. Only by finding out the true picture of the physical mechanisms and the specifics of the formation of these deposits, it is possible to successfully develop new or improve existing effective technologies for the prevention and purification of complicated oil wells, contributing to the corrosion resistance of underground well equipment.

Absence of results of full-scale tests on real models that adequately reflect the physical and thermodynamic processes occurring in the annular space during oil production, does not allow to develop effective technological methods and organizational and technical measures for the prevention and reduction of the probability of hydration, corrosion-active against the surface of casing, and tubing in the annular space of oil wells.

The purpose of the work is to simulate experimentally the process of hydrate formations under conditions close to oil wells.

2 Methodology, Equipment, and Materials

For a detailed study of hydrate formations of iron (Fe) and its oxides in the conditions of the annular space of the oil well, special experiments were performed on the laboratory unit presented in Fig. 1. It permits to study hydration formations over

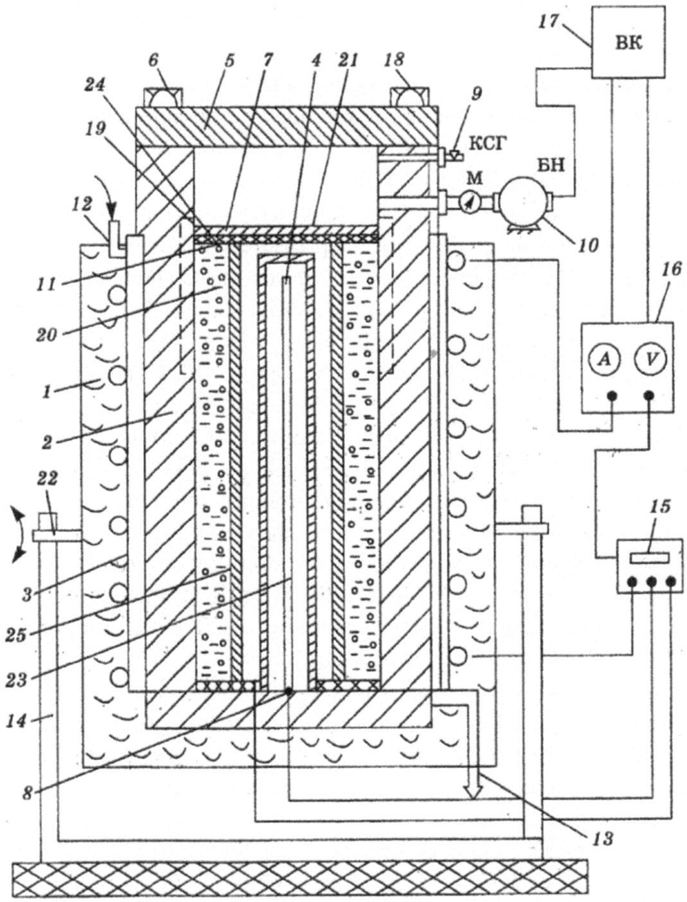

Fig. 1 Schematic diagram of the experimental installation: 1—furnace; 2—chamber of cylindrical shape; 3—cooling jacket; 4—tube; 5—means of cover; 6—squeeze bolts; 7—insulation (textolite) gasket; 8, 15—electronic potentiometers; 9—nipple; 10—pump; 11—grooves; 12—special device; 13—crane; 14—frame; 16—autotransformer with an ammeter and a voltmeter; 17—computational complex; 18—pulse tube; 19—groove; 20—reservoir water and gas fluid; 21, 25—upper and lower walls; 22—pins; 23—electric heater; 24—inner lid

a wide range of pressure changes (0.1–12 MPa) and temperature (−20 to +80 °C). Schematic diagram of the installation allows you to fully approximate the conditions and modes of hydration of iron and its oxides to the real well processes and characteristics.

The installation, in the design and manufacture of which was directly involved the authors of this work, consists of a chamber of cylindrical shape 2, outside of which is a cooling jacket 3, which if necessary pours liquid nitrogen through a special device 12. The discharge of liquid nitrogen at the end of the experiment is carried out through the crane 13. A chamber 2 with a cooling jacket 3 is placed in a furnace

1 with thermal and electrical insulation. Inside the pressure chamber, there are tube 4 with an electric heater 23 asymmetrically welded to the bottom, which is activated when liquid coolant is poured into the cooling jacket (at this time the oven is switched off). The temperature in the chamber is provided by an automatic electric heating system, which includes an autotransformer with an ammeter and a voltmeter 16, a set of thermocouples TP1–TP3, and electronic potentiometers 8 and 15.

The barometric mode in the chamber is provided by the pump BN10, connected to it by a pulse tube 18 through a nipple 9; the pressure is controlled by pressure gauge M. The test objects (1.4 m tubing coils) are placed in the chamber, which is sealed by means of cover 5 and squeeze bolts 6. To eliminate the influence of the upper and lower walls 21, 25 of the chamber on the measurement results, they are closed by insulation (textolite) gasket 7. Loading and unloading of research objects—samples of tubing and reservoir water and gas fluid 20 is carried out in an inclined position of the chamber, for which it is mounted on the frame 14, which allows rotating the camera around the axis of the pins 22. To exclude the possibility of air entrapment in the water–oil mixture under the increased pressure, the inner lid 24 is housed in the chamber, which under pressure moves up and down four grooves 11 cut along the shell of the chamber. The end portion of the auxiliary cover also has a groove 19 into which a seal ring is inserted. The results of testing of the methodology showed that no significant increase in oxygen content in the oil–water environment due to the ingress of air into the chamber was observed.

Data on the measured parameters (temperature and pressure) are fed into the computational complex OK-17. Other information like the volume of the liquid phase being investigated, data of gravimetric measurements, etc., are added thereafter. The computing complex includes a computer, a graphics device, a digital printing device, and peripheral nodes.

In the study, the effect of temperature, pressure, and nature of the oil–water mixture, taken directly from the reservoirs of the oil field, both the products of hydrate formations separately, and the result of their interaction with the steel surface of the tubing and the chamber were investigated. At the same time, as a result of the interaction of the corrosion-active medium with the surface of the tubing and the chamber, sulfides and oxides of iron are formed, which were not taken into account in the experiments on the study of hydrate formations (HF). In addition, we did not specifically investigate the products of asphalt, resin, and paraffin deposits (ARPD) on the walls of tubing and the surface of the chamber, which although in some cases were present, but in a very small amount (less than 5% wt.), Therefore, when processing the results of measurements, their formation was neglected.

The essence of the methodology for the study of hydration of iron and its oxides was as follows. After loading into the chamber, the sample tubes are welded on both ends, to which textolite gaskets were attached, the chamber was filled with a reservoir water–oil mixture, taken directly from the wells of the oil field, characterized by a high degree of hydrate formation of iron and its oxides in the annular space.

After sealing, the chambers were subjected to experiments with a duration of 24 h (as recommended by the International Corrosion Association–NACE). However, when the temperature was kept constant, the pressure was changing from 0.1 to

60 MPa. To study the effect of temperature, similar operations were performed at constant pressure and at different temperatures, which varied in the range from −15 to +60 °C. At the end of each experiment, the tubing sample and the surface of the chamber were cleaned of sediment with a special scraper, and the water–oil mixture was poured through a metal sieve into the tank. Separation of hydrate formations from sulfide and oxide compounds was performed with a magnet wrapped in cellophane. Since the hydrate formations are not sensitive to the magnetic field, after removal of the sulfides and oxides of iron, they were poured into a measuring cup for weighing on electronic scales (measurement accuracy of 0.001 g), and then determined the proportion of hydrate formations in the total mass of deposits. The reference variant was taken as the reference measurement—the pressure of 1 MPa, temperature −15 °C; in this case, the mass of hydrate formations averaged 30% of the total mass of deposits (sulfides and iron oxide, hydrate formations and ARPD). The measurement results were fed to the computer for processing. In total, 120 experiments were performed.

In the course of testing the technique at the laboratory stand, it was found that asphalt, resin, and paraffin deposits (ARPD) are not formed during the experiment (24 h) and therefore do not make an error in the measurement results. The reproducibility of the results of the experiments is 85–95%; the measurement error in the proposed method does not exceed 10%. Data processing was performed using the methods of correlation, dispersion, and regression analysis, which allowed to give a comparative analysis of the processes of formation of iron hydrates or its oxides in the annular space of the oil wells. As experimental specimens, tubing with a diameter of 73 × 5.5 mm, made of 36G2C steel and widely used at Ukrainian deposits, was used. Layered watered oil was collected into the tank directly at the outlet of the well.

3 Results of the Studies and the Discussion

The results of the experiments are shown in Fig. 2 and 3, which make it possible to determine the conditions necessary for the hydration of iron and its oxides. Thus, if the temperature varies from 0 to −15 °C, then at pressures of 0.1–1 MPa, the most favorable conditions for hydration formations are created. If the pressure is reduced (0.1–0.8 MPa), then conditions for the formation of hydrates occur at low temperatures (from −5 to −15 °C). If the pressure is 1 MPa, then only in the range of temperature changes from −15 to +15 °C conditions are created for the maximum formation of hydrates. With an increase in temperature from 20 to 40 °C only at pressures of 5 … 40 MPa, favorable conditions for hydration formations can be realized, although their probability does not exceed 40–50%. In the range of maximum temperatures of 40–60 °C, the formation of hydrates can occur, albeit with a low probability (10–20%), only at elevated pressures of 40–60 MPa.

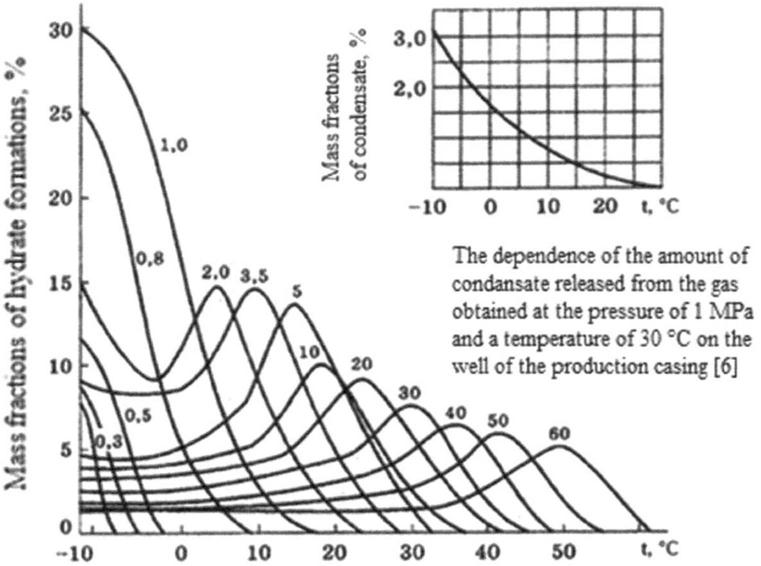

Fig. 2 Graphs of the dependence of the mass fraction of hydrate formations of iron (Fe) and its oxides on temperature: the figures on the curves—the value of pressure in MPa

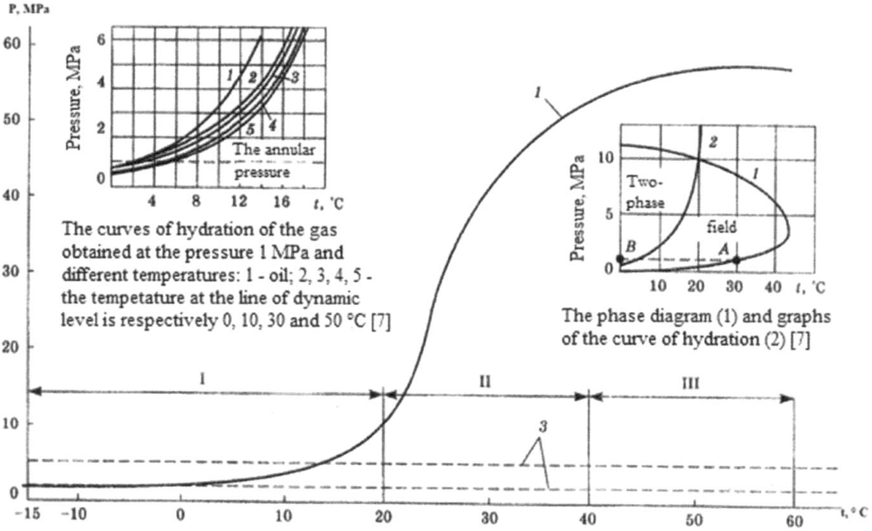

Fig. 3 Graphs of the curve of hydration (1) in the annular space of the oil well: zone I—maximum probability of formation of HF (90–100%), II—average probability of HF formation (40–50%), III—minimum probability of HF (10–20%); 3—lines of thermobaric conditions

The data in Fig. 3 clearly illustrate three zones of hydration formations depending on the temperature–pressure ratios, which are characterized by the different probability of hydrate formation. This is also shown in the graphs of hydration formations of the water–gas mixture and separately the gas released from the oil and water vapor obtained by calculation in [2–4, 18–21]. Comparison of these data in the interval of modes of temperature and pressure changes indicates its convergence, which confirms the adequacy of the developed methodology by the results of thermodynamic calculations of the phase equilibrium "oil–gas–hydrate."

It is noteworthy that in the zone of the line of dynamic (static) level in the well ($t = -5 \ldots +30$ °C, $p = 1 \ldots 5$ MPa), as well as in areas of intense condensation (at t 0 °C—zone of perennially frozen rocks) and t $-5 \ldots -15$ °C—seasonal (in winter season) decrease in temperature at the wellhead zone of the well create favorable conditions and temperature-bar modes for intensive hydration formation (see Fig. 2), which is confirmed by the literature data [2, 7] and the results of our own practical observations. In addition, it is important to note the experimentally confirmed fact known in the literature [6, 7, 11, 12]: as the temperature and pressure increase, the component composition of hydrate formations changes, which is caused by the large content of gas constituents in their structure. This should be taken into account when selecting effective reagents that contribute to the dissolution of hydrate formations and its removal from the surface of the well equipment.

4 Conclusions

1. The information-measuring system and equipment that permit to fully approximate the conditions and modes of hydration of iron and its oxides to the real well processes and characteristics, which make it possible to investigate with a great probability of hydration formation in a wide range of changes in temperature and pressure have been developed.
2. The statistical results of the experiments obtained using the information-measuring system, made it possible to determine the conditions required for the hydrate formations of iron and its oxides in the temperature range from -15 to $+60$ °C and pressures from 0 to 60 MPa, which are confirmed by thermodynamic calculations of phase equilibrium in the annulus of wells known from the literature.

References

1. Zhazyikov, K. T. (1975). K voprosu preduprezhdeniya parafinovyih otlozheniy v nefteprovodah. *Neftyanoe hozyaystvo., 5*, 6–7.

2. Vyazunov, E. V., & Golosovker, V. I. (2001) Issledovanie zakonomernostey parafinizatsii truboprovodov. RNTS «Transport i hranenie nefti i nefteproduktovo. Moskva. VNIINOENG. 12.
3. Mineev B.P. Dva vida parafina, vyipadayuschego na podzemnom oborudovanii skvazhin v protsesse dobyichi nefti / B.P. Mineev, O.V. Boligatova // Neftepromyislovoe delo. – 2004. – # 12. – S. 41–43.
4. V.D.Serediuk, L.O.Knysh. Zapobihannia vidkladanniu asfaltenosmoloparafinovykh vidkladiv u stovburi sverdlovyny // Naftohazova enerhetyka. – 2010. – № 1 (12). – S. 37-40.
5. Bai, Yong. (2001). *Pipelines and risers* (p. 495). Oxford, USA: Elsevier.
6. Ershov, V. A., Chetverkina, V. N., & Shakirova A. H. i dr. (2000). Issledovanie vliyaniya temperaturnyih usloviy na sostav i svoy¬stva smoloparafinovyih oblozheniy. Neftyanoe hozyaystvo. 9, 12–15.
7. Tronov, V. P. (2000). *Mehanizm obrazovaniya smoloparafinovyih otlozheniy i borbyi s nimi*. Moskva: Nedra.
8. Mazepa, B. A. (2002). *Zaschita neftepromyislovogo oborudovaniya ot parafinovyih otlozheniy*. Moskva: Nedra.
9. Shammazov, A. M., Ibraeva, E. M., & Fattahov, M. M. (2000). Issledovanie ravnovesnogo sootnosheniya asfalto-smolistyih veschestv i parafina v anomalnyih sistemah. Izv. vuzov. Neft i gaz. 11, 63–66.
10. Abramzon, A. A., & Chetverkina, V. N. (2003). O primenenii poverhnostno¬aktivnyih veschestv dlya ingibirovaniya parafinootlozheniya. Neftepro¬myislovoe delo. 5, 10–11.
11. Ibragimov N.G. Sovershenstvovanie metodov zaschityi kolonnyi NKT ot asfaltosmoloparafinovyih otlozheniy na promyislah Tatarstana / N.G.Ibragimov // Neftyanoe hozyaystvo. – 2005. – # 6. – S. 110–112.
12. Troyanovskiy, Y. V. (1971) Osobennosti parafinizatsii promyislovogo oborudovaniya i razrabotka mer po borbe s otlozheniyami v usloviyah mes¬torozhdeniy Zapadnoy Sibiri. Avtoref. diss. kand. tehn, nauk. Tyu¬men.
13. Cheremisin, N. A. (2001). *Issledovanie mehanizma obrazovaniya parafinogidratnyih probok v neftyanyih skvazhinah s tselyu sovershenstvovaniya metodov borbyi s nimi*. Tyumen: Avtoref. diss. kand. tehn. nauk.
14. Code of Practice for Pipelines—Part 1. (2004). *Steel Pipelines on Land, PD 41 8010, British Standards Institution*, 44p.
15. Manzhay, V. N., Trufanina, L. M., & Kryilova, O. A. (1999). Termomehanicheskiy sposob udalenie otlozheniy parafina, smol i asfaltenov iz neftepromyislovogo oborudovaniya. *Neftyanoe hozyaystvo, 4*, 35–38.
16. Charonov, V. Y., Muzagitov, M. M., & Ivanov, A. G. i dr. (2001). Sovremennaya tehnologiya ochistki neftyanyih skvazhin ot parafi¬na. Neftyanoe hozyaystvo. 4, 55–57.
17. Charonov, V. Y., Muzagitov M.M., Ivanov A.G. i dr. (2001) Sovremennaya tehnologiya ochistki neftyanyih skvazhin ot parafi¬na. Neftyanoe hozyay¬stvo. 4. 55–57.
18. Ivanova I.K., Koryakina V.V., Semenov M.E. (2018). Issledovanie protsessa gidratoobrazovaniya v emulsiyah asfaltosmoloparafinovyih otlozheniy metodom dsk // Fundamentalnyie issledovaniya. – # 11-2. – S. 143-149.https://doi.org/10.18372/0370-2197.1(82).13488 .
19. Mansoori, G. A. (n.d.). Paraffin / Wax and Waxy Crude Oil. The Role of Temperature on Heavy Organics Deposition from Petroleum Fluids. [Elektronniy resurs]. – Elektoron. dani (1 fayl). — Rezhim dostupu: http://www.uic.edu/~mansoori/Wax.and.Waxy.Crude_html
20. Tertyshna O. V. (2012). Dynamika formuvannia asfaltosmoloparafinovykh vidkladen / O.V. Tertyshna, Yu.Iu. Symonov, K.V. Afanasieva, L.O. Snizhko // Naftova i hazova promyslovist. – №5. – S. 48–50.
21. Vynnykov, Y., Manhura, A., Zimin, O., & Matviienko, A. (2019). Use of thermal and magnetic devices for prevention of asphaltene, resin, and wax deposits on oil equipment surfaces. *Mining of Mineral Deposits, 13*(2), 34–40. https://doi.org/10.33271/mining13.02.034.

Principle of Equireliability at the Internal Water-Supply System Design

Valeriy Novokhatniy, Oleksander Matyash, Sergiy Kostenko, and Stepan Epoian

Abstract The question of reliability increase in construction at the internal engineering networks designing is analyzed. The influence of the houses connection place to the water-supply system from the reliability point of view is established. The acceptance of the basic reliability indicators for calculating the continuity and renewability of water-supply is substantiated. The main indicators of the water-supply systems reliability in the selected direction should be taken as the mean failure time T and the mean recovery time T_R, and the complex indicator—the water-supply system readiness ratio for the direction K_R was set. Mathematical model for calculating the reliability of interior water-supply systems is a consistent combination of renewable elements. It includes a raising pumping station and a water-supply network. The choice of the place of house internal water-supply network connection in the center of equireliability is substantiated, and the calculations of the basic reliability indicators are made. With the optimum connection of the house to the external water-supply network, the faultlessness of the internal network in the least reliability direction was increased by 29%.

Keywords Reliability · The center of equireliability · Inlet of water-supply pipeline to the house

1 Introduction

The water-supply system is a chain of serially connected individual units (structures) from the water source to the consumer. It is necessary to maintain the reliability of all links in this chain for the reliable functioning of the water-supply path. One of these links is the in-house water-supply network—the last link on the water way into the apartment. "The rules for the provision of services to the public for district heating, hot and cold water-supply and sewerage" [1] require that "the permissible period of

V. Novokhatniy · O. Matyash (✉) · S. Kostenko
Poltava National Technical Yuri Kondratyuk University, Poltava, Ukraine
e-mail: matyash19831010@gmail.com

S. Epoian
Kharkov National University of Civil Engineering and Architecture, Kharkiv, Ukraine

V. Onyshchenko et al. (eds.), *Proceedings of the 2nd International Conference on Building Innovations*, Lecture Notes in Civil Engineering 73,
https://doi.org/10.1007/978-3-030-42939-3_65

deviation from uninterrupted water-supply is no more than 6 h per day and no more than two times per month." To ensure regular and renewable water-supply at this last link in the system, it is possible to increase the faultlessness and repairability, both through the use of new technological means and the use of reasonable technical solutions.

2 Problem Statement

Analysis of recent researches [2–11] shows that both scientists and water professionals are paying attention to solving the problem of water-supply reliability. However, when justifying the choice of water source location in an external water-supply network, usually only technological and economic factors are taken into account and reliability issues are not taken into account. For any building, including a residential unit, the water source for the internal water-supply network should be considered as the node (point) of connection to the external network, and then, the connection location should be chosen based on reliability.

3 The Aim of Research

The choice of the place of house internal water-supply network connection in the center of equireliability is substantiated, and the calculations of the basic reliability indicators are made and to carry out practical calculations of the basic reliability indicators.

4 Improving the Houses Water-Supply Reliability by Choosing a Connection to the Water-Supply System

4.1 Problem Formulation

One of the main issues during the design phase of internal water-supply systems is the choice of the building's connection location to an external water-supply system. Considering reliability, the main factors that influence the choice of the building's connection place to the water-supply system are the purpose of the water-supply system, the type, and length of the networks. If the internal water-supply system of a residential building with a conventional height of up to 26.5 m is designed or reconstructed, such water-supply system does not perform fire-fighting functions and has a branched-type water-supply network [12]. In such case, the main reliability indicators should be **the mean failure time** T and **the mean recovery time** T_R, and the complex indicator—**the readiness ratio K_R for the direction** of water-supply. This direction is determined from the input node to the dictating consumer, but the location of the input, relative to the dictating consumers, affects the reliability of the water-supply to them. This is due to the fact that the faultlessness of the supply pipeline is influenced by two main factors: the diameter and length of the pipeline (with the selected pipe material). The greater the length of the feed pipeline, with the accepted diameter and material of the pipes, the less the failure time T is. The optimum condition of reliability is the location of the internal water-supply system inlet, when the operating failure time T from the point of water-supply input to all dictating consumers is maximally possible, and the parameters of failure flow ω to dictating consumers are minimally possible. If the distances from the input node to the dictating consumers are the same, then we get the scheme of equal reliability, that is, the principle of water-supply equireliability to consumers is maintained. The optimization criterion should be the parameter of water-supply system failure's flow in the direction of water-supply from the inlet to the dictating consumer is equal $\omega = \frac{1}{T}$.

4.2 *The Center of Equireliability*

From reliability point view of, a building's connection is optimal, when the reliability of water-supply to all consumers will be maximally possible [13]. Let us call this point of the network as the center of equireliability. In terms of reliability theory, this means that failure time to dictate consumers should be maximized. It is more convenient to perform calculations through the failure flow parameters of the direction $\omega_d = \frac{1}{T}$. Most often there will be a situation where the equireliability center is located on a section of water-supply network between two nodes. If at this point it is not possible to form an input node, the equireliability center must be moved to the nearest node of the internal water-supply network.

4.3 *Mathematical Justification for the Equireliability Center Search*

Branched-type water-supply networks are networks without cycles that can be displayed as trees with a root at the top where the water source is located, and therefore, there is a path between any two vertices and, moreover, only one. In internal water-supply systems, the water source is an input node for connection to external networks. In [13], the mathematical justification of the equireliability center search for external branched-type networks was performed. It is proved that the equireliability center is located in the middle of the largest chain, which consists of series-connected sections of the external network. Vertical raisers, which are connected in the building's basement by the main road, also represent a branched-type water-supply network (Fig. 1).

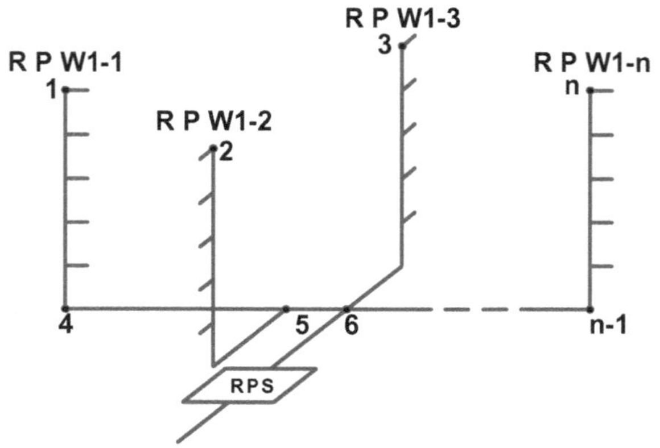

Fig. 1 Scheme of an internal water-supply system of a five-story building

Obviously, the longest chain will be a chain of sequentially connected sections 1–4–5–6–$(n-1)-n$ from node 1 to node n. Next, you need to find the middle of this chain and at this point place the input node and raiser pumping station (RPS). For algorithmically solving the problem, we use an algorithm with the transition to the parameter of the failure flow. First, we find the sum of the failure flow parameters from the first node of the vertex to all others, then from the second—to all others. In the following iterations, we find the values of the failure flow parameters between all the end nodes. In modern water-supply systems, the water pressure in external networks is reduced to 0.1 MPa to reduce water leakage and pipes damage, and in individual residential houses or groups of buildings, RPS is installed. The following example is considered below.

4.4 Mathematical Model of Internal Water-Supply System with RPS Reliability

The elements of the water-supply system will be RPS and network sections of different diameters. We will assume that the pipes and fittings of the network have the same reliability, given the slight difference in diameters. For reliability calculations, we use the following dependencies.

Failure flow parameter for the *i*th section of the network.

$$\omega_{S_i} = \sum_{i=1}^{n} \omega_{0_i} L_i, \tag{1}$$

where

L_i length of the pipeline section of the ith diameter, km;
ω_{0_i} failure flow parameter of ith diameter of 1 km pipeline

Network failure flow parameter in the selected direction

$$\omega_n = \sum_{i=1}^{n} \omega_{s_i}, \tag{2}$$

where

n the number of network sections in the selected direction

Raiser pump station failure flow parameter [14]

$$\omega_{ps} = \sum_{i=1}^{N} \upsilon_i, \tag{3}$$

where

υ_i contribution of the ith enlarged element to the raiser pump station failure parameter;
N the number of enlarged items

Failure flow parameter of the water-supply system in the selected direction

$$\omega_d = \omega_{ps} + \sum_{i=1}^{n} \omega_{0_i} \cdot L_i. \tag{4}$$

Mean failure time of the system in the selected direction

$$T = 1/\left(\omega_{ps} + \sum_{i=1}^{n} \omega_{0_i} \cdot L_i\right). \tag{5}$$

Mean system recovery time in the selected direction

$$T_R = \left(T_R^{ps}\omega_{ps} + T_R^n\omega_n\right)/\left(\omega_{ps} + \omega_n\right), \tag{6}$$

where

T_R^{ps} the mean recovery time of the pumping station;
T_R^n mean time of network recovery in the direction

Mean network recovery time for the selected direction

$$T_R^n = \left(\sum_{i=1}^{n} T_{R_i}^s \omega_{s_i}\right)/\sum_{i=1}^{n} \omega_{s_i}, \tag{7}$$

where

$T_{R_i}^s$ the mean recovery time of the ith network section

Readiness coefficient of the water-supply system in the selected direction

$$K_R = T/(T + T_R). \quad (8)$$

5 Calculation of Water-Supply Reliability

Let us consider, for example, the five-story building with PNS. There are 130 consumers and 300 water dismantling devices installed, pipe material—polypropylene. According to the principle of equireliability, optimal location of water-supply input to the house will be a connection closer to the middle of the house, and not from the end of the house (Fig. 2). Hydraulic calculation for two variants of the house connection was carried out, and diameters of water pipes and distribution network were adopted. The calculations were made in tabular form (Table 1). The most distant from the input of the house water-supply system and the high-mounted water disassembly device was adopted by the dictating consumer.

The calculation of the RPS reliability was performed by the contribution method [2] for two pumping units [14]—working and reserve (Fig. 3).

$$\omega_1 = 2\omega_{зв} = 2 \times 0.1 \times 10^{-4} = 0.2 \times 10^{-4}\, 1/h;.$$

$$\omega_2 = \omega_3 = \omega_p + \omega_v = (1.25 + 0.08) \times 10^{-4} = 1.33 \times 10^{-4}\, 1/\mathrm{h};$$

$$\omega_4 = 2\omega_{dv} = 2 \times 0.6 \times 10^{-4} = 1.2 \times 10^{-4}\, 1/\mathrm{h}.$$

$$T_{R_1} = T_{R_{dv}} = 20\,\mathrm{h};$$

$$T_{R_2} = T_{R_3} = \frac{\omega_p T_{R_p} + \omega_v T_{R_v}}{\omega_p + \omega_v}$$

$$= \frac{(1.25 \times 60 + 0.08 \times 10) \times 10^{-4}}{(1.25 + 0.08) \times 10^{-4}} = 56.99\,\mathrm{h};$$

$$T_{R_4} = T_{R_{dv}} = 10\,\mathrm{h}.$$

$$K_{D_3} = \omega_3 T_{R_3} = 1.33 \times 10^{-4} \times 56.99 = 75.8 \times 10^{-4}.$$

$$\upsilon_1 = \omega_1 = 0.2 \times 10^{-4}\, 1/\mathrm{h};$$

$$\upsilon_2 = \omega_2 K_{D_3} = 1.33 \times 10^{-4} \times 75.8 \times 10^{-4} = 10^{-6}\, 1/\mathrm{h};$$

$$\upsilon_4 = \omega_4 = 1.2 \times 10^{-4}\, 1/\mathrm{h}.$$

$$\omega_{ps} = \upsilon_1 + \upsilon_2 + \upsilon_4 = (0.2 + 0.01 + 1.2) \times 10^{-4}$$
$$= 1.41 \times 10^{-4}\, 1/\mathrm{h}.$$

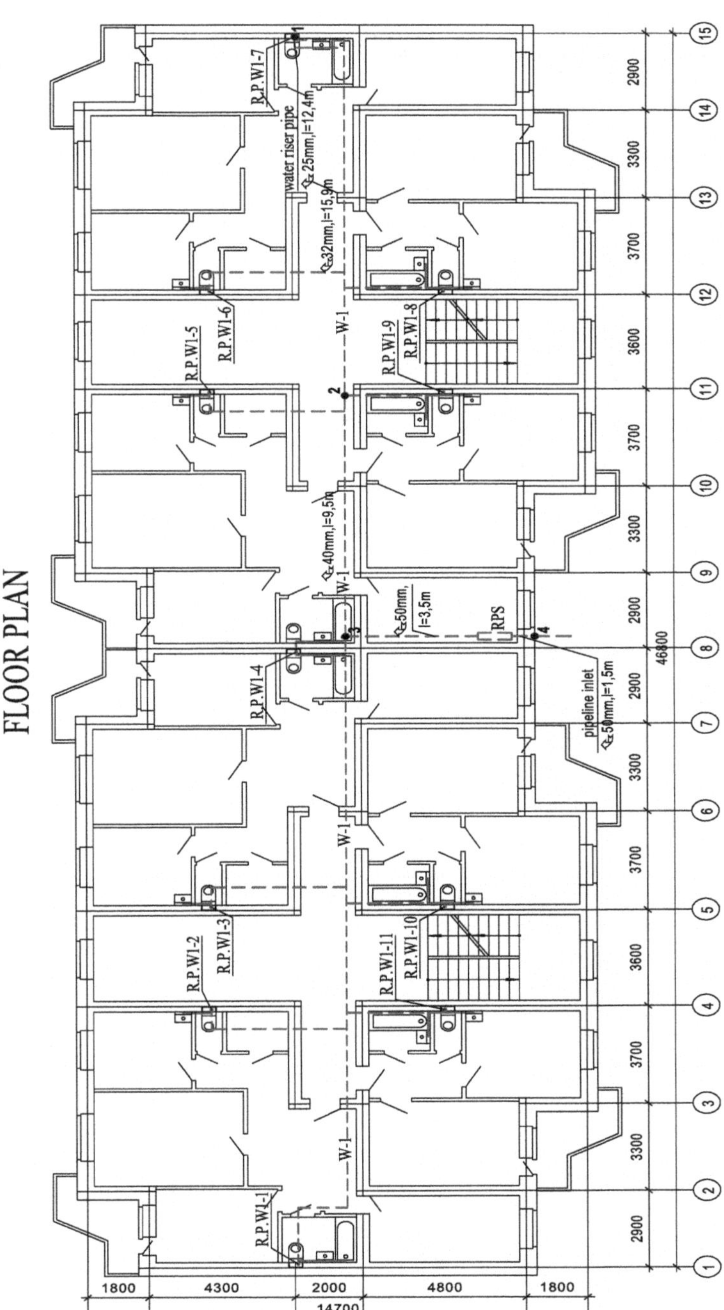

Fig. 2 Network plan for water-supply entering in the middle of the house. *Legend* Rp B1-1—water raiser pipe №1; RPS—raiser pumping station

Table 1 Reliability calculation of the house water-supply system

Internal pipeline sections	Diameters of pipes D (mm)	Lengths of sections L (km)	Failure flow parameter, 1/h				System failure time by the direction, T	Recovery time T_R, (h)	Readiness coefficient K_R
			Network by direction ω_n	RPS Ω_{ps}	Network with fitting by direction ω_s	System by the direction ω_s	Hours/days		
Variant 1 (water input from the end of the house)									
3–4	50	0.0025	2.76×10^{-6}	141×10^{-6}	5.54×10^{-6}	146.54×10^{-6}	6825/284	11.5	0.998318
2–3	40	0.0299							
1–2	32	0.0159							
Raiser pipe	25	0.0124							
Variant 2 (water input in the middle of the house)									
3–4	50	0.005	1.95×10^{-6}	141×10^{-6}	3.90×10^{-6}	144.9×10^{-6}	6901/288	11.6	0.998322
2–3	40	0.0095							
1–2	32	0.0159							
Raiser pipe	25	0.0124							

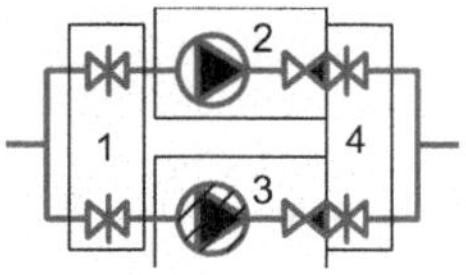

Fig. 3 Scheme of the pumping units

$$T_{PS} = \frac{1}{\omega_{ps}} = \frac{1}{1.41 \times 10^{-4}} = 7092.2\,\text{h}.$$

$$T_{R_{ps}} = \frac{\upsilon_1 T_{R_1} + \upsilon_2 T_{R_2} + \upsilon_4 T_{R_4}}{\omega_{ps}}$$

$$= \frac{(0. \times 20 + 0.01 \times 56.99 + 1.2 \times 10) \times 10^{-4}}{1.41 \times 10^{-4}} = 11.75\,\text{h}.$$

6 Conclusion

1. With optimal house connection to the external water-supply network, the network faultlessness in the direction of the least reliability was increased by 29%
2. The raiser pumping station significantly reduces the reliability of water-supply. Its failure rate is two values more than the network.

References

1. Rules for the provision of services for district heating, hot and cold water supply and sewerage. Resolution of the Cabinet of Ministers of Ukraine No. 630 of 21 July, 2005.
2. Novokhatniy, V. G. (2012). *Reliability of functioning of the supply and distribution complex of water-supply systems* (Doctoral dissertation). Available from National Library of Ukraine named of V. I. Vernadsky. (ДС131058).
3. Tkachuk, O. A. (2008). *Improvement of water supply and distribution systems of settlements*. Rivne: NUVGP.
4. Matyash, A., Novokhatniy, V., & Kostenko, S. (2015). Reliability of water supply of small settlements *MOTROL. Commission of Motorization and Energetics in Agriculture: Polish Academy of Sciences, 17*(6), 95–103.
5. Matyash, A., Usenko, I., Myagkohlib, R., & Kostenko S. (2017). Estimation of reliability of metal water. *Eastern-European Journal of Enterprise Technologies, 3/1*(87), 35–42. https://doi.org/10.15587/1729-4061.2017.101262.
6. Novokhatniy, V., Matyash, A, & Kostenko, S. (2018). Municipal water supply systems of giving-distributive complex reliability with branched networks. *International Journal of Engineering and Technology, 7*(3.2), 653–660. Retrieved from: https://www.sciencepubco.com/index.php/ijet/article/view/14608/5956.

7. Epoyan, S., Karahiaur, A., Volkov, V., & Babenko, S. (2018). Research into the influence of vertical drainage elements on the operational efficiency of rapid filters. *Eastern-European Journal of Enterprise Technologies, 1/90*(91), 62–69.
8. Boryczko, K., Janusz, R., & Tchorzewska-Cieslak, B. (2014). Analysis of risk and failure scenarios in water supply system. *Journal of Polish Safety and Reliability Association Summer Safety and Reliability Seminars, 5*(2), 11–18.
9. Faidaee, M. J., & Tabat, R. (2010). Estimation of failure probability in water pipes network using statistical model. *World Applied Sciences Journal, 11*(9), 1157–1163.
10. Forsyth, P., Robert, D., Rajeev, P., Li C., & Kodikara, J. (2014). Codified methods to analyse the failures of water pipelines: A Review', In R. Das & S. John (Eds.), *Proceedings of the 8th Australasian Congress on Applied Mechanics 2014 (ACAM 8)* (pp. 529–539), Barton, Australia, November 24–28, 2014.
11. UIS 8647. (2016). Reliability engineering. *Evaluation and forecasting reliability test results and (or) use in a small bounce statistics* (p. 54). [Effective as of 2016–05–31], K.: Institute of Mathematical Machines and Systems of National Academy of Ukraine.
12. Internal water supply and sanitation. Part I. Designing. Part II. Construction. DBN B2.5–64:2012 (105p).—K.: Ministry of Regional Development of Ukraine, (2013).
13. Matyash, O. V. (2012). *Improvement of methods of reliability estimation and calculations of branched water-supply networks*. Dissertation abstract. Cand. tech. Sciences. Available from National Library of Ukraine named of V. I. Vernadsky. (PA390290).
14. Kostenko, S. O. (2017). *Reliability of production and group water-supply systems with branched-type networks*. Dissertation abstract. Cand. tech. Sciences. Available from National Library of Ukraine named of V. I. Vernadsky. (PA428713).

Buildings Reconstruction Within the New Educational Space Project in Ukraine

Volodymyr Onyshchenko, Svitlana Sivitska, and Anna Cherviak

Abstract Since 2017, educational system has been reorganized in Ukraine, so the New Educational Space project was implemented within it. The aim of quality assurance in education has always been a topical issue, due to the fast world developing and changing. It is necessary to take into account all the aspects of technical, information, technological, architectural and artistic development of surrounding world and in accordance with these requirements to introduce new systems of educational space. The main components and recommendations for the development of educational space related to the educational environment modernization are the following: Reconstruction of buildings and offices, artistic and energy-efficient component should be considered in details.

Keywords New ukrainian school · New educational space · Quality of education · Reconstruction · Schools · Construction of educational space · Energy efficiency

1 First Section

1.1 A Subsection Sample

Education is of great importance and advantage over other public spheres. It cooperates with the youngest members of society, helps to shape world outlook and develops personality. The Law of Ukraine "On Education" [1] defines the purpose of general secondary education as the comprehensive development of a man as a person and the highest value of society. That is why, the new school implemented in Ukraine should become an area of such development, a space of learning, communication, interaction and joint activity of students, teachers and the local community.

Educational institutions provide education. Their network in Ukraine is quite extensive and is of different forms of ownership. Thus, whatever a school is, the basis of its activities is an effective educational space, a network of educational

V. Onyshchenko · S. Sivitska · A. Cherviak (✉)
Poltava National Technical Yuri Kondratyuk University, Poltava, Ukraine
e-mail: anncherviak@gmail.com

V. Onyshchenko et al. (eds.), *Proceedings of the 2nd International Conference on Building Innovations*, Lecture Notes in Civil Engineering 73,
https://doi.org/10.1007/978-3-030-42939-3_66

institutions operating on the basis of territorial location, demographics and the quality of education in the area.

Since 2017, the process of educational space reforming has begun in Ukraine. This is envisaged by the concept of general secondary education system reforming in Ukraine for 2017–2019, "New Ukrainian School," which was presented in December 2016.

The New Ukrainian School (NUS) is a key reform of the Ministry of Education and Science of Ukraine. It envisages the creation of such environment in schools where children will develop not only physically but also spiritually will create their identity and strive for knowledge that they can learn theoretically and put into practice immediately [2].

From now, the main task for pupils of Ukrainian schools is to master not only separate subjects, but the so-called competences for 12 years of study. Each of these competencies is a combination of knowledge, skills, abilities, ways of thinking, as well as views and values. At the same time, pupils can acquire several competences at the same time, depending on submission form of one or another material [3].

The quality of education in schools depends not only on the system of knowledge and competence of teachers, but also on the environment in which the child learns. Within the NUS project, the New Educational Space project was initiated.

The reform of the New Ukrainian School (NUS) should begin in today's New Educational Space. It is not only the modernization of schools on the principle—new walls, roof and windows. These are completely different approaches to safety levels, energy efficiency, internal motivating design, inclusivity and barrier-free, modern equipment. The Ministry of Regional Development is working on the implementation of the New Educational Space and the Ministry of Education and Science is implementing the New Ukrainian School [4].

New Educational Space (NES) is a project initiated by the Ministry of Regional Development, Construction and Housing and Communal Services of Ukraine, implemented jointly with the Ministry of Education and Science of Ukraine and local governments in order to modernize school infrastructure, create conditions for providing quality and affordable educational services in decentralization reform and reform of the New Ukrainian School [5].

The new educational space provides a set of conditions, techniques and technologies that create safe and accessible educational environment with the use of modern information and communication techniques and technologies in the educational process, including the latest energy efficiency technologies, the latest design technologies, architecture of buildings, structures and territories of institutions education [5].

The purpose of the NES is to modernize school infrastructure, to create conditions in order to provide high-quality and accessible educational services in the context of decentralization and the New Ukrainian School reform [6].

The New Educational Space (NES) is created in educational institutions of Ukraine on the four basic principles [7]:

- energy efficiency;
- modern equipment;
- motivating space; and
- inclusion and absence of barriers.

Artistic decision to renovate school building (motivating space) includes both exterior (appearance of the building) and interior (interior space).

According to the information guide on NES implementation [8], the architecture and design of a school building should be created on the basis of holistic artistic concept, with a single color range and a certain set of finishing agents. Naturally, the overall concept of the building takes into account the differences in perception of external and interior space and the purpose of a certain element, room, group of rooms, section or block. Location and purpose of various parts and elements of the building influence the peculiarities of shapes, color surroundings and approach to using details and to decoration.

While developing the artistic concept of educational institution, the following recommendations should be taken into account:

- Building facade design should be compatible with the design of the surrounding environment, but at the same time, the school building should stand out against it, having a distinctive appearance that makes it clear that it is a school building, but not, for example, a residential building or an entertainment center.
- Within the general concept, the exterior of a building may be more bright and expressive than the interior. The building in the surrounding environment is only a fragment of the landscape that needs to be attractive, while the interior is the place of long-term stay for students and teachers. This place should be comfortable for prolonged activity, not distracting student's attention and not annoying them).
- Groups of rooms intended for students of different ages should be different in decoration. The elementary school block should be warmer and cozier and provide opportunities for both physical activity and quiet communication or privacy. The rooms for senior students are more businesslike, help concentration and create a work moral.
- The group of the main school rooms—the hall, the forum, the assembly hall—is the character of the school. Its décor should be the most expressive, visually reflect the unique character of the school and must correspond with peculiarities of spatial perception, accentuate but not destroy the structure and functions of the rooms. Elements that distort spatial perception, disorient or substitute the reality should not be used.
- School interior decoration must stimulate students' imagination but not substitute it with the designer's imagination. Interesting combinations of space, colors, shapes, materials and textures give space for creative imagination, while narrative or symbolic images rather impose perceptive pattern. In school design, the priority should be given to the students' creativity—you may create opportunities for exhibitions of student works, allocate areas for painting on the walls or allow students design separate spaces themselves during educational projects. Professionally made design elements (murals, prints, lettering, etc.) should be used in

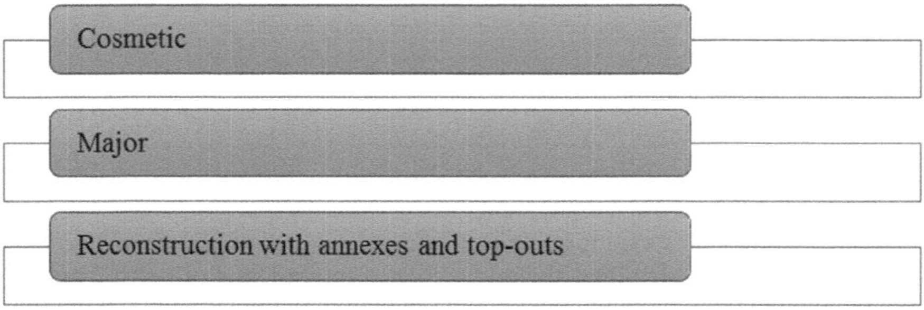

Fig. 1 Types of building repairs. *made on the basis of [8]

a limited way, mainly in general school spaces, as an accent. It is better if such elements are not permanent but regularly changed.

- Communications (ways on the area, the building entrance, the door, corridors, passages, staircases, etc.) should help spatial orientation. They should be noticeable, may be quite bright, but not dark. Small details and complex decorative pattern are not recommended. Navigation elements (signs of entrances, sections, floors and separate rooms) should be included into the overall system of the interior decoration.

School building renovation may be made in several ways. Depending on the seriousness of modification of present structure and building constructions, there may be defined three main types of repairs (see Fig. 1).

Cosmetic renovation [8] requires only the décor renovation of the premises and building façade and also furniture or equipment replacement. It allows to change the color of the facade and interiors, to change the rooms' equipment, to install modern furniture and more. In the process of cosmetic repairs, it is possible to change the designation of some rooms, which allows to improve in a certain way the conditions of educational process organization. But cosmetic repairs do not allow redevelopment of space and adding new opportunities for the organization of the educational process.

Major repairs [8] of the main volume of a building, as a rule, involve, first of all, upgrading of the utility systems (heating, ventilation, etc.), energy system modernization and elimination of technical defects of the basic structures. During major repairs, it is possible not only to change the color scheme of the facades and interiors, but also to redecorate, and, in a certain way, change the relative layout, shape and size of individual rooms in accordance with the modern standard requirements:

- Reconstruction with annexes and top-outs allows adapting the building to the updated requirements of educational process organization. During such reconstruction, it is possible to radically change the spatial organization of a school, to create new rooms in accordance with modern standards and to fully implement new approaches to the educational space [8].

Creation of new educational space at schools provides for updating the main components of the architectural and artistic building structure:

- planning and arrangement of the schoolyard;
- renewal of the building exterior design;
- interior design planning, furnishing and equipment.

According to the information guide [8], depending on the nature of the project, the needs and resources of the community, these changes may be more or less profound. Comprehensive reconstruction of the building is bound to include all these components. Various types of minor and major repairs, as a rule, affect only one or two aspects, which anyway enables to create educational space that meets modern requirements. Improvements of the schoolyard include

- general planning and space zoning of the schoolyard;
- organizing pedestrian, bicycle and car traffic in the schoolyard and nearby;
- arranging, equipping and greening particular schoolyard zones (educational and play space, sports area, utility zone) according to their purpose.

The refurbishment of the school building exterior design in the process of reconstruction with additional buildings and heightening is carried out by means of the architectural volumetric space composition. If the reconstruction does not involve volumetric changes of the building, the following aspects are taken into account in the process of facade renovation [8]:

- facade silhouette and composition;
- facade finishing materials and coloristic;
- solutions for individual elements of the facade: entrance space, windows, etc.;
- decorative elements of the facade, including murals.

School building interior is to be upgraded depending on whether additional building or replanning is set out, or whether only refurbishment and reequipment are possible. While making an interior solution, the following aspects are taken into consideration [8]:

- composition, number and parameters of sections, blocks and separate rooms;
- general layout of the building, relative location of the premises and convenient connections between them;
- interior artistic solutions, decoration and coloristic of the building interior space;
- furnishing and equipment of the premises.

Within the framework of funding from the State Regional Development Fund (SRDF), the Ministry of Regional Development, Construction, Housing and Communal Services of Ukraine regional projects continue to be implemented.

In 2019, funding to the amount of over 2.5 billion UAH was approved for 277 construction projects, reconstructions and thorough overhauls of schools and kindergartens [7].

Within the framework of the funds realization, the project of New Educational Space (NES) is being implemented, by which motivating, creative and unusual designs of educational institutions are created, and the works on energy modernization are carried out.

As part of the educational environment upgrade, schools were provided with the following:

- computer equipment (9500 interactive projectors, 19,200 laptops, 8000 tablets/netbooks, 16,500 multifunctional devices, etc.);
- furniture (364,223 single-seater student kits, 9200 double-seater student kits, 6900 teacher-kits, 77,100 leisure area furniture units);
- didactic materials (361,800 printed educational materials, 57,800 natural objects, 164,300 models and layout sheets, 16,800 musical instruments, etc.).

According to the Ministry of Education and Science of Ukraine, funding from the State Regional Development Fund (SRDF) has been already overally approved for the implementation of 707 regional development projects in all the regions of Ukraine. These include the construction and reconstruction of schools, kindergartens, health care, culture, sports institutions as well as other important regional projects.

Figure 2 shows the allotment of projects by regions.

According to the Law of Ukraine "On Energy Performance of Buildings," thermal modernization of buildings is a set of works aimed at increasing the thermal performance of building envelope structures as well as consumption indicators of energy resources by engineering systems and ensuring the energy efficiency of buildings at

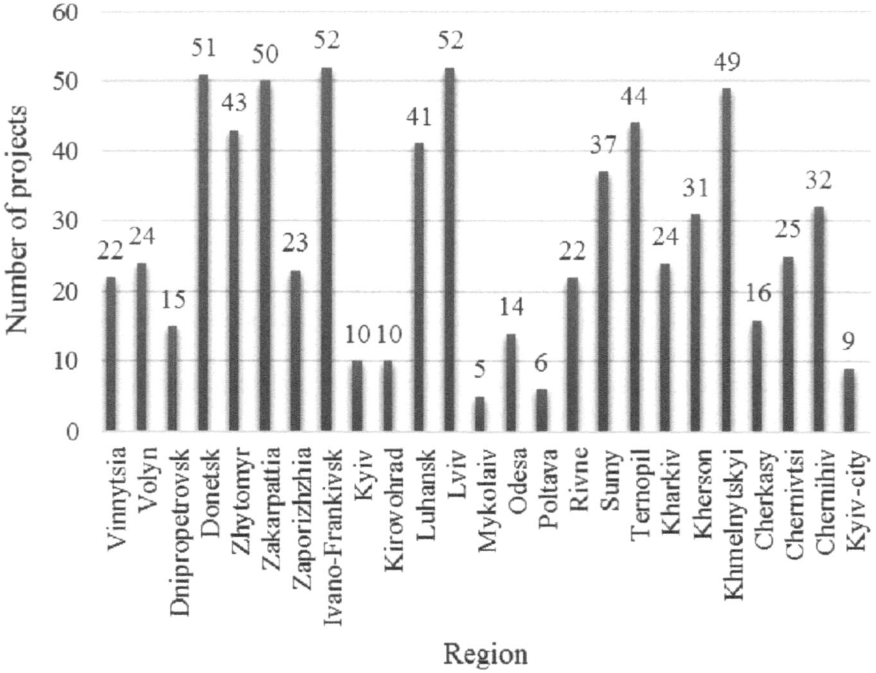

Fig. 2 Allotment of regional development projects by regions

a level not lower than established by the minimum requirements for the energy efficiency of buildings carried out during the reconstruction, overhaul or current repairs of buildings or works that do not require documents that entitle them to completion and after which the object is not subject to be accepted into service.

Thermal modernization of buildings is carried out without developing project documentation as well as obtaining documents that give the right to perform construction works and accepting such an object into service only when performing works on:

1. existing window, door and balcony assemblies or engineering systems (except the works on reconstruction or overhaul of engineering systems).
2. building envelopes with low consequences (CC1).
3. replacement of roof covering of buildings that do not involve interference with enclosures and/or load-enclosing carrying structures.
4. connection and linking-up of individual (manor) dwelling houses, garden cottages, country houses to the engineering networks [9–11].

In 2017, 115 projects of New Educational Space (NES) were created, by the end of 2018—300 modern NES projects. These are remarkable modern institutions all over Ukraine, with more than 120 thousand children studying [6].

According to the Ministry of Education and Science of Ukraine [2], new educational space (reconstruction and renovation of buildings) was first successfully implemented in the following 17 schools:

1. Gorbanivka village, Rivne region.
2. Navariia village, Lviv region.
3. Volnovakha town, Donetsk region.
4. Pisky village, Rivne region.
5. Putrivka village, Kyiv region.
6. Kyiv city, "Erudite" educational complex.
7. Solone urban-type settlement, Dnipropetrovsk region.
8. Korny village, Rivne region.
9. Hryshyne village, Donetsk region.
10. Kyiv city, "Kyivska Rus" ("Kievan Rus") primary school.
11. Illinivka village, Donetsk region.
12. Lviv, "Ecoland" lyceum school.
13. Petrykivka urban-type settlement, Dnipropetrovsk region.
14. Pisochyn urban-type settlement, Kharkiv region.
15. Kyiv city, primary school №. 333.
16. Pokrovsk town, Donetsk region.
17. Yurivka urban-type settlement, Dnipropetrovsk region.

Out of the 115 NES development projects, 60 have been implemented in Ukraine so far. This is not just a bright design, painted walls and murals on school walls: this is an essential educational ecosystem. School stops being a "disciplinary" space, it starts being a space for development and interaction. And that is precisely what the Standards developed by the Ministry of Regional Development for typical projects on New Educational Space are intent on. Hub school is to provide equally qualitative education both in the city and in the countryside [10].

Educational reform is aimed at developing and updating the educational space not only in terms of teaching methods and teachers' qualifications, but also with regard to buildings, classes and the adjacent territory, used for schooling. New Ukrainian School concerns, among other things, the issues of development of pre-school and vocational (vocational-technical) education, training and advanced training of teachers in institutions of higher pedagogical education.

Interior outfitting of classrooms is designed for students' comfortable learning environment. This creates positive emotional atmosphere and an opportunity for applying various teaching methods. Formation of creative learning space is just as necessary as providing schools with skilled staff. Every year, more and more educational institutions that meet New Educational Space requirements are created, which, in turn, increases the level of the educational process.

References

1. The Law of Ukraine "On Education" Homepage. https://zakon.rada.gov.ua/laws/show/2145-19.
2. The Ministry of Education and Science of Ukraine Homepage. https://mon.gov.ua/ua/tag/nova-ukrainska-shkola.
3. To the Day of Knowledge. What is important to know about the new Ukrainian school Homepage. https://nv.ua/ukr/ukraine/events/nova-ukrajinska-shkola-shcho-ce-take-i-shcho-potribno-znati-novini-ukrajini-50039251.html.
4. The Tasks of New Educational Space—Motivation to learn Homepage. https://decentralization.gov.ua/news/9530.
5. The Ministry for Communities and Territories Development of Ukraine Homepage. http://www.minregion.gov.ua/?s=%D0%BD%D0%BE%D0%B2%D0%B8%D0%B9+%D0%BE%D1%81%D0%B2%D1%96%D1%82%D0%BD%D1%96%D0%B9+%D0%BF%D1%80%D0%BE%D1%81%D1%82%D1%96%D1%80.
6. New Educational Space Homepage. https://decentralization.gov.ua/education/novyi-osvitnii-prostir.
7. Government Portal Homepage. https://www.kmu.gov.ua/ua/news/dfrr-profinansuye-budivnictvo-abo-rekonstrukciyu-majzhe-280-shkil-ta-ditsadkiv-z-derzhfondu-regionalnogo-rozvitku-u-2019-roci.
8. New educational space. Information Guide Homepage. https://storage.decentralization.gov.ua/uploads/library/file/407/NOP_Motivuyuchiy-prostir.pdf.
9. The Law of Ukraine "On Energy Performance of Buildings" Homepage. https://zakon.rada.gov.ua/laws/show/2118-19.
10. Vinnytsia Regional State Administration Homepage. http://www.vin.gov.ua/news/ostanni-novyny/10298-aktualni-pytannia-reformuvannia-sfery-osvity-v-umovakh-detsentralizatsii-obhovoryly-v-khodi-vseukrainskoi-selektornoi-narady.
11. Varnaliy, Z., Onishchenko, S., & Masliy, A. (2016). Threat prevention mechanisms of ukraine's economic security. *Economic Annals-XXI, 159*(5–6), 20–24. https://doi.org/10.21003/ea.V159-04.

A New Agent for Removing Concrete Residues from the Surfaces of Polypropylene Molds in the Manufacture of Paving Slabs and Its Advantages

V. O. Onyshchenko, O. M. Filonych, N. V. Bunyakina, and N. B. Senenko

Abstract Washing the working surfaces of propylene molds to remove residues of the concrete mixture is an important stage in the technological process of pavement slab production. Hardened concrete adheres strongly to polypropylene, and molds with residues of hardened concrete can no longer be used. They may be washed in solutions of strong inorganic acids of fairly high concentration. This is dangerous for both the equipment and the polypropylene surfaces of template molds. The purpose of this work was to create a new washing agent of satisfactory quality, enabling the non-destructive washing of polypropylene surfaces without mechanical interference. Another important objective was to reduce the cost of this agent relative to existing counterparts, so as to maintain the profitability of the process. The effectiveness and advantages of the newly developed agent were proved by the weight analysis method. This represents a scientific innovation in that a new effective means for removing hardened concrete residues from the working surfaces of polypropylene molds has been created, and a method for its preparation has been developed. The ability to wash the surfaces of polypropylene molds with maximum preservation of the surface quality, which significantly prolongs their use, is a valuable practical benefit. In addition, the mixture is shown to offer other significant advantages, including the lack of destructive impact on the polypropylene surface of the mold, availability and cheapness of the constituent components, ease of preparation, and absence of sharp and unpleasant odors. Preparation of the new mixture can be carried out both at specialized industrial sites and directly at paving slab production plants. The proposed agent has an unlimited shelf life over a wide temperature range.

Keywords Hardened cement · Polypropylene surface · Dissolving · Paving slabs · Molds

V. O. Onyshchenko · O. M. Filonych · N. V. Bunyakina · N. B. Senenko (✉)
National University "Yuri Kondratyuk Poltava Polytechnic", Poltava, Ukraine
e-mail: natalinasenenko@gmail.com

V. Onyshchenko et al. (eds.), *Proceedings of the 2nd International Conference on Building Innovations*, Lecture Notes in Civil Engineering 73,
https://doi.org/10.1007/978-3-030-42939-3_67

1 Introduction

Concrete pavement tiles (shaped paving elements) are today the most widespread covering of city sidewalks, park paths, floors of shopping malls and exhibition pavilions, and other infrastructure. Their performance properties, namely durability, low wear, frost resistance, low water absorption, abrasion resistance, and resistance to many chemicals, represent indisputable advantages of this type of covering. They provide many years of pavement service life even in the modern metropolis. Eco-friendliness and ease of laying make pavement tiles an optimum method of ground improvement.

Two methods used for the manufacture of such tiles [1, 2] are based on molds made of polypropylene. Technical parameters of products manufactured by both technologies are listed in Table 1 [2].

As noted in [3], the cement-concrete mixture hardens on any surface. After the cycle has been completed, the surface must be cleared of residual concrete. Removing fresh residual concrete is a relatively easy procedure. The problem is that if the mixture is hardened, then the removal process is not easy and requires the use of special solutions. Removal techniques should be safe for both the equipment and the molds. Otherwise, molds with hardened concrete residues will be unusable in the future, as surface defects will form on the products. Moreover, if the residues are removed using mechanical processes or aggressive solutions, the surfaces of the molds will be damaged. This will make them unsuitable for further use. In addition, the reagents used to remove the concrete must ensure maximum safety for workers.

The task of removing residual concrete is complicated by the fact that concrete is a mineral substance, which can be removed only with acid. The desired effect is given by concentrated acid, which penetrates into deep layers of dried cement mortar and destroys it.

Table 1 Technical parameters of paving slabs

Technical requirements	Unit of measurement	According to GOST	Vibropressed pavement tile	Vibrocast paving slabs
Frost resistance	Cycles	200	200	350
Water absorption	%	6	5	3
Abrasion	g/cm^2	0.7	0.7	0.7
Compression strength	g/cm^2	400	450	600
Bending strength	g/cm^2	50	50	60
Release strength of concrete	%	70	70	70–90

Any concrete remover [1–3] contains concentrated acid that destroys the structure of concrete, inhibitors that protect the surface from the action of acids, and substances performing protective functions.

A list of commercial agents currently available in Ukraine and worldwide is given in the works [3, 4]. Unfortunately, they do not represent a satisfactory solution to the problem. Good-quality agents contain a large number of expensive components and are difficult to prepare. This greatly increases the cost of the tiles produced. Cheap agents usually have insufficient washing capacity, resulting in a need for additional machining that destroys the surface of the polypropylene mold.

The authors studied the interaction of concrete with polypropylene, namely its adhesive properties. Modern technologies for the manufacture of construction products use methods based on attracting polypropylene fibers to strengthen concrete mixtures [5–7]. Thus, in the works [8, 9], modification approaches that may improve coverage or polypropylene adhesion are presented. Polypropylene is a synthetic thermoplastic non-polar polymer belonging to the class of polyolefins. At room temperature, it does not dissolve in organic liquids; at high temperatures, it swells and dissolves in some solvents, for example in benzene, carbon tetrachloride, and ether. Polypropylene exhibits low moisture absorption and offers good electrical insulation properties over a wide temperature range. It also has good mechanical properties. Polymeric materials, including polypropylene, are widely used and provide economic advantages. They increase product competitiveness in consumer industries by replacing expensive materials, reducing material consumption, and driving the development of new generations of advanced materials processing technologies [7]. It has been proved that concrete products reinforced with polypropylene exhibit high durability and considerable endurance [7]. Products of pure polypropylene lose their durability properties over time. Their characteristics include high abrasion and a propensity to ignite when exposed to open flames [7].

As the adhesion of concrete to polypropylene surfaces is considerable, the aim of the first part of our work was to develop a washing agent in which the concrete mixture will be highly soluble, while ensuring the protection of the mold surface. Thus, the authors set out to reduce the adhesion between polypropylene and concrete. In [10], the results of investigations into methods of processing polypropylene surfaces and polypropylene products for reducing adhesion are presented. Usually, however, tile manufacturers do not have the capacity to treat and process the polypropylene surfaces of molds in this way. The authors therefore set out to create an agent with properties enabling the complete removal of concrete without additional mechanical intervention.

The authors carried out this work at the request of pavement tile manufacturers, following studies of the properties of mixtures currently in use. The work is therefore of considerable practical importance.

The properties of an agent for removing hardened concrete from the surfaces of trucks, mixers, and other equipment, as presented in [11], were investigated. This mixture contains (by weight): 17.5% hydroxyacetic acid, 4% soap, 2% degreasing agent, and 76.5% water. To remove the hardened concrete, the agent is evenly applied to the surface and kept in contact with the concrete for five minutes. This converts the hardened concrete into soft putty pieces, which are then mechanically removed from the surface. Disadvantages of this agent include the insufficient softening of residues of dried concrete and the need for additional mechanical intervention.

Another known substance is used for cleaning and disinfecting the surfaces of building materials [12]. It is supplied in the form of a solution containing, by weight, 0.1–20% surfactants, 1–50% hypochlorite of alkali or alkaline earth metal, 0.1–50% alkali metal or ammonium silicate, 0.01–2% alkali metal or ammonium fluorosilicate, and the remainder water. The significant number of components is a major drawback of this agent, as it greatly increases its cost and complicates the preparation process. In addition, surfactants adversely affect the environment, and alkali or alkaline earth hypochlorites are unstable compounds that decompose easily, releasing active chlorine with a sharp and unpleasant odor. This complicates work with the agent and makes it dangerous for use in pavement manufacture.

The authors also studied a known washing agent for the removal of solid or semi-solid cement, described in [13]. In one of its forms, it contains, by weight, 60% anhydrous citric acid, 0.01% xanthan as thickener, 0.1% triethanolamine lauryl sulfate as wetting agent, and the remainder water. The mixture is applied to the surface and left there until the cement substance is transformed into solid components and calcium citrate. The remaining solids and calcium citrate either fall off under gravity or can be removed from the surface by water or mechanically. The use of expensive reagents (citric acid, xanthan, triethanolamine, lauryl sulfate), the long exposure time of the applied mixture on the surface, and again the need for additional mechanical cleaning, which can damage the polypropylene surface, are disadvantages of this agent.

In a study of another known agent [14] for the chemical removal of cement mortar residues from solid mineral surfaces and other facing materials, containing (by weight) 5–10% potassium hydrofluoride, 50–57% triple-substituted potassium citrate, 3–10% citric acid, and the remainder water, the authors also identified a number of disadvantages. They include the presence of toxic potassium hydrofluoride and the high cost of the ingredients, meaning that problems arise with the need for environmental protection and personal protection of production workers, in addition to the high cost of the product.

The authors identified another composition [15] as the most suitable for the task. This consists of, by weight, 40–60% HCl, 42% urea, 0.067% complex substituted ketoamine hydrochloride, 0.067% isopropyl alcohol, 0.022% ethoxylated nonylphenol, 0.022% propargyl alcohol, 0.022% methyl vinylketone, 0.022% acetone, and 0.0022% acetophenone. However, a number of shortcomings were also identified: the large number of components, the complexity of preparation, the high cost of the mixture, as well as its ability to destroy machining surfaces.

Taking into account all of the above-mentioned disadvantages, requirements, and features, the authors have developed a cheap agent for removing residues of hardened concrete from polypropylene molds used to form paving slabs. This is a solution containing an acid, a moisturizing agent, and water, in the following proportions (by weight): 13% hydrochloric acid, 0–0.1% moisturizing agent, and the remainder water [4].

The aim of the second part of the work was to improve the properties of the developed agent, maintaining the high ability to dissolve cement mortar while having a minimal impact on the surface of the polypropylene molds, to ensure their reusability. The objective of the work was to develop a new agent [16], to investigate its properties, and to prove its advantages over existing alternatives.

2 Research Methodology and Results

Various methods are used to remove contaminants from surfaces. They may be classified as physical or chemical, depending on the mechanism of the process. Chemical methods are based on the transformation of the contamination by a chemical reaction into new compounds that are easily removed [17]. In the composites developed and presented by the authors [16], as in [4], the hardened concrete residues are dissolved as a result of a chemical reaction between hydrochloric acid and the constituent parts of the concrete mixture.

To improve the properties of the agent, urea and glycerol were added to its composition. Urea is used in detergents to increase detergency without adversely affecting the surface being treated [15]. Glycerol is used to reduce the adhesion forces between the concrete mortar and the surface of the mold material. The proposed improved mixture has the following composition (by weight): 5–15% hydrochloric acid, 5.1–5.9% urea, 30–35% glycerol, and the remainder water.

3 Method of Preparation of the Agent

The method of preparation and use of the mixture developed and proposed by the authors is as follows: Hydrochloric acid, urea, glycerol, and water in the prescribed quantities are added to a reactor and stirred for 30–40 min. The finished product is bottled and packed. The agent is used after dilution with water in a ratio of 1:1.

4 Investigation of the Properties of the Agent

Evaluation of the ability of the improved agent to remove hardened residues of concrete mixture from molds was performed on molds contaminated with material as a result of an actual paving slab production process. For this purpose, samples 2 × 2 cm in size were cut from polypropylene molds. The samples were dried in a drying oven to constant weight at a temperature of 100–105 °C, cooled in a desiccator, and weighed on an analytical balance. They were then immersed for 30 min in the agent to remove residual concrete mixture in stationary mode, washed with distilled water, dried in a drying cabinet, cooled, and weighed with concrete residues as described above. Based on the reduction in the weight of the samples, the ability of the agent to dissolve the hardened residues of the concrete mixture was determined. Each sample was subjected to three sequential treatments of 30 min each. The reductions in the weight of the samples were as follows: 3.00% after the first treatment, 1.74% after the second treatment, and 0.95% after the third treatment.

From these results, we may conclude that the maximum dissolution of the hardened concrete occurs during the first treatment. In addition, the reduction of residual hardened concrete on the surface of the samples was determined visually.

As the paving slab mold has to withstand up to 500 production cycles, it is important to evaluate the degree of destruction of the polypropylene mold surface under the action of the improved agent. Therefore, at the same time, control samples without residual concrete were tested to determine the effect of the chemical agent on the polypropylene surface.

For this purpose, samples of size 2 × 2 cm without concrete mixture were kept in the agent for different times. The degree of destruction of their surface was evaluated based on the reduction in the mass of the samples.

In addition, the authors investigated the effect of the agent on the polypropylene mold surfaces when the components were used in different proportions. At the same time, the authors studied the mixture [4] under similar conditions to compare the effect of the components. To study the effect of the composition on the mold material, mixtures with different ratios of chemicals were prepared:

- solution no. 1—13% hydrochloric acid;
- solution no. 2—13% hydrochloric acid and water in a ratio of 1: 1;
- solution no. 3—100 ml of 13% hydrochloric acid and 50 ml of glycerol;
- solution no. 4—100 ml of 13% hydrochloric acid and 10 g of urea;
- solution no. 5—100 ml of 13% hydrochloric acid, 50 ml of glycerol, 10 g of urea.

The changes in the weights of the samples were determined by treating them in the above solutions for 30 min, and then in a stepwise manner for 60 min (30 min + 30 min), 90 min (30 min + 30 min + 30 min), 120 min (30 min + 30 min + 30 min + 30 min), 1 day, 6 days, and 23 days.

The results of the measurements of weight loss of polypropylene samples without concrete residues (wt%) are given in Table 2.

From the above results, it can be concluded that the samples treated with solution no. 5 [16] and solution no. 2 [4] undergo the smallest mass loss during prolonged treatment. That is, a composition consisting of 13% hydrochloric acid, glycerol, urea, and water has less effect on the polypropylene material of the mold than the individual components of the mixture. Hydrochloric acid solution with a concentration of 13% is recommended, because it is not a precursor in this process. This ensures that the proposed agent will be widely available for production and use.

Changing the proportions of the components, the authors more thoroughly investigated solution no. 5 in terms of its destructive effect on the polypropylene surface. The methodology was similar to that described above. Initially, the effect of the solution with different contents of hydrochloric acid was investigated. The content of hydrochloric acid was calculated in terms of the quantity of active substance,

Table 2 Sample weight loss after processing with different solutions, wt%

Duration of processing						
30 min	60 min	90 min	120 min	1 day	6 days	23 days
Solution no. 1						
0	0	0	0	0.212	0.276	0.286
Solution no. 2						
0	0	0	0	0.237	0.266	0.273
Solution no. 3						
0	0	0	0	0.215	0.289	0.292
Solution no. 4						
0	0	0	0	0.221	0.293	0.299
Solution no. 5						
0	0	0	0	0.209	0.299	0.257

which means that solutions of any concentration can be used to form the mixture. The results of the tests are given in Table 3.

It is clear that the effect of the agent remains almost the same when the content of hydrochloric acid lies in the range 5–15%. With lower acid content, the solubility of hardened concrete in the agent decreases. With higher acid content, the destructive effect on the surface increases. Thus, the optimal range of content of the main active ingredient was determined.

Table 3 Sample weight loss after processing with solution no. 5 (hydrochloric acid from 4 to 16%; figures in wt%)

Hydrochloric acid, wt%	Duration of processing						
	30 min	60 min	90 min	120 min	1 day	6 days	23 days
4	0	0	0	0.364	0.367	0.381	0.387
5	0	0	0	0.232	0.286	0.288	0.254
6	0	0	0	0.287	0.226	0.244	0.259
7	0	0	0	0.236	0.240	0.257	0.255
8	0	0	0	0.229	0.251	0.254	0.260
9	0	0	0	0.239	0.249	0.254	0.254
10	0	0	0	0.239	0.250	0.249	0.253
11	0	0	0	0.241	0.262	0.259	0.258
12	0	0	0	0.244	0.258	0.256	0.249
13	0	0	0	0.236	0.241	0.248	0.254
14	0	0	0	0.240	0.247	0.256	0.258
15	0	0	0	0.239	0.246	0.247	0.249
16	0	0	0	0.433	0.441	0.419	0.469

The next step in the experimental studies was to determine the effect of the excipients on the polypropylene surface at the optimum content of hydrochloric acid. Results obtained by varying the content of glycerol are given in Table 4.

Thus, the optimal range of concentrations of glycerol in the agent was selected. Similar tests were carried out to determine the effect on the polypropylene surface of various concentrations of urea. The results are given in Table 5.

The results confirm the good washing action of the agent, accompanied by a minimal destructive effect on the polypropylene surface of the mold over long processing times.

According to the results of testing, the optimum ranges of content of each component in the agent for dissolving residues of hardened concrete mixture on the surface

Table 4 Sample weight loss after processing with solution no. 5 (glycerol content from 29 to 36%; figures in wt%)

Glycerol, wt%	Duration of processing						
	30 min	60 min	90 min	120 min	1 day	6 days	23 days
29	0	0	0	0.437	0.451	0.459	0.449
30	0	0	0	0.236	0.247	0.246	0.250
31	0	0	0	0.251	0.253	0.251	0.259
32	0	0	0	0.249	0.252	0.253	0.256
33	0	0	0	0.231	0.244	0.256	0.259
34	0	0	0	0.246	0.247	0.249	0.254
35	0	0	0	0.243	0.245	0.255	0.256
36	0	0	0	0.380	0.393	0.397	0.399

Table 5 Sample weight loss after processing with solution no. 5 (urea content from 5.0 to 6.0%; figures in wt%)

Urea, wt%	Duration of processing						
	30 min	60 min	90 min	120 min	1 day	6 days	23 days
5.0	0	0	0	0.381	0.388	0.397	0.403
5.1	0	0	0	0.241	0.243	0.254	0.260
5.2	0	0	0	0.233	0.244	0.248	0.253
5.3	0	0	0	0.231	0.234	0.252	0.253
5.4	0	0	0	0.229	0.242	0.249	0.256
5.5	0	0	0	0.234	0.248	0.239	0.255
5.6	0	0	0	0.226	0.230	0.241	0.252
5.7	0	0	0	0.229	0.234	0.249	0.254
5.8	0	0	0	0.226	0.241	0.251	0.258
5.9	0	0	0	0.234	0.249	0.250	0.257
6.0	0	0	0	0.433	0.439	0.461	0.478

of polypropylene molds used for the manufacture of paving slabs were determined as follows (by weight): hydrochloric acid 5–15%, urea 5.1–5.9%, glycerol 30–35%, and the remainder water. The agent may be prepared from a 13% hydrochloric acid solution.

5 Scientific Novelty

This work represents a scientific innovation in that a new effective means for removing hardened concrete residues from the working surfaces of polypropylene molds has been created, and a method for its preparation has been developed.

6 Practical Importance

The practical value of the work is that it solves the problem of washing the surfaces of polypropylene molds with maximum preservation of the quality of the working surfaces, which significantly prolongs their use. Moreover, significant advantages of the new agent have been proven, such as the lack of destructive impact on the polypropylene surface of the mold, availability and cheapness of the components, ease of preparation, and lack of strong and unpleasant odors. The agent may be prepared both at specialized industrial sites and directly at paving slab production plants. The proposed agent has an unlimited shelf life over a wide temperature range.

7 Conclusions

1. An improved agent has been developed and proposed for dissolving cement mortar from the polypropylene surface of molds used for paving slab manufacture.
2. In practical use of the proposed agent, the quantitative ratios of the components have minimal impact on the material of the mold (polypropylene) over long processing times. The optimal chemical composition and ratio of components have been proposed.
3. The newly developed agent offers several advantages: It is easy to manufacture, does not contain toxic substances, and is cheaper than other washing agents.
4. The agent can be made from a wide range of non-precursor chemical reagents.
5. The proposed substance can be stored indefinitely and does not require special storage conditions.
6. The mixture can be prepared both at specialized industrial sites and directly at paving slab production plants.

References

1. Rajani, B. (2002). Best practices for concrete sidewalk construction. *Construction Technology Update, 54*. ISSN 1206-1220. https://www.nrc-cnrc.gc.ca/ctu-sc/en/ctu_sc_n54/ https://doi.org/10.1139/cjce-24-2-303.2.
2. Momot, I. (2003). Road to the future. *Construction and Reconstruction, 6*, 28–29. https://docplayer.net/55225923-Budivelni-materiali-ta-virobi.html.
3. Onyshchenko, V., Storozhenko, D., Senenko, N., & Bunyakina, N. (2018). Mixes for concrete residues removal from sidewalk tiles production molds invention and analysis. *International Journal of Engineering andTechnology, 7*(4.8), 497–501.
4. Onyshchenko, V., Filonych, O., Storozhenko, D., Bunyakina, N., Senenko, N., & Shulhin, V. (2018). *Patent of Ukraine 129742*. Kyiv: State Patent Office of Ukraine.
5. Aslani, F., & Gedeon, R. (2018). Experimental investigation into the properties of selfcompacting rubberised concrete incorporating polypropylene and steel fibers. *Structural Concrete, 20*(1). https://onlinelibrary.wiley.com/doi/full/10.1002/suco.201800182 https://doi.org/10.1002/suco.201800182.
6. Nobili, A., Lanzoni, L., & Tarantino, A. M. (2013). Experimental investigation and monitoring of a polypropylene-based fiber reinforced concrete road pavement. *Construction and Building Materials, 47,* 888–895. https://doi.org/10.1016/j.conbuildmat.2013.05.077.
7. Trofimov, O., & Savchenko. B. (2014). Fiber concrete technology development. *Visnyk of the KNUTD, 6*(80), 149–156. https://knutd.edu.ua/publications/pdf/Visnyk/2015-3/152_156_2.pdf.
8. Zhang, C. (2014). The approaches for promoting PP adhesion based on the surface modification. *Jornal of Adhesion Science and Technology, 28*(5), 454–465. https://doi.org/10.1080/01694243.2013.838826.
9. Zeiler, T., et al. (2000). Different surface treatments to improve the adhesion of polypropylene. *Jornal of Adhesion Science and Technology, 14*(5), 619–634. https://doi.org/10.1163/156856100742799.
10. Brown, P. S., & Bhushan, B. (2016). Durable superoleophobic polypropylene surfaces. *Philosophical Transactions of the Royal Society a Mathematical, Physical and Engineering Sciencees*. https://doi.org/10.1098/rsta.2016.0193.
11. James Jr, J. M. (1995). *Patent USA SU 5451264 A*. Washington, DC: U.S. Patent and Trademark Office.
12. Panteleimonov, A. V., & Drum, A. Y. (2012). *Patent of Ukraine 72581*. Kyiv: State Patent Office of Ukraine.
13. Gairdner, J. (2003). *Patent USA SU 6592658 B1*. Washington, DC: U.S. Patent and Trademark Office.
14. Chirkov, Y. V. (2018). *Patent EA 028996B1 20180131*. Moscow, Russia: Eurasian Patent Organization, EAPO.
15. MacDonald, J. (2011). *Patent USA SU 7938912 B1*. Washington, DC: U.S. Patent and Trademark Office.
16. Onyshchenko, V., Filonych, O., Storozhenko, D., Bunyakina, N., Senenko, N., Akhmednabiev, R., & Zavora, T. (2020). *Patent of Ukraine 139429*. Kyiv: State Patent Office of Ukraine.
17. Contamination removal methods. Chemicalnow. http://www.chemicalnow.ru/chemies-6183-1.html.

Ukraine Construction Complex Innovation-Oriented Development Management

S. Onyshchenko, S. Yehorycheva, O. Furmanchuk, and O. Maslii

Abstract Organizational and economic principles of the formation and implementation of the construction complex innovation-oriented development management have been studied and determined. The reasons for the low level of the construction complex innovation development have been identified. It is emphasized that regional inter-sectoral complexes, in particular construction, are affected by many factors, threats and dangers of the socio-economic life of the region and require secure functioning. It is proved that the solution of this problem requires the introduction of organizational and economic construction complex innovation-oriented development management, a significant component of which is the cost optimization mechanism. The components of the Ukraine construction complex innovation-oriented development management mechanism have been put forward and developed, which reflects all the constituent parts of the organizational and economic innovative management mechanism

Keywords Construction complex · Innovation-oriented management · Mechanism

1 Introduction

Present-day dynamics of the socio-economic and social life leads to a change in managerial approaches to the functioning and development of the priority industries and economic complexes, where, undoubtedly, the construction complex refers to. Given the strategic importance of the construction complex for the state, its role

S. Onyshchenko · S. Yehorycheva · O. Furmanchuk · O. Maslii (✉)
Poltava National Technical Yuri Kondratyuk University, Poltava, Ukraine
e-mail: pugachaleksa@gmail.com

S. Onyshchenko
e-mail: s07onyshchenko@gmail.com

S. Yehorycheva
e-mail: yehorycheva.sb@gmail.com

O. Furmanchuk
e-mail: o.s.furmanchuk@gmail.com

V. Onyshchenko et al. (eds.), *Proceedings of the 2nd International Conference on Building Innovations*, Lecture Notes in Civil Engineering 73,
https://doi.org/10.1007/978-3-030-42939-3_68

in regional reproduction processes associated with the renewal of the fixed assets, reconstruction, modernization, technical re-equipment of the material goods production, there appears a need to transfer the complex to innovative foundations of functioning, which is the key to the integration of science and production, market initiative stimulation in the process of arising of a new regional development economic model.

2 Main Body

Present-day regional inter-sectoral complexes, in particular the construction one, are affected by many factors of the socio-economic life of the region, which raises a wide range of issues that need to be solved to create favorable conditions for the development of the construction complex, which, in turn, contributes to the implementation of the state's investment policy, determination of the economic proportions, the scope and the pace of the development of individual industries, scientific and technological progress as well as the efficiency of investments and innovations in all the sectors of the economy [1].

An intensive development path of the country and its regions is possible only on the condition of scientific and technological progress achievements being put into practice. For Ukraine's economy, the most promising appears to be a strategy, within which the principles of innovative development can be implemented to the utmost. That is why the introduction of innovations in the construction is to become a top priority for the further development of the complex.

The innovative development of the construction complex is a strategic task, since it contributes to the effective creation of the fixed assets, both productive and non-productive, for the functioning of all sectors of the national economy, meeting the household, socio-cultural needs of the population, creating a large number of new jobs and selling the products of other sectors of the economy that are consumed in the process of carrying out the main activity—construction. As a result of increasing the innovative development of the construction complex, the investment policy of the state is implemented, the national economy proportions, the scope and the pace of the development of individual industries as well as those of scientific and technological progress are specified, and the efficiency of investments in all the sectors of the economy are determined [2].

Construction complex is a priority and strategic for the economy, though in Ukraine the share of the construction industry in the GDP structure amounts to only 2.3%. At the same time, it can be stated that this industry has enormous potential, which can be proved by the comparison of the Ukrainian indicator with those of some other countries (Fig. 1). According to the statistics, construction industry is the engine of economic development worldwide. The average share of construction in the GDP of European countries is about 6%. In Slovakia, it is 8.7%, in Poland—7.9%, in Lithuania and Latvia it comes up to 7.1%, in Kazakhstan and Germany—to about 6%. Therefore, the construction complex of Ukraine needs investments and

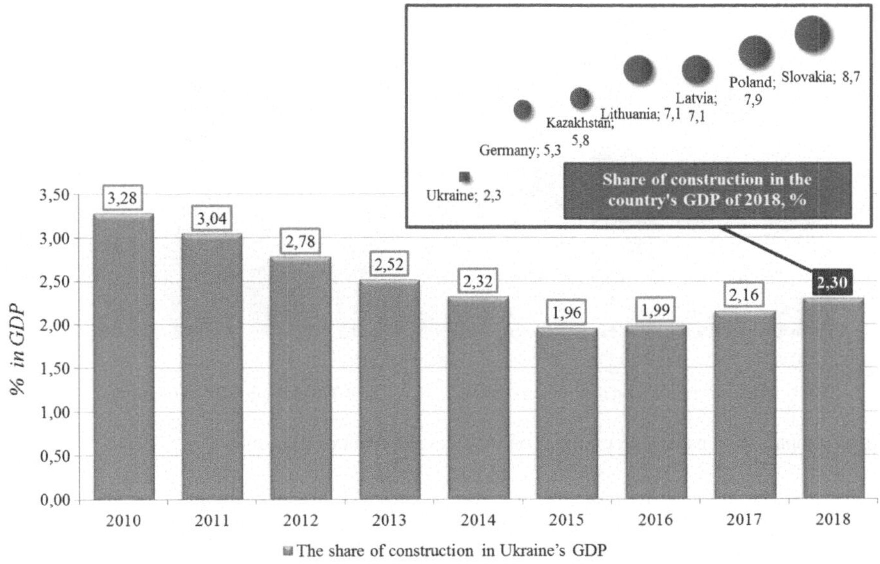

Fig. 1 Level of development of Ukraine construction industry in 2010–2018

innovations that will give impetus for its development. Innovations in the construction complex will enable to enforce production efficiency, the quality of construction products, which will help to save resources, to reduce the costs of operating buildings and structures and to solve a number of social problems by providing the population with affordable housing.

However, currently, scientific research is not always transformed into scientific ideas and the latest building technologies due to the existing obstacles and the imperfect mechanism of introduction of innovations in construction. Innovations in construction in general and in housing in Ukraine in particular are applied spontaneously, without proper scientific substantiation [3].

Consequently, working out an innovative model of the development, with specifying the activity measures of the construction complex structures, provided their strategy being implemented, is necessary for the further development of the construction complex of the country, as the technological backlog of the construction threatens the normal existence of the construction complex of the regions, and the impact of the crisis phenomena can catastrophically affect the existence of the construction complex of Ukraine as a whole [4].

The current economic situation in Ukraine is characterized by an imperfect system of stimulating investment and innovation in all sectors of the economy, increasing the tax burden, significant reduction in state capital investments, the imperfection of the state investment and innovation policy, the abolition of all privileges that stimulate the innovative process of economic development of the country's regions and economic complexes. In recent years, the scientific factor has been virtually excluded from the

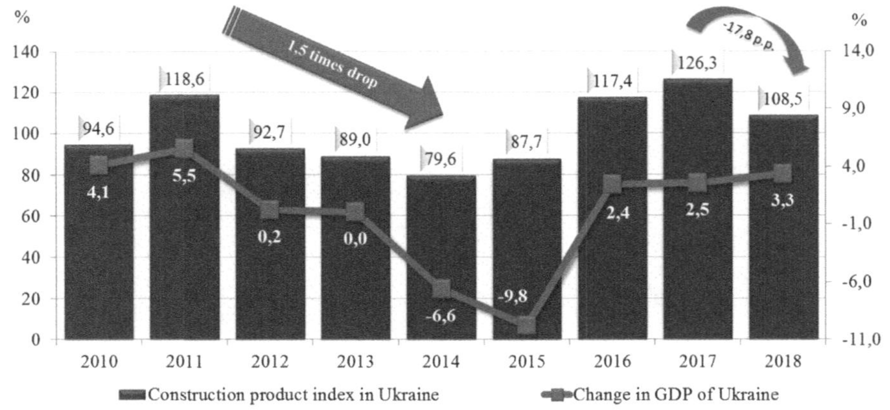

Fig. 2 Impact of construction on the economic development of the country

development of construction, resulting in a significant technological backlog of the construction industry from developed countries [5].

However, today we can talk about a certain growth of Ukraine construction complex in recent years. For example, the construction product index in Ukraine has fallen sharply by 2015, reaching only 87.7% in 2015, which is 1.5 times less than in 2011, when it reached 118.6%. In 2016 and 2017, this indicator began to increase rapidly and reached a level of 126.3% in 2017, which confirms development dynamics of the industry (Fig. 2). However, in 2018, construction product index of Ukraine decreased slightly compared to the previous year. According to experts, each hryvnia invested in the construction brings the state five times more [6–8]. In Ukraine, after some recession, the construction sector has revived over the past three years and continues to develop at a rather active pace. This is also evidenced by the growth of Ukraine's GDP over the last three years, which is confirmed by the statistics in Fig. 2.

The situation in the construction complex has actualized the need to develop a proper mechanism for its innovation-oriented development management, which, in the general sense, implies a close relationship and coherence of interests in the system «education–science–innovation» , aimed at improving development processes, introduction, production and commercialization of innovations, which will actively promote the development of innovative, investment, socio-economic, political activity of a separate construction enterprise as well the construction complex of the region and the country as a whole [9].

Innovation includes not only technical and technological developments, but also the use of new methods of working in the market, new products and services as well as financial instruments. They are the most important factor for the stable functioning of business, financial, credit or any other structures and ensure their economic development and competitiveness [10]. The aim of the innovation development strategy is to create regional innovation infrastructure, which will unite the efforts of state authorities, regional and local self-government bodies, of scientific and technical organizations and the business sector of the region construction complex in order to quickly use the achievements of science and technology for the transition of the region economy to the innovative stage of development.

Thus, the innovative development of the region construction complex is a complex process of transition of a region construction complex to a new, better state through rational use of innovative potential, introduction of scientific developments into construction, effective interaction of science, power structures and production, which will help to reduce the cost of construction projects, to solve the problems of the region construction complex and improve the living conditions of the population [2].

The general tendencies of construction innovative potential of recent years are: decrease of the organizations which carry out scientific and technical works; reduction in the number of employees in scientific organizations; reduction in the number of organizations engaged in innovative activity; almost unchanged number of inventors, designers and innovative proposals; increased funding for scientific and technical works, but this indicator in actual prices, given inflation, is insignificant. All these trends indicate a significant reduction in innovation activity of the country. The main factors that hinder innovation in the construction complex are:

1. Technical and economic ones (low material and scientific base, dominance of traditional production, outdated equipment and technology, orientation to short-term goals, lack of funds for risky projects).
2. Organizational and managerial ones (high centralization and conservativeness of the organizational structure, lack of innovation strategy, slow development and implementation of innovations).
3. Information and communication ones (insufficient information on innovations, lack of protection of ownership of information resources, seclusion and limited interconnection).
4. Socio-psychological ones (stereotypes of behavior, fear, resistance to the new, increasing of uncertainty).
5. Legal (antitrust, tax, patent and credit restrictions) [11].

The consequences of low innovation activity of the state cannot but lead to a sharp decrease in employment in the regions and, consequently, to an increase in social tension in society. Thus, it is critical to revive investor interest in investing in construction projects in general and housing construction in particular. One of the main reasons for these negative phenomena in construction is the increase in the cost of housing construction, which requires optimization and cost reduction that cannot be accomplished without innovative directions for managing the development of the building complex. The high level of costs of construction enterprises is a global economic factor that determines the inefficiency or low efficiency of virtually all market-oriented construction organizations and firms and leads to a decrease in the profitability of the construction complex as a whole.

In order to solve the problem of reducing the cost of construction projects, which, in the author's vision, is one of the most important reasons that inhibits the investment activity of enterprises and leads to insolvency of buyers, a mechanism of the region's construction complex innovation-oriented development management is proposed. It is based on such basic principles as: accessibility and sociality, economic efficiency, competitiveness, systematicity and adaptability, meeting the needs of the regions and distinguishing the elements of organizational and economic mechanisms of innovation management (Fig. 3).

The object of construction complex innovation-oriented development management can be innovations themselves, innovative processes implemented at particular construction enterprises, as well as economic relations arising between participants in innovation activity and the innovation market. The main purpose of managing the innovative development of construction enterprises and the construction complex of the region as a whole is to ensure the effective functioning of a certain innovative process of construction activity and, as a result, to ensure a stable competitive position in the market and profit in the long term. However, the goals of the innovative activity itself can be different: from meeting the new need of the buyer to reducing costs and time in the course of any construction process. Any management system must be built and implemented on certain principles.

For the construction complex, in turn, we offer the following mandatory principles for managing innovative development:

1. Systemacity of business processes—all the elements of the management mechanism and business processes of all segments of the construction complex should be interdependent as in a single system.
2. Complexity of innovations—working out innovative development management system as a whole.
3. Balance of interests of the subjects of the innovation process in the economic chain—the use of innovations should benefit all participants of the innovation process: research enterprises—innovators → construction enterprise as owner → construction enterprise as personnel → manufacturers and wholesale suppliers of construction products → consumers.
4. Innovation orientation toward customer needs—that is, the activity of construction companies should be customer oriented.
5. Rational use of building materials—the principle is connected with limited resources.
6. Alternative development options—taking into account the variability and rapid change of the external environment, it is necessary to have alternative development options for a construction company and the possibility to choose innovations in the market, which is the most optimal for a particular company at a certain point in time.
7. Flexible response—the ability to quickly and effectively adjust the developed plans of innovation in response to changes in the internal and external environment.
8. Risk justification—maintaining a balance between possible losses and expected results.

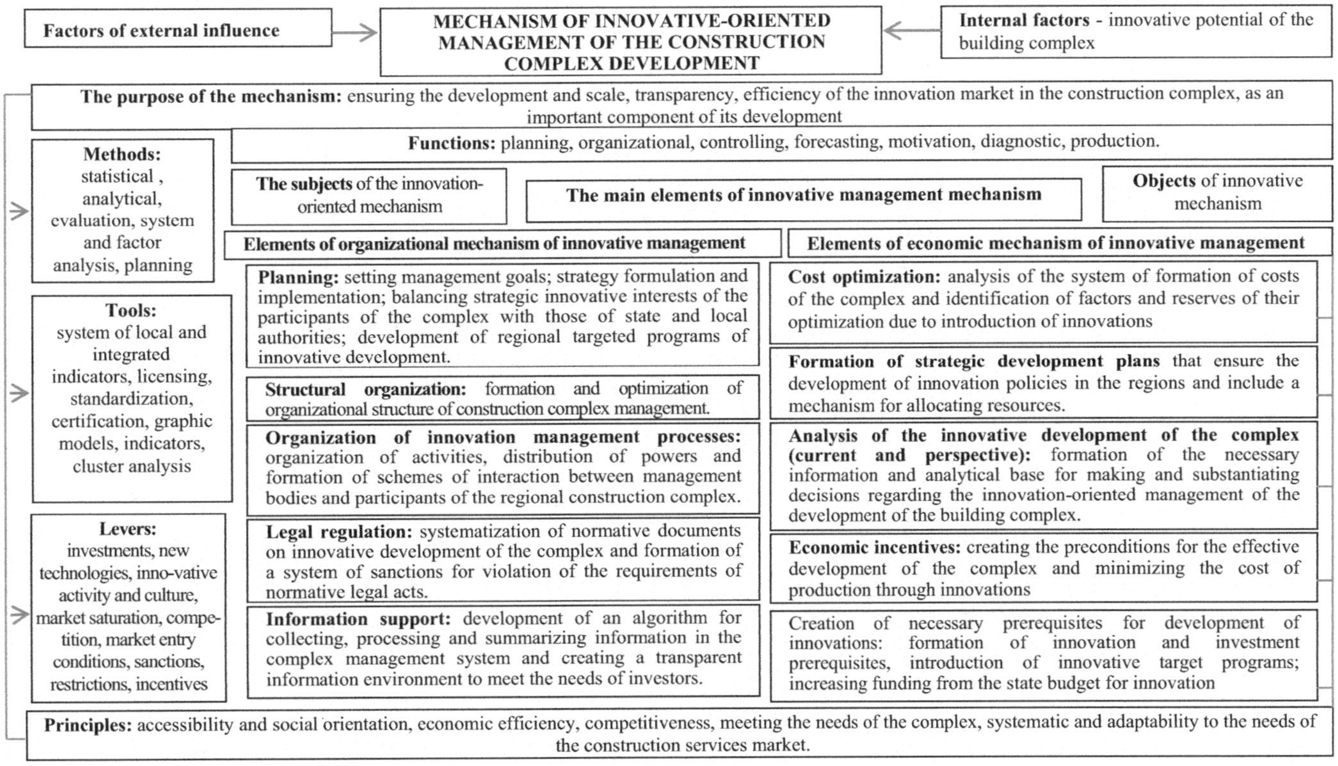

Fig. 3 Components of an innovation-oriented mechanism for managing the development of a construction complex

9. Awareness of market positions—when designing an innovation strategy and goals, construction companies should take into account the market niche they occupy.
10. Sustainability of innovation development—the focus on the constant search and use of innovation by construction companies in their activities.

In general, innovation management involves: developing innovative development goals; creation of innovative strategies system; environmental analysis taking into account uncertainty and risk; innovative potential analysis; market situation estimation; search for innovative ideas, licenses, know-how; market situation forecast; innovative and investment portfolio formation, projects development; planning and organization of scientific developments, their introduction into production; organizational structures management improvement; HR; evaluation innovation performance evaluation; procedure of making managerial decisions; market conditions study, competitors innovative activity; market research for new products and technologies (market capacity, conditions and elasticity of demand, etc.); research of resources necessary for carrying out innovative processes; innovation risks analysis, methods for their minimization determination.

Improving the innovative activity of the region's construction complex, to the author's vision, should start in the scientific sector of higher education institutions in the region and be based on the search and new products development, new technology that will reduce the costs. The uniqueness of the current situation is that the quality assurance of engineering training, on the one hand, and the accelerated development of priority and high-tech branches of science and technology, on the other hand, is possible only by combining the efforts of the intellectual potential and resources at technical universities, academic and industry science and high-tech industry enterprises. It means that Ukrainian technical universities and the high-tech science sector are considered and, in fact, strategic partners who must be objectively interested in improving the quality of higher vocational education and research.

As mentioned above, the system of innovation-oriented governance in the general sense implies close interconnection and coordination of interests the chain «education–science–innovation» . Education involves the transfer of systematic knowledge and skills and is the main mechanism of training qualified specialists who can ensure state economic development. Science provides basic knowledge generation, applied research implementation and new development emergence, while the innovation process enables scientific and educational spheres integration by training specialists and putting into operation the results of scientific and innovation activities.

Thus, based on the above mentioned, the main goals of innovation-oriented management should be considered: (1) ensuring the long-term functioning the innovation process based on the effective organization of all its constituent elements and systems; (2) creation of competitive innovative products, technologies in the most efficient and optimal way.

The allocation of functions in innovation-oriented management is driven by a variety of management activities in the chain: idea—research—development—design—production—innovation [12–14]. The functions of innovation-oriented management are understood as the kind of activity that is necessary for the implementation of the general tasks in innovation management.

The criteria for the effectiveness of innovation processes are economic indicators, which can be used to determine the growth of economic result relative to costs. In this case, the profit and profitability of innovations act not as a goal, but as an important condition and result of innovative activity and creation of new products, technologies, services that affect the standard of living in society. At present, the main disadvantages of constructing complex management are the following: imperfect cost-management mechanism in constructing complex (there is lack of single management entity, which functions are control over the level of expenses and resource allocation at constructing enterprises), low information interaction between constructing enterprises and regional executive bodies (is mainly limited to enterprises obtaining permission for construction, leaving aside any issues of the complex perspective development), imperfect pricing system at constructing that requires constant revising and renovation of state constructing norms, still, there is no their clear structuring, some types of regulating documents are controversial to others, and certain norms are out of date.

The main directions of construction complex innovation-oriented development management can be distinguished, namely development of construction complex innovation development strategy; application of new construction technologies, reconstruction of old buildings with low energy-efficiency indices, construction of new multistoried passive houses and premises; re-equipment of constructing enterprises with new equipment and technologies; development of new organizational-legal and economic mechanisms that stimulate innovative activity development; organization of expertise submitted for application of scientific and technical developments for compliance with advanced world standards; development of new technologies for obtaining, processing and application of structural and functional materials in construction complex, increasing the region export potential; preservation of personnel potential of the scientific and technical sphere; training and retraining of relevant specialists to work in the field of innovation.

According to author's vision, development of new organizational-legal and economic mechanisms that stimulate development of innovative activity; organization of expertise submitted for application of scientific and technical developments for compliance with advanced world standards; development of new technologies for obtaining, processing and application of structural and functional materials in the construction complex, increasing the export potential of the region; preservation of personnel potential of the scientific and technical sphere; training and retraining of relevant specialists to work in the field of innovation. Global and domestic experience of creating and operating cluster structures shows that their organization promotes productivity growth, increases innovation and investment potential of regions, promotes competitiveness of different sectors of economy without attracting additional financial resources. The main direction of constructing complex clustering regional strategy is to reduce costs and full cost of construction sites and their high cost recovery compared to other state measures aimed at creating basic infrastructure and providing services to the population.

Cost optimization of the region's construction complex is a complicated, multidimensional and dynamic process that involves management actions aimed at identifying potential cost reduction reserves in order to increase the level of profitability of the region's construction complex [11].

Global reduction in the cost and value of construction sites or services is a complex and objectively lengthy process that requires considerable investment and innovation. That is why, while enhancing practical value of university research for the ultimate consumer, it is necessary to ensure the full cycle of intellectual property development. By creating scientific and technical products within universities, it is possible to carry out a full research cycle that goes into the development of design and technical documentation, technological processing of the product, possibly to the production of models, samples or batches of goods. This approach is indisputably attractive to mass-market construction companies and those without their own R&D departments and enables businesses to reduce costs when implementing new technologies and in the initial stages of an innovative product's life cycle. Close relationship between education, science and production will result in optimizing the resources and time spent on each participant's goals.

Generalization of scientific research of the problem has made it possible to argue that for efficient innovation-oriented management of construction complex development in Ukraine it is advisable to: create a concept of national innovation system development; improve management and pricing system in construction; to introduce support and development of innovative entrepreneurship in the construction sector at the national and regional levels; to provide state funding for the most important basic technologies; review and introduce tax incentive schemes for innovation activities and promote the development of innovative infrastructure to commercialize research results, bring them into economic turnover, improve the relations between science and production.

The peculiarities of the innovation-oriented management organization in the construction complex can be explained by the fact that the development of the construction industry depends on effective functioning of industrial enterprises, which are the main investors during capital investments. This means that the construction is characterized by an increased level of risk of crisis events, because in the case of financial and economic problems in industry, reduction of investment activity in construction takes place, and it is not possible to exit it without innovation [15].

Every innovation-oriented management system, like any other system, needs its own specific assessment and determination of the effectiveness of its operation. Management effectiveness is understood as its purposefulness to create necessary, useful things, which can satisfy certain needs to ensure the final results achievement, adequate to the set goals. In such an interpretation, the concept of «innovative management performance» is characterized by the result, the effect achieved by the management entity due to its impact on the object [16]. The content of the concept of «innovation management effectiveness» is somewhat different. It is related, first of all, to the interpretation of the terms «effect» and «efficiency» . Thus, the «effect» is the output, the result of the activity, while the «efficiency» is characterized by the ratio of the effect to the cost of resources that helped to achieve the effect [17].

If we assume that the effect of innovation-oriented management is its efficiency, and the costs are the costs of management, we will come to the following logical formula of innovative management effectiveness. In Eq. (1), E_{iy} is the effectiveness of innovation-oriented management; E_y—management effect; B_y—management costs.

$$E_{iy} = E_y/B_y \tag{1}$$

The application of this qualitative dependence for quantitative assessment of innovative management system effectiveness is hampered by a number of circumstances related to the concept of «efficiency» , such as there is a problem of evaluating results that do not come down to a single measure; it is difficult to attribute the results to a certain entity or type of management; time factor should be taken into account—many management actions give effect after some time (recruitment, training, etc.); it is wrong to consider the results of managed processes to be the result of management activity only.

Most of the product is created by manufacturers, but not by managers. Therefore, it is incorrect to compare the result of all the economical activity with the costs of management alone. It is advisable to put the cost effectiveness of all the activities, not just innovation management in the denominator of the previous formula [3].

As a result, we will get an efficiency formula not only for innovative management, but for the whole managed object or process. In Eq. (2), E is efficiency, P is the result, B is the cost of obtaining the result.

$$E = P/B \tag{2}$$

The effectiveness of an innovative management system can be measured by the results of managed objects and processes. Management efficiency can be defined by another use of the original logic formula. For example, management methods that enable achieving a fixed result at the lowest cost for management correspond to a higher level of efficiency. At the same time, to achieve the maximum result under the conditions of limited management costs indicates the highest efficiency criterion of innovation-oriented management.

In every region of Ukraine, there is a relationship between state authorities and society, and the quality and level of which is determined by the effectiveness of government policy. This problem is immediate and fundamental because society will suffer from state organizations inefficiency. In order to determine the effectiveness of innovation-oriented development management of a region construction complex, it is necessary to consider basic approaches to efficiency evaluation.

There are three most common approaches in management effectiveness evaluation in theory and practice: integral, level and temporal [5].

Level approach to management effectiveness assessing identifies three levels of effectiveness in the evaluation process: (1) individual; (2) group; (3) organizational and relevant factors that influence them. Management efficiency is formed as an integrated result of individual, group and organizational efficiency, taking into account the synergistic effect. Accordingly, the level approach can be used at the micro level. Time-based approach to management performance evaluation identifies short-, medium- and long-term assessment processes for each of them specific management performance evaluation criteria can be identified. The time approach cannot be applied to a building complex, as its specificity is in long-term projects, and therefore, the most optimal is to use an integrated approach.

The integral approach to evaluating the effectiveness of innovation-oriented management is based on synthetic (integral) indicator construction that covers several partial (not directly comparable) indicators of management efficiency. It has emerged as one of the options to overcome the major drawback of the vast majority of governance performance indicators—the inability to reflect the multifaceted effectiveness of governance in general. It is an attempt to evaluate the effectiveness of management by means of synthetic (generalized) indicators, covering several major aspects of innovation-oriented management activity of Poltava region construction complex.

Thus, in the author's opinion, the formula for calculating the performance index of innovative-oriented management (W) of a building complex may be as follows:

$$W = P_1 \times P_2 \times P_3 \times P_4 \times P_5, \tag{3}$$

where P_1—growth rate of housing stock, P_2—growth rate of fixed assets, P_3—growth rate of housing commissioning, P_4—growth rate of sold innovative products, P_5—growth rate of innovation spending in the regions.

This indicator is integral because it covers all the main indicators that characterize the state of innovation-oriented management of the construction complex development of the region. The higher this indicator is—the better indicators of the building complex development will be.

The relative growth index of innovation-oriented management efficiency is the ratio of management effectiveness indicators:

$$I = W_1 / W_0, \tag{4}$$

where W_1—management efficiency in the building complex at the end of the forecast period and W_0—management efficiency in the construction complex at the beginning of the forecast period.

If $I > 1$, then there is a positive effect of innovation-oriented management, which is the evidence that management contributes to the building complex development and the system of innovation-oriented management of the building complex is satisfactory.

3 Conclusions

The necessity of using mechanism of innovation-oriented management of construction complex of regions development used in foreign countries and adapted into Ukrainian practice is proved, namely the creation of scientific-production innovative cluster formations and technopark structures that are promising because their main purpose and main idea is the formation on the basis of scientifically grounded and technologically implemented complex solutions, as well as commercial mechanisms, conditions for re-equipment of construction complex of the region (development of new generation construction technologies, new materials) and preparation of complex production and technological packages for profitable investment.

References

1. Varnaliy, Z., Onyshchenko, S., & Zavora, T. (2018). Construction complex development influence on region social and economic security. *International Journal of Engineering and Technology, 7* (3.2), 469–472. http://dx.doi.org/10.14419/ijet.v7i3.2.14574.
2. Mishchenko, R. A., & Furmanchuk, O. S. (2011). Formation pricing mechanism construction sector in the region and ways of its improvement [Formuvannia mekhanizmu tsinoutvorennia budivelnoho kompleksu rehionu ta shliakhy yoho udoskonalennia]. *Economics and Region, 30,* 152–156.
3. Costs for innovation [Vytraty na innovatsiiu]. http://uk.wikipedia.org/wiki. Last Accessed August 14, 2019.
4. Onyshchenko, S. V., & Maslii, O. A. (2017). Organizational and Economic Mechanism of Prevention the Threats to Budget Security of Ukrainian Economy. *Scientific Bulletin of Polissia., 1*(9), 176–184.
5. Demianenko, I. V., & Buriak, A. V. (2009). Innovation and technological progress as driving factors civilization progress society. *Visnyk DDFA: Ekonomichni nauky, 21,* 21–30.
6. Seniv, B. H. (2011). Current state and prospects of the construction industry in Ukraine [Suchasnyi stan ta perspektyvy rozvytku budivelnoi haluzi Ukrainy]. *Innovative Economy, 7,* 19–24.
7. Varnaliy, Z., Onishchenko, S., & Masliy, A. (2016). Threat prevention mechanisms of Ukraine's economic security. *Economic Annals-XXI, 159*(5–6), 20–24. http://dx.doi.org/10.21003/ea.V159-04.
8. Shevchenko, V. O. Status and prospects of development of building materials in Ukraine. Efektyvna ekonomika. Available at: http://economy.nayka.com.ua. Last Accessed July 08, 2019.
9. Furmanchuk, O. S. (2009). Innovation and investment cluster model of development of construction industry in the region [Innovatsiino-investytsiina klasterna model rozvytku budivelnoi haluzi rehionu]. *Productive Forces and Regional Economy, 2,* 206–213. RVPS Ukrainy NAN Ukrainy, Kyiv, Ukraine.
10. Onyshchenko, S. V., & Furmanchuk, O. S. (2012). Technopark as an innovative model of the Poltava region construction complex development [Tekhnopark yak innovatsiina model rozvytku budivelnoho kompleksu Poltavskoho rehionu]. *Business Inform, 7,* 40–43.
11. Bezzubko, L. V. (2008). Innovative construction potential [Innovatsiinyi potentsial budivnytstva]. *Building Ukraine, 7,* 8–11.
12. Lesakova, L. (2009). Innovations in Small and Medium Enterprises in Slovakia. *Acta Polytehnica Hungarica, 6,* 23–25.
13. Vytvytska, O. D. (2011). *The mechanism of realization of motivations of innovative entrepreneurship, monograph*. Kyiv, Ukraine: Ahrar Media Hrup.
14. Kozachenko, H., Onyshchenko, S., & Masliy, O. (2018). Region building complex social and economic security threats. *International Journal of Engineering and Technology, 7*(3.2), 79–85. http://dx.doi.org/10.14419/ijet.v7i3.2.14405.
15. Soldatenkov, V. V. (2012). Investment-innovative model of economic development of Ukraine [Investytsiino-innovatsiina model rozbudovy ekonomiky Ukrainy]. *Building Ukraine, 6,* 22–26.

16. Onyshchenko, S. V., Matkovskyi, A. V., & Puhach, A. A. (2014). Analysis of threats to economic security of ukraine in conditions of innovative economic development [Analiz zahroz ekonomichnii bezpetsi Ukrainy v umovakh innovatsiinoho rozvytku ekonomiky]. *Economic Annals-XXI, 1–2*(2), 8–11.
17. Verkhohliadova, N. I., & Kononova, I. V. (2011). A scorecard for managing the sustainability of a construction company [Systema pokaznykiv dlia upravlinnia stiikistiu funktsionuvannia budivelnoho pidpryiemstva]. *Herald of Khmelnytskyi National University, 1,* 7–10.

Increase of Thermal Resistance of the Gas-Filled Shell and Pneumatic Building for Use as Natural Gas Storages in Gas-Hydrated Form

M. M. Pedchenko, L. O. Pedchenko, and N. M. Pedchenko

Abstract The problems of accumulation, transport, and storage of gases and gas mixtures exist in many cases. Often, the existent technologies appear ineffective for transporting it with pipelines, as condensate or compressed gas. The transport and storage of gases in gas hydrate form can be a serious alternative to traditional technologies. However, capital structures made of reinforced concrete, metal structures, and bricks cannot provide the necessary thermal insulation of the cooled gas hydrate and sufficient tightness. The possibility of using the gas-filled shell and pneumatic building as gas hydrate storage facilities is substantiated in the work. Improvement of such constructions is proposed by using as a material for their thermal insulation of liquid non-solidification foam. The construction of these building and the main parameters of their operation are substantiated. The design of a ground-based storage facility is proposed for the accumulation and storage of natural gas in the composition of a gas hydrate. It is a frameless gas-support structure in the form of two domed, gas-tight soft shells on a thermally insulated basis, and the space between which is filled with liquid foam. The storage facility is equipped with systems for foam generation, collection of foam fracture products, gas and melt water withdrawal, and conditioning of the gas hydrate blocks in the storage facility (which includes a tube heat exchanger at the base of the storage, air cooler apparatus, solar collector for heating of the coolant, and refrigerating unit).

Keywords Gas hydrates · Storage · Gas-filled shell building · Foam

1 Introduction

In recent years, technologies, based on the ability of gas and water molecules to form gas hydrates under certain thermobaric conditions, have been actively developing. Since 1 m^3 of gas hydrate contains to 160 m^3 of methane, large gas volumes can be transported and stored for a long time [1]. Methane clathrates, known as methane hydrates or gas hydrates, are crystalline solid hydrocarbon compounds formed when

M. M. Pedchenko (✉) · L. O. Pedchenko · N. M. Pedchenko
Poltava National Technical Yuri Kondratyuk Universit, Poltava, Ukraine
e-mail: pedchenkomm@ukr.net

V. Onyshchenko et al. (eds.), *Proceedings of the 2nd International Conference on Building Innovations*, Lecture Notes in Civil Engineering 73,
https://doi.org/10.1007/978-3-030-42939-3_69

methane gas is trapped within the crystalline water structure at low temperatures (5–15 °C) and high pressures (2–3 MPa) [2].

The storage of natural gas in the gas hydrate form was first proposed by Benesh (1938) (produced at 283 K and the pressure of 35 MPa, stored at 241 K, i.e., in conditions close to the equilibrium) [3].

Natural gas transport projects in the form of pellets and gas hydrate suspensions developed in Japan and Norway [4] deserve special attention.

To date, different storage technologies for gas hydrate, both in the form of its mixture with water (for example, patents US 6082118, RU 2529928 C2) and dry gas hydrate (e.g. patent US 5964093) are proposed. Of course, dry gas hydrate is a more concentrated form of gas storage compared to a water hydrate mixture.

Until recently, granulated gas hydrate was the main form of transportation and storage. However, it has a lot of disadvantages (freezing of granules, gas loss due to the porosity between the granules, filling the storage volume only at the 70%) [5, 6]. Therefore, it is advisable to form the gas hydrate into monolithic blocks.

Storage of the gas hydrate requires maintaining the thermobaric parameters of its stable existence (to prevent dissociation into gas and water). However, under certain conditions, the gas hydrate can be stored in a metastable state due to the self-conservation effect or forced conservation by an ice layer [7–9]. Also, ensuring its good sealing to prevent gas loss and the formation of the explosive gas–air mixture are compulsory conditions for gas hydrate storage.

The technology of producing gas hydrate blocks preserved by an ice layer is proposed in the patent [10] and work [11]. They are suitable for transportation and storage at atmospheric pressure and slightly subfreezing temperatures (up to −3 °C). Also, the blocks, cooled in the production process, remain stable for a long time without additional cooling. However, a method for their further storage and melting technology for gas consumption is not disclosed in this patent.

2 Review of Information Sources

Therefore, sufficient storage conditions for a gas hydrate, which can remain in a stable state for a long time at atmospheric pressure (for example, cooled and preserved by an ice layer gas-hydrated blocks according to the patent [10]), are to be provided for thermal insulation and sealing.

A method for storing and regasification gas hydrates is disclosed in Patent No. US5964093. It foresees their placement in a cavity formed from the sides and from below by sealed walls, lined with thermal insulation material. The cavity has a roof made of transparent material. If necessary, the sun's rays can flow into the gas hydrate storage through it. For this purpose, a system that regulates the radiation level and therefore the dissociation of the gas hydrate is assumed.

However, the option of melting the surface of the gas hydrate mass has a significant disadvantage. Part of the released water under the influence of gravity will penetrate to some depth into the pores and space between the particles of the gas hydrate. This water will freeze as a result of heat exchange with the cooled gas hydrate. However, the temperature of the wetted layer will increase. Therefore, the level of the hydrate stability will decrease. In the future, this will complicate the operation of the storage facility in the storage mode of the gas hydrate.

In addition, the method of gas hydrate storage, according to patent No. US5964093, involves the use as a storage facility of ground or partially recessed capital structure of traditional building materials (for load-bearing structural elements).

However, capital structures made of reinforced concrete, metal structures, and bricks cannot provide the necessary thermal insulation of the cooled gas hydrate and sufficient tightness. Therefore, they will mainly perform the function of load-bearing structures and the fastening of insulation elements. In addition, this type structures cannot satisfy other requirements of modern energy projects, such as efficiency (their construction time is quite long) and mobility (since the structures are capital).

However, it is known that an effective thermal insulation of the object can be provided by a relatively small layer of porous insulation (foam, mineral wool, etc.), and its high-quality sealing at a pressure that is slightly higher than atmospheric (for example, 0.2–0.3 MPa)—polymeric film of minimum thickness.

In addition, the object can be protected from the weather by, for example, the soft cover of pneumatic structures [12].

Considering the low specific gravity of porous thermal insulation materials and the sealing film coating, the mass of the unit surface area of the storage elements, made of these materials, will be insignificant. In this case, the use of capital structures as supporting elements for their attachment is unjustified.

The purpose of the work is to select the optimal structure of the construction for the arrangement of the natural gas ground storage in the form of gas hydrates (gas hydrate blocks), as well as the method of structure operation (process organization and provision of gas hydrate storage, melting and gas selection).

3 Basic Material and Results

In view of the above, gas hydrates (gas hydrate blocks according to the [10] patent) are proposed to store in ground-bases gas-support constructions, covered soft shells. Such storage facilities are closed structures that “lie” on the gas cushion.

The pressure of gas cushion exceeds the atmospheric pressure only to overcome the force of the shape (bending) formation and compensation of the shell’s own weight (excessive pressure—0.01–1.0 MPa) [13–15].

The analysis of design features and characteristics of shell gas-bearing and air-chamber structures from the point of view of use as gas hydrate storages made possible to highlight their advantages and disadvantages. On the one hand, they can provide quality sealing of the object and protection against atmospheric phenomena.

Such storages can also be mobile—they can relatively easily disassemble, move (transport) and reassemble elsewhere. The main element of gas-supporting structures is the soft shell. However, it is known that the thermal resistance of such structures is very small [13]. Therefore, pneumatic structures need their conditioning to maintain the required temperature (in the case of gas hydrate storage—cooling). Considering the level of costs for the cold production, the operation of such structures as gas hydrate storage is impractical without additional thermal insulation.

Covering these structures with two- and three-layer coatings is one among the solutions. However, the thermal resistance of the gas shut-off layer, according to the method of calculating the heat-protective shell characteristics of gas-supporting structures, cited in [16], significantly increases to only 0.3 m. In [13], it is stated that the overall coefficient of thermal conductivity of the two-layer coverings, considered in it, was 2.8–3.4 W/(m^2 K).

Therefore, regardless from the time of year, significant heat flow into the storage will flow even through the two-layer coating. In [17], foam-polyurethane is proposed to use to reduce gas storage costs. After laying in storage, it is proposed to cover the gas hydrate with a layer of foam-polyurethane 0.5–0.7 mm thick. This will increase the thermal resistance of the storage facility covering to 17.0 (m^2 K)/Вт.

The main characteristic of thermal insulation materials, which determines their useful properties, is porosity. The pores contain gas with low thermal conductivity.

Among others, materials that are foamed solidified polymeric material are used as thermal insulation. In this case, solidification is necessary for the manufacture of products (thermal insulation panels) of a certain size and shape for ease of transportation and use. (Although the technology of insulating objects by applying a liquid foam polymer composition, which then solidify, is also known).

However, the physical characteristics of the bubble shells (liquid or solidified) do not affect the thermal insulation characteristics of the foam. Therefore, the use of solidified porous insulation materials is conditioned by the ease of use.

At the same time, the use of solid thermal insulation materials will be unacceptable in the case of gas-bearing shell and air-chamber structures. First, the construction of structures will be significantly complicated, and therefore their value will increase. Secondly, the functional characteristics and range of possible applications (they will not longer are mobile) will sharply narrow.

In view of the above, the thermal insulation of shell gas-bearing and air-chamber structures must be carried out with foams, which do not solidify but have a sufficient degree of stability. Today, technologies of preparation and foam use are enough advanced. Foams are widely used as a means of extinguishing fires, disinfection, intensification of oil and gas production, etc. A layer of such foam (of a certain thickness) may have a thermal resistance which is sufficient for a high-quality thermal insulation of the structure. In addition, in the case of a transparent cover material, this layer may pass a portion of the scattered sunlight into the storage. In this case, the solar energy input to the structure can be adjusted within a certain range.

The increase of the solar radiation (energy) flow inside the structure is supposed to regulate with the thickness of the foam layer in the respective area of transparent covering until it is completely absent. Adjustment is suggested to carry out by destroying a portion of the foam with a suitable reagent or adding it with a foam generator.

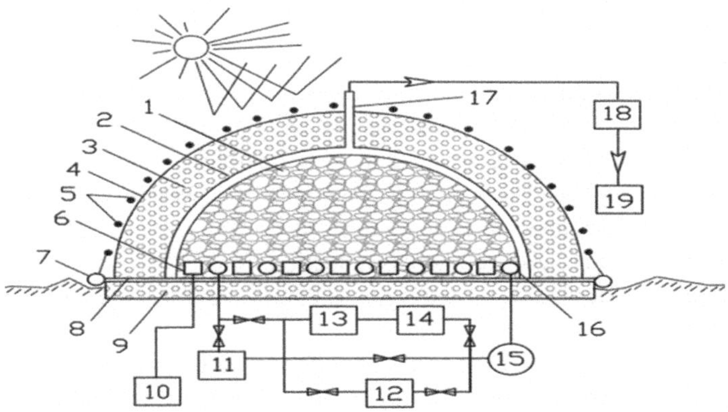

Fig. 1 Schematic diagram of the design of the gas hydrate storage during gas hydrate storage: 1—gas hydrate; 2—the bottom cloth from soft (elastic) material, impermeable to water and gas; 3, 9—a layer of thermal insulation material (liquid foam); 4—a top cloth from soft, water and gas-impermeable, sun-reflective material; 5—ropes mesh; 6—a system of perforated pipes for the removal of water and gas from under the gas hydrate stack; 7—sealing connection of the shell cloth and the base; 8—covering the base of material impermeable to water and gas; 10—water tank; 11—solar collector; 12—coolant heater; 13—refrigeration unit; 14—air cooling apparatus; 15—pump; 16—heat exchanger in the form a pipes system; 17—gas line; 18—compressor; 19—gas consumer

Thus, a schematic diagram of the gas hydrates storage is proposed after taking into account the level of construction technologies development and experience with the use of liquid foam. This scheme is presented in Fig. 1. The storage is a ground-based gas-support structure thermally insulated with a layer of liquid foam. It consists of shell, base, and addition equipments.

The shell for storage facility consists (at least) from two gas-tight soft shells of different diameters. The space between them is filled with the maximal stable liquid foam. Also, the storage facility is equipped with systems for foam generation and collection of the fracture products, air-conditioning (cooling/heating) storage, and gas and water extraction systems. The foam generation system provides foam feed to the upper part of the space between the outer and inner shells. The system for collecting foam fracture products is located between the shells at the level of the storage base.

Selecting the location of a gas hydrate storage facility is the initial step in organizing the gas hydrate storage process. It should be maximal as possible close to gas consumption. Calculation of the thermal balances the process—the level of cooling of the gas hydrate in the production process, taking into account the time and conditions of transportation to the storage location, as well as the duration and storage mode will be the next step.

The heat flux from the earth at an average value of 17 W/m^2 [18] will be 0.03 MW. Insulation of the base will reduce its flow to 9 kW.

Therefore, arrangement of the storage facility base implies the consistent deposition of thermal insulation layer 9 on the prepared site (Fig. 1), a coating from a waterproof material 8, a heat exchanger in the form of a pipe system 16, a system of perforated pipes, and gutters for the removal of water and gas from under the gas hydrate stack 6. The stack of pre-cooled gas hydrate 1 is placed on the prepared base.

The shell design provides at least one layer of thermal insulation material 3 between (at least) two solid cloths 2 and 4 of flexible gas and watertight material. The sun reflecting layer is applied to the upper cloth. Watertight cloth of sheells and bases are tightly connected to seal the storage. The insulated space formed by the gas line is connected to the compressor suction lines.

The shell of the storage to increase gas selection pressure, wind resistance, and mechanical strength is fixed by a ropes mesh 5.

For checking the sealing quality and for prevent the formation of the gas–air mixture, the air components from the storage are removed by evacuating the internal volume of the storage through the gas-lead line 17 by means of a compressor 18.

In order to increase the competitiveness of gas storage technology in gas hydrate form, it is advisable to provide mechanisms for maximizing the use of alternative energy sources to maintain the stability of the hydrate in storage during storage and its melting (regasification) and during the natural gas selection. The temperature of the storage is suggested to be provided by the cold accumulated in the gas hydrate blocks and the system of additional air-conditioning. It is advisable to melt the hydrate directly in storage (without wasting resources on moving it) through solar energy. In addition, for keeping the hydrate in a stable state, it is advisable to maximize the use of natural cold.

The temperature control of the gas hydrate (cooling during storage and heating during gas selection) is carried out as a result of the circulation of the coolant by the heat exchanger 16 under the gas hydrate stack at the base of the storage. Accordingly, the cooled or heated coolant circulates in the heat exchanger. The coolant is cooled by means of a refrigeration unit 13 or an air cooling apparatus 14.

For heating or cooling the coolant, a complex of equipment should include: a solar collector 11, a hydrocarbon fuel heater 12, air cooling apparatus 14, a refrigeration unit 13, a pump 15 is located outside the storage facility.

The gas is selected from the storage facility as a result of the controlled melting of the gas hydrate by the heated coolant. Its heating is mainly due to the sun energy in the solar collector 11 or coolant heaters 12.

At a pressure that depends on the strength of the shell, the gas through the discharge line 17 enters the compressor 18 and then goes to the gas distribution network for consumption. The design features of the storage facility and its mobility will allow the storage of gas hydrates near gas consumers. Therefore, the compression pressure will be at the pressure level of the distribution networks. The water formed after the melting of the gas hydrate, by collector 6 (a system of perforated pipes) is removed from under the gas hydrate stack and go into the water tank 10. Given that water is fresh, it should be considered as an additional product of technology.

4 Conclusions

Gas in gas hydrate form is proposed to store in improved gas-support shell structures. The basis for their improvement is the use of non-hardening liquid foams as thermal insulation. The proposed storage facility design makes possible to significantly improve the technical and economic indicators of the gas storage process in gas hydrate form (and therefore competitiveness), as well as to expand the scope of services for providing natural gas to consumers.

Ground-base storage facility design is proposed for the accumulation and storage of natural gas as part of a gas hydrate. It is a frameless gas-support cover in the form of two domed, gas-tight soft shells on a thermally insulated basis, the space between which is filled with liquid foam.

Equipped storage:

- a foam generation system fixed at the top of the outer shell;
- a system for collecting foam fracture products located at the level of the storage base between the shells;
- a system of perforated pipes, mounted in a thermally insulated storage base, for gas and melt water;
- air-conditioning system (cooling or heating) of the hydrate in the storage, which consists of the tube heat exchanger, the air cooler of the coolant, the solar collector for heating the coolant, the refrigeration unit, and the boiler installed in the base.

In this paper, we consider the use of gas-support shells constructions as gas hydrate storage. However, the scope of their application, based on its excellent characteristics, can be significantly expanded.

References

1. Dawe, R. A., Thomas, M. S., & Kromah, M. (2003). Hydrate Technology for Transporting Natural Gas. *Engineering Journal of the University of Qatar, 16,* 11–18.
2. Sloan, E. D., Jr. (2003). Fundamental principles and applications of natural gas hydrates. *Nature, 426*, 353–359. https://www.ncbi.nlm.nih.gov/pubmed/14628065.
3. Khokhar, A. A. (1998). Storage Properties of Natural Gas Hydrates. PhD Thesis. Trondheim.
4. Rodgers, R. E., Zbong, Y., & Arunkumar et al. (2005). Gas hydrate storage process for natural gas. *GasTIPS, 12*, 34–39.
5. Kanda, H. (2006). Economic study on natural gas transportation with natural gas hydrate (NGH) pellets. In *23rd World Gas Conference*, Amsterdam.
6. Gudmundsson, J. S., Graff, O. F. (2003). Hydrate non-pipeline technology for transport of natural gas. http://www.igu.org/html/wgc2003/WGCpdffiles/10056_1046347297_14776_1.pdf.
7. Gudmundsson, J., Parlaktuna, M., & Khokhar, A. (1994). Storing natural gas as frozen hydrate. *SPE Production and Facilities*, 69–73.
8. Gudmundsson, J. S., & Parlaktuna, M. (1991). *Gas-in-ice: Concept evaluation.* (Technical report), Department of Petroleum Engineering and Applied Geophysics, Norwegian University of Science and Technology.

9. Takeya, S., Ebinuma, T., Uchida, T., et al. (2002). Self-preservation effect and dissociation rates of CH_4 hydrate. *Crystal Growth, 237,* 379–382.
10. Pedchenko, L., & Pedchenko, M. (2013). Method of production of associated oil gas hydrates for the purpose of transportation and storage. Patent Ukraine No.101882.
11. Pedchenko, L., Pedchenko, M. (2012). Substantiation of method of formation of ice hydrate blocks with the purpose of transporting and storage of hydrate gas. Naukovyi Visnyk Natsionalnoho Hirnychoho Universytetu (1), 28–34. http://nv.nmu.org.ua/index.php/uk/component/jdownloads/finish/31-01/460-2012-01-pedch/0.
12. Ermolov, V. V., Byrd, V. U., Bubner, E., et al. (1983). *Pneumatic building structures*. Moscow: Stroizdat.
13. Ermolov, V. V. (1980). *Air-bearing buildings and structures*. Moscow: Stroyizdat.
14. Orsa, Y. N. (1983). *Features of the architecture of pneumatic air support structures*. Stroyizdat, Moscow: Pneumatic building structures.
15. Shikhirin, V. N. (2006). *Elastic mechanisms and designs*. Irkutsk: IrSTU.
16. Aleynikov, A. E., Fedorov, A. B., & Tyutunnikov, A. I. (2004). Methods of heat losses calculation and heat inputs through fencing structures of frame-awning and inflatable structures. *StroyPROFIL, 8*(38), 58–61.
17. Pedchenko, L. A. (2014). Technological and theoretical bases of portage of gas hydrates in the thermal energy systems. PhD Thesis. Poltava. https://drive.google.com/file/d/0Bwbz8yU2lkF9RjUyYUdpYjF5WGs/view.
18. Budyko, M. I.(1956). Thermal balance of the earth's surface. Leningrad: Hydrometeorological Publishing House.

Managing the Field of Reconstruction and Preservation of Historical and Cultural Complexes in Ukraine and Europe

Tetiana Pulina, Tetiana But, Olena Khrystenko, and Valentyna Zaytseva

Abstract The paper analyzes the role and place of the defense architecture monuments—fortresses, castles, defense monasteries, and temples in the historical and architectural heritage of Ukraine—and identifies significant problems of restoration and involvement of castle complexes in modern life. The protection of monuments through their restoration and subsequent adaptation to modern needs, as a rule, concerns well-preserved Ukrainian castle complexes, while a considerable number of demolished castles are practically not restored and closed for inspection, destroyed physically and morally, thus remaining unknown to sightseers and tourists. The paper takes into account the experience of European countries that have long evaluated economic impact of the attendance of fortifications and palaces on the increasing demand in the tourism market of their own country. The paper determines the necessity to enlist and to structure all the monuments at the state level, i.e., to develop an effective mechanism for their inventory, control, reconstruction, and preservation. In addition, it is recommended to use European countries' experience in providing concessions for cultural monuments and at the legislative level to simplify the granting of permits to different ministries and departments to issue such concessions. The financing of preserving historical and cultural complexes in Ukraine is substantiated, which should be:

- budget financing: from the State Budget of Ukraine, from local budgets;
- using extra budgetary funds: bank loans and payments for the use of natural tourist resources, travel businesses' own funds, investors, local fees and charges, grants from international foundations and organizations, charitable contributions, and sponsorship.

Keywords Management · Historical and cultural complexes · Reconstruction · Concessions · Castles · Palaces · Fortresses · Cultural heritage

---

T. Pulina · T. But · V. Zaytseva
National University Zaporizhzhia Polytechnic, Zaporizhzhia, Ukraine
e-mail: tatyanabut1979@ukt.net

O. Khrystenko (✉)
Poltava National Technical Yuri Kondratyuk University, Poltava, Ukraine
e-mail: khrystenko_lena@ukr.net

V. Onyshchenko et al. (eds.), *Proceedings of the 2nd International Conference on Building Innovations*, Lecture Notes in Civil Engineering 73,
https://doi.org/10.1007/978-3-030-42939-3_70

1 Introduction

The development and prosperity of any state is impossible without preserving its cultural heritage. There have long been hundreds of unique castles and palaces in Ukraine that preserve the stories and secrets of many generations. Their appearance was facilitated by the country's geopolitical location at the crossroads of trade routes. There has always been a tremendous tourist interest in such sites.

Compared to Europe where castles and palaces had permanent owners and were inherited, in Ukraine, all castles were nationalized and transferred to various enterprises and organizations that used them as warehouses, sanatoriums, dispensaries, as well as economic and administrative premises, etc.

Today, in Europe, there is a gradual "modernization" of historical buildings, which are increasingly being converted into commercial and residential real estates, while many Ukrainian castles and fortresses are turning into piles of stones.

The study of issues related to the preservation of historical and cultural complexes, and world heritage has gained wide coverage in the scientific literature, in particular, in the works of foreign researchers such as Norros [1], Banfy et al., Chow and Faj [2, 3], as well as domestic scientists: Kot [4], Gorbik [5], Davydova [6], But et al. [7, 8], Timchenko [9], Zamyatin [10], Zhukov [11], Varnaliy [12], and others [13–17]. However, the problems of managing the field of reconstruction and preservation of historical and cultural complexes in Ukraine and Europe are not researched enough.

2 Objective

To identify directions for improving the management of the field of historical and cultural complexes' reconstruction and preservation in Ukraine and Europe.

3 Research Methods

Analysis of statistical data and literature sources according to the paper's objective, comparative analysis, and method of expert assessments.

4 Results

Ukraine is a country that has always been between the west and the east, the north and the south. Such a "buffer" status made the statesmen and residents of Ukrainian lands care constantly about their native land's defense from numerous invaders. Thus, century after century, magnificent castles, mighty fortresses, and defensive monasteries stood on the Ukrainian soil. Major trade routes ran through most of the fortified cities and towns. Due to the fortified territories, trade links between cities in different regions of Ukraine contributed not only to the development of the internal market, but also to the international trade.

Castles and monasteries are the largest historic buildings in the development of settlements. Due to their scale, they are still the compositional centers of many historical cities and villages of Ukraine (Kyiv, Pereyaslav, Chyhyryn, Poltava, Putivl, Chernihiv, Glukhov, Baturin, Okhtyrka, Novgorod-Siversky, Lviv, Kamianets-Podilsky, Bilgorod-Dnistrovsky, Zhovkva, Drohobych, Belz, Sudak, etc.). Some of our cities—Kyiv, Lviv, Volodymyr-Volynsky, Lutsk, Putivl, Novgorod-Siversky, Belgorod-Dnistrovsky—had an extremely developed and complex system of fortifications that had been formed over half a millennium [9].

The castle and hotel business is recognized as the most profitable way of using old fortifications. The prospect of a museum-animated revival of the country's grandest fortifications, which is a nationwide asset and must be accessible to a wide range of visitors, especially young people, is of even greater significance. A priority direction of most Ukrainian castles' and fortresses' tourist renovation is to restore fortifications, place expositions that will present the paintings of the past, and declare these fortifications open-air museums. The revival of tourist interest in them should be facilitated by high-quality information and promotional campaigns, as well as arrangement of various historical–theatrical festivals, competitions, and animation shows.

More than two-thirds of European Union castles (that is more than 1500 fortifications) are now being successfully used in the hotel industry. Spain, Italy, France, the Czech Republic, and other countries have long evaluated the economic impact of the attendance of fortifications and palaces on the increasing demand in the tourism market of their own country. Monuments of the past in these countries are protected by the state and are well preserved.

European tourists have a desire to spend a few days in a real medieval castle, to taste the dishes of ancient ethnic cuisine, and to take part in the animated knightly programs of the castle entertainments. Even Europe's most titled owners of ancestral castles dedicate part of their possessions, along with private apartments, to hotel rooms for guests and tourists.

Table 1 shows the number of fortifications and palaces in Ukraine [13].

Due to the peculiarities of historical and regional development, most fortifications were preserved in western Ukraine. It was in this region of greatest prosperity that

Table 1 Number of fortifications and palaces in Ukraine

No.	Number of fortifications and palaces	Figures
1	Number of fortifications and palaces	3000
2	Number of stately registered architectural monuments	1700
3	Number of castles and fortresses of national importance in the Ministry of Culture's register	75

the Old Rus state reached (the days of Danylo Halytskyi who managed to unite the lands from the Carpathians to the Dnipro) and kept the ancient Ukrainian statehood for the longest period. Subsequently, it was here at the site of Old Rus strongholds that the Polish conquerors built their first stone castles in order to keep the newly captured Rus's lands in submission [8].

Ukraine's most visited castles and fortresses are given in Table 2.

All objects of historical and cultural heritage in Ukraine are protected by the state. The State System of Preserving Historical and Cultural Monuments has been established in Ukraine. The state administration in the field of cultural heritage protection is entrusted to the Cabinet of Ministers of Ukraine, as well as to specially authorized bodies of cultural heritage protection.

The special Law of Ukraine "On Protection of Cultural Heritage" of 8.06.2000 No. 1805-III is in force. It regulates legal, organizational, social, and economic relations in the field of protection of national cultural heritage in order to preserve it, the use of cultural heritage objects in public life, and the protection of the environment's traditional character for the sake of present and future generations [8].

Art. 2 of the Law provides the following official classification of national cultural heritage objects (see Fig. 1).

In addition, there is a great number of public organizations and partnerships, associations and groups that handle cultural heritage protection issues in Ukraine. Among the public groups, the Ukrainian Society for Protection of Historical and Cultural Monuments plays a leading role in our country.

The main activity of the Ukrainian Society for Protection of Historical and Cultural Monuments was to involve the general public in the discovery, protection, research, and promotion of historical and cultural monuments. It is proposed to preserve castles and fortresses in two directions: by turning monuments into museums; by providing excursions to the castle complex with its obligatory inclusion in excursion and tourist routes without creating a museum on the territory of the monument.

Historic and architectural reserves and museums should use the most valuable castle and monastery ensembles in combination with other historical and cultural monuments, picturesque surrounding landscapes. These architectural and territorial complexes are assigned to national historical and cultural reserves. In some architectural monuments, various museums can be arranged, and some part of the territory and architectural heritage can be allocated to establishments of animation and crafts, souvenir-shopping, lodging, and gastronomic purposes.

Since castle complexes need to be used for the benefit of society, taking into account their economic profitability and introducing active technical measures to preserve and restore them, carrying out preventive measures in order to separate objects from aggressive factors that accelerate their damage and destruction, the organization of museums on their basis will be the most effective means of the whole complex's physical and spiritual preservation. In addition, the museum increases the actual importance of a particular monument, more fully discloses its informative value, and stimulates the study of the castle complex's history by carrying out a proper scientific activity.

Table 2 Top ten most visited castles in Ukraine [14]

Region	Castle name	Attendance features
Lviv region	Oleskyi Castle	The castle is one of the most famous in Lviv region, one of the smallest, but best-kept castles in Ukraine. In addition to the museum, there is a restaurant with reconstructed medieval interiors designed for Ukrainian, Polish guests
	Svirzkyi Castle	Has become well known due to filming. It stands vacant, though there were plans to open the Holiday House of the Architects' Union
	Pidgoretskyi Castle	The great castle-palace is a branch of the Lviv Art Gallery. It is still under reconstruction
Transcarpathian region	Castle-palace of the counts of Schenborn	This castle-palace is unique in its kind. It has exactly 365 windows, 52 rooms, 12 front doors
Transcarpathian region	Palanok Castle	The best-kept and reconstructed castle of Ukraine. Located on a hill 70 meters high, in the eighteenth century it was the strongest fortress in the east of the Austrian Empire
Chernivtsi region	Khotin	Inaccessible fortress built by the Moldovans. The movie star castle is the main medieval castle of Soviet films
Volyn region	Lutsk Castle	The only preserved architectural monument of the Grand Duchy of Lithuania in Ukraine. Two impressive Lutsk monuments—the Jesuit Monastery and the Church—are visible from the towers
Ternopil region	Zbarazkyi Castle	The castle was built by the Zbarazky princes, who were Galileo Galileo's students. During the battle near the walls of the fortress, Ivan Bohunwas fatally wounded, and Nestor Morozenko was killed. These events were later described in the novel "By Fire and Sword"
Khmelnytskyi region	Kamianets-Podilskyi Castle	The most picturesque castle of Ukraine. It is located in a very beautiful place, on a rock, along which lies the canyon of the river Smotrich
Odessa region	Bilgorod-Dnistrovska Fortress	The largest fortification building in Ukraine

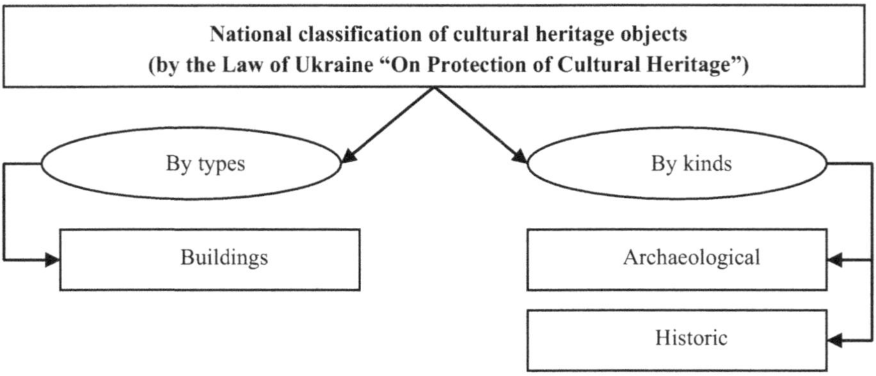

Fig. 1 Current national classification of historical and cultural heritage objects in Ukraine

Some scientific institutes of Ukraine (Ukrproektrestavratsiya, Ukrzakhidproektrestavratsiya, the Research Institute of Theory and History of Architecture and City building) have been conducting research and project development of restoration and further adaptation of castle complexes of Ukraine for many years [11].

There are projects of restoring and further including of the castle complexes in Korets, Novomalin, Starokostiantyniv, Izyaslav (Volyn), Mykulyntsi, Buchach, Pidzamochek (Podilya) developed by Ukrproektrestavratsiya and Ukrzakhidproektrestavratsiya institutes. Much of the ruined castles (almost 40% of the total number of such monuments) have already been provided with the project documentation for reconstruction and further adaptation [10].

The most successful are the projects of preserving and turning castles into museums. Oleskyi, Zolochiv, Lutsk, Ostroh, Zbarazh, Khotin, and Kamianets-Podilsky castle-museums are examples. The cost of priority restoration works for fortifications and palaces of Ukraine is shown in Table 3.

The experience of restoration [7] shows that at all stages—from the project development to its implementation—the problems of the new functional content of the fortification architecture monument, creation of protection zones around the

Table 3 Cost of priority restoration works for fortifications and palaces of Ukraine

No.	Fortifications and palaces	Restoration costs
1	Svirzky Castle	3035 mln. UAH
2	Pidhirtsi Castle	150 mln. UAH
3	Khotin Fortress	150 mln. UAH
4	Chervonohirsky Castle	10 mln. UAH
5	Zhovkva Castle	25 mln. UAH
6	Klevansky Castle	150 mln. UAH
7	Popov estate	35 mln. UAH
8	Pomoryansky Castle	1,5 mln. UAH

objects being restored, the preservation of the natural environment landscape, etc., remain unsolved. Restoration work alone does not solve the problem. There are cases when the restored building is left unused, which leads to its decline and further re-restoration (castles in Skalat, Kryvchy in Ternopil region, church and carriage buildings on the territory of the castle complex in Medzhibozh (Khmelnytsky region), castle in Svirzh in Lviv region). The complete restoration can only guarantee the adaptation of the monuments to the society's contemporary needs. Not having received new functions, not having acquired a true owner, and not being engaged in active social life, they are doomed to moral and physical decline.

The protection of monuments by their restoration and further adaptation to modern needs usually concerns a well-preserved castle complex. On the other hand, apart from being included in the society's active life, there are demolished castles. The buildings are practically not restored, closed for inspection, destroyed physically and morally, while remaining unknown to sightseeing tourists. These are castles in BilyKamen, Stare Selo, Dobromyl, Rakovka (Halychyna), Gubkov, Korka, Novomalin, Taikury, Derman, Dubrovytsya, Izyaslav (Volyn), Vinogradove, Korolevo, Serednyi, Khust, Kvasovo (Transcarpathia) and the vast majority of Podilya castles. The examples cited show that half-ruined castles are not used at all, despite the fact that many objects are important in the history of the Ukrainian people and the development of architectural and building traditions [11].

Let us consider the possibility of developing projects for the "tourist revival" of Ukraine's most famous historical castles, which are in a rather neglected state.

In the village of Urych, there are the remains of the Old Rus rock fortress Tustan (IX–XV centuries)—without exaggeration, the unique architectural monument in ancient architecture. Tustan was the administrative center of the parish, where in the times of KievanRus, Halych, and Halych-Volyn principalities charged a fee for salt transportation. The garrisons of the Tustan Rock Castle and a number of other fortifications guarded the roads leading to the main road to Veretsky Pass (the old name—"the Rusky Way"). The first record of the town fortress Tustan is in the Halych-Volyn chronicle of 1255.

For over 20 years, Tustan was explored by the Carpathian architectural and archeological expedition led by M. F. Rozhko. Back in 1978, he developed a project to reconstruct the Rock Fortress Tustan. But, unfortunately, this project has not been implemented yet (however, investors are still being searched for).

The XVII century Brodivsky Castle of bastion type was built on the best models of the Dutch and Italian defense systems in 1630–1635 under the leadership of the talented architect Andre del Aqua and the French military engineer Guillaume Le Vasseur de Beauplan.

Tourists are now able to view the system of earth shafts with casemates and defensive bastions of the castle, as well as the castle courtyard with barracks.

On the territory of Brodivsky Castle, the baroque palace of S. Potocki is preserved—a spacious two-storied brick building with side risalites extending along the northern side of the castle. Both the castle and the palace are still in a very neglected state, although the architectural features of these objects are ideally suitable for a picturesque castle-hotel with a knightly restaurant and beer cellars on the basis of fortifications.

Uzhgorod Castle of XIII–XVIII centuries is the oldest of all the fortresses of Transcarpathia. Uzhgorod Castle has a maze of underground passages, now is neglected and abandoned. Tourists are only allowed to view the stone 32-m-long well in the castle courtyard which, according to the legend, turns into a horizontal ledge leading into the valley of the River Uzh.

There is an international investment project to revive the castle as a cultural and leisure tourist complex. Investors are still being searched for.

Nevitsky Castle of the XIII–XVII centuries is located to the north of Uzhgorod, upstream of the Uzh, on a steep slope.

Nowadays, Nevitsky Castle is in an extremely attractive half-demolished state.

13 km from the princely capital of Terebovlya in Mykulyntsi over the Seret stands another monument of defense architecture—Mykulynetsky Castle of the XVI–XVII centuries (more precisely, its ruins).

Up to now, about half of the castle fortifications have been preserved: two corner towers with loopholes and wall sections between them. A contemporary researcher of the monument I. Pustynnikova notes alarmed that the territory of Mykulynetsky Castle has now been turned into a landfill, despite the fact that the castle is best suited to become a trade brand of beer popular in Ternopil region.

Berezhansky Castle is the center of the State Historical and Architectural Reserve. It became famous throughout Europe due to the fact that no army could take this stronghold for two centuries.

Berezhansky Castle and Palace were badly damaged during the shelling of the First World War and have not been rebuilt since then. However, it still impresses with its greatness. The thickness of the southwestern wall fragments is 6 m; the angular towers of a pentagonal plan of the stronghold are of the same huge dimensions. The towers have a complex system of loopholes with two or three crossed strokes, which made it possible to aim at different angles.

The revitalized complex will be able to provide lodging, gastronomic, exposition–excursion, and animation-tourist services, and most importantly, it will become a true tourist decoration of Podilya.

Skala-Podilska is a medieval border fortress of Rzeczpospolita located over the Zbruch only 40 km from Kamianets-Podilskyi—the center of the Podolsk district of the Ottoman Empire.

Skala-Podilska Castle should be completely rebuilt by placing a museum exhibit in one part of the palace and a tourist hotel in the other. The perimeter of the internal wooden combat galleries, castle dungeons, gun turrets, and tourist viewing platforms should be restored. The implementation of this project will immediately allow the castle in Skala to be included in a commercial mass tourist route along with the neighboring fortresses in Kamianets-Podilskyi and Khotin.

The presence of a tourist hotel will allow organizing horseback tours over the cliffs of the Zbruch from the castle to the Medobory reserve to karst lakes, reef rocks, the Pearl Tovtry cave, the Medobory water springs, to the location of the Zbrutsky idol and to the cult pagan center of the XIII century Podilskyi tribes consisting of three hillfort-sanctuaries on the three Tovtry mountains Bogit, Govda, and Zvenigorod, as well as tours to the Dniester Canyon, to Kamenets-Podilsky, to the royal fortifications of the "Trinity Trench," to the Atlanta labyrinthine cave and a number of other gypsum caves in the Borshchiv district (Krystalova, Mlynky, Ozerna, etc.).

Buchatsky Castle is an ancestral nest of the Buchastkis, Ukrainian feudal lords (later Polonized), built on a high rocky promontory above the Strypa at the site of the fortifications of a wooden Old Rus castle. The first record of the castle dates back to 1379. According to archaeologists, the north wall of the castle made of blocks of red sandstone dates back to the very XIV century.

The state of the monument preservation gives grounds to start the project of its complete restoration and give it the status of a state historical and cultural reserve. The renaissance palace of the Russian governor Jan Buchatsky should be rebuilt placing in one part a tourist hotel and restaurant, and in the other—a historical museum, the exposition of which will acquaint with the history of the glorious Ukrainian families of Buchatski and Potocki, countless sieges and Polish-Ukrainian and Polish-Ukrainian-Turkish wars of the XIV–XVIII centuries.

Even in a dilapidated state, Medzhibizh Fortress still stuns with its powerful inaccessible stone walls. The walls are 3.5 m thick and 17 m high. They have three tiers of complex loopholes to fire cannon and shotguns; wooden combat galleries led to the second and third tiers. A variety of household and residential buildings lines the walls.

There is a museum on the territory of the fortress. However, nowadays the magnificent Medzhibizh stronghold is in dire need of an investor who would complete its reconstruction. The advent of the historic animation complex, the restoration of respectable palaces, and the organization of a hotel and restaurant complex in one of them will allow Medzhibizh to become a real tourist mecca not only within comfortable Balaklava bay, for which Chembalo castle was built (one of its towers has remained till present).

Once magnificent and inaccessible was Lviv Fortress—the High Castle. Archaeological excavations have shown that the fortress, which originally had wooden and earthen fortifications, occupied the summit of Castle Hill (417 m). At the end of XII –beginning of XIV century, the fortress was rebuilt from wooden to stone. The castle had the shape of an elongated rectangle with four towers at the corners. Barracks and a deep well were located in the castle courtyard.

In the XVIII century, the fortress lost its strategic importance. Its towers and walls are gradually being dismantled for construction material. Up to our time, a fragment of the southern wall with loopholes has been preserved from the High Castle.

We suppose that a significant change in the modern legislation in the field of protection and preservation of cultural heritage is needed in Ukraine, since part of the castles was destroyed even in western Ukraine (Kolomyia, Snyatynsky, Stanislavsky, etc.).

For a system of cultural monuments preservation and restoration to function effectively in our country, the necessary funding must be carried out:

- budget financing from the State Budget of Ukraine, from local budgets;

- extra budgetary funds: bank loans, payments for the use of natural tourist resources, travel businesses' own funds, investors, local fees and charges, grants from international foundations and organizations, charitable contributions, and sponsorship.

It is necessary to use European countries' experience in providing concessions for cultural monuments. Today in Ukraine the legal conditions of the concession mechanism are stipulated by the Law of Ukraine "On Concessions" of the Verkhovna Rada of Ukraine No. 997-XIV of July 16, 1999.

The submission of concessions for memorials should be in accordance with a clear, effective system of state control over compliance with concession terms and tenant-friendly conditions (including simplification of various permits), with the provision of privileges and long-term loans. After all, many monuments, including castles, have neither sewer system, nor electricity, nor other modern conveniences. Moreover, the tenant has to get a number of special permits in different ministries and agencies, which is not easy to apply for.

By creating favorable conditions for investors, Ukraine will not only be able to restore cultural sites, but also to develop the tourism industry by creating the right infrastructure. Investors should be interested in the possibility of building commercial tourist hotels, boarding houses, sanatoria, museum complexes on the basis of castles. The investor can be attracted only on other conditions declared in the investment contract: ownership share after the project implementation; future profit percent; places quotas (in the construction of tourist accommodation facilities), etc.

5 Scientific Novelty

In order to preserve the historical and cultural complexes in Ukraine, a system of management, storage, restoration, and renovation of tourist resources based on the foreign experience has been proposed.

6 Practical Value

Therefore, in order to solve all the above-mentioned problems of Ukraine's historical and cultural complexes reconstruction and preservation, an effective system of management, storage, restoration, and renovation of tourist resources should be formed, using the foreign experience gained. It is necessary to carry out a complete inventory structuring of the monuments at the state level, i.e., to develop a mechanism for their inventory, control, reconstruction, and preservation. In addition, it is recommended to use European countries' experience in providing concessions for cultural monuments and at the legislative level to simplify the process of obtaining permits from different ministries and agencies that issue such concessions.

7 Conclusions

Historical and cultural complexes (fortresses, castles) performed various functions,: protected independence of the state or of a certain territory, acted as composing centers of many historical cities of Ukraine, and guarded trade routes passing through the fortified cities.

The analysis shows that at the present stage of historical and cultural complexes development, the use of castle buildings, some successes in their adaptation, as well as the problem of restoration and incorporation of castle complexes in modern life, remain unsolved.

Therefore, in order to solve all the above problems of Ukraine's historical and cultural complexes reconstruction and preservation, an effective system of management, storage, restoration, and renovation of tourist resources should be formed, using the foreign experience gained. It is necessary to carry out a complete inventory structuring of the monuments at the state level, i.e., to develop a mechanism for their inventory, control, reconstruction, and preservation. In addition, it is recommended to use European countries' experience in providing concessions for cultural monuments and at the legislative level to simplify the process of obtaining permits from different ministries and agencies that issue such concessions.

References

1. Norros, L. (2018). Ponymanyedejstvyia v slozhnykhuslovyiakh: sozdanyesynerhyykul'turnoystorycheskojteoryydeiatel'nosty, Peirce y ekofunktsyonal'nosty. *Mind, Culture, and Activity, 25*(1), 68–85. Retrieved from: https://doi.org/10.1080/10749039.2017.1350714.
2. Banfy, F., Chou, L., RejnaOrtys, M., Ujmet, S., & Faj, S. (2018). Ynformatsyonnoemodelyrovanyezdanyjdliakul'turnohonasledyia: upravlenyeheneratyvnymprotsessomdliaslozhnykhystorycheskykhzdanyj. Springer International Publishing AG, part of Springer Nature M. Ioannides. 119–130, Retrieved from: https://doi.org/10.1007/978-3-319-75826-8_10.
3. Chou, L., & Faj, S. (2017). Razrabotka system veryfykatsyydliapostroenyiaynformatsyonnykhmodelejzdanyjnasledyia s neodnorodnymynaboramydannykh. ArkhypelahFotohrammetrycheskye. *Remote Sensing and Spatial Information Sciences, XLII-2*(5), 125–128. Retrieved from: https://doi.org/10.5194/isprs-archives-XLII-2-W5-125-2017.
4. Kot, S. I. (2018). Pro stanzberezhenniakul'turnoispadschynyUkrainy. Informatsijnoanalitychnimaterialy do parlaments'kykhslukhan' «Stan, problemy ta perspektyvyokhoronykul'turnoispadschyny v Ukraini» 18 kvitnia 2018, Kyiv, 42.
5. Horbyk, V. O., & Denysenko, H. H. (2003). Problemydoslidzhenniaizberezhenniapam'iatokistorii ta kul'tury v Ukraini. Ukrains'kyjistorychnyjzhurnal. 3, 143–151.
6. Davydova, O. G. (2015). Metodyotsiniuvanniaturystychnoipryvablyvostirehionivkrainy. *Ukrains'kyjsotsium, 4*(55), 97–107.
7. But, T. V., Zajtseva, V. M., & Hurova, D. D. (2018). Turysts'kiresursyUkrainy, Zaporizhzhia: TOV RVA «Prosvita» .
8. But, T. V., Zajtseva, V. M., & Pulina, T. V. (2019). RynokturystychnykhposluhUkrainy, Navchal'nyjposibnykdliastudentivvyschykhnavchal'nykhzakladiv, Zaporizhzhia: TOV RVA «Prosvita» .

9. Tymchenko, A. V. (2014). Rozvytokzamkovohoturyzmu v Ukraini. Mozhlyvosti ta rozvytoksuchasnohoturyzmu: svitovyj ta natsional'nyjdosvid: materialyVseukrains'koinaukovopraktychnoikonferentsii 2014. ZNTU, Zaporizhzhia: TOV «LIPS» LTD.
10. Zamiatyn, D. N. (2004). *Meta-heohrafyia: prostranstvoobrazov y obrazyprostranstva*. Moskva: Ahraf.
11. Zhukova, O. V. (2004). MuzeiefikatsiiazamkovykhkompleksivUkrainyiakodyn z zasobivikhzberezhennia. Retrieved from https://www.myslenedrevo.com.ua/uk/Sci/Archeology/Archeometry/Heritage/CastlesMuseums.html.
12. Varnaliy, Z., Onishchenko, S., & Masliy, A. (2016). Threat prevention mechanisms of ukraine's economic security. *Economic Annals-XXI, 159*(5–6), 20–24. https://doi.org/10.21003/ea.V159-04.
13. Krapyva, S., & Pavlov, K. (2016). Epokhavozrozhdenyia. Ynvestytsyonnaiapryvlekatel'nost' y buduscheeukraynskykhzamkov y dvortsov. Byznes, 25–26 (1220–1221), 8–12.
14. Ukrainaincognita.com. (2019). Informatsijnyj portal. 10zamkivUkrainy, iakimaiepobachytykozhen. Retrieved from http://ukrainaincognita.com/pamyatky/zamky-ta-fortetsi/10-zamkiv-ukrainy-yaki-mae-pobachyty-kozhen.
15. Petranivs'kyj, V. L., & Rutynovs'kyj, M. J. (2006). Turystychnekraieznavstvo. (Rozd. 5.). Kyiv, Znannia, Fortetsi,zamkyimonastyriiakkraieznavcho-turystychniob'iekty.
16. Rutyns'kyj, M. J. (2007). Zamkovyjturyzm v Ukraini. Navch. posibnykdliastudentivvysch. navch. zakl. K.: Tsentrnavch. literatury.
17. Chen', L. Y. (2016). Osnovynaukovykhdoslidzhen' u restavratsiipam'iatokarkhitektury. Navch. posibnyk. (2rd ed.). L'viv: L'vivs'kapolitekhnika.

Calculation of Lifetime of Steel Oil Pipelines with the Account of Corrosive Environment Affect

Olena Stepova, Inna Rassoha, Lyudmila Blazhko, and Olena Hanoshenko

Abstract The paper analyzes the operation status of steel pipelines. Existing methods of estimation of residual life, the durability of steel oil pipelines are considered. The purpose of the work is to develop dependencies for estimating the residual life of steel pipelines under the conditions of the corrosive environment with the observance of the requirements of safe operation. Based on the mathematical model of the electrochemical corrosion of the pipeline, in the crack of the insulation coating, under the action of an aggressive electrolytic medium with respect to the metal of the pipeline, dependence was obtained and investigated, which allows calculating the residual depth of corrosion of the pipeline wall during the operation of macro galvanic corrosion pairs under conditions of finding the aggressive solution in the damaged zone. The advantage of this model is the ability to predict the development of corrosion over time, regardless of the chemical composition of the aggressive electrolyte, as well as the ability to obtain the required calculation parameters on structures used. The developed dependence of the estimation of the residual depth of corrosion of the section of the pipeline-trench allows to rationally plan the repair works, to forecast the real terms of construction, to review the mode of operation, etc. The results obtained allow us to reliably estimate the load-bearing capacity of structures working under the conditions of an aggressive crack environment.

Keywords Steel oil pipeline · Electrochemical corrosion · Galvanic element · Corrosion velocity · Corrosion depth · Remaining lifetime

1 Introduction

Oil pipeline transmission of Ukraine started its development in 1950s. The first mighty oil pipeline on the territory of Ukraine was built in late 1960s. Pipes of the main pipelines are made from steel. Since then, many pipelines of big expansion have been built from the given material in the world. Ukraine has a developed network of pipelines that operate in the mid- and strongly aggressive environment that cause

O. Stepova (✉) · I. Rassoha · L. Blazhko · O. Hanoshenko
Poltava National Technical Yuri Kondratyuk University, Poltava, Ukraine
e-mail: alenastepovaja@gmail.com

V. Onyshchenko et al. (eds.), *Proceedings of the 2nd International Conference on Building Innovations*, Lecture Notes in Civil Engineering 73,
https://doi.org/10.1007/978-3-030-42939-3_71

intensive corrosive construction rupture. The first built pipelines have operated more than 48 years while the mid-term of maintenance that amounts more than 35 years.

Among the main ecological problems connected with operating of oil pipe transmission constructions, there is a problem of ecological safety which is based on preventing of environmental pollution by oil products in case of accidental situation.

The term of steel oil pipelines maintenance of Ukraine is close to the planned one. With the increase of their service life, the problem of efficient and continuous operation of pipeline transport, which is ensured by the organization of periodic technical diagnostics of the status of pipeline elements and repair in the areas of unacceptable defects, becomes more urgent.

For such objects, there is a task of assessment and enlargement of the structural behavior remaining lifetime. When planning the repairs and structural maintaining, there is a necessity of determining of the remaining lifetime of structural operating.

The specific tasks that are connected with the problem of the assessment of steel oil pipeline were solved at Physical and Mathematical Institute named after G.V. Karpenko National Academy of Sciences of Ukraine [1], IFNTU of Oil and Gas [2, 3], E. O. Paton Electric Welding Institute of the National Academy of Sciences of Ukraine [4, 5].

Some aspects were considered in the works of the national scientific schools: IFNTU of Oil and Gas [6, 7]. Famous works in lifetime and remaining lifetime assessment of steel pipelines of such scientists as Andreikiv et al. [8–11].

A number of foreign scientists made the research of this issue. Those are Choi et al. [12], Shuai et al. [13], He et al. [14].

The known methods of assessment of the remaining lifetime, lifetime of steel pipeline with the corrosive defects are based on: breaking of lifetime into life cycles according to degrading characteristics; criteria value that characterizes changes in the load-bearing capacity; criteria of «survivability» as a function of accumulation of defection; the development of probable models of structural damage. Thus, corrosive processes of steel of an oil pipeline are taken into account approximately and are not based on the laws of the electrochemical corrosion.

The considered methods reflect the general approaches to the determining of the remaining lifetime of oil pipelines in the aggressive environment. In determining the residual life of structures that work in an aggressive cracked environment in insulating coatings, a clear understanding of the physical and physicochemical processes occurring in the fracture zone of the structures is required, and the calculations should be based on reliable data on the kinetics and features of the corrosion of the steel in the crack.

2 Aim and Objectives of the Research

The aim of this work is to develop correlation for assessment of the remaining lifetime of steel oil pipelines in the corrosive environment with corresponding requirements to safety operation. For achieving this purpose, the following tasks are set:

- To improve and develop the new physical and mathematical models of local electrochemical corrosion of steel oil pipelines in the cavities of coating when there is aggressive electromechanical liquid affect that is based on the real parameters found out by the undisruptive method when examining the constructions;
- Based on mathematical model of local electrochemical corrosion of steel oil pipeline, correlation was found out that enables calculating thickness of the wall and the remaining lifetime of the pipeline sector in the cavity of coating in operating of macro galvanic corrosive couples and permanent state of aggressive liquid in the cavity.

3 Materials and Methods of the Research

Main oil pipelines are used in natural conditions, mainly, below the ground. For protection of steel oil pipelines from aggressive attack of the environment, different coatings are used as a rule. However, during operation, insulating coatings break the continuity and the aggressive external environment has to access to the pipeline steel. In such case, the question of the remaining lifetime of the oil pipeline is very urgent. In the below ground with the sectors, where the coating is broken, anode and cathode polar characteristics of steel essentially change, and as a result, the potentials are steady in these places. Due to the fact that the operation of the pipeline with areas where the insulation is broken due to the electrochemical corrosion of the pipeline metal, attention should be paid to the characteristics of the corrosion process when inspecting the pipeline. Electric potential of the given galvanic couples is a universal indicator for calculating the expenses for metal damage in the cavity.

Coating, as capillary-porous material, is a conductor of 2-d type that is why the process of steel corrosion in it is possible to consider from ordinary electrochemical corrosion of metals in the electrolytes. The authors [15–17] consider the electrical field near the heterogeneous electrode, model of which consists of two sectors of arbitrary width, which differ in steady potentials. Local corrosive element is in the sector of oil pipeline under the coating (cathode) and sector of the pipeline in the cavity under the electrolyte.

Electrical boundary of the electrical field potential is determined in this case on the basis of the two-dimensional solution of Laplace's equation.

$$\begin{aligned}\phi(x, y) &= \frac{a(E_a - E_K) + cE_K}{c} \\ &\quad + \sum_{K=1}^{\infty} \frac{2(E_a - E_K)}{\pi k\left(1 + \frac{\pi k}{c} L\right)} \sin \frac{\pi k}{c} a \cos \frac{\pi k}{c} x e^{-\frac{\pi k}{c} y} \\ &= \frac{a(E_a - E_K) + cE_K}{c} + \frac{2(E_a - E_K)}{\pi} \\ &\sum_{K=1}^{\infty} \frac{\sin \frac{\pi k}{c} a}{\left(1 + \frac{\pi k}{c} L\right)k} \cos \frac{\pi k}{c} x e^{-\frac{\pi k}{c} y},\end{aligned} \tag{1}$$

where φ—potential; a—width of anode selector, m; $E_{a,}$ E_K—current less potentials of anode and cathode, mB; c—width of cathode selector, m; x, y—flow coordinates; k =1, 2, 3; L—coefficient that depends on electro productivity of electrolyte and coefficient of polarization, $L = \gamma \cdot b$; γ—electric conductivity of electrolyte; b—polarization efficiency.

Taking into account Ohm's law in the differential form in the Eq. (1), expression for determining current density on the surface of one local element is found out.

$$i(x) = \frac{2(E_a - E_k)\gamma}{c} \sum_{k=1}^{\infty} \frac{\sin \frac{\pi k a}{c} \cos \frac{\pi k x}{c}}{k\left(1 + \frac{\pi k L}{c}\right)} \tag{2}$$

Quantitative assessment of corrosive damages, therefore, the remaining lifetime of safety operation of an oil pipeline depends on the remaining wall pipe gauge of the external surface of an oil pipeline.

For the calculation of wall pipe gauge in the aggressive electrolyte liquid in the zone of damaged coating, the dynamics of pit depth corrosion of an oil pipeline with operating galvanic element "oil pipeline with the damaged coating—oil pipeline coated" are considered.

According to Faraday's law, the corrosive metal damage is counted according to the formula:

$$M = K \times I \times t, \tag{3}$$

where M—weight removal value, hour; I—current of electrochemical corrosion A; t—time, h; K—electrochemical coefficient of metal that is counted according to the formula $K = \frac{A}{F \times U}$, g/A h); M—chemical equivalent, h/mol; A—atomic weight metal, for metal $A = 55.847$ h/mol; U—metal valency, *valency*, for metal $n = 2$; F—Faraday's constant, $F = 96.485$ A s/mol = 26.80139 A h/mol.

Supposing local corrosive pit depth is round. Let us consider corrosive pit depth on the pipe in equilateral cone that is spreading equally. Mass of the damaged metal of an oil pipeline quantitatively will amount to the volume of this cone with the account of absolute weight of metal.

Let on the last phase of the external diameter of the corrosive pit depth amounts to the gauge wall pipe and depth h gauge pit depth. Gauge pit depth:

$$h = \frac{K \times i \times t}{0.33 \times D}, \tag{4}$$

where D—absolute weight of metal oil pipeline (metal), g/cm^3, $D = 7.874$ g/cm^3; i—current intensity that flows through the density of cross section surface depth pits, A/cm^2

Taking into account the current of electrochemical corrosion of macro galvanic vapor, calculated by the formula (2), expression (4) takes the form:

$$h = \frac{K}{0.33 \times D}\left(\frac{2(E_a - E_k)\gamma}{\pi} \times \sum_{K=1}^{\infty} \frac{\frac{1-\cos 2\frac{\pi k a}{c}}{2}}{K\left(1 + \frac{K\pi L}{c}\right)}\right)t. \tag{5}$$

Then, the remaining thickness of pipe wall of an oil pipeline is counted according to the formula

$$\nu = \delta - h, \tag{6}$$

where δ—nominal thickness of an oil pipeline wall sector, mm

For the assessment of operational status of an oil pipeline, it is rational to count the thickness of the oil pipeline wall which corresponds to the full wear of operational status of the construction and compare it with the remaining thickness of the pipe wall. For providing accident-free operation of an oil pipeline sector, the factual absolute remaining wall thickness has to be less than admissible: $\delta \leq [h]$. Formularization of the remaining lifetime of an oil pipeline according to the factor of steel corrosion in the coating cavity is counted according to the reducing of oil pipeline wall thickness that is taken onto account. The period before the first leakage from oil pipeline is counted according to the formula (5)

$$t = \frac{h \times 0.33 \times \pi \times D}{K \times 2\gamma(E_a - E_k)\sum_{k=1}^{3} \frac{\sin^2 \frac{k\pi a}{c}}{k\left(1+\frac{\pi k L}{c}\right)}} \tag{7}$$

In case of several progressive depth pits, it is necessary to take into account corrosion prediction of their quantity.

Crash time before the first leakage of oil pipeline is counted according to the formula:

$$T = t - t_{в}, \tag{8}$$

where t—The time to the first leak from the pipeline was obtained by converting Eq. (5); $t_{в}$—time of oil pipe operation in the given conditions, years.

The scientific novelty of the findings is that on the basis of the mathematical model of local corrosive element in the oil pipeline sector, correlation has been worked out that enables to count the depth of corrosive oil pipe wall pit and the remaining thickness of the pipe wall of the operating macro galvanic corrosive couples. Practical value of the results of the work is as follows: the developed methods of assessment of the remaining lifetime give an opportunity to rationally plan the repairs, to predict the real terms of operational status of the construction, to reconsider the regime of operating. The found out results enable a more valid way to assess the load-bearing capacity of the oil capacity transmission of pipelines that operate in the aggressive environment with the cavities in the coating.

4 Conclusion

Based on mathematical model of electrochemical pipeline, corrosion in the coating cavity correlation is found out and researched which enables counting the remaining lifetime of the oil pipeline of operating macro galvanic corrosive couples in the conditions of aggressive liquid in the damaged coating. The developed correlation enables possibility of rational planning of the repairs, to predict the real terms of construction operation, to reconsider operating regime, etc. The results enable valid assessment of the load-bearing capacity and the remaining lifetime of the constructions with the cavities in the coating that operate in the aggressive environment.

References

1. Andreikiv, O. Y., & Hembara, O. V. (2013). Vplyv gruntovoi korozii i transportovanykh produktiv na dovhovichnist zvarnykh ziednan naftohazoprovodiv. *Physicochemical Mechanics of Materials, 2,* 52–58.
2. Poberezhnyi, L. I., Yavorskyi, A. V., Tsykh, V. S., Stanetskyi, A. I., & Hrytsanchuk, A. V. (2017). Pidvyshchennia rivnia ekolohichnoi bezpeky truboprovidnykh merezh naftohazovoho kompleksu Ukrainy. *Naukovo-tekhnichnyi zhurnal «Tekhnohenno-ekolohichna bezpeka», 1,* 24–31.
3. Hrabovskyi, R. S. (2010). Otsinka Mitsnosti naftoprovodu z koroziino-mekhanichnymy defektamy. *Rozvidka ta rozrobka naftovykh i hazovykh rodovyshch, 3*(36), 38–42.
4. Andreikiv, O. Y., Kushnir R. M., & Tsyrulnyk, O. T. (2006). Vyznachennia zalyshkovoho resursu truby naftoprovodu z urakhuvanniam naiavnykh defektiv u yii stintsi i realnykh umov ekspluatatsii. *Problemy resursu i bezpeky ekspluatatsii konstruktsii, sporud ta mashyn/ Pid zah. red. B.Ie. Patona. Kyiv: Instytutelektrozvariuvannia im. Ye.O. Patona NAN Ukrainy*, 328–331.
5. Bekker, M. V. (2007). Obespechenye nadёzhnoi rabotы naftotransportnoi systemы DK «Ukratransneft» : *sbornyk dokladov nauchno-praktycheskoho semynara K.: Ynstytut эlektrosvarky ym. E. O. Patona,* 3–5.
6. Kryzhanivskyi, Y. I., & Nykyforchyn, H. M. (2011). Osoblyvosti koroziino-vodnevoi dehradatsii stalei naftohazoprovodiv i rezervuariv zberihannia nafty. *Fizyko-khimichna mekhanika materialiv, T, 47*(2), 11–20. http://nbuv.gov.ua/UJRN/PHKhMM_2011_47_2_4.
7. Kryzhanivskyi, Y. I., Hrabovskyi, R. S., Fedorovych, I. Y., & Barna, R. A. (2015). Otsiniuvannia kinetyky ruinuvannia elementiv ekspluatovanoho hazoprovodu. *Fizyko-khimichna mekhanika materialiv, T, 51*(1), 13–19. http://nbuv.gov.ua/UJRN/PHKhMM_2015_51_1_4.
8. Andreikiv, O., Ivanytskyi Ya, L., Terletska, Z. O., & Kit, M. B. (2004). Otsinka dovhovichnosti truby naftohonu z poverkhnevoiutrishchynoiu pid dvovisnym blochnym navantazhenniam. *Fizyko-khimichna mekhanika materialiv, 3,* 103–108.
9. Osadchuk, V. A., Andreikiv, O. Y., Banakhevych, Y. V., Drahiliev, A. V., & Kychma, A. O. (2014). *Zalyshkova mitsnist ta dovhovichnist dilianok naftohazoprovodiv z defektamy* (p. 264). Lviv: Vydavnytstvo Lvivskoi politekhniky.
10. Pichuhin, S. F., Zyma, O. I., & Vynnykov, P. I. (2015). Nadiinist liniinoi chastyny pidzemnykh mahistralnykh truboprovodiv. *Zb. nauk. prats (haluzeve mashynobuduvannia, bud-vo)/ PoltNTU im.. Yu. Kondratiuka., Poltava:PNTU, 1*(43), 17–29.
11. Pichuhin, S. F., Pashynskyi, V. A., Zyma, O. I., Vynnykov, P. I., & Bila, Z. I. (2018). Nadiinist liniinykh chastyn mahistralnykh truboprovodiv: monohrafiia. *Poltava:PP «Astraia»*, 439.
12. Choi, J. B., Goo, B. K., & Kima, J. C. (2003). Development of limit load solutions for corroded gas pipelines. *International Journal of Pressure Vessels and Piping, 80*(2), 121–128.

13. Shuai, J., Ma, B., Liu, D. X., & Xu, K. (2013). Assessment on failure pressure of high strength pipeline with corrosion defects. *Journal Engineering Failure Analysis, 32*(1), 209–219.
14. He, D. S., Guo, J., & Zhang, P. (2007). Assessment method for remaining strength of corroded pipeline and its application. *Journal of Acta Petrolei Sinica, 28*(6), 125–128.
15. Stepova, O., & Paraschienko, I. (2017). Modeling of the corrosion process in steel oil pipelines in order to improve environmental safety. *Eastern-european journal of enterprise technologies, industrial and technology systems, 2, 1*(86), 15–20. http://journals.uran.ua/eejet/article/view/96425.
16. Stepovaja, E., Holik, Y. Y., & Fraňa, K. (2018). Methods for precautionary management of environmental safety at energy enterprises. *Naukovyi visnyk Natsionalnoho hirnychoho universytetu. Naukovo-tekhnichnyi zhurnal, 6*(168), 173–177.
17. Varnaliy, Z., Onishchenko, S., & Masliy, A. (2016). Threat prevention mechanisms of ukraine's economic security. *Economic Annals-XXI, 159*(5–6), 20–24. https://doi.org/10.21003/ea.V159-04.

Building a System of Diagnosis Technical Condition of Buildings on the Example of Floor Beams Using Methods of Fuzzy Sets

O. O. Terentyev, P. E. Grigorovskiy, A. A. Tugaj, and O. V. Dubynka

Abstract The article discusses the example of analysis of the technical condition of structural elements, for example, beams using the methods and apparatus of fuzzy set theory. Fuzzy technology is gaining more supporters among professionals working in the field of decision making, where the constraints and consequences of possible actions are not known precisely. Took start in 1965 from the work of Lotfi Zadeh, fuzzy logic has found wide application in various fields of technology in the late nineties. In basic provisions, definitions and terms, the theory of fuzzy (vague, not clearly defined, vague) sets is formulated by L. Zadeh and developed further in the works. In this case, the basis of diagnostics of states of the considered subsystem, the proposed algorithm of diagnosis. When making a diagnosis, the parameters are taken—nine diagnostic features with their digits, presented in the form of ranges of change. Diagnosis is to match one of the solutions to each combination of parameter values.

Keywords Fuzzy sets · Diagnostics of condition · Strength test · Diagnostic matrix · Information technology · Combination of parameter values

1 Introduction

The urgency of the work lies in the effectiveness of the proposed models and methods of diagnosis and prediction of technical condition of buildings sold analytical apparatus of fuzzy sets, which makes it possible to create an information platform to collect real, accurate and minimally unprofitable media as a powerful tool for accurate and reliable diagnosis of temporal phases of buildings within the life cycle [1]. The main purpose of development is to build a system of diagnosis technical condition of buildings on the example of floor beams using the methods of the theory of not clear sets [2, 3].

O. O. Terentyev · P. E. Grigorovskiy · A. A. Tugaj · O. V. Dubynka (✉)
Kyiv National University of Construction and Architecture, Kiev, Ukraine
e-mail: dainfo28@gmail.com

O. O. Terentyev
e-mail: knuba@knuba.edu.ua

V. Onyshchenko et al. (eds.), *Proceedings of the 2nd International Conference on Building Innovations*, Lecture Notes in Civil Engineering 73,
https://doi.org/10.1007/978-3-030-42939-3_72

Formulated states S_1, S_2, S_3, S_4, S_5 types of diagnoses are considered to be recognized. In the diagnosis, parameters are set—nine diagnostic features of their discharges, in the form of ranges change:

k_1—damaged concrete, reducing its protective properties against fittings (1 or 0);
k_2—longitudinal cracks in the protective layer of concrete reinforcing rods along, corrosion products along the reinforcing rods (1 or 0);
k_3—the presence and width of the opening of normal cracks (0–1 mm);
k_4—availability inclined cracks (1 or 0);
k_5—strength concrete (0–31%);
k_6—reinforcement corrosion (0–21%);
k_7—deflection (0–31%);
k_8—strength test conditions for normal sections (1 or 0);
k_9—strength test conditional on inclined sections (1 or 0).

2 Strength Problem Solving

The task of diagnosis is to put each combination of parameter values in line one solution: $s_i (i = \overline{1.5})$; see Tables 1 and 2.

Parameters k_1–k_9 are regarded as linguistic variables. For further analysis, it should be divided into groups that are linguistic variables: k_1, k_2, k_4—group of options identified during the examination visually; k_3, k_5, k_6, k_7—a group of parameters are determined by the results of instrumental measurements; k_8, k_9—group parameters determined by the results of test calculations [4–6].

Also, introduced linguistic variable *S*, which is measured by a range of conditions, diagnoses S_1–S_5. Structure model of diagnostics of considered beams is shown in a tree inference (Fig. 1)

That meets the following equation:

$$s = f_s(x, y, z); \quad x = f_x(k_1, k_2, k_4);$$
$$y = f_y(k_3, k_5, k_6, k_7); \; z = f_z(k_8, k_9). \tag{1}$$

For evaluation of linguistic variables k_1–k_9, and *x*, *y* and *z*, there are single scale qualitative terms introduced:

L low;
BA below average;
A average;
AA above average;
H high

Table 1 Diagnostic matrix for beams overlap in general

№ п/п	**Диагностические признаки**	K_{js}	Разряды признаков	$p(k_j)$	Сост. S_1	Сост. S_2	Сост. S_3	Сост. S_4	Сост. S_5
					$P(S_1)$	$P(S_2)$	$P(S_3)$	$P(S_4)$	$P(S_5)$
1	Повреждения бетона, снижающие его свойства по отношению к арматуре	k_{11}	да	$p(k_{11})$	$p(k_{11}/S_1)$	$p(k_{11}/S_2)$	$p(k_{11}/S_3)$	$p(k_{11}/S_4)$	$p(k_{11}/S_5)$
		k_{12}	нет	$p(k_{12})$	$p(k_{12}/S_1)$	$p(k_{12}/S_2)$	$p(k_{12}/S_3)$	$p(k_{12}/S_4)$	$p(k_{12}/S_5)$
2	Продольные трещины в защитном слое вдоль арматурных стержней	k_{21}	да	$p(k_{21})$	$p(k_{21}/S_1)$	$p(k_{21}/S_2)$	$p(k_{21}/S_3)$	$p(k_{21}/S_4)$	$p(k_{21}/S_5)$
		k_{22}	нет	$p(k_{22})$	$p(k_{22}/S_1)$	$p(k_{22}/S_2)$	$p(k_{22}/S_3)$	$p(k_{22}/S_4)$	$p(k_{22}/S_5)$
3	Нормальные трещины (ширина раскрытия)	k_{31}	< 0,4 мм	$p(k_{31})$	$p(k_{31}/S_1)$	$p(k_{31}/S_2)$	$p(k_{31}/S_3)$	$p(k_{31}/S_4)$	$p(k_{31}/S_5)$
		k_{32}	до 1,0 мм	$p(k_{32})$	$p(k_{32}/S_1)$	$p(k_{32}/S_2)$	$p(k_{32}/S_3)$	$p(k_{32}/S_4)$	$p(k_{32}/S_5)$
		k_{33}	≥ 1,0 мм	$p(k_{33})$	$p(k_{33}/S_1)$	$p(k_{33}/S_2)$	$p(k_{33}/S_3)$	$p(k_{33}/S_4)$	$p(k_{33}/S_5)$
4	Наклонные трещины (наличие)	k_{41}	да	$p(k_{41})$	$p(k_{41}/S_1)$	$p(k_{41}/S_2)$	$p(k_{41}/S_3)$	$p(k_{41}/S_4)$	$p(k_{41}/S_5)$
		k_{42}	нет	$p(k_{42})$	$p(k_{42}/S_1)$	$p(k_{42}/S_2)$	$p(k_{42}/S_3)$	$p(k_{42}/S_4)$	$p(k_{42}/S_5)$
5	Прочность бетона	k_{51}	проектная	$p(k_{51})$	$p(k_{51}/S_1)$	$p(k_{51}/S_2)$	$p(k_{51}/S_3)$	$p(k_{51}/S_4)$	$p(k_{51}/S_5)$
		k_{52}	≤ 30%	$p(k_{52})$	$p(k_{52}/S_1)$	$p(k_{52}/S_2)$	$p(k_{52}/S_3)$	$p(k_{52}/S_4)$	$p(k_{52}/S_5)$
		k_{53}	> 30%	$p(k_{53})$	$p(k_{53}/S_1)$	$p(k_{53}/S_2)$	$p(k_{53}/S_3)$	$p(k_{53}/S_4)$	$p(k_{53}/S_5)$
6	Коррозия арматуры	k_{61}	< 5%	$p(k_{61})$	$p(k_{61}/S_1)$	$p(k_{61}/S_2)$	$p(k_{61}/S_3)$	$p(k_{61}/S_4)$	$p(k_{61}/S_5)$
		k_{62}	5 – 20	$p(k_{62})$	$p(k_{62}/S_1)$	$p(k_{62}/S_2)$	$p(k_{62}/S_3)$	$p(k_{62}/S_4)$	$p(k_{62}/S_5)$
		k_{63}	> 20%	$p(k_{63})$	$p(k_{63}/S_1)$	$p(k_{63}/S_2)$	$p(k_{63}/S_3)$	$p(k_{63}/S_4)$	$p(k_{63}/S_5)$
7	Прогиб	k_{71}	допускаемый	$p(k_{71})$	$p(k_{71}/S_1)$	$p\ k_{71}/S_2)$	$p(k_{71}/S_3)$	$p(k_{71}/S_4)$	$p(k_{71}/S_5)$
		k_{72}	≤ 30%	$p(k_{72})$	$p(k_{72}/S_1)$	$p(k_{72}/S_2)$	$p(k_{72}/S_3)$	$p(k_{72}/S_4)$	$p(k_{72}/S_5)$
		k_{73}	> 30%	$p(k_{73})$	$p(k_{73}/S_1)$	$p(k_{73}/S_2)$	$p(k_{73}/S_3)$	$p(k_{73}/S_4)$	$p(k_{73}/S_5)$
8	Условие прочности по нормальным сечениям	k_{81}	да	$p(k_{81})$	$p(k_{81}/S_1)$	$p(k_{81}/S_2)$	$p(k_{81}/S_3)$	$p(k_{81}/S_4)$	$p(k_{81}/S_5)$
		k_{82}	нет	$p(k_{82})$	$p(k_{82}/S_1)$	$p(k_{82}/S_2)$	$p(k_{82}/S_3)$	$p(k_{82}/S_4)$	$p(k_{82}/S_5)$
9	Условие прочности по наклонным сечениям	k_{91}	да	$p(k_{91})$	$p(k_{91}/S_1)$	$p(k_{91}/S_2)$	$p(k_{91}/S_3)$	$p(k_{91}/S_4)$	$p(k_{91}/S_5)$
		k_{92}	нет	$p(k_{92})$	$p(k_{92}/S_1)$	$p(k_{92}/S_2)$	$p(k_{92}/S_3)$	$p(k_{92}/S_4)$	$p(k_{92}/S_5)$

Each of these terms is fuzzy sets, given by protecting supplies. Using qualitative terms imposed intervals and analyzing the changes in the numerical values of probability of realization signs $p = (k_{ij}/S_i)$ in diagnostic matrix (Table 2), Eq. (1) are given in Tables 3, 4, 5 and 6.

Using Tables 3, 4, 5 and 6 and logical operations • (AND—min) and V (OR—max), we can write the system of fuzzy logic equations linking the membership function diagnoses and linguistic input variables:

Table 2 Diagnostic matrix for beams overlap with numerical values and probabilities of the signs

№ п/п	Диагностические признаки	k_{ij}	Разряды признаков	$p(k_{ij})$	Сост. S_1	Сост. S_2	Сост. S_3	Сост. S_4	Сост. S_5
					$P(S_1)$	$P(S_2)$	$P(S_3)$	$P(S_4)$	$P(S_5)$
					0,18	**0,29**	**0,35**	**0,13**	**0,05**
1	Повреждения бетона, снижающие его свойства по отношению к арматуре	k_{11}	да	$p(k_{11})$	0,14	0,29	0,40	0,56	0,76
		k_{12}	нет	$p(k_{12})$	0,86	0,71	0,60	0,44	0,24
2	Продольные трещины в защитном слое вдоль арматурных стержней	k_{21}	да	$p(k_{21})$	0,05	0,27	0,38	0,59	0,88
		k_{22}	нет	$p(k_{22})$	0,95	0,73	0,62	0,41	0,12
3	Нормальные трещины (ширина раскрытия)	k_{31}	< 0,4 мм	$p(k_{31})$	0,92	0,78	0,67	0,54	0,22
		k_{32}	до 1,0 мм	$p(k_{32})$	0,07	0,21	0,28	0,34	0,48
		k_{33}	≥ 1,0 мм	$p(k_{33})$	0,01	0,01	0,05	0,12	0,30
4	Наклонные трещины (наличие)	k_{41}	да	$p(k_{41})$	0,03	0,05	0,28	0,30	0,35
		k_{42}	нет	$p(k_{42})$	0,97	0,95	0,72	0,70	0,65
5	Прочность бетона	k_{51}	проектная	$p(k_{51})$	0,79	0,49	0,33	0,28	0,20
		k_{52}	≤ 30%	$p(k_{52})$	0,17	0,33	0,36	0,39	0,42
		k_{53}	> 30%	$p(k_{53})$	0,04	0,18	0,31	0,33	0,38
6	Коррозия арматуры	k_{61}	< 5%	$p(k_{61})$	0,73	0,52	0,33	0,26	0,13
		k_{62}	5 – 20	$p(k_{62})$	0,23	0,28	0,34	0,39	0,40
		k_{63}	> 20%	$p(k_{63})$	0,04	0,20	0,33	0,35	0,47
7	Прогиб	k_{71}	допускаемый	$p(k_{71})$	0,93	0,82	0,59	0,35	0,11
		k_{72}	≤ 30%	$p(k_{72})$	0,05	0,11	0,28	0,39	0,43
		k_{73}	> 30%	$p(k_{73})$	0,02	0,07	0,13	0,26	0,46
8	Условие прочности по нормальным сечениям	k_{81}	да	$p(k_{81})$	0,93	0,75	0,53	0,29	0,09
		k_{82}	нет	$p(k_{82})$	0,07	0,25	0,47	0,71	0,91
9	Условие прочности по наклонным сечениям	k_{91}	да	$p(k_{91})$	0,94	0,83	0,75	0,55	0,33
		k_{92}	нет	$p(k_{92})$	0,06	0,17	0,25	0,45	0,67

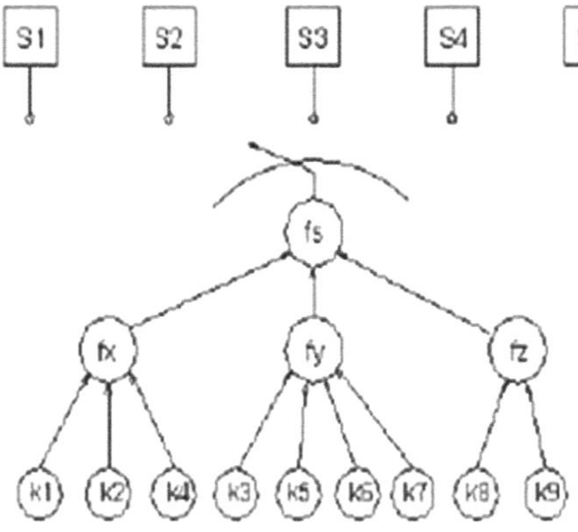

Рисунок 1. Дерево логического вывода

Fig. 1 Tree inference

Table 3 Knowledge of the interrelation *s*

s	s_1			s_2			s_3			s_4			s_5		
X	H	B	B	HC	C	BC	HC	HC	C	HC	C	C	HC	HC	BC
Y	H	H	B	H	HC	C	H	HC	HC	HC	HC	C	H	HC	C
Z	H	C	B	HC	C	BC	C	C	BC	HC	C	BC	HC	C	B

Table 4 Knowledge of the interrelation x

X	Н			НС			С			вС			В		
k_1	НС	С	ВС	С	НС	С	Н	ВС	ВС	НС	С	ВС	НС	НС	С
k_2	Н	НС	ВС	НС	С	С	Н	В	В	НС	С	ВС	НС	НС	С
k_4	НС	НС	С	НС	С	С	Н	В	В	Н	НС	ВС	НС	С	ВС

Table 5 Knowledge of the interrelation y

Y	Н			НС			С			вС			В		
k_3	НС	НС	С	Н	НС	С	Н	Н	В	Н	НС	ВС	Н	НС	С
k_5	Н	НС	НС	НС	НС	НС	Н	Н	ВС	Н	НС	С	НС	НС	НС
k_6	Н	НС	С	НС	НС	НС	Н	НС	ВС	НС	НС	С	НС	НС	НС
k_7	Н	НС	НС	Н	НС	НС	Н	Н	В	Н	Н	ВС	Н	НС	С

Table 6 Knowledge of the interrelation z

Z	Н			НС			С			вС			В		
k_8	Н	С	В	НС	С	ВС	Н	С	В	НС	С	ВС	С	С	С
k_9	НС	С	ВС	С	С	С	Н	С	В	Н	НС	ВС	НС	С	ВС

$$\begin{gathered}
\mu^{S_1}(s)=\left[\mu^{\text{н}}(x)\cdot\mu^{\text{н}}(y)\cdot\mu^{\text{н}}(z)\right]\vee\left[\mu^{\text{в}}(x)\cdot\mu^{\text{н}}(y)\cdot\mu^{\text{с}}(z)\right]\vee\left[\mu^{\text{в}}(x)\cdot\mu^{\text{вс}}(y)\cdot\mu^{\text{в}}(z)\right];\\
\mu^{S_2}(s)=\left[\mu^{\text{нс}}(x)\cdot\mu^{\text{н}}(y)\cdot\mu^{\text{нс}}(z)\right]\vee\left[\mu^{\text{с}}(x)\cdot\mu^{\text{нс}}(y)\cdot\mu^{\text{с}}(z)\right]\vee\left[\mu^{\text{вс}}(x)\cdot\mu^{\text{с}}(y)\cdot\mu^{\text{вс}}(z)\right];\\
\mu^{S_3}(s)=\left[\mu^{\text{нс}}(x)\cdot\mu^{\text{нс}}(y)\cdot\mu^{\text{с}}(z)\right]\vee\left[\mu^{\text{нс}}(x)\cdot\mu^{\text{нс}}(y)\cdot\mu^{\text{с}}(z)\right]\vee\left[\mu^{\text{с}}(x)\cdot\mu^{\text{с}}(y)\cdot\mu^{\text{вс}}(z)\right];\\
\mu^{S_4}(s)=\left[\mu^{\text{нс}}(x)\cdot\mu^{\text{нс}}(y)\cdot\mu^{\text{нс}}(z)\right]\vee\left[\mu^{\text{с}}(x)\cdot\mu^{\text{нс}}(y)\cdot\mu^{\text{с}}(z)\right]\vee\left[\mu^{\text{с}}(x)\cdot\mu^{\text{нс}}(y)\cdot\mu^{\text{вс}}(z)\right];\\
\mu^{S_5}(s)=\left[\mu^{\text{нс}}(x)\cdot\mu^{\text{н}}(y)\cdot\mu^{\text{нс}}(z)\right]\vee\left[\mu^{\text{нс}}(x)\cdot\mu^{\text{нс}}(y)\cdot\mu^{\text{с}}(z)\right]\vee\left[\mu^{\text{вс}}(x)\cdot\mu^{\text{нс}}(y)\cdot\mu^{\text{в}}(z)\right];\\
\mu^{\text{н}}(x)=\left[\mu^{\text{нс}}(k_1)\cdot\mu^{\text{н}}(k_2)\cdot\mu^{\text{нс}}(k_4)\right]\vee\left[\mu^{\text{с}}(k_1)\cdot\mu^{\text{нс}}(k_2)\cdot\mu^{\text{с}}(k_4)\right]\vee\left[\mu^{\text{вс}}(k_1)\cdot\mu^{\text{вс}}(k_2)\cdot\mu^{\text{с}}(k_4)\right];\\
\mu^{\text{нс}}(x)=\left[\mu^{\text{нс}}(k_1)\cdot\mu^{\text{нс}}(k_2)\cdot\mu^{\text{нс}}(k_4)\right]\vee\left[\mu^{\text{нс}}(k_1)\cdot\mu^{\text{с}}(k_2)\cdot\mu^{\text{с}}(k_4)\right]\vee\left[\mu^{\text{с}}(k_1)\cdot\mu^{\text{с}}(k_2)\cdot\mu^{\text{с}}(k_4)\right];
\end{gathered}\tag{2}$$

$$\begin{gathered}
\mu^{\text{с}}(x)=\left[\mu^{\text{н}}(k_1)\cdot\mu^{\text{н}}(k_2)\cdot\mu^{\text{н}}(k_4)\right]\vee\left[\mu^{\text{вс}}(k_1)\cdot\mu^{\text{в}}(k_2)\cdot\mu^{\text{в}}(k_4)\right]\vee\left[\mu^{\text{вс}}(k_1)\cdot\mu^{\text{в}}(k_2)\cdot\mu^{\text{в}}(k_4)\right];\\
\mu^{\text{вс}}(x)=\left[\mu^{\text{нс}}(k_1)\cdot\mu^{\text{нс}}(k_2)\cdot\mu^{\text{н}}(k_4)\right]\vee\left[\mu^{\text{с}}(k_1)\cdot\mu^{\text{с}}(k_2)\cdot\mu^{\text{нс}}(k_4)\right]\vee\left[\mu^{\text{вс}}(k_1)\cdot\mu^{\text{вс}}(k_2)\cdot\mu^{\text{вс}}(k_4)\right];\\
\mu^{\text{в}}(x)=\left[\mu^{\text{нс}}(k_1)\cdot\mu^{\text{нс}}(k_2)\cdot\mu^{\text{нс}}(k_4)\right]\vee\left[\mu^{\text{нс}}(k_1)\cdot\mu^{\text{нс}}(k_2)\cdot\mu^{\text{с}}(k_4)\right]\vee\left[\mu^{\text{с}}(k_1)\cdot\mu^{\text{с}}(k_2)\cdot\mu^{\text{вс}}(k_4)\right];
\end{gathered}\tag{3}$$

$$
\begin{aligned}
\mu^{\text{н}}(y) &= \left[\mu^{\text{нс}}(k_3)\cdot\mu^{\text{н}}(k_5)\cdot\mu^{\text{н}}(k_6)\cdot\mu^{\text{н}}(k_7)\right]\vee\left[\mu^{\text{нс}}(k_3)\cdot\mu^{\text{нс}}(k_5)\cdot\mu^{\text{нс}}(k_6)\cdot\mu^{\text{нс}}(k_7)\right]\vee\\
&\vee\left[\mu^{\text{с}}(k_3)\cdot\mu^{\text{нс}}(k_5)\cdot\mu^{\text{с}}(k_6)\cdot\mu^{\text{нс}}(k_7)\right];\\
\mu^{\text{нс}}(y) &= \left[\mu^{\text{н}}(k_3)\cdot\mu^{\text{нс}}(k_5)\cdot\mu^{\text{нс}}(k_6)\cdot\mu^{\text{н}}(k_7)\right]\vee\left[\mu^{\text{нс}}(k_3)\cdot\mu^{\text{нс}}(k_5)\cdot\mu^{\text{нс}}(k_6)\cdot\mu^{\text{нс}}(k_7)\right]\vee\\
&\vee\left[\mu^{\text{с}}(k_3)\cdot\mu^{\text{нс}}(k_5)\cdot\mu^{\text{нс}}(k_6)\cdot\mu^{\text{нс}}(k_7)\right];\\
\mu^{\text{с}}(y) &= \left[\mu^{\text{н}}(k_3)\cdot\mu^{\text{н}}(k_5)\cdot\mu^{\text{н}}(k_6)\cdot\mu^{\text{н}}(k_7)\right]\vee\left[\mu^{\text{н}}(k_3)\cdot\mu^{\text{н}}(k_5)\cdot\mu^{\text{нс}}(k_6)\cdot\mu^{\text{н}}(k_7)\right]\vee\\
&\vee\left[\mu^{\text{в}}(k_3)\cdot\mu^{\text{вс}}(k_5)\cdot\mu^{\text{вс}}(k_6)\cdot\mu^{\text{в}}(k_7)\right];\\
\mu^{\text{вс}}(y) &= \left[\mu^{\text{н}}(k_3)\cdot\mu^{\text{н}}(k_5)\cdot\mu^{\text{нс}}(k_6)\cdot\mu^{\text{н}}(k_7)\right]\vee\left[\mu^{\text{нс}}(k_3)\cdot\mu^{\text{нс}}(k_5)\cdot\mu^{\text{нс}}(k_6)\cdot\mu^{\text{н}}(k_7)\right]\vee\\
&\vee\left[\mu^{\text{вс}}(k_3)\cdot\mu^{\text{с}}(k_5)\cdot\mu^{\text{с}}(k_6)\cdot\mu^{\text{вс}}(k_7)\right];\\
\mu^{\text{в}}(y) &= \left[\mu^{\text{н}}(k_3)\cdot\mu^{\text{нс}}(k_5)\cdot\mu^{\text{нс}}(k_6)\cdot\mu^{\text{н}}(k_7)\right]\vee\left[\mu^{\text{нс}}(k_3)\cdot\mu^{\text{нс}}(k_5)\cdot\mu^{\text{нс}}(k_6)\cdot\mu^{\text{нс}}(k_7)\right]\vee\\
&\vee\left[\mu^{\text{с}}(k_3)\cdot\mu^{\text{нс}}(k_5)\cdot\mu^{\text{нс}}(k_6)\cdot\mu^{\text{с}}(k_7)\right].
\end{aligned}
\tag{4}
$$

$$
\begin{aligned}
\mu^{\text{н}}(z) &= \left[\mu^{\text{н}}(k_8)\cdot\mu^{\text{нс}}(k_9)\right]\vee\left[\mu^{\text{с}}(k_8)\cdot\mu^{\text{с}}(k_9)\right]\vee\left[\mu^{\text{в}}(k_8)\cdot\mu^{\text{вс}}(k_9)\right];\\
\mu^{\text{нс}}(z) &= \left[\mu^{\text{нс}}(k_8)\cdot\mu^{\text{с}}(k_9)\right]\vee\left[\mu^{\text{с}}(k_8)\cdot\mu^{\text{с}}(k_9)\right]\vee\left[\mu^{\text{вс}}(k_8)\cdot\mu^{\text{с}}(k_9)\right];\\
\mu^{\text{с}}(z) &= \left[\mu^{\text{н}}(k_8)\cdot\mu^{\text{н}}(k_9)\right]\vee\left[\mu^{\text{с}}(k_8)\cdot\mu^{\text{с}}(k_9)\right]\vee\left[\mu^{\text{в}}(k_8)\cdot\mu^{\text{в}}(k_9)\right];\\
\mu^{\text{вс}}(z) &= \left[\mu^{\text{нс}}(k_8)\cdot\mu^{\text{н}}(k_9)\right]\vee\left[\mu^{\text{с}}(k_8)\cdot\mu^{\text{нс}}(k_9)\right]\vee\left[\mu^{\text{вс}}(k_8)\cdot\mu^{\text{вс}}(k_9)\right];\\
\mu^{\text{в}}(z) &= \left[\mu^{\text{с}}(k_8)\cdot\mu^{\text{нс}}(k_9)\right]\vee\left[\mu^{\text{с}}(k_8)\cdot\mu^{\text{с}}(k_9)\right]\vee\left[\mu^{\text{с}}(k_8)\cdot\mu^{\text{вс}}(k_9)\right].
\end{aligned}
\tag{5}
$$

The total number of fuzzy logic equations is 20. In general, each input variable k_1–k_9 has its own membership function, fuzzy terms (L, BA, A, AA, H) used in these twenty Eqs. (2)–(5). However, to facilitate further construction, modeling within the fuzzy knowledge base for all variables k_1–k_9 is only one form of the membership function [7–9]. To do this, each variable spacing reduced to one universal interval (0, 4), carried out using the following ratios:

$$\mu^{j}(k_i) = \tilde{\mu}^{j}(u);\ u = 4\frac{k_i - \underline{k_i}}{k_i - \underline{k_i}},\quad j = H, \text{н}C, C, \text{в}C, B, \tag{6}$$

where (k_i)—change interval variable k_i, $i = 1.9$.

The final analytical model of the membership function is as follows:

$$\tilde{\mu}^{j}(u) = \frac{1}{1 + \left(\frac{u-b}{c}\right)^2}. \tag{7}$$

Option b for terms L, BA, A, AA, H is set to 0, 1, 2, 3, 4, respectively. Option *c* for terms with all the same equals to 0.923. The choice of these relationships (6) and options *b* and *c* in that form due to the fact that they are proven approximation membership functions is obtained by pairwise comparisons. Further, the fuzzy logic Eqs. (2)–(5) with functions of fuzzy terms (7) allow us to decide on the specific diagnosis of the following algorithm [10, 11].

1. fixed values for the condition of the subject element is implementation-defined attributes in the form of quantitative values that fall within the boundaries of these ranges of change

$$K^* = \left(k_1^*, k_2^*, k_3^*, \ldots, k_9^*\right). \tag{8}$$

2. Using relations (7) and options *b* and *c*, determine the value of membership functions for fixed values of the parameters k_i*.
3. Using the logic Eqs. (2)–(5), values are calculated at fixed membership functions of the state vector $K*$ for all diagnoses S_1, S_2, S_3, S_4, S_5. It is important to note that logical operations AND (Λ) and OR (V) of the membership functions are replaced by operations min and max:

$$\mu(a) \wedge \mu(b) = \min[\mu(a), \mu(b)] : \mu(a) \vee \mu(b) = \max[\mu(a), \mu(b)]; \tag{9}$$

4. Finally, the final decision for which

$$\mu^{s_j^*}\left(k_1^*, k_2^*, \ldots, k_9^*\right) = \max\left[\mu^{s_j}\left(k_1^*, k_2^*, \ldots, k_9^*\right)\right], \quad j = \overline{1.9}. \tag{10}$$

An example of the proposed algorithm can be reduced by building a state vector $K*$ in the form that corresponds to the implementation of signs of observed beams. In this case, follow the same conditions diagnosis:

(1) damage to the outer surface is that $K_1 = 1.0$ mm;
(2) longitudinal cracks in the protective layer are that $K_2 = 1.0$ mm;
(3) normal cracks in the tension zone opening to 0.4 mm are accepted that $K_3 = 0.2$ mm;
(4) inclined crack is that $K_4 = 1.0$;
(5) concrete strength was 10% below the design, i.e., $K_5 = 10\%$;
(6) there is exposure fittings, 5% of which section affected by corrosion, that is $K_6 = 5\%$;
(7) deflection does not exceed normative, i.e., $K_7 = 0$;
(8) Conditions strength calculation of normal intersections satisfied, that $K_8 = 1.0$;
(9) Conditions strength calculation on inclined sections satisfied, that $K_9 = 1.0$.

Using the model (8) and the accepted value parameters *b* and *c*, calculate the value of membership functions in points $*k_i$ ($i = 1.9$) for all fuzzy terms [12-14]. The results are shown in Table 7.

These values obtained are presented in Eq. (2):

Table 7 Value of membership functions

№ п/п	k_i^*	u_i^*	$\mu^{H}(k_i^*)$	$\mu^{HC}(k_i^*)$	$\mu^{C}(k_i^*)$	$\mu^{BC}(k_i^*)$	$\mu^{B}(k_i^*)$
1	1,0	4,0	0,0506	0,086	0,176	0,460	1,0
2	1,0	4,0	0,0506	0,086	0,176	0,460	1,0
3	0,2	2,0	0,176	0,460	1,0	0,460	0,176
4	1,0	4,0	0,0506	0,086	0,176	0,460	1,0
5	10,0	1,29	0,339	0,910	0,628	0,226	0,104
6	5,0	0,95	0,485	0,997	0,437	0,169	0,084
7	0,0	0,0	1,0	0,460	0,176	0,086	0,0506
8	1,0	4,0	0,0506	0,460	0,176	0,460	1,0
9	1,0	4,0	0,0506	0,460	0,176	0,460	1,0

$$\mu \mathrm{H}(x) = [0.086 \bullet 0.0506 \bullet 0.086] \vee [0.76 \bullet 0.086 \bullet 0.176] \vee [0.460 \bullet 0.460 \bullet 0.176] = 0.176.$$

Similarly,

$$\mu \mathrm{HC}\,(x) = 0.176;\ \mu \mathrm{C}(x) = 0.460;\ \mu \mathrm{BC}\,(x) = 0.460;\ \mathrm{MKB}\,(x) = 0.176.$$

According to Eq. (10),

$$\mu \mathrm{H}(y) = [0.460 \bullet 0.339 \bullet 0.485 \bullet 1.0] \vee [0.460 \bullet 0.910 \bullet 0.997 \bullet 0.460] \vee [1.0 \bullet 0.910 \bullet 0.437 \bullet 0.460] = 0.460.$$

Similarly,

$$\mu \mathrm{HC}(y) = 0.460;\ \mu \mathrm{C}(y) = 0.176;\ \mu \mathrm{BC}(y) = 0.460;\ \mathrm{MKB}(y) = 0.176.$$

According to (2.13),

$$\mu \mathrm{H}(z) = [0.0506 \bullet 0.086] \vee [0.176 \bullet 0.176] \vee [1.0 \bullet 0.460] = 0.460.$$

Similarly,

$$\mu \mathrm{HC}(\mathrm{z}) = 0.176;\ \mu \mathrm{C}(z) = 0.176;\ \mu \mathrm{BC}(z) = 0.460;\ \mathrm{MKB}(z) = 0.460.$$

Finally, according to Eq. (2.6),

$$\mu S_1(s) = [0.176 \bullet 0.176 \bullet 0.460] \vee [0.176 \bullet 0.460 \bullet 0.176] \vee [0.176 \bullet 0.460 \bullet 0.460] = 0.176.$$

Similarly,

$$\mu S_2(s) = 0.176; \mu S_3(s) = 0.460; \mu S_4(s) = 0.176; \mu S_5(s) = 0.176.$$

3 Conclusion

1. Based on the results of fuzzy sets of the data, it is formulated as follows: since the most important decision membership functions are corresponding to S_3, then the state is taken as unsuitable technical condition (category III) element floor beams.
2. Thus, the considered approach promotes certainty in recognizing states of buildings with limited and inaccurate input.
3. Together with probabilistic approaches and methods of information theory of fuzzy sets considered approach add confidence to the expert in justifying the necessary decisions on the extent and depth of engineering interventions in order to bring it in normal technical condition.

References

1. DSTU B V.2.7-214: 2009. (2010). Building materials. *Concrete. Methods of determination of control samples*. Kyiv: Minregionstroy of Ukraine.
2. Mikhailenko, V. M., Terentyev, O. O., Eremenko, B. M., & Bolshakov, V. I. (2013). Information technology assessment of technical condition of building structures using fuzzy models. *Construction, Materials, Engineering, Scientific Collection, 70*, 133–141. (Works Under the General editorship of Professor V. I. Bolshakov edition).
3. Mikhailenko, V. M., Terentyev, O. O., Eremenko, B. M., & Bolshakov, V. I. (2014). Treatment of experimental results of the expert system for diagnostics of technical condition of buildings. *Construction, Materials, Engineering, Scientific Collection, 78*, 190–195. (Works Under the General editorship of Professor V. I. Bolshakov edition).
4. Terentyev, O. O., Sabala, Y. Y., Malyna, B. S. (2015). Fundamentals of the organization of fuzzy inference for the task of diagnosing the technical condition of buildings and structures. *Managing the Development of Complex Systems, Collection of Scientific Papers, 22*, 138–143). (Kiev).
5. Terentyev, O., & Tsiutsiura, M. (2015). The method of direct grading and the generalized method of assessment of buildings technical condition. *International Journal of Science and Research (IJSR), 4*(7), 827–829.
6. Normative documents W issues of abstieg, pasportist, bezpechno Nadine exploits of virobnichih budwell first sparud. Kyiv, p. 144 (2003).

7. DSTU B V.2.7-224: 2009. (2010). Building materials. Concretes. *Strength control rules*. Kyiv: Minregionstroy of Ukraine.
8. DSTU B V.2.6-7-95. (2008). Construction of buildings and structures. Concrete and reinforced concrete construction products. In *Methods of load testing*. Kyiv: Minregionstroy of Ukraine.
9. DSTU B V.2.6-62: 2008. (2009). Construction of buildings and structures. Marches and stairs are reinforced concrete. *Specifications*. Kyiv: Minregionstroy of Ukraine.
10. Catalog of instruments for non-destructive testing of concrete.
11. Varnaliy, Z., Onishchenko, S., & Masliy, A. (2016). Threat prevention mechanisms of Ukraine's economic security. *Economic Annals-XXI, 159*(5–6), 20–24. https://doi.org/10.21003/ea.V159-04.
12. Cherniha, R., & Serov, M. (2006). Symmetries, ansätze and exact solutions of nonlinear second-order evolution equations with convection terms, II. *European Journal of Applied Mathematics, 17*(5), 597–605. https://doi.org/10.1017/S0956792506006681.
13. Yakovlev, S. V., & Valuiskaya, O. A. (2001). Optimization of linear functions at the vertices of a permutation polyhedron with additional linear constraints. *Ukrainian Mathematical Journal, 53*(9), 1535–1545. https://doi.org/10.1023/A:1014374926840.
14. Cherniha, R., & Pliukhin, O. (2013). New conditional symmetries and exact solutions of reaction-diffusion-convection equations with exponential nonlinearities. *Journal of Mathematical Analysis and Applications, 403*(1), 23–37. https://doi.org/10.1016/j.jmaa.2013.02.010.

Features of Investment Regulation of Construction in Agriculture

O. O. Tomilin, V. V. Gryschko, and S. A. Kolomiyets

Abstract The article deals with the foundations of construction industry formation and its economic development and the features of investment regulation of construction in agriculture. The basic key problems of construction and ways of their solution in agriculture are determined. The features of capital investments of modern production are revealed. The role of direct foreign investments (share capital) from the countries of the world in the national economy of Ukraine is substantiated. It was established that an important feature of the proportional interaction of industries and spheres of production is the structural reorganization of the sphere of the economy. Improvement of structural policy in the construction industry through the development of economic and organizational mechanisms for regulation of inter-sector economic relations will contribute to improving the efficiency of the functioning of all sectors and spheres of agro-industrial production. It is proved that the main ways of structural policy in the construction industry are: firstly—structural (formation of branch development); and secondly—budget (effective definition of market regulators of the gross domestic product). The simulation of economic development of the main branches of the national economy was carried out. It is proved that the growth rate of investments in the studied branches of the economy was less than the growth of the cost of fixed assets. The main problems of economic development of Ukraine, which lead to intensification of the structural adjustment of the national economy, are revealed. The hierarchical levels of structural policy are shown in dynamics. The mechanism of state regulation of investment processes in building activity is offered. It is proved that the main goal of the state structural policy in construction activity is the formation of inter-sectoral proportions, financial planning, development strategies, and the provision of incentives to stimulate the industries.

O. O. Tomilin
Poltava State Agrarian Academy, Poltava, Ukraine

V. V. Gryschko (✉)
Poltava National Technical University Named After YuriiKondratiuk, Poltava, Ukraine
e-mail: gvito@ukr.net

S. A. Kolomiyets
Kharkiv National Agrarian University Named After V.V. Dokuchaev, Kharkiv Oblast, Ukraine
e-mail: sk1974@i.ua

V. Onyshchenko et al. (eds.), *Proceedings of the 2nd International Conference on Building Innovations*, Lecture Notes in Civil Engineering 73,
https://doi.org/10.1007/978-3-030-42939-3_73

Keywords Construction · Investment · Investment regulation · Agriculture · Agro-industrial production · Inter-industry relations · Structural policy · Mechanism of state regulation

1 Introduction

The research focuses on the construction industry, which is considered one of the strategic, profitable, and investment-attractive in the external and domestic markets, because it depends on the effectiveness of the operation of the entire system of management in the country, ensuring the implementation of expanded reproduction of production potential.

Based on world experience, it can be convincingly stated that foreign capital attracted to its territory is determined on the basis of national interests in order to build the construction industry.

Companies that avoid sending short-term messages have a better chance of attracting investors with long-term perspectives. If the aim of responsible investment is to produce long-term change, then a consideration of whether it aligns with extant practices is critical [1].

For the effective functioning and development of the national economy and the construction industry, it is necessary to study the theoretical essence and forms of manifestation of structural policy.

The investment activity of the construction industry is actively influenced by the state through lending and depreciation policy, providing financial assistance in the form of grants, subsidies and subventions, and tax policy. In order to effectively formulate and develop the construction, it is necessary to carry out a deep analysis of the development of the construction industry, to identify specific problems, and most importantly to find effective solutions through the interference of the state regulatory bodies.

The feature of capital investment in modern production is that the structure of capital investments depends on the specialization of the economy. The stock of equity capital is formed at the expense of the founders (owners), reinvestment of profits, and other sources. The study of the in-depth processes of interbranch economic relations is an important element for understanding the trends and directions of the development of the national economy, the formation of an adequate structural policy. The need for a methodological elucidation of the nature of the inequivalence of interdisciplinary exchange, its origins, and justification of ways to reduce the negative effects in the branches and areas of the national economy arises precisely from these processes.

2 An Overview of the Latest Sources of Research and Publications

At the present stage of the market economy, the construction industry needs to address many of the internal as well as external problems.

Construction includes construction management, administration, and engineering services, all of them requiring the construction industry in order to provide their professional services. This reflects both the increasing number of domestic construction projects and the market demand for professional construction practitioners [2]. Researches show that the main problems in the development of the construction industry are low efficiency of state programs in housing construction and investment attraction. It was established that the main obstacles to the development of housing construction in the country are related to the low efficiency of state development programs in this area and mortgage lending, as well as the process of investment and guaranteed housing. Particular attention of scientists deserves problems of energy efficiency of housing construction, its advantages and disadvantages in the context of macroeconomic development of the country, and the transformation of trends of influence [3].

The leading role of public authorities in improving the quality and environmental friendliness of the housing construction is not in contrasting the methods of state regulation with the methods of market self-regulation [4].

The worst housing conditions with a low level of comfort is in the rural population. Researchers draw attention to the fact that the population of the country positively assesses housing conditions against the background of a lower self-esteem of property and social status.

The original study is considering the main problematic questions related to construction, which are connected to implementation of marketing plans. They are as follows: studying of a state and dynamics of demand for construction products at all stages of its lifecycle; providing high degree of civil production to market requirements; the maximum possible using of production capacities and active impact on buyers credit worthiness [5].

Some authors have proved that an increase in investments in the construction and transport sectors will allow to increase the GRP by 3404.74 million UAH. In the formation conception of the spatial development of Ukraine, it was advisable to ignore the development potential and the nucleus of economic growth [6].

The analysis also shows that a high level of investment in fixed assets in a region stimulates employment reallocation. Job creation is also associated with the growth of the housing stock. Perhaps surprisingly, industries tend to experience less employment growth in regions where they initially had high employment. It could be that where conditions conducive to growth are experienced by a relatively large industry that the competition for local resources is more intense within that industry and it hinders growth [7].

The policy environment for cross-border investment is subjected to constant change. At the national level, governments continue to adopt investment policy measures that influence the overall business environment for investors. At the international level, new investment agreements have been concluded at a rate of more than one per week for the past few years [8].

Research focuses on the fact that technology progress is crucial for the structural transformation and economic development. However, reaching o the frontier of world technology requires intensification of research and higher and technical education to implement institutional arrangements and policies facilitating entry and exit of firms and investments.

When it comes to switching jobs, residential location is more important than the location of the previous job, especially in the case of cross-industry switches. The location of firms is not sufficient when defining the labor pool. Growth differentials are also an important dimension of laborpooling, as there are substantial fluctuations in employment growth rates in region–industry combinations that greatly affect the availability of labor with specific skills [7].

The class of the «new rich» distinguishing themselves in ultrahigh quality requirements of life, in particular, big requests concerning prestigious housing (elite houses and estates) which arose in the last decades can become a new perspective segment of construction products market. This market is characterized by the fact that the circle of these consumers is very narrow, but it has industry is characterized by some almost unlimited requests supported with financial potential [9].

3 Main Body

One of the unresolved problems that impede the effective development of agriculture in the process of inter-sectoral interaction is not the elaboration of the economic mechanism of management, the inequality of inter-sectoral exchange, and the discrepancy between incoming and outgoing financial flows with other sectors of national economy. The lack of necessary structural changes in agro-industrial production and scientifically based organizational and economic mechanisms for their regulation led to the unmanageability of these processes.

A distinct geographical pattern can be observed in which new jobs are created in the residential locations of people who have lost their jobs in other industries rather than where the jobs were lost. This pattern of job reallocation indicates that these are people rather than other released resources who drive job creation elsewhere [7].

Formation and development of construction production in the country lead to the need for foreign investment, in order to meet social needs in society. The share of investment in financing construction production is rather high.

Studies found that general definition of the construction industry based on world trends include needs of increasingly sophisticated economy, client demands, technological and social change, and globalization leading to competitive pressures. In the result, the construction industry represents most of every nation's savings [10].

The construction industry, as one of the main strategic industries of the country, requires the use of the mechanism of state regulation of investment processes in construction activities. Foreign investments are used when domestic production is not able to provide the proper level of production of goods by own means. In a difficult economic situation experienced by many countries, including Ukraine, external financial resources are often the most important means of developing modern production. In today's market conditions of management, a significant part of machines, equipment and technologies necessary for production, entering the domestic producer of industrially developed countries, is provided mainly by external financing. Actual value of investments of enterprises depends on its investment potential.

Therefore, studies point to construction problems that are related to the economic and social spheres. Construction economics deals with methods that enable one to make economic decisions toward minimizing costs and/or maximizing benefits to business organizations.

Decisions within the construction are linked with economic and social fields that are associated with project development, its location, and suitability of a particular type of project. This industry is characterized by some features, such as: unique character of the construction outcome (investment result), a large number of small companies, and the general state of the economy influencing demand, terms of construction projects' scale, and expenses [10].

According to the Statistical Yearbook of Ukraine, the main investors of the national economy of Ukraine are the following countries: Cyprus ($317 million), Great Britain ($133 million), Denmark ($56.8 million), the United States of America (51.5 million dollars), Germany (41.1 million dollars), Austria (28.9 million dollars, France (20.7 million dollars), and the Russian Federation (18 million) [11].

The direct investments (share capital) from the countries of the world in the Ukrainian economy for the period from 2014 to 2017 are presented in Table 1.

According to the Statistical Yearbook of Ukraine, the following countries have become the main investors in the national economy of Ukraine: Cyprus (9690 million USD), the Netherlands ($ 5948 million), the Russian Federation (4317 million USD); Great Britain (2 million USD), Germany (1585 million USD), Virgin Islands (BR.)—1686 million USD), Switzerland (1452 million USD), France (1294 million USD), and Austria (1268 million USD) [11].

The industries of the national economy include construction, industry, commerce, agriculture, management, and others, which are subdivided into industries and types of production.

The main problems of the economic development of Ukraine, which lead to intensification of the structural adjustment of the national economy, are:

- structural deformation in the branches of production;
- low-level technology of modern production;
- inefficiency of interbranch connections of branches of the national economy;
- not the welfare of the economic mechanism of management for all sectors and spheres of production.

Table 1 Direct investments (share capital) from the countries of the world in the economy of Ukraine (2014–2017), million USD

Countries	Volume of direct investments as of January 1 of the respective year				On December 31, 2017
	2014	2015	2016	2017	
Cyprus	17725.6	12769.4	10239.5	9690.1	10008.6
Netherlands	9007.5	6986.7	6184.7	5948.4	6292.9
Russian Federation	3525.9	2338.9	3036.9	4317.4	4598.4
Great Britain	2768.2	2153.4	1790.3	2008.7	2169.0
Germany	2908.4	2105.2	1598.2	1584.6	1792.6
Virgin Islands (Brit.)	2275.9	1988.3	1715.0	1683.1	1601.8
Switzerland	1351.0	1391.5	1390.8	1451.9	1537.6
France	1520.5	1394.6	1305.4	1294.5	1346.6
Austria	2314.0	1648.7	1559.8	1268.2	1265.9
Luxembourg	555.8	398.8	363.9	964.2	942.3
Poland	819.8	808.6	758.3	764.4	815.5
Hungary	685.9	593.2	614.9	770.1	796.0
Belize	1026.6	652.5	535.1	604.2	600.6
USA	934.7	701.6	634.1	576.0	538.7
Other countries	6284.2	4794.0	4427.6	4587.8	4837.5
Total	53704.0	40725.4	36154.5	37513.6	39144.0

Source Developed by the Authors

The investment process is a complex economic category that characterizes the socio-economic state of the state and allows you to model its future.

In the sphere of investment activity, a special place is occupied by financial investments. Investing in funds in financial instruments (corporate rights, securities, derivatives, etc.) aims at obtaining future profits, establishing control over an enterprise, and other purposes [12].

Financial investments can be carried out in the following areas:

- investment activity in investing in financial instruments and in particular in securities. In today's conditions of economic development, the tendency of intensification of investment operations in the securities market, the aggregate of which is about 90% of the total volume of financial investments of enterprises, is manifested.
- the acquisition of derivatives (derivative securities) involves the purchase of these financial instruments in order to hedge operations in the spot market or receive speculative income;
- depositary operations, which is a form of short-term investment of temporarily free funds for the purpose of generating investment profits;
- introduction of shares in joint ventures.

Table 2 Direct foreign investment in Ukraine (at the beginning of the year, million USD)

Years	Direct foreign investment in Ukraine	Direct investment from Ukraine
2000	3281.8	98.5
2005	9047.0	198.6
2010	39824.5	6226.3
2011	46293.5	6424.8
2012	49494.4	6456.2
2013	53178.1	6588.7
2014	53704.0	6702.9
2015	40725.4	6456.2
2016	36154.5	6315.2
2017	37513.6	6346.3
2017–2000 (%)	11.4 times more	64.4 times more

Source Developed by the Authors

This form of financial investment is less capital-intensive, but more operational. It allows the establishment of control over the economic activities of the enterprise.

The processes of decentralization, which are declared by Ukraine, enable to distribute investments in the economy of certain sectors. The specificity of the construction is that, when implementing large-scale investment projects for new objects, reconstruction, and modernization, social needs of local communities and the impact on the ecological situation must be taken into account [13].

Foreign direct investment (equity) from the countries of the world into the national economy of Ukraine is given in Table 2.

It should be emphasized that in the last 17 years of research, the largest volume of foreign direct investment in Ukraine was determined by us in 2014. Direct foreign investments in 2014 came from the economies of the world to Ukraine at 53,704.0 million USD, while direct investments from Ukraine totaled 6702.9 million USD.

Thus, one of the reasons for the small inflow of foreign investments into the country's economy is the imperfect regulation of the regulatory framework in the field of investment legal relations, as well as to develop a clear strategy and tactics for attracting foreign investment.

According to the State Statistics Committee of Ukraine, the total volume of direct foreign investments as of January 1, 2017 amounted to $37513.6 million, which is $1359.1 thousand more than the volume of investments at the beginning of 2016 (Table 2). The estimated amount of the program's funding is 501,284,68 thousand UAH, including: 210,142.1 thousand UAH—at the expense of the state budget, 277,882.58 thousand UAH—local budgets, and 13,260 thousand UAH—at the expense of other sources. The amount of expenditures necessary for the implementation of the program is determined annually, taking into account the possibilities of state and local budgets during the formation of their indicators.

Table 3 State target social program «Youth of Ukraine» for 2017–2020, thousand UAH

Sources of funding	Amount of funding	Years			
		2017	2018	2019	2020
The state budget	210142	43109	48057	52363	56267
Local budgets	277883	49201	54643	60187	66488
Other sources	13260	2510	2750	3000	3250
Total	501285	94820	105450	115550	126005

Source Developed by the Authors

Construction is a branch of material production, whose activity consists in the construction and repair of buildings and structures or their separate parts, and creates production and non-productive fixed assets.

The construction productivity depends on several factors such as the built environment efficiency and nature. The availability of the properly configured and located resources within construction investment is affected by flexibility, mobility, the workforce, and the productivity of companies involved in the investment [14].

Construction is based on a powerful construction industry, which is a collection of construction and installation organizations.

General definition of the construction industry based on world trends includes needs of increasingly sophisticated economy, client demands, technological and social change, and globalization leading to competitive pressures. In the result, the construction industry represents most of every nation's savings [10].

Investments and construction activities in Ukraine are presented in Table 3.

In the state target social program «Youth of Ukraine» for 2016–2020, the section «Creating conditions for the promotion of youth to work and residence» is set out, which provides:

- housing for youth—by creating conditions for providing youth with housing;
- employment of young people—by creating conditions and implementing measures aimed at youth employment (provision of primary and secondary employment and youth self-employment);
- provide support for youth in employment and facilitate the creation of new jobs;
- provide effective implementation of state and regional programs in order to provide youth with housing;
- providing young plots of land for self-management and individual construction;
- optimization of the network of vocational schools for the training of mass workers in various fields of activity;
- restoration of work of intercity vocational training combines;
- increase of the quota for rural youth admission to higher educational institutions on a state commission basis;
- provision of concessional long-term loans to young families and single young citizens for the construction (reconstruction) and purchase of housing with a maturity of up to 30 years.

A distinct geographical pattern can be observed in which new jobs are created in the residential locations of people who have lost their jobs in other industries rather than where the jobs were lost. This pattern of job reallocation indicates that these are people rather than other released resources who drive job creation elsewhere [7].

Most social institutions provide services in the non-profit sector of the economies, and their activities are not profitable, but aimed at obtaining a social effect. At the same time, for the activation of construction and successful operation, social facilities require significant capital investment [15]. Subjects of investment process enter into various social relations, which also have cross-sectoral character.

The indisputable fact is that interbranch communications have an active influence on the economic and technical situation in the manufacturing sectors due to the active influence on other sectors of the national economy.

Immediately, the process of inter-sectoral exchange of individual spheres and industries is carried out through direct and reciprocal interbranch links. The solution of the problems of the system of interbranch communications will contribute to improving the efficiency of the functioning of all sectors and sectors of production.

Investments and construction activities in Ukraine are presented in Table 4.

Indicators of economic development have been investigated in three areas: agriculture, industry, and trade, and are reflected in the descriptive model in three main indicators: fixed assets, investments, and gross value added.

We investigated the effect of two factors, the growth rate of fixed assets and the growth rate of investment costs, that affect the gross domestic product and presented by us as a two-factor model.

The following factors are included in the two-factor model of the growth rate of gross value added:

x_1—rate of growth of the cost of fixed assets (percent);

x_2—the growth rate of investment value (percent).

Table 4 Investments and construction activity in Ukraine (2000–2017)

Name	Years				2017–2000 (%)
	2000	2015	2016	2017	
Capital investments, total, (UAH million)	23629	273116	359216	448462	1.2 times more
Capital investment indices (up to the previous year)	114,4	98,3	118,0	122.1	107.0
Investments in housing construction (UAH million)	3404	45610	44610	53372	1.7 times more
Housing fund (million sq. m of total area)	1021.6	973.8	977.9	984.8	96.4
Put into operation houses (thousand sq. m)	5558	11044	9367	10206	1.8 times more

Source Developed by the Authors

The effective indicator (Y) represents the growth rate of gross value added. The obtained equation of communication has an analytical form:

$$Y = 82,388 + 0.00091 \times 12 - 0.00131 \times 22 \quad (1)$$

The value of the correlation coefficient (0.72383013) indicates a high degree of correlation between the productive and the factor characteristics.

The value of the determination coefficient (0.52393005) exceeds the critical one for this population (0.130) and indicates that the variation of the effective indicator by 52.4% is due to the influence of these factor indicators.

The value of t-criterion (71.7) significantly exceeds the critical for this population (2.84), which confirms the reliability of this model. Consequently, the parameters of the correlation–regression model indicate that the growth of the cost of fixed assets by 1% will lead to an increase in the growth of gross value added by 0.000912% on the average for the studied population.

An increase in the volume of investments by 1% will lead to an increase in the growth rate of gross value added by 0.001312% on the average for the studied population. The value of the criterion of significance (p-level) for this model is well below 0.05, which confirms the reliability of the model.

According to our calculations, the growth rate of investment in all three of the studied sectors of the economy was less than the growth of the cost of fixed assets. Investment growth should exceed the value of fixed assets.

The set of state measures, structural parameters, and proportions that are oriented toward the most effective market instruments are the main features of structural policy.

The main methods of adjusting the structure of the national economy can be attributed to:

- definition of the priorities of the structural transformation of the national economy;
- development of state programs of structural adjustment of production;
- financial support of structural shifts in the economy through state investment;
- formation of special funds for the financing of programs of structural adjustment of the economy;
- attraction of foreign investments for the purposes of structural policy; use of system of state orders and purchases;
- production and technological modernization of the processing industry.

Ensuring the equivalence of inter-sectoral economic relations in proportion to invested capital, taking into account its turnover, will enable an increase in the efficiency of the construction industry. An important feature of the proportional interaction of industries and spheres of production is the structural reorganization of the economic sphere. This goal can only be achieved through the regulation of structural policy in production. In our opinion, politics is a set of tools and methods for realizing certain interests in order to achieve a certain goal in a certain social environment. The structure is the interconnection of the components of an object that are in a stable interaction. The structure of the national economy is a proportional relationship between the spheres of production. Structural policy is a multifaceted concept that reflects the equivalent ratio of various elements of economic phenomena.

In order to balance the socio-economic development of the national economy, its effective functioning, and stable growth, it is necessary to select promising and priority directions of structural policy. Consequently, the improvement of structural policy in production through the development of economic and organizational mechanisms for regulation of inter-industry economic relations will contribute to improving the efficiency of the functioning of all sectors and sectors of production.

Thus, structural policy is a state policy aimed at creating inter-sectoral macroeconomic proportions in the areas of production, distribution of goods, and capital and labor in the sectors of the national economy [12]. The mechanism of structural transformations can be called transformations in the structure of indicators of inter-industry economic relations between the state, households, holding structures, corporations, etc., which can overcome the disproportions that arise between individual spheres, the sectoral structure of the national economy, stabilization, and production growth. In our opinion, the main mechanisms for adjusting structural policy can be attributed to: the implementation of integrated target programs, free pricing, competition accumulation, and transfer of capital from industry to industry. Violations of the mechanism, respectively, and the change in the structural elements of regulation of interbranch economic relations in production lead to a decrease in the efficiency of industries, a decline in production, and a negative impact on the development of the entire national economy. In our opinion, the main ways of structural policy in specific market conditions can be attributed:

(a) structural—the formation of sectoral development;
(b) budget—effective definition of market regulators of GDP.

Structural policy should be based on the optimal growth rates of various sectors and sectors of the national economy, while respecting the macroeconomic proportions of agricultural enterprises. Consequently, the improvement of structural policy in production through the development of economic and organizational mechanisms for regulation of interbranch relations will contribute to improving the efficiency of the functioning of all sectors and sectors of production.

We believe that economic instruments for regulation of inter-sectoral economic relations should be considered as structural–organizational; financial; prices; tax. The main objectives of the structural and organizational regulatory instrument in the construction industry are customs and tariff; socio-psychological; market pricing; financial and credit; innovation-investment.

As a result of our research, we propose to consider the levels of structural policy in dynamics and at the lowered hierarchical levels, in particular:

Level 1—industry in the world economy;
Level 2—industry in the national economy;
Level 3—the structure of industries (subsectors);
Level 4—the structure of the consumer market;
Level 5—the structure of producers by category;

Level 6—is the structure of the cost of production.

The hierarchical model of the levels of organizational policy in dynamics is presented in Fig. 1.

The levels of structural factors we propose to extend to the coverage from the structure of material and non-material costs in the cost price at the micro level, then on the mezzanine—changes in the number of producers of gross output by the list of industries and subsectors and then to the macro level, i.e., the share of gross output and gross value added of individual industries in the structure of the national and world economy. In addition, the implementation of structural adjustment should be directed to the following types of economic structural components: reproductive, sectoral, territorial, sectoral, institutional, technological and technical, and economic. The restructuring of the economy is realized due to adaption of the national economy to new requirements and modern realities help to eliminate it from a deep crisis. It

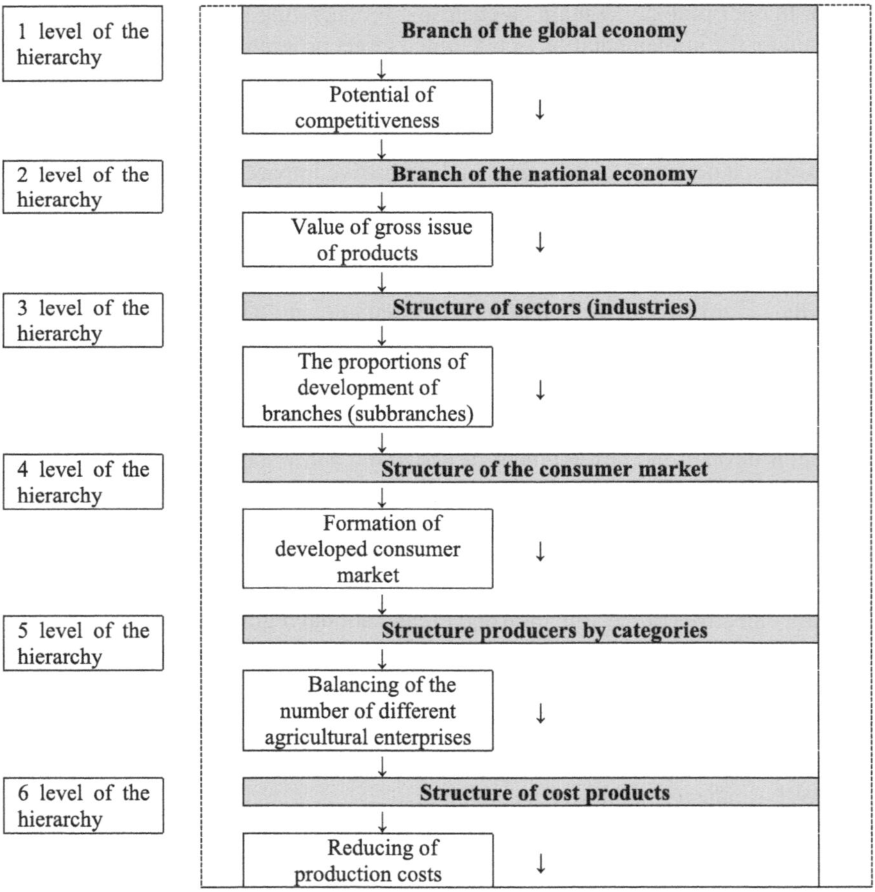

Fig. 1 Hierarchical model of the levels of organizational policy. *Source* Developed by the Authors

directly affects the stabilization and growth of production, market balance, other most important macroeconomic indicators, enables the national producer to enter the process of international division of labor, to create an adequate social base. According to the authors, the main strategy of structural policy is the development of market inter-industry economic relations, the formation of equivalent relations between the spheres of production. In order to establish optimal macroeconomic proportions in agriculture and the construction industry, it is necessary to intensify the support of the state, to implement structural restructuring of the economy and regulate inter-sector economic relations.

4 Conclusion

The development of the national economy is associated with constant fluctuations, which correct the structural macroeconomic proportions and change structural equilibrium. In order to ensure proportionality in the construction industry, it is necessary to take into account the main interrelated aspects: the proportionality between the spheres, within the spheres, the sectoral structure of production, and the types of economic structure. Successful socio-economic development of the construction industry is possible under the condition of state regulation and scientifically grounded structural changes in the national economy of the country. The structural policy should be based on the optimal growth rates of various sectors and sectors of the national economy, with macroeconomic proportions in the fields related to agro-industrial production. State regulation of investment activity should be aimed at increasing the inflow of foreign direct investment in the country's economy in order to implement investment programs in order to provide citizens with housing and profit. In the absence of internal resources, investments can prove to be a very effective tool for updating existing production and serve as an effective means of economic growth, but the volume of investment is much less than necessary, which requires optimization of the structure of the sources of its financing [16–21]. When evaluating the results of our study, it should be noted that the problems of formation and regulation of non-equivalent inter-sectoral economic relations require detailed study.

Unresolved problems in this area necessitate some changes in the methodological approach to the consideration of inter-sectoral economic relations: from theoretical foundations to the design of appropriate measures and an optimal model that will combine the economic background with the forms of organizational support. Relying on the world's European standards on the provision of citizens with the standard of living, our research shows that our country has an extraordinary potential for capital investment and construction activities.

Acknowledgements Further research will be needed to find methodological approaches to the study of investment activity in Ukraine, further development of a scientific discussion on the problems of formation, and development in various spheres of production, including the construction industry in Ukraine. A promising direction to improve the formation of interbranch economic relations is the creation of an optimal model that combines the system of economy and agro-industrial production.

References

1. Himick, D., & Audousset-Coulier, S. (2016). Responsible investing of pension assets: Links between framing and practices for evaluation. *Journal of Business Ethics, 136*, 539–556. https://doi.org/10.1007/s10551-014-2530-z.
2. Liu, C., & He, S. (2016). Input-output structures of the Australian construction industry. *Construction Economics and Building, 16*(2), 56–70. https://doi.org/10.5130/AJCEB.v16i2.4819.
3. Brunko, P. (2016). Financing of energy-efficient construction. *Skhid. Economic Sciences*, №1, pp. 5–13. https://doi.org/10.21847/17289343.2016.1(141).61374.
4. Aggogeri, F., Borboni, A., Merlo, A., Pellegrini, N., & Ricatto, R. (2017). *Materials 10*(3), art. no. 297 https://doi.org/10.3390/ma10030297.
5. Moon, S., & Chi, S. (2018). Predicting construction cost index using the autoregressive fractionally integrated moving average model. *Journal of Management in Engineering, 34*(2), Article Number: 04017063. https://doi.org/10.1061/(asce)me.1943-5479.0000571.
6. Melnyk, M., Synyutka, O., & Kushniretska, O. (2016). Spatial policy of regional metropolis development in Ukraine: Conceptual principles of formation. *Economic Annals-XXI, 159*(5–6), 43–47. doi: http://dx.doi.org/10.21003/ea.V159-09.
7. Morkute, G., Koster, S., & van Dijk, J. (2017). Employment growth and inter-industry job reallocation: Spatial patterns and relatedness. *Regional Studies, 51*(6), 958–971. https://doi.org/10.1080/00343404.2016.1153800.
8. UNCTAD-IPFSD. (2015). Investment policy framework for sustainable development. https://unctad.org/en/PublicationsLibrary/diaepcb2015d5_en.pdf.
9. Said, H. M., & Bartusiak, J. (2018). Regional competition analysis of industrialized homebuilding industry. *Journal of Construction Engineering and Management, 144*(2), Article Number: 04017108. https://doi.org/10.1061/(asce)co.19437862.0001424.
10. Rangelova, F. (2015). *Fundamentals of economics in sustainable construction*. Bulgaria: Bultest Standard Ltd.
11. Werner, I. E. (Ed.). (2017). Statistical yearbook of Ukraine for 2016. K.: State Statistics Service. 541p.
12. Tomilin, O., Galych, O., & Kalinichenko, O. (2016). Economic aspects of development of inter-branch relations in the agrarian sector: Monograph. Opole: University of Opole, Poland, 2016. 171p. http://dspace.pdaa.edu.ua:8080/handle/123456789/1568.
13. Luiz, J. M. (2003). The relevance, practicality and viability of spatial development initiatives: A South African case study. *Journal of Theoretical Social Psychology, 23*(5), 433–443. https://doi.org/10.1002/pad.282.
14. Lawson, R. (2013). The importance of construction sector to the overall economy. Walesbusiness.org. The Business Blog for Wales.
15. Sitar, L., & Skrenkovsky, R. (2015). Current state and problems of social sphere of Ukraine. *Effective Economy*. https://www.economy.nayka.com.ua/?op=1&z=3954.
16. Bukhalo, O. V., & Kolomiets, S. A. (2018). Features of attraction of capital investments in agricultural production. *Bulletin of Kharkiv National Agrarian University named after V. V. Dokuchaev. Series "EconomicSciences", 1*, 169–183 (Kharkiv. Vip).
17. Resolution of the Cabinet of Ministers of Ukraine dated 18.02.2016 № 148/ On Approval of the State Target Social Program "Youth of Ukraine" for 2016–2020 and amending certain resolutions of the Cabinet of Ministers of Ukraine. https://www.kmu.gov.ua/ua/npas/248881l.

18. Zhuk, I. M. (Ed.). (2016). Statistical yearbook of Ukraine for 2015. K.: State Statistics Service. 575p.
19. Varnaliy, Z., Onishchenko, S., & Masliy, A. (2016). Threat prevention mechanisms of ukraine's economic security. *Economic Annals-XXI, 159*(5–6), 20–24. https://doi.org/10.21003/ea.V159-04.
20. Cherniha, R., & Pliukhin, O. (2013). New conditional symmetries and exact solutions of reaction-diffusion-convection equations with exponential nonlinearities. *Journal of Mathematical Analysis and Applications, 403*(1), 23–37. https://doi.org/10.1016/j.jmaa.2013.02.010.
21. Cherniha, R., & Serov, M. (2006). Symmetries, ansätze and exact solutions of nonlinear second-order evolution equations with convection terms. II. *European Journal of Applied Mathematics, 17*(5), 597–605. https://doi.org/10.1017/S0956792506006681.

The manufacturer's authorised representative in the EU is Springer Nature Customer Service Centre GmbH, Europaplatz 3, 69115 Heidelberg, Germany. If you have any concerns regarding our products, please contact ProductSafety@springernature.com

Printed and bound by CPI Group (UK) Ltd, Croydon, CR0 4YY
05/10/2026
02228121-0020